The Flying Circus of Physics

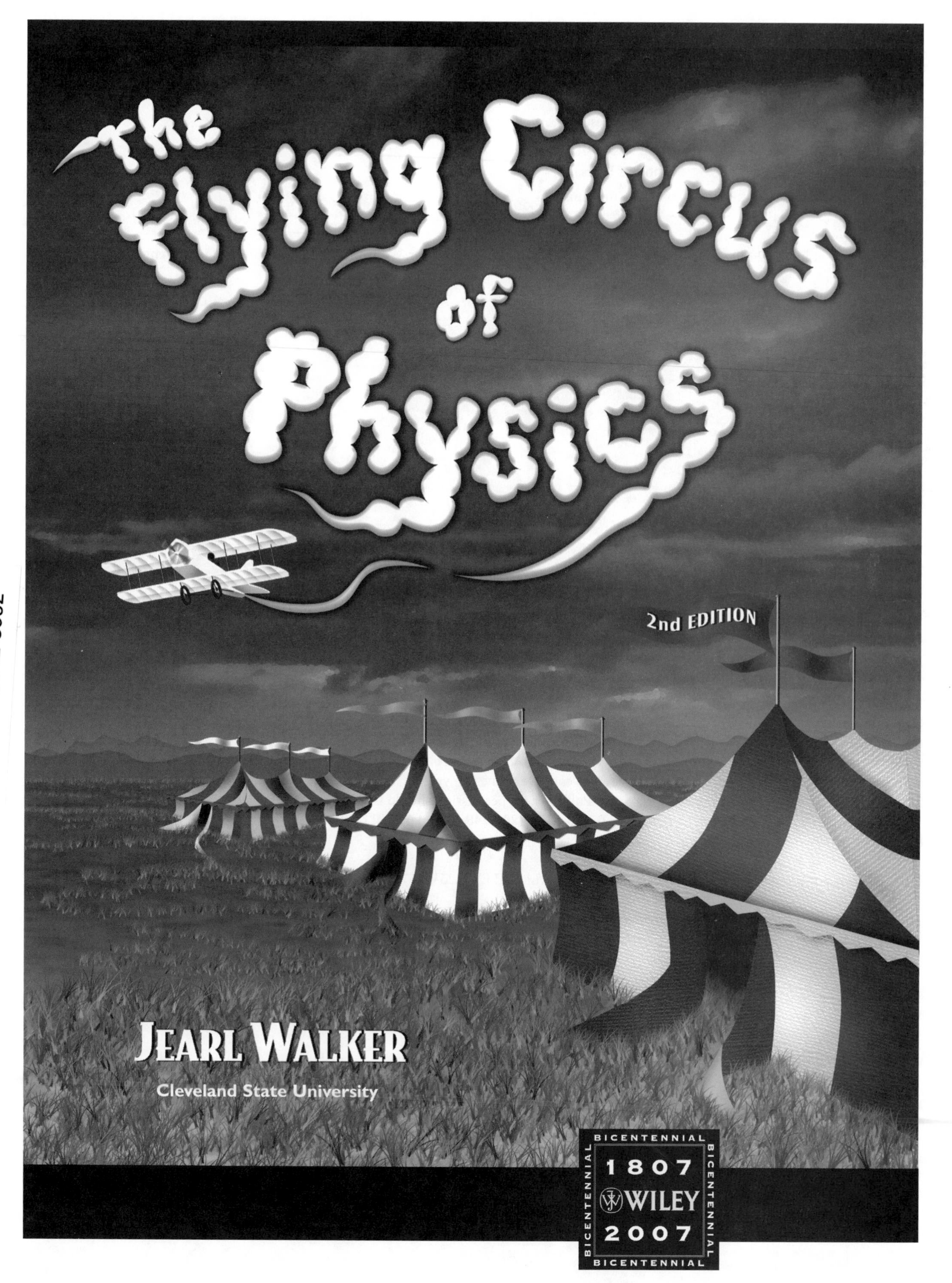

The Flying Circus of Physics
2nd EDITION
JEARL WALKER
Cleveland State University
BICENTENNIAL
1807
WILEY
2007

ACQUISITION EDITOR	Stuart Johnson
SENIOR PRODUCTION EDITOR	Elizabeth Swain
MARKETING MANAGER	Amanda Wygal
DESIGNER	Madelyn Lesure
SENIOR ILLUSTRATION EDITOR	Anna Melhorn
MEDIA EDITOR	Thomas Kulesa
COVER ILLUSTRATOR	Norm Christiansen

This book was set in 10/12 Minion by Preparé and printed and bound by Courier/Westford. The cover was printed by Lehigh Press.

This book is printed on acid free paper. ∞

To order books or for customer service please, call 1-800-CALL WILEY (225-5945).

Library of Congress Cataloging in Publication Data:
Walker, Jearl
The flying circus of physics / Jearl Walker.-- 2nd ed.
p. cm.
Includes index.
ISBN-13: 978-0-471-76273-7 (pbk.: acid-free paper)
ISBN-10: 0-471-76273-3 (pbk.: acid-free paper)
1. Physics--Problems, exercises, etc. I. Title.
QC32.W2 2007
530--dc22

2006008029

Printed in the United States of America

10 9 8 7 6 5 4 3 2 1

This book is dedicated to my wife

Mary Golrick

who sat with me for the 13 years I wrote "The Amateur Scientist" for *Scientific American*, the 16 years (and still counting) I spent writing *Fundamentals of Physics*, and the 200 years (seemingly) I spent on developing and writing this edition of *The Flying Circus of Physics*. Without her encouragement, support, love, and tolerance, I would have ended up staring at the wall instead of a word-processing screen.

The Flying Circus of Physics began one dark and dreary night in 1968 while I was a graduate student at the University of Maryland. Well, actually, to most graduate students nearly all nights are dark and dreary, but I mean that particular night was really dark and dreary. I was a full-time teaching assistant, and earlier in the day I had given a quiz to Sharon, one of my students. She did badly and at the end turned to me with the challenge, "What has anything of this to do with my life?"

I jumped to respond, "Sharon, this is physics! This has everything to do with your life!"

As she turned more to face me, with eyes and voice both tightened, she said in measured pace, "Give me some examples."

I thought and thought but could not come up with a single one. I had spent at least six years studying physics and I could not come up with even a single example.

That night I realized that the trouble with Sharon was actually the trouble with me: This thing called physics was something people did in a physics building, not something that was connected with the real world of Sharon (or me). So, I decided to collect some real-world examples and, to catch her attention, I called the collection *The Flying Circus of Physics*. Gradually I added to the collection.

Soon other people wanted copies of the *Flying Circus* material, first students in Sharon's class, then my fellow graduate students, and then some of the faculty members. After the material was printed as a "technical report" by the Physics Department at Maryland, I landed a book contract with John Wiley & Sons.

The book was published in 1975, a few years after I became a physics professor at Cleveland State University; it was revised in 1977. Since then it has been translated into 11 languages for publication around the world. This is the second edition of the book, which is completely rewritten and redesigned.

When I began writing *Flying Circus* material, I searched through only a few dozen research journals, page by page, and discovered few relevant papers. Indeed, my metaphor for the project was that I was digging for gold in an almost barren mountainside—the gold nuggets were few and hard to find.

The world has changed: Now, many hundreds of research papers with potential *Flying Circus* material are published every year and, in terms of my metaphor, I find huge gold veins. And now I don't dig through just a few dozen journals; I look through about 400 journals directly and use search engines to sort through hundreds more. On many days my fingers just fly over my computer keyboard. I wish Sharon could look over my shoulder at all the really curious things I find. With this book you get that chance: Come look over my shoulder and you'll see that physics "has everything to do with your life."

Web site for The Flying Circus of Physics

The web site associated with this book can be found at **www.flyingcircusofphysics.com** and contains:

- Over 10 000 citations to journals and books of science, engineering, math, medicine, and law. The citations are grouped according to the items in the book and marked as to difficulty.
- Bonus items.
- Corrections, updates, and additional comments.
- An extended index.

Origin of the Flying Circus name

I named my original collection of problems after the very early airshows in which daredevil pilots performed hair-raising stunts. I thought such an airshow was generically known as a "flying circus" and hoped that the image of daredevil pilots would entice someone to read my words.

I've since learned that a flying circus was originally a circus that moved about on a train and then later the name given to German aircraft that were moved in that way. The term came to be associated with the famous German pilot Red Baron, who in World War I painted his airplane blood-red to scare the pilots he fought in the air.

The comedy troupe known as Monty Python's Flying Circus first appeared in England about a year after I had begun using the *Flying Circus* name. The name must have just been in the air on both sides of the Atlantic Ocean that year. (The "dead parrot routine" is, however, entirely Monty Python's.)

Bibliography

All citations are listed in the *Flying Circus of Physics* web site, grouped according to the items here in the book, and marked as to mathematical difficulty. The site contains over 10 000 citations.

Sending stuff to me

I would very much enjoy receiving corrections, comments, new ideas, and citations. If the latter, I would appreciate if you would send the full citation without abbreviations and with the full page numbers, but if that is not possible, even a scrap will interest me. If you can send me a photocopy of a paper or a web site address, that would be great.

I generally do not list web sites in the citations because I cannot frequently check whether the web sites remain active.

I teach full time, work on this book full time, and work on the textbook *Fundamentals of Physics* twice full time. That is a lot of full times, and yet I am only finite. So, please understand why I cannot answer every letter or note.

Cleveland State University

If you want to attend a solid, middle-size university, come to Cleveland State University (www.csuohio.edu) in Cleveland, Ohio. I have been teaching here for over 30 years and have no intention of stopping (although I hear that Nature will eventually slow me down). I'm the fellow in the small office, surrounded by research papers, with my fingers flying over a keyboard desperately trying to meet yet another publication deadline.

Textbooks

The material in this book assumes that you took elementary physics or physical science in grade school. If you want a good textbook to go with this book, here are some suggestions:

• *How Things Work: The Physics of Everyday Life*, Louis A. Bloomfield (John Wiley & Sons), non-mathematical introduction to physics

• *Physics*, John D. Cutnell and Kenneth W. Johnson (John Wiley & Sons), algebra-based introduction to physics

• *Fundamentals of Physics*, David Halliday, Robert Resnick, and Jearl Walker (John Wiley & Sons), calculus-based introduction to physics

Acknowledgments

I have many people to thank because they encouraged me at times when I thought, "All hope is lost!" Well, ok, that is part of the reason. The rest is that many people tolerated me when I became completely obsessed and thought, "I need to work as if there is no tomorrow!"

Jearl and Martha Walker (my parents, who, when I was a teenager, surely spent many sleepless nights worrying whether I would end up successful or incarcerated), Bob Phillips (my high school math and physics teacher who opened new worlds for me), Phil DiLavore (who got me started in teaching), Joe Reddish (who was instrumental in getting the original notes of *The Flying Circus of Physics* published as a technical report by the University of Maryland Physics Department), Phil Morrison (who was the first to encourage me to publish the technical report as a book and who then wrote a nice review about the book in *Scientific American*, which probably got me the 13-year job of writing the magazine's "Amateur Scientist" section), Dennis Flanagan (the editor at *Scientific American*, who hired me and then guided me for years), Donald Deneck (the physics editor at John Wiley & Sons in the early 1970s, who offered me the first book contract for *The Flying Circus of Physics*), Karl Casper and Bernard Hammermesh (who thought enough of the book work to hire me as an Assistant Professor at Cleveland State University), David Halliday and Robert Resnick (who allowed me to take over their textbook *Fundamentals of Physics* in 1990), Ed Millman (who tutored me on how to write textbooks), Mary Jane Saunders (the Dean of the College of Science at CSU, who set up such a positive atmosphere that this edition of *The Flying Circus of Physics* was possible and who critically reviewed many of the manuscript pages), Stuart Johnson (the physics editor at John Wiley & Sons who guided me through this book and multiple editions of *Fundamentals of Physics*), Carol Seitzer (who read through the manuscripts for this book, making many solid changes), Madelyn Lesure (the designer of this book), Elizabeth Swain (the production editor at John Wiley & Sons who managed the production of this book), Chris Walker, Heather Walker, and Claire Walker (my grown children who tolerated my obsession with writing and teaching their entire lives), Patrick Walker (my growing child—not only did he tolerate the many years I spent working in the basement, but he also taught me how to climb the overhang at the rock-climbing gym), and (most of all) Mary Golrick (my wife, who contributed many ideas to this edition and who kept me going whenever I exclaimed, "All hope is lost!").

Physics for...

• **a first date:** 1.57, 1.75, 1.122, 1.124, 2.51, 2.90, 4.78, 5.17, 5.19, 6.98, 6.122, 7.15, 7.16, 7.50

• **a pub:** 1.110, 1.122, 1.149, 2.10, 2.24, 2.25, 2.51, 2.76–2.78, 2.87–2.91, 2.96, 2.108, 2.120, 3.27, 3.40, 4.24, 4.42, 4.60, 4.78, 6.98, 6.113, 6.130, 6.136, 6.138

• **an airplane trip:** 1.17, 1.18, 4.53, 4.69, 5.34, 5.35, 6.10, 6.34, 6.35, 6.37, 6.44, 6.63, 6.91, 6.100, 6.105, 6.129

• **a bathroom and toilet:** 1.93, 1.193, 2.21, 2.23, 2.41, 2.60, 2.150, 3.67, 4.65, 4.66, 6.88, 6.99, 6.110

• **a garden:** 1.132, 2.11, 2.80, 2.93, 2.94, 2.99, 3.25, 4.29, 4.57, 4.84, 5.32, 6.84, 6.92, 6.115, 6.118, 6.120, 6.121, 6.126, 7.38

I invite you to create other groupings of problems for certain occasions and locations.

Jearl Walker
Department of Physics
College of Science
Cleveland State University
2121 Euclid Avenue
Cleveland, Ohio USA 44115
Fax: USA 216.687.2424

C • O • N • T • E • N • T • S

Brief Contents

Contents

CHAPTER 7

Armadillos Dancing Against a Swollen Moon
(VISION) **305**

Slipping Between Falling Drops

Figure 1-1 / *Item 1.1*

1.1 • Run or walk in the rain?

Should you run or walk when crossing a street in the rain without an umbrella? Running certainly means that you spend less time in the rain, but it also means that you may be intercepting more of the raindrops. Does the answer change if a wind blows the drops either toward or away from you?

If you drive a car through rain, what speed should you choose to minimize the amount of water hitting the front windshield so that you might maintain visibility?

Answer If the rain falls straight down or if the wind blows it toward you, you should run as fast as possible. Although you run into raindrops, the decreased time in the rain leaves you less wet than if you move slower. To decrease the number of drops you run into, you should minimize your vertical cross-sectional area by leaning forward as you run. To move rapidly while also bending over, you might, as one researcher suggested, ride a skateboard through the rain, but that is certain to attract attention and, besides, a skateboard is more trouble to tote than an umbrella.

If the wind is at your back, the best strategy is to run at a speed that matches the horizontal speed of the falling drops. In that way, you still get wet on the top of your head and shoulders, but you do not run into drops along your front surface, nor do they run into you along your back surface. However, this strategy does not work for an object being moved through the rain if the object has a much larger horizontal cross-sectional area than you do. Such an object will collect an appreciable amount of water on its top surface even if its speed matches the horizontal speed of the raindrops. To minimize the wetting, such an object should be moved as quickly as possible.

If you drive through rain, you are concerned with maintaining visibility rather than minimizing wetness. If the drops fall straight down or if they are being blown toward you, you should drive slowly. If the rain is being blown in the direction you are driving, you ideally should match the car's speed to the horizontal speed of the drops, but that may not be practical.

1.2 • Traffic platoons and gridlock

If heavy traffic is to flow smoothly along a street without stopping, how should the light sequence at the intersections be timed? Should the timing be varied when rush hour begins? Why does the scheme sometimes fail, such as in a snow storm, in which case *gridlock* sets in and the traffic effectively freezes in place?

Answer The cars move in groups called *platoons.* Suppose that a platoon has been stopped at a red light at intersection 1. When the light turns green, the platoon leaders first accelerate and then travel at a certain cruising speed. Before they reach intersection 2, the light there should turn green so that they are not cowed into slowing. If you know the distance between the intersections, the typical acceleration of the leaders, and the time spent at cruising speed, you can calculate when the light at intersection 2 should turn green.

The motion of the cars farther back in the platoon is delayed from the onset of the green because a *start-up wave* must travel to them (the drivers do not begin to move simultaneously). Several tens of seconds may be required. If the tail of the platoon is delayed too much, it is stopped by the next red light at intersection 2. Suppose that the next platoon traveling down the street is as long or longer than the previous one. Then the number of cars stopped by the next red light at intersection 2 increases.

The situation worsens if the platoons continue to be long. The line of cars stopped at intersection 2 might lengthen until it extends back into intersection 1 and blocks the cross traffic there. Gridlock then begins. To relieve the problem, the light sequence of intersections 1 and 2 must be reversed: The light at intersection 2 must now turn green *before* the light at intersection 1 does, so that the cars stopped at intersection 2 can clear out before the next platoon arrives. The change in sequence could be made manually or by a computer that monitors the number of cars stopped at intersection 2.

Platoons can also be found in tunnel traffic (especially where lane changing is forbidden) and on two-lane rural roads. In each case, a platoon begins when faster cars encounter a slower vehicle, such as a truck. In the rural situation, the platoon dissipates if drivers manage to pass the slow vehicle.

1.3 • Shock waves on the freeway

When the traffic density increases on a freeway or highway, why do "waves," in which the drivers slow or speed up, propagate through the traffic? Sometimes they are created when an accident or a stalled car blocks a lane, and sometimes they are caused by *phantom accidents* in which traffic slows because of a relatively minor reason, such as a car changing lanes. Do the waves move in the direction the cars move or in the opposite direction? Why might a wave persist long after the accident or stalled car has been removed?

Answer When the density of vehicles is quite low, the actions of one driver have little effect on other drivers, especially if passing is possible. When the density is somewhat greater, the drivers begin to interact in the sense that they slow, partially because of safety concerns but also because the possibility of passing is diminished. Suppose that you are driving in such traffic. If the driver ahead of you slows or speeds up, you will do the same after a response time of about a second. The driver behind you follows suit after another response time of a second. And so on. This action of speeding up travels back through the lane of cars as a wave. Such a wave is probably invisible to anyone alongside the road because the adjustments to speed are usually slight.

Now suppose that the driver in front of you abruptly brakes hard. You and the drivers behind you will also brake hard, with each requiring about a second for response. The sudden braking also travels back through the lane of cars as a wave, but now the action is apparent to a roadside observer. Such a wave is a shock wave. Depending on the concentration of the cars before and after the wave has passed, the wave can travel in the direction of traffic (downstream) or in the opposite direction (upstream), or it can even be stationary.

Suppose that a shock wave is created when a car stalls in moderately heavy traffic and that 15 minutes are required by the driver to push the car off the road. As cars then begin to accelerate back to the normal cruising speed, a *release wave* travels back through the long line of waiting cars. It may be much later before the release wave catches up with the shock wave that still travels back through the traffic. Only then is the full traffic flow restored to normal.

1.4 • Minimum trailing distance for a car

If one car trails another, what is the minimum separation that will allow the trailing car to stop before colliding with the leading car should the leading driver suddenly brake to a stop? Conventional advice is that there should be at least one car length of separation for each 10 miles per hour (about 16 kilometers per hour) of speed for the cars. Is the advice sound?

Answer The advice is not sound because it is based on two shaky assumptions. One is that the drivers have identical response times to an emergency. If the trailing driver is slower to respond than the leading driver, more separation is required. The other, more subtle, assumption is that the cars slow at the same rate. If they are not put into a full slide, the assumption is likely to be off. The dangerous situation is, of course, when the leading car slows more rapidly than the trailing car.

Suppose that there is only a small difference in the slowing rates. Is there a simple rule for calculating the minimum separation that avoids an accident? Surprisingly there is not, because the minimum separation depends on the square of the speed and thus it is not easily calculated mentally for a given situation. So, if you travel fast behind another car, you had best allow much more separation than suggested by the conventional advice.

1.5 • Running a yellow light

Suppose that the light at an intersection turns yellow shortly before you reach the intersection. Should you brake to a stop before entering the intersection, continue to travel at your current speed, or accelerate? You might make a decision based on experience by judging your speed, the distance to the intersection, the intersection's width, and an assumption about the duration of the yellow light. Is there a possibility that you would violate the law with any of the choices even if you do not exceed the speed limit?

Answer Local law may influence the answer because in some regions you violate the law if you are in the intersection when the light turns red, while in other regions you can be there legally as long as you have entered the intersection before the light is red. In the first situation, you might very well be in a no-win situation because you may not be able to stop in time or accelerate sufficiently (and still not exceed the speed limit) to clear the intersection. In such a situation, there is a range of distances from the intersection in which any strategy fails to avoid violating the law. The problem is worse when the duration of the yellow light happens to be short and the legal speed is low, but the danger of a collision is lessened if the green light for the perpendicular traffic is delayed for one or two seconds after your light has turned red.

1.6 • Spinout during hard braking

When some types of cars without antilock braking systems are braked hard, they begin to spin and may even end up traveling backwards down the road (Fig. 1-2*a*). What produces the spin, and why don't all types of cars spin? If your car begins to spin, what is the best strategy for regaining control of its motion: Should you turn the front wheels in the direction of the skid or in the direction you wish to go?

Answer Reversal is common to cars with front-mounted engines because they have more weight riding on the front wheels than the rear wheels. That means that the rear wheels are likely to *lock up* and begin to skid before the front wheels, and then any chance spin given to the car by, say, some irregular feature in the road quickly leads to reversal.

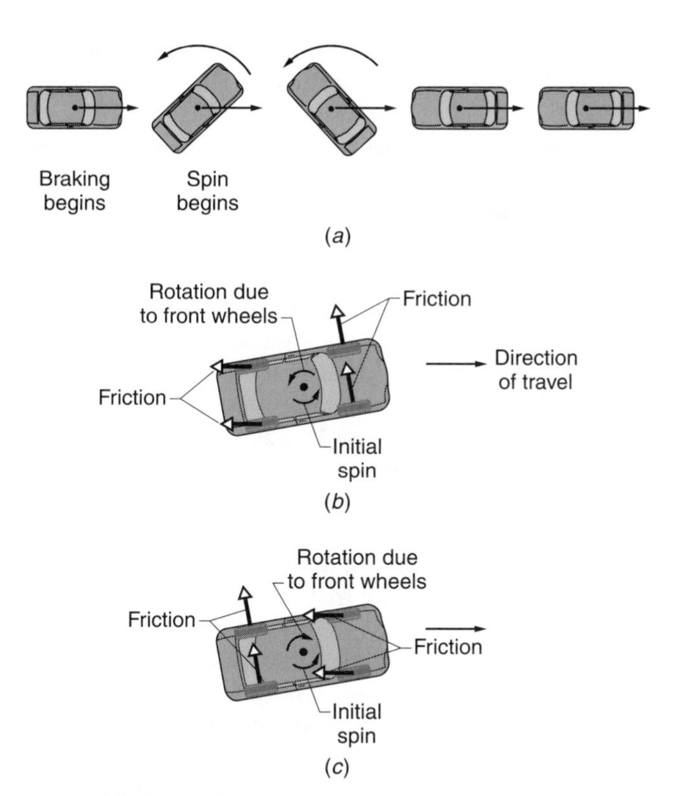

Figure 1-2 / *Item 1.6* (*a*) Reversal of car due to hard braking. Frictional forces on tires for (*b*) front-mounted and (*c*) rear-mounted engines.

To picture the reversal, consider the friction on the tires when a car begins to spin to the left of its intended direction of travel (Fig. 1-2*b*). The rear tires, which are sliding, have frictional forces on them directly toward the rear. The front tires, which are still rolling, have frictional forces on them that are parallel to the front axle and point partially toward the left rear. All the forces create torques that attempt to rotate the car horizontally around its center of mass. The torques from the friction on the front wheels dominate because they attempt rotation in the same direction, which is in the direction that the car has begun to spin. So, spin is enhanced and the car is reversed.

If the engine is in the rear of the car, the roles of the frictional forces on the front and rear wheels are interchanged, and the torques from the rear wheels dominate—they counter the initial spin (Fig. 1-2*c*).

According to conventional advice, if your car begins to spin, you should turn the front wheels toward the intended direction of travel. In doing so, you create a torque on the front wheels that counters the spin, but unless you are skilled, you may overshoot and spin out of control in the opposite direction.

1.7 • To slide or not to slide

Suppose that you are driving down the road when a large moose jumps into your path some distance in front of you. And also suppose that your car lacks an antilock braking system. Should you lock the wheels by braking as hard as you can, or should you apply the brakes as much as possible without locking the wheels? If your car goes into a full slide, why is the end of the slide so abrupt?

Answer Textbooks have traditionally argued for the second choice, correctly pointing out that it is the friction on the tires that stops the car. When the wheels are rolling, the friction can be increased to some maximum value by an appropriate amount of braking. If you brake even harder, the wheels lock and the tires skid. The friction is then smaller and, with less friction, the stopping distance must be longer.

The best choice is to brake just hard enough to put the wheels on the verge of sliding, and then you stop in the least distance, right? Well, actually, maybe not, because that choice might give a stopping distance that is 25% longer than if you lock the wheels and fully slide.

The textbook argument may be unwise in an emergency situation for two reasons. One is that you scarcely have time to experiment with the braking. The other has to do with the torques put on the car by the frictional forces on the wheels: Those torques tend to pitch the car forward by attempting to rotate the car around a horizontal axis through its center of mass (Fig. 1-3). The attempt at rotation decreases the load on the rear wheels and increases the load on the front wheels.

Suppose that you brake just hard enough to put the car on the verge of sliding. Because all the wheels are still turning and because the load on the rear wheels is decreased, the rear wheels are the ones on the verge of sliding (not the front wheels with the extra load) and the friction on the rear wheels is small. If the front and rear brakes are identical, the friction on the front wheels is the same small amount, and so the total friction on the car is small, and the stopping distance for the car is large.

Now suppose that you brake hard enough to lock all the wheels and fully slide. With the wheels sliding, the friction on them depends on the load on them. Since the load on the front wheels is increased, the friction on them is large. Even though the friction on the rear wheels is small, this increased friction on the front wheels means that the total friction on the car is greater than in the previous situation, and thus the stopping distance for the car is shorter. However, locking up the wheels is still not desirable because the sliding eliminates your control of the car; you may easily begin to spin (see the

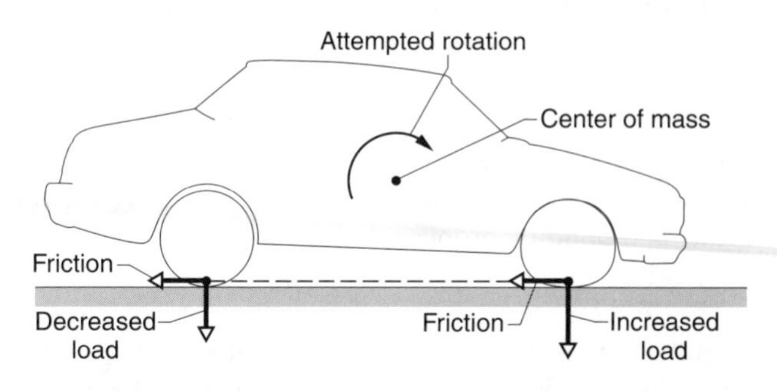

Figure 1-3 / *Item 1.7* A car pitches forward during braking.

preceding item), colliding with adjacent cars or cars in the opposing lane of travel.

The abrupt end to a full slide is due to a sudden increase in the friction on the tires. During the slide the region of contact between the tires and the road is lubricated with melted tar and rubber (see the next item). But as the car slows, less material is melted and the lubrication decreases, which suddenly increases the friction.

1.8 • Skidding to a stop

If a car's wheels are locked during emergency braking, the tires will slide along the pavement and leave skid marks. Suppose that a car slides to a stop from a certain speed. Does the car's weight influence the length of the skid marks? How about the tread design and width of the tire? What if the tire is smooth?

Why is stopping a car more difficult when the road is only somewhat wet than when it is covered with flowing water?

Answer In emergency braking the friction on the tires from the road first increases to some maximum value and then it drops when the wheels lock and the tires begin to slip. The slippage rips off bits of the tire and heats the tires and road. The tire may melt and if the road consists of bituminous material, it too may melt. If either one melts, it produces a fluid that lubricates the slide, which further decreases the friction.

The melted material soon re-solidifies, but the trail—the skid mark—persists, perhaps for months. The trail often has striations that run along its length; they are due to the ribs on a tire or to loose gravel that is ground into the road. On concrete surfaces skid marks are rarer and almost invisible, consisting primarily of the bits of tire that have been melted or torn free.

When a car slides to a stop without colliding with anything, the length of its skid marks allows an investigator to estimate the speed of the car when the sliding began. However, so many variables are involved that the calculation can be only an estimate. One of the variables is the mass (or weight) of the car—a heavy car requires a slightly longer distance to stop than a lighter car, due primarily to the increased lubrication that the greater weight produces. (In traffic court and many physics books, this effect is generally neglected.)

The length of a skid mark also depends on the conditions of the roadway—the marks are usually shorter when the road surface is pebbled and longer when it has been polished by extensive use. The stopping distance does not depend on the width of a tire, because generally the frictional force on a tire depends on only the weight bearing down on the tire and the texture and bonding properties of the tire and road surface, not the tire width.

The tread on a tire has little effect on the stopping distance when the road is dry, but it may be crucial when the road is wet. If the water is substantial, as during a heavy downpour, the tires may tend to glide (or *hydroplane*) over a thin layer of water that offers almost no friction. That is, the tire does not touch the roadway because the water cannot get out of the way or out from under the tire. Hydroplaning becomes even worse when the road is dirty and rain has just begun to fall, because the water and dirt mix to form a very viscous lubricant, much like a clay slurry. Thus the friction between tire and road drops significantly, which can take many drivers by surprise in an emergency stop because they think that as rain begins, the road is not yet wet enough to cause hydroplaning. After the rain has cleaned the road and the road has dried, the friction between tire and road is higher than before the rain because the contamination has been removed. Tires that are designed to minimize hydroplaning have tread that channels or throws water from beneath the tire off to one side.

If the water is not enough to make the car hydroplane, it can still significantly reduce friction on the tires. A tire grips a dry road surface because the weight bearing down on it momentarily seals the bottom of the tire against the surface. That sealing allows the tire to mesh with the irregular surface, molding down into the slight valleys and snagging against the slight upward projections. That tight fit of the tire into the irregular road surface can produce much of the friction the tire requires during emergency stopping. However, when the road surface is wet, those valleys are full of water. Then when the tire is momentarily sealed against the road, it traps the water in those valleys, making the road surface relatively smooth and effectively eliminating the upward projectiles. So, the tire can no longer be snagged against those projectiles.

If the car begins to spin during an emergency stop, the marks left on the road will be curved. Such spin can be initiated if the rear wheels lock up before the front ones, or it can be due to a slant of the road. (Often the crown is higher than the sides to drain rainwater.)

If a wheel is still rolling during a spin, it scrapes sideways on the road and leaves a *scuff mark* that lacks the striations which characterize a skid mark. Either type of mark may be intermittent if the road is uneven enough to make the car bounce or if the braking is nonuniform. Short gaps in the marks are usually due to bouncing, while longer gaps might indicate that the driver pumped the brakes.

SHORT STORY

1.9 • Some records for skid marks

The record for skid marks on a public road appears to have been set in 1960 by the driver of a Jaguar on the M1 in England—the marks were 290 meters long. In court the speed was alleged to have been in excess of 160 kilometers per hour (or 100 miles per hour), when the wheels were first locked. If we assume that the coefficient of friction between tires and roadway was 0.7, we can calculate that the car's speed was about 225 kilometers per hour (or 140 miles per hour).

The Jaguar's skid marks were impressive, but they pale next to the ones left by Craig Breedlove in October 1964 at the Bonneville Salt Flats in Utah. Attempting to topple the land speed record and break through the 500-mile-per-hour "barrier" (805 kilometers per hour), Breedlove drove his rocket-powered *Spirit of America* through a measured mile, first in one direction and then in the opposite direction, so that the effects of wind could be averaged out. When he made his second pass through the mile, he was traveling about 540 miles per hour.

To slow, he released a parachute, but its cord snapped under the strain; the secondary chute also failed. Next he applied his brakes, pressing the pedal to the floor, but they did little more than leave skid marks that were almost six miles long before burning out. The vehicle was then traveling at about 500 miles per hour as it narrowly passed through two lines of telephone poles without crashing. It finally stopped when it rode up and over an embankment and then plummeted, nose down and at 160 miles per hour, into a pool of brine that was 5 meters deep. Since Breedlove was firmly strapped into his seat, he nearly drowned in the submerged compartment. Still, Breedlove's runs through the mile set a speed record and broke the 500-mile-per-hour barrier with an average speed of 526 miles per hour.

1.10 • Woodpeckers, bighorn sheep, and concussion

A woodpecker hammers its beak into the limb of a tree to search for insects to eat, to create storage space, or to audibly advertise for a mate. During the impact, the rate at which the head slows is about 1000 *g*s (1000 times gravitational acceleration). Such a deceleration rate would be fatal to a human or at best severely damage the brain and leave the person with a concussion. Why then doesn't a woodpecker fall from a tree either dead or unconscious every time it slams its beak into a tree?

To determine dominance during the mating season, male bighorn sheep charge one another to slam their horns and heads together in a violent collision. Yet, they don't drop to the ground unconscious (it is hard to be a female's choice if sprawled unconscious on the ground). Some types of horned dinosaurs (such as the *Triceratops*) may have had similar collisions. Why don't the collisions hurt the sheep?

Answer The ability of a woodpecker to withstand the huge deceleration when it hammers at a tree limb is not well understood, but there are two main arguments. (1) The woodpecker's motion is almost along a straight line. Some researchers believe that concussion can occur in humans and animals when the head is rapidly rotated around the neck (and brain stem), but that it is less likely in straight-line motion. (2) The woodpecker's brain is attached so well to the skull that there is little residual movement or oscillation of the brain just after the impact and no chance for the tissue connecting the skull and brain to tear.

Head-banging sheep are usually protected by three features. (1) Their horns bend so as to prolong the duration of the collision and thereby reduce the force in the collision. (2) The skull bones (cranial bones) also shift or rotate about their junctions (sutures) in a spring-like or hinge manner in order to cushion the blow to the head. (3) Most of the energy of a collision ends up in the strong neck muscles of the animals. Although the collisions look terribly violent, the muscles and horns of the animals have evolved to the point where breaking a horn or hurting the brain is unlikely. The *Triceratops* probably also benefited from an extensive sinus system that overlaid the brain case and that could have acted as a shock absorber.

SHORT STORY

1.11 • The game of *g*s

In July 1977, at El Mirage Dry Lake, California, Kitty O'Neil set two records for a dragster on a 440 yard run. From a standstill, she reached the greatest *terminal speed* (speed at the end of the run) ever recorded and also broke the record of the lowest elapsed time with her mark of 3.72 seconds. Her speed was an astounding 392.54 miles per hour (about 632.1 kilometers per hour). Her average acceleration during the run was 47.1 meters per second-squared, which is 4.81 times the acceleration of gravity, or 4.81 *g*s for short.

In December 1954, at Holloman Air Force Base, New Mexico, Dr. John Stapp, an Air Force colonel, was strapped in the seat of a rocket sled with nine rockets behind it. When they were fired, Stapp and the sled were propelled along a track for 5 seconds, reaching a speed of 632 miles per hour, about 1018 kilometers per hour. His acceleration during the propulsion stage was approximately 56.4 meters per second-squared, or 5.76 *g*s. The numbers are certainly impressive, but the real test of Colonel Stapp was the stop by the water brakes, which took only 1.4 second—he slowed (decelerated) at the rate of 20.6 *g*s.

In May 1958, in a similar sled at Holloman, Eli L. Beeding Jr. reached a speed of about 72.5 miles per hour, or 117 kilometers per hour. The speed hardly seems noteworthy because it is common on some highways, but it commands respect when the time for the acceleration is noted. The time was 0.04 second, less than the blink of an eye. Beeding's acceleration of 82.6 *g*s remains the record in a controlled situation.

In July 1977, in Northamptonshire, England, David Purley's race car crashed and his speed dropped from 108 miles per hour to zero while he moved through a distance of only 26 inches. (The speed was 174 kilometers per hour; the distance was $\frac{2}{3}$ meter.) His deceleration was a seemingly lethal 179.8 *g*s but, although he had 29 fractures, three dislocations, and underwent six heart stoppages, Purley survived.

1.12 • Head-on car collision

Suddenly you realize that a car is headed toward your car the wrong way in a one-way tunnel. To minimize your danger from the impending accident, should you match your speed to that of the other car, go even faster, or slow to a stop?

A head-on collision is the most dangerous type of car collision. Surprisingly, the data collected about head-on collisions suggest that the risk (or probability) of fatality to a driver is less if that driver has a passenger in the car. Why is that?

Answer The best advice is to stop and, if possible, put your car in reverse. You can get a measure of the severity of the collision by considering the total kinetic energy or the total momentum of the cars before the collision. If you do not decrease your speed toward the other car, both quantities are large, and so the collision will be severe.

The situation is unlike football, where one player may choose to speed up when running head-on into another player. The difference is that a player may want the collision to be violent, and by properly orienting his body he can shift the collision to his opponent's vulnerable area or cause his opponent to become unbalanced and slam into the field.

Data collected about head-on car collisions indicate that adding a passenger to your car reduces your risk of fatality. That risk depends on the change in your velocity during the collision: A large change means that you undergo a severe acceleration due to a severe force. For example, if your car has a small mass and the other car has a large mass, your velocity may be changed so much that you end up going backward. Additional mass in your car, from a passenger or even a sandbag in the trunk, can decrease your change in velocity and thus also your risk. Here is one numerical result: Suppose your car and the other car are identical and that your mass and the other driver's mass are identical. Your fatality risk is reduced by about 9% if you have an 80 kilogram passenger in your car.

SHORT STORY

1.13 • Playing with locomotives

Waco, Texas, September 15, 1896: William Crush of the Missouri, Kansas, and Texas Railroads dreamed up a surefire idea for a show. He arranged for two obsolete locomotives to face each other at opposite ends of a 4 mile track. One was painted red, the other green. The idea was to crash the locomotives into each other at full speed.

Well, nothing sells quite like violence, and 50 000 spectators paid to see the crash. After the engines were fueled and their throttles fixed open, the locomotives accelerated toward each other. When they met, they were going about 90 miles per hour, which is 145 kilometers per hour.

Several of the spectators were killed by the scattered debris and hundreds were hurt. The rest of the crowd probably got their money's worth. Being near the collision, with its transformation of kinetic energy of the trains into kinetic energy of flying debris, was like being near a moderate explosion.

1.14 • Rear-end collision and whiplash injury

In a rear-end collision, a car is hit from behind by a second car. For decades, engineers and medical researchers sought to explain why the neck of an occupant of the front car is injured in such a collision. By the 1970s, they concluded that the injury was due to the occupant's head being whipped back over the top of the seat as the car was slammed forward, hence the common name "whiplash injury." The neck was apparently extended too far by the head's motion. As a result of this finding, head restraints were built into cars, yet neck injuries in rear-end collisions continued to occur. What actually causes these injuries?

Answer The primary cause of a whiplash injury is the fact that the onset of the forward acceleration of the victim's head is delayed from that of the torso. Thus, when the head finally begins to move forward, the torso already has a significant forward speed. This difference in forward motions puts a huge strain on the neck, injuring it. The backward whipping of the head happens later in the collision and could, especially if there is no head restraint, increase the injury.

1.15 • Race-car turns

High-speed motor races are often won by the performance of car and driver on the turns, which is where the speed is lowest. Consider a 90° turn on flat track, as in Formula One racing. Obviously, the best way to take the turn depends on the handling characteristics of the car, the skill and experience of the driver, and the conditions of the track. However, in general, should the driver follow a circular path around the turn? That choice usually guarantees the least time spent in the turn, but why might it not be the best choice?

Why do drivers who are experienced on the flat tracks of Formula One courses have a difficult time if they switch to Indy car racing, which usually has banked turns? In particular, why is such a driver prone to spin out as the car enters the turn?

Answer A novice driver takes a turn along a circular path. A skilled driver brakes while turning a little, then turns sharply, and then follows a less sharply turned path while accelerating. The procedure takes more time in the turn but allows the driver to enter the straightaway at greater speed than the novice driver. That greater speed on the straightaway more than makes up for the time lost in the turn.

The procedure has another advantage. If the turn is taken too fast, the limit of the frictional forces on the tires will be exceeded and the car will slide out of control. To maintain the friction, the skilled driver first brakes and only then takes a sharp turn. Because the rest of the turn is gradual, the driver can accelerate without overwhelming the friction.

A skilled Formula One driver has an intuitive feel for the sensations of force and motion during a flat turn. The sensations on a banked turn are quite different, and a Formula One driver will probably be too late in making the sharp-turn part of the turn procedure.

1.16 • Sprint tracks

Why is a race on a straight track generally faster than one of the same distance on a curved track? When the track is flat and oval, why does a runner in the outside lane generally have an advantage over one in the inside lane, even though the distances in the two lanes are the same? Why does the speed of a race on such a track depend on the shape of the oval?

Answer Entering a curve, a runner slows; leaving the curve, the runner accelerates back to the straightaway speed. In any turn, a centripetal force toward the center of the turn is required. Here, the centripetal force is provided by the friction force on the runner's shoes. During that inward force on the shoes, the runner's body tends to lean outward in the turn, as if it is being thrown outward. So, to maintain balance, the runner slows to decrease the forces and leans inward to offset the outward leaning tendency. The sharper the turn is, the more the runner must slow and lean. Thus, someone running in an outside lane (which has less curvature) will generally have an advantage over someone running in the inside lane (which has more curvature).

When the track is flat and oval, the amount of the race along the curved portions partially determines the speed of the race. In general, a wide oval provides a faster race than a narrow oval because the curvature of the curved portions of a wide oval is smaller than that of the sharp turns of a narrow oval. The best geometry (other than a straight track, of course) is a circle; its curvature is least.

1.17 • Takeoff illusion

A jet plane taking off from an aircraft carrier is propelled by its powerful engines while being thrown forward by a catapult mechanism installed in the carrier deck. The resulting high acceleration allows the plane to reach takeoff speed in a short distance on the deck. However, that high acceleration also compels the pilot to angle the plane sharply nose-down as it leaves the deck. Pilots are trained to ignore this compulsion, but occasionally a plane is flown straight into the ocean. What is responsible for the compulsion?

Answer Your sense of vertical depends on visual clues and on the vestibular system located in your inner ear. That system contains tiny hair cells in a fluid. When you hold your head upright, the hairs are vertically in line with the gravitational force on you, and the system signals your brain that your head is upright. When you tilt your head backward, the hairs are bent and the system signals your brain about the tilt. The hairs are also bent when you are accelerated forward by an applied horizontal force. The signal sent to your brain then indicates, erroneously, that your head is tilted back. However, the erroneous signal is ignored when visual clues clearly indicate no tilt, such as when you are accelerated in a car.

A pilot being hurled along the deck of an aircraft carrier at night has almost no visual clues. The illusion of tilt is strong and very convincing, with the result that the pilot feels as though the plane leaves the deck headed sharply upward. Without proper training, a pilot will attempt to level the plane by bringing its nose sharply down, sending the plane into the ocean.

SHORT STORY

1.18 • Air Canada Flight 143

On July 23, 1983, Air Canada Flight 143 was being readied for its long trip from Montreal to Edmonton when the flight crew asked the ground crew to determine how much fuel was already on board. The flight crew knew they needed to begin the trip with 11 300 kilograms of fuel. They knew that amount in kilograms because Canada had recently switched to the metric system; previously fuel had been measured in pounds. The ground crew could measure the onboard fuel only in liters, which they reported as 7682 liters. Thus, to determine how much fuel was on board and how much additional fuel was needed, the flight crew asked the ground crew for the conversion factor from liters to kilograms of fuel. The response was 1.77, which the flight crew used (1.77 kilograms corresponds to 1 liter) to calculate that 13 597 kilograms of fuel was on board and that 4917 liters was to be added.

Unfortunately, the response from the ground crew was based on pre-metric habits—1.77 was the conversion factor not from liters to kilograms but rather from liters to *pounds* of fuel (1.77 pounds corresponds to 1 liter). In fact, only 6172 kilograms of fuel were on board and 20 075 liters should have been added. This meant that when Flight 143 left Montreal, it had only 45% of the fuel required for the flight.

On route to Edmonton, at an altitude of 7.9 kilometers, the airplane ran out of fuel and began to fall. Although the airplane had no power, the pilot managed to put it into a downward glide. Because the nearest working airport was too far to reach by gliding, the pilot angled the glide toward an old, nonworking airport.

Unfortunately, the runway at that airport had been converted to a track for race cars, and a steel barrier had been constructed across it. Fortunately, as the airplane hit the runway, the front landing gear collapsed, dropping the nose of the airplane onto the runway. The skidding slowed the airplane so that it stopped just short of the steel barrier, with the stunned race drivers and fans looking on. All on board the airplane emerged safely. The point here is this: Quantities without proper units are meaningless numbers.

1.19 • Fear and trembling at the amusement park

What accounts for the thrill of a ride on a roller coaster? Surely, the heights, speeds, and illusions of falling are factors, but all those sensations can be had in a fast, exterior elevator that is glass encased. No one queues up and pays for elevator rides.

And how about the rides that sling you about? Why do you clutch, and maybe even scream, during the rides?

Roller coasters are designed to give the illusion of danger (that is part of their fun), but in fact engineers go to extreme lengths to make them exceedingly safe for riders. In spite of this attention to passenger safety, an unlucky few of the millions of people who ride roller coasters each year end up with a medical condition called *roller-coaster headache.* Symptoms, which might not appear for several days, include vertigo and headache, both severe enough to require medical treatment. What causes roller-coaster headache?

Answer Some rides are exciting because of heights, high speeds, or large accelerations (up to 4 gs on a roller coaster), or because rapid rotation creates an amusing sensation of centrifugal (outward-directed) force, but the most frightening rides are usually those that produce rapidly changing and unexpected forces on you. When you feel a constant force and undergo a constant acceleration, things seem under control, but when the force suddenly changes size or direction and you accelerate unexpectedly, you subconsciously sense danger. The element of surprise on a subconscious level generates an existential flirt with death.

Standard roller coaster: The heights and the high speeds are enthralling, as is the clatter on an old wooden coaster. When you travel quickly through a curved low section, an apparent centrifugal force on you seemingly presses you into the seat; when you travel over a short but highly curved hill, the force seems to throw you out of the seat. When you go over the edge of the first and largest hill, you have a distinct feeling of falling. The illusion is best when you sit in the front car so that little of the coaster is in front of you. However, I think that sitting in the rear car is even more frightening. As you approach the edge and more of the coaster begins to descend, the force on your back builds, gradually at first and then ever more rapidly (the rate is exponential), and just as you reach the edge, the force disappears. The experience is as if some diabolical agent shoves you toward the edge in a frenzy and then hurls you into free fall.

Mouse roller coasters: The cars are sent separately along the track. The compartment in which you sit pivots above a wheeled framework that follows the track, with the pivot located near the back of the car. When you reach a sharp turn, the framework faithfully follows the curved track but the compartment continues to travel forward for a moment before it too turns. In that moment you have the illusion that the compartment is flying off the track.

Modern roller coasters: Vertical loops and corkscrews produce sensations of centrifugal forces that rapidly change size and direction, and you are also turned upside down. Both factors produce fear. As you ascend a vertical loop, the apparent centrifugal force should decrease as you slow, but the curvature of the track sharply increases so as to maintain that apparent force. On some coasters you might move along the track while facing backwards so that you cannot foresee any of the changes in force, speed, or acceleration that you are about to undergo. Riding a coaster in darkness also eliminates anticipation and enhances fear.

Rotor: When you stand next to the interior wall of the large spinning cylinder, you feel pinned by a powerful centrifugal force (Fig. 1-4*a*). The force may alter your perception of the downward direction and create the illusion that you are tilted backward. If the force is large enough, the floor can be dropped away while you are held in place by a frictional force between you and the wall. Although the idea of an outward force may then be quite convincing, the force that actually pins you is an inward force—the wall pushes on you toward the center of the cylinder in order to keep you going

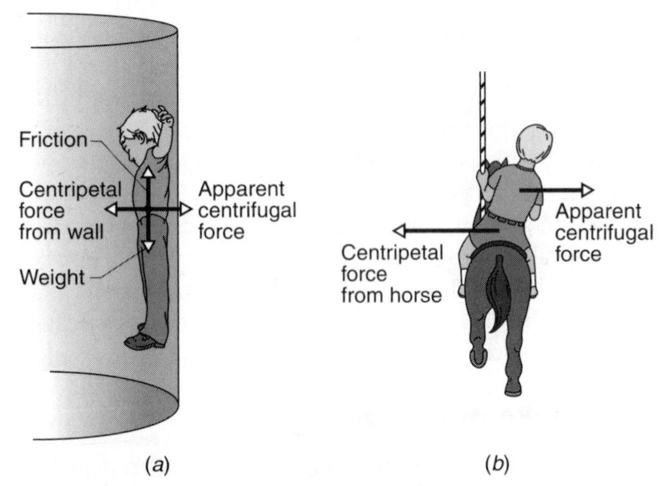

Figure 1-4 / *Item 1.19* Forces involved in (*a*) a rotor and (*b*) a merry-go-round.

in a circle. Since you do not slide down the wall, the frictional force on you must be upward and equal to your weight.

Ferris wheel, merry-go-round, and rotating swings: These rides offer milder sensations of a centrifugal force. When your cage on a Ferris wheel rotates through the top of the circle, you feel as though you are being lifted by the force. At the bottom of the circle, you feel as though you are being shoved downward into the seat. On a merry-go-round, the centrifugal force seems to throw you outward (Fig. 1-4*b*), especially if you ride an outside horse, which moves faster around the circle than horses closer to the center. When you ride in a swing that is rotated around a central hub, the chains move off vertical as if a centrifugal force is pushing you outward. In each of these three rides, there really is no centrifugal force. Rather there is a centripetal force (from the seat in the Ferris wheel, the horse in the merry-go-round, and the chains in the swing), and that force is what keeps you circling.

Rides with rotating arms: You sit in a compartment that is at the outer end of an arm which pivots around the outer end of another, more central arm. If the arms rotate around their pivots in the same direction, you feel the greatest centrifugal force and have the greatest speed when you pass through a point that is farthest from the center of the apparatus. When the directions of rotation are opposite, your speed is least at the far point (due to the opposing rotations), but the force on you varies most rapidly there because you are being whipped through a highly curved path.

Vertical falls: You sit in a compartment that is some 40 meters high when it is suddenly released and allowed to drop in almost free fall. You have a sense of weightlessness because you and the seat below you fall at almost the same rate, and so you no longer feel any support from the seat. Some riders think that the sensation is fun.

Roller-coaster headache can result from any amusement park ride in which the acceleration is large and rapidly changing in direction. The large acceleration puts a strain on the brain and any abrupt change in direction can then cause the brain to move relative to the skull, tearing the veins that bridge the brain and skull.

SHORT STORY

1.20 • Circus loop-the-loop acts

The modern amusement parks may be packed with thrills, but they pale compared to some of the circus stunts involving bicycles that were performed between 1900 and 1912. As one circus attempted to outshine another, daring acts were devised and performed, some more than once if the performers escaped injury. One of the early stunts was demonstrated in 1901 by Adam Forepaugh & Sells Bros. Circus. A man known as "Starr" rode a bicycle down from a height of 18 meters along a 52° ramp. That may not sound too challenging, but the ramp consisted of three sections of extension ladders, which meant that the ride was quite rough, especially near the bottom portion.

The next year at New York's Madison Square Garden, Forepaugh & Sells introduced Diavolo and a bicycle–loop act. With an ambulance standing by, Diavolo began his ride down a ramp from just beneath the ceiling's incandescent lamps and then passed along (inside) a vertical loop with a diameter of 11 meters and into nets to stop the motion. In 1904 the same circus presented the "Prodigious Porthos" in another bicycle act. The ramp was similar but the top of the loop was excluded, requiring Porthos to fly 15 meters through the air, while inverted, to reach the second portion of the loop.

Perhaps the most daring bicycle stunt took place in 1905 when Barnum & Bailey Circus played Madison Square Garden. The act began with Ugo Ancillotti on a bicycle high on one ramp and his brother Ferdinand similarly mounted even higher on a second, facing ramp (Fig. 1-5). On signal, the brothers began their descents. Upon reaching the sharply curved lower end of his ramp, Ugo was projected 14 meters to land on another ramp, and then he repeated the performance across a second gap of 9 meters. Meanwhile, Ferdinand was sent into a curved path by the bottom portion of his ramp so that he soared upside down to reach another curved ramp. The most gripping aspect of the performance was when Ferdinand soared upside down and only few feet below Ugo, who was crossing his first gap. The danger in the performance was quite real—when the act was attempted again in the evening show, Ferdinand took a bad fall during the "loop the gap," and the act was apparently canceled.

Circuses began to substitute "autos," partially because of the novelty of automobiles at the time. One or two occupants would ride an auto down a ramp and flip once or twice in the air before reaching a second ramp. However, these types of circus stunts waned after 1912, probably because audiences grew accustomed to the danger involved. The associated physics did not receive another injection of theatrics until more modern times when Evel Knievel, his son Robbie Knievel, and other stunt people began to ride a motorcycle up or down a ramp and soar over cars and trucks.

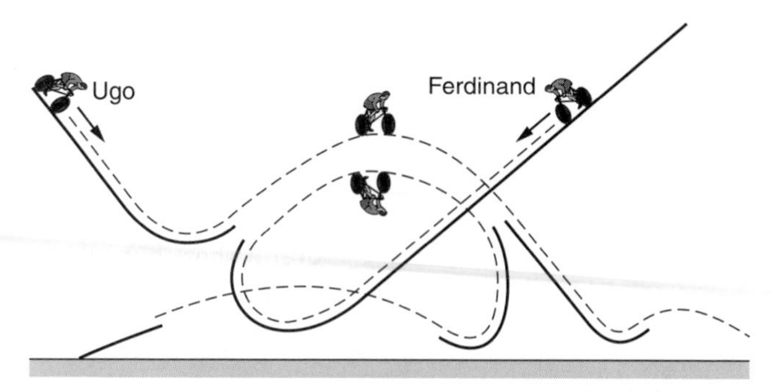

Figure 1-5 / *Item 1.20* The bicycle act of Ugo and Ferdinand Ancillotti.

1.21 • Catching a fly ball

When a high fly ball is hit to the outfield, how does the player in the area know where to be to catch it? The outfielder may run to the proper point and wait for the ball. Or the outfielder may run at a measured rate and arrive at the proper point just as the ball arrives. Either way, playing experience surely helps, but are there clues hidden in the ball's motion that can guide the outfielder?

As an example of an outfielder's skill, Robert Weinstock of Oberlin College relates how Babe Ruth once caught a high fly from Jimmy Foxx of the Philadelphia Athletics. Ruth was waiting deep in left field, expecting a long fly ball from Foxx, but Foxx hit the ball askew and it went high and short. As soon as the sound of the hit reached Ruth, he ran to the precise spot on the field, waited, and then snared the ball with his glove.

Answer Although an outfielder uses many clues to catch a fly ball, two angles appear to be important. One is the vertical angle α through which the ball moves in the player's view as the ball travels toward the outfield (Fig. 1-6*a*). If the player is already at the proper point to catch the ball, this angle increases but at a decreasing rate (at first, it increases rapidly and then it increases less rapidly). If the player is too close (and must retreat), the vertical angle increases at an increasing rate; if the player is too far (and must advance), the vertical angle first increases and then begins to decrease. The player knows from experience to move until, in the later part of the ball's flight, the vertical angle increases at the proper decreasing rate.

The other important angle comes into play when the ball is hit off to the left or right of the player. As the ball travels toward the outfield, it moves horizontally through angle θ in the player's view (Fig. 1-6*b*). The player runs so that this angle increases at a constant rate. This allows the player to run to the proper catching point at a fairly steady rate instead of making a dash at the last second. Doing all this well takes practice but it must also come naturally because dogs, such as those who catch a thrown Frisbee with their mouth, use the same procedure (as revealed by video cameras attached to them).

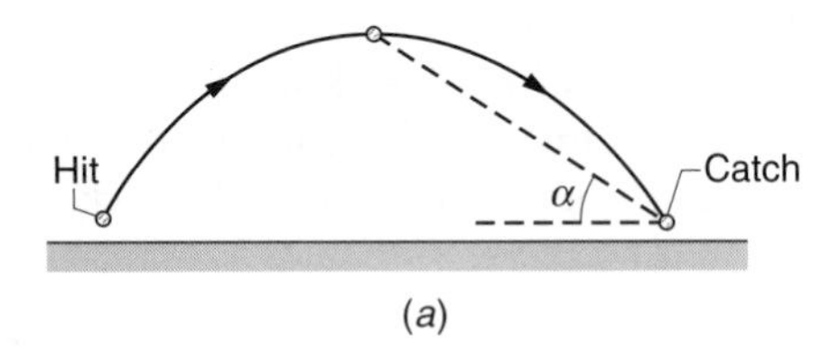

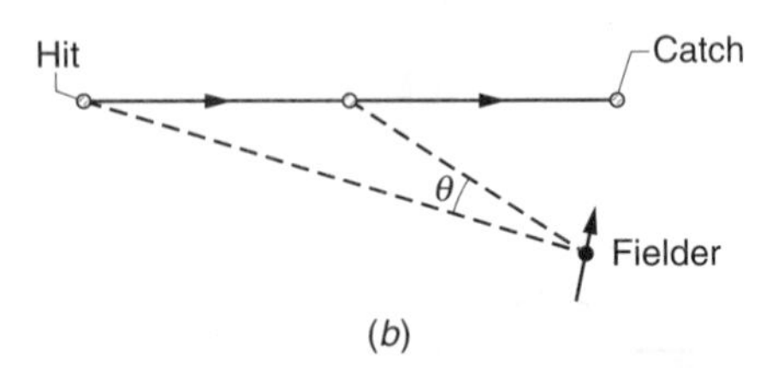

Figure 1-6 / *Item 1.21* (*a*) Side view of fly ball's path. (*b*) Overhead view of the path.

SHORT STORY

1.22 • High ball

In August 1938, Frankie Pytlak and Hank Helf, two catchers from the Cleveland Indians, set out to capture the world's record for catching the longest dropped baseball. While they waited at street level beside Terminal Tower in Cleveland, Ken Keltner, the third baseman, prepared to toss the balls from the top of the building, about 700 feet (or 213 meters) high. The previous record of 555 feet had been set in 1908 by two catchers from another team who caught baseballs tossed off the Washington Monument in Washington, D.C.

Keltner had no way of seeing his fellow players on the street and so he tossed the balls out blindly. Pytlak and Helf wore steel helmets to guard against injury by the balls, which reached estimated speeds of almost 140 miles per hour (or 225 kilometers per hour). Helf made the first catch, claiming with a grin that there was nothing to it, but the next five balls for Pytlak went astray. One bounded up to the 13th floor and was fielded by a police sergeant after its third bounce. On the sixth try, Pytlak made his catch and the shared record.

The next year Joe Sprinz of the San Francisco Baseball Club attempted to catch a baseball dropped 800 feet from a blimp. (Some reports claim the fall was much higher.) On his fifth attempt he got a ball in his glove, but the impact drove hand, glove, and ball into his face, fracturing his upper jaw in 12 places, breaking five teeth, and knocking him unconscious—and he dropped the ball.

Even more ludicrous was an attempt in 1916 to catch a baseball tossed from a small airplane. Wilbert Robinson, the manager of the Brooklyn Dodgers and a former catcher, arranged for the Dodger trainer Frank Kelly to toss the ball from the airplane while at a height of 400 feet. But unknown to Robinson, Kelly substituted a red grapefruit for the ball. When the fruit exploded on impact, its red contents drenched Robinson, who cried, "It broke me open! I'm covered with blood!"

1.23 • Hitting a baseball

If you are right-handed, why do you hold a baseball bat with your right hand higher than the left one and turn your left side to the pitcher? How long does a baseball take to reach home plate? How much time do you have to execute a swing? How much error can you have in your swing and still hit the ball?

Some home-run sluggers prefer heavy bats, claiming that the extra weight in the collision results in a longer hit. Other players choose a light- or moderate-weight bat with a similar claim. (Occasionally, when a wood bat is used, a player will illegally install a cork core to lessen the weight.) Who is correct in the argument about weight? Should a player warm up with a standard bat with a lead doughnut slipped over the outer end or a bat that is much lighter or much heavier than the bat that will be used in the game?

Where should the ball hit the bat to give it the greatest speed? Why does the bat sometimes sting your hands and attempt to jerk out of your grip during its collision with the ball?

Pitchers so feared the power of the legendary hitter Babe Ruth that they sometimes threw him a slow ball instead of a faster one. They figured that if the ball hit the bat with a slow speed, it would rebound with a slow speed and not go as far. Was their reasoning sound?

Answer If you are right-handed, you generally use the right hand in tasks that demand control, such as writing. Swinging a bat is such a task because to hit the ball you must swing the bat almost without error. When you swing, you push on the bat with your right hand and arm while pulling on it with your left hand and arm. The left side does most of the work; the right side does most of the guiding. You can guide the bat better if the right hand is high, and you can pull it better if the left hand is low. In the conventional stance with your left side toward the pitcher, you can turn into the pitch with your controlling hand behind the bat, where it can more easily guide the bat's motion.

Even a slow ball takes less than a second to reach home plate, while a fast ball might take as little as 0.4 second. (A record speed for a fast ball, 100.9 miles per hour, was set on August 20, 1974, by Nolan Ryan, then playing for the California Angels.) You actually have less than 0.4 second to swing, because you must first size up the pitch and mentally extrapolate the ball's flight across home plate. Professional players can swing in about 0.28 second, but some of the highly skilled batters manage to swing in as little as 0.23 second. The faster swing gives a player the advantage of studying the ball's flight a bit longer before beginning the swing.

To hit the ball out of the park, your guidance of the bat must be accurate within a few millimeters. If the bat is slightly low, the ball pops up. If it is slightly high, the ball hits the ground before it goes very far. In addition, your timing of the swing must be accurate within a few milliseconds. And, to make the task even more demanding, you must do all this without seeing the ball as it nears the bat, because your visual system cannot track it during the latter part of its flight. It is a wonder that some players are so consistently successful at hitting the ball.

Experiments have shown that the speed of a batted ball improves with an increase in the weight of the bat until the weight exceeds about 35 or 40 ounces. A bat of moderate weight (about 32 ounces) is better than a heavier bat for at least three reasons. Two are obvious to most players: The moderate weight bat is easier to swing and to control than a heavy bat. Both factors are due to the smaller *rotational inertia* of the bat—that is, the distribution of mass with respect to the center (or centers) about which the bat is rotated during a swing. The third reason has to do with the energy transfer during the bat–ball collision. In general, the transfer of energy in a collision of two objects improves the closer those objects are matched in mass (or weight). Thus, in the bat–ball collision, more energy is transferred from the bat to the ball with a moderate-weight bat than a heavy bat.

So, why do some batters still prefer a heavy bat? The choice might be based on the associated length of the bat. A light bat is short, requiring that the player stand near the plate. If the ball travels through the near part of the *strike zone*, the player might have to hit it with the section of the bat near the hands. As explained below, such a collision greatly diminishes the chance of a good hit. To avoid the problem, players might choose a heavier bat because of its additional length. They then can stand farther from the plate, and the collisions occur in a better region on the bat.

Experiments reveal that a player will swing a bat with *less* speed if the player first warms up with a heavier or lighter bat or the same bat weighted with a lead doughnut on the outer end. The reason appears to be that when warming up with a bat, the player sets up a certain mental program (the procedure of using the muscles) to swing that bat. If the warm-up bat is significantly different than the one actually used in play, then the mental program will not be exactly appropriate and the bat used in play will not be swung well.

The forces you feel during the collision with the ball depend on where it hits the side of the bat. The collision usually shoves and rotates the handle, but not if the ball happens to hit the *sweet spot* that is known as the *center of percussion* (COP). If the collision is between the center of mass and the COP, the handle is shoved in the direction of the pitch. If it is outside the COP, the handle is jerked toward the pitcher.

Another sweet spot involves the oscillations a collision can set up in the bat, which can sting your hands. For most cases, two types of oscillations appear on the bat. The simplest type, called the *fundamental*, is one in which the far end of the bat oscillates the maximum amount. You probably won't notice this oscillation because of its low frequency.

The other oscillation, called the first overtone, is quite perceptible and can even hurt your hands a little. In it the free end of the bat oscillates vigorously, but there is a point, called a *node*, somewhat closer to you that does not move at all. The node also carries the nickname of sweet spot, because if the ball hits there, the first overtone is not produced and so there is no perceptible oscillation on the hands.

You can find the node on a bat by dangling it from your fingers and tapping on its side. When you strike the node, you will notice little or no oscillations. But when you strike at other points, especially nearer the center of the bat, the oscillations can be both felt and heard.

To give the ball the greatest speed, you generally should hit it at a point between the sweet spots and the center of mass, but the exact location depends on the ball's initial speed and on the ratio of the bat's mass to the ball's mass. The faster the ball or the lighter the bat, the nearer to you the ball should hit the bat.

I can imagine that when Ruth saw a slow ball coming his way, he began to grin. Hitting the ball out of the park depends primarily on the control of the bat during the swing and an accurate assessment of where the ball will travel across the plate. A slow ball gave Ruth ample study and the chance to position and time his swing.

1.24 • Legal passes in rugby

In rugby, one player can legally pass the ball to a teammate if it is not forward. If the player with the ball is running toward the opponents' goal, what direction of throw is allowed? Can a toss be toward his rear and still be illegally forward?

Answer The problem has to do with the velocity of the player. When he throws the ball toward the rear, the ball's velocity might actually be forward relative to the playing field. For example, in Fig. 1-7*a* the velocity of the ball relative to the player is toward the rear, but once it is added to his running velocity, it is angled forward (Fig. 1-7*b*).

If the umpire is running while watching the pass, he will see the velocity of the ball angled in yet another direction, because of his own velocity. Only the stationary spectators will properly see whether the ball's flight is illegally forward or not.

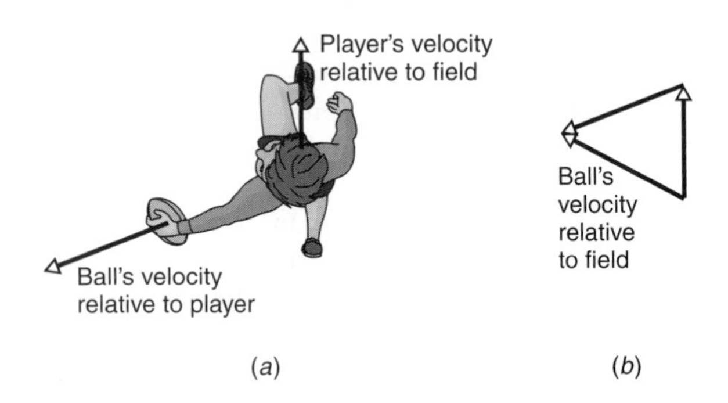

Figure 1-7 */ Item 1.24* A rugby pass back to the left may seem legal relative to the player (*a*) but actually be forward to the field (*b*).

1.25 • Juggling

The world's record for juggling rings is currently 11; records for other objects involve fewer numbers. Obviously juggling requires good eye–hand coordination and practiced tossing and catching, but is there any other factor that limits the number of objects that can be juggled?

Answer Gravity, of course, imposes a limit. If you want to add to the number of objects being juggled, you must toss the objects higher so that you have some extra time for the additional objects. However, the gain in time is always small. If you throw an object twice as high as previously, you gain only about 40% more time for its flight. Plus, you would have to throw it with 40% more speed, which means that the toss is more likely to be erratic.

1.26 • Pole vaulting

Fiberglass poles revolutionized pole vaulting in the early 1960s. Earlier in the sport, poles were bamboo. Steel and aluminum poles became popular in the 1950s. But nothing could beat out fiberglass poles, and once they were introduced, the record jump quickly rose from 4.8 meters to over 5.8 meters. Some say that eventually the record should be well above 6.0 meters. Why was the fiberglass pole so instrumental in raising the record?

Answer The fiberglass pole was much more flexible than the previous poles of bamboo, steel, or aluminum. This flexibility gives two advantages to a pole vaulter. The athlete can better convert the kinetic energy of the run toward the jump into elastic potential energy of the pole as it is bent. (That stored energy comes from the run, not a muscular effort by the athlete in bending the pole.)

Perhaps that much is obvious. More subtle is that the flexibility of the pole delays the conversion of the elastic potential energy back into the kinetic energy of the then rising athlete. This delay allows the athlete to reposition the body so that the gain of energy from the then straightening pole results in upward motion rather than forward motion.

To make a good jump, an athlete must not only run fast toward the jump to ensure that there is plentiful kinetic energy to be used but must also measure the stride so as to place the far end of the pole properly in the *box* on the ground. As the pole catches in the box, the athlete must jump forward so as to maintain the forward motion and bend the pole properly. As the pole bends, it stores some of the athlete's initial kinetic energy. During this bending and the eventual unbending, the athlete tucks the legs and leans backward so as to rotate the legs and body into a vertical orientation. To help unbend the pole so as to gain back more energy, and to help the reorientation of the body, the athlete pushes forward with the upper hand while pulling backward with the lower hand. If all is timed correctly, the unbending pole feeds back its stored energy to send the athlete upward.

1.27 • Launch of an atlatl and a toad tongue

Several ancient cultures, such as the Aztecs and tribes in the far north of North America, developed a launching mechanism in which a spear (or dart) is propelled by means of a wood stick that is rapidly brought forward until the spear flies free of the stick (Fig. 1-8). Why does the launching device, now called an *atlatl*, give a greater spear speed than if the spear is simply thrown forward? The speed was large enough that the spear could fly through about 100 meters and then rip through, say, the armor on a Spanish conquistador confronting the Aztecs. Why was a stone often attached to the launching device?

How can a toad propel its tongue outward at a surprising speed and for a surprising distance to catch a fly?

Answer In a conventional launch of a spear, you provide the spear's kinetic energy through the work your hand does in moving the spear forward through a certain distance. The launching device that ancient cultures discovered adds to the length through which the spear is propelled and so also to the energy that is given it. The advantage of attaching a stone to the launching device is not understood. Indeed, experiments indicate that the added mass results in a slightly slower launching speed of the spear.

A toad appears to snare its prey with its tongue by a mechanism similar to that of an atlatl. When it spots the prey, the toad rapidly propels its tongue toward the prey, but the soft outer portion of the tongue remains folded back on the (now stiffening) rest of the tongue. As the tongue nears the prey, the outer portion is suddenly rotated forward to plop down on the prey. By thus rotating the outer portion forward while the rest of the tongue is still moving forward, the toad adds to the kinetic energy of the outer portion. This extra energy increases the chance that the prey will stick to the outer portion even if the prey lies on a surface (such as a leaf) that yields when the prey is hit. Once the prey is stuck, the toad rapidly pulls the tongue and prey back into its mouth.

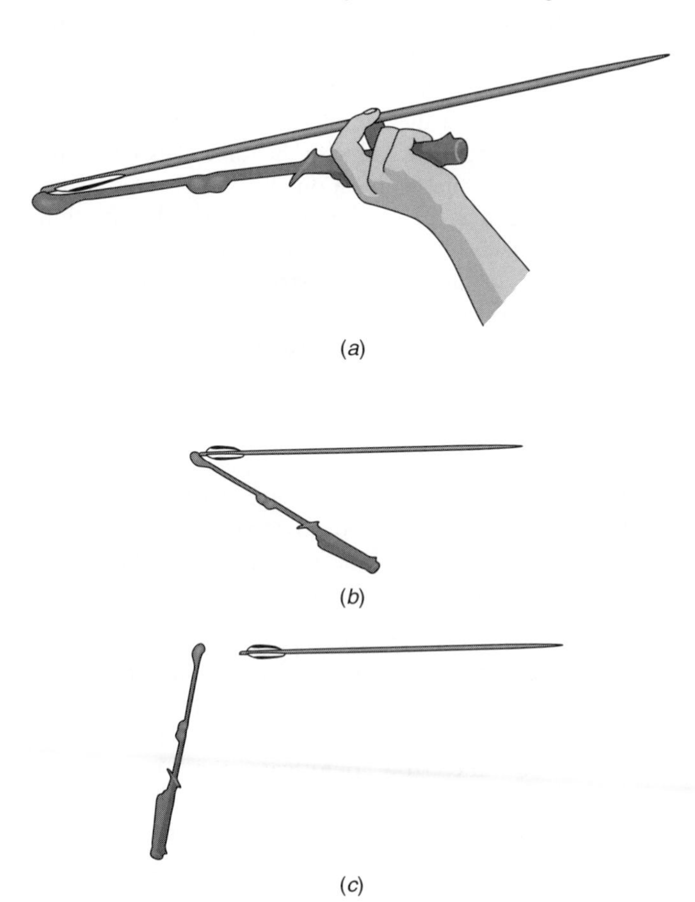

Figure 1-8 / *Item 1.27* The launch of a spear by an atlatl.

1.28 • Slings

Someone moderately skilled with a sling can hurl a stone of 25 grams at a speed of 100 kilometers per hour (about 60 miles per hour) to hit a target 200 meters or more away. How is the stone given such a large speed, or more to the point, such a large momentum? In some battles of the past the weapon proved more valuable than an arrow, for even if an enemy soldier wore leather armor, the stone's collision could inflict lethal internal damage whereas an arrow might just be deflected. When the soldier lacked any armor, the stone could easily penetrate the body. A sling was also more accurate than an arrow and could often travel farther. For this reason, slingers were often grouped behind the archers, who needed to be closer to the enemy to be effective.

The most famous battle involving a sling was, of course, the brief one between David and Goliath. For 40 days the giant Philistine had challenged the Israelis, but none had dared take up the fight until David. He chose five smooth stones from a brook and then walked into range of Goliath. David kept the situation safe because Goliath's sword was useless at such a large separation. David retrieved the first stone from his carrying pouch and slung it at the giant. The stone hit with such momentum that it burrowed into the giant's forehead.

Answer The stone, which may be a real stone or one that is made from clay or metal, is placed in a flexible pocket to which two straps are attached. The opposite ends of the straps are held in the hand—the right hand if the person is right-handed. One of the straps is fastened around several fingers, while the other one has a knot that rests against the thumb and forefinger.

The straps are made taut by the left hand as all is lifted above the person's head. There the left hand lets go and the right hand does work on the stone by pulling the pocket toward the rear, then down and toward the front. This motion is accomplished largely with the wrist rather than the whole arm. The stone is then pulled around in a vertical circle three or four times to build up its kinetic energy. Just as the stone reaches the bottom of the last circle, the knotted strap is released, loosening the stone, which then flies toward the target.

The advantage in the weapon is that work can be done on the stone for a longer distance and time than if the stone is merely thrown forward like a baseball. The radius of the circle also plays a role, because the larger it is, the greater the launch speed of the stone is, and so also the range. In times past some soldiers carried several slings with different strap lengths in order to sling stones at different ranges.

1.29 • Tomahawks

Someone skilled at burying the sharp edge of a tomahawk in a target may just be experienced, but is there any scientific basis to the skill? Knowing that basis, would you be able to hit a target on the first try?

Answer To launch a tomahawk, you hold its handle perpendicular to your forearm, pull your arm back past your head, and then rotate the forearm and tomahawk forward around the elbow, releasing the tomahawk so that its velocity is horizontal and forward. The weapon then spins about its center of mass (located in the heavy head) as it flies through the air.

Unless you are well practiced with throwing one, a tomahawk will probably have a different launch speed and spin rate every time you throw it. That seemingly means that hitting a target at a certain distance will take luck. However, a curious feature of the launch is that the ratio of the launch speed to the spin rate is independent of how rapidly you bring your forearm forward. That independence means that, regardless of your launch, the tomahawk will turn and be in a striking orientation at certain distances from you. So, to hit a target, all you must do is stand at one of those certain distances (which you would determine by observation or by calculation) and throw the tomahawk. You probably could accomplish this on the first try.

Of course, when tomahawks were really used as weapons in the early days of the United States, a warrior could not afford to adjust his distance from a target before he threw his tomahawk. Instead, he would quickly adjust the distance between his hand and the head of the weapon. That hand–head distance determines the values of the target distances at which the weapon will be in striking orientation. To make this adjustment for any target distance in a fighting situation, the handle of the tomahawk must be long; indeed, early tomahawks were made with long handles.

1.30 • Bolas

A bola consists of three heavy balls connected to a common point by identical lengths of sturdy string (Fig. 1-9*a*). To launch this native South American weapon, you hold one of the balls overhead and then rotate that hand about its wrist so as to rotate the other two balls in a horizontal path about the hand. Once you manage sufficient rotation, you cast the weapon at a target. During the weapon's flight, its rotation rate increases, and when it reaches the target, the string rapidly wraps around the target until the balls crash into the target. Why does the rotation rate of the balls increase during the flight?

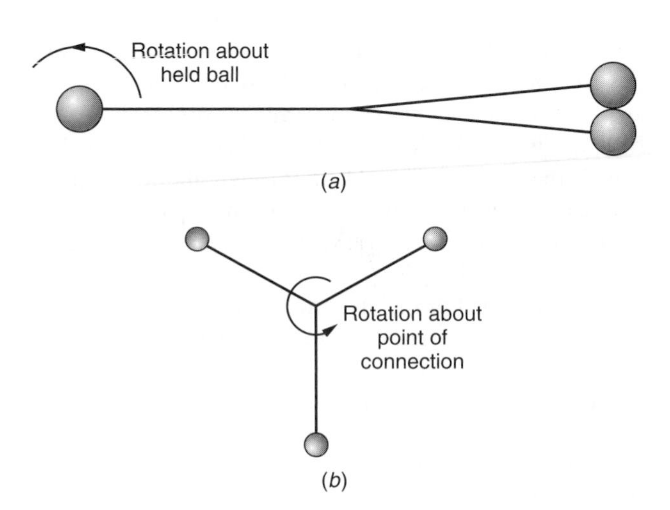

Figure 1-9 / *Item 1.30* A bola as (*a*) it is thrown and (*b*) it flies through the air.

Answer Let L be the length of the string from any one of the balls to the common point to which the balls are attached. As you spin up the bola with your hand holding one of the balls, the other two balls begin to orbit (together) about the held ball at a distance of $2L$. But once you cast the bola and it flies freely through the air, this configuration of two balls orbiting the third ball is unstable, and the bola soon begins to orbit about the common connection of the three strings, at a distance of L and with the three balls symmetrically placed about that connection (Fig. 1-9*b*). This change in configuration reduces the bola's mass distribution. Because the bola is flying freely, its angular momentum cannot change. So, with the mass distribution decreasing, the rotation rate must increase. The situation is similar to an ice skater spinning on point while bringing in the arms to reduce the mass distribution and thus increase the rotation rate.

1.31 • Siege machine

Suppose that you are in a medieval siege of a heavily fortified castle. You don't want to get too close to the castle because of the archers on the fortress walls. From a distance, how could you attack the walls?

Answer Two main types of siege machines were used to attack fortified walls: the catapult and the trebuchet. The catapult was effectively a bow that fired an arrow or a stone (perhaps 25 kilograms). The machine was much bigger than an archer's bow, the arrow could have been 2 meters long, and the string was ratcheted back so that far more energy could be stored and then transferred to the arrow during launch. Still, the arrows could do little damage against a stone wall because both the energy and momentum of the arrow were not large.

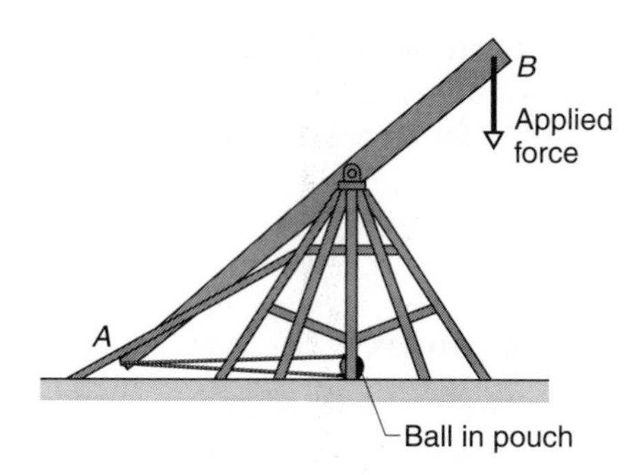

Figure 1-10 / *Item 1.31* Trebuchet.

The trebuchet was far more destructive, and some models could hurl 1300 kilogram stones. They could also hurl dead horses or even bundles of human corpses. The latter was used when an attacking army was ravished by the Black Death and they wished to send the disease into the fortress to infect the defenders. In more humorous situations, modern trebuchets have been used to hurl pianos and even small cars.

Figure 1-10 gives the essential trebuchet design. A projectile lies in a pouch that is attached to end *A* of a long wood beam. A large downward force is suddenly applied to end *B* so that the beam is rotated around an axle and the pouch is rapidly brought up and then over the machine. As the pouch and projectile pass over the machine, the pouch's tie on the beam slips off a hook and then they fly through the air. The energy given to the projectile thus comes from the work done by the force applied at end *B*.

That force could simply be the coordinated downward pull by several men. However, the trebuchets that could hurl large objects significant distances used a heavy counterweight at *B*; then the applied force was the gravitational force acting on the counterweight. The counterweight was first gradually lifted by men using a ratchet. Then the counterweight was allowed to fall so that some of the gravitational potential energy stored in it by the men could be transferred to the kinetic energy of the projectile. The kinetic energy and momentum of the projectile were very large and if the projectile was stone, it could knock a hole into a fortress wall. Once trebuchet use became widespread, castle walls were redesigned so that they could better withstand the impacts. For example, some walls were slanted instead of vertical so that the projectile might move somewhat along the wall instead of directly into it.

1.32 • Human cannonball

The circus stunt in which a person is propelled into the air from a cannon or some other contraption began in the early 1870s when a human cannonball was sent up only a short distance and was caught by an assistant on a trapeze bar. When the Zacchini family revived the stunt in 1922, they decided to make more daring flights by having the performer fly through the air and land in a net. Their first cannons depended on springs to propel the performer, but by 1927 compressed air was put to work.

Striving to increase the excitement of the stunt, the family began to send the performer over Ferris wheels. They started with one Ferris wheel, but by 1939 or 1940 they reached the limit of even unreasonable safety when Emanuel Zacchini soared over three Ferris wheels and through a horizontal distance of 70 meters.

The human cannonball act is probably one of the most impressive acts of projectile motion, for it obviously involves the chance that the performer might miss the net. Are more subtle dangers involved?

Answer To get ready for a shot, the performer would slip his or her legs down inside "metal trousers" on the piston inside the barrel of the cannon. The trousers were fitted closely to the shape of the legs and were needed to supply support when the piston was suddenly shoved upward. The subtle danger involved that shove, because the acceleration required for a long flight was so severe that the performer would momentarily black out. Part of a performer's training was to regain awareness during the flight so that a controlled roll could be made on the net. If the landing were uncontrolled, then the collision and rebound on the net could easily break the limbs or neck of the performer. The family claimed that the muzzle speed of a performer was as much as 600 kilometers per hour, but a speed of less than 160 kilometers per hour seems more credible.

Another subtle danger lay in the air drag encountered by a performer. The size of the air drag depended on the orientation of the body as it flew through the air: It was smaller if the body was oriented along the direction of travel, and larger if the body was oriented perpendicular to that direction (which might happen during the descent). A smaller air drag increased the range of the shot; a larger air drag reduced the range. Because the performer's orientation varied from shot to shot, someone had to calculate (or guess) approximately how far the performer would go and then make the net wide enough to account for the possible variations due to the air drag.

1.33 • Basketball shots

Basketball is, of course, a game of both skill and chance. Is there some best way in which to throw the ball to increase the probability of making a basket? For example, is it better to toss the ball in a high arc or to throw it along a flatter trajectory? When might spin be beneficial, and when is it undesirable?

In a *free throw* (where a player gets an uncontested shot at the basket from about 4.3 meters), a player might employ the *overhand push shot*, in which the ball is pushed away from about shoulder height and then released. Instead, the player might use an *underhand loop shot*, in which the ball is brought upward from about the belt-line level and released. The first technique is the overwhelming choice among professional players, but the legendary Rick Barry set the record for free-throw shooting with the underhand technique. Does one technique actually provide a better chance at making a shot?

Answer From any position on the court, there is a wide range of angles at which you can launch the ball to send it through the basket, provided that you give the ball the proper speed. However, the fact that the ball is smaller in diameter than the basket allows a certain margin of error in the launch speed. If you choose a low angle, the margin for error is small and you must be quite accurate. You also must give the ball a large speed, which requires more force from you and which works against accuracy. If, instead, you choose an intermediate angle, the margin for error in the speed is larger, and the speed and force are smaller. So, you have a better chance at making the shot. For even larger angles, the margin for error is approximately the same, but the required speed and necessary force are larger, which makes larger angles less desirable.

Novice players usually shoot the ball along too flat a trajectory, but seasoned players learn through practice to arc the ball into the basket. The higher the shot is released, the slower the required launch must be, which gives an advantage to a tall player. The height advantage is so strong that some players elect to release the ball during a jump even when unchallenged by an opponent. If you put backspin on the ball and happen to hit the backboard instead of the basket, the spin creates friction that may cause the ball to rebound into the basket. When the shot is taken from one side, sidespin on the ball may also help.

The underhand free throw has a greater chance of success than the overhand throw, but the reasons are still debated. The success might be because the underhand throw is easier to execute, but a greater advantage seems to lie in the fact that the throw allows a player to put more backspin on the ball, which can make up for an errant toss onto the backboard.

SHORT STORY

1.34 • Records in free throws

In 1977, Ted St. Martin set the world's record for consecutive baskets while standing at the free-throw line—he made the shot 2036 times. The next year Fred L. Newman made a stranger record. While blindfolded, he made 88 consecutive baskets from the line. During a 24 hour period several years later, and with his eyes open, Newman managed to score 12 874 baskets out of 13 116 attempts.

1.35 • Hang time in basketball and ballet

Some skilled basketball players seem to hang in midair during a jump at the basket, allowing them more time to shift the ball from hand to hand and then into the basket. Similarly, some skilled ballet performers seem to float across the stage during the leap known as a *grand jeté*. Obviously no one can turn off gravitation during a jump or a leap, so what accounts for these two examples of apparent hanging in midair?

Answer The hang-in-midair of both basketball player and ballet performer is an illusion. In basketball the illusion is primarily due to a player's agility to perform so many maneuvers during the jump. In ballet's grand jeté, the illusion comes from a shift of the performer's arms and legs during the leap: She raises her arms and stretches her legs out horizontally as soon as her feet leave the stage. These actions shift her center of mass upward through her body (Fig. 1-11). Although the center of mass faithfully follows a

Figure 1-11 / Item 1.35 Path of the center of mass during a grand jeté.

(curved) parabolic path across the stage as required by gravitation, its movement relative to the body decreases the height that would have been reached by the head and torso in a normal leap. The result is that the head and torso follow a nearly horizontal path during the middle of the leap. This path seems strange to the audience who, out of normal experience, expect a parabolic path even if they do not even know the term.

A basketball player can similarly flatten the path taken by the head during a jump across the floor if the player pulls up the legs and raises up the arms and ball. However, I don't think that this technique is commonly planned by players. Although a player raises the arms and ball toward a basket during a jump near the basket, a player rarely lifts the legs, and the resulting slight flattening of the path taken by the head hardly seems to fool a defensive player who jumps alongside the shooting player.

1.36 • Golfing

How should you swing a golf club to best hit a golf ball during a drive? For example, should you swing down as hard as you can, somewhat like striking an assailant with a club in a fight? If, instead, you should increase or decrease your effort sometime during the swing, will the flexibility of the shaft of the golf club affect when you make that change?

Why is hitting a 1 meter putt considerably harder than a half meter putt? Is a 3.5 meter putt considerably harder to hit than a 3.0 meter putt? Why can the ball be rolling directly toward the cup and yet still not go in?

Answer When you swing the golf club down during a drive, you begin the swing with your wrists cocked so that the club is at an angle of about 90° with your arms. If you swing the club as in a fight, you will automatically allow the wrists to uncock during the swing. The club head will actually have a greater speed when it hits the ball if you resist that uncocking by reducing the torque you apply to the club somewhere during the swing. Just when this uncocking should be done is learned through experience. Once the wrists are uncocked, the club swings around the wrists as they swing around the shoulders, resulting in the increased speed of the club head.

Many players believe that the flexibility of the club shaft affects the flight of the ball because it determines the angle at which the club's head meets the ball. The argument has been that a more flexible shaft first bends backward during the swing and then springs forward more just before impact with the ball than does a stiffer shaft and thus delivers more energy to the ball. However, studies show that the club's flexibility has little effect on the ball's flight—indeed, greater flexibility might result in a decrease in the energy transferred to the ball because the impact sets the club oscillating. Thus, a stiff club is more desirable because it gives greater control in hitting the ball squarely.

One measure of the difficulty of a putt is the angle occupied by the cup in the ball's point of view. If you move the ball away from the cup, the angle initially decreases rapidly, which means that the difficulty of making the shot increases rapidly. However, beyond a distance of about a meter, the angle begins to decrease rather slowly, which means that the associated difficulty begins to increase rather slowly. Of course, this simple analysis overlooks other difficulties with a long putt, such as the increased number of variations in the grass texture and in the slope of the ground along the path to the cup.

If a ball is rolling directly toward the cup, it will not score if its speed is above some critical value when it leaves the near side of the cup's rim. Such a ball travels across the mouth of the cup, falling during its passage, but the fall is insufficient to keep the ball from rolling out of the cup once it hits the wall on the far side.

SHORT STORY

1.37 • Curtain of death of a meteor strike

Whenever a metallic asteroid reaches the ground (instead of burning up in the atmosphere), it digs a crater by throwing rock into the air. However, the *ejecta material*, as it is called, does not come out haphazardly. Rather, the faster moving rocks tend to be ejected at steeper angles to the ground. If you were to witness this ejecta fly toward you, you would see that at any instant it forms a thin, curved curtain (Fig. 1-12): Particles higher in the curtain are ejected at greater speeds and angles than the particles lower in the curtain. The slower rocks hit the ground earlier than the higher rocks; thus you see and hear a steady pounding of the ground as the curtain moves toward you.

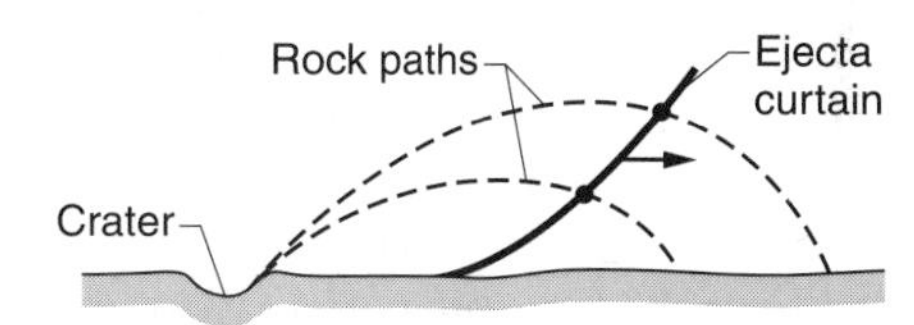

***Figure 1-12** / Item 1.37* Rocks ejected from a meteor strike.

1.38 • The high jump and the long jump

A novice in the sport of high jumping might be tempted to hurdle over the bar by throwing one leg up over it and then dragging the other leg over, while bent forward at the waist. A more successful jump is made with the *straddle*, in which the person essentially rolls over the bar face down and with the length of the body parallel to the bar.

When Dick Fosbury won the high-jump contest in the 1968 Olympics in Mexico City, he introduced what appeared to be a bizarre way to jump. The technique is now known as the *Fosbury flop* and is used almost universally by high jumpers. To flop, a competitor runs with a measured pace up to the bar and then twists at the last moment, going over the bar backwards and face up. What advantage does such a style have? Why is the approach to the bar at a measured pace? Surely a faster pace would give the athlete more energy to jump higher.

One of the most stunning events in the history of track and field sports also occurred at the Mexico City Olympics. In mid-afternoon on October 18, Bob Beamon prepared for the first of three allowed attempts at the long jump by measuring off his steps along the approach path. Then he turned, ran back along the path, hit the takeoff board, and soared through the air. The jump was so long that the optical sighting equipment for measuring the jumps could not handle it, and a measuring tape had to be brought out. One judge said to Beamon, who then sat dazed off to one side, "Fantastic, fantastic." The jump was an astounding 8.90 meters, easily beating the previous record of 8.10 meters (a difference of nearly two feet!).

Beamon was certainly aided somewhat by the wind at his back, because it was just at its allowed upper limit of 2.0 meters per second. Did he also benefit from the high altitude and low latitude of Mexico City; that is, did matters of air density and the strength of gravity account for his astonishing jump?

The length of a long jump is measured to where the jumper's heels dig out sand upon landing, unless the jumper's buttocks then land on and erase the heel marks. If those marks are erased, the length of the jump is only to the near edge of the hole left in the sand by the buttocks. Thus, landing in the proper orientation is important in the long jump.

When a long jumper takes off, with the final footfall on a takeoff board, the torso is approximately vertical, the launching leg is behind the torso, and the other leg is extended forward. When the long jumper lands, the legs should be together and extended forward at an angle so that the heels will mark the sand at the greatest distance but still disallow the buttocks from erasing that mark. How does the jumper manage to go from the launching orientation to the landing orientation during the flight?

In the standing long jump in the ancient Olympiad, why would some of the athletes jump with handheld objects called *halteres* that were several kilograms in mass?

Answer The height that is recorded in high jumping is, of course, the height of the bar, not the maximum height of the head or some other part of the jumper. Suppose that during a jump, the athlete can raise the center of mass (com) to a height L. If the athlete hurdles over the bar, the bar must be considerably lower than L if the body is not to touch it, and so the height of the jump is not very much (Fig. 1-13*a*). In a straddle jump, the body is laid out horizontally and can pass over the bar with the bar much closer to the center of mass, and so the bar can be higher (Fig. 1-13*b*). In a flop, the curvature of the body around the bar lowers the center of mass to a point below the body, and the athlete can pass over an even higher bar than with a straddle jump (Fig. 1-13*c*). The last-moment twist and backward leap in a flop also gives a stronger launch.

The approach to the jump is slow compared to, say, a sprint, because the key to winning is a flawless execution, and so timing is essential. At the end of the approach, the athlete plants the launching foot well ahead of the body's center of mass, and then, as the launching leg flexes, the body is twisted around that foot. This procedure allows some of the kinetic energy of the run to be stored in the flexing leg. As the leg then pushes against the ground, it propels the athlete upward, transferring some of the stored energy, and also additional energy gained from muscular effort, into the flight of the athlete.

Beamon's long jump was aided only slightly by the wind and the location. Mexico City is at an altitude of 2300 meters, which is considerably higher than the altitudes of many other locations for the Olympics. The high altitude meant that the air density was low, and so the air drag retarding the jump was smaller than if the jump at been at a lower altitude. The high altitude also meant that the gravitational acceleration was smaller, and so the gravitational pull that opposed Beamon's launch and that eventually pulled him back to the ground was smaller. The acceleration and pull were further reduced because of the *effective* centrifugal force on Beamon due to Earth's rotation. That effective force is larger at lower latitudes, because such places travel faster during the rotation.

However, all of these factors played only a small role in the jump. So, why then did Beamon travel so far? The primary reason is that he hit the launch board while running rapidly. Most long jumpers approach more slowly so as to avoid placing their last step just past the board, which would disqualify the jump. They also want to avoid taking off before the board and losing the solid support it gives during the launch while also losing distance in the jump since the jump is measured from the board. Because the board is only 20 centimeters long, the final step must be planned.

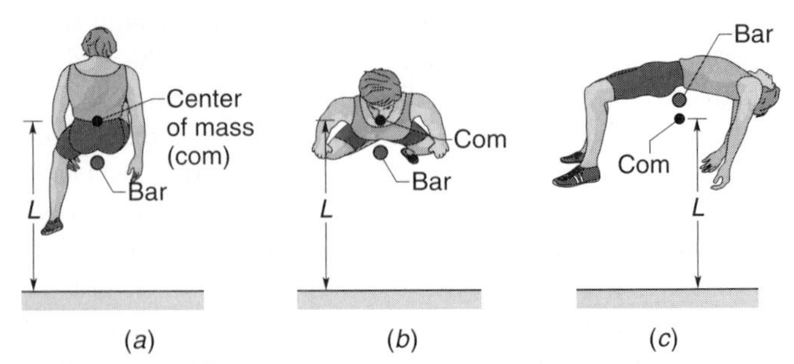

Figure 1-13 */ Item 1.38* The (*a*) hurdle, (*b*) straddle, and (*c*) flop styles of high jumping.

Beamon, who was known for disqualified jumps, apparently decided to gamble on his first try and sprinted to the board. His last step barely avoided extending beyond the board. Had he gone beyond the board, he presumably would have made his next two jumps with more concern about the board and less speed.

No one jumped as far as Beamon, including Beamon himself, for the next 23 years. Then, finally, at the 1991 World Track and Field Championship, Mike Powell jumped 8.95 meters—2.0 inches farther than Beamon. He did it in Tokyo and thus without any benefit of higher altitude, and he did it with only a mild wind of 0.3 meter per second at his back. Powell stunningly demonstrated that the effects of altitude and wind are secondary to athletic ability.

To consider the reorientation of a long jumper during flight, suppose that the jump is to the right in your perspective. During the launch from the board, the force on the launching foot from the board produces a clockwise rotation of the body, which tends to bring the trunk of the body forward and the forward leg rearward. This tendency of clockwise rotation is increased as the trailing leg is brought forward to ready for the landing. The reason is that the jumper is then free of the ground, and so the angular momentum of the body must remain constant. So, when the trailing leg is rotated counterclockwise to be forward, the rest of the body tends to rotate clockwise.

To decrease the clockwise rotation, so that the jumper is in the proper orientation for landing, the arms are rapidly swung clockwise about the shoulders. In addition, the legs might continue to move as in running, with a leg outstretched when rotated clockwise to the rear and pulled in when rotated counterclockwise to the front. (None of this motion alters how far the jumper goes; it only alters the orientation of the body.) Novice jumpers often fail to swing the arms sufficiently or, worse, they swing one or both arms in the wrong direction. The trunk and legs are then not in the best orientation, and the jump is short because the heel marks are short or the buttocks erase the heel marks.

The halteres used by jumpers in the ancient Olympiads could increase the length of the jump. An athlete would swing the handheld objects forward and backward in preparation for a jump, then swing them forward during the first part of the jump, and finally swing them backward in preparation for the landing. Properly used, this technique could add 10 or 20 centimeters to the length of the jump for two reasons. (1) As the center of mass of the athlete–halteres system moved through the air, the last backward swing shifted the halteres backward relative to the center of mass and thus shifted the athlete forward relative to the center of mass. (2) During the launch, the forward swing of the halteres increased the downward force on the launch point, thereby giving a greater launch force on the athlete. (In effect, the athlete was using shoulder and arm muscles in addition to the leg muscles during the launch.) A jump could have been increased a bit more if the athlete would have hurled the halteres backwards during the last part of the flight, effectively rocketing the body forward. The center of mass of the athlete–halteres system still lands at the same point, but the athlete is now a bit forward of that point.

1.39 • Jumping beans

If a young girl who is sitting on a blanket gathers up the blanket's four corners and then pulls up very hard on them, can she lift herself? Well, of course not, although I know of one girl who tried with all her might. How, then, do jumping beans manage to jump up into the air?

Answer A bean contains a small worm that first pushes off from the bottom of the bean and then collides with the top, propelling the bean upward. The external force (the force outside the worm–bean system) that is responsible for the motion is the upward force on the worm as it initiates the jump.

1.40 • Somersault of a click beetle, attack of a mantis shrimp

If you poke a click beetle when it is lying on its back, it throws itself up into the air as high as 25 centimeters, with a noticeable click. During the leap it may turn over so that it lands right-side up. The launch involves an acceleration that is as large as 400 *g*s (that is, 400 times the acceleration of gravity) and requires a power that may be 100 times the possible power of any one muscle in the beetle. How does the beetle produce such enormous power, which, of course, cannot be due to its legs because it begins on its back? One clue is the click and another is the fact that the beetle cannot immediately repeat the performance.

The peacock mantis shrimp (*Odontodactylus scyllarus*) attacks its prey by quickly rotating a feeding appendage toward it. The appendage does not strike the prey but instead produces air bubbles that produce a destructive sound wave when they suddenly collapse. The acceleration at the outer tip of the appendage can be as much as 10 000 *g*s. How can a shrimp achieve such high acceleration?

Answer The beetle's leap is somewhat like a mousetrap that is triggered—both jackknife upward. In the beetle a muscle in the front of the body slowly contracts and moves a peg-like section over the *mesosternum* until a notch in the peg (the peghold) catches on the *lip* of the mesosternum, arching the beetle (Fig. 1-14*a*). After tension builds in the muscle, the peg suddenly slips over the lip and slides

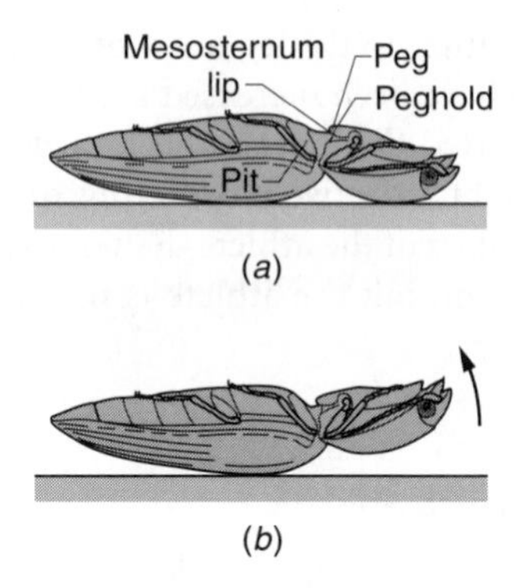

***Figure 1-14** / Item 1.40* (*a*) Click beetle on its back, with peghold caught and muscles under tension. (*b*) Peg has slipped over the catch and beetle jackknifes upward.

down into a pit. The abrupt slip forces the front end of the beetle to jackknife upward and causes the hind end to push down on the ground (Fig. 1-14*b*). The push hurls the beetle upward, and the rotation initiated by the slip of the peg allows the beetle to spin around its center of mass while in the air. It may spin around enough so that it lands on its feet. The click emitted by the beetle is produced either by the slip of the peghold over the lip or the abrupt stop of the peg after it enters the pit.

The initial slow contraction of the muscle allows the beetle to store energy. The sudden release of that energy is responsible for the high power of the jump. Before the jump can be repeated, energy must again be stored, which takes some time. This type of energy storage and sudden release is used by many animals for abrupt motion, either to catch lunch or to avoid becoming lunch.

A similar process is used by mantis shrimp. The appendage used in an attack is held tightly against the body while a saddle-shaped element is slowly put under tension like a spring can be compressed. The appendage is held in place by a latch. Once the saddle-shaped element is under maximum tension, the latch is released and the element drives the rapid rotation of the appendage.

SHORT STORY

1.41 • Some record lifts

In the sport of weight lifting, records are frequently broken, but the record for the greatest lift of any kind was firmly set in 1957 by Paul Anderson. He employed a *back lift* in which he stooped beneath a reinforced wood platform that was supported by sturdy trestles. In front of him was a short stool against which he could both steady himself and push downward. On the platform were auto parts and a safe filled with lead. With an astonishing effort of both arms and legs, he lifted the platform—the composite weight was 6270 pounds (27 900 newtons)!

Perhaps equally impressive was a reported lift by Mrs. Maxwell Rogers of Tampa, Florida, in April 1960. Discovering that a car had fallen off a bumper jack and onto her son who was working underneath the car, she lifted one end of the car so that her son could be rescued by a neighbor. The car weighed 3600 pounds (16 000 newtons), of which she presumably lifted at least 25%. She suffered several cracked vertebrae. (Accounts of this sort appear occasionally in newspapers. In a panic, an untrained person can manage to lift something that has a weight greatly exceeding the person's body weight and that could not possibly be lifted under calmer circumstances.)

1.42 • Chain collisions

If one ball runs into a stationary ball, under what conditions does the second ball receive the most energy? Are the same conditions needed if the second ball is to receive the greatest speed? What are the answers if a ball runs into a chain of initially stationary balls?

Suppose that initially there is a large ball that is moving and a smaller ball that is stationary. Can you increase the energy that is given to the smaller ball by inserting additional balls between the two? If so, what should the masses of the intermediate balls be?

A golf ball is flying toward your head. If you wish to decrease the energy that will be transferred to your head, should you guard your head with a hand so that the hand is knocked into your head?

A popular toy consists of a row of adjacent balls that can each swing as a pendulum (Fig. 1-15*a*). The balls are elastic; that is, only a little energy is wasted when they collide with other objects. You draw back an end ball and then release it

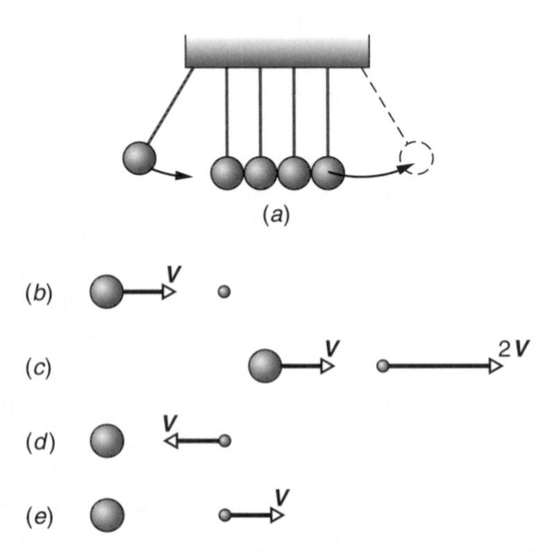

***Figure 1-15** / Item 1.42* (*a*) The first ball is released; the last ball is knocked aside. (*b*) Before and (*c*) after a collision of a very large ball with a very small ball. (*d*) Before and (*e*) after the collision from the perspective of the large ball.

so that it crashes into the next ball. Why does only the ball at the opposite end of the row move?

Rehang the balls so that there is a small space between them, and then send the first ball into the second at a slight angle to the row. Although the initial collisions are skewed, the misalignment gradually disappears as the collisions proceed. However, if you increase the space between the balls enough and repeat the demonstration, the misalignment increases with each collision. The collisions may even stop if one ball is knocked to the side so much that it fails to hit the next ball. Why does the alignment–misalignment behavior depend on the spacing between the balls?

Answer The second ball receives the greatest energy when its mass matches that of the first ball. If the balls are highly elastic, almost the full energy is transferred, in which case the final speed of the second ball almost equals the initial speed of the first ball, and the first ball stops.

The second ball receives the greatest speed when its mass is much less than the mass of the first ball. Let V represent the speed of the first ball (Fig. 1-15*b*). If the mass ratio is very large and the collision is very elastic, the second ball may receive a speed that is $2V$ (Fig. 1-15*c*). That may seem incorrect, but for a moment take the perspective of the first ball, as if you were that ball. The second ball seemingly approaches you with a speed of V (Fig. 1-15*d*), bounces elastically, and then heads away from you with a speed of V (Fig. 1-15*e*). Now go back to your original perspective. The second ball moves away from the first ball with a relative speed of V. What is the first ball doing? Since the second ball has such little mass, the collision does not appreciably alter the speed of the first ball and it is (approximately) still V. So, the speed of the second ball must be $V + V$, or $2V$. If there is a chain of such collisions, then the speed imparted by each collision is (approximately) double that imparted by the preceding collision.

When the end balls are already chosen and you want to improve the transfer of energy to the smaller ball, insert intermediate balls such that each one has a mass that is the geometric mean of the masses on the opposite sides of it. (The geometric mean of the masses is the square root of the product of the two masses.) Other choices of intermediate mass also improve the transfer of energy but not as much.

This conclusion figures into the question about the golf ball. If you guard your head with a hand, you may actually increase the transfer of energy to your head, because your hand has a mass that is intermediate to the masses of the ball and your head. Still, inserting a hand is wise because it is broad and will spread the force you will receive on your head.

The toy with a series of pendulum-like balls is usually explained in terms of the momentum and kinetic energy of the initially moving ball. The only way for those quantities to go unchanged as they are relayed through the series is for the final ball to end up with all of the momentum and kinetic energy. So, in the end, it alone moves. However, the explanation is misleadingly simple, because the actual behavior of the intermediate balls can be quite complex if they are initially touching.

In the demonstration where the first ball hits the second one at a skewed angle, the ratio of the separation D between the balls to their radius R is important. If D/R is smaller than 4, the misalignment decreases during the collisions because the collisions gradually shift inward and become more direct. If the ratio is greater than 4, the misalignment increases because the collisions shift gradually outward on the curved surfaces of the balls.

1.43 • Dropping a stack of balls

Hold a baseball just above a basketball with a slight separation and then drop the balls from about waist height (Fig. 1-16*a*). Although neither ball bounces particularly well on its own, the combination of the two gives a surprising result: The basketball goes almost dead on the floor and the

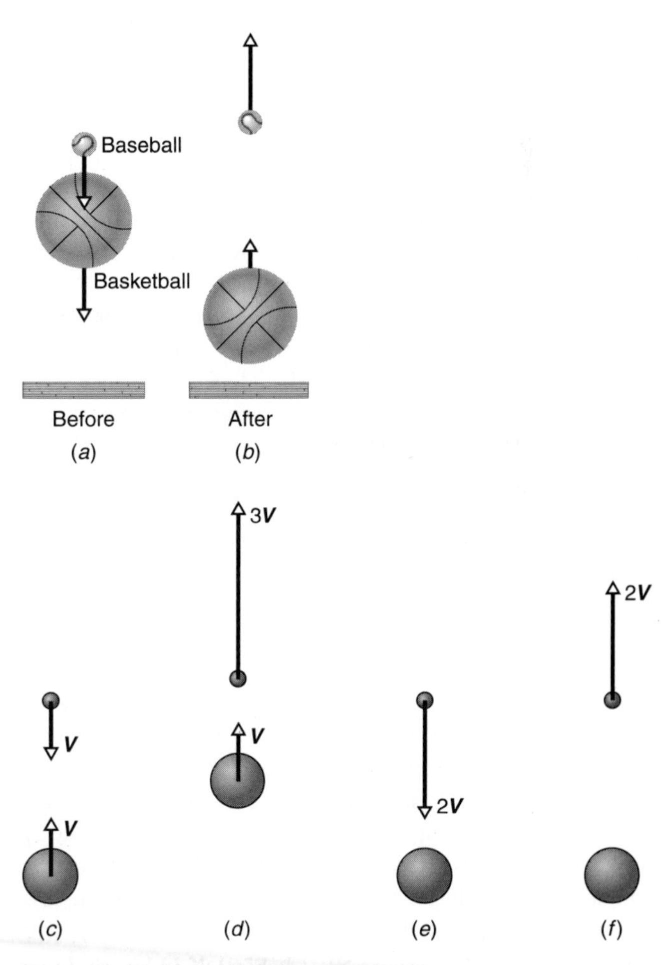

Figure 1-16 / *Item 1.43* (*a*) Before and (*b*) after a baseball and basketball are dropped together on a hard floor. (*c*) Before and (*d*) after a collision of a very large ball with a very small ball. (*e*) Before and (*f*) after the collision from the perspective of the large ball.

baseball may bounce to the ceiling (Fig. 1-16*b*). The height reached by the baseball is greater than the sum of the heights that the balls would bounce individually. (Be careful. If the alignment is off, the baseball shoots out sideways with such speed that it can injure you or someone nearby.) If you repeat the demonstration but add a small elastic ball to the top of the stack, the third ball takes off like a rocket and can go even higher than the baseball did, although it receives less energy.

In theory and if the balls are chosen appropriately, the top ball in a dropped stack of two balls can reach a height that is nine times the height from which the stack is dropped. With three balls, again appropriately chosen and under ideal conditions, the top ball can reach a height that is 49 times the drop height.

You might like to experiment with a variety of different balls, such as a Ping-Pong ball, a "Super Ball" (a highly elastic ball, under trademark by Wham-O Inc.), or a tennis ball. How should the balls in a stack be chosen to launch the top ball to a great height, and why does it go so far?

Answer When a stack of two balls is dropped, the bottom ball rebounds from the floor and then collides with the still falling second ball. The collision transfers energy to the top ball and gives it an upward velocity. Suppose that you want to maximize the transfer of energy so that the bottom ball goes dead. If the balls are elastic, then the best transfer of energy occurs when the bottom ball is about three or four times as massive as the top ball, as is the case with a basketball and a baseball.

If, instead, you want the top ball to go as high as possible, you should choose one that is much lighter than the bottom ball. The height reached by the top ball depends on the square of the velocity it receives from the collision. If the mass of the top ball is much smaller than the mass of the bottom ball, the top ball gets a large velocity-squared and can reach a height that is nine times the drop height.

To see the result, first examine the speeds of the balls just before their collision. The top ball falls with a speed V while the bottom ball heads upward with the same speed V (Fig. 1-16*c*). If the collision is very elastic, the second ball may receive a speed that is $3V$ (Fig. 1-16*d*). That may seem wrong, but for a moment take the perspective of the first ball, as if you were that ball. The second ball seemingly approaches you with a speed of $2V$ (Fig. 1-16*e*), bounces elastically, and then heads away from you with a speed of $2V$ (Fig. 1-16*f*). Now go back to your original perspective. The second ball moves away from the first ball with a relative speed of $2V$. What is the first ball doing? Since the second ball has such little mass, the collision does not appreciably alter the speed of the first ball and it is (approximately) still V. So, the speed of the second ball must be $V + 2V$, or $3V$.

If you play with a larger stack of balls, you need to arrange for the masses of the balls to decrease upward in the stack. When the bottom ball rebounds, it hits the second ball and transfers some of its energy. Once the second ball is redirected upward, it runs into the descending third ball, and transfers some of its energy. The third ball then reverses direction and runs into the fourth ball, and so on. If the stack were large enough, you could, theoretically, launch the top ball into orbit.

SHORT STORY

1.44 • A crashing demonstration

When he was a student in the 1970s, John McBryde of Houston and two other students experimented with the physics of dropped balls by releasing a softball and basketball from a third story walkway that linked two dorms. Repeatedly, the basketball went dead on the ground and the softball was hurled up well over their heads, at least 10 meters above the ground. The stunt was great fun until on the last drop the alignment of the balls was skewed, and the softball shot through the window of the resident assistant, hurling glass everywhere within the room. The cost of the repair was $250, but the penalty may have been considerably higher had the resident assistant been in his room at the time.

1.45 • Karate

Consider a forward punch in which the closed fist begins palm up near the belt and is then thrust forward and turned palm down. Why is this procedure taught with two precautions: Go out to a full arm's length but no farther (you don't lean forward), and make contact with your opponent when the fist has traveled about 90% of the way out (so you aim the punch about 10% of the way into the opponent's body)? Why are the hips and torso swiveled during the early stage of the punch?

Why are a punch, chop, kick, and other maneuvers usually made with a small area of contact? How fast can an expert move fist or foot, and how much force and energy can be delivered? When a karate expert breaks the bone in an opponent, why isn't bone in the expert also broken? When a stack of objects such as wood slabs is broken, why are the objects set up with separators such as pencils?

I never broke boards in karate class, but when I began teaching I thought board breaking would make a vivid demonstration of the forces involved in a collision. So, one day as I raced off for lecture, I hastily grabbed two pine boards that I found in the lab. In class I chose a burly student to hold the boards vertically so that I might punch into them, striking with the first two knuckles on my right fist. Unfortunately, the student flinched when I struck the boards, and they did not break. I struck again and again but with no better luck. After partially covering the front board with blood and swelling my first two knuckles by several millimeters, I quit and shuffled from the classroom. These days I use a "patio brick" that rests on rigid supports at each end, and I strike the block with the bottom of my clenched fist. Why is the new strategy more successful than the previous one?

Answer You should not lean forward into the punch for at least two reasons. You want to be stable so that you can immediately deliver another strike, and you want proper stance so that the force you experience does not break one of your bones. Karate experts can throw a barrage of strikes that are so fast that you cannot see them clearly. Ron McNair, one of the astronauts killed in the explosion of the *Challenger* space shuttle, was such an expert. He could deliver a multitude of strikes with hands, feet, knees, and elbows so rapidly that he appeared to be a fluid flowing around his opponent.

When you fight in karate, you want to make contact with your opponent when your fist is traveling its fastest, because it then has the greatest momentum and you will deliver the greatest force and energy. That optimum point is when the fist is about 90% of its way out, and so you mentally space the punch as if the fist will come to a full arm's extension about 10% into the opponent's body. If you make contact too early or late, the force and energy of the collision are less.

You should strike with a small part of your body so that the force per unit area on your opponent is largest and you transfer energy to only a small section of the opponent's body. The strike might then bend and break bone in the opponent. The technique also serves to protect you. When you strike properly, such as with the first two knuckles, the side of an open and rigid hand, or the edge of a foot, and also orient yourself correctly, the force in the collision does not break any of your bones.

The fact that bending is important in breaking an object is demonstrated when a karate expert strikes a board or concrete block that spans two supports. Each support is positioned at one end of the object so that when the strike is delivered to the center of the span, the force creates a large torque around each support point. The torques rotate the left and right halves of the object around the support points, and the object bends downward. If it bends enough, a crack begins at the lower surface and races upward, and the object fully breaks.

When a stack of separated objects is broken, the karate expert breaks the first object, and its pieces then break the second object, and so on. The breaking travels through the stack faster than does the expert. Dry white pine boards and concrete patio blocks are typical props for such demonstrations. The pine is cut and mounted with the grain running across the short width; such a board is weaker to a strike than if the grain runs lengthwise. The patio blocks are usually dried in an oven beforehand so that internal water is eliminated, because the water can add to the block's strength.

Collisions with board or block typically last for 0.005 second. The speed of a fist in a forward punch can be up to 10 meters per second. Kicks and downward strikes can be even faster. A strike with the fist can deliver a force up to 4000 newtons when a typical board breaks. The force is larger when the board does not break because the hand then does not penetrate through the board with some residual momentum. Instead the hand must stop or even ricochet, either of which requires that the force in the collision be greater than with the board breaking.

When my student flinched, he allowed the boards to move toward him. The action increased the duration of the collision, and because my force in the collision depended inversely on that duration, he decreased my force and it was then insufficient to break the boards. The demonstration with a brick is more dramatic but is also more trustworthy because the brick is rigidly mounted and the duration of the collision is short. It is also safer because the fleshy bottom of the fist hits instead of the knuckles, which are rather vulnerable, as anyone who has hit an opponent in the chin with bare knuckles can attest.

1.46 • Boxing

Why, exactly, do boxing gloves make boxing safer? In spite of the measure, why does the sport still lead to long-term brain damage and an occasional death?

Answer In earlier days when men fought with bare knuckles, injuries and deaths were more likely. A glove serves to spread the force out over a larger area, making injury to both fighters less probable. The glove also softens the blow because its material must be compressed during the impact. That action increases the duration of the collision and thus decreases the force in the collision. Still, especially in heavyweight boxing, the force delivered by a powerful fighter can be severe, even lethal.

A skilled fighter knows how to *roll with a punch* directed to his head; that is, he moves his head backwards. Were he to keep his head stationary or, worse, move into the punch, the force in the collision would be greater. The most dangerous time in a fight comes during the later rounds when both men are tired and unable to anticipate a punch and to respond to it by moving backward.

The most dangerous punch is one delivered to the chin or forehead, especially when the punch is skewed, because it rotates the head backward, compressing the brain stem and shearing the brain (attempting to make part of the brain slide past another part). Even if a fighter is not knocked out by a punch, the brain inevitably undergoes damage from a punch because the skull crashes into it to initiate the backward motion. The collision disrupts the blood flow in the area of the collision and abrades the surface of the brain. Shearing from the backward rotation damages the interior of the brain. Additional damage occurs on the side of the brain opposite the punch, because when the skull begins to move backward, pulling away from the brain, the fluid pressure in the space separating the skull and brain decreases, causing capillaries to rupture.

With repeated damage, the fighter's ability to think, remember, and speak diminishes and he is then irreversibly *punch-drunk*. The sport may be an adult's game, but the game reduces the capabilities of a participant to those of an infant.

1.47 • Skywalk collapse

July 17, 1981, Kansas City: The newly opened Hyatt Regency was packed with people listening and dancing to a band playing favorites from the 1940s. Many of the people were crowded onto the walkways that hung like bridges across the wide atrium. Suddenly two of the walkways collapsed, falling onto the merrymakers on the main floor, killing 114 people and injuring almost 200 others.

What caused the collapse? Certainly the weight of the throng on the walkways was a factor, but was there a structural fault in the design of the walkways? After several days of investigation, a Kansas City newspaper pointed out that a detail in the original design had been altered during construction. In the original the end of three walkways was to be suspended from a single rod hanging from the ceiling. A washer and nut threaded onto the rod just below a skywalk would bear the skywalk's weight (Fig. 1-17*a*).

Apparently someone responsible for the actual construction realized that such a suspension system would be nearly impossible to build, and so where a single rod was to run through a skywalk, two rods extending from the skywalk were used instead (Fig. 1-17*b*). How would such a simple and reasonable change lead to the tragic death and injury during the celebrations that Friday evening?

Answer Consider the way in which weight was supported at the end of the highest walkway. In the original design, the weight of the walkway and the people on it would have been supported by the nut that was to be threaded onto the rod there. How about in the altered design in which two nuts were used? On the top walkway, the nut on the rod that extended downward had to support the weight of both lower walkways and people on them. More dangerous, the nut on the rod that extended upward had to support the weight of all three walkways and all the people on them. Apparently when the walkways became crowded, the combined weight ripped or broke some of these nuts and caused the structure to collapse. A simple change—a tragic difference.

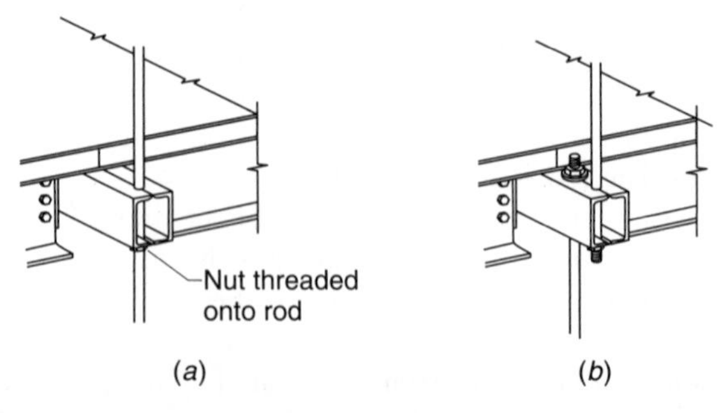

Figure 1-17 / Item 1.47 (*a*) The original design and (*b*) the actually used altered design for the skywalk support.

1.48 • World Trade Center collapse

Why, physically, did the WTC Twin Towers eventually collapse after being struck by airplanes on September 11, 2001?

Answer There have been two major explanations of the collapses of the Twin Towers.

(1) The collision and the ignition of an airplane's fuel led to a fire with a temperature exceeding 800° Celsius. Because the collision removed the thermal insulation on the vertical steel columns, the high temperature caused the columns to soften and then buckle under the weight of all the floors higher in the building. Then, suddenly, many of those vertical columns failed and the higher part of the building collapsed onto a lower floor. Even if the columns of the lower floor were not heated, this sudden and huge impact caused its supporting columns to buckle. Therefore, floors *pancaked* downward.

(2) The collision and the ignition of the fuel led to a fire, but the temperature was less than the amount needed to soften the vertical supporting columns. (As some researchers have reasoned, the airplane-damaged floors did not have enough ventilation for a large fire and the smoke escaping through the hole created by an airplane did not indicate a large fire.) Instead, the fire caused one or more floors and their supporting horizontal beams (trusses) to expand. Because these floors and horizontal beams were constrained, they could expand only by bowing, which then pulled the vertical supporting columns inward. This inward pull could have been enhanced if the columns and horizontal beams were softened by the fire. Once the vertical columns were pulled inward, they could no longer support the higher part of the building, which then collapsed.

1.49 • Falls from record heights

February 1955: A paratrooper fell 370 meters (1200 feet) from a C-119 airplane without managing to deploy his parachute. He landed on his back in snow, creating a crater that was a meter deep. Air-evacuated to a hospital, he was found to have only several minor bone fractures and a few bruises.

March 1944: Flight Sergeant Nicholas Alkemade, an RAF rear gunner on board a Lancaster bomber on a bombing raid over Germany, discovered that his plane was on fire and that he was unable to reach his parachute. After jumping from 5.5 kilometers, he hit a tree and then snow, and yet he suffered only scratches and bruises.

World War II: I. M. Chissov, a lieutenant in the Soviet air force, decided to bail out of his airplane when attacked by a dozen Messerschmitts. Since he did not want to be a "sitting duck" for the German pilots, he decided to delay the deployment of his parachute until he was well below them.

Unfortunately, he lost consciousness during the 7 kilometer fall. Fortunately, he happened to hit a snowy ravine. Although hurt by the impact, he was back in military service in less than four months.

Perhaps even more bizarre is the stunt that was long performed by Henri LaMothe. He would dive from 12 meters to belly-flop into a pool of water that was only 30 centimeters deep, hitting with a force that was about 70 times his body weight. (The stunt is quite dangerous and should not be repeated. I heard of one foolish young man who attempted it and ended up being paralyzed from the neck down.)

The news media commonly report stories of other survivors of long falls (and plenty of stories of nonsurvivors). Why do the survivors survive?

Answer The lethal factor in a fall is, of course, the force the victim experiences during the collision with the ground (or some other solid surface). The force depends directly on the momentum of the victim just before the collision and inversely on the duration of the collision. The momentum depends on the victim's speed and mass. When the fall is from a great height, the victim reaches a *terminal speed* sometime during the fall. Although gravity certainly continues to pull downward, the victim's acceleration is eliminated by air drag that then matches the pull of gravity. The size of the terminal speed depends on the victim's orientation: Being spread-eagled creates more air drag than a feet-down or head-down orientation and so has a smaller terminal speed. However, landing spread-eagle after falling a great distance is hardly an advantage.

The time the collision takes is a more crucial factor. If the collision is "hard," it may last for 0.001 to 0.01 second, and the force stopping the victim is certain to be lethal. But if the collision is "softer" (the victim takes longer to stop), then the force is smaller and the victim might survive. A fall into deep snow may prolong the collision enough to reduce the force to a survivable level. Apparently the 30 centimeters of water was sufficient for La Mothe to survive his dives.

A victim that falls head-down is far more likely to be killed than with any other orientation, because of the great vulnerability of the spine, brain stem, and brain.

1.50 • A daring parachuting rescue

In April 1987, during a jump, parachutist Gregory Robertson noticed that fellow parachutist Debbie Williams had been knocked unconscious in a collision with a third sky diver and was unable to open her parachute. Robertson, who was well above Williams at the time and who had not yet opened his parachute for the 4 kilometer plunge, managed to catch up with Williams and, after matching her speed, grab her. He opened her parachute and then, after releasing her, his own, with a scant 10 seconds before impact. Williams received extensive internal injuries due to her lack of control on landing but survived. How did Robertson manage to catch Williams?

Answer Robertson was able to saved Williams by manipulating the (upward) air drag he experienced as he fell. As a sky diver begins to fall and the downward speed increases, that force, which opposes the gravitational force pulling the diver downward, builds in strength until the air drag matches the gravitational force. Once this match is made, the diver falls with a constant speed, said to be the *terminal speed*. The size of the terminal speed depends on the cross-sectional area the diver has with the passing air. A diver has less cross-sectional area and a greater terminal speed when either head-down or feet-down than when horizontally spread-eagle.

When Robertson first noticed the danger to Williams, he reoriented his body head-down so as to minimize the air drag on him and thus maximize his downward speed. Williams, falling uncontrollably with much more air drag on her, had reached a terminal speed of about 190 kilometers per hour. Robertson, with his streamlined orientation, reached an estimated speed of 300 kilometers per hour, caught up with Williams and, as he neared her, went into a horizontal spread eagle to increase the air drag on him and slow him to Williams's speed.

1.51 • Cats in long falls

Humans rarely survive long falls, but cats apparently have much better luck. A study published in 1987 considered 132 cats who had accidentally fallen from heights of two to 32 floors (6 to 98 meters), most of them landing on concrete. About 90% survived, and about 60% even escaped injury. Strangely, the extent of injury (such as the number of fractured bones or the certainty of death) decreased with height if the fall was more than seven or eight floors. (The cat that fell 32 floors had only slight damage to its thorax and one tooth and was released after 48 hours of observation.) Why might a cat have a better chance of survival in a longer fall? (The survival is by no means guaranteed, so if you live in a high-rise apartment, be sure to keep your cat away from any open window.)

Answer If a drowsy cat accidentally topples from a window sill, it quickly and instinctively reorients its body until its legs are underneath. The cat then uses the flexibility of its legs to absorb the shock of the landing: The flexibility lengthens the time of the landing and thereby reduces the force on the cat.

As a cat falls, the force of air drag that pushes upward on the cat increases. If the fall is from the sill onto the floor, the air drag is never very much. But if the fall is longer, then the air drag may become large enough to reduce the cat's downward acceleration. In fact, if the fall is more than about six floors, the air drag can become large enough to match the gravitational force pulling downward on the cat. The cat then falls without acceleration and with a constant speed called *terminal speed*.

Unless terminal speed is reached, the cat is frightened by its acceleration and keeps its legs beneath its body, ready for the landing. (Your body is also sensitive to accelerations rather than speeds.) But if terminal speed is reached, the acceleration disappears, and the cat relaxes somewhat, instinctively spreading its legs outward (in order to increase the air drag on it) until it must finally get ready for the landing.

Once the cat spreads out, the air drag automatically increases, which reduces the speed of the cat. The longer the fall, the more the speed is reduced, until a new and reduced terminal speed of about 100 kilometers per hour is reached. Thus, a cat falling from, say, 10 floors will land with a speed that is less than that of a cat falling from five floors and will have a better chance of escaping serious injury.

1.52 • Land diving and bungee jumping

On Pentecost Island in New Hebrides, one native test of manhood is to dive from a high platform and toward the ground, trusting that a length of liana vines tied around the ankles and secured to the top of the platform will halt the fall before the ground is reached. In May 1982, a young man made such a leap from more than 81 feet. Just before he was stopped by the vines, his speed was reported to be about 55 kilometers per hour. The acceleration that he then underwent while being stopped was estimated to be 110 *g*s (110 times the free-fall acceleration). There are no reports on how well he could walk thereafter.

A tamer version of vine jumping, but one that still occasionally leads to injury and death, is bungee jumping, in which a person leaps from a high platform with an elastic band attached to the legs and to the platform. This practice began on April Fool's Day (of course!) in 1979 when members of the Dangerous Sports Club leaped from a bridge in Bristol, England. Suppose that you bungee-jump from a bridge (and, of course, stop short of whatever lies below you, which does not always happen). Where do you experience the greatest force and acceleration? If you are frightened of the experience and decide to use only half the length of the bungee cord, will the greatest force and acceleration be half as much?

Answer You feel the greatest force and undergo the greatest acceleration, both upward, when you reach your lowest point, where the bungee cord is reversing your motion and you are momentarily stopped. If we can treat the cord as being exactly like an idealized spring as used in textbooks, then the values of the greatest force and acceleration are independent of the length of the cord and thus also the distance that you fall. Although a shorter fall gives you less downward speed for the bungee cord to arrest, the correspondingly shorter cord that you would use will be stiffer (like a shorter and thus stiffer spring) and will reduce your smaller speed to zero with the same acceleration as a less stiff cord would reduce a larger speed to zero.

The upward acceleration stopping the jumper is sometimes great enough to injure the jumper. The eyes are especially vulnerable because, with the head downward during the stopping, the increased blood pressure in the eyes can cause hemorrhage.

1.53 • Trapped in a falling elevator cab

Suddenly it happens—you are in an old elevator with no safety backup system when the cable snaps, and the elevator cab falls. What should you do to optimize your chance of survival, as slim as that might be? For example, should you jump up just before the cab collides with the bottom of the shaft?

Answer The best advice is probably that you should lie down. You might think that the action is impossible since both you and the floor are falling, but there is sure to be some drag on the cab from the guide rails along which it slides and from the air through which it falls. So, you can drop to the floor. There you should spread out, preferably on your back. The idea is to spread the force you are about to experience over as much surface area as possible.

Standing is ill advised, because then the force is spread over a smaller area, such as the cross section of your ankles. If the collision is severe, your ankles will collapse and the trunk of your body will then crash to the floor.

Jumping up at the last moment (surely that is impossible to time from an enclosed cab) may be the worst thing to do. If you jump up sometime during the fall, you will probably only reduce your downward speed. Suppose that the cab ricochets from the bottom of the shaft. You are then traveling downward as the floor of the cab is traveling upward, and shortly later . . . well, no need for the gory details.

SHORT STORY

1.54 • Bomber crashes into Empire State Building

At 9:45 A.M. on Saturday, July 28, 1945, a U.S. Army B-25 bomber crashed into the 78th and 79th floors of the Empire State Building in New York City while flying through dense fog. The airplane's three occupants and ten workers inside the building were killed, and 26 others were injured. Had the day been a regular work day, the toll could have been much higher.

The collision ripped off the airplane's wings and sent the fuselage and the two motors into the interior of the building, where the fuel then burst into flames that were so bright that onlookers from the street could see them in spite of the fog. One motor went completely through the building and out the other side to fall onto the roof of a 12-story building, where it started another fire.

As the airplane crashed through the Empire State Building, it struck one of the girders in the elevator region, damaging it and some of the elevator cables. An elevator operator, who had just opened her door on the 75th floor, was blown out of the elevator by the explosion of the airplane and then set on fire by burning fuel that had been propelled down through the shaft. Her flames were extinguished by two nearby office workers. After giving her first aid, they escorted her to another elevator where a fellow operator agreed to take her to the first floor for further medical help. Just as the door closed, the cables on the elevator were heard to "snap with a crack like a rifle shot," and the elevator cab then fell to the building's subbasement.

Rescuers who reached the subbasement shortly later expected to find both occupants of the cab dead. However, after they cut a hole in the basement wall to reach the cab, the rescuers found both women alive, although badly hurt. The women had fallen more than 75 stories, but the safety devices on the elevator had apparently slowed the descent sufficiently to diminish the crash at the bottom of the shaft. There is no report on what the women did during the fall, but because of fear and the jostling, I doubt that they remained standing.

1.55 • Falls in fighting, landing during parachuting

When someone is thrown down in judo or aikido, how should the person land to minimize the chance of being hurt? How do professional wrestlers manage to go unhurt when they throw themselves or each other down onto the mat of the wrestling ring? In any of the cases, if a person does not land properly, there is a good chance bones will be broken or internal organs hurt.

How should a parachutist land so as to lessen the danger of injury? Although the parachute greatly reduces the downward speed, the speed is still appreciable, being equivalent to a jump from a second-story window.

Answer You should land so as to maximize the region of contact. The technique reduces the force per unit area on the part of the body hitting the floor and lessens the chance that a bone will bend or twist to the point of breaking or an internal organ will be stressed to the point of rupturing. If you are thrown in judo or aikido, you should slap the mat as the trunk of your body hits. The arm adds to the area making contact, and the slap also helps lift the body and reduce the collision force on the rib cage. Professional wrestlers are usually in superb shape and can withstand long falls (such as when they jump from the top of the ropes onto an opponent lying on the mat). They also fight on a floor that is highly flexible. When they land on it, the collision duration is lengthened by its give, and the force in the collision is thereby reduced.

A parachutist is trained to collapse and roll by first making contact with the balls of the feet, then bending the knees and turning so as to come down on the side of the leg and finally the back side of the chest. The procedure has two advantages: It prolongs the collision (and so it reduces the force on the parachutist) and it spreads the force of the collision out over a large area. If the parachutist were to land standing up, the compression on the bones in the ankles would likely rupture the bones.

1.56 • Beds of nails

I introduced the beds-of-nails demonstration to physics education after seeing it as part of a theatrical karate demonstration. My version comes in two parts: In the first part, I am sandwiched shirtless between two beds of nails with one or two persons standing on top of the sandwich. Although the nails hurt a great deal, I am rarely punctured by them. What factor decreases the risk of puncture?

In the second part, I am again sandwiched between the two beds of nails when an assistant places a concrete cinder block on the top bed and then smashes it with a long, heavy sledgehammer. This part is dangerous for many reasons, one of which is the debris that can hit eyes and teeth. (Once when I gave the "Flying Circus Show" with the beds-of-nails demonstration as a finale, my regular assistant was unable to make the trip, and I enlisted the aid of the professor who had invited me. He swung the sledgehammer hard but came in at such an angle that most of the concrete chunks were propelled across my face. One of the chunks cut deeply into my chin, and when I staggered up to deliver my closing remarks, I bled profusely over body, pants, and shoes. I have never again had such a dramatic end to a talk, or such audience response.) Why is a large block somewhat safer to use than a small block?

Answer When people stand on me, their weight is spread over enough nails in the top bed that the force from each nail is usually insufficient to pierce my skin. The force from the nails on my back is larger, because they must also support my weight. By experimenting I discovered how much weight the people standing on me can have before I am pierced. (Don't think that I go without pain, because the demonstration hurts a lot.)

The large block that is smashed not only adds a theatrical flare to the demonstration but it also increases the safety in three subtle ways. (1) If I am to be squeezed hard, then the block and top bed must accelerate rapidly downward; a larger block diminishes the acceleration because of its greater mass. (2) Much of the energy in the sledgehammer goes into rupturing the block rather than into the bed's motion. (3) The fact that the block disintegrates means that the collision time is longer than if the block were not present, and so the force in the collision is smaller than it would be otherwise. When I first gave the beds-of-nails demonstration in class, I used a small brick instead of a large block. The impact of my assistant's sledgehammer left me lying stunned on the floor for several minutes.

1.57 • Hanging spoons

Clean a lightweight spoon and the skin on your nose, breathe lightly onto the interior surface of the spoon's bowl and then hold it so that the surface rests against your nose. Test for adherence by repositioning the spoon and partially releasing it. When you feel it hold, let it go. There, just what you've always wanted: A spoon dangles from your nose. Who can resist you now?

Why does the spoon hang? How does breathing on it first help? Can you hang spoons from other parts of your face, or, if you're into that kind of thing, from other parts of your body?

How long can you hang a spoon from your nose? I have long claimed that my record is 1 hour and 15 minutes, set in a French restaurant in Toronto. However, the truth is that it was actually in a truck stop in Youngstown, Ohio, where a burly member of a motorcycle gang suggested that the spoon would hang better if he reshaped my nose.

Answer If the spoon and your nose are free of oil, there can be enough friction between the spoon and the skin to hold the spoon in place. The spoon is stable provided that the center of its mass distribution lies along a vertical line through the region where it sticks to your nose. Otherwise, gravity rotates the spoon when you release it, and the motion may cause the spoon to slide off. Condensation from your moist breath helps glue the spoon to your nose. Although a water layer acts like a lubricant when it is relatively thick, a very thin layer acts like a glue because of the electrical attraction between water molecules and the nearby surfaces of spoon and skin.

1.58 • Trails of migrating rocks

Stones in the dry lake beds dotting California and Nevada sometimes have long trails extending from them across the hard-baked desert floor. The trails might be tens of meters long, and the mass of the stones can range up to 300 kilograms. What causes the trails? Are the rocks trying to make a break for the gambling casinos in Las Vegas? Is some weirdo pushing them about? Whatever the cause, the trails must be difficult to make because the friction between a rock and the desert ground is certainly large.

Answer Many theories attempt to explain how the stones leave trails. One involves the rare freezing of rainwater. Stones trapped in a thin ice sheet catch in chance wind gusts and scrape trails in the underlying desert floor when the gusts are strong enough to move the stones and ice sheet.

Another theory is that a stone leaves a trail when pushed by wind during one of the rare rainstorms in the region. Once the water lubricates the ground, the wind in the storm can push or roll a stone over the ground so that it leaves a trail. The friction between the stone and ground is least when the water forms a thin layer of mud lying over a still-firm base. A gust of wind might then abruptly shove a stone from its sitting position. Once moving, the stone would require less force to keep moving.

1.59 • Hitches

The clove hitch that is illustrated in Fig. 1-18*a* has one free end and one end under load. If the load increases, can the hitch slip; that is, can the free end be pulled through the knot so that the knot fails? Or is the knot self-tightening?

Answer The friction forces and the tension in a knot can be mathematically analyzed to determine whether the knot will hold or fail under some arbitrarily large load. Here, let's just do a simple analysis, beginning with the free end, which is under no tension (Fig. 1-18*b*). The cord passes beneath another section in a *wrap-over*—the top section presses down on the lower section. If the free end is not to slip through the wrap-over, the friction created by that pressure must not be smaller than the tension attempting to pull the free end through the wrap-over.

Next, in the hitch the cord winds around the rod in two *wrap-arounds*. The end of this coiled section that is nearer the free end is under small tension, while the other end of the section is under larger tension. If the section is to be stationary, the friction between the cord and the rod must be large enough to withstand the difference in tension between the two ends.

Finally, the cord passes through another wrap-over. On the other side, the cord is under whatever tension is created by the load. If the top section in the wrap-over presses on the bottom section hard enough, the wrap-over is stable.

So, there are three requirements of the friction at points along the cord in the clove hitch. If either the wrap-overs or

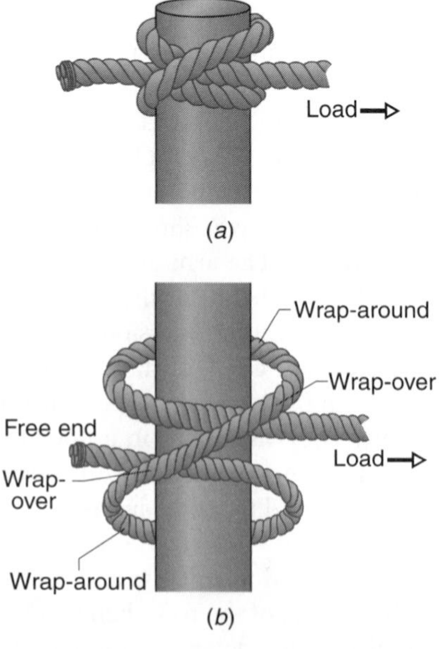

Figure 1-18 / *Item 1.59* (*a*) Clove hitch. (*b*) The elements of a clove hitch.

wrap-arounds are especially strong, the hitch will hold under any load. But if either are weak, then the hitch will fail if the load is made large enough. Other types of knots will fail under a large load even though the wrap-overs and wrap-arounds are both strong, while some knots will automatically tighten to meet any load, and then failure can occur only if the rope between the knot and load breaks.

1.60 • Rock climbing

When climbing a wide crack on a mountainside you might be able to *chimney climb* by pressing your shoulders against one wall and your feet against the opposite wall (Fig. 1-19). You are stable as long as the pressure on the rock is large enough, but the procedure is tiring. Is there a particular vertical distance between feet and shoulders that minimizes the pressure you need to apply?

A narrow, vertical crack where the rock extends out more on one side than the other can be climbed with a *lie-back*. You get on the side opposite where the rock juts out, grip the near side of the crack with your hands, and press your feet on the exposed opposite side. The technique is quite tiring because of the tension in your arms. If you want to minimize the tension, how far down from your hands should you place your feet?

Here are a few more of many possible questions:

(a) If, while climbing a nearly vertical rock face, you find a narrow ledge at foot level, should you press the toe of your climbing shoe or the side of the shoe onto it?

(b) Suppose that you are confronted with a steeply slanted rock slab on which you can stand upright. Are you more stable if you bend over and put your hands on the slab to get some friction on the hands?

(c) If two tilted slabs join at an acute angle, is it safer to climb directly up one of the slabs or along their juncture?

(d) How can you gain purchase with vertical cracks in the rock without having to employ a lie-back?

(e) Why do climbers frequently dip their fingers into a bag on their belt to coat the fingers with chalk?

Figure 1-19 / *Item 1.60* Chimney climb.

(f) As you climb on a rope, the rope runs back down through one or more *runners* (metal loops anchored into the rock) to a climbing partner. Should you use a rope with a lot of stretch or one with almost no stretch?

(g) The advantage of using runners is that a climber can fall only a certain distance below the highest runner. A subtle danger, however, is that the rope may snap as it is stretched during the fall. Many novice rock climbers reason that this danger depends on the climber's height above the last runner just before the fall: the greater this height, the greater the stretch and thus the greater the danger of snapping the rope. Why is that reasoning wrong?

(h) Some types of spiders climb with a safety line of silk, called a drag-line, that would help arrest a fall. Surprisingly, the drag-line has little stretch and should snap even if the spider goes through a moderate fall. Why then does the spider produce a drag-line?

(i) Many skilled rock climbers suffer from chronic pain that runs along their fingers, and some climbers also display a noticeable bulge along the palm side of an injured finger when they draw the finger in toward the palm. What is the connection between the bulge, the pain, and the physics of climbing?

Answer First, a serious caution: None of the rock-climbing examples discussed should be tested without expert instruction, because there are so many variables and assumptions involved that the explanations are only approximate.

In a chimney climb there is a best location of the feet if you wish to minimize the push that is required against the rock at feet and shoulders. In principle you can find it by first putting your feet at some low position and then decreasing your push until the feet are on the verge of slipping. If you then raise your feet while continuing to keep your feet close to slipping, you further diminish the required push. However, the action increases the friction required at the shoulders because the friction at the feet is now less, and the sum of the frictional forces must always equal your weight if you are not to fall. If you continue to shift your feet upward until your shoulders are also about to slip, you are then in the position that requires the least push against the rock.

A lie-back also has a best location for the feet, at which the tension is minimized in the arms. Start with the feet high and then gradually lower them as you decrease the tension. When they are low enough that they are on the verge of slipping, the tension is least.

Answers for the rest of the questions, in order:

(a) The least effort is gained by using the side of the shoe. To stabilize the foot, the leg muscles must counter a torque by the force from the ledge. The torque is larger when the toe is used because the distance between toe and leg bone is larger than the distance between the side of the foot and the leg bone.

(b) As a rule, you are more stable if you stand upright. Leaning over can easily require too much friction at the feet, and so they might slip. Also, you gain little friction from the hands and, if you lean too far forward, the friction on them can actually be down the slope and work against your stability.

(c) Climb the juncture because it is necessarily less tilted than either slab.

(d) Many vertical cracks can offer support if you can jam fingers, hand, arm, foot, or leg into them and then press against the sides.

(e) Chalk is used by climbers to absorb moisture on the fingertips, to give a firmer grip on the rock face. The wide belief is that the moisture decreases the static friction between fingers and rock, and so chalk should bring the friction back up to the dry-skin value. However, one study revealed that chalk actually *decreases* the friction for two reasons: (1) In drying the skin, the chalk decreases the compliance of the fingertips. (2) The chalk particles form a slippery layer between the fingertips and the rock. Still, chalk is an overwhelming favorite among rock climbers; more study is needed here.

(f) Rock climbers (as opposed to spelunkers) use rope that gives considerably under stress, so that if you fall, your stop at the end of the fall is not sudden and the force stopping you is not huge. As the rope begins to stretch, the rope materials rub against one another and become warmer; most of the potential energy and kinetic energy that you lose during the fall ends up as thermal energy within the rope.

(g) Skilled climbers know that the danger of snapping a rope depends on the *fall factor* $2H/L$, where H is the climber's height above the highest runner and L is the length of rope between the climber and where the rope is secured, probably at the *belayer* handling the rope. Depending on the values of H and L, the fall factor may be dangerously high when H is small if L is also small. As a climber ascends and L increases, the same value of H is not as dangerous.

(h) As the spider reaches the end of the drag-line during a fall, the force on it from the drag-line pulls more line from the spider's spinnerets. The force on the spider from the drag-line is not so severe as to snap the drag-line as the spider is brought to a stop.

(i) Many rock climbers have injured their fingers when hanging by them in a *crimp hold*, in which a climber presses down with four fingers to gain purchase on a narrow, overhead ledge. When the full weight of the climber is supported in this way, the fingers can be damaged. Specifically, the fingers are held in place by means of tendons passing through guiding sheaths, called *pulleys*, that are attached to the finger bones. When the full weight is supported by the fingers, the forces required of those tendons can rip the tendons through the pulleys. Thereafter, the climber has not only pain in the fingers but also a noticeable bulge when the fingers are clinched because the tendons are no longer constrained to be next to the bones.

1.61 • Rock climbing by bighorn sheep

Rock climbers wear shoes with special soles to get large frictional forces between the shoes and the rocks on which they climb. If the rocks are wet, the climbing can be treacherous. Indeed, many of us have trouble walking across a wet floor without slipping. Bighorn sheep don't wear shoes with special soles but still manage to scamper up rocky slopes without an obvious care and even when the rocks are wet or covered with moss. How do the sheep cling to the rocks?

Answer A walking person makes first contact on a floor with the heel of the descending foot. If the floor is wet, the heel encounters little frictional force to stop it at the first point of contact and may slip forward, causing the person to fall. A bighorn sheep makes first contact with a rock with the rear section of a cloven hoof, at the point where the two digits of the hoof join. This section is narrow enough that it penetrates moss or anything else coating the rock. As weight then comes down on the hoof, more of the two digits make contact with the rock and slide away from each other to form a V-shaped region of contact with the rock. By sliding in this way, the two digits scrape the rock to remove material that might be slippery and jam themselves into any rough regions on the rock, thereby preventing the hoof from sliding forward as the weight is brought down on the hoof.

1.62 • Pulling statues across Easter Island

The prehistoric people of Easter Island carved hundreds of giant stone statues in their quarries and then moved them to sites all over the island. How could they do this using only primitive means?

Answer The giant stone statues of Easter Island were most likely moved by the prehistoric islanders by cradling each statue in a wooden sled and then pulling the sled over a "runway" consisting of almost identical logs acting as rollers. Although pulling the sled required a tremendous effort by the islanders (a tremendous amount of energy), it was far easier than pulling a statue across the ground, which would require overcoming the friction from the ground. In a modern reenactment of the roller technique, 25 men were able to move a 9000 kilogram Easter Island–type statue 45 meters over level ground in 2 minutes.

1.63 • Erecting Stonehenge

How were the stone blocks for Stonehenge, the megalithic construction on the Salisbury Plain in England, transported to the site and lifted into position? The *sarsens* are the huge upright stone blocks; the *lintels* are the somewhat smaller stone blocks that straddle pairs of sarsens.

Answer The stone blocks were unlikely to have been transported for more than 5 to 10 kilometers in spite of the romantic tales and ingenious schemes that have been attributed to them. The blocks were all available to the ancient builders, perhaps after having been moved from the stone source by ice flow during early ice ages, long before Stonehenge was built. To move a block, the builders may have made a rolling rig out of it by binding logs and smaller blocks around the main block so as to form (roughly) a cylinder. Then, with teams pulling on ropes, the cylinder could be rolled over level ground and even up gentle slopes. Modern-day enthusiasts have moved blocks in this way.

A more likely procedure for the ancient builders was to leverage a block onto a sled constructed from lashed-together logs. The sled then was dragged by teams of people or work animals pulling on ropes, and the progress was eased by grease slopped onto the ground in front of the sled's runners. Modern-day enthusiasts have also moved blocks in this way.

Uprighting a sarsen at the building site was likely done by pulling the sled up onto a mound that ended abruptly at a hole (Fig. 1-20*a*). A counterweight block was probably placed on the top rear of the sarsen as the sarsen was pulled past the mound's edge overlooking the hole. The counterweight served as a control to the motion of the sarsen and also allowed the center point of the sarsen to be pulled past the edge. With the sarsen thus poised, the counterweight was pulled forward until the sarsen rotated and fell into the hole. Ropes around the top of the tilted sarsen were pulled to make it vertical.

One possible way to lift a lintel to the top of a pair of adjacent sarsens was tested in modern times in a small Czech town. A concrete block (5124 kilograms) was pulled along two oak beams with surfaces that had been debarked and lubricated with fat (Fig. 1-20*b*). Each of these 10 meter beams extended from the ground to the top of one of two upright pillars onto which the block was to be raised. The pull on the block was via ropes wrapped around it and around the top ends of two spruce logs. A platform was strung at the opposite end of each log. When enough workers sat or stood on a platform, the attached spruce log would pivot about the top of its upright pillar and pull one end of the block a short distance up a beam. Once the block was moved, stops were positioned at its lower end to prevent it sliding back down as the platform was repositioned for another pull on the block. By duck-walking the block up the oak beams (moving one side and then the other side), only eight or nine people were needed on the platform.

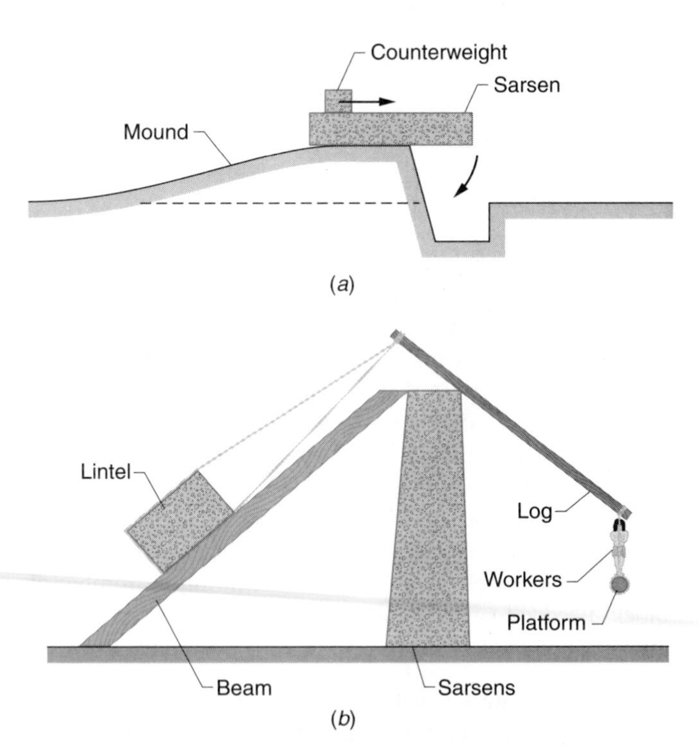

***Figure 1-20** / Item 1.63* (*a*) Righting a sarsen at Stonehenge. (*b*) Raising a lintel.

1.64 • Lifting the blocks for the Egyptian pyramids

At the quarry, the builders of the Egyptian pyramids had to lift the stones (which averaged 2300 kilograms and were as massive as 14 000 kilograms) from the quarry onto sleds, which were then moved out of the quarry. How could the stones be lifted without machines, pulley systems, or any wheeled devices?

The following method may have played a role: A block is wedged upward to allow a number of flexible poles to be positioned underneath it, extending out on opposite sides of the block. Then the exposed ends of one or more of the poles are lifted up slightly (say, half a centimeter) and held in place by sturdy material wedged under the ends. The procedure is next repeated for more of the poles, until all of the poles have been raised the same amount. The block is then higher. How does the technique allow a tremendous weight to be lifted by a few people, and why is the flexibility of the poles important?

At the pyramid site, how were the workers able to pull the blocks up into place on the pyramid? In particular, were earthen ramps used?

Answer Raising a large stone block with flexible poles is considerably easier than with rigid poles. Suppose the rigid poles are in place. To raise the exposed ends of one of them, say one at the end of the block, workers would have to apply upward forces on the ends that almost match the weight of the stone. The reason is that as the stone is lifted by that one pole, it would lose contact with (and thus support from) all but one of the other poles. So, the workers would have to supply that tremendous support.

However, with flexible poles in place, you alone can raise the end of any one pole with a force that is considerably less than the weight of the block. The reason is that as an end is lifted, the block does not lose contact with the other poles, which continue to help support it.

To get the blocks into place on a pyramid, the workers may have used earthen ramps, either directly up a side of the pyramid or spiraling around the pyramid. Presumably teams of men would pull a stone up such a ramp with ropes, using water to decrease the friction between the stone and its path. The gradual climb of a ramp would decrease the required force and thus also the required number of men in a team. However, as appealing as that fact is, the ramps would have been enormous (up to 1.5 kilometers long) and pulling a huge block around the corner of a spiraling ramp would have been slow and daunting.

More likely the blocks were pulled directly up the side of the pyramid on sleds, using the side as a ramp (Fig. 1-21*a*). As each layer of the pyramid was finished, workers mounted blocks on the outside surface and then dressed them (smoothed them). A sled pulled along the dressed stone faces, with water lubricating the runners, would encounter surprisingly little friction. Calculations suggest that a team of 50 men could raise an average block within several minutes, a pace that would allow a pyramid to be built in the time intervals as historically recorded. Even fewer men would be needed if the ropes looped over the construction site to a sled on the opposite side (Fig. 1-21*b*). That sled and the men that could have ridden within it would have acted as a counterweight. Once the men on top of the construction site got the upward-directed sled moving, the downward-directed sled would have helped drag it to the top. This scheme had the advantage of getting empty sleds back down to the ground where they could be reloaded.

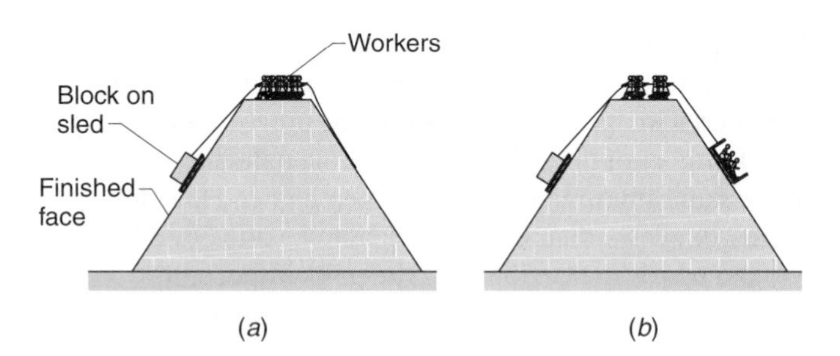

***Figure 1-21** / Item 1.64* Two arrangements to pull a stone block up a pyramid.

1.65 • A Slinky

A Slinky is the well-known spring toy of Poof-Slinky, Incorporated, that can be made to climb down (somersault down might be a better description) a flight of stairs. You place the spring on the highest step, pull the top of the spring up and then down onto the second step, and then let go. Provided the step dimensions are appropriate, the Slinky then climbs down the steps until it reaches the bottom of the flight. The time the Slinky takes to climb down a flight depends on the number of steps it takes (you might arrange it to take the steps two at a time) but is independent of the height of each step. (A Slinky climbs down a tall step and a short step in the same time.) How does the Slinky accomplish this motion?

Answer When you pull the coils up and then down onto the lower, second step, you send a wave through the length of the coil. As the wave travels, more coils move onto the second step by first moving upward, then along the arc of the spring, and then down onto the second step. When the wave reaches the last coils on the first step, those coils are pulled up with enough speed along the arc that they overshoot the second step and (provided the step dimensions are appropriate) land on the third step. The whole process is then repeated.

A Slinky's success in climbing down stairs (and slowly enough that you can see the climbing) is due to the rectangular cross section of its wire. That design, patented by Richard T. James in 1947, reduces the ratio of the spring's stiffness to its mass compared to wire with a circular cross section. The smaller ratio results in a slower speed for the wave you set up along the length of the spring. A plastic Slinky, with a different ratio and thus a different wave speed, climbs about half as fast as the original steel-wire Slinky.

With either type, the time required for a Slinky to climb down one step is set by the ratio of stiffness to mass, not the height of the step. On a short step, the wave travels slowly; on a tall step, the wave travels faster; and the time required by the wave to travel the length of the Slinky is the same for the two steps.

1.66 • Leaning tower of blocks

Using blocks, books, dominoes, cards, coins, or any other set of identical objects, construct a stack that extends from the edge of a table. For a given number of objects, what arrangement gives the maximum overhang (the horizontal distance from the table's edge to the farthest point on the stack)? Suppose that the objects are dominoes with a length L. How many are needed to give an overhang of L? How about $3L$?

With a full set of 28 dominoes, build an arch that spans the space between two tables with identical heights. What arrangement gives the maximum span?

Lego bricks (under copyright by Wham-O, Inc.) are small, plastic, toy blocks. On one of the broad sides of a block there are four holes and on the opposite side four short projections. One block can be stacked directly over another so that four connections are made, or the top block can be shifted by half a block's length so that two connections are made. Let x be half the length of a block, and n be the number of blocks you have. How many different stable (free-standing) towers can you build with all n blocks?

Consider a tower in which each block except for the lowest one is stacked either directly above or shifted to the right of the block just below it. What is the minimum number of blocks needed to give an overhang of, say, $4x$? Is there a more efficient way of stacking to get the same overhang?

Answer A stack is stable if a vertical line through its center of mass extends through the table. So, of course, to get a large overhang, you want the line to pass through the edge of the table. One popular way to build a large overhang is based

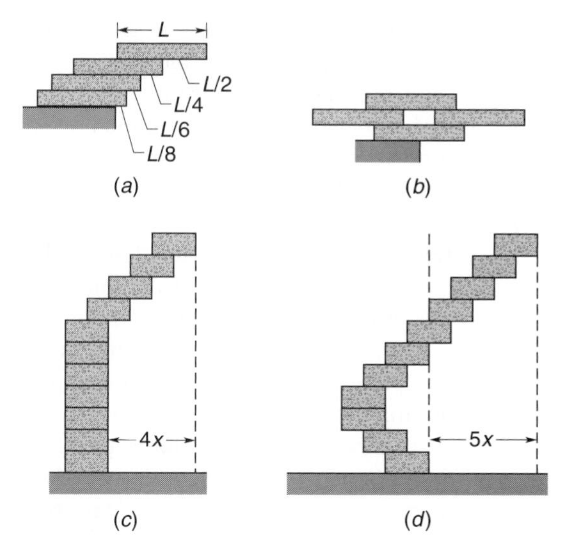

Figure 1-22 / *Item 1.66* Stacking schemes for (*a*)–(*b*) blocks and (*c*)–(*d*) Lego blocks.

on the *harmonic series* (Fig. 1-22*a*). Suppose that you use dominoes. To balance one domino, you would put its center over the edge and get an overhang of $L/2$. Then you can replace the edge of the table with the edge of another domino and arrange for the center of mass of the two dominoes to be over the edge of the table. The overhang is now $(L/2)(1 + 1/2)$. Next you replace the edge of the table with the edge of a third domino and arrange for the combined center of mass of the three dominoes to be over the edge of the table. The overhang is now $(L/2)(1 + 1/2 + 1/3)$. With n dominoes played in this fashion you produce an overhang of $(L/2)(1 + 1/2 + 1/3 + \cdots + 1/n)$, where the expression in parentheses is the harmonic series. Here are some results:

Overhang	Number of dominoes required
L	4
$2L$	31
$3L$	227
$4L$	1674

There is no theoretical limit to the scheme, only practical ones.

More economical schemes use dominoes to counterbalance the ones extending outward from the edge. For example, in one stacking scheme four dominoes give an overhang that is slightly larger than L (Fig. 1-22*b*), and another uses only 63 dominoes to get an overhang of $3L$.

Counterbalancing also helps if you want to build an arch with a full set of 28 dominoes. If the left and right sides are self-supporting, the span can be about $3.97L$, but there is at least one design in which the sides are not individually self-supporting that gives a span of about $4.35L$.

All of the overhangs and arches can be improved if you arrange for the diagonals of the dominoes, rather than their long sides, to be perpendicular to the table's edge.

With three Lego blocks, you can build five different towers (mirror-image arrangements excluded) and four of them are quite stable. One tower is marginally stable—the slightest disturbance will topple it because the center of mass is on a line that passes through an edge of the lowest block. The maximum overhang is $2x$ (a block's length) for the marginally stable tower, x for three of the other towers, and zero for the most stable tower (which is built straight up).

The rules under which a leaning tower can be built determine the proper strategy for gaining the maximum overhang. Suppose that you are to avoid any marginally stable tower and must either put one block directly over another or shift it only to the right. Then the most economical plan is to build a tower straight up except for the last blocks, which are fastened to make a stair-like structure to the right. For example, to get an overhang of $4x$, you need a minimum of 11 blocks positioned with the top four forming stairs (Fig. 1-22*c*). To get an overhang of nx, you need a minimum of $0.5n(n + 1) + 1$, with the top n forming stairs. For a marginally stable tower, leave off the lowest block.

Fewer blocks are needed to get a given overhang if you build first to the left and then to the right. For example, 11 blocks can give a stable overhang of $5x$ (Fig. 1-22*d*).

1.67 • Leaning tower of Pisa

The famous tower in Pisa, Italy, began to lean toward the south even during its construction, which spanned two centuries. Indeed, when the bell chamber was finally added at the top, it was made vertical in the hope of arresting the lean of the rest of the tower. (If you see the tower or a photograph of it, you will notice that the bell chamber gives a banana-shape to the tower.)

The tower was closed to tourists for many years after a tower in Pavia collapsed, killing four people. But was the tower in Pisa on the verge of collapse? After all, it was leaning toward the south by only a little more than 5°, and though the leaning had increased yearly, the increase amounted to only a little more than 0.001° per year. If the tower was to collapse, didn't the tower's center of mass have to move out beyond the base of the tower? That would not have happened for quite some time.

Answer Although the lean of the tower was always small and the center of mass was well over the support area of the base of the tower, before modern repair work was done on the tower, the lean had shifted the support of the tower's weight onto the outside wall at the south side. This shift put the lower section of the south wall under tremendous compression, which threatened to buckle the wall outward in an explosive failure. The danger was increased by a staircase spiraling around the outside of the tower, weakening the structural strength of the wall. From the start, the leaning was due to the compressible soil beneath the tower and the situation was worsened with each heavy rainfall. To stabilize the tower and partially reverse the leaning, engineers have put in an underground drainage system to decrease the water content of the soil and have excavated soil beneath the north side of the tower.

1.68 • Falling dominoes

Once the first domino in a long line of upright, regularly spaced dominoes is toppled against the second one, the toppling moves like a wave along the line. After the wave begins, how many dominoes are moving at any given instant and what determines the speed of the wave? Obviously the dominoes should be spaced no farther apart than the length of one domino, but is there also some minimum spacing? Why doesn't a line of children's blocks topple in domino fashion? Can you send a chain reaction through a line of dominoes where the first one is quite small and each of the others is scaled up from the preceding domino by some factor?

Answer An upright domino has two positions of stability, or *equilibrium*. It reaches one when it stands flat on its bottom side (Fig. 1-23*a*) and the other when it is angled so that its center is directly above its supporting edge (Fig. 1-23*b*). In both positions, the gravitational force, which we assume acts at the domino's center of mass, pulls downward through a supporting point. However, the second position is said to be one of *unstable equilibrium* because the slightest disturbance will upset the domino, shifting the downward gravitational force to the left or right of the support edge. If it is to the right as in Fig. 1-23*c*, the domino topples over.

When you topple the first domino in a line of dominos, it rotates through the position of unstable equilibrium and then falls over to crash into the second domino. If you barely nudge the first domino, then the energy in the crash comes from its fall from the position of unstable equilibrium. When the dominos are too close, the fall is too short to provide enough energy to topple the second domino. Toppling is more likely with a larger spacing, provided that it does not exceed a domino's length. The story is similar for dominos farther down the line. (Of course, you could just wallop the first domino and not worry about the spacing, but that is hardly sporting.)

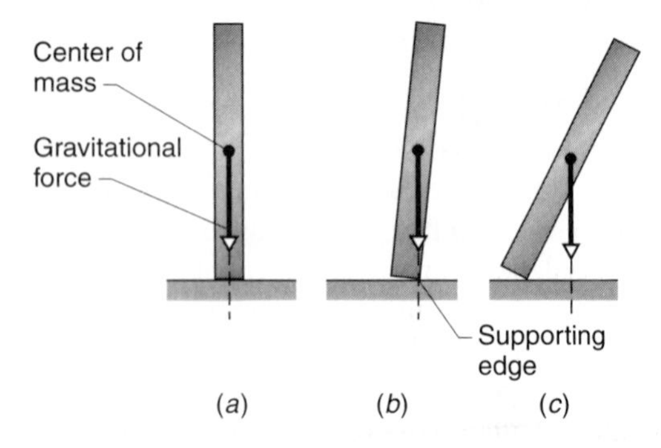

Figure 1-23 / *Item 1.68* Domino passing through position of unstable equilibrium.

At any one instant there may be five or six dominoes in motion. The wave picks up speed as it moves along the line, with the speed approaching a certain value that depends on the spacing, the friction between dominoes, and how well the dominoes bounce from one another. When the spacing is smaller, the wave travels faster and the clatter from the collisions has a higher pitch.

Lorne Whitehead of Vancouver once described how a chain reaction sweeps through a line of dominoes that are scaled up by a factor of 1.5 (on all sides) from one domino to the next. When he topples over the first by "nudging it with a long wispy piece of cotton baton," the energy in the chain reaction is amplified by about 2 billion when the 13th and last domino is knocked over. He figures that given the proper set of dominoes, it would take a line of only 32 to knock over a domino as tall as the Empire State Building. (That's more than King Kong could do.)

1.69 • Falling chimneys, pencils, and trees

When a tall chimney falls, it likely will rupture somewhere along its length. What causes the rupture, where is it located, and which way does the chimney bend after the rupture (Fig. 1-24*a*)? You can check your answer by toppling a stack of children's blocks and noting which way the stack curves during the fall. You might also set up a stack of short, hollow cylinders that are held together internally with elastic bands.

If you stand a pencil upright on its point and then let it fall, does the pointed end move in the direction of the fall or in the opposite direction?

When a tree falls, which way does the lower end move and what shape does the tree take during the fall? Can a tree break like a chimney? Why does a tree sometimes seem to float just before it hits the ground? Why does the butt end sometimes strike the stump so hard that it might almost uproot it?

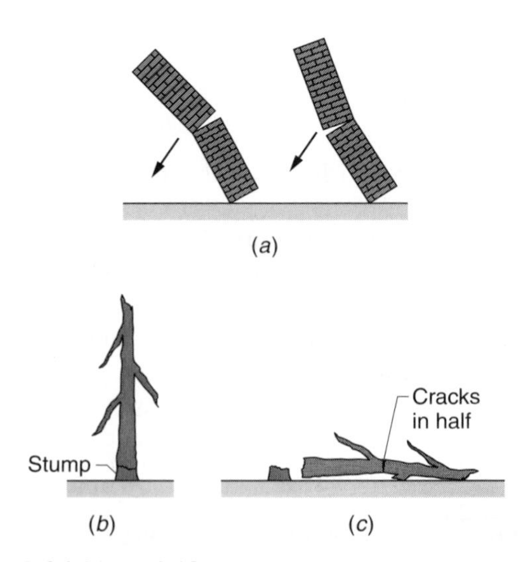

Figure 1-24 / *Item 1.69* (*a*) Which way will a chimney break? An old tree (*b*) initially and (*c*) when the top hits the ground and cracks the tree in half.

(There you are, out in the woods, playing at being a lumberjack, and watching your first big tree fall. You're no dummy—you figure out which way the tree will fall and stand on the opposite side. But just after the tree hits the ground, it comes roaring back at you with revenge, hitting you in the chest and breaking three ribs. It's time to put away the axe.)

Answer As the chimney rotates around its base, the lower part attempts to rotate more quickly than the upper part, and the chimney begins to bend backwards. If the chimney is a uniform cylinder, the greatest attempt at bending is at a height of $\frac{1}{3}$ the chimney's height, and so the chimney is most likely to rupture there. If a chimney has some other shape, the rupture point will be elsewhere. The rupture begins to travel across the width of the chimney from the front of the fall, but the compression on the back side drives the crack downward somewhat. A second break point sometimes occurs lower on the chimney as the top part attempts to slide backwards over the bottom part, thus pulling against the fall on the upper surface of the bottom part.

The direction in which the pointed end of a pencil moves when the pencil topples depends on the amount of friction between the tip and the surface it touches. If the friction is small, the pointed end moves opposite the direction of the fall. With larger friction, the pointed end moves in the direction of the fall, even if it might first move opposite that direction.

A felled tree will bend backward like a chimney but will break only if it is dead and rotted. If the break occurs early in the fall, the top part may fall in the opposite direction of the lower part, making the situation dangerous if you are nearby. If you cut a notch in one side of a live tree and then cut a horizontal slice almost through on the opposite side, the tree will fall toward the notched side, snap the hinge and hurl the butt end upward and then pull it in the direction of the fall. If the tree has plentiful branches, they will be compressed when the tree hits the ground, and their recoil might propel the butt back toward the stump. The impression of floating comes from the air resistance that a tree with full foliage meets as it nears the ground.

Some trees end up in pieces on the ground because of the way they hit the ground. If the initial break is due to, say, strong winds and is at the top of a short stump (Fig. 1-24*b*), then the top of the tree can hit the ground first. In that case, the falling section can snap in half (Fig. 1-24*c*). That leaves a shorter section that hits shortly later; it too snaps in half. Before the final piece hits the ground, tree sections may snap in half several times.

1.70 • Breaking pencil points

The point of a wood pencil often breaks when I write with enthusiasm. Where exactly does the break take place? Why is a break more likely if the point is sharp and less likely if it has been dulled from use?

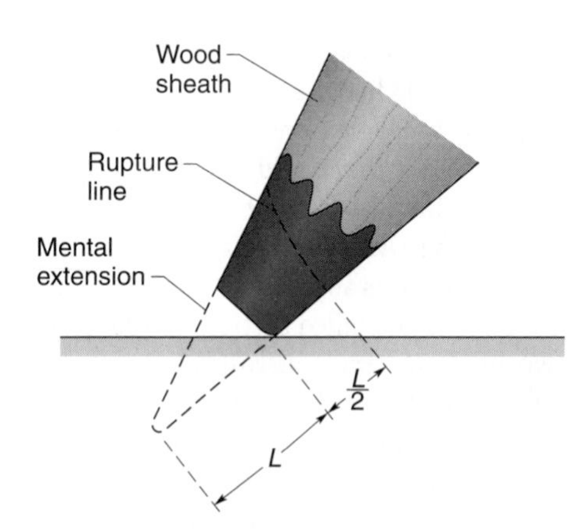

***Figure 1-25** / Item 1.70* Rupture line in a pencil point.

Answer When you write, you press down on the pencil point with the pencil held at a tilt. The action creates forces that attempt to bend the exposed length of lead by elongating the lower side (the side facing the paper) and compressing the top side. The lead is weaker to the attempted elongation, and so the fracture begins on the lower side. While the fracture races up across the width of the lead, it is driven back toward the wood sheath as one side of the rupture attempts to slide past the opposite side.

The rupture begins at the spot where the attempt at elongation is greatest. To find that spot, mentally complete the cone of which the pencil point is part (Fig. 1-25). If the missing length is L, then the rupture begins $L/2$ up from the actual writing tip, or $3L/2$ up from the imaginary tip of the completed cone. That fact means that the rupture begins at the spot where the diameter of the lead is $\frac{3}{2}$ times the diameter of the writing tip, a result that can be tested if you wish to sacrifice a number of pencils. (You should do this in private, because repeatedly breaking off pencil points is probably a sign of abnormal behavior—a pencil-breaking syndrome, or something like that.)

When a point is freshly sharpened, the fracture occurs at a narrow section, and so it requires only a small force to initiate it. If the point is blunter, the maximum extent of bending occurs farther up the lead and at a wider section, and the required force is larger. In that case breaking is less likely under normal writing conditions. If the point is so blunt that the spot of maximum bending is within the wood sheath, the analysis here is inappropriate, and the point can break only if you slam the pencil point down on the writing surface (which is certainly a sign of abnormal behavior).

1.71 • Failure of a bridge section

June 28, 1983, Greenwich, Connecticut, USA: At 1:28 A.M., a 30 meter length of the bridge spanning the Mianus River on Route I-95 collapsed. In the dark the occupants of two cars, a tractor-trailer, and another truck failed to spot the missing section in time, drove over the exposed edge, and fell 20 meters into the river. Three people were killed and three others were hurt.

Bridges sometimes collapse because of age or disrepair, but that bridge on I-95 had seemed to be in good condition. Was there something odd about its design or the way in which traffic crossed it that could have led to the tragedy?

Here are some clues. Because of the angular approach the highway takes to the river, the bridge sections are diamond shaped. Each section was supported along two edges. Along the southern edge of the failed section, the support was provided by two *pin and hanger assemblies*, one at each corner (Fig. 1-26*a*). Each assembly consisted of two steel bars through which steel pins passed. At both ends of each pin a large nut had been tightened and welded to secure the pin.

The assemblies allowed some flexibility in the bridge section so that it could respond to vibrations from the traffic load and to any variation in length from a temperature change. Apparently one of the nuts at the corner farther from the center of the section fatigued and its pin worked its way free, causing the section to fall into the river. What sideways force freed the pin? The answer proves to be a worthy study if such catastrophes are to be avoided.

Answer Consider a truck in an outside lane as it crossed a section of the bridge. For the truck to maintain speed, its tires had to push back continuously against the section, creating a torque that attempted to rotate the section around its center (Fig. 1-26*b*). The attempted rotation produced a sideways force on both sets of support pins and nuts on the southern end, but the force was largest at the farther corner because of its greater distance from the center.

After considerable vibration and stress, one of the nuts at that corner failed and its pin slipped out of place, allowing the corner to drop. The diminished support of the section overloaded the rest of the support points and the section fell. Had the section been square instead of diamond, the resistance to rotation would have been shared uniformly by all four corners, and so the failure at one corner would have been less likely.

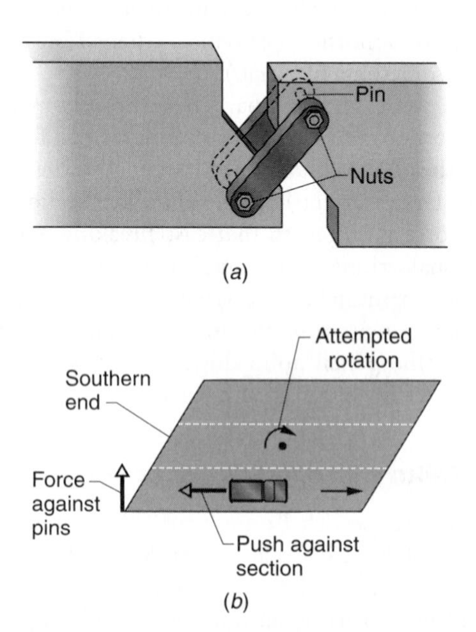

Figure 1-26 / *Item 1.71* (*a*) A pin and hanger assembly holds the span. (*b*) Tendency of rotation due to truck.

1.72 • Jackknifing of a train

When a train engine happens to collide with a massive object and derail, why do the engine and cars usually jackknife instead of all being deflected to one side of the track? Why doesn't the jackknifing extend past the first few cars?

Answer Suppose that the front end of a train engine strikes a massive object that partially straddles the track. Consider the force on the engine in two parts: The force parallel to the track acts to slow the train. The force perpendicular to the track causes the engine to move off to one side of the track. That perpendicular force also tends to rotate the engine around the engine's center of mass. Suppose the front end of the engine is deflected to the right of the track. Then the rotation tends to bring the rear of the engine to the left of the track. Since that rear is attached to the first car, the leftward deflection is not as great as the rightward deflection of the front of the engine.

As the front of the first car is deflected to the left, the car tends to rotate around its center of mass, which brings the rear of the first car to the right of the track. And because of the attachment between the first and second cars, the front of the second car is also deflected to the right. However, this deflection is less than that of the engine or front of the first car. And so it goes.

1.73 • Bowling strikes

In ten-pin bowling (Fig. 1-27) how should you play the ball so as to maximize the chances of a *strike*, in which all the pins are knocked down? Novice bowlers aim for the headpin (the central and foremost one) from the middle of the alley, but seasoned bowlers throw the ball from one side of the alley while putting sidespin on it. The ball seems to *break* or *hook* (that is, changes its course abruptly) at some point down the alley and then head toward the pins along an oblique path. Ideally, the ball should enter the array just to one side of the headpin in what is called the *pocket* (usually the right side if the ball is released on the right side of the alley).

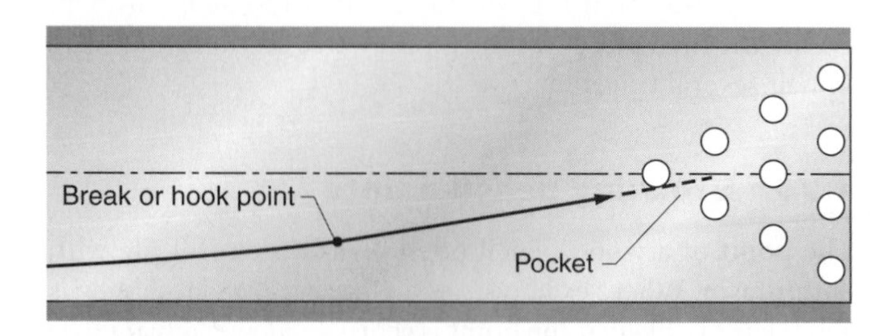

Figure 1-27 / *Item 1.73* Path of a bowling ball.

Is the break real or illusionary? And is the seasoned bowler's strategy of angling the ball into the array actually warranted?

Answer Getting strikes with the novice's scheme is difficult for at least two reasons. The ball may penetrate the array but the pins on the extreme left and right are likely to be left standing. If the ball is somewhat off the mark, its collision with the headpin may knock the ball to the side so severely that it misses the rest of the pins.

If the ball enters the array along a slanted path through the pocket, a wide bounce is much less likely, and so more pins will be downed. If the path is angled several degrees with respect to the central axis of the alley, and if the ball hits the side of the headpin properly, the outside pins along both sides of the triangular array fall in domino fashion and the ball crashes through two of the interior pins, causing one to fall against the other.

The angle of the ball's approach to the pocket depends on the initial ratio of sidespin to forward speed and also on the increase in the friction the ball encounters as it moves down the lane. Normally the first 50% or so of the lane is oiled to decrease the friction. Just after it is thrown, the ball slips over the oiled lane and travels in a curved path toward the pins. When it suddenly begins to roll without slipping, somewhere in the dry (oil-free) region of the lane, its path straightens. The hook is the tightly curved path taken by the ball just before rolling begins. The ability of a bowler to hook a ball depends primarily on the change in the friction along the ball's path, but it also depends somewhat on the fact that the ball is not a uniform sphere because of the finger holes.

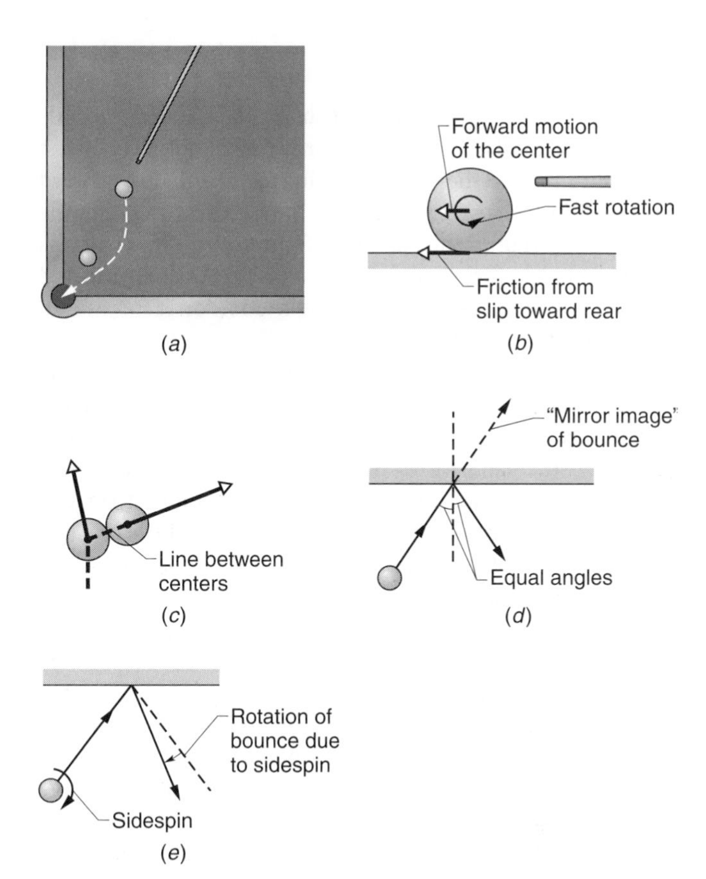

Figure 1-28 / *Item 1.74* (*a*) Masse shot. (*b*) High strike produces a friction force in the forward direction. (*c*) A glancing collision. Bounce from a rail (*d*) without English and (*e*) with left English.

1.74 • Shots in pool and billiards

Where should the cue stick strike the cue ball for the following results, and why do they occur?

(1) The ball immediately rolls without slipping.

(2) The ball runs into a stationary object ball, and then, shortly afterwards, follows after that ball—a *follow shot.*

(3) The ball similarly runs into an object ball but comes back toward you afterwards—a *draw shot.*

(4) The ball similarly runs into an object ball but then stops after moving only briefly.

When the cue ball is hit by the stick anywhere along a vertical plane through its center and then runs into an object ball, what is the angle between the paths then taken by the balls? If the cue ball collides with the rail (the elevated side of the table) at a certain angle, in what direction does it bounce? If it has been struck by the stick off to either side of the central vertical plane and then runs into the rail, how does the direction of the bounce differ?

A *masse shot* can send the cue ball around an object ball that is an obstacle along the direct path to the target ball (Fig. 1-28*a*). How is the shot executed and what produces the curved path that is taken? (The shot is outlawed in most pool-rooms because it runs the chance of ripping the felt covering on the table.)

Why is the height of the rail always $\frac{7}{5}$ of a ball's radius R?

Answer Situations 1 through 4 involve hitting the cue ball somewhere along the central vertical plane through the ball. For 1 and 4, hit the ball at a height of $\frac{7}{5}R$ (that is, $\frac{2}{5}R$ above the center). For 2, hit it anywhere else above the center, and for 3, anywhere below the center.

The answers involve the way the stick puts spin on the ball. If the ball is struck at a height $\frac{7}{5}R$, the impact produces just enough *topspin* for the ball to roll forward without initially slipping over the table. If the ball then hits an object ball, the energy associated with the forward motion is transferred to the object ball, and the cue ball spins briefly in place until friction from the rubbing drains its rotational energy. (The friction is in the forward direction and may drive the ball a short distance in that direction before the ball stops spinning.)

If the ball is hit anywhere else above its center, its spin is in the proper direction for the ball to roll forward, but the spin rate is either too large or too small, and so the ball initially slips. The slippage generates friction, which brings the spin and forward motion into synchrony, whereupon the ball rolls smoothly forward.

For example, suppose you hit the ball higher than $\frac{7}{5}R$. Its spin is then too large for its forward speed, and so its lowest point slips toward the rear, producing friction that is forward (Fig. 1-28*b*). The friction decreases the spin and increases the forward speed until the ball can roll smoothly. If the ball hits an object ball before then, it transfers its forward motion and briefly spins in place, but the strong friction on it forces it to chase after the object ball.

If the ball is hit below the center, its *backspin* is in the wrong direction for smooth rolling and the friction is large and toward the rear. The friction soon reverses the spin and also slows the ball's forward motion, and then the ball rolls smoothly. If the ball hits an object ball before that stage is reached, the forward motion is transferred, and the cue ball briefly spins in place before the strong friction on it causes it to roll back toward you.

When a cue ball strikes an object ball along a glancing path, the object ball is propelled off to the side along a line that extends through the centers of the balls at the moment of impact (Fig. 1-28*c*). The cue ball bounces to the opposite side. The angle between the final paths is often cited as being 90°, but it has that value only when the collision is on the extreme side of the object ball. (The initial path of the cue ball is actually curved because the cue ball slips on the table just after the collision, but the curved portion is usually too short to perceive.)

If the cue ball rolls smoothly into the rail, its angle of approach matches the angle at which it bounces (it is much like a light ray reflecting from a mirror). One way to visualize the bounce is to imagine that the target (pocket or object ball) lies on the opposite side of the rail, just as distant from the rail as it truly is on the near side (Fig. 1-28*d*). It is then like an image "inside" a mirror. Shoot the cue ball toward that image and the ball will take the appropriate bounce on the rail to hit the target.

However, if the ball has *sidespin* or *English* (it spins around a vertical or tilted axis in addition to spinning around a horizontal axis for rolling), the angle of the bounce is altered. Such sidespin is created when the ball is hit to the left or right of the centerline. From your view, left English (the ball is hit on the left side) rotates the bounce clockwise (Fig. 1-28*e*), and right English rotates it counterclockwise.

A masse shot is made by striking downward on the side of the cue ball. The strike forces the ball to spin as if it has a combination of draw and English. The strike also propels the ball in one direction, but the friction that is produced by the spin curves the path.

The height of the rail is chosen so that a ball's collision with the rail does not cause the ball to slip over the table and lose energy to friction. Instead, the ball rolls smoothly immediately after the collision.

1.75 • Miniature golf

In a game of putt-putt, or miniature golf, a golf ball is hit along a small course enclosed by short walls. The idea is, of course, to get the ball into a cup, or hole, with the fewest strokes. Often the cup lies beyond some barrier or corner of the course, and to reach it economically, a player must bounce the ball off the wall. How should the ball be played to put it in the cup with only one stroke?

Answer When a golf ball bounces from a wall, it reflects much like a ray reflects from a mirror: The angle of its reflection equals the angle of its approach. That fact allows you to picture a tricky shot as if it involves reflecting a beam of light off a mirror. Figure 1-29 shows an example in which a ball is to be hit off a wall and into a cup. Pretend that the wall is a mirror producing an image of the cup. That image, which appears to be behind the wall, has the same distance from the wall as the cup. If you hit the ball toward the image position of the cup, it will reflect from the wall and into the cup.

Players that are skilled at miniature golf (and pool, in which similar bounces occur) can mentally picture a sequence of such bounces. Of course, several practical matters, such as rough and tilted terrain and details of an actual collision with a wall, spoil this simple analysis, and so miniature golf still requires some measure of luck.

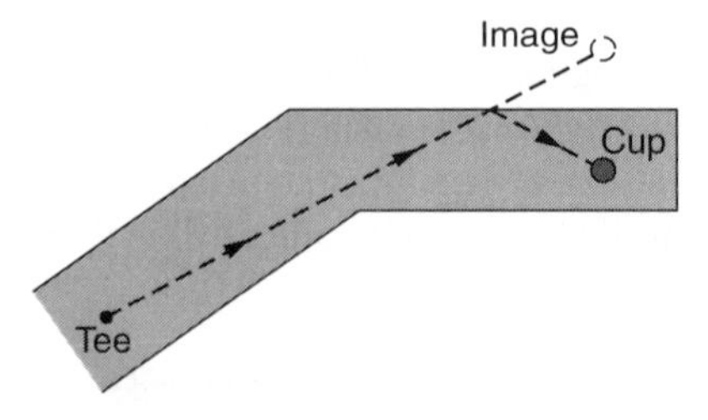

Figure 1-29 / *Item 1.75* Overhead view of a miniature golf course hole.

1.76 • Super Ball tricks

If you drop a Super Ball (the highly elastic rubber ball from Wham-O, Inc.), it bounces almost back to your hand. Suppose that you throw it straight down and also put spin on it. Where then does it go?

If you throw the ball down at an angle and with *backspin*, it bounces back and forth between two spots on the floor (Fig. 1-30*a*). If instead you give it *topspin*, it alternates between long and short bounces while moving away from you (Fig. 1-30*b*). (The height of the bounces may seem to alternate between being low and high, but the impression is an illusion.) If you put backspin on the ball while throwing it beneath a flat table, it may refuse to continue beneath the table and fly out toward you (Fig. 1-30*c*). If you throw it onto one wall of two parallel vertical walls that are fairly close together, the ball will probably bounce back to you (Fig. 1-30*e*). What accounts for such strange and headstrong behavior, and why

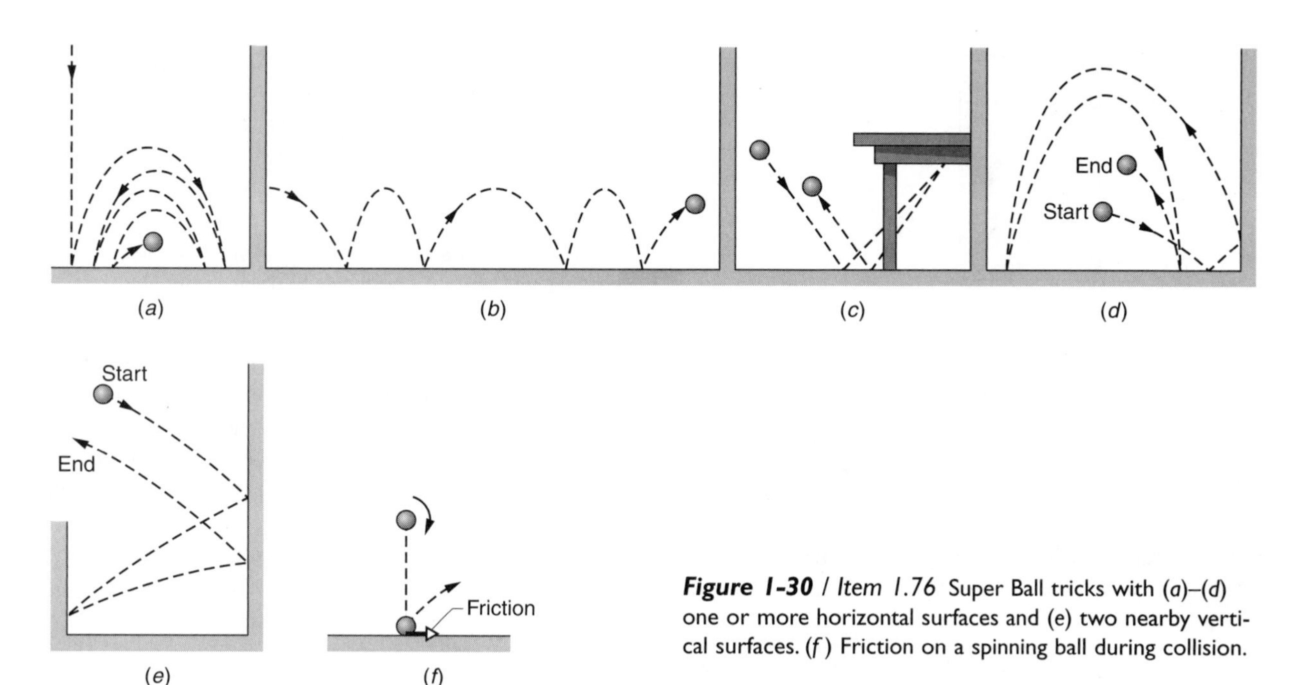

Figure 1-30 */ Item 1.76* Super Ball tricks with (*a*)–(*d*) one or more horizontal surfaces and (e) two nearby vertical surfaces. (*f*) Friction on a spinning ball during collision.

does a Super Ball bounce so much better than a normal rubber ball?

Answer When the ball has spin, its rough surface momentarily catches on the floor, and the friction that is generated sends the ball off in a surprising direction. The friction also alters the spin on the ball, and so the next bounce may be quite different.

For example, if the ball is thrown downward with clockwise spin as seen from one side, the friction is toward the right (Fig. 1-30*f*). The ball also experiences an upward force from the floor during the collision. The combination of the two forces directs the ball upward to the right. When the ball is thrown down at an angle and with spin, it may bounce away from you, straight up, or even back toward you depending on the direction and size of the spin, which determines the direction and size of the friction.

The illusion that the bounce height alternates comes from the variation in steepness of the path taken by the ball. As the ball alternates between short and long hops, the angle of the bounce also alternates. (The illusion is so seductive that I twice gave it credence in my writings even though I had just argued how the height cannot vary.)

A Super Ball bounces so well because of the way the collision sets up oscillations within it. When a normal rubber ball hits, the sudden compression on its lower side causes the ball to oscillate. The time for one oscillation depends on the material makeup of the ball. Chances are that the time differs from the time required for the full collision, in which case the ball continues to oscillate after it has left the floor. The oscillations require energy, and so the ball then has less energy for its upward travel and does not go particularly high.

A Super Ball consists of a core that is surrounded by a shell of different material. The construction alters the oscillations so that the time taken by the first one matches the time the ball is on the floor. Just as the bottom of the ball is reversing its compression and shoving off from the floor, the oscillation is outward against the floor, and so it helps launch the ball. As a result, the oscillation's energy is put back into the upward motion of the ball, allowing the ball to go high.

To determine the direction in which a spinning Super Ball bounces, here is a recipe that comes from the need to keep the ball's kinetic energy and angular momentum constant during the bounce. The vertical velocity is merely reversed. The horizontal velocity of the lowest point on the ball is also reversed, but it is harder to picture because it consists of both the ball's spin and the horizontal velocity of its center. If you combine the directions of the vertical and horizontal velocities just after the collision, you then have the direction in which the ball bounces.

1.77 • Racquetball shots

The bounce of a racquetball, which is a reasonably elastic ball, is partially determined by the spin on the ball. You can give it spin by stroking the racquet over the top or bottom of the ball as you hit it. Or you can hit the ball into a wall or ceiling so that the collision produces spin. Once created, the spin might give the ball a bounce that will perplex your opponent. For example, what does the ball do if it hits the front wall horizontally with either topspin or backspin?

One of the finest plays in racquetball is the Z-shot, which was discovered in the 1970s. As shown in Fig. 1-31*a*, the ball is hit from the right side of the court. After it strikes high on

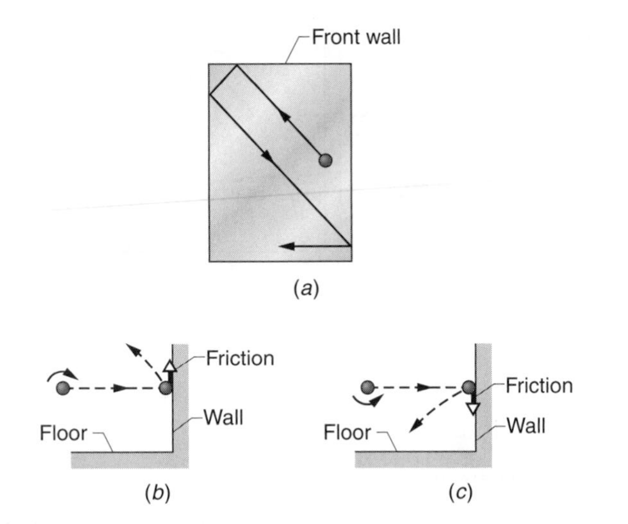

Figure 1-31 */ Item 1.77* (*a*) Racquetball Z-shot. Bounce from a wall with (*b*) topspin and (*c*) backspin on the ball.

the left side of the front wall and next on the forward part of the left wall, it bounces low on the rear part of the right wall. It then travels parallel and so close to the rear wall that your opponent has great difficulty in returning the shot. One reason is that the ball travels across the width of the court, an unusual situation in the game. The other reason is that the ball is so near the wall that your opponent cannot get behind it for a forward return. The only hope is to slam the ball into the rear wall hard enough that it rebounds to the front wall.

What accounts for the path taken by the ball in a Z-shot?

Answer The role of spin and friction on a bouncing ball is explained in the preceding item. If you slam the ball horizontally into the front wall with topspin, the ball leaps high and then descends far down the court (Fig. 1-31*b*). If, instead, you give it backspin, it bounces down to a spot near the front wall (Fig. 1-31*c*). (Thus, by using spin you can have your opponent running all over the court.)

In a Z-shot you hit the ball without giving it spin, but it picks up clockwise spin (as seen from overhead) with its first two bounces. When it undergoes the third bounce, the friction generated by the spin arrests the ball's rearward motion, and the collision propels the ball along a path that is perpendicular to the right wall. The player who first discovered this shot stunned his opponents because the ball's final trajectory was so novel that their playing experience could not anticipate it.

SHORT STORY

1.78 • A controversial goal

In the 1975 World Cup final for field hockey, India won a goal with a play in which an umpire ruled that the ball crossed the goal line, struck the upright wood member on the right of the goal (which was scarcely inside the goal line), and then bounced back into the field of play (Fig. 1-32, which is an overhead view and not to scale). Although such a bounce is highly unlikely and certainly uncommon to the sport, it might occur if the ball happens to take a certain angle toward the goal while also having spin. Less spin is required if the shot comes from the near left side of the goal. If the angle to the goal (between the incoming path and the goal line) exceeds 25°, the backward bounce of the ball is impossible. No one remembers the exact details of the shot in the contest, but the umpire's ruling was at least plausible.

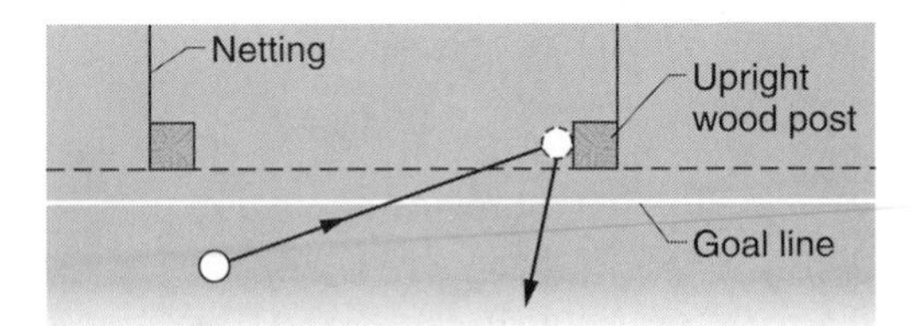

Figure 1-32 */ Item 1.78* Overhead view of a field hockey ball striking goal post and bouncing back into play.

1.79 • Tennis

Where on the head of a tennis racket should you hit the ball if you want (a) the greatest speed for the ball, (b) the least force on your hand due to the collision, or the least attempt of the racket handle to rotate out of your grip during the collision, or (c) the least oscillations of the racket due to the collision (and thus the least oscillations of the handle against your hand)?

Does the firmness of your grip affect the rebound speed of the ball? Are there really such things as a fast court and a slow court?

Answer When you hit the ball, you should strive to have the collision along the long axis of the racket; you will not only give the ball more speed, but you will also avoid the racket twisting in your grip. But just where along that axis you should hit the ball depends on the type of racket and exactly which goal of those listed in the question you value most. Each of the regions on the strings meeting any one of those goals is said to be a *sweet spot*. Thus the term is often confusing unless the goal is defined with it.

Sweet spot 1 is a region where a collision will give the ball the largest speed. This sweet spot is close to the throat of the racket and not, as you might think, at the center of the head. The position has to do with the energy lost in the collision. During the collision, both the racket and the ball deform and then spring back to their original shape. The energy going into the racket's deformation is not returned to the ball because the ball leaves the strings before the racket springs back. So, to minimize this loss of energy, the ball should hit very near the throat where the frame of the racket is stiffest because of the nearby handle. However, the energy lost in the ball's deformation shifts the sweet spot up from the throat somewhat. This loss is greater very near the throat where the strings are tighter and thus present a more rigid structure to the ball than those nearer the center. So, sweet spot 1 is located near the throat because of the racket's stiffness but somewhat above the throat because of the looser strings there.

Sweet spot 2 is a region where the collision produces no force on the hand at the handle. Although the collision tends to push both racket and hand backward, it also tends to rotate the racket. When the collision is at sweet spot 2, the backward push on the hand is canceled by a forward motion of the handle due to the rotation. If the ball hits farther from the hand than sweet spot 2, the rotation of the racket causes the handle to pull outward from the hand. If it hits nearer the hand than the sweet spot, the rotation causes the handle to move into the hand.

Sweet spot 3 is a region where the collision produces little oscillation of the racket (and thus little oscillation against the hand at the handle). If the racket is hit elsewhere, it briefly and perhaps strongly oscillates, somewhat like a plate on a xylophone oscillates when struck.

There is also an ill-defined sweet spot 4 where a player subjectively judges the collision to be best, for any number of reasons.

Although some tennis instructors advise a player to hold the racket very firmly during the ball–racket collision so as to increase the rebound speed of the ball, research shows that the rebound speed is unaffected by the firmness of the grip. The main advantage of a firmer grip seems to be the better control against the resulting twisting of the racket when the collision is off the long axis of the racket. The main disadvantage of a firmer grip is that the impact force and the resulting oscillations of the racket are transmitted more to the arm, which may contribute to the ailment known as *tennis elbow*. Perhaps to lessen this transfer, a skilled player partially relaxes the grip on the racket just prior to impact with the ball by ceasing to accelerate the racket just then.

The material along a court (clay, wood, grass, carpet, and other coverings) can affect the horizontal speed of a ball that is hit low over the net and then strikes the court, skidding across the court before it rebounds. Just how much of the ball's horizontal speed is retained after the strike determines whether the court is fast or slow: On a fast court, the friction is low and more of the horizontal speed is retained. On a slow court, the friction is high and less of the horizontal speed is retained. When the ball is sent into a high lob, it comes down at a steep enough angle that it rolls (rather than skids) over the court, and the ball always loses about 40% of its horizontal speed for any of the traditional court materials.

1.80 • Bicycles and motorcycles

Why is a moving bicycle or motorcycle fairly stable, even if you ride without using the handlebars? How do you initiate a turn? Can you turn a bicycle without using the handlebars? Why is the modern design of a bicycle considerably more stable than earlier designs? In particular, why does the modern bicycle have a front wheel fork that curves away from the rider? What advantage would a bicycle with a low center of mass have in a race?

Answer The question of why a moving bicycle or motorcycle is stable has long been debated. Some investigators have championed the idea that the wheels act like a gyroscope—they tend to resist any chance tilt because of their angular momentum. However, research has demonstrated that the effect is small, especially for a bicycle. Another argument is that you turn the wheel in the direction of the deflection, and the motion of you and the bicycle in the forward direction rights the bicycle. But that cannot be the whole story, as anyone who has ridden a bicycle without touching the handlebars knows. Both explanations also fail to explain how a rider can keep a bicycle upright even when the bicycle is stationary.

The best explanation appears to be one that involves the *trail* of the front wheel—that is, the distance along the ground between where a vertical line through the front axle touches and where a mental projection of the steering axis touches. If the trail extends forward from the tire (as is the case with most—perhaps all—bicycles), then when you undergo some chance tilt, the front wheel automatically steers into the tilt, reducing it. If you help turn the wheel, you can aid the correction, but you do not have to help. If the bicycle had a trail that extended toward the rear instead of forward, the front wheel would not automatically steer to correct a chance lean, and so you would have to make the correction yourself, which makes such a bicycle difficult to ride.

The question of how you initiate a turn on a bicycle or motorcycle has also been long debated, partly because the correct explanation seems wrong. If you want to turn a bicycle, say, to the right, you must turn the front wheel to the left in what is called *countersteering*. You, the bicycle frame, and the front wheel then automatically lean to the right, that is, into the intended turn. This lean causes a torque that opposes the countersteering, turning you, the bicycle frame, and the front wheel to the right. The bicycle is then brought upright.

In a bicycle race where the rider is upright and pumping rapidly on the pedals, the bicycle is thrown vigorously left and right, pivoting around the contact points on the roadway or ground. The lower the center of mass of the bicycle is, the closer it is to the pivot points, and the easier the left and right oscillations are for the rider.

1.81 • Motorcycle long jumps

The stuntman Evel Knievel made numerous stunning jumps in the 1960s and 1970s in which he rode a motorcycle up a ramp, flew through the air over numerous cars or trucks, and then landed on another ramp on the far side. He often made the jumps successfully but on occasion lost control of the motorcycle during the landing and was seriously injured. In

1978, a young man attempted a similar jump over the wings of a DC 3 aircraft but made the fatal mistake of keeping the motorcycle's throttle wide open during the flight. Why did that mistake lead to his death?

Answer When the rear wheel clears the first ramp, the friction retarding its motion suddenly disappears. If the throttle is still open so that the engine continues to drive the wheel, the wheel then turns faster than when it was in contact with the ramp. Since the motorcycle and rider are airborne and free of any external torques, their angular momentum cannot change. So, when the rear wheel begins to turn faster, the motorcycle and rider must rotate in the opposite direction to maintain the initial angular momentum. The rotation brings the front of the motorcycle upward, perhaps by as much as 90°, which makes landing on the far ramp almost impossible. Closing off the throttle at the instant of launch would prevent the dangerous rotation. Slowing the wheel somehow would be even better because it would pitch the front of the motorcycle downward, readying it for the landing.

1.82 • Skateboards

Why do you have an easier time of balancing on a skateboard when you get it moving than when it is stationary? How can you get a skateboard (and yourself) to jump over an obstacle, a maneuver known in the street as an *ollie*?

Answer Your instability comes from an inevitable tilt to the left or right. One investigator showed that in a simple model of a skateboard, the tilt is automatically corrected by the forward motion of the board provided that your speed exceeds a critical value, about 0.8 meter per second. Any chance tilt then turns the front and rear wheels and sets up a small oscillation to the left and right without spilling you off the board. The frequency of the oscillation increases with speed.

In a more complicated model the investigator discovered that when the speed exceeds a second critical value, the board is again unstable to chance tilts and requires agility by the rider. However, stability seems to be reinstated if the speed exceeds a third critical value, but such a speed is uncommonly high in the sport.

To execute an ollie as you ride the skateboard on a sidewalk toward an obstacle, you go through the following routine. At the appropriate moment, you slide your forward foot toward the rear, lower yourself, and then push downward hard on the skateboard to propel yourself upward. Because your rear foot is on the rear of the skateboard, behind the rear wheels, your downward push slams the rear of the skateboard onto the sidewalk. The collision propels the skateboard upward and also begins to rotate it around its center of mass. As the skateboard rises and rotates, you contract both legs so as not to impede the rise, but you also slide your forward foot toward the front to control the rotation. If you are successful, your forward foot will level the skateboard near the top of the skateboard's rise. You then ready yourself for the landing, allowing your legs to flex upon impact so as to cushion the shock.

1.83 • Pitching horseshoes

In the game of horseshoes you toss a metal shoe (it resembles a horseshoe) at a metal stake 12 meters away. In the toss, you bring your arm down and to the rear, and then you rapidly swing it forward, releasing the shoe once your arm is about horizontal. When the shoe lands on the ground, you want the stake to be within its arms. It might end up there if it skips over the ground, but your chance is better if it hits the stake during its flight and then falls into place.

If you are unschooled in the game, you might be tempted to toss the shoe in what is called a *flip*, holding the midpoint of the shoe as shown in Fig. 1-33*a*. When you release the shoe, its plane is horizontal and its arms point toward the stake. During the release you flip the shoe so that it then rotates end over end during the flight.

A flip was originally the most common tossing technique, but then skilled players adopted other ways of holding, orienting, and spinning the shoe. In one technique the grip is up on an arm, with the plane of the shoe extending away from you and at a tilt to the vertical, and with the arms pointed upward (Fig. 1-33*b*). Depending on the speed of the flip you give the shoe, it turns through $\frac{3}{4}$, $1\frac{3}{4}$, or even $2\frac{3}{4}$ revolutions before it hits the stake. In another technique the grip is also off-center, but the arms point downward and the shoe is rotated through $\frac{1}{4}$, $1\frac{1}{4}$, or $2\frac{1}{4}$ revolutions. Why might these more modern techniques produce more *ringers* (the shoe hits the stake, circles it, and then drops into place) than does a flip?

Answer If the interior of the shoe hits the stake in a traditional flip throw, the shoe will likely rebound toward you and land away from the stake (Fig. 1-33*c*). In the modern tosses, part of the shoe's rotation is around a vertical axis. When the interior of the shoe hits the stake, that part of the rotation continues, causing the shoe to spin around the stake. Quickly the flared section on one arm catches on the stake, trapping the shoe, and the shoe eventually falls into place (Fig. 1-33*d*). The name *ringer* probably comes from the ringing of the shoe around the stake or from the clatter that the rotation makes.

1.84 • Spinning hula hoops and lassos

How does one keep a hula hoop (the ring under trademark by Wham-O, Inc.) up and spinning around the body in an almost horizontal plane? How does a cowboy accomplish a similar motion with a lasso?

Answer Both types of motion are due to the force on the spinning object that is provided at a support point. With the hula hoop, the force is at the hoop's point of contact with

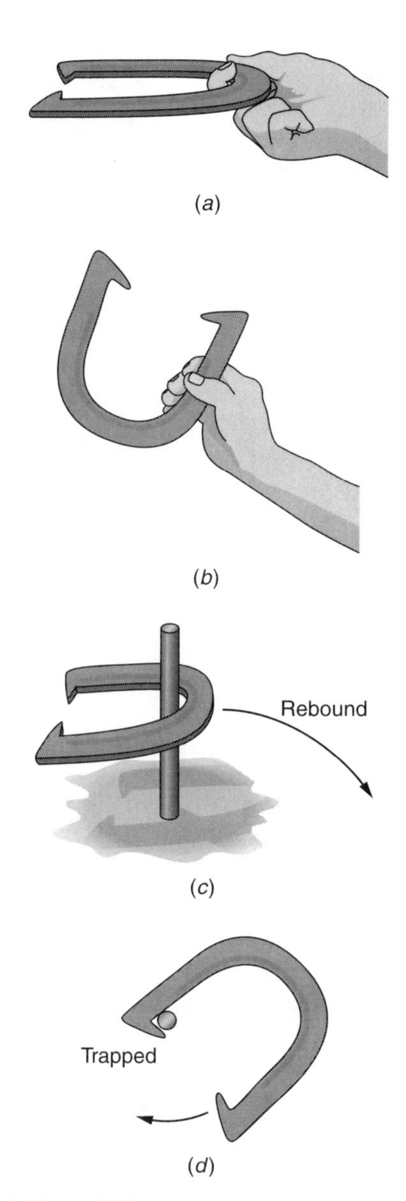

Figure 1-33 / *Item 1.83* (*a*) Release for horseshoe flip throw. (*b*) A better release. (*c*) Rebound from the stake. (*d*) A ringer.

the body. With the lasso, the force is due to the hand pulling on the short length of rope between the hand and the loop of rope. In each case the support point moves in a small circle and pushes or pulls outward on the spinning hoop or lasso, and the force tends to bring the plane of the object close to horizontal. To maintain the spin, the circling by the support point must be somewhat ahead of the circling by the object.

1.85 • Yo-yo

When a yo-yo is tossed downward in the normal manner, how does its rotation gain energy? Why does its downward speed first increase and then decrease? Why can some types of yo-yos *sleep*—that is, remain spinning at the end of the string—while others immediately begin to climb the string once they have reached its end? How do you awaken a sleeping yo-yo to initiate a climb? Why does it climb more poorly, or not at all, if you let it sleep too long? When a yo-yo is near the hand, why does its plane turn around the string (a motion that is called precession)? Why is a sleeping yo-yo much less likely to precess?

A number of tricks can be performed with a yo-yo, including *around the world* and *walking the dog*. In the first one, the spinning yo-yo is made to swing around a large vertical circle while at the end of the string. In the second, a sleeping yo-yo is lowered to the floor, where it then rolls. If the string is kept taut and is also made horizontal, which way will the yo-yo move if the string is suddenly jerked?

Yo-yos come in a variety of styles but among the most impressive must have been the one that was constructed at MIT in 1977. The string (actually, nylon cord) was 81 meters long; the yo-yo structure consisted of two 66 centimeter bicycle wheels coupled with a steel shaft; and the yo-yo was released over the side of a 21-story building.

Even more striking was a 116 kilogram yo-yo that in 1979 Thomas Kuhn yo-yoed about 30 meters from a crane to set the record for the heaviest yo-yo. The yo-yo, 1.3 meters tall and almost 0.80 meter wide, was proportional to a standard yo-yo.

Yo-yos in space: On occasion an orbiting astronaut may play with a yo-yo. Why is it difficult to make a yo-yo sleep in such an environment?

Answer Suppose that you drop the yo-yo rather than throw it downward. Normally when you drop an object, its potential energy is transformed into kinetic energy and the object travels progressively faster with descent. A yo-yo is different for two reasons: It rotates and the rotation rate depends on the thickness of the layer of wound-up string on the yo-yos shaft. As the yo-yo descends and unwraps that string layer turn by turn, the yo-yo spins faster and faster. That leaves less energy for the descent itself. As a result, the rate at which the yo-yo descends first increases and then, about halfway down, it decreases. When the yo-yo reaches the end of its descent and the string is completely unwound, the yo-yo bounces.

If the string is attached to the shaft (usually through a hole in the shaft), the yo-yo immediately begins to wrap back up on the shaft, with the direction of the yo-yo's rotation unchanged. If, instead, the string is looped around the shaft and the bounce is not severe, the yo-yo will sleep. You can awaken it by jerking upward on the string. The jerk yanks the yo-yo upward and momentarily relieves the tension in the string. Since the yo-yo is turning, it catches up some of the slack string on the shaft. Provided there is enough friction, the captured section of string holds, and then the yo-yo is forced to wrap up more string, which makes it climb. If you wait too long to awaken a sleeping yo-yo, too much energy is lost to the rubbing between the shaft and the loop around it, and the yo-yo will be unable to climb back to your hand.

In space, gravity is effectively turned off because both the astronaut and the yo-yo are in free fall. To yo-yo, an astronaut must throw the yo-yo—it won't fall away on its own. When it reaches the end of the string, the abrupt stop makes it bounce, and it most likely will catch up on slack string and be forced to return. To make it sleep, the astronaut must pull gently on the string during the bounce so that the tension prevents slack string. The astronaut could instead swing the yo-yo in a circle to maintain the tension.

Chance disturbances tend to make the yo-yo precess, but the precession is normally appreciable only when the yo-yo is near the hand, where it spins slowly. When it is sleeping, the high speed of rotation creates a large angular momentum that stabilizes the yo-yo against the disturbances. The yo-yo is then much like a gyroscope.

I leave the tricks for your analysis, but for walking the dog, you might want to consider some variations in the orientation of the string as suggested by the next item.

1.86 • Unwinding a yo-yo

Suppose that you unwind a short length of the string on a yo-yo, place the yo-yo on a table such that the string unwraps from the bottom of the central axle, and then pull the string horizontally toward you. Will the yo-yo move toward you or away from you, or will it turn in place? What will it do if you pull upward at an angle to the tabletop? How does it behave if you turn it over so that the string unwraps from the top of the central axle? Before you test a yo-yo, guess at the answers. If you do not have a yo-yo, you can substitute many types of spools, such as a spool of thread.

Stand a bicycle against a table, arrange for a pedal to be in its lowest position, and then pull that pedal toward the rear of the bicycle. Does the bicycle move, and if so, in what direction?

Answer The analysis of the yo-yo is easiest if you consider the point of contact between the yo-yo and the table as being the point around which a torque is to be calculated. Since the friction on the yo-yo from the table acts at that point, friction does not create any torque to turn the yo-yo. To determine the direction in which the yo-yo moves, you then need only consider the torque from the string. If the torque is clockwise (see the figures), the center of the yo-yo must move past the contact point in a clockwise direction and so toward you. If the torque is counterclockwise, the motion is the reverse.

Suppose that the string unwraps from the bottom of the axle. When you pull the string horizontally, the torque it creates with respect to the point of contact is clockwise and the yo-yo moves toward you (Fig. 1-34*a*). To see what happens when you pull somewhat upward, extend the force vector from the string backward until the extension reaches the table. If the extension is to the left of the contact point as in Fig. 1-34*b*, the torque is still clockwise and the yo-yo still moves toward you. If the extension passes through the contact point (your pull is at a larger tilt), the torque is eliminated and the yo-yo spins in place (Fig. 1-34*c*). If the extension is to the right of the contact point (your pull is at an even larger tilt), the torque is counterclockwise and the yo-yo moves away from you (Fig. 1-34*d*).

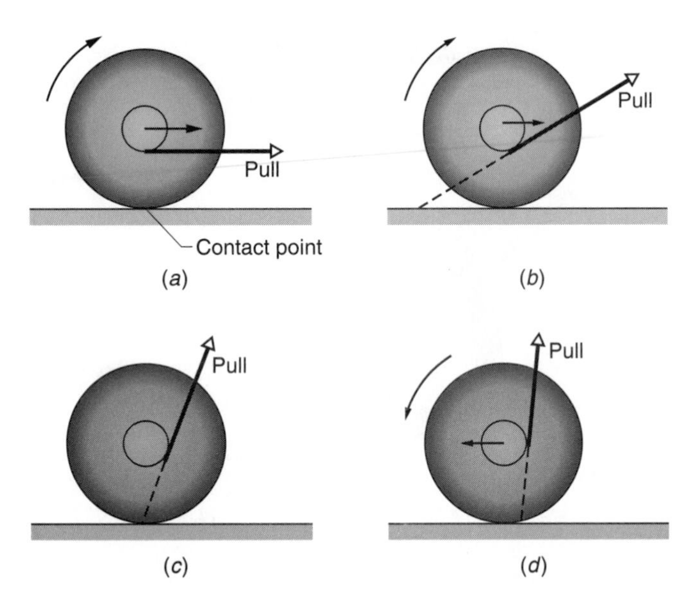

Figure 1-34 / *Item 1.86* (*a*)–(*d*) The direction of pull determines the direction of the yo-yo's roll.

When the string unwraps from the top, the yo-yo rolls toward you for any angle of pull because the extension is always to the left of the contact point.

For the bicycle setup, the bicycle rolls to the rear because of your pull. The forward-pointing frictional forces on the tires, though smaller than your pull, act at a large radius and dominate the turning, rotating the pedal forward against your pull.

1.87 • Driving through the sound barrier

The current land-speed record was set in the Black Rock Desert of Nevada in 1997 by the jet-powered car Thrust SSC. The car's speed was 1222 kilometers per hour (or 759 miles per hour) in one direction and 1233 kilometers per hour in the opposite direction. Both speeds exceeded the speed of sound at that location (1207 kilometers per hour), and the car sent shock waves (sonic booms) across the desert floor to the observers. Setting the land-speed record was very dangerous for many obvious reasons, such as the chance that the air pressure under the car's nose could lift the nose and cause the car to flip over backwards (while traveling faster than sound!). A more subtle danger had to do with the car's wheels. Can you find that danger?

Answer With the car traveling faster than the speed of sound there on the desert floor, each wheel rotated in excess of 6800 revolutions per minute, with a huge centripetal acceleration of 35 000 *g*s (35 000 times the gravitational acceleration) on the rim. Although the wheels were cast aluminum, the radial acceleration put the material of the wheel on the edge of what it could withstand without rupturing. The unknown factor was how that material would fare as the wheels rolled over the desert. Had a wheel hit even a small object, the shock would have caused the wheel to explode and the car to crash. Because that part of the desert had once been used for artillery practice, the ground crew had to walk the route to carefully inspect for partially buried artillery shells and similar debris before the car could be run along its course.

SHORT STORY

1.88 • Spin test explosion

Large machine components that undergo prolonged, high-speed rotation are first examined for the possibility of failure in a *spin test system.* In this system, a component is *spun up* (brought to high speed) while inside a cylindrical arrangement of lead bricks and containment liner, all within a steel shell that is closed by a lid clamped into place. If the rotation causes the component to shatter, the soft lead bricks are supposed to catch the pieces for later analysis.

In early 1985, spin testing was being conducted on a solid steel rotor (a disk) of mass 272 kilograms and radius 38 centimeters. When the sample reached an angular speed of 14 000 revolutions per minute, the test engineers heard a dull thump from the test system, which was located one floor down and one room over from them. Investigating, they found that lead bricks had been thrown out in the hallway leading to the test room, a door to the room had been hurled into the adjacent parking lot, one lead brick had shot from the test site through the wall of a neighbor's kitchen, the structural beams of the test building had been damaged, the concrete floor beneath the spin chamber had been shoved downward by about 0.5 centimeter, and the 900 kilogram lid had been blown upward through the ceiling and had then crashed back onto the test equipment. The exploding pieces had not penetrated the room of the test engineers only by luck.

1.89 • Kayak roll

You are riding the rapids in white water when you and your kayak are up-ended. Realizing that it would be unwise to continue the trip upside down, you try to right the kayak without leaving the cockpit so that you are still somewhat protected. How do you do this?

Answer Here is one strategy. As you roll through the inverted orientation, bend over and extend your paddle toward the water surface that lies in the direction of the roll. Then pull down sharply on the paddle so that the upward resistance it meets provides a torque that will continue the roll and bring you up to the surface. Instead, you might tilt the plane of the paddle and pull parallel to the length of the kayak. In that case, the upward force on the paddle comes from the deflected path the water is forced to take by the paddle.

Until your body breaks the surface, it experiences buoyancy that effectively cancels your weight. However, as your body clears the surface, your weight becomes important and can easily stop the rotation. To avoid stopping, keep your body in the water as long as possible by bending over to the side and let the kayak continue to roll to the upright position while you continue to pull down or backward on the paddle. Just as the kayak becomes upright, sit up.

Some kayakers also employ a *hip snap* during the inverted phase. By snapping the hips in the direction opposite the intended roll, the kayak is forced into the roll. This procedure is most helpful when the paddle has been lost and only outstretched arms can be used in its place.

1.90 • Curling

In the sport of curling, a rotating *stone* is sent sliding along an ice rink toward a target region. The stone, a heavy object, is supported by a narrow circular band. The path the stone takes is initially straight but gradually begins to curve to one side, with the curvature increasing as the stone nears the end of the path. For example, if the stone is launched with a clockwise spin as seen from above, the path curves off to the right. Skilled curlers employ the curvature to send their stone around another one that might shield the target. Why does a stone's path curve?

Curling is often played on pebbled ice (ice with small upward protrusions) that is formed when water is sprayed onto the rink, possibly because such a surface gives more deflection. Vigorously sweeping the ice just in front of the stone is thought by many players to add length to the path and also to increase the deflection. What might account for these effects?

Answer The sideways deflection of the stone (curving the stone's path) is due to the friction on the stone's narrow supporting band. The friction is not a *dry friction* between the band and the ice. Rather, it is a *wet friction* between the band and a thin layer of liquid water melted from the ice by the rubbing of the band. The amount of the friction is not uniform around the band because the friction at any point depends on the speed of the point. If the stone were sent sliding with no rotation, every point would have the same speed and would experience the same amount of friction. However, in play, the stone is sent sliding with some rotation. The combination of the forward motion with this rotation means that different points on the band move at different speeds and thus experience different amounts of friction. The result of this uneven distribution of friction on the band is a net force to the side, deflecting the stone. If the stone is rotating clockwise, the net force and the deflection are rightward. The uneven distribution of friction is also responsible for the stone's behavior at the end of its path: For a while after its forward motion ceases, it spins around one side as if pinned at that side.

The matter of pebbled ice is not understood, and the practice of sweeping is sometimes unfairly scoffed at. A pebbled surface may enhance the friction's dependence on speed. Sweeping certainly removes grit and loose ice that would hinder the stone, but it may also lubricate the stone's motion by partially melting the ice.

1.91 • Tightrope walk

How does a long, heavy bar help a tightrope walker maintain balance, especially if the performance is outdoors and in a moderately gusty wind?

Some tightrope performances have seemed incredibly dangerous. In 1981, Steven McPeak walked a wire strung peak to peak at the Zugspitze, which lies on the border between Austria and Germany. During part of the traverse he was a kilometer above the ground. On the same day he walked up the cable that is normally used by the cable cars on the mountain. He somehow managed slopes that exceeded 30°.

In 1974, Philippe Petit walked across a wire that was between the twin towers of the World Trade Center in New York City and 400 meters above street level. He had shot the wire from one tower to the other with a bow and arrow. After at least seven passages, he was arrested by the police for criminal trespassing. Presumably, they could think of no other reason to stop him, because lawmakers had not foreseen the possibility of criminal wire walking.

Answer Balance is maintained if the center of mass is kept, on the average, over the rope. When the performer leans too far in one direction, the body must bend back in the opposite direction to correct the problem. A heavy bar helps: If the performer leans, say, to the left, the bar is shoved to the right so that the combined center of mass of the performer and bar is kept over the rope. The procedure must be executed quickly before the performer leans too far. A light bar is of little help—with little mass, it would have to be shifted too far to be practical.

1.92 • Bull riding

Why is riding a wild bull or a bucking bronco in a rodeo (or a mechanical bull in a bar, as was sometimes popular in the 1970s) so difficult? Is there anything that a seasoned rider does to help stay on the bull other than just hold on to the strap wrapped around the animal's chest?

Answer The rider's perch depends on the location of the bull beneath her, but the bull suddenly twists, pitches, runs, and stops. With each sudden move of the bull, the rider's momentum and angular momentum tends to send or rotate her from her perch. If she just holds on to the strap with both hands, she must use her strength to arrest the motion of the upper portion of her body off her perch.

She can do better if she throws one arm high while holding onto the strap with the hand of the other arm. She can then throw the free arm in a direction that counters any sudden rotation given her by the bull. The free arm must be held high instead of low so that its mass is far from the center about which the rider tends to rotate at any given instant. Only then can the motion of the free arm effectively counter the rotation of the more massive upper body. If the rider holds a large hat with the free hand, the air drag on the hat as it is waved might give an extra measure of resistance to the rotation of the upper body.

A first-time skater, either on blades or wheels, does something similar to partially correct a problem with imbalance. During my first time on roller skates, when the skates tended to roll out in front of me, I automatically rotated my arms in vertical circles back over my shoulders (like a windmill) to keep my center of mass positioned over the skates and thus to maintain my balance and what little was left of my pride.

1.93 • Tearing toilet paper

One of the frequent, albeit minor, frustrations of life is pulling on a roll of perforated toilet paper, only to pull off a single square, which, of course, is useless to your purpose. The problem is characteristic of fresh rolls and rarer with ones that have been almost depleted. Why are fresh rolls so troublesome? Does the angle at which you pull matter? Is the problem worse if the paper is pulled off the top of the roll or, with the roll reversed, off the bottom?

Answer Your force on the loose end of the toilet paper creates a torque that attempts to rotate the roll. Countering your torque is a torque from the friction between the cardboard interior of the roll and the dispenser rod. When your pull is small, the friction is also and just enough to prevent the roll from turning. As you increase your pull, the friction increases until it reaches some upper limit. Any stronger pull forces the roll to turn, and once there is slippage, the friction is suddenly diminished. But if the required pull is too much, the paper rips.

When the roll is fresh, its weight bears down on the rod and makes the upper limit to the friction large, which means the pull that is required to turn the roll is certain to rip the paper. When the roll is almost depleted and weighs less, the upper limit is smaller and then you can overwhelm the friction with a smaller pull, probably without ripping the paper. If your pull is upward, as is usually the case if the loose end is on the bottom of the roll, you help support the roll, and the upper limit to the friction is thereby smaller. You are then less likely to rip the paper. (In this explanation I have ignored the role played by the lever arms of the torques. You might want to re-examine my conclusions by considering how the lever arm of your pull changes as the roll is depleted.)

Alas, there is no escape from physics, not even in the bathroom.

1.94 • Skipping stones and bombs

How can you manage to skip a flat stone over water? Can you increase the number of skips by increasing the speed or spin that you give the stone? How does a stone skip over wet sand, and why is its path marked with widely spaced pairs of closely spaced nicks?

During World War II, the skip of stone, over water inspired one of the weapons of the British Royal Air Force. The RAF was determined to demolish several of the vital dams in Germany, but the dams were so sturdy that they could be broken only if explosives were placed near the bottom. Bombing the top surfaces would have been useless, and torpedoes dropped by airplanes into the water would have just been snared by the nets that had been deployed near the dams. The difficulties of the task were augmented by the facts that the dams were located in narrow, deep valleys that would confine an air attack, and any attack would have to be made on a dark night if the aircraft were to avoid the artillery that protected the valleys.

To solve the problem, the RAF developed a cylindrical bomb with a length of about 1.5 meters and a slightly smaller diameter. As an airplane approached a dam, a motor gave a bomb a large backspin (top moving opposite the airplane's motion), and the bomb was then released 20 meters above the water surface. (The airplane was equipped with two bright lamps whose beams were angled so that they crossed 20 meters below the airplane. By finding the altitude that gave the smallest spot of light on the water, the pilot could put the airplane at the right height.)

What did the bomb do when it reached the water? Was its rotation of any further use at the dam?

Answer To get a good skip you need to skim the stone over the water so that its plane and its approach path are both nearly horizontal. You should also give it as much spin as possible because the spin stabilizes the stone's orientation, much like spin stabilizes a gyroscope. When the stone hits the water properly, a small wave springs up in front of the leading edge, and the stone then ricochets from it in the forward direction. The initial speed of the stone determines the distance between skips. The number of skips is set by the loss of energy at each skip. The stone not only gives up energy to make the wave but it also briefly rubs against the water surface.

Skipping stones is an ancient pastime, but in recent years a manufactured "stone" consisting of sand and plaster has been introduced. Its bottom surface is concave so as to reduce the rub with the water and so also the energy loss. While the world's record with a natural stone is currently about 30 skips, the artificial stones give 30 or 40 skips.

To explain a stone's pockmarked path over sand, suppose that it hits first along its trailing edge. A shallow pit is dug into the sand, and the collision rapidly flips the front edge down to dig out another nearby pit. The second collision then propels the stone through the air and also reorients the stone so that another pair of pits is dug out farther along the beach.

When the RAF bomb hit the water, its backspin forced it to skip because of the rapid motion of the bottom surface against the water. The gradual loss of energy during the skips reduced the length of each hop, but the hops were still large enough to pass over the tops of the torpedo nets. When the bomb hit the dam wall, the backspin forced the cylinder to roll down the wall. A hydrostatic charge, set for a depth of 10 meters, then ignited the bomb. One reviewer commented, "It was a beautifully simple idea for positioning a bomb weighing almost 10,000 pounds to within a few feet."

A similar bomb, but smaller and spherical, was developed to sink ships. Two of the weapons were to be given backspins of 1000 revolutions per minute and then dropped from 8 meters about 1.5 kilometers from the target. The thought was that as they went leaping over the water surface like flying fish, they might possibly avoid the nets and booms that guarded the target. Once they collided with the hull, they would roll down it until at some preselected depth the 600 pound charge would be detonated. The weapons could also be used to penetrate long tunnels: Released into a tunnel's opening, they would skip their way deep in the tunnel before exploding. For various reasons, the smaller bombs were never used for either purpose. (Physics, though always interesting, can sometimes be horrible in its application.)

1.95 • Spinning ice-skater

An ice-skater spinning on point is a standard device to demonstrate the conservation of angular momentum. When he pulls in his arms, the skater spins faster. The increase in spin is due to the fact that there are no external torques on him, and so his action cannot change his angular momentum. Thus, because he moves some of his mass (arms and possibly one leg) toward the axis around which he spins, his spin rate must increase. This argument is certainly correct, but what force makes him spin faster, and why exactly does his kinetic energy increase?

Answer Both questions can be answered in terms of two *fictional forces* experienced by the skater. The forces are said to be fictional because, although quite apparent to the skater from his rotating perspective, the forces do not actually exist from our stationary perspective—they are not real pushes or pulls. Instead, they are his interpretation of what he feels. One of these interpreted forces is radially outward, a *centrifugal force*. When he brings in his arms and a leg, he must work hard against that apparent outward force. His work goes into his increased kinetic energy. The other interpreted force, a *Coriolis force*, seemingly pushes him around the axis about which he spins. As he brings in his arms and legs, he feels as though an invisible agent is pushing on him with that force, making him spin faster.

1.96 • Spinning a book

Fasten a rubber band around a book to keep it closed and then toss it into the air while spinning it around one of the three basic axes that are shown in Fig. 1-35*a*. For two choices of axes, the spin is stable. Why does the book noticeably wobble when spun around the other axis? Similar instabilities can be seen when a hammer, tennis racket, or a variety of other objects are flipped into the air.

Answer The axes through the book are characterized by the rotational inertias associated with them. The rotational inertia has to do with the way the mass is distributed with respect to the axis around which the book rotates. With one axis the mass is distributed far from the axis (the *rotational inertia* is largest), while with another the mass is close to the axis (the rotational inertia is smallest). (See Fig. 1-35*b*.) When you spin the book around either of these axes, the spin is stable.

The troublesome axis is the one for which the mass distribution and the rotational inertia are intermediate. Were you to toss the book perfectly around that axis, it would spin stably. The problem is that you cannot execute such an ideal toss. Inevitably you err, and the error then produces a wobble that quickly grows. In one interpretation, the error in the initial alignment produces an effective centrifugal force (a fictional force that is radially outward) on the book that causes it to rotate around the axis with the largest rotational inertia. The wobble you see is the combination of the spin you intended and the extra spin produced by the centrifugal force.

The problematic axis with an intermediate distribution of mass shows up in all sorts of objects. However, if any two of the axes have equal rotational inertias, rotation around either axis is unstable, and the rotation may amount to a slow roll around an axis rather than an obvious wobble. In addition, if air drag on a spinning object is significant, rotation about the axis with the largest mass distribution and rotational inertia is also unstable. To show this feature, toss a rectangular card into the air while making it spin around that axis; chances are that the card ends up spinning around the axis with the smallest rotational inertia.

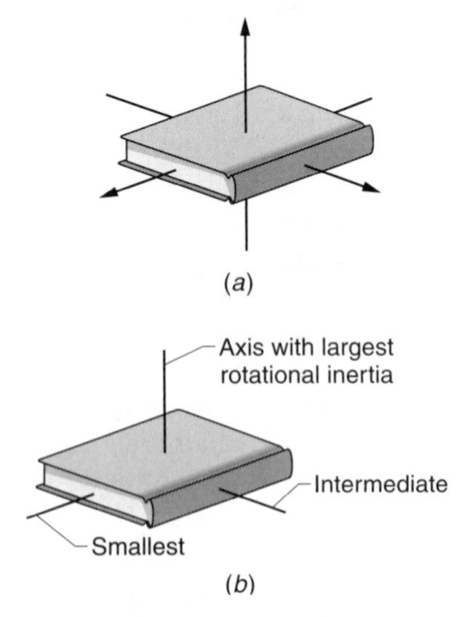

Figure 1-35 / *Item 1.96* (*a*) Three axes through the book. (*b*) The rotational inertias associated with the axes.

1.97 • Falling cat, astronaut antics, and fancy diving

If a cat is released upside down from a height of a meter or more, it quickly rights itself so that it lands on its paws. The action seems to violate a firm rule in physics: When there is no torque acting on an object, the object's angular momentum cannot change. A cat is such an object. It begins the fall with no rotation and so has zero angular momentum, and there is no torque acting on it. Yet its rotation seems to imply that its angular momentum does not remain zero. Does the cat violate the rule?

In an orbiting spacecraft, how can an astronaut undergo *yaw*, turn left or right, without touching anything? How might the astronaut *pitch*, which is to rotate forward or backward around a horizontal axis that runs left and right? Is a *roll*, which is a rotation around a horizontal axis that runs forward and rearward, possible? (Here again is an object that has zero angular momentum and no torque acting on it, and yet it somehow turns.)

A diver jumping from a board or platform is different because the fall usually begins with some angular momentum when the diver pushes off from the diving surface during the jump. In the simplest dive the diver turns over so that the hands are the first to enter the water. Why does the rate of rotation increase when the diver *pikes* or *tucks* before straightening out to enter the water? The rapid rotation is needed if the diver is to turn over several times before reaching the water.

How does a diver manage to add twist to a dive? For example, a diver might include three twists in a forward one-and-a-half somersault. Must the twisting motion come from a certain push from the diving surface, or can the diver leave the surface in a purely somersault motion and then initiate the twist while in midair? Many of the physics techniques employed by divers are also employed by skiers *hot dogging* (performing fancy maneuvers while airborne), gymnasts, skateboarders, and BMX bikers.

Some dives and some jumps from a trampoline are similar to the fall of a cat in that they begin with no angular momentum. But somehow, without benefit of a torque from a push on a surface, the diver or trampoline performer manages to create rotations while in the air.

Answer Explanations for how a cat turns over have been offered for about a century and still there is controversy. I'll give two of the explanations (each supported by photographic evidence), but keep in mind that since cats do not study physics, they may not all use the same technique.

Explanation 1: Think of the cat as consisting of two halves connected by a flexible joint that is midway along the spine. An axis runs through each half, and the two axes initially meet at an angle because the body is convex downward. Once the cat is released, both halves rotate around their individual axes in the *same* direction, while the joint rotates around a horizontal axis in the *opposite* direction. For example, if from an end view the halves both rotate clockwise, then the joint rotates counterclockwise. (Notice that since the two halves rotate together, the body of the cat does not become twisted.) Each rotation involves angular momentum, but the sign of the clockwise angular momentum is negative while the sign of the counterclockwise angular momentum is positive. So, the net angular momentum of the cat during its turning continues to be zero, which is the value with which the cat began its fall.

Explanation 2: Again take an end view. The cat pulls in its front legs, keeps its rear legs extended, and whips its tail around counterclockwise. The action forces a clockwise rotation of both head and body, but with the front legs pulled in, the front half of the cat turns more than the rear half. (Notice that in this explanation, the body of the cat becomes twisted.) As the tail continues to turn, the cat then extends its front legs and draws in its rear legs. The adjustment causes the rear half to turn clockwise more rapidly than the front half, and so the twist of the body diminishes. Eventually the cat is righted and lands, catching the floor with its front paws. (If the cat lacks a tail, one of its rear legs takes on the role of the tail.) As in the first explanation, the net angular momentum remains zero throughout the fall.

If you are the astronaut mentioned in the item, here is one way for you to create yaw. Extend your right leg forward and your left leg rearward. Then bring the legs together again after sweeping the right leg to the right and rear and the left leg to the left and front. As seen from above, the legs move in a clockwise direction. During their motion, your torso must turn in the counterclockwise direction so that your net angular momentum continues to be zero.

To undergo pitch, raise your arms off to the left and right and then move them in circles in the same direction, as if swimming. Your torso rotates in the opposite direction, and again, your net angular momentum remains zero. A roll comes from a combination of pitch and yaw. (Where do you end up if you undergo a sequence of left yaw, forward pitch, and then right yaw? How about a sequence of forward pitch, right yaw, and then backward pitch? Surprisingly, you end up in the same orientation, although you resemble one of the Three Stooges with either sequence.)

If you pike or tuck during a somersault dive, your rate of rotation increases because you pull mass in toward the axis around which you spin. (You are like an ice-skater who pulls in arms and a leg while spinning on point.) The inward pull reduces your distribution of mass. Your angular momentum, which is the product of that distribution and your spin rate, goes unchanged.

If during a somersault you move your right arm upward and your left arm downward, their motion forces your torso to rotate, with your head moving rightward. The action does not alter your angular momentum but it does misalign the axis around which you somersault from the direction of the angular momentum. The result is a twist. So, you need not initiate a twist by some special push from the diving surface but can instead produce it in midair.

1.98 • Quadruple somersault

July 10, 1982, Tucson, Arizona, USA: The aerial acrobat, Miguel Vazquez, released his grip on the bar of his swing during a performance of the Ringling Brothers and Barnum & Bailey Circus, tucked, rotated a full four times, and then was caught in the hands of his brother, Juan, who was inverted on another swing. It was the first time that a *quad* had been performed before a circus audience, although the attempt had been made since 1897 when the first triple somersault was accomplished. What made a quadruple jump so difficult (and thus probably makes a *quad and a half* impossible)?

Answer To set up the jump, the aerialist and his partner each swing on a trapeze. When the aerialist moves upward and toward his partner, he releases from his trapeze, immediately pulls into the tuck position, and then somersaults. As he completes his fourth rotation, he must stretch out so that his partner can catch his arms. Thus the jump has two primary requirements: (1) The aerialist must turn rapidly enough to complete four rotations in the time he takes to fly to his partner. (2) He must pull out of the rotation just as he reaches his partner, or he will be rotating too quickly to be caught.

To meet the first requirement, the aerialist goes into the tuck position to bring his mass closer to his center of mass about which he rotates. This move increases the rotation rate just as the rate is increased when a spinning ice-skater pulls in arms and a leg. However, most aerialists cannot pull themselves in tightly enough to get the required rotation rate for a quadruple jump.

To meet the second requirement, the aerialist must see his surroundings well enough to know how many times he has rotated, so that he comes out of the rotations just in time to be caught. However, the rotation speed for a quad (and thus also a quad and a half) is so large that the surroundings are too blurred for the aerialist to judge the rotation correctly. Thus, the catch is almost never made.

1.99 • Tumbling toast

A piece of toast lies butter side up across the edge of a kitchen table when the table is accidentally bumped, sending the toast tumbling to the floor. Is there any truth to the common thought that the toast always lands butter side down (as an example of Murphy's law, which here might be stated, if a mess *can* happen, it *will* happen)?

Answer If the toast is nudged (rather than hit hard) from the table, the face that lands on the floor can be predicted if we know three quantities: the height of the table, the amount of friction between the toast and the table edge, and the initial overhang of the toast (how far the center of the toast lies beyond the table edge when the tumbling begins). When the table is bumped, the center of the toast is moved out beyond the table edge and the toast begins to rotate around that edge. It also slides along the edge. Both the rotation and the sliding determine the rate at which the toast rotates as it falls through the table's height to the floor. If the rate is enough to turn the toast between 90° and 270° during the fall, then the toast lands butter side down. For typical table height and friction and for common toasted bread, a range of small overhang values and a range of large overhang values yield a butter-side-down landing, while intermediate overhang values yield a butter-side-up landing. Here you can do your own experimentation.

1.100 • Ballet

The grace and beauty of ballet are partly due to a subtle, hidden play of physics. If the performer is skilled, you will never notice the physics. Instead you will see motions that seem oddly wrong, as though they defy some physical law, and yet you may not be able to pinpoint what is odd about them. Here are two examples:

In a *tour jeté* (or turning jump), the performer leaps from the floor with no apparent spin and then somehow turns on the spin while in midair. (The performer does not go through the gyrations of an astronaut as described in a preceding item—they would hardly be regarded as graceful and probably would require too much time.) Just before the performer lands, the spinning is turned off.

A *fouetté turn* is a series of continuous *pirouettes* in which the performer turns on one foot while periodically extending the opposite leg outward and then drawing it inward. One of the most demanding examples of *fouetté* in classical ballet occurs in Act III of *Swan Lake* when the Black Swan must execute 32 turns.

In both of these examples, how is rotation accomplished?

Answer In a tour jeté, the illusion that the spin is turned on and then off in midair is due to how the performer moves her arms and legs inward and then outward during the jump. That shift changes her *rotational inertia*, which has to do with the performer's mass and how it is distributed relative to the axis around which she spins. The performer's *angular momentum* is the product of her rotational inertia and the rate at which she spins. While she is in the jump, she cannot change her angular momentum. She begins the jump with her arms and one leg extended and with a small rate of spin, too small for the audience to perceive. Once she is in the air, she gracefully moves her arms and leg inward to decrease her rotational inertia. Because her angular momentum cannot change, her rate of spin increases and is then perceptible to the audience, so to them it seems to magically turn on after she leaves the stage. As she prepares to land, she re-extends her arms and a leg and regains her initial rotational inertia. Her rate of spin is then again too small for the audience to perceive, and she has seemingly turned off the spin while still in the air.

In a fouetté turn, the performer pushes on the floor to initiate the turn and then comes up *on point* on one foot. She next brings the opposite leg in toward the body axis to increase the spin. As she turns back toward the audience, she extends the free leg so that it gradually takes up the angular momentum of the rest of the body, and for a moment the leg continues to turn around the body axis while the rest of the body does not. The pause allows her to drop momentarily off point and push on the floor with the foot for another revolution.

1.101 • Skiing

There are a variety of ways you can turn while skiing down a slope, but what exactly makes you turn? In the *Austrian turn*, you bring your body down toward the skis and then quickly lift it while also rotating the upper part in the opposite direction of the intended turn.

Another technique requires that you keep the skis flat on the snow while you shift your weight forward or to the rear. Which way you turn depends on how your path is angled down the hillside. The path directly down the slope is the *fall line*. If you travel to the left of it and bring your weight forward, you turn clockwise as seen from above. A backward shift of your weight produces an opposite turn. The results are reversed if you ski to the right of the fall line.

Turns can also be produced if you *edge* your skis—that is, tilt them so that the uphill edge bites into the snow. For example, if you shift your weight forward while edging as you ski to the left of the fall line, you turn counterclockwise. Notice that with edging the shift creates a direction of turn that is opposite to the one created when you keep the skis flat.

Why is the outside edge on a racing ski curved from front to rear? Why do some skiers prefer long skis instead of short ones? When you ski down the fall line, why must you lean forward so that your body is perpendicular to the slope? If you unwisely choose to remain upright, why do you fall?

A novel way of turning while skiing was reported in 1971 by Derek Swinson of the University of New Mexico. Instead of ski poles, Swinson carried a heavy rotating bicycle wheel, holding it by an axle equipped with handles. The plane of the wheel was vertical and the section at the top of the wheel rotated away from him. When he wanted to turn right, he lowered his right hand and raised his left hand. Turning to the left required just the opposite adjustments. What caused the turns?

Answer The Austrian turn is similar to the rotations discussed in the preceding items. When you quickly lift your body, you lessen the contact between the skis and the snow, momentarily reducing or eliminating the friction on the skis. Your angular momentum just then is zero, and since friction no longer acts, it cannot produce a torque on you and the angular momentum cannot change. So, if you rotate the upper part of your body to the left, the lower part of your body and the skis must rotate to the right. When your weight is again felt by the skis and friction returns, the friction allows you to turn the upper part of your body to face the new direction of travel.

To see how the other turning techniques work, consider the case where you ski to the left of the fall line and assume that your normal stance places your weight over the center of the ski. Also assume that the friction on the ski is uniformly distributed along its length. The friction along the front of the ski is partially uphill and creates a torque that attempts to rotate you to the left around your center of mass (Fig. 1-36*a*). The friction at the rear counters with a torque that attempts to rotate you to the right. In each case the torque size depends on how much friction there is and also how it is distributed with respect to your center of mass (*com*). Friction at a point far from your com creates a greater torque than friction at a point nearby. With both amount and distribution evenly matched between front and rear of the ski, you do not rotate.

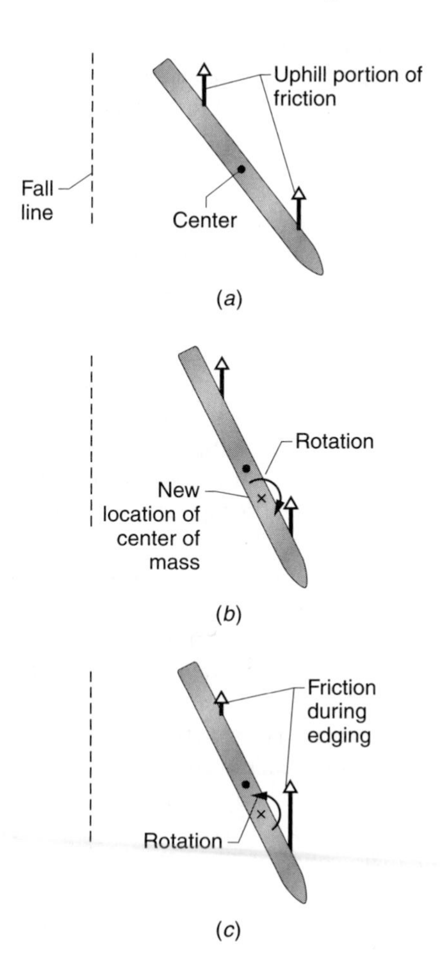

***Figure 1-36** / Item 1.101* Forces on a ski with (*a*) normal stance, (*b*) forward stance, and (*c*) rear stance.

If you shift your com forward, you upset the balance of torques (Fig. 1-36*b*). There is now more ski behind your com and less in front, and so the total friction behind is larger than the total friction in front. Also, the friction at many of the points along the rear is now far from your com, while most of the friction at the front is now near it. So, the torque from the rear wins out and you turn to the right.

If you edge the ski while shifting forward, the bite into the snow increases the size of the friction at the front and decreases it at the rear (Fig. 1-36*c*). In this case, the torque from the front wins out, and you turn to the left.

The edge of a racing ski is mildly curved so as to allow easier turns. When you push the edge down onto the snow, the ski meets the least resistance by gliding along a path that is a continuation of the curve.

Short skis are so easily vibrated by a rough path that you may lose your balance. Although long skis are more difficult to maneuver, they vibrate less.

To see why you must lean forward while skiing down the fall line, imagine that your weight is represented by a vector at your center of mass. The vector can be broken up into two *components*, or parts. One component points down the slope and is responsible for your motion, and the second component points directly toward the slope. If you are to be stable, the second component must also point toward your feet. If you decided to ski while being vertical, the second component would create a torque around your feet and rotate you rearward to the snow.

In Swinson's demonstration, assume that the friction on the skis can be neglected—he and the wheel are isolated from any outside torques. Next, consider an overhead view. Since the wheel initially rotates around a horizontal axis, there is no rotation of either the wheel or Swinson around your line of sight. That means there is no angular momentum of the wheel and Swinson around the vertical, a condition that cannot change because of the lack of any outside torques. If Swinson lowers the right handle and raises the left one, you would then see the wheel rotate counterclockwise, which means that it now has some vertical angular momentum. To keep the total angular momentum zero as it was initially, Swinson must be turned clockwise in your view. So, his maneuver turns him to his right.

1.102 • Abandoned on the ice

You wake up to find that you are deserted in the middle of a large frozen pond with ice that is so very slippery that you can neither walk nor crawl over it. How can you escape?

Suppose that you happened to be lying face down on the ice. As you consider your escape, you decide that you must turn over onto your back to keep from freezing to death. How can you turn over?

Answer Throw a shoe or any other item in one direction; you will move (albeit, slowly) in the opposite direction. Since there is no force on you from the ice, the total momentum of you and the tossed item must remain zero. When you give momentum to the item, you also give just as much momentum to your body in the opposite direction.

Similar physics occurs if someone attempts to roll a bowling ball while on in-line skates with little friction on the wheel rotation. I attempted this. Although the skates began to move backward, my torso did not and I managed to avoid a face-down fall only by grabbing the closest person.

To roll over while on very slippery ice, raise one arm and then, with it outstretched, strike it smartly against the ice. Although there can be no friction on your hand from the slippery ice, there is a force on your hand vertically upward from the ice. That force allows you to rotate the trunk of your body so that you are then on your back.

SHORT STORY

1.103 • Rotation sequence matters

If you are to walk 3 meters north, 3 meters east, and 3 meters south, you end up at the same point no matter how you choose the sequence of those three short walks. Rotation can be different. Hold your right arm downward, palm toward your thigh. Keeping your wrist rigid, (1) lift the arm until it is horizontal and forward, (2) move it horizontally until it is pointed toward the right, and (3) then bring it down to your side. Your palm faces forward. If you start over, but reverse the motion, which way does your palm end up facing?

1.104 • Personalities of tops

Why does a spinning top stay up, even when it tilts appreciably from the vertical? Why do some tops initially *sleep*—that is, stay vertical—while others undergo *precession* (the top's central axis rotates around the vertical as in Fig. 1-37*a*)? Why does the precession often involve *nutation*, a nodding up and down of the top's central axis? Are there distinct types of nutation? Why do some spinning tops die rapidly while others linger?

Answer Usually when a force pulls on an object, the object moves in the direction of the force. But if the object is spinning, the force can make it move perpendicularly to the direction of the force. Such motion seems all wrong, and that is one reason why tops are so fascinating. Even if a child knows nothing about the laws of physics, he or she still knows that a tilted top should just fall over, not precess around in a circle.

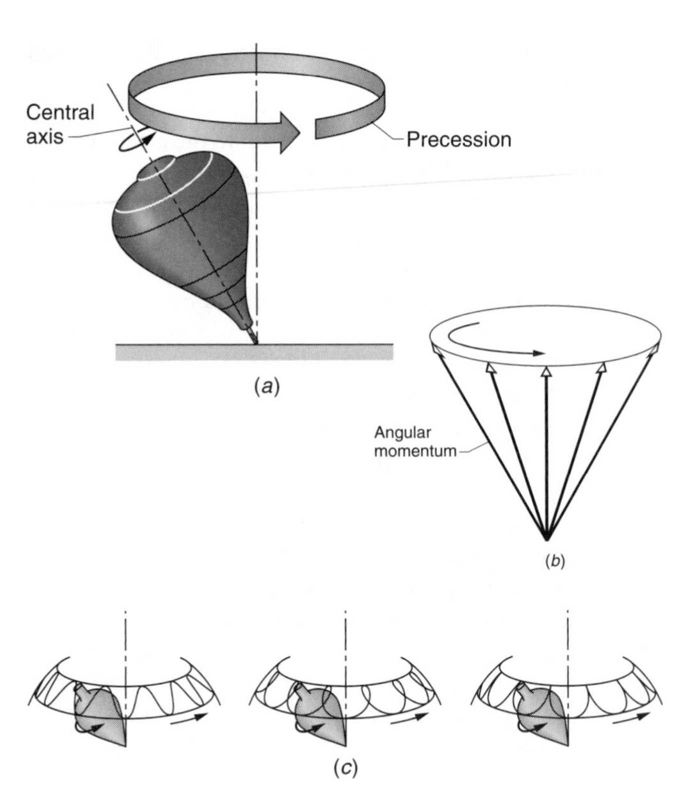

Figure 1-37 / *Item 1.104* (*a*) Precession of a top around a vertical axis through the contact point. (*b*) The top's angular momentum vector moves around the vertical. (*c*) Nutation during the precession.

The traditional explanation of the precession involves the top's angular momentum. This quantity involves the rate at which the top spins around the axis along its length. Moreover, it is a vector quantity that points along that axis. Consider a snapshot of a top that is tilted somewhat and that has a fast counterclockwise spin as seen from above. In Fig. 1-37*b*, the top's angular momentum is represented by a vector that points upward along its central axis.

Because the gravitational force pulls downward on the top, it creates a torque on the top, tending to rotate it around its point on the floor and thus causing it to fall. In fact, if the top were not spinning, it would fall. However, because the top is spinning and already has angular momentum, the torque merely changes the orientation of that angular momentum, rotating the vector around its tail so that its head traces out a cone. Because the angular momentum is along the top's central axis, the central axis also traces out a cone.

Once a top is launched, its center of mass falls somewhat, as the top leans over, and two rules must be obeyed: Its angular momentum around the vertical axis and its total energy must each remain constant. Since the fall tilts the top's spin away from the vertical, the precession must be fast enough to keep the total angular momentum around the vertical constant. The kinetic energy for the precession comes from the fall of the top's center of mass and consequent decrease in the potential energy.

The top cannot continue to fall and still obey both rules, and so the center of mass comes to some lowest point. Then it begins to rise again and the precession slows. The up and down oscillations between the extreme points allowed by the rules is the nutation that is superimposed on the precession. The nutation comes in three types that are characterized by what the center of mass does at the highest point. The top might momentarily stop its precession, continue to move in the same direction as it does at the lowest point, or briefly move opposite that direction (Fig. 1-37*c*). Which occurs depends on the initial precession you give the top in the launch—it might be in the same direction as the precession generated by gravity or opposite it, or maybe you give the top no precession.

If you launch a top with enough spin, it will stay vertical without precession and nutation. But as air drag and the friction on the point gradually steal energy, the spin drops below some critical value, and then the top begins to fall, precess, and nutate. With further drain of energy, the top tilts more, precesses faster, and nutates more wildly, until it finally hits the floor.

A *sleeper* is a top with a design that allows it to spin above the critical value long enough for friction on the point to rotate the top vertically so that it stands upright. Typically such a top is wide and has a blunt point, but the floor surface is also a factor. The friction occurs because the point slips as it spins while also moving through a circle on the floor due to the precession.

SHORT STORY

1.105 • A headstrong suitcase

Robert Wood, the noted physicist from Johns Hopkins, was said to have once played a joke on an unsuspecting porter at a hotel. According to the story, Wood spun up a massive flywheel and then closed it up in his suitcase before the porter arrived. When the porter walked the suitcase down a straight hallway, he noticed only the weight. But when he attempted to turn the corner, the suitcase mysteriously refused to turn. The porter was reportedly so frightened that he dropped the "possessed" suitcase and ran from the scene.

1.106 • Tippy tops

A peculiar type of top, called a *tippy top* or a *tippe top*, consists of part of a sphere with a stem substituting for the missing section. You spin the top by twirling the stem between thumb and finger, releasing it with the spherical (and heavier) side down. Provided there is enough friction between the top and the floor, the top rights itself and then spins on the stem. Relative to you the direction of spin goes unchanged, but relative to the top it reverses.

You can see the same sort of standing up if you spin a football, hard-boiled egg, or the type of school ring that has a smooth stone. In each case, why does the object's center of mass move upward against gravitation?

Answer There has been no simple explanation of a tippy top, only mathematically difficult ones. However, the key element is the friction on the part of the top that touches the floor. Somehow the friction creates a torque that leads to the righting but the details of the process remain elusive. Here is a simple possibility: The friction acts to increase the precession (see above), which then causes the center of mass to move upward, as happens with other types of tops.

1.107 • Spinning eggs

You can tell whether an egg is fresh or hard-boiled without cracking it open if you spin it on its side. A fresh egg spins poorly while a hard-boiled egg spins well. If you spin a hard-boiled egg fast enough, it rises up on one end. If you briefly touch the top center of a fresh egg that is spinning on its side, the rotation starts up again after the touch, but with a hard-boiled egg, the touch eliminates any subsequent motion. Can you account for these behaviors?

Answer The difference between the two types of eggs is, of course, that one is filled with a fluid that sloshes while the other is rigid. The sloshing interferes with the spin of a fresh egg and restarts the rotation when you briefly touch and stop the egg. When a hard-boiled egg is spun fast enough, it behaves like a tippy top (see the preceding item) and stands up.

1.108 • Diabolos

A diabolo is an ancient toy that consists of a spool with conical ends that join at a narrow waist (Fig. 1-38). It is spun by means of a string that runs under the waist and is tied to handles. You begin with the toy on the floor and (if you are right-handed) your right hand low and your left hand high. You then tighten the string by bringing your right hand up smartly and letting the string drag the left hand down. The friction between the string and waist rotates the diabolo.

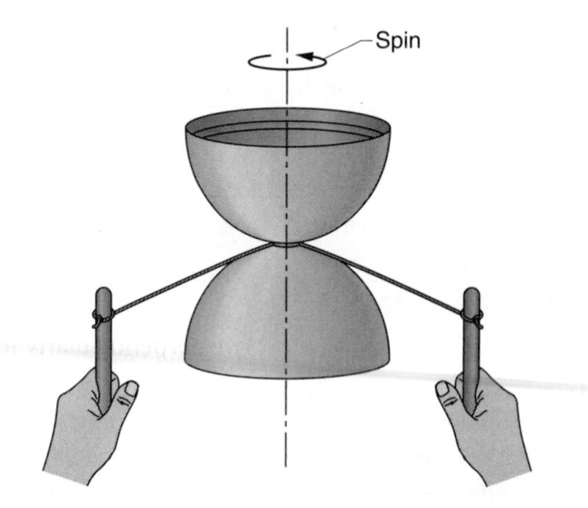

Figure 1-38 / *Item 1.108* Overhead view of a spinning diabolo.

You increase the speed by loosening the string somewhat, allowing the diabolo to descend, repositioning your hands, and then repeating the procedure. If you produce enough speed, the diabolo spins stably on the string. By rapidly raising both hands, you can even toss the diabolo up into the air and then recapture it on the string when it descends.

Why does spin stabilize the diabolo? (Without it the toy merely topples off the string.) If it begins to tilt, how can you stabilize it? For example, if the far end begins to droop, how can you bring it back up? How can you turn the diabolo to your left or right? (Search the Web under "diabolo" for sites that list and demonstrate diabolo tricks.)

Answer If you pick up the diabolo with the string without making the diabolo spin, it is unlikely to be balanced on the string and will fall off. If, instead, you rapidly spin the diabolo, you give it angular momentum, which stabilizes its perch. The angular momentum is a vector that lies along the main axis of the toy. With the launch described in the item, the vector is horizontal and points toward you. A spinning diabolo is stable, because only a torque can change the direction of its angular momentum. If the diabolo is balanced on the string, the diabolo's center is just above the string, and the pull of gravity on the diabolo is down through the string and it does not cause any torque about the string. Thus, the angular momentum cannot change.

If the diabolo is almost balanced, the pull of gravity on the heavier side creates a small torque and gives the diabolo a small, additional angular momentum vector that points to the left or right. As a result the diabolo does not topple over because of gravity but *precesses* left or right; that is, its central axis rotates left or right. (Friction from the string also creates a torque, but if the string is centered or nearly so, this torque will only gradually slow the spin.)

If the far end begins to droop, you can use the string to create a torque to bring the end back up. Pull the string in your right hand toward you and against the right side of the diabolo. The press against the side produces a torque that is downward and that brings the angular momentum vector of the toy back to the horizontal.

To make the diabolo turn to the right, move your hands apart and then pull them toward you. Either the string pulls against the underside of the diabolo or it actually slips toward you, which makes the far end of the diabolo heavier than the closer end. If it does not slip, the pressure on the underside creates a torque that turns the diabolo. If it does slip, the torque from gravity on the heavier side makes the diabolo turn.

1.109 • Rattlebacks

A *rattleback* (also called a *celt stone* or a *wobblestone*) is a curious type of top that has a skewed ellipsoidal bottom surface. The ones sold as a toy insist on spinning in only one direction. If you spin one in the other direction, it quickly stops, rattles up and down, and then spins in the direction it desires. Some stream-polished rocks behave similarly, but you might find a rare type that will reverse its spin several times before its energy dies out. Why does a rattleback reverse direction?

Answer A rattleback is quite difficult to explain in detail but its spin reversal is due to the fact that its bottom surface is an ellipsoid that is misaligned with the general shape of the stone. That is, the long and short axes of the ellipsoid are not aligned with the length and width of the stone. When the stone is spun around the vertical in the "wrong" direction, the misalignment destabilizes the spin and the stone begins to wobble. The friction on the stone from the tabletop transfers energy from the spinning to the wobble. When the transfer is almost complete, the friction reverses the transfer, but this time the stone spins in the opposite direction. With some rattlebacks, the spin in the "correct" direction is also somewhat unstable, in which case wobbling again appears and the spin is again reversed.

1.110 • Wobbling coins and bottles

Flick a coin with your finger to send it spinning on a tabletop, and then both watch and listen to it. As it begins to lie down, the pitch of its clatter first drops and then rises. Is it simply spinning faster? No, if you look down on it, its face is initially blurred by the motion and then later becomes clear enough to recognize.

Balance a bottle on an edge, and then by pulling in opposite directions with a hand on each side, spin it. As it spins, it gradually moves toward the vertical and the clatter increases in pitch. You can also spin a bottle that is almost horizontal, but the launch is more difficult. If you can manage the launch, the bottle will gradually become horizontal during its spin, but unlike the coin, the clatter only decreases in pitch during the descent.

Can you account for these behaviors?

Answer The coin spins around its central axis but the axis is also driven around the vertical, a motion that is called *precession.* The precession comes from a torque that is created by the coin's weight, which can be considered to act at the coin's center. As friction and air drag gradually drain energy from it, the coin begins to lie down and also spin around its central axis slower, which makes the face easier to see. Initially the drain slows the precession but then later the descent of the center of mass actually begins to convert potential energy into addition kinetic energy for the precession. The clatter you hear is made by the precession as it slaps the edge of the coin on the table. The pitch of the clatter increases as the rate of precession increases.

When a bottle is spun in a nearly upright orientation, it too precesses. As its central axis gradually moves toward the vertical, its center of mass descends, and again energy is fed into the precession and the pitch increases. When a bottle is spun in a nearly horizontal orientation, its precession decreases as the bottle falls until the precession reaches some final small value. Then the bottle lies down and rolls on the table.

1.111 • Judo, aikido, and Olympic wrestling

Karate often relies on strength and on collisions with large forces, but judo, aikido, and Olympic wrestling usually employ techniques by which you make your opponent unstable enough to fall. Most familiar is the basic hip throw in judo—you somehow cause your opponent to rotate over your hip from your rear to fall to the mat. You might be surprised to learn that unless you think through the physics, the technique is likely to fail, especially if your opponent is larger and stronger than you. How should you properly execute a hip throw?

Consider also the following example from aikido. An opponent grabs you from the rear, with his arms wrapped around yours and his hands tightly on your wrists. How can you throw him to the mat?

Aikido includes stick fighting, in which the following might occur. An opponent thrusts at you with the end of a long stick. The opponent is too near for you to grab the stick and pull it forward even farther, and besides, that plan would pit strength against strength. Is there a better way to down your opponent?

Answer To perform a hip throw, wait for your opponent to step forward on his right foot, and then step forward with your right foot between his feet, pull down toward the right on his uniform to curve his body forward and move his center of mass out to his navel, and simultaneously turn around toward your left and bring your hips up next to him.

His center of mass is then approximately on your right hip (Fig. 1-39*a*). By pulling on the right shoulder of his uniform, you can easily rotate him around your right hip and to the mat. A key element here is to bend him over in the initial move. If you do not, his center of mass remains buried in his body (Fig. 1-39*b*). If you then twist about and attempt to rotate him over your hip, you must fight against his weight, which creates a torque that counters your torque and so also your attempt at rotation. Your throw then requires strength because you must essentially lift him; if he is heavy, you will probably fail.

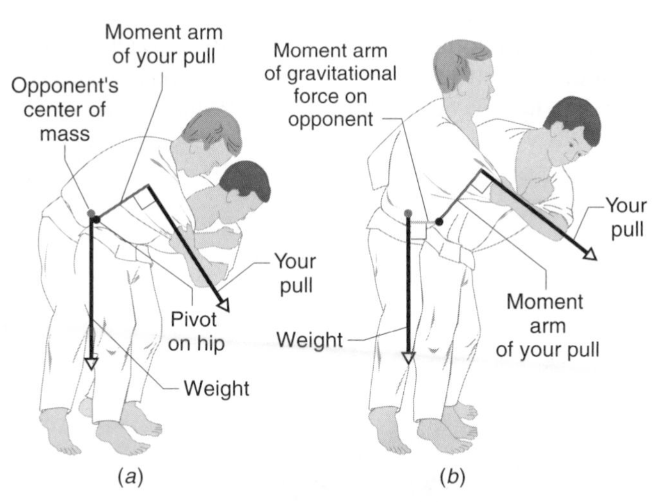

Figure 1-39 / *Item 1.111* A judo hip throw (*a*) correctly executed and (*b*) incorrectly executed.

In the first aikido question, you should bring your hands smartly to your chest (to trap your opponent's arms) while also sliding your right foot forward, dropping downward and rotating your body to the right. In doing so, you bend him and move his center of mass to a point of rotation on your back. He then cannot prevent being thrown over you to the mat.

Stick fighting is difficult to master and my answer here is too brief to explain the art. When your opponent thrusts forward, you should step to the right of the stick, turning so that your left hand can grab its outer portion and your right hand can grab the portion between his hands. Then rapidly bring the stick up and backwards over his head so that he topples backwards. It is important that you begin the technique while the stick is moving forward, because then your opponent is committed to the forward momentum he has created and cannot counter your upward deflection of the stick.

1.112 • Bullet spin and long passes

Why does a rifle come with *rifling* (spiral grooves along the bore's interior) that imparts a spin to a bullet? When the shot is long and arced, what causes the bullet to nose over so that it hits the target front first?

In the American version of football, why must the quarterback put a strong spin on the ball if it is to fly smoothly and then nose down during the latter half of the flight? This procedure not only makes the ball go farther but also makes the receiver's catch easier. A punter kicks the ball with a bit of spin in order to get the same smooth trajectory, but why? Doesn't that just make the ball easier to catch by someone on the opposing team?

Answer If the bullet or the thrown or punted football is given sufficient spin around its long axis, it acts like a gyroscope in that it tends to maintain its orientation instead of tumbling, which would disrupt and thereby shorten its flight. As it travels along a curved path due to the pull of gravity, it encounters air drag on its underside. The drag can be considered to be concentrated at a point somewhat in front of the center of the object. If the spin is large enough, the object is then like a top and tries to align itself with the force it feels—namely, the air drag. So, as it passes through the arc, it gradually noses down.

Some quarterbacks can manage only a wobbly pass because they fail to spin the ball only around the long axis. The additional spin around a short axis through the football produces the wobble, which is an example of precession—the long axis around which the football spins rotates around a circle. The spin and precession are in the same direction (for example, clockwise if the quarterback is right-handed), and the precession is about $\frac{3}{5}$ the rate of the spin.

If a quarterback successfully spins the ball during a pass, not only will it travel farther because of its more streamlined orientation but also the receiver can determine far better just where the ball will come down. When a punter puts spin on a ball, usually the intent is to get the ball to travel farther, but a second intent is to keep the ball aloft longer so that the punter's team can get downfield before the ball comes down. The time aloft is the so-called *hang time*. When the ball is kicked without spin or made to fly in an erratic manner, air drag more quickly removes kinetic energy from the ball and the hang time is decreased.

When bullets are fired directly upward, they sometimes maintain their stability throughout the flight, returning to the ground base-first. Although they probably would then not be lethal, they could still injure someone. If they tumble while falling back down, they will come in much slower than their muzzle speed and the chance of injury is decreased. Still, if someone near you starts shooting in the air, you best hide instead of standing in the open in admiration.

1.113 • Pumping playground swings

How do you *pump* a swing to get it to go higher? And if the swing is initially at rest, how do you start it without shoving off from the ground or having someone push you?

Answer One method is to stand in a swing and pump it by squatting at the high points of the arc and standing up at the lowest point. Standing increases your speed. You can explain the increase by arguments of either energy or angular momentum. By standing you lift your center of mass and do work against the centrifugal force you feel. The work goes into your kinetic energy and increases your speed. By standing you also shift your center of mass toward the point about which you rotate. The action is similar to that of an ice-skater who spins on point while pulling in her arms: The fact that her angular momentum cannot change means that her speed of rotation must increase. On a swing, your rotational speed also increases. With either argument, the increased speed at the low point adds height to the arc. Although the height of your body influences the rate at which you put energy into the swinging, your mass does not.

You can also pump on a swing by pulling on the ropes when you swing forward and pushing on them when you swing back. The distortion you create in the ropes produces forces on your hands that propel you—forward when you pull and rearward when you push.

One way to start a swing is to stand or sit upright with your hands on the ropes and your arms bent, and then fall backward until your arms are fully extended. Your center of mass rotates around the seat of the swing, while the seat rotates around the bar that supports the swing. Your brief fall supplies the kinetic energy for the motion and also its angular momentum.

1.114 • Incense swing

For the last 700 years, ceremonies at the Cathedral of Santiago de Compostela in northwest Spain have been marked by a dramatic swinging of a large censer that hangs some 20 meters from its support. The censer, which weighs about as much as a thin man, is held by rope that wraps around a support and extends down to floor level where it is controlled by a team of volunteers (Fig. 1-40).

After someone begins the pendulum motion with a push, the team pumps the swinging by pulling hard on the rope when the censer passes through its lowest point and then relaxing their pull when it reaches its highest point. The hard pull reduces the length of the pendulum by about three meters, and the reduced pull restores the length. After 17 pulls, which takes less than two minutes, the censer swings up by almost 90°, nearing the ceiling. Its rapid travel through the lowest point fans the coals and incense that burn within it. Why does the team's timed action add energy to the pendulum?

Answer Energy is added to the incense swing by the same mechanics behind the stand-and-squat procedure in the preceding item. When the team reduces the length of the pendulum, the censer is moving rapidly through the low point of its circular arc and they must pull very hard. Thus they do a lot of work on the censer in reducing the pendu-

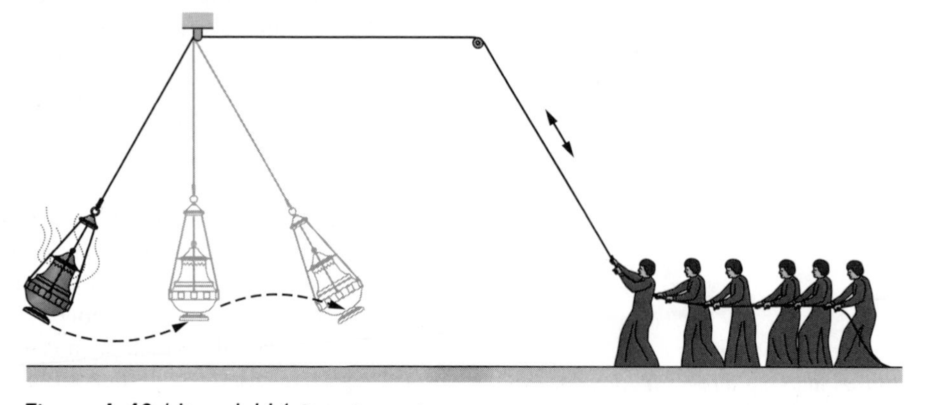

Figure 1-40 / *Item 1.114* Pumping an incense censer.

lum's length, and that energy goes into the kinetic energy of the censer. When the team restores the length at the pendulum's high point, the censer is moving slowly or is momentarily stopped.

1.115 • The pendulum in the pit

In Edgar Allan Poe's masterpiece of terror, "The Pit and the Pendulum," a prisoner finds himself strapped flat on a floor above which a pendulum hangs some 30 to 40 feet. Initially the pendulum seems to be motionless, but later when the prisoner looks up again, he discovers that it is swinging through a distance of a yard and that it appears to have descended. And then to his horror, he realizes that the lower end consists of "a crescent of glittering steel, . . . the under edge as keen as that of a razor . . . ."

As hours go by, the pendulum's motion becomes mesmerizing—the crescent gradually descends and the swing increases in distance, becoming "thirty feet or more." Its intent becomes clear: It is to sweep directly across the prisoner's heart. "Down—steadily down it crept. I took a frenzied pleasure in contrasting its downward with its lateral velocity. To the right, to the left—far and wide—with the shriek of a damned spirit! . . . Down—certainly, relentlessly down!"

Suppose that the crescent is suspended by a rope that is gradually let out. Why would the extent of swinging increase with the crescent's descent?

Answer The extent of the swinging increases because as the crescent descends, its potential energy is gradually converted into kinetic energy. However, calculations reveal that given the initial height and swing as graphically described by Poe, the crescent is unlikely to swing left and right by more than 10 feet when it reaches the prisoner, not the 30 feet or more as reported in the narrative. (The mathematical inconsistency would hardly be of any comfort to the prisoner in Poe's tale.)

1.116 • Inverted pendulums, unicycle riders

If a standard pendulum is inverted, it is, of course, unstable and will easily topple. However, if its support is oscillated vertically and quickly and if there is a bit of friction between the pendulum and the support, why does the pendulum stand upright? It is so stable that were you to nudge it sideways, it would quickly regain its upright stand.

If the pendulum's support is, instead, rapidly oscillated horizontally, the pendulum will swing about the vertical while upside down, as if gravity were reversed in direction. A unicycle rider employs similar physics. When the rider begins to fall—say, forward—stability is momentarily regained by driving the wheel somewhat forward. As the rider then begins to fall backward, the wheel is driven backward.

Can multiple rods, connected in series, be made to stand upright as a series of inverted pendulums if the lower one is oscillated vertically? Can a long wire be made to stand upright like this? And the biggest question of all is, can a rope be made to stand upright like in the classic Indian rope trick where a rope extends upward with no means of support at the upper end?

Answer During vertical oscillations, the pendulum stands approximately upright if the acceleration produced by the oscillations exceeds the acceleration of gravity. In a sense, the pendulum has no chance to fall over, because it is periodically pulled rapidly downward and thereby righted. If the support is horizontally oscillated rapidly enough, the pendulum is also unable to fall. As in the unicyclist's strategy of support, as soon as the pendulum begins to fall one way, the support is brought underneath it in that direction and the fall is arrested.

Several rods connected in series can be made to stand upright if the lower one is oscillated vertically and rapidly enough. A wire that is too long to stand up on its own (it bends over under its own weight) can be made to stand upright if oscillated. However, a rope cannot be made to stand upright because it is too flexible, and thus the Indian rope trick remains just an illusion.

1.117 • Carrying loads on the head

In some cultures, such as in Kenya, people (especially women) can carry enormous loads on their heads. They may have strong neck muscles and an acute sense of balance, but the really surprising feature is how little effort is required of them. For example, a woman might be able to carry a load up to 20% of her body weight without having to breathe heavily (in fact, without any extra effort on her part), whereas a European or American woman of comparable health and strength would find carrying such a load very difficult. What is the secret of the elite load bearers?

Answer During walking, a person's center of mass moves up and down in a periodic fashion. The high point occurs when the body is over one foot while the other foot is moving past that foot, toward the front. The low point occurs when both feet are on the ground and her weight is being shifted from the rear foot to the front foot. This periodic vertical motion of the center of mass, with the support point periodically moving horizontally beneath the center of mass, is similar to the motion of a unicyclist who moves back and forth to maintain balance. In particular, part of the woman's energy is periodically shifted between potential energy (related to the height of her center of mass) and kinetic energy (the speed at which her center of mass moves forward). Normally a person is inefficient in the energy transfer for about 15 milliseconds just after the high point is reached. That is, as the center of mass begins to descend, not all of the potential energy is transferred to kinetic energy, and muscles are used to propel the person forward.

An elite load bearer in, say, Kenya walks in this normal and slightly inefficient way when she is *not* carrying a load.

However, when she carries a load, the interval of inefficiency just after the high point is reached is less. In fact, carrying a moderate load (20% of body weight) may require no more effort than carrying no load at all, presumably because the load causes the woman to shift potential energy to kinetic energy more efficiently than normally. Only if the load exceeds 20% of body weight does a woman have to expend more energy than when unloaded, but even then she expends less energy than, say, a European woman who walks differently.

1.118 • Carrying loads with oscillating poles

In Asia, some people carry light to moderately heavy loads by tying them to opposite ends of a springy pole such as a bamboo pole (Fig. 1-41). As the person walks or runs, the load and pole oscillate vertically. Does this arrangement offer any advantage in carrying the loads?

Answer The vertical oscillations of the person's torso cause the pole and the loads to oscillate vertically. Suppose a rigid pole is used across a shoulder. Then when the torso moves upward, the shoulder must apply a large force to lift the pole and its load. And when the torso moves downward, the shoulder applies little force because the pole and its load fall with the shoulder. So, there can be a considerable variation in the force on the shoulder as the person walks or runs.

The primary purpose of a springy pole is to smooth out the variation in the force on the shoulder. The key is that once oscillations are set up in the pole, the loads oscillate out of step with the pole's center—when the loads move upward, the center moves downward, and vice versa. The center also oscillates out of step with the shoulder—when the shoulder moves upward, the center moves downward. Thus, the shoulder is in step with the loads, which results in a nearly constant force required of the shoulder. When the shoulder moves upward, the spring of the pole is sending the loads upward. When the shoulder moves downward, the upward motion of the center of the pole helps support the downward-moving loads.

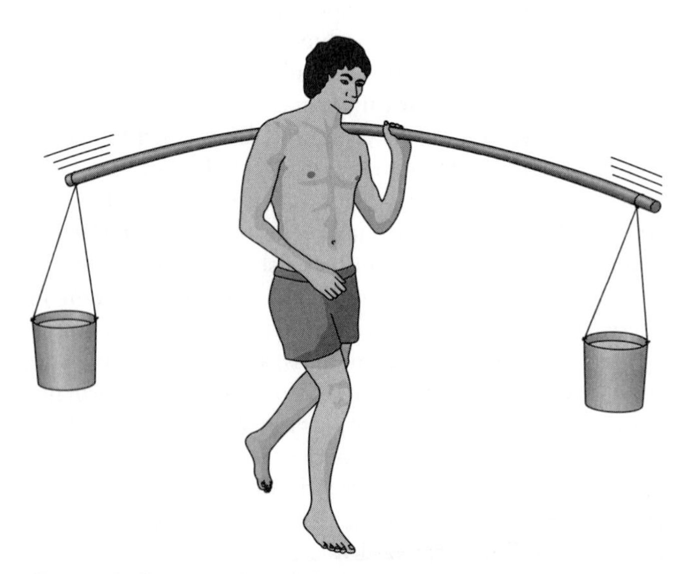

Figure 1-41 / *Item 1.118* Heavy loads are carried on poles that oscillate.

1.119 • Coupled pendulums

Make a system of pendulums by attaching two equal lengths of thread to a support and then wrapping each thread once around a horizontal rod (Fig. 1-42*a*). Add an identical object to the lower end of each thread, and position the rod about a third of the way from the top. Hold one of the objects, move the other one to one side parallel to the rod, and then release both objects. You might think that the displaced pendulum would be the only one to swing, but the motion is gradually transferred to the second pendulum. Once the transfer is complete and the first pendulum is stationary, the transfer is reversed. Thereafter, the motion periodically shifts between the two pendulums.

Similar behavior is shown by the other systems in Fig. 1-42. In Fig. 1-42*b*, a spring connects two pendulums. In the third system (Fig. 1-42*c*), the pendulums are tied to a narrow tube that can rotate around a horizontal string, and the pendulums swing perpendicularly to the tube. In the fourth (Fig. 1-42*d*), they swing perpendicularly to a short string that connects them.

Perhaps surprisingly, the exchange of oscillations can be seen with two identical toy compasses. Place one on a table and then place the other next to it after shaking it to make the needle oscillate. The oscillations are shuttled back and forth between the compasses.

What accounts for the behavior?

Answer Let's consider only the first system described. The transfer of motion comes from a transfer of energy as the pendulums push and pull on each other by means of the rod. If you were to swing the pendulums in either of two special ways, called *normal modes*, there would be no transfer. In one of these modes, the pendulums are made to swing in step (Fig. 1-42*e*), in which the full length of the threads participate in the motion and the swinging has a low frequency. In the other normal mode, the pendulums are made to swing exactly out of step (Fig. 1-42*f*). The opposing motions prevent the thread above the rod from participating, and so the effective length of the pendulums is now smaller than in the first normal mode and the swinging has a higher frequency.

If you start only one pendulum, both modes appear and compete with each other. The pendulums then swing with a frequency that is the average of the frequencies associated with each mode. Their amplitude (the extent of each swing) varies at a rate that is equal to the difference of the frequencies of the modes. As the amplitude for one pendulum decreases, the amplitude for the other pendulum increases, and then the changes are reversed. Similar exchange of

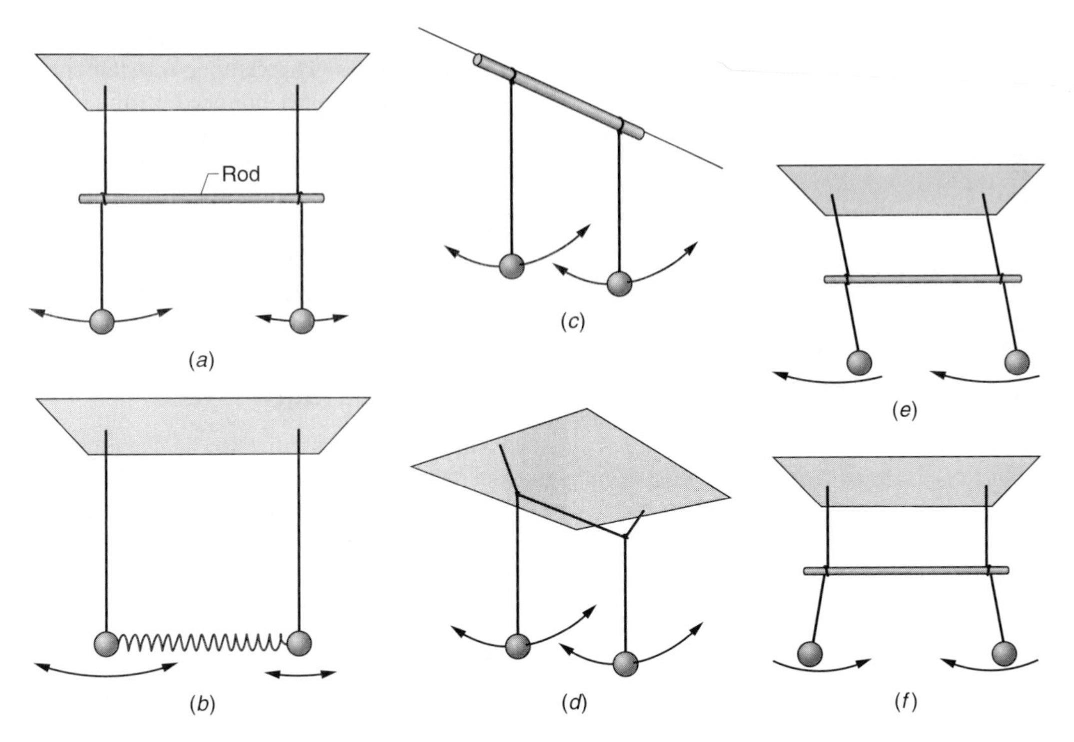

Figure 1-42 / *Item 1.119* (*a*)–(*d*) Coupled pendulums. (*e*)–(*f*) Normal modes.

motion occurs with the compasses because their needles oscillate around the direction of magnetic north just like pendulums oscillate around the direction of gravity.

1.120 • Spring pendulum

Hang a fairly stiff spring by one end and then attach an object at its lower end so that the spring is stretched to about $\frac{4}{3}$ of its initial length. Pull down on the object and then release it. The object initially bobs vertically (Fig. 1-43*a*) but soon the bobbing is replaced with a pendulum motion (Fig. 1-43*b*). Once the bobbing has disappeared, the pendulum motion begins to die out and the bobbing reappears. Thereafter, the motion periodically shifts back and forth between the two types. You can also set up the bimodal behavior if you start the pendulum motion instead of the bobbing.

A similar exchange of motion is displayed by the apparatus shown in Fig. 1-43*c*). The pendulums are connected by a flexible beam that happens to oscillate at twice the frequency of either of the pendulums were they free. In this case energy is periodically exchanged between the pendulum motion and the oscillations of the beam.

An equally complicated example is shown in Fig. 1-43*d*. The horizontal bar can pivot around the support rod. At one end of the bar a vertical bar is fixed in place, while at the other end another vertical bar is free to swing around a pivot. There are two pendulums here: Pendulum A is the second vertical bar and pendulum B is the combination of the horizontal bar and the fixed vertical bar. If the lengths of the bars are adjusted so that the frequency of swinging of A is twice that of B, there will be a periodic exchange of motion once the swinging of A is begun by hand (as in Fig. 1–42*a*).

In these examples, what accounts for the periodic exchange of motion?

Answer Let's consider just the first arrangement. If you could pull down and then release the object perfectly vertically, the object might just bob, but such perfection is unlikely as you are sure to give the object some slight sideways motion. When you choose the object as described, a purely bobbing motion has a frequency twice that of purely pendulum motion.

Suppose that at one moment the object is primarily bobbing. Energy then begins to transfer from the bobbing to pendulum motion. The transfer is due to the fact that the length of the pendulum changes during the bobbing. The situation is similar to a child standing and squatting twice during each complete oscillation of a playground swing. The child changes the effective length of the swing, and the action feeds energy into the swing's motion so that it goes higher.

Once the transfer is complete, it is reversed because of the pull of the object on the spring each time the object swings to an extreme point. The pull comes twice during each complete pendulum swing, and so its frequency matches the frequency of the purely bobbing motion and the bobbing reappears. When it again dominates, the energy is transferred back to the pendulum motion. And so on.

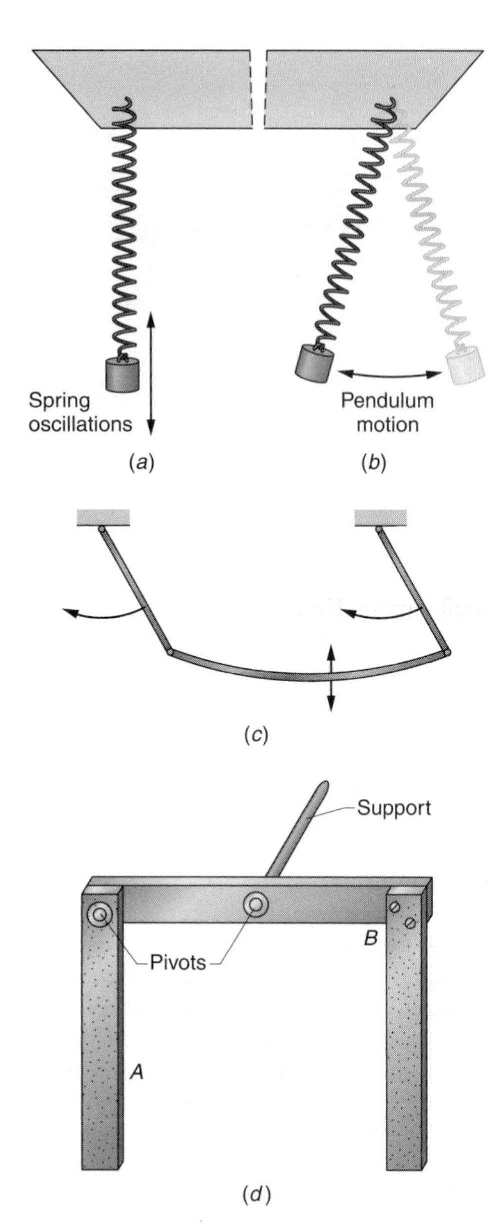

Figure 1-43 / *Item 1.120* Oscillations alternate between (*a*) spring oscillations and (*b*) pendulum oscillations. (*c*) Oscillations alternate between pendulum oscillations and vertical oscillations of the interconnecting beam. (*d*) Oscillations alternate between pendulum oscillations of part *A* and part *B*.

1.121 • The bell that would not ring

When a bell was once installed at the Cathedral of Cologne, it would not ring when swung because it and its clapper swung in step and so the clapper never collided with the bell's interior. What can be done about the problem, short of throwing the bell out of the belfry?

Answer When two pendulums are hinged together and one is both shorter and less massive than the other, they can swing in step. At the cathedral the bell was the longer, more massive pendulum and the clapper was the shorter, less massive pendulum. The clapper was too short. After the bell struck it, the clapper bounced off the bell and matched the bell's motion. So, the two swung together, with no additional collisions. To eliminate the synchronous motion, the clapper was lengthened and thus also made more massive. Then when the bell struck it, the clapper was slower to move and did not keep up with the bell. Thus, as the bell swung back and forth, it would run into the clapper.

1.122 • Spaghetti effect

Why do you sling sauce wildly if you slurp a long strand of spaghetti into your mouth? The effect is not only great fun at the dinner table but is also of concern to engineers designing equipment that either pulls in sheets of paper (which can exhibit the *spaghetti effect*) or ejects sheets of paper (which can exhibit the *reverse spaghetti effect*).

Answer Here is one explanation. Suppose that when the strand is free of the dish, it has some sideways motion. As you suck the strand into your mouth at a constant rate and decrease the length that is still free, the kinetic energy associated with the sideways motion becomes concentrated in a smaller amount of mass. If the amount of kinetic energy is not to change, the speed of the sideways motion must increase. As the end of the strand nears your mouth, the speed becomes large enough that sauce on the strand is slung outward.

A compatible explanation involves angular momentum. If the free end of the strand initially rotates around the entrance point of your mouth, it must rotate faster as it nears that point. It is somewhat like an ice-skater who first spins on point with outstretched arms and then brings them inward.

The spaghetti effect can also be seen with a metal tape measure that is automatically drawn into its case when a button is pressed. As the end of the tape nears the case, the end may lash out sideways and injure you. Instructions advise that you draw in the last part slowly to avoid the problem.

1.123 • The spider and the fly

How does a spider that sits at the center of an orb web know where a fly has become entangled or stuck to the web? Why doesn't a web just break when the fly runs into it? After hitting a web, why doesn't a fly simply fly away?

Answer Because of its thrashing, the fly sends waves along the threads, including some of the radial threads on which the spider sits. The waves on the radial threads can be divided into three types according to the direction of the oscillations of the threads. In two types, the oscillations are perpendicular to a thread, either in the plane of the web or perpendicular to the plane. In the third type the oscillations are parallel to the thread. It is the third type that alerts the spider. If the spider samples that oscillation on two or three adjacent threads, it can quickly determine the direction of the fly because the thread that runs out toward the fly carries the strongest oscillations. Even if an ensnared prey does not thrash about long enough to be detected, the spider can still locate it by plucking the radial lines with its legs. Any line weighted down by the prey will oscillate differently than a free line, cluing the spider about the direction and maybe even the distance to the prey. (There is some experimental evidence that a human can also determine the distance to a load attached to a taut string—without looking—by simply oscillating the string.)

Some spiders tune their webs in the sense that they adjust the tension in the threads. When they are very hungry, they increase the tension so that even the thrashing of a small prey sends noticeable waves through the web. When they are less hungry, they decrease the tension so that the thrashing of only a large prey sends noticeable waves.

In 1880, C. V. Boys (well remembered for his popular book on soap films) described how he could attract the attention of a garden spider by touching a vibrating tuning fork (tone A) to the edge of a web or whatever supported the web. Provided the spider was at the center of the web, it could easily find the fork. However, if the spider was not centered, it had to go to the center first before being able to find the fork. When Boys brought the fork near the spider rather than to a portion of the web away from the spider, the spider took the vibrations as a danger and quickly dropped from the web on a safety thread.

A certain type of tropical spiders is said to be kleptoparasitic because it does not weave its own web but steals prey from a host spider who does weave a web. To monitor the web, the kleptoparasitic spider runs threads (20 or 30 centimeters long) from its resting place to the hub and radial lines of the web. Whenever the host spider's web ensnares, say, a fly, oscillations are sent along the monitoring threads. From the pattern of oscillations, the kleptoparasitic spider can even tell whether the fly was wrapped by the host for later eating. If so, then it will soon sneak onto the web to steal the wrapped food.

A web functions as a filter to catch flying prey that are approximately the size of the spider or smaller by absorbing the prey's kinetic energy and momentum. The web is designed to fail (break) if the prey is larger than the spider because then the prey could hurt the spider.

When a prey hits the web, the threads stretch but they act like a viscous liquid in that they retain most of the energy of the collision internally. Thus, the prey cannot simply bounce from the web. In addition, adhesive drops (which look like microscopic beads) are positioned along some of the threads (the *capture threads*) to trap the prey. (The beads are spaced far enough apart that the spider can pick its way along a thread without itself being stuck to the thread.) A prey will struggle but because the thread is so easily stretched, the prey cannot find anything against which it can push to free itself from the drops.

1.124 • Footbridge and dance floor oscillations

In 1831, cavalry troops were traversing a suspension bridge near Manchester, England, while supposedly marching in time to the oscillations they had created on the bridge. The bridge collapsed when one of the bolts that anchored it failed, and most of the men fell into the water below the bridge. Ever since, troops have been ordered to *break step* when marching across any lightweight bridge. How can marching in step cause a bridge to collapse?

In 2001, a sleek, low-slung footbridge across the Thames was opened in London to connect the Tate Modern art gallery with the vicinity of St. Paul's Cathedral and to mark the new millennium. However, as the first surge of pedestrians began to walk over it, the Millennium Bridge, as it is called, began to oscillate so much that some of the pedestrians kept their balance only by hanging on to the handrail. What caused the oscillations?

Why can similar oscillations occur on a dance floor or at a lively rock concert?

Answer The danger is that if the troops march in step to the oscillations they set up in the bridge, the oscillations might grow to the extent that they rupture part of the bridge's support. (I cannot tell if such was actually the case in the Manchester example.) By breaking step, the pounding on the bridge by the troops was no longer coordinated (synchronized) and the oscillations could not grow.

In walking over the Millennium Bridge, each pedestrian produced forces on the bridge that were not only downward but also left or right. Such forces occur because a person normally walks by swinging the body left and right. These left–right forces are small but on the bridge they happened to occur at a frequency (0.5 hertz, or 0.5 times per second) that approximately matched the frequency at which the bridge could sway left and right. Such a match in frequency is said to be a condition of *resonance*, and the oscillations tend to grow much like the swinging of a child in a playground swing grows if you provide a push with a frequency that matches the swinging frequency.

Initially the pedestrians were largely out of step with one another, the forces were largely unsynchronized, and thus the oscillations grew only slowly. Soon, however, the oscillations were large enough that some of the pedestrians kept their balance by walking in step with them. As more pedestrians fell into step, the oscillations grew even more, which made walking even more difficult and caused even more pedestrians to fall into step. Eventually about 40% of the pedestrians on the bridge were walking in step and the left–right oscillations were appreciable and had even led to up–down oscillations. To fix the bridge, engineers installed devices to drain energy from any left–right swaying of the bridge and thus to prevent pedestrians from falling into step.

Similar oscillations, but largely from vertical impacts, can occur in the floors of offices, gymnasiums, and dance halls. It is especially noticeable when spectators jump in unison as in some forms of dancing such as *pogoing*. The oscillations can also occur in the spectator seating area at a concert if the spectators stamp their feet or even vigorously clap their hands in time with the music. Such spectator activity usually has a frequency of 1 to 3 hertz. If that value is close to the lowest resonant frequency of the dance floor or seating area, resonance can be set up, and then the amplitude and acceleration of the spectators can be not only noticeable but even frightening. To avoid resonance and possible damage or collapse to the structure, building codes commonly recommend that the lowest resonant frequency of the structure be no less than 5 hertz.

1.125 • Precariously balanced structures and rocks

During some earthquakes, seemingly stable block-like structures have been toppled over by the ground shake while seemingly unstable column-like structures have been left standing. Even structures such as community water tanks in the shape of a golf ball balanced on a tee have survived earthquakes while cylindrical water tanks have not. What accounts for the stability of the seemingly unstable structures? This question is obviously important in the design of modern structures in regions where there is seismic activity. It is also important in the preservation of ancient structures such as the classic statues and columns in regions such as Greece.

In rock-strewn terrain, where rocks have been exposed by weathering, the rocks can reveal whether appreciable seismic activity has occurred there. For example, the rocks in some regions in California, even as close as 30 kilometers from the notorious San Andreas fault, indicate that there has been no appreciable seismic activity there any time in the last several thousand years. What simple evidence in the rocks can indicate that lack of activity?

Answer Ground shake (a single pulse, a series of pulses, or to-and-fro oscillation) can cause an unanchored structure to rock on its edges (Fig. 1-44*a*). If the center of mass of the structure moves over an edge, the structure will topple. If you attempt to topple the structure with a push at its top (as you might topple an upright domino), then the structure is more unstable the taller it is. However, toppling by ground shake is a very different mechanism because the push is at the bottom of the structure. Now, the stability of a structure depends on the distance R from the structure's center of mass to an edge (Fig. 1-44*b*); greater R generally means greater stability. Although the effect of ground shake depends on a great many variables, a tall column with a large R may be more stable than a short column with a small R when each is set rocking by the shaking.

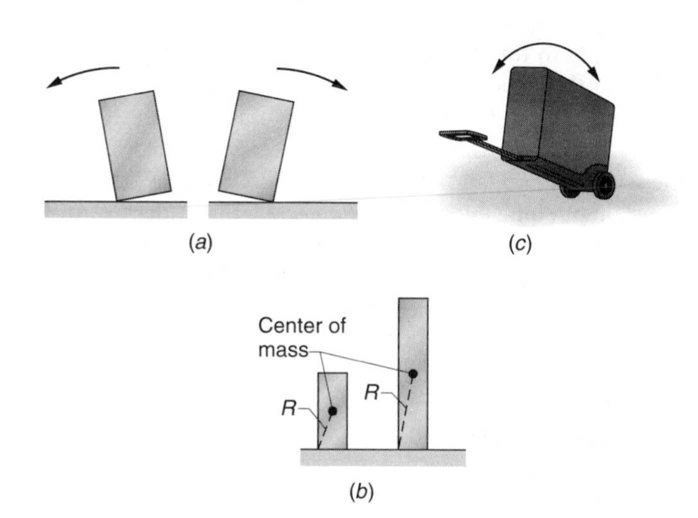

Figure 1-44 / *Item 1.125* (*a*) Block set rocking by ground shake. (*b*) Danger of toppling depends on distance *R*. (*c*) Two-wheeled suitcase can rock and then topple.

You may have seen similar rocking if you have ever pulled a two-wheeled suitcase through an airport (Fig. 1-44*c*). If you walk slowly so as to pull steadily on the suitcase handle, the suitcase is stable (remains upright). But if you walk quickly, periodically pulling on the handle, the suitcase can rock left and right on its wheels. If the rocking is large enough, the suitcase will topple over, even if you attempt to arrest the fall by twisting the handle in the opposite direction.

In some rock-strewn regions, weathering of rocks has left some of them balanced on a narrow pedestal of remaining material. Such *precariously balanced rocks* (as they are called) can usually be toppled by hand and would be toppled by even moderate ground-shaking during seismic activity. Thus, the fact that the rocks have been standing for thousands of years means that the region has not seen appreciable seismic activity during that period.

1.126 • Sinking of the nuclear submarine *Kursk*

In August 2000, as Russia's Northern Fleet conducted exercises in the Barents Sea north of Russia, the nuclear submarine *Kursk* mysteriously sank. As word of the loss spread, seismologists from around the Northern Hemisphere realized that on the day the *Kursk* sank, they had recorded unusual seismic waves originating in the Barents Sea. Analysis of the data suggested the reason the submarine sank, and—more surprising—it also revealed the depth. How could the submarine's depth be determined from measurements made very far away?

Answer Seismic waves are waves that travel either through Earth's interior or along the ground. Seismology stations are set up mainly to record seismic waves generated by earthquakes, but they also record seismic waves generated by any large release of energy near Earth's surface, such as an explosion. As the seismic waves travel past a station, they oscillate a recording pen and the pen traces out a graph. The traces attributed to the *Kursk* consisted of a first set of small-amplitude oscillations; 134 seconds later, oscillations with much greater amplitudes began.

From these tracings, analysts concluded that the first seismic waves were generated by an onboard explosion, possibly a torpedo that failed to launch when fired. The explosion presumably breached the hull, started a fire, and sank the submarine. Then later, much stronger seismic waves were generated after the submarine was sunk and were possibly generated when the fire caused several (probably five) of the powerful missiles on board to explode simultaneously. These stronger waves arrived at seismology stations as pulses separated by a time interval of about 0.11 second.

From that time interval, analysts could calculate the depth of the sunken submarine. The stronger explosion occurred when the submarine was sitting on the ocean floor (the seabed). It sent a pulse into the seabed and a pulse upward through the water. The pulse traveling through the water "bounced" several times between the water surface and the seabed. Each time it hit the seabed, it sent another pulse into the ground, and seismology stations detected those ground pulses as they arrived one after another. Thus, the time of 0.11 second between any two successive ground pulses was equal to the round-trip time for the water pulse to travel up to the water surface and back to the seabed. Using that time interval, analysts calculated the submarine was at a depth of approximately 80 meters; in fact, it was later discovered at a depth of about 115 meters, remarkably close to the calculated depth.

Seismologists have recorded other major explosions, such as the truck-bomb blast in Nairobi, Kenya, in 1998 in a terrorist attack on the American embassy. In 1989, they recorded the seismic waves produced by the (acoustic) shock wave generated by the space shuttle *Columbia* as it flew over Los Angeles on its (successful) return to Edwards Air Force Base. And on September 11, 2001, seismologists recorded the collisions of the hijacked airplanes with the towers of the World Trade Center and the subsequent collapse of the towers.

1.127 • Detection by sand scorpion

When a beetle moves along the sand within a few tens of centimeters of a sand scorpion, the scorpion immediately turns toward the beetle and dashes to it (for lunch). The scorpion can do this without seeing (it is nocturnal) or hearing the beetle. How can the scorpion so precisely locate its prey?

Answer A sand scorpion determines the direction and distance of its prey from the waves the prey's motion sends along the sand surface. With one type of wave, transverse waves, the sand on the surface moves vertically and thus perpendicularly to the wave's travel direction. With the other type, longitudinal waves, the sand moves parallel to the wave's travel direction. The longitudinal waves travel three times faster than the transverse waves. The scorpion, with its eight legs spread roughly in a circle about 5 centimeters in diameter, intercepts the faster longitudinal waves first and learns the direction of the beetle; it is in the direction of whichever leg is disturbed earliest by the waves (Fig. 1-45). The scorpion then senses the time interval between the first interception and the interception of the slower transverse waves and uses it to determine the distance to the beetle. For example, a time interval of 0.004 second between the arrivals of the two types of waves means that the waves originated 30 centimeters from the beetle. In this way, the scorpion immediately determines the direction and the distance to its prey.

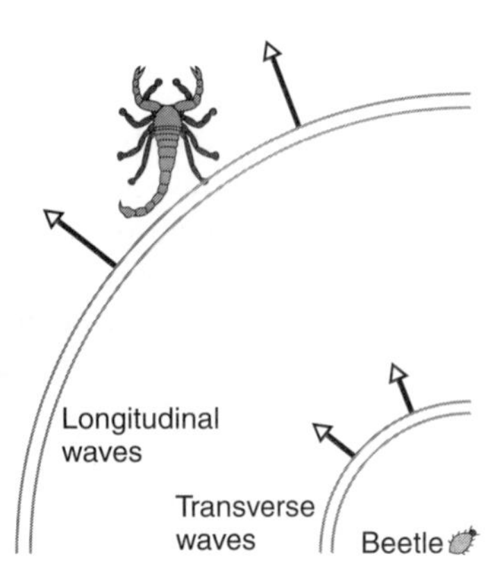

Figure 1-45 / *Item 1.127* Waves alert a scorpion as to the beetle's motion.

1.128 • Snow waves

Why, in apparently rare circumstances, can a footstep in a field of snow set off a *snowquake* that travels away from the site, usually with a low-frequency plopping sound?

Answer A snowquake is probably the progressive lowering of the snow surface due to the collapse of a structurally weak layer of hoarfrost below the snow (and thus hidden). The footstep causes the collapse of the hoarfrost just below it, and that collapse tugs and shakes on the surrounding hoarfrost, which then collapses, and so on. As the hoarfrost collapses, the snow slumps down with a plop, making a sound much like snow does when it falls from a tree branch onto a bed of snow.

1.129 • Football-stadium wave

A football-stadium wave is a spectator-created pulse that travels around large, crowded stadiums during sporting events. The effect first gained widespread attention at the 1986 World Cup (soccer to the United States, football to the rest of the world) in Mexico and thus is often called the *Mexican wave* or *La Ola.* As the pulse travels around the stadium, spectators stand with raised arms and then sit back down. How does the wave get started (there is no signal from, say, an announcer) and how fast does it travel?

Answer The wave can begin only if it is noticeable. One or even a few people standing and sitting back down is insufficient because such action would be lost in the normal motion of spectators. Instead, a wave needs a sizable number of participants to stand and sit in unison. Thus, the wave can begin only if one or more initiators can organize the first group of, say, 20 or 30 participants. The initiators could turn to face the group with perhaps a flag to gather attention. The simultaneous motion of the first group would then be noticed by the adjacent group of people, who would then stand and sit, and so on. Studies show that the wave usually travels clockwise around the stadium (as seen from above), but I cannot explain why. The speed is approximately 12 meters per second, which seemingly depends on the time required for a spectator to react by standing after seeing an adjacent group of spectators stand.

1.130 • Body armor

How does the fabric type of body armor stop small-arms munitions (handgun bullets and fragments from bombs or grenades)? Why doesn't it stop a knife?

Answer When a high-speed projectile strikes body armor, the fabric stops the projectile and prevents penetration by quickly spreading the projectile's energy over a large area. This spreading is done by longitudinal and transverse pulses that move radially from the impact point, where the projectile pushes a cone-shaped dent into the fabric. The longitudinal pulse, racing along the fibers of the fabric ahead of the denting, causes the fibers to thin and stretch, with material flowing inward into the dent. One such radial fiber is shown in Fig. 1-46. Part of the projectile's energy goes into this motion and stretching. The transverse pulse, moving at a slower speed, is due to the denting. As the projectile increases the dent's depth, the dent increases in radius, causing the material in the fibers to move in the same direction as the projectile (perpendicular to the transverse pulse's direction of travel). Part of the projectile's energy goes into this motion. Some of the energy is dissipated by the rubbing of the fibers where they cross one another in the weave or, in body armor consisting of multiple layers, by the stretching and breaking of the fibers.

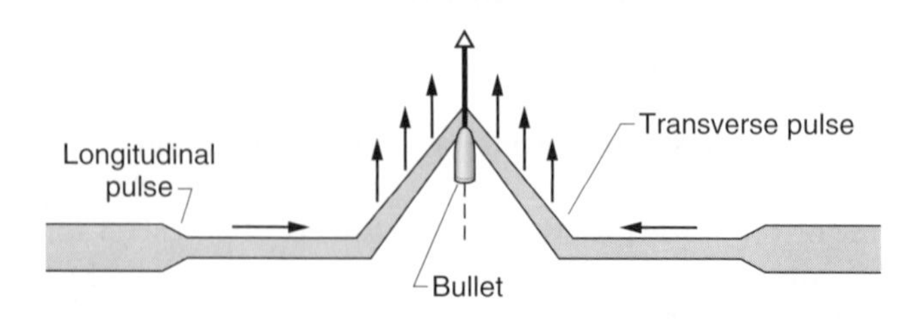

Figure 1-46 / Item 1.130 Dent in body armor due to projectile.

Fabric body armor does not stop a knife because the knife point can easily penetrate the fabric between the fibers, and the sharp edge can then cut the fibers as the knife continues to move forward. You might think that mail, the flexible armor worn in the days of fighting knights, would work better, but it was designed to stop the broad slash of a sword, not the concentrated projection of a knife point.

1.131 • Archer's paradox

No matter how well aimed an arrow is, once it is released and begins to move past the bow's stock, it will point off the target by as much as 7°. Yet it still heads toward the point of aim. The deviation of the arrow is even stranger if the arrow is followed in slow motion. Although it rests against the stock when aimed, the arrow never again touches the stock once released. Instead of sliding along the stock, it snakes around it. What accounts for the behavior, and how does the arrow still find its mark?

When the longbow was used in combat, why was the arrow prepared with a ball of beeswax formed on its point?

Answer Suppose that the arrow is on the left side of the stock. Just as it is released, both the string and the stock push its ends leftward, and the arrow buckles and then begins to oscillate left and right. The oscillations allow the arrow to snake around the stock without losing any energy through rubbing and without the feathered end striking the stock. Although the tip of the arrow does not always point toward the target during the oscillations, the flight is still in that direction. Soon after the arrow clears the bow, the oscillations die out, and then the tip points in the intended direction.

For an arrow to snake properly around the stock, it should be able to undergo one full oscillation by the time it leaves the string. The requirement demands a certain flexibility of the arrow. If it is too flexible, the oscillations are too slow and the feathered end hits the stock. If it is too stiff, the oscillations are too fast or the extent of the sideways motion is too small, and then the arrow fails to leave the bow with its full energy due to rubbing or a collision of the feathered end.

Reportedly, a ball of beeswax was placed on the tip of an arrow so that the arrow might better pierce the armor of a soldier. The reason given is that because the ball strikes the armor first, the collision causes the arrow to become more perpendicular to the armor just as the point of the arrow reaches the armor. Thus, the arrow is less likely to glance off the armor and more likely to penetrate it.

1.132 • Oscillating plants

A tree can be broken or uprooted if it is bent over by the gale of a hurricane or typhoon. How can it be in just as much danger in a substantially smaller wind?

Answer Any given tree will sway with what is called its *natural frequency*, in which the base is fixed in place, the top sways the most, and intermediate points sway by intermediate amounts. The value of the natural frequency depends on the length of the tree, the strength of the tree material (its ability to bend), and the air drag on its branches and leaves. Although a single gust of wind can set the tree swaying, the motion soon dies out and is unlikely to bend the tree over enough to break or uproot it. Those dangers come when a series of gusts hit the tree at a rate that is close to the tree's natural frequency, a condition said to be *resonance*. The situation is then like you pushing with a modest force on a playground swing carrying a child. If you push with the natural frequency of the swing, you can gradually build up the extent of the swinging. For gusts and trees, the swaying can also be built up.

Of course, wind gusts do not occur at a fixed rate but if their average frequency falls close to the resonant frequency of a tree, the swaying may be enough to break or uproot the tree. However, if the tree is surrounded by other trees, not only is the tree partially shielded from the gusts but also the energy of its motion is gradually lost to the rubbing of limbs with those other trees. Any tree, whether clustered or isolated, will also lose energy to the air drag on the foliage and to the stretching and compression of its stems.

Crop plants are also subject to resonant swaying by wind gusts, and so they too can be broken or uprooted by persistent gusts coming at a rate roughly at the resonance frequency. For corn stalks, that is a frequency of 1 to 2 hertz, somewhat higher than for trees.

1.133 • Oscillating tall buildings

The impact of wind on a tall building can cause the building to oscillate, which can be irritating or even nauseating to the building occupants. Making the building stiffer to decrease the deflection due to the wind is not practical or economical. How then can you reduce the oscillations to an acceptable level?

Answer One way to decrease the oscillations is to mount a spring–block device on the roof, with the spring aligned with the prevailing direction of the wind. One end of the spring is attached to the roof; the other end is attached to a block that can move along a track parallel to the spring. The frequency at which the block would naturally oscillate on the end of the spring is adjusted until it matches the frequency at which the building naturally sways. Then, when the building sways, the spring is stretched, causing the block to oscillate at the same frequency. However, the block's oscillation is delayed from the building's oscillation so that the two oscillations are exactly out of step. For example, when the building sways to the left, the block is oscillating to the right and thus tends to offset the building's sway.

Some buildings have double spring–block oscillators, one spring–block oscillator mounted on the block of a larger spring–block oscillator. The oscillations of the smaller oscillator are fine-tuned by an electronic circuit monitoring the building oscillations. Other buildings have a water oscillator in which water oscillates from side to side out of step with the building. To diminish the sway of the Petronas Tower in Kuala Lampur, Malaysia, which is 101 stories (508 meters) tall, a pendulum with a 680 000 kilogram ball is mounted on the 92nd floor.

1.134 • Diving from a springboard

In springboard diving, a skilled diver knows how to make a running dive: The diver first takes three quick steps along the board to start the board oscillating and then leaps to the free end of the board so as to be catapulted high into the air. A novice diver can imitate that procedure but fail miserably to be catapulted upward and might even be knocked off the board. What is the "secret" of a skilled diver's high catapult?

Answer A competition diving board sits on a fulcrum about one-third of the way out from the fixed end of the board. In a running dive, a diver takes three quick steps along the board, out past the fulcrum so as to rotate the board's free end downward. As the board rebounds back through the horizontal, the diver leaps upward and toward the board's free end. A skilled diver trains to land on the free end just as the board has completed 2.5 oscillations during the leap. With such timing, the diver lands as the free end is moving downward with greatest speed. The landing then drives the free end down substantially, and the rebound catapults the diver high into the air.

1.135 • Fly casting

If you throw a fishing fly as hard as you can, it travels only a short distance because of air drag. How, then, can you cast a fly a great distance with a rod and reel? Surely, the task is even more difficult because the line also meets air resistance, and yet the procedure gives a large speed to the fly.

Answer To cast, you bring the rod up and somewhat past the vertical to toss the fly and line to the rear, and then you pull the rod sharply forward to hurl them forward. Your force on the fly and line is effectively at the tip of the rod. Were you to throw them by hand with an identical force, you would do little work and give them little kinetic energy because the distance your hand moves is small. Since the tip of the rod moves through a larger distance, your work and the energy you give the fly and line are larger.

Once the tip of the rod is forward and stationary (Fig. 1-47*a*), the kinetic energy and speed of the fly increase even though you no longer do any work. To see the increases, first note the shape of the line just then (Fig. 1-47*b*): It extends forward from the tip of the rod, curves upward and toward the rear, and then extends back almost horizontally

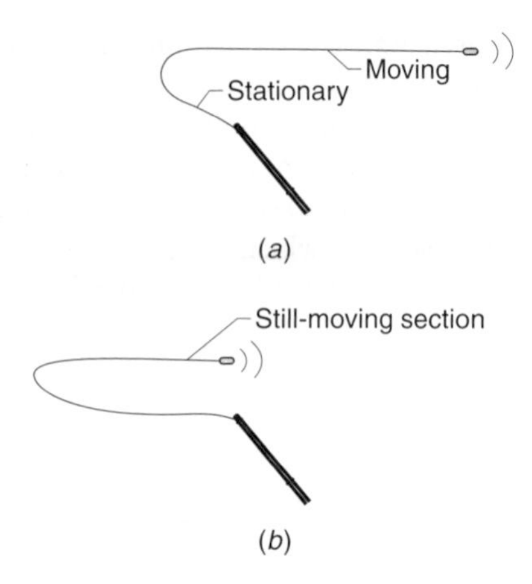

Figure 1-47 / *Item 1.135* A fishing line cast forward. (*a*) Most of the line is moving. (*b*) Less of the line is moving.

to the fly. The first section is stationary because the rod is stationary, while the last section is moving along with the fly. As the fly moves forward, more of the line is in the stationary section, and so the kinetic energy becomes concentrated in the fly and the length of line that is still moving. When the fly reaches the forward-most point, it has all of the kinetic energy and moves quickly, much faster than if you threw it by hand. If you are holding some of the line loosely just then, the fly can pull it out and thus travel far out over the water.

Air drag limits the travel distance of the line. That is the reason why anglers strive for a small loop in the line so that the loop runs into less air. They also strive for an asymmetric loop with a forward-pointing top and less-curved bottom. The air drag on the bottom part of such a loop produces a vertical lift on the line that allows a longer cast. This is the technique used by anglers in competition fly casting.

Some anglers believe that the bending of the rod on the preliminary backward cast of the line is the primary source of energy for the fly during the forward cast, but studies reveal that such a contribution is small. However, the flexibility of the rod is important in the placement of a cast and in the handling of a fish. The stiffness of a rod is measured by the amount of the load required on the tip of the rod to bend the rod by a certain amount. Anglers usually choose a stiffer rod when fishing for large fish because they don't want the pole to bend over. The frequency of a rod is measured by the rate at which it will oscillate when the butt end is clamped in place and the tip is first deflected by a small amount and then released. High-frequency rods, said to be "lively," are often preferred for long casts. Low-frequency rods offer more control and are used to place a fly accurately.

1.136 • The Falkland Islands battle, Big Bertha

During World War I, British and German navies battled near the Falkland Islands, which is at a latitude of about 50° S. Although the British shots were well aimed, they mysteriously landed about a hundred meters to the left of their targets. Were the gun sights off? Seemingly not, for they had been precisely set back in England. What was wrong?

During the German shelling of Paris in that war, a huge artillery piece nicknamed Big Bertha hurled shells into the city from 110 kilometers away. Had not the Germans taken scientific principles into account, their shots would have missed the mark by almost 2 kilometers.

When the Germans first began to test long-range artillery, they were surprised to find that if a shell is fired at a large angle, larger than 45°, it went much farther, perhaps twice as far as when 45° was the angle. Since in many common situations a launch at 45° produces the maximum range for a shot, why were their shells going farther with a larger angle?

Answer We usually invent a fictitious force, the *Coriolis force*, to account for the apparent deflection of a long-range shell that is actually due to Earth's rotation during the shell's flight. The apparent deflection is rightward in the Northern Hemisphere and leftward in the Southern Hemisphere, and it is larger at larger latitudes and zero at the equator.

When a long-range shell is launched, it has not only the velocity due to the launch but also a certain velocity due to the rotation of Earth at the location of the launch. During the shell's flight, the target continues to circle around Earth's axis because of the rotation. If the target's motion is not taken into account, the shell will miss the target. For an example in the Northern Hemisphere, suppose the target lies due north of the shell's launch site. Both the target and the launch site circle eastward around Earth's axis, but the target, being at a higher latitude, follows a smaller circle than does the launch site. Since both sites must complete a full circle in a day, the target moves slower than does the launch site. When the shell is launched due north, it also has the same eastward speed as the launch site. During its flight, it travels eastward faster than does the target and so it ends up to the east of the target. In the perspective of someone at the launch site, the shell is deflected eastward—that is, toward the right of the target.

Gunners allow for the deflection through trial and error but the correction to the gun sights depends on latitude and is in the opposite directions in the two hemispheres. The British guns were set well for the latitude of England but woefully wrong for the southern latitude of the Falkland Islands. With the long flight of the shells from Big Bertha, the Germans knew to correct for the Coriolis deflection—during a shell's flight, Paris moved.

When the Germans fired long-range shells at an angle larger than 45°, the shells traveled through the thin reaches of the atmosphere, which reduced the air drag on them, and so the shells went surprisingly farther.

1.137 • Jack and the beanstalk to space

Is there any way to put a satellite into orbit and then drop a line from it to the ground so that material can be hoisted to the satellite? Is there any way that the satellite could be moved away and the disconnected line then remain in place? (You would then have a "beanstalk" but no giant.)

Answer If the satellite is in an equatorial orbit and at the proper altitude so that it rotates around Earth just as fast as Earth turns, then in principle a line could be lowered to Earth and an elevator system could even be rigged. If the satellite is higher, the effective centrifugal force on it would pull on the line—the arrangement would then be a *skyhook* that could lift materials on the line without need of an elevator system. A strong, lightweight line might actually be left standing free, like the fabled beanstalk, if the effective centrifugal force balanced the line's weight, but calculations show that the line would need to be about 143 million meters long, a bit much.

If the satellite is in an orbit that leaves the lower end of the line skimming over Earth, and if the line is elastic, then it might provide a virtually free means of transportation. A passenger compartment could be added to the line's lower end and, as the line stretched out because of the pull by the satellite, the compartment would hop into the atmosphere and then come down again after traveling a great distance. Although during the ascent the compartment's pull against the satellite would reduce the satellite's energy, most of the energy would be given back during the descent when the compartment pulls the satellite along its orbit. To allow for inevitable energy losses, the satellite might be equipped with a small rocket.

1.138 • Spring fever and the standing of eggs

Try standing a raw egg on its end. Chances are that it will just topple over. Do you have any better chance of standing it up on the vernal (spring) equinox as some people believe?

Answer To understand the vernal equinox, imagine a plane that extends through Earth's equator and out to the Sun. Also imagine that the Sun orbits Earth instead of the other way around. In this arrangement, the Sun's orbit is slanted with respect to the extended plane, and the Sun passes through the plane twice a year. One of those days is the vernal equinox. According to rumor, the gravitational pull by the Sun on objects on Earth, particularly an egg, is somehow different on that day. The rumor is just untrue.

So why then does the rumor continue? One reason may be that a few people make a concerted effort on the vernal equinox, and on that day only, to stand up eggs. (Those people must be pretty bored.) If they gain moderate success on that day, then they claim there is something special about gravity then. Were the idea true, surely you will feel the difference—your mass is larger than an egg's and so you should feel the stabilizing pull from the Sun all that much more. Needless to say, you feel no different on the vernal equinox and probably do not even know when it passes.

If you find an egg that will stand on end, that end will likely be slightly flat, though perhaps only in a small region. A sneaky way to stand up most any egg is the following: Make a small mound of salt, lightly press the blunt end of the egg down into the mound, adjust the egg so that it is upright, and then carefully blow salt away from the egg. The few salt crystals that remain are wedged between the egg and the table and provide enough support to keep the egg upright. Someone unaware of what you have done might not even see the remaining crystals; you could then attribute the egg-standing to, say, an increased flux of cosmic rays. (Why not? It makes as much sense as attributing the stunt to the vernal equinox.) You could also cheat by flattening the end of an egg with sandpaper.

Here's another way that sometimes works. Shake the egg so as to rupture the membrane holding the yolk. Then hold the egg upright on a table for a few minutes to allow the yolk to settle into the bottom end and thereby lower the center of mass of the egg. The bottom-heavy egg might then continue to stand up when you release it.

The habit of standing up eggs on the first day of spring appears to have begun in China thousands of years ago. Since then, an uncountable number of eggs have been stood on that particular day. You might think that such success proves that the day is special. Well, no. The first day of spring on the Chinese calendar is about 90 days before the vernal equinox.

1.139 • Moon madness

Most people believe that the number of births, car accidents, admissions at hospital emergency rooms, aggressive acts, and a whole host of other human activities increase during a full Moon. Just how would the Moon cause this lunar effect—is the effect due to the gravitational force of the Moon, is it psychological, or is nonexistent?

Answer Can gravitation be the cause? No, the gravitational pull on you due to the Moon is imperceptibly small. Were it noticeably large, then you would feel the effect as the Moon rose in the sky and you thereby move somewhat closer to it, increasing the Moon's gravitational pull on you. Do you feel lighter as the Moon climbs the sky? No, of course not.

Could a tidal effect due to gravitation be the cause? The Moon certainly has an appreciable and easily seen effect on oceans by causing the tides. Do people somehow respond to the same effect? No, the tides are due to the variation of the Moon's (and Sun's) gravitational force across the width of Earth. That variation over such a large distance produces a bunching of water. As Earth turns, some ocean regions turn through this bunching and experience high tide. The variation of the Moon's gravitational force across the width (or length) of a person is too small to produce any similar tidal effect. So, this too is not the answer.

But why even consider gravitation? The term full Moon means that the full face of the Moon (in our perspective) is illuminated by the Sun. That extent of illumination would in no way alter the gravitational force on us due to the Moon.

So, one might guess that the lunar effect is psychological—people are somehow driven into a frenzy because of the added illumination at night, even if they happen to live in bright city lights or not go out at night.

Alas, if you plot the number of births, car accidents, admissions at hospital emergency rooms, aggressive acts, or other human activities versus the phase of the Moon, you actually will find no peak at full Moons. The lunar effect is just a legend, even among professional health workers who should know better.

1.140 • Gravity hill

Scattered around the world are places where gravity seems to pull a car up a hillside. One location is just outside Mentor, Ohio. When I coast down the hillside with my car in neutral, the car gradually slows to a stop and then begins to travel back toward the crest of the hill. Can gravity actually pull upward in such places? (If you visit one of these hills, be extremely cautious that you are not hit by another car—a driver will not expect to find a stopped or slowly moving car.)

Answer The effect is an illusion, but it can be so convincing that the experience is unnerving. (When I first sampled the illusion near Mentor, one of my daughters, then a child, was in the car. Although she knew little about gravity, she knew enough to burst into tears as the car rolled uphill.) If you sight along the surface of the road, the illusion disappears and you see the true inclines of the road. You may find that there is a shallow dip in the overall moderate slope of the hill. (Be careful of other cars.) When the car rolls backwards and toward the crest, it is actually rolling back into the dip. When you sit in the car, the dip is imperceptible and the illusion of rolling up the hillside is strong. If the trees alongside the road are tilted appropriately, they can enhance the illusion.

The illusion of a wrong tilt of a section is sometimes due to a much stronger tilt of the road before and after the section. In one example, if the near and far road sections are strongly downhill and the middle section is only moderately downhill, that middle section may appear to be uphill. The apparent horizontal can also affect the perception of tilt: For example, imagine a horizontal road that curves off to the left just in front of a hillside that hides the true horizon. You get the impression that as the road approaches the hill, it curves downward because the apparent horizon is at the hilltop and thus high.

1.141 • Falling through the center of Earth

Imagine a hole that extends along the rotation axis of Earth, from one pole to the other. If you were to fall into such a hole, how long would you take to reach the opposite end, and if you did not somehow escape then, what would happen to you? Would matters be any different if the hole went through Earth elsewhere?

A shorter version of such a tunnel has been suggested for transportation in heavily traveled regions, such as between New York and Washington, D.C. A straight tunnel would be dug between the two cities and track laid. When a train is released at one end of the track, it would require almost no energy from an engine to reach the opposite end of the track. What would propel it, and how much time would the trip require?

In *Pole to Pole*, an early science fiction story by George Griffith, three people attempt to make a trip through Earth by using a naturally formed (and of course fictional) hole that extends through the North and South Poles. Beginning at the South Pole, their capsule first falls toward the center of Earth while being slowed by balloons that are filled with helium or hydrogen. As the story goes, the gravitational force on the passengers becomes alarmingly large as they approach the center of Earth and then, exactly at the center, it momentarily disappears.

The later ascent toward the North Pole proves slower than anticipated, and the balloons are again employed to give lift, but calculations by the onboard scientist reveal that the capsule will rise only to a certain height and there it will slow to a stop, trapping the passengers. Even ditching heavy machines fails to lighten the load sufficiently. In desperation, the scientist lowers himself through a bottom hatch, briefly clings by his hands, and then drops away from the capsule. The loss of his mass is enough to allow the capsule to reach the end of the tunnel where the remaining two passengers escape. (Scientists are often wont to sacrifice themselves for the benefit of others.) Is the story sound?

Answer Suppose that you fall into a straight tunnel connecting the poles. Freeze the picture after you have fallen to a certain radius from the center. Imagine a sphere with that same radius and centered on Earth's center. The mass inside the sphere pulls on you but the mass outside the sphere effectively does not, because for every outward pull by an outside section on your side of Earth, there is a matching inward pull by some other outside section on the opposite side of Earth.

Now continue the fall. As you approach the center, the radius of your location and the size of the remaining mass contained in a sphere of that radius shrink, and so the gravitational pull on you also shrinks. When you pass through the center, the pull is momentarily zero. The ascent through the rest of the tunnel is a reverse of the descent. Assuming ideal conditions, such as no air drag, equal distances for descent and ascent, and the miraculous idea that you could somehow survive the heat and other lethal conditions in Earth's core, you would come to a stop just as you reached the opening at the end of the ascent.

The time of full passage would be about 42 minutes. (The result assumes that the density of Earth is uniform. If the core is denser than the rest of Earth, the result is several minutes less.) If you did not escape, you would bob back and forth in the tunnel forever.

If the tunnel were located somewhere else, it would have to be curved if you were not to crash into its sides. The trouble is that you begin the descent with whatever rotational speed the ground has at the opening to the tunnel. As you fall toward the center, you pass into sections with less rotational speed and will run into the side of the tunnel.

The straight-line tunnel between cities is nearest the center of Earth at the tunnel's midpoint. A train would essentially fall down the first half of the track and then climb back up the second half. Extra energy would be needed only to overcome friction and air drag. The trip would take 42 minutes, the same as in the pole-to-pole tunnel.

I leave the details of the science fiction story for your analysis.

1.142 • Stretching of plastic shopping bags

When you load up a plastic shopping bag with groceries and then carry the bag by the loops at the top of the bag, why will the loops initially withstand the load but then, several minutes later, begin to stretch, perhaps to the point of tearing?

Answer If you suspend a load from the lower end of a spring hanging from a ceiling, the spring will stretch by a certain amount and then stay stretched. Plastic, which consists of polymers, is different. If you suspend a load from the lower end of a plastic strip, the strip will initially stretch like the spring but thereafter it will gradually stretch more in what is called *viscoelastic creep*. The mechanism of this creep can vary from polymer to polymer but a simplistic explanation is this: The polymer consists of many long and entangled molecules, somewhat like a pile of spaghetti strands. When the polymer is put under load, these molecules gradually disentangle somewhat because they are pulled in the direction of the load. The gradual reorientation of the molecules allows the plastic to gradually stretch. If the plastic stretches enough, it may also narrow perpendicular to the direction of the load in what is called *necking*. You can easily see necking in the plastic retainer for a beverage six-pack. Pull the retainer off the beverages and then pull it in opposite directions with your hands until it necks.

1.143 • Giant's Causeway and starch columns

Giant's Causeway in Northern Ireland is an ancient lava (basalt) bed that now consists of basalt columns of various heights. The columns are stunning because in cross section they are polygonal, with many being hexagonal. How could the once-fluid lava break up into vertical, polygonal columns? You can produce similar columns with a mixture of water and cornstarch dried by a heat lamp.

Answer As lava slowly cools, randomly positioned cracks (fractures) develop at the top surface and then extend into the bulk of the lava. The cracks occur because as the lava cools, it tends to contract, which puts the lava under *stress* (a tendency to pull apart). When the stress is so large that it overwhelms the strength of the lava, the lava ruptures into a crack, reducing the stress. Where a developing crack happens to extend toward an existing crack, the stress along the existing crack steers the developing crack to make a perpendicular intersection.

After this first stage of crack formation, a secondary system of cracks develops in the lava. These cracks may each start in a straight line but as they extend into the bulk of the lava, they tend to split (*bifurcate*). Depending on the rate at which the lava cools, the intersection of the secondary cracks with the first-stage cracks tends to split the lava into columns that are pentagonal or hexagonal in cross section.

You can see similar first-stage cracks and secondary cracks in many situations, such as drying mud layers. You can also study the crack formation in a controlled way with a mixture of water and cornstarch. As water diffuses (spreads) through the mixture and then evaporates, the mixture attempts to contract and thus is under stress and tends to form cracks. Depending on the rate at which water evaporates, the secondary cracks can produce pentagonal and hexagonal columns of dried cornstarch.

1.144 • Broken fingernails

If you tear a fingernail, why does the crack tend to veer left or right across the nail rather than travel down the nail?

Answer After you tear a nail at its exposed edge, the crack tends to travel in a direction that requires the least energy to separate cells from one another. The nail consists of three layers: the lower layer is moderately hard keratin, the thicker middle layer is harder keratin, and the upper layer is softer keratin. The strength of the nail is largely determined by that middle layer, which consists of long, narrow cells that run left and right across the nail. Less energy (about half as much) is needed to separate two lines of those cells from each other than to break across many lines of the cells. Thus, the crack tends to veer left or right rather than traveling down the nail.

1.145 • Crumpling paper into a ball

Take a sheet of paper and crumple it between your hands, squeezing it into a ball. Quickly you reach the point where you cannot collapse the ball further. Yet, 75% of the ball is just air. What stops you from crushing the ball further?

Answer As you crumple the paper, you form *curved ridges* (folds) and *conical points* (peaks). Energy from you is required to rearrange the fibers of the paper into these new configurations, and force from you is needed to overcome the friction between the fibers and between pieces of the paper that rub together. Here's another way of saying all this: Energy is stored in the places where the paper is stressed. If you unfold the sheet, you can see the lines and regions of permanent distortion due to the stress.

To collapse a crumpled ball more, you must collapse the existing ridges and also create new ones, which requires more energy from you. The rearrangement of the fibers now becomes more difficult. Eventually you reach the stage where further collapse requires more energy and force than you can supply. Still, if you were to put the ball under a heavy load, it would gradually collapse further over the next few weeks or even the next few years. The fibers gradually move in a *plastic flow* as if they were in a hot plastic that was somewhat fluid.

1.146 • Playful to tragic examples of explosive expansion

One day R. V. Jones of the University of Aberdeen happened upon a beaker of water outside a lab in Oxford while carrying a pistol. For amusement he fired at the beaker, expecting it to shatter into a heap of fragments as a beaker should when hit by a bullet. Instead, it disappeared. He later lectured on why.

Years afterward the Royal Engineers of Aberdeen set out to topple a tall industrial chimney with the physics of his lesson plan. They placed an explosive charge inside the bottom of the brick chimney and then filled the chimney with 2 meters of water. They expected that the explosion would blow out the foundation and send the chimney to the ground. Well, they were half right. The lowest 2 meters of the chimney certainly blew away, but so cleanly that the rest of the chimney dropped neatly and stably onto the remains of the old base. The Royal Engineers then had an even worse problem on their hands.

Why did the beaker and the lowest 2 meters of the chimney get blown away so completely?

A series of stunning photographs by "Doc" Edgerton of MIT, some of the early strobe photographs, reveals the response of a common lightbulb penetrated by a bullet. When the bullet enters the bulb, it reduces the glass at the entrance point to a powder and then some of the powder is propelled back toward the weapon. Shouldn't considerations of force and momentum require that the powder be sent solely in the direction the bullet is traveling?

When President John F. Kennedy was assassinated, some of the brain debris was sent back along the rear of his car in the general direction of Lee Harvey Oswald, who is believed by most investigators to have fired the lethal shot. However, some investigators believe that the rearward spray of debris is actually evidence that another shot must have come from a second sniper on a grassy knoll some distance in front of the car. Must that be the case?

Answer When a bullet hits an empty beaker, the glass around the entrance and exit points shatters into a powder while the rest of the glass is broken into larger pieces as fracture lines travel around the sides of the beaker. If the beaker is filled with water, the water cannot expand upward fast enough to accommodate the space required by the bullet and the effect of its shock wave, and so the water pushes outward on the beaker's walls, reducing the glass to a powder all around and hurling the grains away in all directions. So it went with the bricks in the lowest 2 meters of the chimney when the explosive device suddenly demanded additional volume.

The back-splatter of powdered glass in Edgerton's demonstration is also due to the expansion of a fluid, the small amount of gas contained by the bulb. I don't mean to trivialize or even objectify President Kennedy's death, but the spread of brain debris on the back of the car was most likely due to the response of the fluid in the brain to the sudden impact of Oswald's bullet.

1.147 • Why a hanging picture becomes crooked

If you hang a picture with a short length of cord passing over a support such as a nail, chances are that it will eventually be crooked. What makes it unstable? Is there anything you can do to stabilize it other than tying the cord to the nail or using two, widely spaced nails?

Answer When the cord is short, the picture is unstable because any chance disturbance will allow it to lower the center of its mass distribution by hanging crooked. You can eliminate the instability by substituting a long cord. The minimum length involves the angle between the sections of the cord at the nail and the angle at the left and right between diagonals across the picture (Fig. 1-48). When the angle between the diagonals is less than the angle at the nail, the picture is unstable. By substituting a longer cord, you decrease the angle at the nail. When it is smaller than the angle between the diagonals, the picture cannot lower its center of mass by becoming crooked and thus is stable.

1.148 • A two-spring surprise

Find two springs that are approximately equal in length and strength and arrange for them and three lengths of string to support a block as shown in Fig. 1-49. One of the strings con-

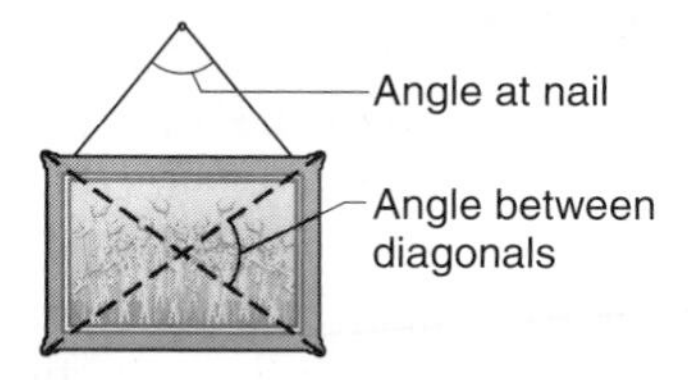

Figure 1-48 / *Item 1.147* Angles are important to the picture's stability.

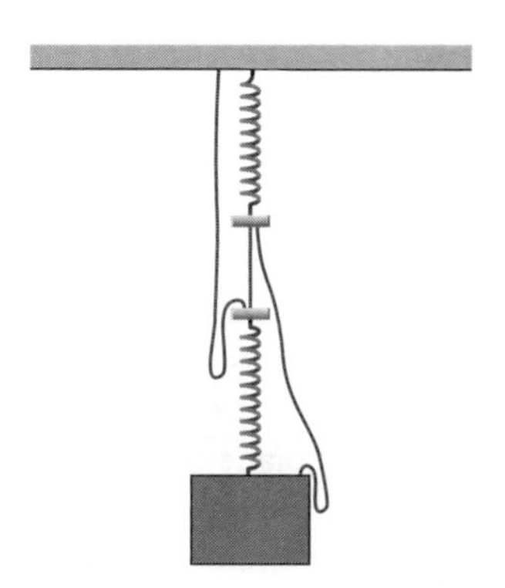

Figure 1-49 / *Item 1.148* Arrangement of two springs and limp strings.

nects the springs and is under tension. The other two strings are of equal lengths but are slightly too long to help support the block and are thus limp.

If you cut the short, interconnecting string so that the longer strings must then help support the block, does the block descend?

Answer When you cut the short string, two factors determine the new level of the block. One factor is that the block now hangs from the two longer strings; since those strings were originally limp and are now under tension, the block tends to be at a lower level. The second factor has to do with how much the springs are stretched. In the original arrangement, each spring supported the full weight of the block, but in the new arrangement, each supports only half that weight. So, in the new arrangement, the springs are now stretched less, which tends to move the block upward. Provided that the longer strings are not too long, this second factor wins out and the block ends up higher than initially.

1.149 • Stability of a pop can

The stability of a can of pop or beer on a table is measured by the energy needed to tilt the can from its normal resting position up to where its center of mass lies directly above the edge still on a table. Is a full can more or less stable than an empty one? Is the can most stable for some particular height of the liquid? The question might be important if the table happens to be on a shaky airplane flight or train ride or if a bartender attempts to slide the can across the surface of a bar.

Answer A completely filled can is more stable than an empty one. Although the center of mass is at mid-height for both cases, the extra mass in a full can means that more energy is needed to tilt the can to the point where it will topple onto its side.

If you slowly drain liquid from a can, three factors influence the can's stability. The center of mass drops until the liquid surface reaches it, and then the center of mass begins to rise. The mass of the liquid decreases as the liquid is removed. And when the can is tilted, the remaining liquid flows so that its top surface remains horizontal. Consideration of these factors shows that a typical can of beer or pop is most stable when the height of the liquid is slightly larger than the radius of the can.

1.150 • Wilberforce pendulum

The strange pendulum shown in Fig. 1-50 is named after L. R. Wilberforce, a British physicist who investigated it in 1894. It consists of a spring attached to a small object with adjustable arms. When the spring is pulled down and released, the object first bobs up and down, but the motion is soon replaced with a rotational motion of the object. Thereafter the motion is periodically exchanged between the spring and rotational motions. The arms on the object are needed because if the apparatus is to display this exchange of motion, the frequency of the purely spring-like oscillations must match the frequency of the purely rotational motion. To make that so, the arm lengths need to be adjusted. Why does the *Wilberforce pendulum* behave so strangely?

Answer The Wilberforce pendulum is similar to the coupled pendulums described in a previous item. Here the normal modes of motion are the oscillations of the spring and the rotation of the object. The modes are coupled because when the spring oscillates and changes length, the coiling and uncoiling requires that it also rotate. The rotation is initially slight but soon it gains all of the energy. As the object then rotates, it coils and uncoils the spring, which changes the spring's length. The variation is initially small but soon it steals all of the energy. Then the process of transfer repeats itself.

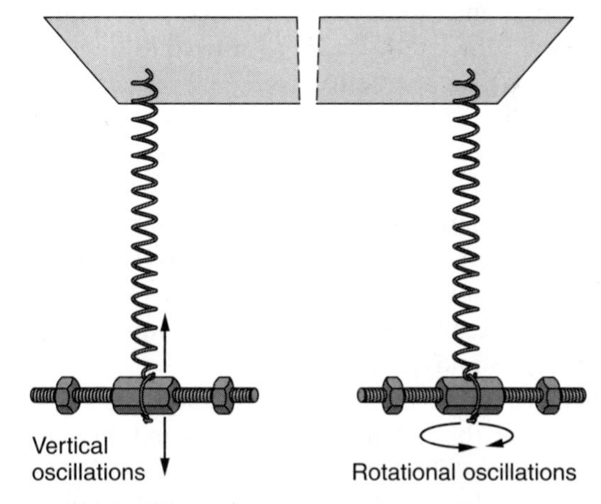

Figure 1-50 / *Item 1.150* Wilberforce pendulum alternates between vertical oscillations and rotational oscillations.

1.151 • Drag racing starts

In drag racing there are two measures of interest on a quarter-mile course: the final speed and the elapsed time. To prepare for the race, why does a driver gun the engine so that the rear wheels spin? Why does the procedure decrease the elapsed time for the run but does not appreciably increase the final speed?

Answer The rear tires are spun so that some of the material melts. After cooling for a few seconds, the material is sticky and so adds to the traction of the wheels when the race begins. The increased traction allows a large initial acceleration and thereby reduces the elapsed time in the run, but the final speed is largely set by the power limitation of the engine—that is, the maximum rate at which the engine can provide energy.

1.152 • Turn or stop

It is difficult to find any physics with more real-world impact than that which involves your possible death. For example, suppose you suddenly discover that you are driving straight for a brick wall at a T-intersection. Should you apply your brakes fully, turn left or right at full speed, or turn while applying the brakes?

Suppose instead that you spot a box ahead of you on a stretch of highway. To avoid hitting the box should you apply your brakes fully or attempt to steer around the box?

If your car and another car are headed toward an intersection along perpendicular streets and with identical speeds, should you and the other driver brake fully without changing direction, or should you each swerve away from each other so that the cars end up leaving the intersection along adjacent paths?

Answer Let's ignore all practical matters such as brake fade, reaction time, and nonuniform road conditions. Then, according to one study, braking while continuing to head toward the wall is the best choice. Consider the situation where the frictional force on the tires is maximum and just barely brings you to a stop in front of the wall. A circular turn onto the side street would require a force on the tires that is twice as large, because extra force is needed to make the car turn from its initial travel. So, if you elect to turn, the force would overwhelm the friction, and you would slide, spin, and eventually hit the wall. Even if you braked while turning, you would still hit the wall.

Whether you can steer around a box depends on the ratio of its width to the distance between you and the box when you begin to act. The marginal case is when the width is about half the distance. For a wider box, studies suggest you should brake fully while headed straight toward the box. For a smaller box you can steer around it.

In the situation where two cars are about to collide in an intersection, it may be best for the drivers to swerve. However, the danger is hardly diminished because the cars then leave the road and plow through whatever is in their way.

1.153 • Slipping past a bus

A bus slows to make a turn at an intersection, but there is enough room in your lane adjacent to the bus's lane for your car to slip by the bus (Fig. 1-51). Is that a wise thing to do?

Answer As the bus turns, its rear effectively pivots around the rear wheels and swings out in a direction that is opposite the turn. Unless the turn is gradual, the rear of the bus might move into your lane by a meter or so and thus strike your car if you are trying to slip by. The sharper the bus is turned, the more its rear encroaches on your lane.

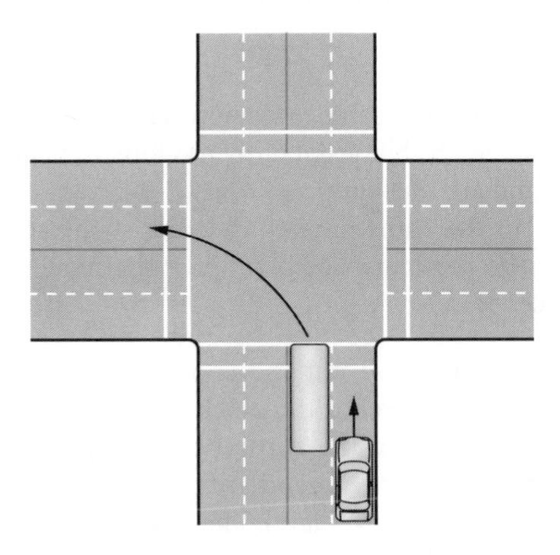

Figure 1-51 / *Item 1.153* Car slipping past a turning bus.

1.154 • Compression region in sticky tape

For many types of sticky tape, as you pull the tape from the roll, a short region of compression (where the tape is noticeably pressed onto the roll) forms just ahead of where the tape leaves the roll. You might see the compression region better if you stick two tape strips together and then slowly peel them apart. What causes the compression line?

Answer As the tape is pulled from the roll, the separated portion rotates away from the roll, about the line where the tape separates from the roll. The tape is stiff enough for this rotation of the separated portion to cause a rotation of the portion that is about to be separated, squeezing the gummy adhesive below it down onto the roll. When you stop separating tape from the roll, you eliminate the rotation and so the region of compression disappears.

1.155 • Bobsled in a curve

In bobsledding, the goal is, of course, to complete the run from the top of the course to the bottom in the least time. Often the winner is decided by a fraction of a second, a margin that may be only one part in 1000. In the straight sections the idea is to slide as smoothly as possible. What strategy should be used in a turn? When you enter the turn, should you steer the sled up high on the slope or keep it as low as possible? Is there a danger of a spill (and so a crash) for either case?

Answer Imagine taking a circular turn on a flat track. In order to turn, a centripetal force must act on you toward the center of the circle. The faster you take the turn, the larger the centripetal force must be. That force is provided by the friction on the sled that counters the sled's tendency to slip sideways. (It is the friction that is perpendicular to the runners on the sled, not the friction along their lengths, which tends to slow the sled.) If you come into the turn with too much speed, the friction is overwhelmed and you slip sideways and crash.

A turn on a bobsled track is banked so that the speed can be high. The bank tilts the supporting force on you and the sled from the ice surface. The tilt is toward the center of the circle so that the support force provides additional centripetal force. Now you might take the turn fast without slipping sideways, provided that you ride up on the bank.

However, you do not want to climb the bank any more than is necessary for three reasons. (1) The higher you go, the farther you must travel to complete the turn, and so you add to the travel time. (2) If you go high, both the friction along the length of the runners and the air drag have longer to act on the sled, and so you come out of the turn with a smaller speed than if you had stayed low. (3) Going high with too small a speed may result in a spill because of the tilt.

1.156 • Too quick to slide

The novel apparatus in Fig. 1-52 consists of a ring that is free to slide along a rod. The upper end of the rod (at the pivot) is forced to oscillate horizontally through a small distance. If the oscillations are slow, the ring slides off the rod, but if they are quick, the ring stays on the rod in spite of its weight. What holds it on the rod?

Answer Were the pivot stationary or oscillated slowly, gravity would, of course, pull the ring off the rod. But when the oscillations are rapid, gravity lacks the opportunity. The pivot moves the slowest near the extreme ends of its oscillations and the fastest in the center. So, most of the time the rod is at a slant. Suppose the slant is leftward (the pivot is leftmost). Although gravity attempts to slide the ring down the rod and toward the right just then, before the ring can move, the pivot shifts to the right and the rod slants rightward. Now gravity tries to move the ring down and to the left, but, again before the ring can move, the rod changes orientation.

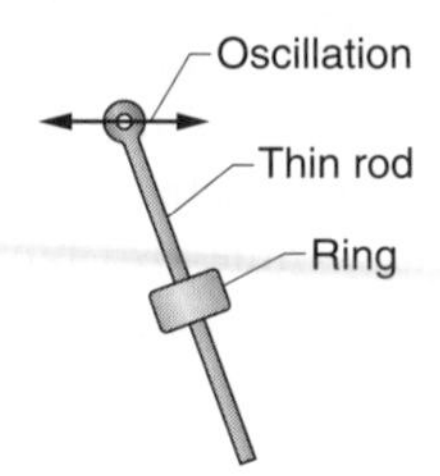

Figure 1-52 / *Item 1.156* Oscillation can keep the ring on the rod.

1.157 • The home of the Little Prince

The mysterious visitor that appears in the book of this enchanting parable, *The Little Prince*, was said to come from a planet that was barely larger than a house. What would life be like on such a planet—for example, could the Little Prince walk about on the planet?

Answer The source of this whimsical item, J. Strnad, considers a planet somewhat larger than the one in the celebrated book and finds that even walking on the planet would be quite difficult because of the tiny gravitational pull. If the Prince were to move faster than 11 centimeters per second, he would be launched into space, unable to return, and if he moved slower but still faster than 80 millimeters per second, he would be sent into orbit around the planet. Someday astronauts will need to cope with such conditions if they mine house-size asteroids.

1.158 • Parachuting with a pumpkin

In 1987, as a Halloween stunt, two skydivers passed a pumpkin back and forth between them while they were in free fall just west of Chicago. The stunt was great fun until the last skydiver with the pumpkin opened his parachute. The action caused the pumpkin to be ripped from his hands. Unfortunately, the pumpkin then plummeted about half a kilometer, ripped through a roof of a house, slammed into a kitchen floor, and splattered all over the newly remodeled kitchen. What caused the skydiver to lose control of the pumpkin?

Answer When the skydiver opened his parachute, the parachute exerted a sudden, large, upward force on him to reduce his downward velocity. The force was more than enough to rip him away from his pumpkin, and the poor pumpkin then fell to its death in that house just west of Chicago.

1.159 • Pulling in a feisty fish

If a fish is small, you might be able to reel it in by simply rotating the handle on the reel and winding up line, but if it is large and full of fight, what should you do to bring it in?

Answer Pulling in a fish involves a battle of torques. When you point the rod toward a strong, feisty fish, you must apply a large force to the reel's handle if you are to generate enough torque to rotate it. The problem is the short lever arm that you work with—it is the distance between the handle and the center around which it turns. You have an easier time if you grab the rod above the reel and pull so as to rotate the rod around its lower end. If the fish is strong, you can prop the lower end on a pivot and then pull with both hands. Either way, you work with a larger lever arm, and so less force is required of you. After you raise the tip of the rod, you gradually lower it again as you wind up line.

If you wish to wear out a fish, holding it in check is easier with a rod that bends because that condition shortens the distance between your hand and the tip of the rod and decreases the torque created by the fish. You then need less torque from your hands to keep the rod in place.

1.160 • Fiddlesticks

Fiddlesticks is a toy consisting of a plastic ring that rotates around a wooden rod. If you hold the rod vertically with the ring at the top and then spin the ring, it will gradually move down the rod. Why does its rate of descent slow and the spin increase during the fall? If you quickly invert the rod before the ring reaches the bottom, you can keep the motion going indefinitely.

Answer If you allowed the ring to roll down an inclined plane, it would progressively turn faster during its descent, with the increased energy for rotation coming from the decreased potential energy. The ring on the rod essentially rolls down the rod in a similar fashion, but on the inside surface of the ring rather than the outside. At any given instant, the ring is at a slant, with part of its inner surface touching the rod. In the next instant, the point of touch has moved around the rod and also down it (Fig. 1-53). The point of touch continues to spiral down the rod. As the ring descends, it converts potential energy into the kinetic energy for the spinning.

The rate of descent is set by the pitch of the spiral, which is fixed by the slanted orientation of the ring. As the ring spins faster, it becomes more horizontal, and the pitch of the spiral and the rate of descent both decrease.

If two rings are set spinning near the top of the rod, the higher ring might happen to catch up with the lower ring. When they touch, the higher ring bounces upward from the collision, spiraling upward.

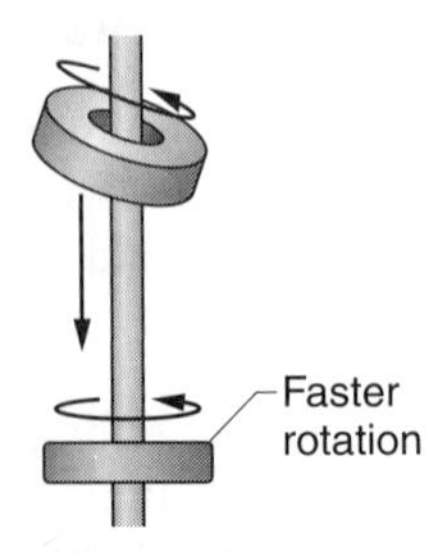

Figure 1-53 / Item 1.160 The ring is initially slanted and spinning slowly. Lower down, it is less slanted and spinning faster.

1.161 • Rotor on a notched stick

A rustic toy that is certain to spark controversy consists of a rotor on a notched stick and a second stick that is rubbed along the notches (Fig. 1-54). The wood rotor is supported by a pin that runs through a hole in the rotor and into one end of the notched stick. Holding your forefinger on one side of the notched stick and your thumb on the opposite side, you stroke the second stick over the notches. If you press hard with your forefinger, the rotor turns in one direction; if instead you press hard with your thumb, the rotor turns in the opposite direction.

When you show the toy to someone unfamiliar with it, you can slyly shift the pressure between thumb and forefinger to reverse the rotation. There is no end to the causes you can attribute to the reversal, such as variation in the cosmic ray intensity.

How does the toy work?

Answer If you apply no pressure on the sides of the notched stick, the vibrations only wiggle the rotor. But when you press against one side, the pressure retards that side's response to the vibrations. The asymmetry in the response of the two sides forces the pin to travel around an elliptical path, and then the friction between the rotor and the pin makes the rotor spin around in the same direction. When you shift the pressure to the opposite side of the stick, the pin travels around the ellipse in the opposite direction, and so the rotor's spin is reversed.

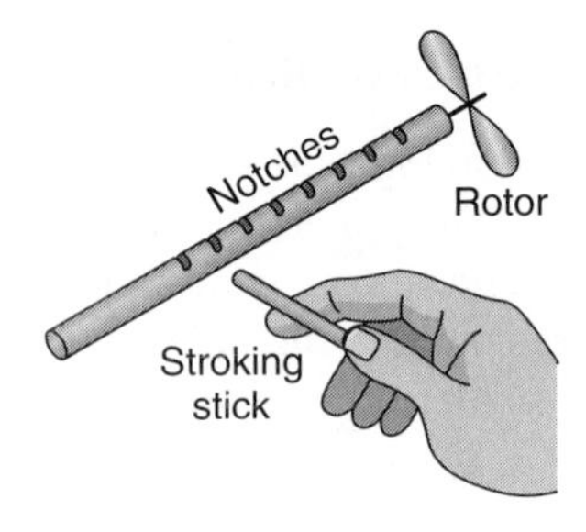

Figure 1-54 / Item 1.161 Rotor spins on the pin after the stick is set oscillating.

1.162 • Shot put and hammer throw

At what angle should a shot be put to maximize the range? Is it 45° as some textbooks claim? If not, is the deviation due to air resistance the shot encounters during the flight?

At what angle should the hammer in a hammer throw be launched? Why does the athlete spin around prior to the release of the hammer, while also moving forward? Just before the release, why is the hammer pulled toward the athlete's body?

Answer If the shot were launched by a machine at ground level, the theoretical optimum angle would be 45°. If it were launched by a machine at the typical release height in shot putting, the theoretical optimum angle would be about 42°. However, most shot putters prefer a smaller angle, perhaps as small as 29°, because the release is then physically more efficient and the shot is launched with greater speed. Although the smaller angle tends to decrease the shot put range, the greater launch speed more than makes up for that decrease. (Air resistance has little effect.)

To build up the kinetic energy of the hammer prior to its release, the athlete twirls the hammer around several times (with feet planted) and then begins to spin several times along with the hammer while also moving across the launching circle in order to add even more speed to the hammer. The hammer's motion is not horizontal. Rather, the hammer rises to a high point when the athlete faces the direction of the intended launch, and then the hammer falls to a low point in just the opposite direction. During the fall, the athlete, with both feet momentarily planted, pulls in the direction of travel of the hammer and thus adds to the kinetic energy.

As the athlete almost completes the last spin and reaches the edge of the launching circle, the hammer is suddenly pulled toward the body to increase its speed. (The situation is like an ice-skater bringing in arms and a leg while spinning on point—the action increases the rate of spinning.) The hammer is then released at about shoulder height. Thus, the hammer should be released at an angle somewhat less than 45° because of that initial launch height.

1.163 • Jumps during downhill ski race

If an experienced skier who is coming down a hill notices that the slope suddenly steepens along the way, the skier will crouch and then jump upward to lift the legs and become airborne before reaching the increase in slope. Why doesn't the skier wait until reaching the increase in slope to use the slope to become airborne?

Answer If the skier does not jump prior to reaching the sudden increase in slope, but instead uses that increase to become airborne, then the skier travels farther through the air and thus falls farther to land on the snow. The longer fall results in a more jarring landing, which might easily topple the skier.

1.164 • Pulling a tablecloth beneath dishes

Pulling a tablecloth out from a set of dishes is a classic classroom demonstration. When I was still dating and faced with trying to start a conversation on a first date, I would often perform the stunt for my date, only I would not use common dishes or glassware, but instead would partially fill lab beakers and flasks with wine. The technique always got the conversation going. However, if you should try the demonstration on one of your dates, be aware that showing it is not enough. To keep the conversation moving, you need to explain how the demonstration works. How does it work?

Answer If you pull smoothly and rapidly on the tablecloth, you immediately reduce the friction between it and the dishes. Such reduction is common. When two surfaces begin to slide over each other, the friction between them is generally smaller than when they are on the verge of sliding. In the case of the tablecloth, all or part of the reduction stems from the skipping the dishes undergo. Since they do not fully touch the cloth at all times, the friction on them is smaller, and the cloth can slide out from underneath. However, that smaller friction will still move the dishes in your direction. The longer you pull, the more the dishes will move, which is another reason for you to pull rapidly so as to reduce that time.

SHORT STORY

1.165 • Pulling with teeth

On April 4, 1974, John Massis of Belgium managed to pull two passenger cars of New York's Long Island Railroad by clamping his teeth down on a bit that was attached to the cars with a rope and then leaning backward and pushing against the railway ties with his legs. The cars weighed about 80 tons, or 700 000 newtons, but since Massis did not lift the cars, it was their mass that was important. Somehow he was able to move a mass of about 71 000 kilograms through a measurable distance. (In a physics classroom the concept of work being the product of the force and the distance an object is moved is sometimes left cold, but Massis instilled new life, albeit bizarre, into the concept.)

1.166 • Jerking chair

If you are in a chair on a common, smooth floor, you can move yourself and the chair over the floor with a series of jerks and without touching the floor with your feet. But an initially stationary object (here, you and the chair) cannot move unless there is some "outside" (external) force acting on it. What is the force that propels you and the chair?

Answer To move yourself and the chair, you first suddenly and strongly push your hands down and toward the rear on the chair. The downward force increases the compression of the chair against the floor, allowing more friction between the chair and the floor, which prevents the chair from sliding backward. Your push against the chair also propels your body forward. Once you are moving, suddenly pull upward and forward on the chair. The upward pull reduces the compression of the chair on the floor and thus also the friction between the chair and the floor, allowing the chair to slip forward. So, the outside force that moves you is actually the friction encountered in the first stage of the procedure.

1.167 • Lifting a person with fingers

You may have seen the following stunt in a magic show: A performer chooses three people from the audience to assist him in lifting a fourth, somewhat heavy person also chosen from the audience. The stunt requires that the magician and his three new assistants lift the fourth person from a chair while each use only a single index finger. The magician places his finger below an armpit of the seated person. Each of the assistants similarly places an index finger: one is below the other armpit, one is below the left knee, and one is below the right knee. With a great effort, the magician and three assistants try to lift the seated person, with no success—he is just too heavy.

To apply some magic, the magician and the assistants place their hands on the head of the seated person and then press down with a little pressure. This pressure presumably reduces the weight of the seated person. The fingers are then repositioned for a lift, which is attempted on a signal from the magician. This time, the seated person is easily lifted.

What's going on? Obviously, if a little downward pressure on my head could reduce my weight, I would never have to worry about being even a bit overweight.

Answer During the first attempt at the lift, the three assistants and the magician happen to lift in an uncoordinated manner, with some forces being applied sooner than others. The unequal forces at the four points on the body of the seated person merely topples that person to one side because of the torque those unequal forces produce, and then there is little chance that the lift can be made. Indeed, the magician can help initiate this toppling.

During the second attempt the four forces are applied simultaneously, due to the coordinating signal from the magician. There is then no torque on the seated person, and with that person's weight now shared equally between the four people, the person can be lifted with a reasonable effort.

1.168 • Rockets, and a problem with an iceboat

Suppose that an initially stationary rocket is ignited while in space. Can it reach a speed that is larger than the speed by which its propellant is issued? Does its final speed depend on whether the fuel burns slowly or quickly? Why are ground-launched rockets made to fire in stages? (The idea originated in China around 1000 A.D.) Is there some optimum number of stages? Could a single-stage rocket be made that is powerful enough to launch a satellite into orbit or send people to the Moon? How feasible was the plan in one of the Jules Verne novels where a manned capsule was fired like a ball from a large cannon sunk into the ground?

You are on a small iceboat on a wide, flat expanse of very slippery ice that you wish to cross. Strewn around the shore of the ice are rocks. You decide to load some of them into the boat so that you can propel the boat by throwing the rocks from the boat and toward the shore, but you have room for only a certain total mass of rocks. To give the iceboat the greatest final speed, should you choose many small rocks or a fewer number of larger rocks? That is, should you throw a large or small amount of mass with each toss? For argument, assume that you always hurl the rocks, large or small, with the same speed relative to you and the boat.

Answer The rocket can be made to go faster than the speed at which material is issued from it provided that the ratio of the rocket's initial mass to its final mass exceeds 2.72 (which is equal to the exponential of 1.0). The rate at which the fuel is burned makes no difference on the rocket's final speed. A single stage rocket cannot launch a payload from Earth into orbit because it cannot reach the required final speed of about 11.2 kilometers per second. So, rockets are built with stages. When the lowest stage has exhausted its fuel, it is dropped off so that its mass need not be lifted any higher, and then the next stage is ignited. There is an optimum number of stages, about four or five for common rockets, due to the expense of additional stages.

The people in Verne's story would have been killed by the acceleration they experienced.

You will give the iceboat a greater final speed if you throw a large number of small rocks instead of a fewer number of large rocks. To see the point, first consider throwing only one large rock and then consider throwing two smaller rocks, each with half the mass of the larger rock. In the second scheme, the first rock gives a certain forward speed to the boat and also to the second rock that you still hold. That means that when you throw the second rock, the increase in the boat's speed is greater than when you threw the first rock.

SHORT STORY

1.169 • Earth to Venus

The first attempt at sending a man to Venus came in Baltimore, Maryland, in 1928. Robert Condit and two assistants built a rocket from angle iron and sailcloth. It was powered by gasoline that was vaporized and sprayed into steel tubes and then ignited by spark plugs.

Condit was to make the trip alone, carrying along some food, water, two flashlights, and a first aid kit. Guidance was not a concern because he planned to aim the craft carefully when he took off. When he got to Venus he was to deploy a 25-foot silk parachute to slow his descent. How exactly he was to get back was not too clear, but if there was no food or water on the planet, he did not intend to stay long.

On the day of the test firing, Condit climbed into the craft and turned on the engine to go up a quarter-mile, just to check it out. Great billows of fire and smoke erupted from the steel tubes, but there was no liftoff. Condit increased the flow of gasoline, and the fire became so arresting that it stopped the traffic on the street. But there was still no liftoff. Condit kept at it until he finally ran out of fuel.

He never reached Venus, or otherwise you would already know his story.

1.170 • A choice of hammers

To cut into wood or soft stone with a chisel, should you use a wood or steel mallet? Which is best if you are to cut into something much harder, such as granite? Why does a steel hammer serve better than a wood one to drive a nail into wood?

Answer When the material is soft and requires only a small force to penetrate, the idea is to transfer as much energy as possible to the cutting. In that case a wood mallet serves best because, although it delivers only a moderate force to the chisel and so to the material, it transfers much of its energy. When the material is hard, the cut is more difficult to make, and so force is more important. A steel hammer delivers a large force because it is massive and because it bounces from the chisel. However, the strike of the hammer against the chisel is elastic; that is, little energy is transferred and most is kept by the hammer.

A steel hammer is, of course, also the choice for driving a nail. A wood mallet deforms when it strikes the head of a nail, wasting some of the mallet's energy.

1.171 • Pressure regulator

A conventional pressure cooker consists of a pot that is tightly sealed except for a central tube that is mounted with a loosely fitted cylinder. The cylinder has three holes drilled in its side, each with a different diameter. I set the pressure in the pot by choosing which of the holes is to rest on the tube. But how does the procedure work? After all, the weight of the cylinder is not changed by a different choice of hole.

Answer The steam that is generated inside the pot pushes upward against the weight of the cylinder. The pressure that is maintained in the pot roughly matches the pressure needed to support the cylinder. When the pressure of the steam becomes too large, it lifts the cylinder and allows steam to escape so that the pressure inside the pot is brought back to the desired level. If you choose a wide hole in the cylinder, the weight of the cylinder is spread over the large cross-sectional area of the hole, and so the pressure needed to lift the cylinder is small. A smaller hole gives a larger pressure in the pot.

1.172 • Sliding a stick across fingers

Hold a meter stick horizontally on your index fingers, with the fingers at opposite ends of the stick, and then move your fingers uniformly toward each other. Does the stick slide uniformly? No, it alternates between sliding on one finger and then the other, changing several times before the fingers reach the center of the stick. Why?

Answer In spite of appearances, the initial conditions on the fingers are not symmetric. You inevitably pull slightly harder with one finger—say, the right one—and overcome the static friction on it from the stick, and so it begins to slide beneath the stick. The friction on it is then *kinetic friction*, which is initially smaller than the *static friction* on the left finger. But as the right finger moves toward the center, the portion of the stick's weight that it supports increases, and so does the sliding friction, until the friction there exceeds the friction on the left finger. Then the right finger stops and the left finger begins to slide. Soon the left finger supports so much weight that it stops and the right finger begins to move again. The cycle is repeated until your fingers get near the stick's center, and then the stick tends to topple off your fingers.

SHORT STORY

1.173 • Giant tug-of-war

Harrisburg, Pennsylvania, June 13, 1978: Some 2200 students and teachers attempted to set the world's record for a tug-of-war. The braided-nylon rope was 600 meters long, 2.5 millimeters thick, and built to withstand a force of 57 000 newtons (13 000 pounds). However, soon after the contest began, the rope suddenly snapped. The contestants near the center then relaxed their grip but the ones farther away continued to pull, and so the rope quickly slid through some of the hands. At least four students lost fingers or fingertips from the friction.

1.174 • Shooting along a slope

Suppose that you set the sights on a rifle for a certain distance while at a shooting range. If you then shoot at a target at the same distance but either up or down a slope, will the shots land true, too high, or too low?

Answer Perhaps surprisingly, the shots land high when you shoot either up or down the slope. To correct the sights, you must multiply the distance to the target by the cosine of the slope's angle with respect to the horizontal.

1.175 • Starting a car on a slippery road

When the road is slick and a car has a manual shift, should you start the car's motion in first or second gear?

Answer Since the road is slick, the friction on the tires is easily overwhelmed, in which case the tires will slip. To avoid slippage, you initially want only a small torque applied to the wheels. You might be able to use first gear if you let out the clutch smoothly and gingerly. Otherwise, you should shift to second gear to reduce the torque.

1.176 • Balancing a tire

When a new tire is mounted on a wheel, it must be *balanced*, which is a procedure in which a small lead weight is attached to the rim. If the tire is not balanced, it will not roll smoothly but will shimmy or bump against its mount. Both problems are due to the fact that the wheel's mass is not uniformly spread around the center—the wheel acts as though it has an extra mass at some spot inside it. When the wheel is balanced, the lead mass offsets that extra mass, and then the wheel runs smoother.

One way to balance the wheel is to lay it down on a tilt-table stand with a bubble balance. The wheel and stand are then like a playground see-saw in that the extra mass creates a tilt in one direction. You place a lead mass on the opposite side of the tilt and then trim the weight with cutters until the stand is level, which is indicated by the bubble being centered in the balance. This technique is called a *static balance*.

In a *dynamic balance* the wheel is spun horizontally around its center. The extra mass on one side makes the wheel wobble, but when a lead mass is added to the rim and trimmed appropriately, the wobble disappears.

Are the two techniques of balancing equivalent? That is, do they each eliminate both bump and shimmy?

Answer The two techniques for balancing are not equivalent. The static balance eliminates bumping; the dynamic balance eliminates shimmy. Although the lead mass may end up in the same location, it will be trimmed to different sizes in the two techniques.

To see the difference, first consider the see-saw arrangement of a static balance. The extra weight on one side of the wheel creates a torque that attempts to rotate the wheel in one direction around its center. The size of the torque depends on the size of the extra mass and on how far horizontally it is from the center. The lead mass creates a torque in the opposite direction. Since it must be on the wheel's rim, its distance from the center is fixed. So, to balance the two torques, you start with a lead mass that is too large and then trim it until its torque matches the other torque. When the wheel is put on the car, it will not bump against its mount.

Shimmy depends on how deeply the extra mass lies buried inside the wheel. Again consider the wheel when it is horizontal. If it is to spin smoothly, it must rotate around the vertical axis through its center. However, the buried extra mass makes it rotate around an axis that is tilted from the vertical—the wheel wobbles. To right the spin axis, a lead mass is attached somewhere on the rim as previously, but it now must have a different size, and its location could also be different. Although it eliminates the wobble, it no longer exactly balances the see-saw play of torques, and there is still some bumping. Because the residual bumping is usually small, a dynamic balance is considered to be the better of the two balancing techniques.

1.177 • Carnival bottle swing

While touring a sideshow, you happen upon a stand in which the game is to knock over a bottle with a pendulum bob that is suspended at the bottle's level. The fellow (the *carnie*) running the game explains that you are not allowed to swing the bob directly at the bottle; instead you must arrange for the bob to hit the bottle on its return swing. Doesn't sound too hard, does it? With a few practice swings, you should win a prize, right?

Answer The game is dishonest because the bob will always orbit the bottle if it clears the bottle on the forward swing. For it to hit the bottle on the return swing only, its angular momentum would have to change during the travel, and yet there is no torque on it to do that. However, you might be sneaky and twist up the string before you release the bob. Then, the bob rotates around its center during its swing and can encounter forces from the passing air that are similar to those that account for a curve ball in baseball. Those forces can alter the return swing so that the bob hits the bottle. (You had best be careful, though, because an angry carnie is an unpleasant sight.)

1.178 • Hanging goblet, ready to crash

Tie a glass goblet or some other somewhat heavy object to a small and lighter object, such as a rubber eraser, by means of a meter length of cord. Hold a pencil horizontally, drape the cord over it, and then pull the lighter object toward your left or right until the goblet is just underneath the pencil and the lighter object is approximately horizontal with the pencil. If you now release the lighter object, what happens? Silly question, I know. The heavy goblet will drag the cord (and eventually the lighter object) over the pencil as it falls, until the goblet smashes on the floor. Right?

Answer Once you release the lighter object, it begins to fall while also being pulled toward the pencil by the cord because of the falling goblet. The combined motion means that the lighter object rotates around the pencil with a decreasing radius. The situation is then somewhat like an ice-skater pulling in the arms while spinning on point—the angular speed (spin) increases in order to maintain the angular momentum. Here, the angular momentum must also be maintained because there is no torque to change it. So, the rotation rate of the lighter object increases, which increases the tension in the cord, slowing the goblet's fall. Once the lighter object has rotated several times around the pencil, the force on the goblet is enough to halt the goblet's descent, and so the goblet never reaches the floor.

1.179 • Breaking a drill bit

If a high-speed drill bit is lowered too firmly onto a work surface, why does the bit break?

Answer The forces on the ends of the bit tend to buckle the bit slightly. If the rotation speed is greater than some critical value, that slight bulge is quickly enhanced to the point that the bit snaps.

1.180 • Swinging watches

A windup pocket watch, popular in earlier times, kept good time when worn but not when the watch was hung from a support by its chain. Then the watch might gain or lose 10 minutes or more per day, while it also mysteriously swung itself like a pendulum. One investigator reported the bizarre sight of a wall full of watches hanging by their chains and swinging merrily. What accounts for such untimely behavior?

Answer The pendulum motion is brought about by the rotational oscillations of a balance wheel (part of the timing mechanism) when the frequency of the wheel's oscillations is near the frequency at which the case swings. When frequency of the wheel is somewhat lower than the frequency of the swinging, the two motions are out of step and the watch gains time. When the frequency of the wheel is somewhat higher than the frequency of the swinging, it loses time.

SHORT STORY

1.181 • Flattening the Golden Gate Bridge

On its 50th anniversary in 1987, the Golden Gate Bridge was opened to pedestrians who walked across it in celebration of the magnificent bridge. A surprising number of people showed up. When 250 000 were packed rather tightly on the bridge, its midsection became flat instead of forming the normal arch, and some of the supporting cables became slack. The bridge also began to sway sideways (as experienced with London's Millennium Bridge in 2001). That day of celebration turned out to be an unscheduled integrity test of the Golden Gate Bridge. Thankfully it passed the test.

1.182 • Hunting by railway vehicles

In a traditional design for a train, the wheels are *coned* (slanted), constrained to remain on a rail by an interior flange, and are linked in pairs by an axle. The rails, which have a round top, usually lean slightly inward. When the train moves along straight track, why do the compartments or cars sway from side to side, a motion that is called *hunting*?

Hunting not only limits the speed of a train, but it also tends to deform both rail and roadbed. Because the resulting wear is not even on the left and right sides, the cars pulled by an engine are occasionally turned around so as to even out the wear on the two sides.

Answer If the car shifts, say, to the right, then a wheel on the right rides on a large radius while one on the left turns on a smaller radius, due to the slanted shape of each. Since the wheels are rigidly connected to each other, they turn at the same rate, but the difference in the turning radii means that the wheel on the right travels farther along the track than the wheel on the left in a given amount of time. The difference in the speed along the track skews the axle and the car runs crooked along the track until it is left of center. The situation is then reversed. If the oscillations continue, the train is said to "hunt" in the sense that it "seeks" a stable location.

A chance deflection can initiate the hunting, but frictional forces created by distortion on the rail and wheel due to the train's weight can also start the oscillation. If the train's speed is below some critical value, the oscillations from each deflection die out. But if the speed is higher, the oscillations build and only the flanges can save the train from derailment. Sometimes the oscillations are so extreme that the wheel climbs off the rail in spite of the flanges.

1.183 • Oscillating car antenna

Some types of vertical car antenna, especially the whip type, may begin to oscillate as you drive. Why does the antenna oscillate in the pattern of Fig. 1-55*a* for low to moderate speeds and in the pattern of Fig. 1-55*b* for higher speeds?

Answer If we mounted the antenna in a vise and somehow made it oscillate, it would oscillate in what are called *resonant modes* (or patterns) and at *resonant frequencies.* We are said to set up *resonance* when our oscillations set up one

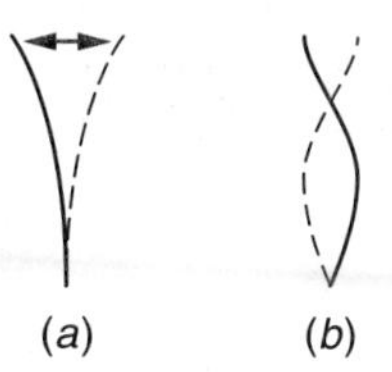

Figure 1-55 / *Item 1.183* Oscillating car antenna at (*a*) low speed and (*b*) higher speed.

of the patterns. The simplest pattern is called the *fundamental mode*, which oscillates at the lowest resonant frequency (Fig. 1-55*a*). In this pattern, the bottom of the antenna does not move (because it is fixed in place), the top moves the most, and the intermediate points move at intermediate distances. The next more complicated pattern, the *first overtone mode*, has a point of no oscillation located somewhat down from the top. When the antenna is on a moving car, the passing air tends to create vortexes on the back side of the antenna. The variations in air pressure due to the vortexes tend to make the antenna oscillate. At low to moderate speeds, the fundamental mode is set up. At greater speeds, with vortexes being shed by the antenna at a greater rate, the first overtone mode is set up.

1.184 • A ship's antiroll tank

A ship's rolling is often just unsettling, but if the waves strike the side of the ship with the same frequency as the roll that is introduced, the rolling can be built up to a dangerous extent. (Such a match of frequency is an example of *resonance*; a similar buildup of oscillations can be seen if you push a child on a swing each time the child swings back to you.) To lessen the danger, some ships in the past were outfitted with a tank that ran across the ship's width and was partially filled with water. The dimensions of a tank were chosen so that the sloshing of the tank water had the same frequency as the rolling of the ship. Doesn't that seem precisely wrong, because won't the sloshing then augment the rolling?

Answer Suppose that the waves push on the right side of a ship at the ship's resonant frequency. The roll of the ship is not instantaneous but, because of the ship's mass, lags behind the push by about a quarter step—that is, a fourth of a full roll to the left and right. Similarly, the sloshing in the tank lags behind the roll of the ship by another quarter-step, which makes it a half-step behind the push on the ship. So, when the waves try to rock the ship to the left, the sloshing tries to rock the ship to the right, and the ship tends to remain upright.

Antiroll tanks were used primarily on German ships around the 1900s. Although they worked well on a regular seaway, they proved unworthy when the waves were irregular, and in some cases they actually enhanced the rocking.

1.185 • Road corrugation

Many unpaved road surfaces are initially made smooth but soon develop patterns of shallow hills and valleys, with separations of 0.5 to 1 meter, along the paths taken by the wheels of cars. The periodic pattern cannot be due to weather erosion as common chuckholes are. What produces the patterns? Why aren't the patterns eliminated and the road smoothed by the impact of the wheels on the hills? Similar corrugations can be found on train and trolley rails and on the downhill paths taken by skiers. Do the patterns migrate along the road, rails, and paths?

Answer A corrugation pattern on a road begins once an irregularity first appears in the initial smoothness. When a tire hits the irregularity with sufficient speed, it might hop slightly and then dig into the road when it comes down. Even if the tire does not actually leave the road surface, the tendency to hop first momentarily relieves the weight on the tire and next drives the wheel down extra hard. The downward push digs out a shallow valley from which the tire then must climb, and so it tends to hop again. With additional cars traveling over the surface, the pattern is enhanced and extended down the road but does not migrate.

1.186 • Seeing only one side of the Moon

Why do we see only one side of the Moon? (There is a variation in what we see, but not by much.) Since the Moon orbits Earth, shouldn't we see its entire surface?

Answer The strength of Earth's gravitational field varies with distance from Earth. That means that the gravitational field on the far side of the Moon is weaker than on the near side. This variation in the field has created slight tidal bulges in the Moon, one on the far side and one on the near side, so that the Moon is not spherical. Because of these bulges, Earth's gravitational field causes the Moon to rotate around its center as it orbits around Earth. The result is that the Moon always presents (roughly) the same face toward Earth. Many other natural satellites in the solar system also point the same face toward the planet they orbit.

1.187 • Intelligence satellites

When activities at some region on Earth's surface are to be monitored from space, intelligence satellites photograph the region. The satellites are timed so that when one satellite no longer looks down on the region, another one takes over. Wouldn't it be easier to keep one satellite over the region, traveling around in orbit just as fast as the interesting region turns around Earth's axis of rotation? That may seem a better strategy, but for most regions on Earth it is impossible to arrange. Why, and where does the easier strategy work?

Answer An orbiting satellite is held in orbit by the gravitational pull on it from Earth. That pull is always directed toward the center of Earth, and so the orbit must be around the center. That fact eliminates the possibility of a satellite staying over, say, New York City, because the orbit would then be around the northern half of Earth rather than Earth's center. However, a satellite, said to be a *geostationary satellite*, could stay over a point on Earth's equator because the orbit would then be around the center. The satellite would have to be placed at the proper altitude (about $\frac{1}{10}$ of the distance to the Moon) so that it orbited just as fast as the point on the equator turned around Earth's axis of rotation. For any other point on Earth's surface, an intelligence satellite must photograph along a slanted line.

1.188 • Air drag speeds up satellite

Most satellites orbit Earth in the thin reaches of the atmosphere and experience a small amount of drag. The drag should slow a satellite just as air drag on a coasting car slows the car. However, in the satellite's case, the drag increases the speed. How can a retarding force result in an increase in speed and thus kinetic energy?

Answer The drag reduces the satellite's total energy, which consists of both kinetic energy and potential energy, and the satellite gradually drops into a smaller orbit. With the fall, the satellite's potential energy decreases, but only half of the decrease is converted into thermal energy by the friction from the atmosphere. The other half goes into the kinetic energy, the increase in speed being necessary because of the smaller orbit. Upon inspection, the result may not be so surprising—normally when things fall toward Earth, their speed increases.

1.189 • Moon trip figure eight

When a spaceship is sent to the Moon, why is its path in the form of a distorted figure eight instead of an ellipse that encompasses Earth and the Moon?

Answer The figure eight path requires less energy by the ship because for much of the trip the ship stays close to the line between the centers of Earth and the Moon. Since along that line the gravitational pulls from Earth and the Moon compete, the net force on the ship is smaller than if the ship is in an elliptical orbit. So, less energy is required to overcome the net force.

1.190 • Earth and Sun pull on Moon

Since the Moon is captured in orbit around Earth, the gravitational pull on it from Earth must dominate the pull from the Sun, right? Well, actually, no, because the Sun's pull is more than twice as large as Earth's pull. Why then don't we lose the Moon?

Answer The Sun's force does dominate the Moon's motion: The Moon orbits the Sun. The smaller force from Earth acts as a perturbation to the main motion and produces loops in the orbit. We can make sense of the motion by simply saying, "The Moon goes around Earth while Earth goes around the Sun."

1.191 • Gravitational slingshot effect

If a space capsule moves near enough a planet, it might undergo a *gravity assist* or *slingshot effect* in which it gains energy. But isn't the idea faulty? Imagine monitoring the capsule from the planet. As the capsule approaches, it certainly should gain energy due to the gravitational pull from the planet, but isn't the gain nullified as the capsule recedes?

Answer The problem in the interpretation presented lies in your location—the planet that you are on is moving. From that view, the capsule will seem to gain no energy. But take the view of someone at rest with respect to the Sun. Such an observer would see the capsule gravitationally attracted to the planet. If the capsule passes near the planet on the rear of the planet's orbit, the capsule is effectively dragged along the orbit by the planet and so the capsule gains energy. The planet loses an equal amount of energy but the change is immeasurable because of the planet's huge mass, while the energy increase of the capsule is appreciable because of the capsule's much smaller mass.

1.192 • Making a map of India

In the past when India was surveyed, the measures were reportedly slightly inaccurate because the plumb line did not hang exactly along the vertical, especially in the northern part of the country. Why might the story be credible?

Answer The mass at the lower end of the plumb line can be pulled toward the Himalayas through several arc seconds by the gravitational pull from the mass of the mountains. In other regions, a nonuniform distribution of mass introduces similar errors.

1.193 • Shaving with twin blades

If one shaves with a razor having twin blades, is there an optimum speed at which the blades should be drawn across the skin, or should the blades be drawn as quickly as possible or as slowly as possible?

Answer When the first blade encounters a hair that extends from the skin, it snags the hair at the skin's surface and then drags the hair along the skin in the direction of the blade's motion, pulling the buried base of the hair upward from its original position. At some point during this dragging, the first blade slices off the length of hair that initially extended above the skin.

The remaining length of hair then springs back to its original orientation and next begins to retract into the skin. If the second blade catches the hair after it has sprung back and before it is retracted, then the blade can remove even more of the hair, delaying the next need of a shave. To get such a close shave, the razor should not be moved so quickly that the spring-back does not occur or so slowly that the retraction is completed. The optimum speed is about 4 centimeters per second, but the value will vary between shavers because of different properties of the skin and hair (especially the elasticity).

1.194 • The handedness of river erosion

There are arguments that on the average the right bank of a river in the Northern Hemisphere suffers more erosion than the left bank, while in the Southern Hemisphere just the opposite is true. Although the effect is certain to be small and masked by other factors, why might the idea be correct?

Answer Earth's rotation can produce an apparent deflection of flowing streams, rightward in the Northern Hemisphere and leftward in the Southern Hemisphere. The deflections are not true deflections because we watch the streams from a rotating surface. However, they can be quite apparent in large-scale motions, such as the air flow around weather systems, giving the familiar counterclockwise rotation around a hurricane in the Northern Hemisphere. The flow of a large river, such as the Mississippi River, might also display the apparent deflection.

Racing on the Ceiling, Swimming Through Syrup

2.1 • Race cars on the ceiling

A car traveling through a flat turn in a Grand Prix race depends on friction to stay in the turn. However, if the car is going too fast, friction fails and the car slides out of the turn. In earlier times, a car had to take the flat turns rather slowly. But modern race cars are designed so that they are literally pushed down onto the track to give the tires good grip. In fact, that push down, called *negative lift*, is so strong that some drivers boast they could drive their cars upside down on a long ceiling. What causes negative lift, and can a race car actually be driven upside down as done fictionally by a sedan in the first *Men in Black* movie?

Negative lift is dependable when a car is the only one taking a turn as in, say, a time trial, but a skilled driver knows that negative lift can disappear during a race. What causes it to disappear?

Answer About 70% of negative lift on a car is due to one or more wings that deflect the passing air upward. The rest of the negative lift is called *ground effect* and has to do with the airflow beneath the car. The faster a car moves, the greater both aspects of negative lift are. At the high speeds typical in a Grand Prix, the negative lift is larger than the gravitational force on a car. Thus, if the car were to move from a normal track to a ceiling (without slowing), the now upward negative lift would more than offset the downward gravitational force. Thus, the car could indeed be driven upside down as in *Men in Black.*

Ground effect is due to the constricted flow of air below a car. As air is squeezed into a small passage beneath the car, its speed increases at the expense of its pressure. So, there is lower air pressure below the car than above it, and the pressure difference presses the car onto the track. In a race, a driver can reduce air drag on the car by closely following another car, a procedure known as *drafting.* However, the leading car disrupts the steady flow of air under the trailing car, eliminating the ground effect on it. If the trailing driver does not anticipate that elimination and slow accordingly, sliding into the track wall may be unavoidable.

The Chaparral 2J was an early race car using ground effect. It had two fans at the rear to suck air under the car from openings at the front. A road-hugging skirt along each side prevented air from the sides from entering the flow. The

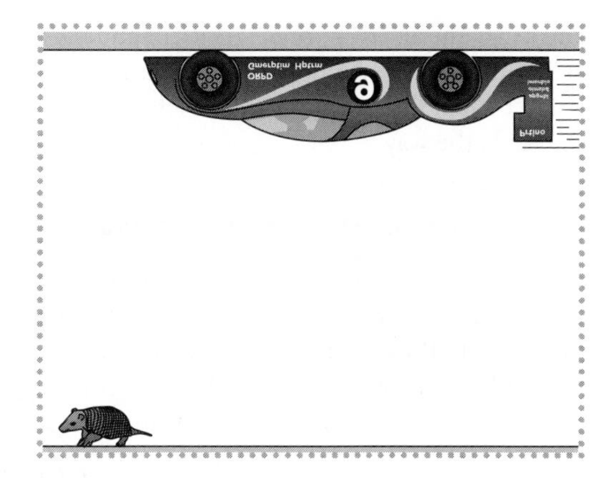

Figure 2-1 / *Item 2.1*

low pressure beneath the car held the car on the track in a fast turn, and the outflow from its fans diminished the normal vortex formation behind the car, reducing the air drag on the car. As a result, the car was reasonably fast on the straight-track sections and unbeatable on the turns. It was so good, in fact, that it was banned from races.

2.2 • Drafting

Race car drivers in many styles of racing take advantage of one another by *drafting,* in which a trailing car is positioned almost directly behind a leading car. This is obviously dangerous. What advantage does it offer?

Answer In spite of its aerodynamic design, a race car still meets a lot of drag. One source of drag is the pressure difference between the front and rear of the car. At the front, the air impact creates high pressure. At the rear, the airflow breaks up into vortexes, which have reduced air pressure. So, the pressure difference between front and rear acts to slow the car, requiring greater fuel consumption for the car to maintain its high speed.

If a trailing car pulls behind a lead car, both cars have an advantage. The trailing car disrupts the vortex formation at the rear of the leading car, and the leading car has a smaller front-to-rear pressure difference. The trailing car has less air impact on the front, and so it too has a smaller front-to-rear pressure difference.

The trailing driver can use a *slingshot pass* to go around the leading driver: The trailing driver pulls back somewhat from the leading car to allow vortex formation to be set up behind the leading car. The low-pressure vortexes act to slow the leading car and pull the trailing car forward. Timing this carefully, the trailing driver can accelerate into the vortex region and then snap around the side of the leading car.

Reportedly, Junior Johnson was the first to use these aerodynamic techniques in the 1960 Daytona 500 NASCAR race, which he won in spite of his car being rated as slower than other cars in the race.

Drafting is used in other sports, notably in bicycle racing. It is also practiced by animals, as when a mother duck leads her ducklings in a line across a pond. Ducks, of course, do not move fast enough to worry about aerodynamics, but the ducklings can benefit from the calm wake left by the mother duck as she leads the way.

2.3 • Aerodynamics of passing trains

A high-speed train, moving at 270 kilometers per hour or even faster, produces a compression wave as it slams into the air, forcing the air to flow around the sides and top of the train. What happens when the train travels into a tunnel? What happens when two such trains pass close to each other in opposite directions? If a high-speed train passes a person standing near the track (as could happen at a station where the train does not stop), is the person in any danger?

Answer When a train speeds through a tunnel, we can simplify the picture by assuming the train is stationary and the air flows past it. As air is forced into the confining space between the train and the tunnel wall, its speed increases. The energy required for that increase must come from the store of energy associated with air pressure, and so the pressure decreases. A passenger inside the train might feel this pressure decrease in the ears as air trapped on the inside of the ear pushes outward on the ear drum. (The sensation is similar to what you feel when rapidly ascending in an airplane.)

When two trains pass each other, the pressure in the air between them is also decreased. If the trains are inside a tunnel, the decrease can be even greater. In earlier days when the train speeds were being increased, windows were sometimes sucked out of the trains during passing.

Whether in a tunnel or out in the open, airflow around trains approaching, passing, and leaving each other is very complicated and requires computer modeling. However, we can use a simplified explanation for the pressure reduction. Each train drags air from the space between them, and so, with less air there, the pressure is reduced.

As a high-speed train passes a person, the compression wave shed by the front of the train and the highly turbulent airflow that follows can knock the person down or (worse) onto the train or track.

2.4 • Collapse of the old Tacoma Narrows Bridge

One of the most dramatic physics videos shows the violent twisting of the old bridge at Tacoma Narrows on November 7, 1941. The wind was only moderate that morning (about 68 kilometers per hour, which is about 42 miles per hour), yet this very sturdy bridge was ripped apart within a few hours after the twisting oscillations began.

While building the bridge, the construction workers dubbed it "Galloping Gertie" because of its tendency to oscillate, making the bridge resemble a roller-coaster track. Indeed, after the bridge was officially opened, drivers would converge on the bridge for the novelty of the oscillations, which were sometimes enough to cause cars to disappear from the view of a driver. Although many people have attributed the bridge's collapse to that galloping tendency, galloping apparently had little or nothing to do with the collapse. What did cause the collapse?

Answer The bridge's girder had the form of a squat H, with a reinforcing beam running along each side. As the wind encountered the blunt face on the windward side of the bridge, vortexes were formed above and below the bridge's horizontal section. As these vortexes swept along the horizontal section, they caused the bridge to flutter by oscillating vertically like a flag flaps in a wind. The bridge design was flawed (although no one could have known at the time) in that it was structurally weak in resisting both the flutter and the torsional oscillations (the twisting) that can be seen in the video of the collapse.

When the oscillations became violent (and frightening), two people were forced to crawl on all fours to get off the bridge. A professor went out onto the bridge to rescue a dog left in an abandoned car but gave up when the frightened dog tried to bite him. The video shows him coming back from the car, trying to walk along the relatively stable central line around which the bridge was twisting. Shortly later, a section of the bridge fell and the fluttering died out, but then the fluttering restarted and much of the rest of the bridge's span across the river fell.

Although many physics instructors used the bridge collapse as a dramatic example of resonance, the collapse was due to fluttering and twisting, not resonance and galloping. Indeed, the driving force was a fairly steady wind, not a wind somehow pulsing at a resonance frequency of the bridge. The bridge did shed vortexes much like a cable in a wind. Such vortexes can cause galloping if the frequency at which they appear matches a resonant frequency of the cable's oscillation. However, the galloping in the bridge would have never been strong enough to rupture the bridge.

2.5 • Aerodynamics of buildings

On a windy day, why is the wind especially gusty for people walking near a building? If you want to avoid the gusts but must be near a building, where should you stand? Why do buildings sway in a wind? Some buildings have an open area at ground level, either for traffic or pedestrians. Why can the wind through such an open area be especially strong?

Answer Wind veers around a building and breaks into vortexes (or swirls) upon passing the building's corners (Fig. 2-2*a*). So, a sidewalk pedestrian will find the wind to be most gusty at or just beyond those corners. The wind will be the least gusty behind the building, where there can be relative

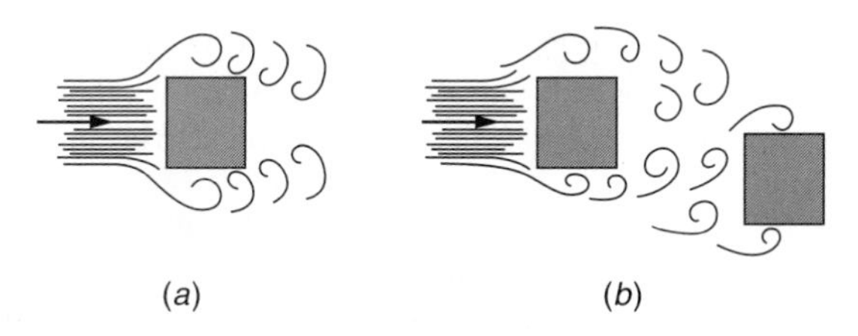

***Figure 2-2** / Item 2.5* (*a*) Overhead view of wind breaking into vortexes as it passes corners of a building. (*b*) The region between staggered buildings can be very gusty.

calm. The air pressure in that area is likely to be lower, which can cause outward bulging of windows. In extreme cases, windows can be sucked out.

Gusts will also be less at the point on the windward side where the wind splits in half, so that half goes around the building on one side and the other half goes around on the opposite side.

If a building has an open passage through which wind can travel, the wind will speed up because it is funneled into the passage. This increase in speed has two consequences: First, the wind can bend over (and even knock down) pedestrians, and doors along the passage can be difficult to open if they must swing into the oncoming wind. Second, the air pressure in the passage is reduced because energy is shifted to meet the required increase in wind speed. So, windows and doors along the passage bulge toward the passage. In some cases, windows break and doors cannot shut.

Where several tall buildings are clustered, the separation of the wind and the formation of vortexes can be complex. For example, if two buildings are positioned in a wind with the downwind building staggered from the upwind building, the low-pressure region behind the upwind building can pull the wind off the face of the downwind building, creating very gusty vortexes between the two buildings (Fig. 2-2*b*). In other situations, where tall buildings are clustered in rectangles with streets between the clusters, streets parallel to the wind can effectively become wind tunnels. As you walk from the leeward shelter of a building into one of these wind-tunnel streets, you can be knocked down. Moreover, because the wind is funneled into the street, the air pressure there is reduced, causing street-facing windows to bulge outward.

The variation in wind pressure on the windward side of the building can cause the building to sway or oscillate, with the top of the building moving most. This swaying can nauseate some building occupants. They can also be nauseated by the infrasound and audible howling that a strong wind makes when producing vortexes at the corners of the building. Tall buildings that are prone to sway in strong winds are usually equipped with anti-sway devices, such as a roof-mounted block–spring arrangement in which a heavy block moves opposite the building's motion.

Extreme winds produced by a hurricane or tornado can blow down a house or a somewhat larger building. They can also peel off the roof either by catching it under the windward edge or by substantially reducing the air pressure above the roof until portions of it are pulled up into the wind and blown away. Also, these extreme winds not only can blow windows into a building on the windward side, but they can also suck windows out on the leeward side or along the sides where vortexes form.

2.6 • Kites

What keeps a kite aloft and what determines a stable flight, as opposed to a chaotic one in which the kite constantly loops and flutters?

Answer A triangular kite is a flexible surface that faces into the wind while tilted up at an angle, said to be the *angle of attack*. Four forces act on the kite. (1) The gravitational force, of course, pulls downward. (2) Because the wind is deflected downward by the kite's face, the kite experiences an upward *lift*. (3) The wind also produces a drag force in the direction of the wind. (4) The string produces a force that is downward and into the wind.

If the kite is not in stable flight, the torques due to the forces automatically rotate the kite around the *bridle point* (the point where the main, long string branches into separate strings running to various points on the kite frame). The rotation changes the kite's angle of attack and thus alters the lift and drag. As a result, the kite not only rotates but also moves vertically. The vertical motion changes the angle at which the string pulls on the bridle point and thus also the horizontal and vertical pulls of the string.

Stable flight occurs if three quantities vanish: (1) the torques, (2) the net vertical force, and (3) the net horizontal force. For these to vanish, not only must the kite have the proper orientation but also the string must pull at the proper angle and with the proper force. The kite is then said to be in an *equilibrium state*. For a given wind speed, there might be more than one equilibrium state. If the wind speed changes, both the orientation of the kite and the angle of the string must change for the kite to find a new equilibrium state.

2.7 • Ski jumping

Why can a ski jumper jump for some 200 meters by using good form but a drastically smaller distance by using bad form? Why do some of the jumps end in dangerous tumbling, and how does a jumper avoid tumbling?

Answer The long flight of a ski jumper is due to the lift produced on the body and the skis, which are held in a V opened toward the oncoming air. If the jump is done correctly, the ski jumper glides through the air like a paper airplane. However, the force of the passing air also presents a serious danger, because it can suddenly produce more lift on the front of the skis than on the rear. The imbalance of forces creates a torque that rotates the jumper, and then very quickly the imbalance becomes much worse and the jumper loses control and begins to tumble. The uncontrolled landing can be fatal.

A skilled jumper knows how to quickly get body and skis into the proper orientation to maximize the lift early in the gliding path. The secret lies in the upward jump at the end of the takeoff ramp. That jump must produce a forward rotation to bring the jumper and skis into the proper downward orientation, so that the skis and body make the correct angle with the passing air. In addition, the jumper must time this forward rotation so that the torque on the skis and body vanishes just as the jumper reaches the proper orientation. All this maneuvering is essential for a good, safe jump but is complicated by the dependence on air density, which partially determines the force of the passing air on the jumper. If a jumper is accustomed to the air density at, say, a low altitude and then attempts to jump at a high altitude where the air is less dense, the timing and jump can be wrong.

2.8 • Speed of a downhill skier

Speed is the goal of many downhill skiing events, especially for skiers focused on beating the world's speed skiing record (faster than 240 kilometers per hour). Air drag is the main obstacle in these events. Indeed it is more important than the friction on the skis. How can a skier minimize air drag?

Answer Here are a few things a skier does to reduce the air drag. The suit is skintight to eliminate flapping. The helmet is not only smoothly shaped to part the passing air but it also fits over the shoulders so that the air meets no sudden jump at the shoulders or back and so it does not break up into swirls under the back of the helmet. (A small skier may be able to hide most of the shoulders under the helmet.) The legs meet the air with fairings designed to part the air and avoid vortex breakup behind the legs. Such vortexes are points of low pressure. So, with high pressure on the front of the legs and low pressure behind them, the pressure difference can be a significant drag. The ski poles bend around the body instead of jutting into the passing air. The skier squats in an *egg position* to minimize the front area presented to the oncoming air.

One of many difficulties in a downhill speed race is the effort required to keep the legs in their proper positions. Because air is funneled between the thighs, the air speed there is greater than around the outside of the thighs. The energy required for the speed increase comes from the air pressure. Thus, the air pressure is less between the thighs than on the outside, and the thighs tend to be pulled together. The skier must constantly fight this tendency.

2.9 • Boomerangs

Why do boomerangs boomerang—that is, return? Some boomerangs might make a round trip of as much as 200 meters, and some can make several round trips before they land. You throw a boomerang with its plane almost vertical. Why does the plane generally lie down during the flight? Boomerangs come in a variety of shapes besides the classic bent-banana look. Can a straight stick be made to boomerang?

Answer Each arm of a boomerang resembles a classic *airfoil* (the shape attributed to an old-style airplane wing). It has a blunt front edge that cuts through the air and a thinner trailing edge; the top surface is curved and the bottom surface is flat. When the boomerang is in flight, this airfoil shape deflects the passing air and as a result the boomerang experiences lift in the opposite direction.

To throw the boomerang with your right hand, you first hold it near your head with the curved surface facing you but the plane of the boomerang slanted slightly rightward from the vertical. Then you bring your throwing arm forward quickly while snapping your hand around your wrist. The lift on the boomerang is then up and to your left; the upward portion of the force is what keeps the boomerang in flight.

The size of the lift on an arm of the boomerang depends on the speed at which air passes the arm. Since at any given instant the top arm rotates forward (the same direction as the boomerang's flight) and the bottom arm rotates backward, there is more lift on the top arm than on the bottom one.

Since the lift is up on the top arm, at some distance from the center of the boomerang, it creates a torque that attempts to turn the plane of the boomerang. Because the boomerang is spinning like a top, the torque turns the axis around which the boomerang spins so that the axis points more toward you and you see more of the top face. As the boomerang turns, its flight path curves. The result is a curved path that brings the boomerang back to you.

A straight stick can be made to boomerang if thrown like a boomerang. The initial spin around the short axis through the center is unstable, and the rotation shifts to the long axis that runs through the length of the stick. The switch reorients the stick, but the direction around which the stick spins is not changed. During the return flight, the spin deflects the passing airstream downward, which results in lift on the stick.

2.10 • Throwing cards

Release a credit card (or any other stiff card) with its long edge down and horizontal (and its faces left and right). Why doesn't the card merely slip through the air to hit the floor directly below the release point?

A common trick is to throw playing cards at an open-top box. I use a plastic credit card. If I throw it randomly, it almost immediately flaps about, stalls, and then drops to the floor. Is there any way to stabilize the card while it flies through the air so that I have a good chance of hitting a target?

Answer The flight of a card released long edge down is very challenging to explain, and mathematical attempts have been made to do so since 1854. The flight can be chaotic but instead it might show the following patterns: (1) *Flutter* is where the card slides through the air, alternating between sliding leftward and sliding rightward. (2) *Tumbling* occurs when the card rotates around an axis while it also glides either leftward or rightward. Which behavior occurs depends on the dimensions of the card. A standard playing card usually develops a steady tumbling while drifting at a certain angle to the vertical. Starting from the initial vertical orientation, the release deflects the lower end either left- or rightward. Then as the card slides at an angle to the vertical, the airflow past it creates a high-pressure point below the leading edge and above the trailing edge. These high-pressure regions rotate the card around the central axis along its length. As the card reaches the face-down orientation, its fall slows but the rotation continues until the card approaches being vertical again. As it does this, it slides through the air more easily and so its downward speed increases. The process is then repeated.

The trick in throwing a card is to stabilize it so that it does not undergo flutter and tumbling. One popular way is to hold the card face down with the thumb on top, the first finger on the long edge, and the second finger on the bottom. With a bend of the wrist, the card is brought back until it touches the base of the palm. Then, with a forward snap of the wrist, the card is launched while spinning around a vertical axis. Forces on it from the air rotate the card so that it becomes vertical, spinning around a horizontal axis. The flight can then be remarkably straight, and the card can hit a target forcefully. In fact, care must be taken that the card does not hit someone in the eye.

Some performers are remarkably skilled at throwing playing cards across an audience, up into the balcony section, or even around a boomerang path.

2.11 • Seeds that spin

How do seeds from ash, elm, and maple trees manage to stay in the air long enough that a breeze might carry them away from the parent tree?

Answer A seed from one of these trees is winged and prolongs its fall by spinning. For example, a single-wing samara from a maple tree spins around its center of mass (the center of its mass distribution), which lies between the bulge portion and the flat-wing portion. The plane of the wing may be tilted by as much as 45°. As the wing spins around during the seed's fall, it propels air *downward*, and so the seed undergoes a force that is *upward*. The force can also push the seed off to the side so that it takes a helical path to the ground (Fig. 2-3).

Probably the action is easier to picture if you ride along with the seed. As the air comes up past you, it pushes against the underside of the wing portion. The component (or part) of the push that is perpendicular to the wing is *lift*, the force that helps support the seed. The push of the air makes the wing spin like a helicopter blade and also allows the seed to glide off to one side. Often the combination of spin and glide allows the seed to come down on a spiral while also spinning around its center of mass.

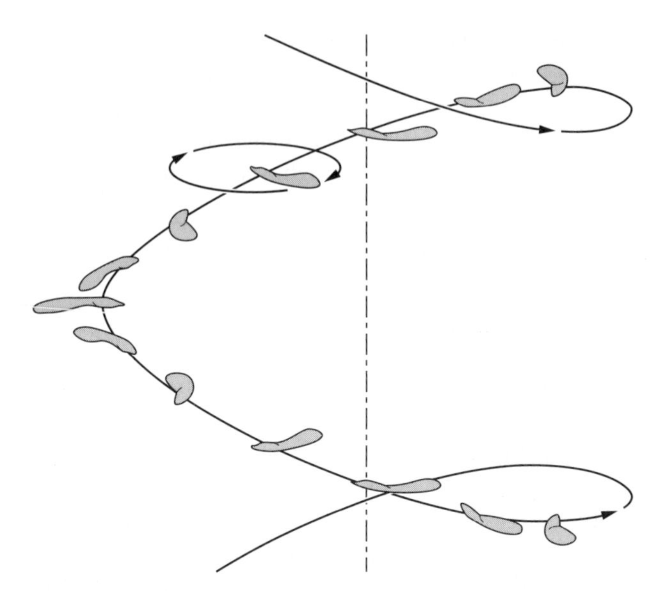

Figure 2-3 / *Item 2.11* Possible path of a winged seed, spinning opposite the direction of rotation along a helical path.

2.12 • Flying snakes

For those who fear snakes, there is a snake that can produce nightmares for a lifetime. The paradise tree snake (*Chrysopelea paradise*) can climb a tree, jump from a high point, and then glide toward the ground. It can even alter its glide path during flight to take advantage of a new target, such as another tree. How can a snake glide through the air?

Answer The snake, hanging from a branch, jumps upward and away from the branch. As its body straightens, the bottom side flattens, starting at the head and proceeding toward the tail (but the tail itself does not participate). In addition, the bottom side along the latter half of the snake becomes somewhat concave, with most of it flat but with downward lips at the left and right sides. The horizontal width of the snake is doubled at midlength.

The flattened region serves as an airfoil to give the snake lift. So, the snake's gliding is something like that of a paper airplane. However, after gaining speed, the snake does something quite different: It takes on an S shape and then begins to oscillate horizontally at a frequency of about 1.3 times per second. Immediately its glide path becomes shallower. So, the oscillation must somehow increase the lift on the snake. The snake has an airspeed of about 8 meters per second, descends at a speed of about 5 meters per second, and glides at an angle of about 30°. It can change directions by tilting the latter half of its body just as its head is moving in the new direction during the head's oscillation.

The lift created by the snake's oscillation is not well understood. However, we can speculate that during the left–right motion of the concave latter half of the body, the orientation of the bottom side may change. If it alternates by tilting to the left and tilting to the right, it might increase the lift.

2.13 • Air drag on tennis balls

Why will a used tennis ball generally reach the receiver faster than a new tennis ball that is hit exactly the same way?

Answer The flight time of a tennis ball is determined by air drag on the ball. If a given shot (speed and angle) is repeated many times, starting with a new ball, the air drag first increases and then gradually decreases to some minimum value. Presumably the reason lies in the nap (the surface fuzz). Initial play raises the nap, which then "catches" more air and thus increases the air drag. However, eventually the nap becomes torn away or flattened, and the air drag decreases. Thus, the server has a slight advantage when playing with a well-worn ball because the ball encounters less drag than a new one and reaches the receiver in less time, making a return more difficult.

2.14 • Veering a football around a defensive wall

How can a football (soccer) player use a free kick to send the ball along a *curved* path around a defensive wall of players and into the goal? Such a kick, dubbed a *banana kick* because of the shape of the flight path, seems unearthly and can often take a goalkeeper completely by surprise, especially if the wall blocks his view of the first phase of the ball's flight.

Answer Figure 2-4*a* shows an overhead view of a ball in flight through stationary air. Let's ride along with the ball so that the air passes us, as in Fig. 2-4*b*. If the ball does not spin, the air passes symmetrically on the sides of the ball, and then somewhere on the back side the two airstreams break free of the ball and form vortexes behind the ball. If, instead, the ball spins—say, clockwise as in Fig. 2-4*c*—the airstreams are not symmetrical. Instead, the stream moving against the spinning surface breaks up into vortexes early and the stream moving with the spinning surface clings to the surface and breaks free of it late. We can think of the airstream as being slung off the spinning ball, much as mud is slung off a tire. Because the spin on the ball forces the airstreams to be deflected, the ball is forced off in the opposite direction. That is, the spinning ball's deflection of the air causes the ball to veer off to one side. The effect is commonly called the *Magnus effect* after an earlier investigator.

In the soccer free kick, let's assume that the ball is kicked toward the left side of the defensive wall and with a clockwise side spin (Fig. 2-4*d*). The ball should be launched at an angle of about 17° with the ground and directed out of arm's reach

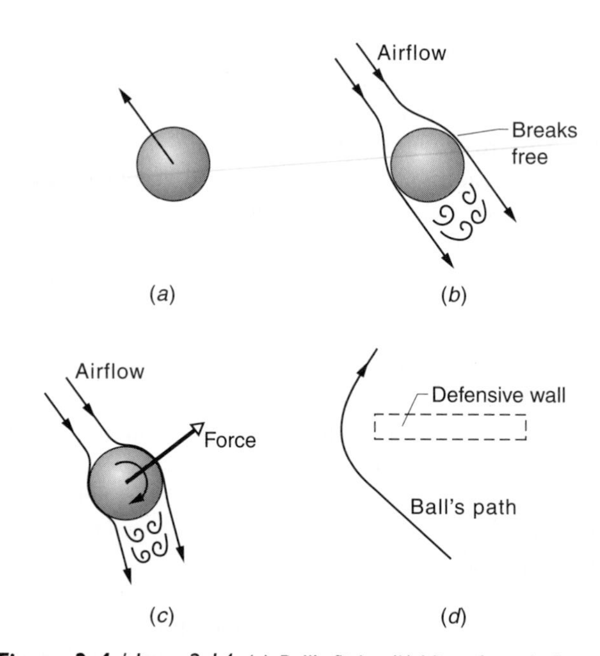

Figure 2-4 */ Item 2.14* (*a*) Ball's flight. (*b*) View from ball. (*c*) Spinning ball deflects passing air. Ball deflected to one side. (*d*) Deflection takes ball around a defensive wall and toward the goal.

at the end player on the wall. As the ball travels through the air, the spin causes the airstream to be deflected to the left and thus the ball to veer to the right. If the kick is executed well, the ball neatly veers around the end of the wall just out of reach and then into the goal.

Part of the magic of the shot may come from the change in the ball's speed during the flight. The air drag on the ball is largely due to the difference in the high-pressure impact of air on its front surface and the low-pressure vortexes on the rear surface. As the ball slows, the extent of the vortex region changes, first increasing and then decreasing, and so the air drag varies in the same way. Thus, the slowing of the ball first increases and then decreases, which can trick a goalkeeper.

Other sports balls will veer during flight if spinning, including tennis balls, Ping-Pong balls, and volleyballs. (Early on, veering was noticed in the flight of spinning cannonballs and rifle bullets.) Of course, sending the ball along a curved path in any direction can confuse an opponent. A spinning ball also has an advantage of bouncing from the field, court, or wall in a surprising direction. However, a smooth beach ball seems to be different in that it can veer first one way and then another, following a path that is more S-shaped than banana-shaped. This perplexing last deflection, called a *reverse Magnus effect*, occurs when the ball's speed and spin rate fall to low values.

2.15 • Golf-ball aerodynamics

Why are golf balls dimpled? If a golf ball is hit with topspin (where the top rotates in the direction of the ball's travel), the ball will roll forward when it hits the green. Is that desirable for a player who normally hits the ball well short of the hole?

Answer Originally golf balls were smooth, but then golfers noticed that once it was nicked and beat up, a ball would generally fly farther with the same drive of the golf club. Eventually the modern dimple designs were perfected, allowing the golf balls to be driven much farther than an identical smooth ball.

The primary function of the dimples is to reduce the air drag on the ball by reducing the pressure difference between the front and rear of the ball. At the front, the pressure is high because the ball is colliding with the air. As the ball slides through the air, the air clings to the ball and then at some point breaks free. When it breaks free, it forms vortexes in which the air pressure is reduced. If the region of vortex formation at the rear of the ball is extensive, the pressure difference between the front and rear of the ball can be large. That is, the air drag is large and the ball does not fly very far. With the ball dimpled (or even scarred), the air along the sides of the ball is turbulent, which allows the air to cling better to the surface and reach farther along the rear surface before it breaks away and forms vortexes. So, a dimpled ball has a smaller region of vortex formation at the rear (and thus less air drag) than a smooth ball.

Although a dimpled ball flies farther down the course, it can also fly farther off the course in a hook or slice. In other words, the dimples increase distance, not control.

A spinning ball experiences *lift*. If the club hits the ball low so that the ball has *backspin* (or *bottom spin*), the lift is positive, meaning upward, and so the lift keeps the ball in the air and the ball flies far. If the club hits the ball high so that the ball has *forward spin* (or *topspin*), the lift is negative, meaning downward, and so the lift brings the ball down early. Thus, topspin may give a long roll on the course but shortens the flight through the air.

The (positive or negative) lift on a golf ball is due to the way air is slung off a spinning ball. With backspin, the air is slung downward by the ball's spin (Fig. 2-5). Because the air is forced downward by the spin, the ball is forced upward—that is, given positive lift. With forward spin, the air is slung upward, and the ball is forced downward—that is, given negative lift.

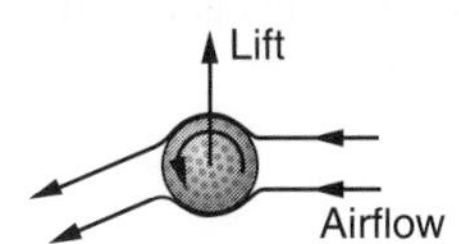

Figure 2-5 / *Item 2.15* Golf ball traveling to the right. Air passes to the left and is deflected downward by the spin on the ball. Ball experiences lift.

2.16 • Baseball aerodynamics

How does a baseball pitcher throw a fastball in which the ball does not fall as much as with gravity alone during its flight to the batter? If the batter does not properly anticipate a fastball, the swing could be too low to make good contact with the ball. How does a pitcher throw a curveball in which the flight path curves toward or away from the batter or toward the ground?

Answer The fastball is thrown by snapping the wrist directly toward the batter after the ball is brought forward with a powerful overhand motion of the arm. This procedure puts backspin on the ball; that is, the top of the ball rotates back toward the pitcher. As the ball moves through the air, the air that passes over the top of the ball is slung off the ball toward the ground because of the spin. This downward deflection of the air produces an upward deflection (*positive lift*) on the ball. Now, the ball does not actually move upward because it is falling during its flight. But the ball does not fall quite as far as it would were it not spinning, which can fool a batter.

In addition to this basic deflection, the seam on the rotating ball encounters air drag, which can deflect the ball's flight and slow the ball. The seam winds around the ball to hold together the two leather pieces making up the ball's surface. The way the ball is held and thrown is usually described in terms of what the batter sees. In one basic throw, dubbed the two-seam fastball, the batter always sees two sections of the seam on the ball as the ball rotates during its flight. In a *four-seam fastball*, the batter sees successive sections of the ball rotate into view as the ball flies through the air. Although both types of fastball seem to give the same positive lift, some pitchers claim one or the other is superior.

A curveball is thrown with either sidespin or topspin. If a sidespin slings air off to the pitcher's left, then the ball is deflected to the right, either toward or away from the batter, depending on whether the batter is right-handed or left-handed. If the ball is given topspin in a pitch known as a *drop*, air is slung off upward, producing a downward deflection (negative lift). A *slider* is a throw with sidespin but with less rotation so that the deflection is less, which can trick a batter.

These are basic throws. A good pitcher can tilt the spin on a ball to get any desired direction of lift and deflection, varying it from pitch to pitch to confuse the batter. Professional baseball players at bat look for clues as to what pitch is on its way, such as the last orientation of the pitcher's hand or the rotating seams on the ball. The job is difficult because the batter can see the ball clearly only in the first part of the flight, and then the ball becomes a blur and the swing must begin.

2.17 • Cricket aerodynamics

In cricket the bowler *bowls* a ball with a straight-arm throw, hurling the ball at the ground so that it bounces in the general direction of the batsman, who then attempts to hit it with a bat. Although that basic description hardly seems to be exciting, the game can be filled with surprises and entire nations sit transfixed watching matches. How can the bowler control (at least approximately) where the ball goes?

Answer The bowler has several ways to confuse a batter. (1) He can make the ball take an unexpected bounce on the ground by throwing the ball with spin. (2) He can make the

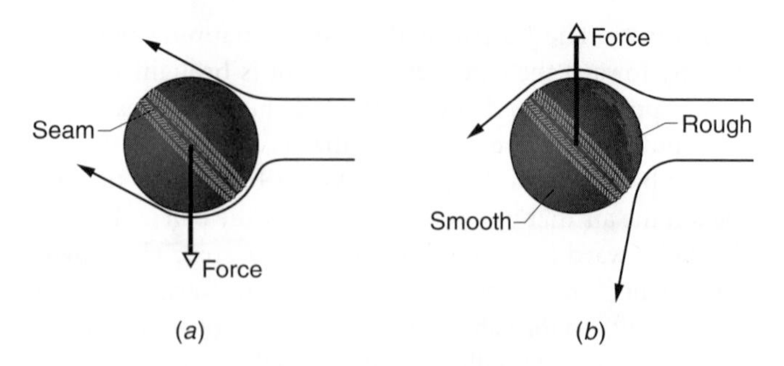

Figure 2-6 / *Item 2.17* Airflow past a bowled cricket ball and the deflection force on the ball. (*a*) New ball. (*b*) Ball roughened by use and polished on one side by the bowler.

ball *swing* (be deflected) as it travels toward the ground by positioning the seam on the ball. (3) He can make the ball swing by putting spin on the ball. Of course, he can also use a combination of these techniques.

In technique 2, the ball presents an approximately constant face toward the air. In the overhead view of Fig. 2-6*a*, the ball is taken to be stationary and the air is taken to be moving leftward. In that view, the seam is in the lower part of the front face and the air passes over and under the ball. On the top, the airflow is fairly smooth, which leaves the layer immediately next to the ball unstable. It breaks free of the ball before reaching the back side. On the bottom, the flow is initially fairly smooth, but the seam *trips* the flow into turbulence. The turbulent mixing of air immediately next to the ball allows the airflow to cling to the ball until it reaches the back, where it breaks free. We can say that the seam causes the airflow to be slung upward in the diagram, which is to the bowler's left. Thus, from the bowler's view, because the air is forced to his left, the ball is forced to his right—that is, the ball swings rightward.

In *reverse swing*, the deflection of the ball is opposite what we see in Fig. 2-6*a*. The bowler repeatedly polishes the ball on his trousers to make one face smooth and to leave the opposite face roughened from repeated use. He then throws the ball with the rough face forward (Fig. 2-6*b*). Now, the airstreams passing both sides of the ball are immediately in turbulent flow and thus tend to cling to the ball. However, the seam on the right side now acts like a launching ramp in that it sends the turbulent flow free of the ball. The result is that the flow coming around the top side (in the figure's view) leaves on the back of the ball and the flow coming around the bottom side leaves at the seam. In the bowler's view, the airflow is forced rightward and thus the ball is forced leftward—the reverse swing.

2.18 • Birds flying in V formation

When birds fly long distance in a flock, why do many flocks fly in a V formation?

Answer When a bird flies by flapping its wings (instead of gliding), each downward push by a wing creates a vertical vortex (swirl) in the air trailing the bird. The vortex circulates downward on the bird side, outward on its bottom side, upward on its far side, and inward on its top side. If a trailing bird positions itself in the up-flow part of the vortex, it receives a free lift. It still must flap to stay aloft, but it does not have to flap quite as hard, and thus its energy expenditure is not quite as much. The savings can be significant over a long journey.

To be in the upflow, a trailing bird should be off to one side of a leading bird, and a V formation is one of the best formations for placing the birds properly. It also allows them visual contact. However, birds are rarely in exactly the best position for saving energy, and the spacing within a V formation is often uneven, suggesting that flying in formation is actually quite difficult.

Although the front bird experiences some of the upflow from the birds just to the left and right of it, the lead position is usually the most tiring. Presumably, many of the birds in a flock take turns as leader. The birds could, instead, fly in a flattened V or a straight line and then the lead position would not be so tiring.

Energy savings can also be one reason why fish swim in schools. The vortex formations by the leading fish can help reduce the energy requirements of fish farther back in the school.

2.19 • Speed swimming in syrup

A swimmer must push or pull against the water to be propelled through it. The water is, of course, fluid and thus the push or pull is not as effective as it would be against a solid object. Suppose we added something to the water to make it more viscous, so that it was less fluid. Could someone swim faster through such water?

Answer In an experiment, guar was added to a swimming pool to produce a viscosity twice that of pure water. Swimmers were then timed while swimming 25 yards. The result was that the increased viscosity did not change their swimming speeds. The viscosity increase gave a swimmer a better push or pull, but it also increased the drag against the swimmer, and the two effects canceled each other.

2.20 • Contrails

Why do aircraft sometimes leave trailing white lines in the sky? Why do the trails sometimes puff up or loop?

Answer When an airplane flies through abundant water moisture at high altitudes, it can form a trailing cloud called a *contrail* (short for condensation trail). Usually a contrail consists of at least two white lines that begin somewhat behind the airplane. As the airplane pushes its way through the air, vortexes (swirls) are left by the wingtips (and other projecting portions). The air in a wingtip vortex moves upward, toward the airplane, downward, and then outward. The engines inject soot into this circulating flow, causing water moisture to form drops or ice crystals, which strongly scatter sunlight, making the vortex trails visible. Because the scattering usually does not depend on wavelength (or color), the contrails are usually white.

These vortexes can be dangerous to other aircraft, especially a smaller and lighter one that can be flipped over by a vortex. So, small-craft pilots take great care to avoid the wake of a much larger aircraft. However, wingtip vortexes were reportedly put to good use in the English skies during World War II: During an attack of V-1 flying bombs, a British pilot would fly alongside one of the bombs and use a wingtip vortex to flip it over so that it would crash.

The length of a contrail consisting of drops is usually short because the drops tend to evaporate. Ice, however, can persist to give a long and long-lasting contrail provided the ice does not become large enough that it simply falls. A long-lasting contrail can expand as water moisture forms additional drops or crystals. In some regions of high-density air traffic, expanding contrails can overlap and cover much of the sky.

Contrails sometimes develop into loops as they begin to dissipate; then only the cores of the vortexes remain visible. Contrails can also develop into a *popcorn* (puffy) structure if portions descend, which causes them to widen.

If a contrail in bright light casts a shadow on underlying aerosols (such as smoke, fog, or mist), the shadow appears as a dark line across the sky. When the Sun is behind the airplane, the dark line might appear to be in front of the airplane, as if a dark extension of the airplane's contrail.

An airplane can also produce a dark line known as a *distrail* when it flies through a fairly thin cloud, eliminating the cloud's water drops and ice crystals by vaporizing them either with the thermal energy of its engines or by mixing warm air from above the cloud down into the cloud. It can also produce a distrail if its engines dump enough moisture in the cloud that the ice crystals grow large enough to fall out of the cloud.

2.21 • Inward flutter of a shower curtain

When showering, the shower curtain invariably flutters inward and brushes against my leg. My curtain is not unusual, because unless they are weighted down or fitted with small magnets, most curtains display this annoying characteristic. What drives them inward?

Answer One popular explanation is that when air is heated by the hot water, it rises over the curtain, and so cooler air from the room must blow into the shower along the bottom of the curtain. Such a chimney-like flow is certainly present when you take a hot shower; however, the curtain blows inward even if you take a shower with water that is cooler than room air.

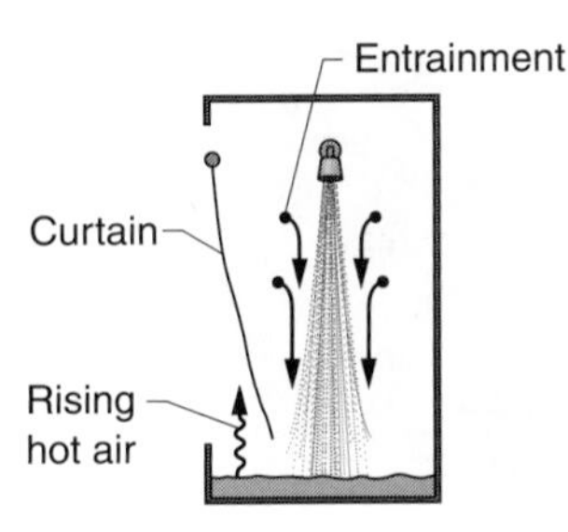

***Figure 2-7** / Item 2.21*
Entrainment of air causes inward flutter of shower curtain.

The primary cause of the curtain's motion is that as water falls, it *entrains* (traps and carries along) the adjacent air (Fig. 2-7). So, there must be a continuous flow of air toward the falling stream to replenish the adjacent air. Part of that flow is beneath the shower curtain, and the flow pushes the curtain inward. If the water is hot, the pool of water on the bottom of the shower heats the air just above it, and that air rises up against the in-blown curtain, helping support the curtain.

Air motion due to the entrainment of air by flowing water can also occur when water flows down into a cave system. The entrainment carries air into the cave along the water's path, which means that an equal amount of air must flow out of the cave. In some systems, a cave explorer can feel the outward flow of air.

2.22 • Prairie dog and giant ant nests

The prairie dog, a rodent that thrives in the open plains of the U. S. Midwest and in many residential areas, builds long burrows between 1 and 5 meters deep, connecting two or more entrances. Wind cannot blow into the burrows to bring oxygen to the prairie dogs. So why don't they suffocate in their burrows?

The leaf-cutting ants build huge nests, to a depth of perhaps 6 meters, to house some five million ants. Not only must the ants breathe within the complex maze of underground passages, but also the fungi they cultivate for their young requires oxygen and cannot withstand temperatures above 30°C. The activity of all those ants in the nest could easily elevate the temperature above this value. How is such a nest ventilated to control both oxygen and temperature?

Answer A prairie dog builds mounds around each of the entrances to a burrow, typically a rounded dome-like mound around one entrance and a steeper cone-like mound around another (Fig. 2-8). The mounds, which are made from the evacuation of the burrow and from surrounding dirt (and which are carefully maintained), can function as a lookout point for the animal, but their primary purpose is to ventilate the burrow. When wind blows across one of these openings, it tends to *entrain* (grab and remove) air molecules at the entrance. Because the mounds have different shapes and different heights, the entrainment is more pronounced at one entrance than another. Thus, air is pulled out through one opening, causing air to enter the other opening and flow

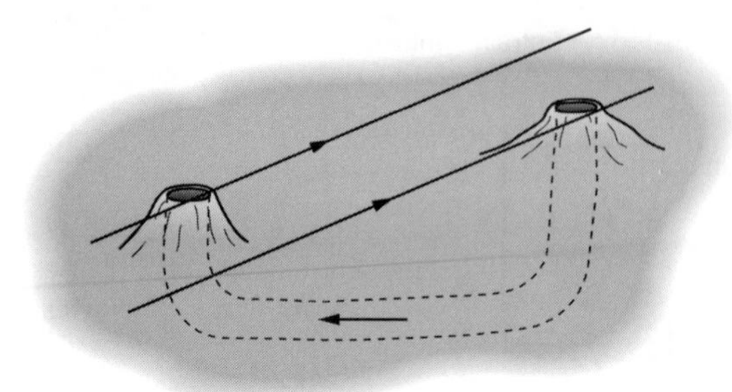

Figure 2-8 / *Item 2.22* Wind passing over two mounds of a prairie dog burrow.

through the burrow. With this dependable supply of oxygen, the animal does not suffocate.

The ants and fungi in the huge nests of leaf-cutting ants generate much thermal energy and thus heat the air in the nests. Although this warmed air tends to rise out of a nest, the nests are too large and complex to be vented this way. Instead, they are vented by entrainment of air by wind moving past the openings on the surface, much like prairie dog burrows are vented.

2.23 • Bathtub vortex

When water drains from a bathtub, why does it swirl above the drain, spiraling into a vortex, and what is the direction of the swirling, clockwise or counterclockwise? If the swirling direction depends on the hemisphere in which the bathtub is located, what is the direction when the bathtub is located near the equator? Does water flow into the swirl primarily from the top surface, as if the vortex is a drainpipe draining water from the top surface? What determines the depth of the vortex? (The vortex could be a mere dimple of air into the water surface, or it could be a column of air extending down into the drain.) Why is the swirling direction sometimes abruptly reversed in the last few minutes of drainage from a tub? Why is sound produced by some bathtub vortexes?

Answer The bathtub legend about the swirling direction is based on the general atmospheric circulations seen in large-scale systems such as hurricanes. When air flows over a large expanse, the rotation of Earth produces an apparent deflection of the winds in what is called the *Coriolis effect.* The deflections give a counterclockwise flow in the Northern Hemisphere and a clockwise flow in the Southern Hemisphere.

The water flow in a draining tub is a very short-scale system that is dominated by much greater effects than the Coriolis effect. The swirling direction is determined primarily by the net direction of swirling of the water when it pours into the tub or when someone makes it swirl. If the water is dominated by, say, clockwise flow, then the water can retain some of that flow direction for an hour or even longer. If the water is drained while it is still swirling clockwise, then the vortex over the drain will be clockwise. Other factors that can determine flow direction include the lack of symmetry in the bathtub (the drain may not be symmetrically located), the disturbance of pulling the plug, and temperature differences between water on one side (say, the side that opens into the rest of the bathroom) and the opposite side (at the wall).

The Coriolis effect has been demonstrated in a special bathtub when certain precautions were taken: The tub was circular, the drain was central, the water was allowed to settle for a long time, the temperature of the water was stabilized, the water was isolated from chance disturbance by people in the room, and the plug was pulled very carefully. With these precautions, the swirling was driven by the Coriolis effect and, because the tub was located in Boston, the water drained counterclockwise.

Most of the water going down a drain moves toward the drain near the bathtub base. When the water reaches the drain, some goes immediately down the drain but much of it spirals *upward* before it goes down the drain. The water going down the very center of the drain comes from the top water surface—that is, from the dimple that is seen over the drain. If the vortex is vigorous, the bottom of the dimple is thin and unstable, with air bubbles breaking off it.

The extent of the vortex (the depth of the air column in the vortex) is determined partly by the diameter of the drain. A wide drain usually produces only a dimple on the water surface. A narrow drain usually produces a narrow, vigorous vortex with an air column extending into the drain. An intermediate drain can produce a vortex that initially grows downward and then retreats upward.

The reason for a last-minute reversal of swirling is not well understood. One explanation is that when the water layer becomes very shallow, the flow into the vortex is suddenly hampered by the friction along the tub base.

A bathtub vortex can produce sound if it is vigorous enough to *entrain* (capture) air as bubbles, which emit sound as they oscillate and collapse. The water surface may also oscillate, sending out variations in air pressure as sound waves.

2.24 • Vortex in a cup of coffee

Carefully stir a cup of black coffee and then remove the spoon. As the coffee rotates around in the cup, slowly and carefully pour cold milk or cream into it at the center. Why does a dimple appear at the center? Why does the dimple fail to appear if the milk is warm or hot?

Answer You leave many small vortexes embedded in the general rotation you see in the coffee. Because the cold milk is denser than the coffee, it sinks down along the central axis of the swirling, pulling some of those vortexes into the center and stretching them downward. This gathering and stretching increases the rotation speed of the liquid near the center. The surface near the center then becomes concave, or dimpled, as swirling liquid normally does, but here the concave shape is more pronounced.

2.25 • Gathering of tea leaves, spinning of olives

If you stir a cup of tea with tea leaves scattered across the bottom (and then remove the spoon), why do the tea leaves gather at the bottom center? Just before they reach the center, why do they form a rough ring around the center and then move inward?

If a martini with an olive is stirred, the olive will move around the center of the glass with the swirling liquid but it will also spin around an internal axis. Why does the direction of spinning tend to be opposite the direction of swirling?

Answer As explained by Albert Einstein, the tea-leaf motion reveals a circulation pattern of tea water in the cup. Because your stirring makes the water rotate around the central vertical axis, the water tends to spiral outward. That is, each parcel of water moves as if on a flat, rotating merry-go-round.

However, in the teacup, the water next to the bottom surface is retarded by friction and thus does not rotate as vigorously as that at the top surface. So, the tendency to spiral outward is strong on the top and weak on the bottom surface. This difference sets up a circulation system called *secondary flow*: While the liquid circles around the central axis, it also moves outward along the top surface, down along the wall, inward along the bottom surface, and then upward along the central axis (Fig. 2-9). The flow along the bottom surface drags tea leaves to the center and then abandons them.

What Einstein did not notice was that tea leaves form a ring soon after the spoon is removed and before they finally reach the center. Tea leaves farther out than this ring are dragged into it by the secondary flow. Tea leaves closer to the center spiral outward to it. As the rotation of the water in the cup dies out, the radius of this ring decreases, and so the leaves gradually move to the center, finally coming to rest there.

If the tea were stirred by placing the cup on a rotating table, such as a record player, the stirring would begin at the bottom of the liquid because of friction between the liquid there and the surface of the cup. The rotation of the liquid would gradually ascend to the top surface. During that ascent, the liquid along the bottom surface tends to spiral outward and the liquid along the top surface has no tendency to spiral. As a result, a secondary flow is established: outward along the bottom, up along the wall, inward along the top, downward along the central axis. This flow is opposite the flow produced by spoon stirring, and now tea leaves will end up just below the wall.

When a martini with an olive is stirred, the olive straddles faster-moving liquid near the center of the glass and slower-moving liquid farther out. Thus, the drag on the olive can be greater at the point closest to the center, causing the olive to spin in a direction opposite the swirling. (Because many variables are involved, including the distribution of mass in the pitted, stuffed olive, the olive can, instead, spin in the direction of the swirling or spin chaotically.)

2.26 • Meandering rivers

Why does a river tend to meander (forming an *ox bow*) instead of following a straight line? From an airplane view, you might find that some rivers meander wildly. What causes the isolated loops of water, called *oxbow lakes*, that lie alongside a wildly meandering river?

Answer Meandering begins by chance in the complex flow of a river, but once even a slight change in direction is established, the water flow can enhance the change, creating a bend and then a loop. The water flow makes these changes by eroding the soil or rock along the shore and along the river bed. The process can be very complicated and can depend on the special circumstances of a river, but here is a simple explanation: Figure 2-10*a* shows an overhead view of a river bend, and Fig. 2-10*b* shows a vertical cross section taken across the bend. When water flows into this bend, it tends to spiral outward, as if thrown outward. The flow along the river bed is retarded by friction with the bed, which decreases the outward motion. The flow along the top surface of the river is not retarded. So, as water flows generally downstream in the bend, it also follows a *secondary flow*, which is outward on the top surface, down along the outward bank, inward along the bed, and upward along the inner bank. This secondary flow carves out material on the outer bank and deposits the material somewhat downstream on the inner bank. Thus, the bend grows outward as the outer bank is gradually removed.

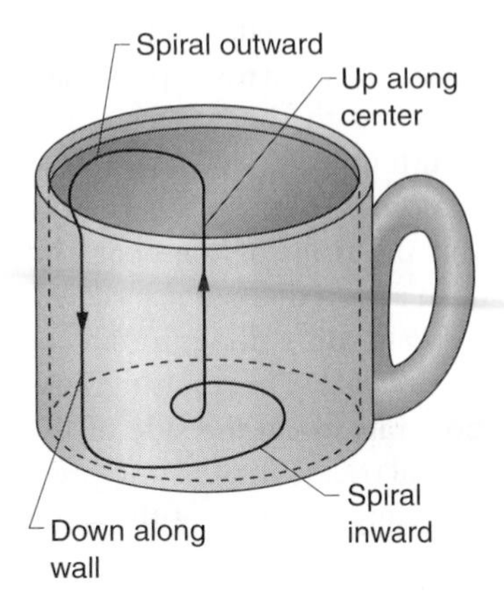

Figure 2-9 / *Item 2.25* Secondary flow in stirred tea.

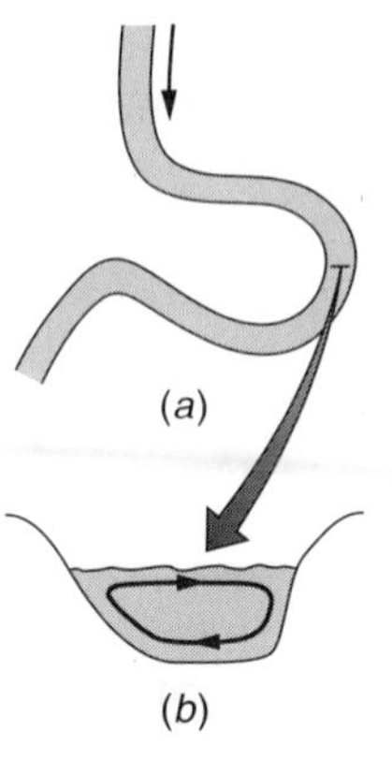

Figure 2-10 / *Item 2.26* (*a*) Overhead view of a river meander (ox bow). (*b*) Secondary flow in a cross section within the meander.

If the loop becomes large, erosion along the turns leading into the loop can cut off the loop, which then is an oxbow lake.

2.27 • Bird spinning in water

Why does a phalarope (a small wading bird) spin vigorously on water while snapping its head down to peck at the surface?

Answer When there aren't enough prey on the water surface, a floating phalarope rapidly spins by pushing hard against the water with its toes outstretched and then folding the toes out of the way during retraction of the foot. Its efforts cause an upwelling of the underlying water, which brings prey to the surface, rotating in the *opposite* direction of the bird. The bird very rapidly pecks at the prey as they reach the surface. The best results probably occur when the upwelling collects prey off a somewhat shallow bottom that otherwise would be out of the bird's reach.

2.28 • Water climbing a spinning egg

If you spin a hard-boiled egg like a top, it will stand up on one end. If you spin it in a shallow pool of water (a few millimeters deep), why does the water climb up the side of the egg before it is thrown off?

Answer Normally when you stir water as in a cup of tea, the water spirals outward, leaving a dimple at the center and thus marking a vortex there. When the egg spins in the water, the water tends to move toward the outside but it also adheres to the egg. By climbing the underside of the curved shell, it can remain clinging while also moving outward. At some point on the curve of the shell, the gravitational force and the general instability cause the water to break free of the shell. It then flies through the air as drops, which land in a circle around the egg.

2.29 • Circular water-flow pattern in a sink

When a smoothly flowing water stream from a faucet hits a flat sink with an open drain, why does a circle form around the impact point, with deeper water on the outside of the circle?

Answer When the water from a faucet hits the sink, it spreads radially at a rate that is said to be *supercritical* because the water moves faster than waves can move over the water. Initially, the flow is stable because any chance disturbance is quickly eliminated. However, as the water spreads outward, the effects of the water's viscosity become important and the flow becomes unstable. In one description, the viscous flow begins along the sink surface and then gradually extends upward. At a certain radius from the impact point, the viscous flow reaches the surface and the water depth suddenly increases, an effect known as a *hydraulic jump* (Fig. 2-11). Beyond this wall, the water speed is slower (*subcritical*). Thus the hydraulic jump is the transition from faster, shallower flow to slower, deeper flow.

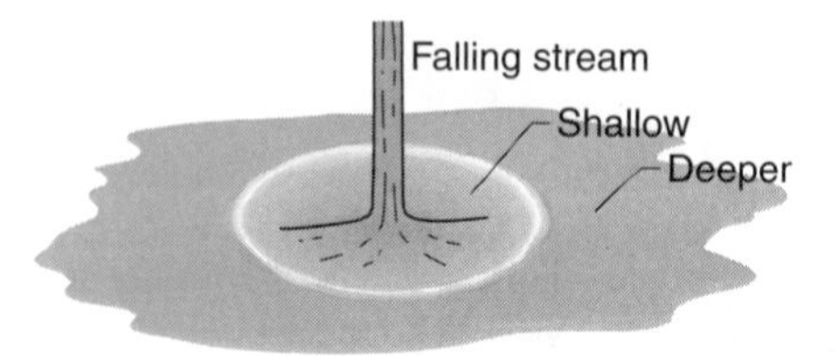

Figure 2-11 / Item 2.29 Circular wall of water around stream's impact on basin.

Hydraulic jumps often appear on many common streams, such as where water flows down a driveway, along a curb, through underground drainage pipes, and along tilted irrigation cannels. Look for a stationary wave on the flow, especially where an obstacle lies in the flow. Waves are created as the water flows over or past the obstacle. Most of those waves merely lose their energy and disappear, but one wave, with a particular wavelength, moves up the stream as fast as the water moves down the stream, and so the wave is stationary. The continuous disturbance of the water by the obstacle continuously feeds energy into the wave, allowing it to persist. You might see a stationary series of crests and valleys instead of a single wall as you do in a sink. Hydraulic jumps on fast-moving streams can be a serious (or even fatal) problem for white-water rafting, because a raft can become trapped at the jump and overturned by the turbulence.

If you carefully place a water drop just upstream of a hydraulic jump in a kitchen sink, the drop can become trapped against the jump's wall and remain floating for a long time (without merely merging into the water) because air is continuously being dragged under it by the flowing water.

A stream of a viscous fluid, such as antifreeze (ethylene glycol) can form a circular hydraulic jump, but it can also spontaneously transform into a polygonal hydraulic jump with straight edges and sharp corners.

2.30 • Water level in canals

Suppose you are in a boat moving along a fairly narrow and shallow canal. As the bow passes a point on the side of the canal, does the water level at that point go up or down?

Answer As the boat moves along the canal, water in front of the boat must be pulled through the narrow open space alongside the boat to end up behind the boat. That water motion is accomplished by a continuous dip in the water at the bow. Because the water pressure is lower in the dip, the dip effectively pulls water from the front of the boat and sends it to the side of the boat, an action dubbed *drawdown*. Thus, as the boat advances, the water dips along the side of the canal in what is called the *canal effect*. The resulting variation in water pressure and flow rate along the side of the boat can make navigation of a canal tricky, and the low-pressure pulse can affect equipment along the sides of the canal or even in adjoining water branches.

2.31 • Solitary waves

In 1834, the British engineer and naval architect John Scott Russell witnessed a curious wave on a water channel near Edinburgh. A boat was being pulled rapidly through the water by horses when suddenly the horses and boat stopped. However, the mound of water at the bow of the boat did not stop; instead, it continued along the channel at about 4 meters per second. On horseback, Russell was able to follow this mound, which was about $\frac{1}{3}$ meter high and about 10 meters long (across the channel), for about 3 kilometers before he lost sight of it in the "windings of the channel." He was astounded that the wave did not seem to diminish as it traveled. If you splash in a stream, the waves you make diminish rather quickly and certainly could not travel several kilometers even on a broad body of water. What was so different about Russell's wave?

Answer If a boat goes through a channel faster than the speed of waves over the water, the bow of the boat tends to bulldoze water into mounds in front of the boat. If the boat speed is only slightly faster than the wave speed, then the water develops several distinct crests and valleys. If the boat goes faster, then the valleys are filled in and the bulldozed water becomes a single prominent mound called a *solitary wave* or *soliton.*

Russell saw a solitary wave that was released from the boat when the boat suddenly stopped. The mathematics of such a wave is notoriously difficult, but the solitary wave is itself simple. Normally, waves sent across water undergo sorting by wavelengths, a process known as dispersion. So, if a splash sends out a bunch of waves with various wavelengths, the waves disperse while also dying out with distance. In a solitary wave, the disturbance to the water level is amplified by the wave itself, preventing dispersion and thus maintaining the wave's shape. In fact, a solitary wave should travel for a very long distance because it loses its energy very gradually due to the low viscosity (internal friction) of the water.

In an ordinary wave, a parcel of the water moves in a circular or elliptical path but is not transported in the direction of the wave's travel. For example, if you splash to send waves across a pond, only the waves and not the water move over the pond. A solitary wave is different because it *does* transport water. To show this, Russell had many horse-dragged boats send solitary waves along a long canal. Russell found that the water depth increased at the far end of the canal and decreased (by as much) at the near end.

2.32 • Tidal bores

When the tide enters and then runs up certain rivers, why does the incoming water develop a turbulent wall, called a *tidal bore*, that can dramatically increase the depth of the water as it sweeps past a given point on the river? On some rivers (such as the Severn River in England) and on some occasions (the tide has to be right), the wave that sweeps up the river can be high enough that surfers ride the wave for several kilometers.

Long before surfing was a sport, fishermen would routinely position their boats at a river's mouth to hitch a ride upstream on a bore. That use was apparently unknown to the captains and crews of the Royal Navy ships surveying the Chien-tang-kiang River in China in 1888. One night, with the ships at anchor in the river, the crews heard a tremendous roar. About 30 minutes later, a bore caught the ships and carried them upstream by about a kilometer even though the ship engines were set on full power to oppose the motion. The roar came from the turbulence in the bore, which could have capsized the ships.

Answer A tidal bore can occur when a large amount of water flows upstream on a river, either forming a turbulent single wall of water or a smooth, short series of crests and valleys. The best conditions for a bore are these: (1) Large tidal variations in water height in the body of water into which the river empties. (2) The river should be shallow with sloping sides and a funnel-shaped estuary. As the long-wavelength waves in the deeper water become funneled into the shallow water in the estuary and then into the river, the water forms a front or wall. In an *unduluar jump*, the crests and valleys of the waves increase in height and depth and the front of the crests become steeper until the crests overrun the valleys. The result is a single wall or pulse of water, called a *solitary wave*, moving upstream against the normal river flow. A bore can be devastating if it catches a boater by surprise. For a historical example, in 1843 the daughter of the famous 19th century writer Victor Hugo was knocked from her boat on the lower Seine by an unexpected, viscous bore and, because she was unable to swim, drowned.

2.33 • Tides

What causes the tides? Why do most shore locations have two high tides per day but others have only one?

Answer Here is a simple answer. The primary cause of tides is the Moon's gravitational pull on Earth's oceans even though that force is not strong enough to lift the water. Because the force varies over Earth's surface (strongest on the side facing the Moon, weakest on the opposite side), the force reshapes the water distribution by stretching it parallel to the line connecting Earth and the Moon. The stretching produces two bulges in the water distribution, one on the side facing the Moon and one on the opposite side. If Earth did not rotate, then a shore location in the bulge facing the Moon would have high water (*high tide*) all day and a shore location in the opposite bulge would as well. However, Earth's rotation means that a shore location rotates through both bulges in about a day and will thus have two high-water intervals.

Here are some complicating factors. The bulges are not exactly positioned on a line through Earth and the Moon because the water motion encounters friction within the water and against shorelines. The friction delays the water's response to the stretching by the Moon. So, the high-water point in a port city may occur an hour or more after the Moon is highest in the sky. For example, the high-water point in the English Channel is delayed by many hours because the water motion encounters considerable resistance.

Another complicating factor is that the gravitational force from the Sun also tends to stretch the water distribution. However, the solar effect is, roughly speaking, less than half the lunar effect. Although the Sun is much larger than the Moon, it is also much farther from Earth. During New Moon and Full Moon, the Sun and Moon are aligned and their tidal effects sum to give larger tides called *spring tides.* When the directions to the Sun and Moon are separated by 90°, the sum gives *neap tides.* Because of these various complications, some shore locations can have only one noticeable tide per day.

2.34 • Tides in the Bay of Fundy

The tides in the Bay of Fundy (Nova Scotia, Canada) are dramatic, sometimes with a 15 meter range of lowest level to highest level occurring over a few hours. Why can the tidal range be so large there and not so large at other places?

Answer You can make the water in a bathtub oscillate (slosh) if you periodically push down on it (pump it) with a paddle. You obtain the strongest oscillations if you time the pushing so that each pump occurs every time the water is highest at one end of the tub. With that timing, you are said to be *resonantly* pumping the water, and the time interval of the pumping is said to match the *oscillation period* of the tub.

The water in a bay can also be made to oscillate if it is resonantly pumped. For example, the tidal effects due to the Moon tend to set up oscillations in a bay, to make the water slosh. However, for most bays the oscillations are small because the timing of the tidal effects does not match the oscillation period of the bay. The Bay of Fundy is different: Its oscillation period is about 13.3 hours, which is reasonably close to the 12.4 hours between tidal bulges. Thus the sloshing in the Bay of Fundy is significant.

Historical records suggest that tidal variations in the Bay of Fundy have gradually increased because the resonance period of the bay has been gradually shifting closer to the tidal variation period. The shift could be due to changes in the bay's shape because of an increase in the sea-level height.

2.35 • Dead water

While on a polar expedition in August 1893, the ship *Fram* encountered what is now called *dead water* on the northern coast of Siberia. The ship was capable of traveling 6 or 7 knots but in the dead water could manage only 1.5 knots even though both the water and the weather were calm. Moreover, control of the ship was marginal; in fact, the captain was forced to travel in loops to escape the dead-water region. The water was not visibly different from any other stretch of ocean water. What caused the speed reduction and loss of helm control?

Answer Dead water occurs when a layer of relatively fresh water overlies salt water, which can happen when a river empties onto ocean water. Two interfaces play a role: the air–fresh water interface and the fresh water–salt water interface. Normally, much of the energy of a ship engine creates waves along the first of those interfaces—think of the wave production as a form of drag on the ship. In dead water, the ship produces two sets of waves, one along each interface, and so the drag is significantly more. The faster the ship tries to go, the faster its energy drains to the *internal waves,* as they are called, on the fresh water–salt water interface.

The ship's bow is located above the first crest in the internal wave. The water just below that crest moves in the direction opposite the ship, opposing the ship's motion. The *Fram's* length was such that the rudder was also above a crest of the internal wave, and so the rudder was of little use in maneuvering the ship.

2.36 • Tornadoes

Tornadoes can occur in many places around the world but they are an almost continuous threat in *tornado alley,* a broad stretch through the center of the United States. They strike both awe and fear in anyone seeing them or hiding from them, because they are a simultaneous combination of beauty and evil.

What causes tornadoes, and why do they occur so frequently in tornado alley? Does a tornado destroy a home by pushing in walls or pulling them outward? Can tornado winds actually drive straw into wood as sometimes claimed in the popular press?

Answer A tornado is a large atmospheric vortex that can form in an extensive storm where moist warm air slides under cool dry air, with the two bodies of air moving in opposite directions. As the warm air rises through the cool air and its water vapor begins to form water drops, a lot of thermal energy is released by the transition from vapor to liquid. This energy release causes the rising warm air to rise even faster. The complex air motion (opposite moving bodies and accelerated rising air) produces *wind shear* where adjacent airflows have very different speeds and directions. In a way that is not well understood, these conditions can lead to swirling and then to a tornado. Although these actions can be simulated on large computers, the simulations do not reveal a simple answer about what causes a tornado or how the large tornadoes derive so much energy.

A tornado is visible only if it picks up dirt and other debris off the ground and brings it up along the funnel or if it causes significant condensation of water into water drops. The larger tornadoes are probably composites of several simultaneous vortexes: Several smaller vortexes circle around a larger central one. Tornadoes come in a variety of shapes, including those that resemble funnels, pillars, or ropes (snakes). Some tornadoes are nearly vertical, while others reach horizontally before dipping downward. All tornadoes appear to move in a haphazard way, leaping over the countryside and dipping down occasionally to leave great scars on the ground.

Contrary to popular belief, the danger to a house from a tornado is not a sudden reduction of air pressure outside the house, which would cause the walls to explode outward. Indeed, the air pressure does not drop by very much. So, if a tornado is approaching, don't spend time opening windows in the hope that the internal air pressure will match the external air pressure. Run! Hide! The basement may offer the best protection but, lacking that, a bathroom with its partial shield of a tub and plumbing might be best.

The danger to a house is from the high-speed winds around a tornado. Once they catch under the eaves, they can lift a roof off a house. Then with the structural integrity of the house lost, the wall facing the wind will be blown inward and the other three walls will be blown outward. In spite of what we see in *The Wizard of Oz*, a house is unlikely to be entirely lifted and transported by a tornado. Rather, it is more likely to be blown apart and its debris transported, perhaps even used as shrapnel in ripping apart a nearby house. If a house is not blown apart, it might end up being rotated around an anchor, probably the plumbing in a bathroom, so that it ends with a new orientation.

The winds of a tornado can be strong enough to drive straw into wood or a wood stick into a steel pipe. In laboratory simulations of tornado winds, splinters, toothpicks, and broom straw have penetrated various types of wood targets after being shot from a pneumatic gun.

SHORT STORY

2.37 • Looking up into a tornado

A few people have survived the experience of looking up into the funnel of a tornado. The most thorough description on record is by Captain Roy S. Hall whose house was attacked by a tornado in May 1948. After the roof had been taken and some of the walls pushed in, Hall was able to see his neighbor's house and was relieved that his house was not flying through the air as he had momentarily feared. However, he then saw something horrible: About 20 meters away something descended to about 6 meters above the ground and hovered with a slow vertical oscillation. That something was curved with a concave surface facing him. With shock he realized that this hovering thing was the inside surface of the tornado funnel, and *so he was inside the funnel*!

As he looked up into the funnel, it seemed to stretch for 1000 feet, swaying and gradually bending. It contained a bright central region that glowed like a fluorescent light fixture. As the funnel bent over, rings formed along its length. Hall saw nothing being pulled up through the funnel's interior, had no trouble breathing (so the air pressure could not have been noticeably low), and marveled at the complete silence (in contrast to the dramatic noise during the tornado's approach). Suddenly the funnel moved away, and Hall's family came out of hiding to find him.

2.38 • Waterspouts and funnel clouds

What causes a waterspout, those large vortexes seen over water? Why can boats sometimes survive an encounter with a waterspout?

Answer A waterspout typically forms over water where there is a strong updraft surrounded by a region of downdraft. Air drawn into the updraft picks up moisture and thermal energy from the underlying water. As it ascends in the funnel, it is warmer and moister than the surrounding air. Because it is warmer, the air accelerates upward, but then the moisture begins to condense as drops. That change releases much thermal energy, which warms the air even more, increasing the upward acceleration. This process is the so-called *heat engine* driving a waterspout. Air in the surrounding region, especially air cooled by rain, descends to take the place of the air lost to the funnel. Although a waterspout resembles a tornado and is often described as being a very weak tornado, the heat engine driving it and the instability of the air that produces the updrafts are more like those of a dust devil.

Although boats often survive weaker waterspouts, larger waterspouts can do considerable damage and easily capsize even a moderate-size boat. The larger waterspouts are probably responsible for stories about fish raining from the sky: A waterspout can pull up a lot of water and fish before it moves onto land, where it loses its heat engine, dissipates, and dumps its load.

The lower end is surrounded by a *spray sheath*, a swirling squat cylinder of spray. The lower third of the funnel is visible largely because of water pulled up into the funnel, and the rest might be visible if water vapor condenses into drops, which can then scatter sunlight.

2.39 • Dust devils, fog devils, and steam devils

Dust devils are whirlwinds that often appear in hot regions but also appear on the cold surface of Mars. They are made visible by dust, dirt, and other debris they pick up from the ground and then carry upward. Many dust devils are small and harmless, but some are a kilometer tall and big enough to carry small animals (or even children). The dust devils on Mars are even larger, perhaps 6 kilometers tall.

Fog devils are whirlwinds that can appear in fog, and steam devils are whirlwinds that can appear over water on cold days. Both are fleeting and harmless.

What causes these types of whirlwinds?

Answer These various types of whirlwinds are due to an unstable arrangement in which cooler air lies over warmer air. For example, a dust devil can occur when bright sunlight intensely heats the ground, which then heats an overlying thin layer of air. That hot (low-density) air should rise away from the ground, but if the region has little or no wind, cooler air sits like a blanket on the hot-air layer. The situation is unstable and even a rabbit scurrying through the region can trigger an upward burst of hot air. Then additional hot air flows across the ground to the bursting point, swirling as it

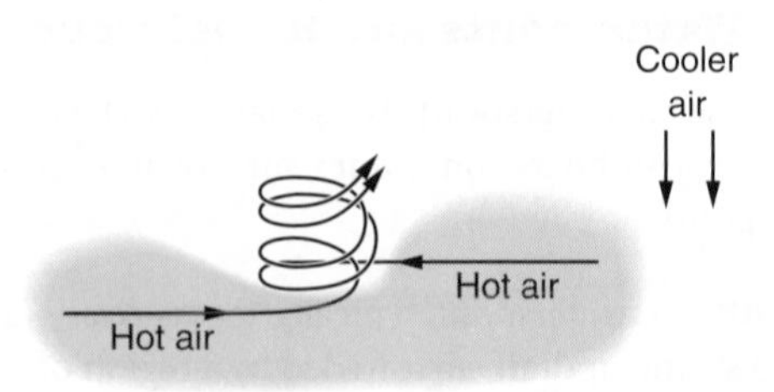

Figure 2-12 / *Item 2.39* Hot air flows along hot ground and then swirls upward. Cooler air descends.

enters the ascending column of hot air—this is a dust devil (Fig. 2-12). The swirling can be either clockwise or counterclockwise, depending on the flow over the ground and the obstacles encountered. In the surrounding region, cooler air descends to replace hot air lost to the dust devil.

A dust devil can move from land to water, but unless it picks up a lot of water, it may be difficult to see and the only evidence might be the circular ridge it produces on the water.

A fog devil can occur when fog rises from brightly lit wet grass. The grass heats the air just above it, which then begins to ascend somewhat like hot air in a dust devil. However, the moisture in the air condenses to form drops, a process that releases a lot of thermal energy and causes the hot air to rise even faster. A steam devil can occur over water when the air temperature is below freezing but the water temperature is above freezing. The air next to the water is then warmer than the air above it, an unstable situation.

You might be able to produce miniature steam devils on a cold day: Place a wide container beneath a window and fill it with very hot water. Next, open the window so that cold dense air flows through the window and across the water. Hot air and water vapor rising from the water are accelerated upward as they enter the cold air because the cold air is denser and because the water vapor begins to condense, releasing thermal energy. They are also being pushed horizontally by the cold airflow. The combined vertical and horizontal motion, along with some turbulence, might produce fleeting vortexes, which are made visible by the condensation of water drops.

2.40 • Ring vortexes

How does a smoker blow a smoke ring? Why does the smoke ring expand if it approaches a wall? How does a dolphin produce a similar ring of air in the water?

Answer A smoke ring is a *ring vortex* blown with a strong puff from a mouth filled with smoke. As the smoke and air leave through the rounded opening of the mouth, the flow near the lips is slowed by friction, and so the flow through the center of the opening tends to outrun it. The tendency causes the flow to curl outward around the lips, thus starting the vortex motion. The smoke merely acts as a *tracer*, making the air motion visible.

If the smoke ring approaches a wall, the friction on the airflow from the wall causes the ring to expand. The rate at which the air swirls decreases, somewhat like the rate at which an ice-skater spins on point decreases when the arms are extended outward.

A dolphin also likes to play with a ring vortex and can produce one in a number of ways. Here is probably the most common way: The dolphin swims on its side while flicking its (then) vertical tail fin from side to side. As the fin moves through the water, the flow next to the fin is retarded by friction, causing a curling motion that develops into a ring vortex in a vertical plane. The dolphin turns back, presents its blow hole to the ring vortex, and blows air into the core of the vortex, where it is quickly distributed throughout the vortex. The air affects the buoyancy of the vortex and also acts as a tracer. The dolphin might play with the vortex by following it, swimming through it, producing a second ring vortex to interact with it, or breaking off a section that will loop to form a secondary, smaller ring vortex.

In a classroom, a vortex can be formed with an *air cannon*, which is a box with a circular opening at the front and a flexible covering (such as a plastic garbage bag) fitted across the otherwise open back. When the flexible covering is pulled to the rear and released, it pushes a stream of air through the circular opening. Just as with a blown smoke ring, the flow forms a ring vortex but without the benefit of a tracer. With an air cannon, you can startle someone across the room with a large ring vortex that approaches with no warning.

A ring vortex can also be produced by allowing a drop to fall into either the same type of fluid or a fluid with which it can mix. As the drop hits and penetrates, it forms into a ring vortex. The formation is easier to see if the drop contains a small amount of dye.

If one ring vortex trails another, with the two roughly centered on the same axis, the trailing vortex might catch the leading one. Depending on the circumstances, the two vortexes may merge to form a single vortex or they can play the following game (Fig. 2-13): The trailing vortex shrinks and rotates faster while the leading vortex expands and rotates slower. The tailing vortex then passes through the leading vortex to become the new leader. This leap-frog exchange can occur several times. You might see similar leap-frog action if you let a second drop fall into a liquid quickly after a first drop enters the liquid. If each drop develops into a ring vortex, the second vortex might pass through the first vortex.

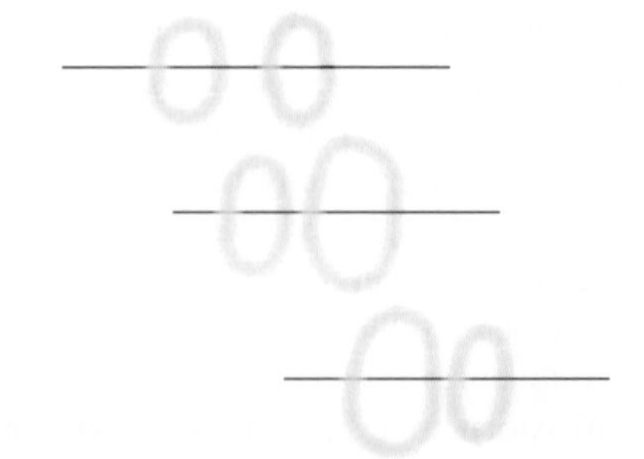

Figure 2-13 / *Item 2.40* Trailing smoke ring passes through leading smoke ring.

2.41 • Siphons and toilets

How can a liquid be siphoned out of a container (Fig. 2-14)? That is, why does the liquid climb up the tube inserted into it? In particular, does the air pressure on the liquid push it up the tube? What limits the height of the climb? Why must the tube's free end be lower than the end inside the container?

Answer To initiate siphoning, the tube draped over the container wall must be entirely filled with liquid. (It may have been filled by a squeeze ball that pulls the liquid up over the top of the tube.) Although liquid flows and does not have the substance of a solid, it is nevertheless cohesive. That is, each portion is attracted to adjacent portions. When the liquid in the tube's downflow section begins to fall out of the tube, portions near the top pull other portions over the top, which pulls still more portions up to the top. The whole action is as if a chain were in the tube. As long as the chain section outside the container is longer than the section inside the container, gravity will pull the chain up, over, and down the tube.

Contrary to common belief, the atmospheric pressure does *not* push liquid up the tube. Indeed, if the atmospheric pressure changes, the siphoning action is unaffected.

When liquid is siphoned, it is said to be under *tensile stress* because any portion of it in the upflow section is being pulled both upward and downward. Surprisingly, water can withstand tensile stress up to a limit, but beyond that limit, the water suddenly forms cavities and vaporizes into them. The siphon height can be increased until that transformation occurs in the top of the tube. Then, with air cavities breaking the continuity of the water, the siphoning action stops and the water merely drains from the tube.

The siphoning action will also stop if air seeps into the upflow section, to gather at the top and break the continuity of the water. This type of breaking occurs in the common toilet. When water is dumped into the toilet bowl from a tank, the increased pressure at the bottom of the bowl pushes water into the drainage pipe, which forms a siphoning tube. Water and whatever is in the water is then siphoned into the pipe until the water in the bowl is nearly completely drained. Then air can bubble into the siphon, breaking the continuity of water at the top of the siphon and stopping the siphoning action. Usually water from the tank continues to enter the bowl for a few more minutes, but it is not enough to restart the siphoning action. However, it does serve to act as a barrier against the odors that would otherwise come into the room from the drainage pipe.

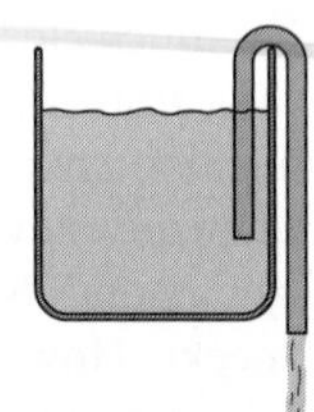

Figure 2-14 / *Item 2.41*
Siphoning tube.

2.42 • Lizards walking on water

How does a basilisk lizard manage to run over water without merely sinking into it? Not only do the young, low-weight lizards escape predators this way, but the older, more weighty adults do also.

Answer During a run, a footfall begins when the lizard slaps its foot against the water. The slap produces an upward, supporting force on the lizard, but because water is a fluid with low viscosity (internal friction), the foot soon begins to sink into the water. As it sinks, it pushes out a cavity of air, first downward and then toward the rear. The rearward push provides a forward force on the lizard, allowing it to run. Because the lizard does not want to fight water drag on its leg, it pulls the leg out of the cavity before water enters and surrounds its foot and leg. By then the other leg has begun a footfall with a slap on the water. Although the lizard does sink somewhat, the average upward force it experiences with the series of foot slaps is enough to support even an adult.

2.43 • Lead bar floating in a boat

Suppose you sit in a boat floating in a small backyard pool. Inside the boat you have a very large cork and a bar of lead. What happens to the water level in the pool if you throw the cork onto the grass, the cork onto the water, the bar onto the grass, and the bar into the water?

What happens to the water level if a hole is made in the bottom of the boat so that water gradually enters to sink it? If the water level changes, does it begin to change when the water first enters the boat?

Answer When an object floats, it displaces water; that is, it occupies space that otherwise the water would occupy. The amount of space is set by this simple rule: The mass of the displaced water equals the mass of the object. So, if a cork with a mass of 1 kilogram floats in water, it sinks into the water until it has displaced a volume of water with a mass of 1 kilogram. The cork displaces that much water regardless of whether it floats directly in the water or via the boat. When you throw the cork from the boat onto the water, the amount of displaced water is unchanged and so is the water level in the pool. If you throw the cork onto the grass, it is no longer displacing water in the pool. So, the water level descends.

When the bar of lead is in the boat, the same rule about matching masses applies. Suppose the bar has a mass of 1 kilogram. Then it displaces an amount of water that has a matching mass of 1 kilogram. That takes a lot of water, the volume being about 11 times the volume of the bar. If you throw the bar onto the grass, it no longer displaces that large amount of water, and the water level in the pool decreases. If, instead, you throw the bar into the water, it completely submerges. Then the amount of water displaced has the volume of the bar. The amount is now $\frac{1}{11}$ times what was displaced when the bar floated via the boat. Thus, the water level decreases.

When a boat begins to take on water, it is still floating and thus displaces the same amount of water. The water level changes only when the boat stops floating—that is, when it is fully submerged. Then the water level abruptly decreases.

2.44 • Floating bars and open containers

Does an open container, such as a food or beverage container, float upright or with a list? If a long bar with a square cross section floats in liquid, which of the orientations in Fig. 2-15 does it take?

Answer In any floating situation, an upward buoyancy force on an object balances its downward gravitational force. That balancing can generally be met with many orientations if the container is suitably submerged. However, most orientations are unstable in that the buoyancy force causes the container to rotate. The resulting orientation is difficult to describe in general—you might explore results in a sink or during a bath. Here are a few results: A squat container will float upright (bottom face down) but a tall, narrow container will list, maybe to the point of capsizing. Perhaps the most curious floating behavior is that of a light container when water is gradually added to it: From its empty orientation of being upright, it lists and the list increases and then it decreases until the container is upright again when it is about to sink.

The orientation of a square bar depends on the ratio of the bar's density to the liquid's density. Because the bar floats, the ratio cannot be more than 1. At a value of almost 0, the bar is so light that it is hardly submerged and floats facedown. If we gradually decrease the density of the liquid, the bar gradually sinks into the liquid but continues to float facedown. However, when the ratio reaches about 0.21, the bar begins to list, and when the ratio reaches about 0.28, it is floating with its sides at 45° to the horizontal.

If we continue to decrease the liquid's density, the orientation does not change until the ratio reaches about 0.72, and then the list decreases until the bar is again facedown when the ratio is about 0.79. When the ratio reaches 1, the bar is fully submerged and still facedown.

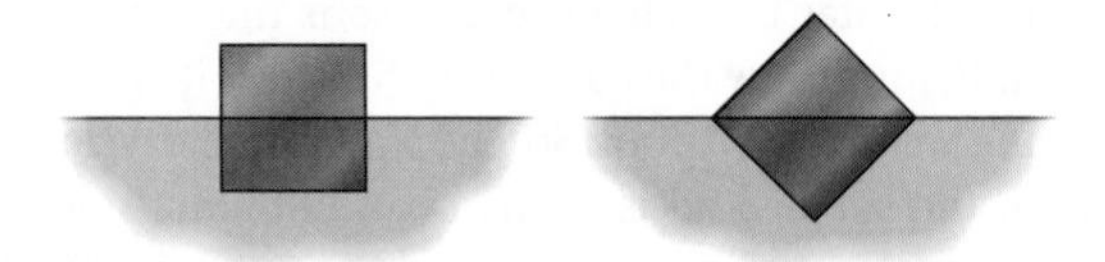

Figure 2-15 / *Item 2.44* Two orientations of a floating square bar.

2.45 • Hole in a dam, ship in dry dock

A popular tale involves a Dutch boy who saves his town from flooding by stuffing his finger into a hole he discovers in the dike holding back the North Sea. How can one boy push back the entire North Sea?

If a boat is put into dry dock, the water is drained while the walls are moved inward until finally the boat is held in place by the walls. During the draining process, what is the least amount of water needed to float the boat?

Answer The water pressure on the boy's finger depends on the depth of the hole from the water surface, not on the width or full depth of the ocean. So, provided the depth of the hole was not significant, perhaps the story could be true.

The dry-dock question lacks a full answer. However, the boat's ability to float does not depend on the depth or width of the body of water. All that matters is the height of the water alongside the boat. In principle, if that height is maintained, the water pressure on the boat always provides the upward buoyancy needed to match the downward gravitational force on the boat. Thus, even a thin layer of water that hugs the hull should be enough. However, if we consider a very thin layer, the layer becomes unstable and a chance disturbance can suddenly cause the boat and the dry-dock wall to touch, ending the floating.

2.46 • g-LOC in pilots

"Top gun" pilots have long worried about taking a turn too tightly because they can undergo what is called *g*-LOC (*g*-induced loss of consciousness). There are several warning signs for a pilot to ease up. When the centripetal acceleration is 2 or 3 *g*s, the pilot feels heavy. At about 4 *g*s, the pilot's vision switches to black and white and narrows to *tunnel vision*, in which the peripheral view disappears and only the straight-ahead view remains (as if looking into a tunnel). If the acceleration is maintained at 4 *g*s or increased, vision ceases and, soon after, the pilot becomes unconscious. What causes these changes in a pilot?

Answer If a pilot takes a turn with the head toward the center of the circle, as is normally done, the blood pressure in the brain drops, impairing vision and ultimately causing unconsciousness. Modern jets are both powerful and highly maneuverable, and so a pilot can easily take a turn too quickly, especially when in a dogfight. The pilot will then fall into *g*-LOC without warning. If the pilot does not regain consciousness in time, the airplane will either stall or fly into the ground.

2.47 • Blood circulation in snakes, giraffes, and tall dinosaurs

Why is the heart at the midpoint in a water snake, somewhat closer to the head in a ground snake, and much closer to the head in a tree-climbing snake? How does a giraffe manage to send blood to its head while not allowing blood to pool in its legs? How does it avoid suffering brain damage or even fainting when it bends over to drink from, say, a pond? Sauropod dinosaurs were huge, with very long necks. How did they manage to send blood to the head and drink water?

Answer If a snake is head up and vertical, the heart must pump blood uphill to the brain, and the blood tends to pool in the lower half of the body. However, neither effect is a problem for a water snake because the water pressure on the snake increases with depth. The higher pressure on the lower half of the snake prevents blood pooling. The heart is located at the snake's midpoint, and so the higher water pressure there and the lower water pressure at the head helps push blood to the brain.

A vertical ground snake lacks the water pressure and thus suffers from blood pooling. However, its heart has a better placement because it is closer to the brain than the snake's midpoint. A tree snake is even more adapted: Its heart is even closer to the brain and the lower half of the snake is tightly constructed to prevent blood pooling. Thus, a tree snake can climb a tree without fainting.

A giraffe has an even more serious blood-flow problem. Because its head is so much higher than its heart, the blood pressure must be very large. For example, for a giraffe 4.0 meters tall, the average blood pressure in the aorta must be about 250 mm Hg (millimeters mercury) if the blood pressure in the brain is to be a reasonable 90 mm Hg. Because the feet are so far below the heart, the very large blood pressure would cause severe pooling in the legs and feet were it not for their construction. The legs are muscular and have tight skin that acts somewhat like pressure stockings. When a giraffe brings its head down to drink, it moves slowly to allow its blood pressure to adjust. It also spreads its front legs to lower its heart. Although a *profusion mesh* supplying the brain with blood helps to protect the brain, a sudden increase in blood pressure could cause the animal to faint or suffer brain damage.

A sauropod had a worse blood-flow problem, even if it never raised its head fully. It probably moved slowly to allow pressure adjustments. It also had a huge heart, up to 5% of its body weight.

2.48 • Did the sauropods swim?

The dinosaurs known as sauropods, including the *Apatosaurus* (once also named *Brontosaurus*) and the surrealistically long-necked *Mamenchisaurus*, were large, even in dinosaur measures. A long-standing question has been how they could manage to walk (let alone, run). One suggestion was that they spent much of their life swimming or wading in water. Could such a huge dinosaur swim?

Answer Because we don't have sauropods to observe, the best way to explore the question is to make scale models to see whether they float. (The tricky part is allowing for the lungs.) This research revealed that the center of the buoyant forces lifting a model is somewhat behind the center of the gravitational forces pulling the model downward. This would have been an unstable arrangement, because the net buoyant force would have rolled a sauropod forward until its neck was at least partially submerged. In fact, a sauropod would probably also have rolled onto one side or the other. In short, sauropods would not have enjoyed a day swimming at the beach.

However, they could have waded through water up to their chests without trouble, and the ones with longer forelegs could have then punted their way through the water like gondoliers in Venice who pole their way through it. Indeed, tracks left by punting dinosaurs have been found: The tracks differ from the whole-foot tracks left by walking dinosaurs because a punting dinosaur digs a claw tip into the mud and then drags it backwards, leaving a narrow channel with some mud thrown up at the rear.

2.49 • Gastroliths in dinosaurs and crocodiles

Why do the stomachs of many tetrapods, both living ones such as crocodiles and fossil ones such as plesiosaurs, contain pebbles and stones, which are called *gastroliths* or *stomach stones*?

Answer Gastroliths were long thought to be needed for digestion, so that the animal can grind food in the stomach. However, a stronger argument is that the stomach stones are used to help offset the animal's buoyancy, so that it might be able to float and swim relatively low in the water. This positioning would allow a crocodile to float with just its eyes and nose out of the water, thus lurking motionlessly and almost invisibly, able to ambush prey. The stones also act as ballast and decrease the energy required to counter rolling in a current. The stones would also aid a crocodile in pulling prey down into the water.

The gastroliths in the plesiosaur also served as stabilizing ballast by allowing the animal to sink well into the water. Because the plesiosaur had a long heavy neck in front of its buoyant lungs, it suffered from rolling when the water was choppy. The rolling would have been decreased if it had stones in the stomach behind the lungs.

2.50 • Coanda effect

If a fluid (either liquid or gas) flows through air near a solid surface, why does the fluid stream tend to curve toward the surface and then attach itself to the surface? You can easily see the effect in a kitchen sink by holding a curved surface in the smoothly flowing stream from the faucet. For example, hold a glass jar horizontally so that the stream hits the top of the curved surface and travels down one side (Fig. 2-16*a*). The stream may adhere to the surface so well that it rounds the bottom and almost climbs back to the top on the other side. If you angle a rod in the stream and adjust the stream speed, the stream may cling to the rod and spiral around the rod several times before it breaks free to fall (Fig. 2-16*b*).

When provoked by ants, bombardier beetles produce froth or spray that is hot (100°C) and toxic. The more common type of bombardier beetle (brachinines) can direct its jet-like spray by rotating the tip of its abdomen like a gun turret. If an ant

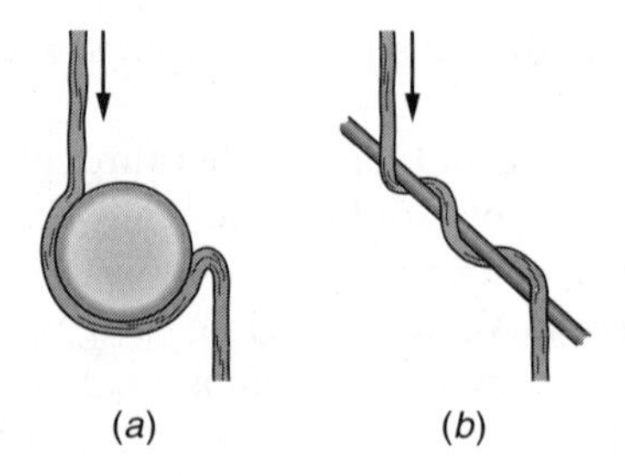

Figure 2-16 / *Item 2.50* A falling water stream wraps around (*a*) a jar and (*b*) a tilted rod.

attacks, say, a front leg, the abdomen tip is aimed down and forward, targeting the leg. Once drenched, the ant quickly scampers off. The paussines, a less common type of bombardier beetle, do not have such a mobile abdomen tip; rather their spray can be issued only toward the rear or to the side. Still, the beetle can deftly target an ant even if the ant is in front of the beetle or on a forward leg. How can the beetle shoot the spray forward when it cannot be issued forward?

Answer This attraction of a stream toward a solid surface and the resulting attachment is called the *Coanda effect* after Henri Coanda, the Romanian engineer who discovered it. Suppose a water stream is reasonably near a solid surface. The stream *entrains* air; that is, it grabs adjacent air molecules, forcing them to move along with the stream. This action removes air molecules, and so other air molecules (farther from the stream) tend to flow in as replacements. However, the solid surface impedes this inflow. With insufficient molecules between the stream and the solid surface, the air pressure between them is reduced. The air on the other side of the stream is still at atmospheric pressure, and so the stream is pushed toward the surface, becoming attached to it. This attachment can persist even if the surface is curved away from the original direction of the stream.

The paussines possess flanges just forward of the gland opening from which the spray issues. To shoot the spray forward, the opening is controlled so that the spray hits a flange. There it can be deflected by as much as 50° as it moves around the curvature of the flange via the Coanda effect. When the spray leaves the flange, it flies through the air as a thin jet. The beetle can control the final direction of the jet by controlling where on the flange the spray hits as it is emitted from the gland.

2.51 • Teapot effect

A properly designed teapot spout (or spout on any other liquid container) allows water to pour freely, so that it ends up where you want, such as in a teacup. An improperly designed teapot spout displays the so-called *teapot effect*: Instead of pouring freely, the water doubles back under the spout and runs down the spout, perhaps for several centimeters, before it breaks free and falls (Fig. 2-17). Even if it does not cling to the spout's underside, the falling stream can arc back toward the teapot. Of course, the clinging or the unpredictable backward arcing can produce a mess. What causes the teapot effect?

Answer If the water is poured quickly enough, the stream probably moves through the air in an expected and familiar path, often called a *projectile path* because a solid object shot from the spout would take the same path. The unexpected behavior occurs when the water leaves the spout more slowly. The stream develops a change in pressure across its thickness, with atmospheric pressure at the air–water surface and a reduced pressure close to where the water moves rapidly around the spout lip. The greater external pressure pushes the stream against the lip. If the stream speed is moderate, the stream is forced around the lip before it breaks free, arcing back toward the teapot.

If the water is flowing more slowly, the point at which the water last touches the spout can be shifted onto the underside of the spout. This adhesion is usually attributed to the mutual attraction of water molecules and spout molecules. It is loosely explained as being due to *surface tension* and is generally described as an example of *wetting*. However, the main reason that the stream turns around the lip and then runs down the spout is that air pressure pushes the stream against the spout. Even if you coat the underside of the spout with butter, to reduce the molecule–molecule attraction and eliminate wetting, the stream still turns around the lip and runs along the spout.

Many factors determine the location of where the stream breaks free of the spout. If you experiment with a certain teapot and a controlled water flow rate, the distance that the water can run along the underside can change from trial to trial. To eliminate the teapot effect, a spout might have a small hole on the underside, near the lip. When the stream reaches the hole, the abrupt change in surface curvature causes the stream to detach (it *trips* the stream). A similar technique is used on windowsills to defeat the teapot effect where rainwater flows under the sill and possibly into an uncaulked point on the wall: The underside of the sill has a narrow vertical cut that causes the water to detach from the sill. To defeat the teapot effect when you pour liquid from, say, a pan, position a vertical knife or rod at the point where the liquid leaves the pan. The liquid adheres to the knife or rod, descending along it instead of flowing down the outside of the pan.

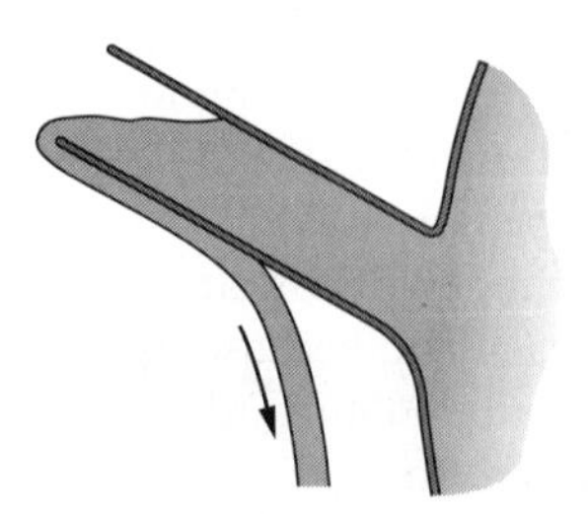

Figure 2-17 / *Item 2.51* Water stream clinging to teapot spout.

The teapot effect can also be seen in some water fountains where the water overflows the edge of a container and either runs along the underside of the edge or arcs back toward the container as a falling sheet. If the edge is circular, the falling sheet may form a closed surface called a *water bell.*

2.52 • Ascents after deep diving

When a scuba diver or deep-sea diver is brought back to the water surface, why must the diver wait at certain depths for set time intervals instead of being brought up continuously? Why do many divers feel fine when they follow this procedure but undergo pain if they fly soon after the dive? (The pain sets in soon after takeoff.) Whales frequently dive to great depths. Do they suffer any damage?

Answer If a diver breathes air under pressure, nitrogen molecules from the air are forced into solution in the bloodstream. When the diver ascends, the reduced water pressure on the body allows the dissolved nitrogen to form bubbles. (Bubble formation in a freshly opened carbonated drink occurs on the container walls, but the nitrogen bubble formation here occurs within the fluid, that is, within the blood.) The bubbles tend to move with the blood flow, collecting into "slugs" if they move into larger veins (toward the heart) or becoming wedged and blocking the blood flow if they move into smaller veins (away from the heart). Such nitrogen poisoning, called the *bends*, can produce terrible pain, long-term disability, and even death. A victim is usually repressurized and forced to breathe air with an elevated oxygen level so that the nitrogen will eventually come out of solution in the blood and be dissipated. To avoid the bends, a diver is brought up to the surface in stages, with a set amount of time spent at each stage to allow the nitrogen to come out of solution.

The ascent schedule is meant to eliminate enough of the dissolved nitrogen that a diver does not have nitrogen bubbles upon surfacing. However, the residual dissolved nitrogen could still form bubbles if the diver ascends in an airplane soon after surfacing. Although modern airplanes have pressurized cabins, the air pressure is lower than the normal atmospheric pressure at ground level, and that reduced pressure could allow nitrogen bubbles to form.

Although whales were thought to be immune to the dangers of deep diving, some evidence suggests that they too can suffer from the bends, especially if they are forced to surface especially quickly.

The term *bends* was first used to describe ailments of workers who constructed tunnels deep enough to require pressurization, such as during the construction of the Brooklyn Bridge piers in the 1860s. Upon ascending to the surface, some of the workers experienced so much muscular pain that they were hunched over, somewhat like high-society women of their times who walked with an exaggerated posture called the *Grecian bends.* Thus the name bends came to be applied to the ailments due to breathing pressurized air.

2.53 • Snorkeling by people and elephants

In snorkeling, a swimmer breathes by means of a tube that extends above the water level. Why is the length of the tube restricted to about 20 centimeters? That is, what is the acute danger in using a longer tube, other than the difficulty in circulating air into and out of it? Using its trunk, an elephant can also snorkel. How can it survive its common snorkeling depth of about 2 meters?

Answer Because the water pressure on a diver increases with depth, the blood pressure also increases. If a diver is swimming by holding the breath, the pressure in the lungs also increases. The match of the blood pressure and the lung's air pressure allows the continued transfer of oxygen to the blood and the removal of carbon dioxide from the blood. However, if the diver began to breathe through a tube, the air pressure in the lungs would drop to atmospheric pressure. This decrease is small if the diver is not far below the water surface, but for greater depths the mismatch between the blood pressure and the lung's air pressure could be fatal, a condition called *lung squeeze.* Then the small blood vessels at the lung surface rupture, and blood seeps into the lungs.

A mature elephant would seemingly undergo lung squeeze with each submerged swim because its lungs are about two meters below the water surface, which means the pressure difference between its blood pressure and the lung's air pressure is large. However, its lungs are protected in a special way. The *pleura* is a membrane that encloses the lungs in any mammal. Unlike in the other mammals, the elephant's pleura is filled with connective tissue that holds and protects the small blood vessels in the walls of the lungs. Thus, the vessels do not rupture during snorkeling.

2.54 • Deep diving, submarine escape

One safety measure used in scuba diving is learning how to get a lungful of air from someone's tank and then ascend to the surface. What is the danger in the ascent? Can someone escape from a disabled submarine by getting a lungful of air in the submarine and then swimming up to the surface?

What is the danger of taking a lungful of air at the water surface and then descending in water, as some people do for sport (free diving) and commercial work (the *ama* divers of the South Pacific)? What is the danger of momentarily losing air pressure when deep-sea diving in a diving suit?

Answer The pressure on a swimmer's body increases significantly with water depth. If someone takes a breath from a scuba tank at the bottom of a swimming pool and then ascends while holding that air, the pressure decreases and the lungs expand until they reach their limit of expansion. Unless the person exhales to avoid expansion, the pressure in the lungs can exceed the blood pressure, and then air is forced into the bloodstream, which can be fatal. Every year a few people die while practicing scuba diving because they neglect to exhale.

In principle, someone could swim from a disabled submarine to the surface if the submarine is not terribly deep and the person exhales during the ascent. However, exhaling takes great discipline because the tendency is to maintain air in the lungs for the terrifying swim through an unknown distance to the surface. The urge to take a breath is even worse. That urge depends on how much of the pressure in the lungs is due to carbon dioxide. If that pressure reaches a certain critical value, the urge to take a breath may be unbearable. If the person is exhaling properly during the ascent, the critical value is reached not just below the surface but farther down. If the person can swim through that point, the rest of the ascent might be relatively easy.

Submarine occupants can also be rescued if a bell-shaped chamber is lowered to the submarine. In fact, such a chamber was used to rescue 33 crewmen from the U.S. submarine *Squalus* in May 1939 when it was disabled at a depth of 80 meters. Divers ran guide cables from a ship on the surface to a hatch on the submarine. Then the chamber was lowered along the guideline. The chamber, which was open at the bottom, did not fill with water because air was released from tanks to increase the air pressure. When the chamber reached the hatch, it made a watertight contact with a ring around the hatch. After the chamber was bolted onto the ring and the air pressure was reduced, the hatch was opened to allow several of the crewmen to climb into the chamber for a trip to the surface.

In free diving, the ability to hold one's breath for a long dive comes from training, the shock of cold water on the face (the so-called *diver's response*), and the willingness to undergo physical damage. Training can increase lung capacity and the time between breaths. The cold-water shock decreases oxygen consumption. The descent is usually aided by a heavy object (perhaps added to a belt) that is dropped at the end of the descent. However, even without that object, the diver develops *negative buoyancy* (net downward force) during the descent. Normal buoyancy (*positive buoyancy*) is upward and is due to the diver occupying a certain volume in the water. However, as the diver descends, the lungs are compressed and the diver occupies less volume. The buoyancy decreases and becomes smaller than the gravitational force. The result is that the net force on the diver is downward, and the diver sinks. The lungs are squeezed down to the size of a soft-drink can, and blood leaks into the space they should be occupying.

These physiological changes occur if the diver starts at the surface with a lungful of air. However, if the diver starts from a submerged chamber, breathing air (or some other oxygen mixture) at the ambient water pressure, the diver may feel no discomfort. Although dives into the deepest parts of the oceans seem unlikely, they are not physiologically impossible.

When a person works in a deep-sea diving suit, air is supplied via a hose to the *hard hat* (the helmet), and a pump at the surface end of the hose increases the air pressure in the suit to match the ambient water pressure. If the pump falters or fails, safety valves snap shut to prevent the pressure inside the suit from dropping to surface pressure. In earlier days when the equipment lacked such valves, pump failure meant that the water pressure would literally ram a diver's body up into the hard hat.

2.55 • Lake Nyos disaster

In August 1986, in the valley below Lake Nyos in Cameroon, Africa, a deadly gas or aerosol cloud killed about 1700 people and an uncounted number of animals. Investigators who reached the site several days later declared that the lake itself was the culprit, rather than some type of toxic volcanic gas. How could a lake emit a deadly gas?

Answer The lake contains a high concentration of dissolved carbon dioxide, especially in the lower reaches because of the larger water pressure. Apparently, something caused the lower water to rise toward the surface, allowing much of the water's carbon dioxide to form bubbles, which were buoyed to the surface. Those bubbles emerged from the water with such vigor that waves were thrown across the lake. The carbon dioxide pooled above the water and overflowed the lake's boundary, pouring down into the valley where it smothered the victims. The victims were submerged in the carbon dioxide and died because of oxygen starvation.

We shall probably never know for sure what triggered the initial rise of lower water. The trigger could have been a combination of rainwater draining into the lake and an especially strong wind blowing over the lake from the end where the rainwater entered. Because the rainwater was somewhat cooler and thus denser than the existing lake water, the arrangement of rainwater over existing lake water was unstable. If a strong wind blew the rainwater layer across the lake and the rainwater then sank there, its sinking could cause deep water to swell up back at the opposite end of the lake. As that deep water rose into progressively less water pressure, the gas would have started to come out of solution.

Lake Nyos is still highly charged with carbon dioxide, and researchers fear that another lethal outbreak could occur. Indeed, they warn everyone away from the lake, especially in the rainy season.

SHORT STORY

2.56 • House-hopping, and riding the skies in a lawn chair

In September 1937, on a golf course in Old Orchard Beach, Maine, Al Mingalone, a newsreel cameraman, spent the latter half of an afternoon trying to shoot footage of a stunt known as *house-hopping*. Wearing a harness to which 27 large hydrogen-filled balloons were attached, he repeatedly ran toward a house and jumped upward, hoping that the buoyancy of the balloons would carry him up and over the structure. But each attempt failed because he could manage to ascend by only 25 feet, which was inadequate.

With the light failing, he exclaimed, "This time let's put on a load for a decent jump and get it over with." Five more balloons were inflated and attached to the harness, and Mingalone made his last jump for the day. But as he rose, the safety line between the harness and a car bumper went taut and then snapped.

In the dimming light and with a storm brewing, Mingalone began to drift toward the Atlantic Ocean. His father and an assistant first watched in horror and then jumped into the car. They were joined by Father Mullen of a local church, who had the foresight to grab a high-powered 22-caliber rifle. The trio drove after Mingalone, but the balloonist was frequently lost from sight as he drifted into rain clouds. Plus, the car was constrained to follow roads that, of course, did not run exactly along Mingalone's path.

After an hour and when Mingalone was about 250 meters off the ground, the trailing motorists happened to catch sight of him. They pulled to a stop and hopped out, and then Father Mullen carefully shot out three of the balloons to bring Mingalone to the ground unhurt. The loss of buoyancy was just enough to allow him to descend slowly; of course, if the Father had shot out too many balloons, the ending would have been tragic. Along his flight, Mingalone dropped his camera, but it was later discovered undamaged in a potato field. The camera had recorded a more thrilling event than Mingalone had originally intended.

In July 1982, Larry Walters of San Pedro, California, made a similar flight but intentionally. Taking off from a driveway in that suburb of Los Angeles, Walters rode in a lawn chair with 42 helium-filled weather balloons attached. He initially raced upward at about 250 meters per minute, soon reaching an altitude of 5 kilometers where pilots of two airliners spotted him. Their report of a man in a lawn chair buoyed by balloons might have sounded bizarre to the air traffic controllers of Los Angeles International Airport had not one of Walters' cohorts previously managed to phone the controllers.

Once up in the thin, cold air, Walters began to decrease the buoyancy by shooting out some of the balloons with a BB gun. But in the excitement and while suffering from oxygen starvation, he accidentally dropped the gun. Although oxygen starvation produces a degree of euphoria, Walters was alarmed when his chair momentarily began to rise again. Once continuous descent prevailed, he controlled his fall by periodically dropping water-filled containers to lighten the load.

As he neared the ground, the balloons became snarled in power lines, but luckily he ended up dangling about 2 meters off the ground. The separation from the ground was enough to eliminate the immediate possibility that the power would short-circuit through him to the ground. To avert the danger of electrocution, rescuers turned off the power in the area before lowering Walters.

His trip lasted 1.5 hours, and extended 5 kilometers up and 16 kilometers horizontally. The Federal Aviation Agency was initially perplexed with how to charge Walters (there were no written rules about flying lawn chairs into airline flight patterns), but after a 6-month deliberation the agency charged him with a large fine for several violations, including one for operating an aircraft without an airworthiness certificate.

2.57 • Flow of medieval cathedral window glass

Some windowpanes installed in cathedrals during medieval times are thicker at the bottom than at the top. Has the glass gradually flowed downward during the passing centuries?

Answer Glass can be considered a viscous fluid that can flow or settle. However, calculations reveal that downward flow is far too gradual to give perceptible changes in the medieval windowpanes. Indeed, a perceptible change might require a million years.

Other explanations of the windowpane shapes involve the manufacturing process. For example, the glass may have first been blown as a cylinder, then split, and finally flattened. The lower portion of the cylinder may have been thicker than the upper portion, and thus one portion of the final windowpane would be thicker. Workmen may have naturally mounted these windowpanes with the thicker portion at the bottom.

2.58 • Strange viscous fluids

Why does ketchup (catsup) flow more easily from a bottle if the bottle is first shaken? You might have noticed this effect if attempting to pour ketchup on your hamburger only to discover that someone else at the table had just shaken the container—you can end up with more ketchup than hamburger on the plate.

Why does ink from a ballpoint pen flow easily from the pen when you use it but not when the pen is in your pocket or bag? Why does one-coat paint flow easily onto a wall but then won't flow off the wall and onto the floor? Why does butter spread over room-temperature bread when a knife is used but not on its own? Why is a thick mixture of water and cornstarch difficult to stir if you attempt to stir it quickly but easy to stir if done more slowly?

Why is silicone putty (such as sold under the trademark Silly Putty) or a mixture derived from polyvinyl alcohol (sold under the trademark Slime) rigid if you slap it, fairly elastic if you bounce it off a floor, and fluid if you drape it over a rod?

Answer The unusual properties of these various fluids are due to their *viscosity*, which is a measure of how easily or with how much difficulty a fluid can flow. For example, cold molasses has high viscosity and flows sluggishly, whereas water has low viscosity and flows easily. The viscosity of most fluids depends on temperature, but at a given temperature, most fluids have a certain viscosity. Such fluids are said to be *Newtonian fluids*.

A second class of fluids is said to be *non-Newtonian* because their viscosity depends on *how* they are made to flow. Ketchup is an example: If it is left undisturbed for a while, it has a high viscosity, making it difficult to pour from a bottle with a narrow opening. However, if it is shaken or stirred for a few seconds, its viscosity noticeably drops. So, to get ketchup flowing easily from a bottle, you can shake the bottle several times. The shaking causes portions of the fluid to slide over other portions, and the relative motion (said to be *shearing*) probably untangles some of the interlocked long-chain molecules in the ketchup mixture, allowing for easier flow. When shearing decreases a fluid's viscosity, the fluid is said to be *thixotropic* or *shear-thinning*.

Ink in a ballpoint pen, a one-coat paint, and butter are all shear-thinning fluids. When you press on them via the pen's ball, a paintbrush, or a knife, the pressure and the attempt at motion decreases the viscosity and the substance flows fairly easily. As soon as the pressure and the attempt at motion stop, the viscosity is again too large to allow flow.

A thick mixture of cornstarch and water is called *shear-thickening* because relative motion in the mixture increases the viscosity. (A dilute water–cornstarch mixture does not show this effect.) If you hit the thick mixture with the palm of your hand, the relative motion immediately increases the viscosity so much that the mixture is almost rigid and certainly does not splash, yet the viscosity and the ability to flow return almost immediately. The momentary increase in viscosity is probably due to the alignment of cornstarch molecules perpendicular to the direction of flow, thereby quickly stopping the flow. As soon as the attempted flow disappears, the alignment disappears. If you throw a handful of the thick mixture at the floor, it is almost rigid during the collision but then flows over the floor just afterward. If you stick a fairly wide rod or fairly large spoon down into the mixture and then yank it upward, you might be able to lift the mixture and its container, at least momentarily.

Silly Putty and Slime are both viscous non-Newtonian fluids. If they are draped over a rod so that gravity gently pulls on them, they will flow downward. If a larger, more sudden force is applied, as in a collision, they react like an elastic ball because the long molecules in the material are coiled and behave like springs. If an even larger force is applied, they break. For example, if you suddenly pull the two ends of a strand of Silly Putty in opposite directions, the strand will break much like a metal rod pulled in opposite directions. You can also cut Silly Putty and Slime with scissors: As the blades suddenly apply large pressures, shearing the fluid, the fluid becomes rigid and brittle.

You can see another curious effect if you push Silly Putty through a hollow tube. When it exits the other end, it expands in what is called *die swell*. The expansion is due to the long molecules recoiling as they emerge from the tube, after being compressed as the material is forced through the tube's confines.

Some types of non-Newtonian fluids can siphon themselves out of a container. If you pull part of the fluid up the side of a beaker and drape that part over the rim, its flow along the outside of the beaker can drag the rest of the material upward and over the rim.

2.59 • Soup reversal

If you stir certain canned soups, such as tomato soup, and then remove the stirring utensil, why does the swirling reverse just before it stops? To see the reversal, first mix a can of condensed tomato soup with a small amount of water (less than the normal amount). Then experiment with the lighting on the soup's surface.

Answer When you stir the soup, you not only force the stirring instrument through the soup but also force various layers of the soup to move relative to one another. The relative motion, said to be *shearing*, unravels the normally coiled long-chain molecules in the soup. As motion and shearing decrease, the molecules suddenly pull themselves back into coils, reversing the swirling as though the soup were an elastic membrane.

2.60 • Bouncing liquid stream

Pour a thin stream of hair shampoo or liquid hand soap onto a flat surface where the stream can form a mound that oozes outward. For certain pouring heights and certain liquids, why does the stream occasionally take a big leap sideways? (I get very nice, frequent leaps with Ivory Hand Soap.)

Answer The type of shampoo that jumps is said to be *viscoelastic* because it is viscous (has internal friction opposing motion) and also elastic (it acts like a rubber membrane). The viscosity of the shampoo is fairly high when the shampoo moves slowly in the falling stream and in the mound. However, when the stream runs into the mound, the collision causes shearing; that is, it causes one viscous layer to move quickly across another viscous layer. The motion decreases the viscosity of that portion of the stream. Because the liquid is elastic, this sudden decrease in viscosity allows the colliding portion to bounce somewhat like a rubber ball, and so the stream forms a wide loop that extends (stretches) off to one side of the stream and mound (Fig. 2-18). The loop is so

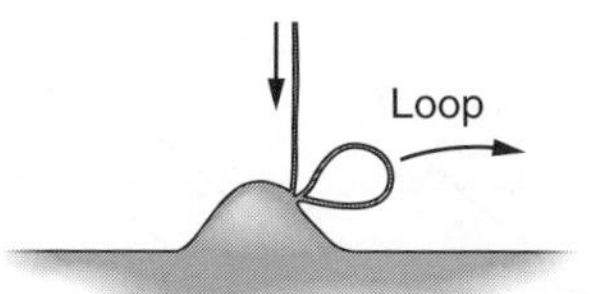

***Figure 2-18** / Item 2.60* Falling shampoo stream appears to bounce from shampoo mound.

***Figure 2-19** / Item 2.62* Honey coils like a rope.

fleeting that we see only the top section and the stream looks like it bounced off the mound.

2.61 • Rod-climbing fluids

If you insert a rotating rod into a bowl of water, the rod makes the water swirl and form a vortex that stretches down the rod. If you replace the water with egg white, STP oil treatment, or certain other fluids, why does the swirling fluid *climb up* the rod, a behavior called the *Weissenberg effect*?

Answer This climbing tendency of certain fluids is due to the way the rod forces the fluid to rotate. To see this shearing effect, imagine the fluid to be in cylindrical shells centered on the rod. The rod's rotation also drags the inner shell into rotation. That shell slides past the next shell, dragging it into rotation. And thus, shell by shell, the fluid is dragged into rotation. Because the motion is set up by dragging and slipping, the shells are said to undergo shearing. When the fluid is water, the shearing is not very successful past the first few shells and so the rotation diminishes farther out. However, with these special fluids the molecules are not only cohesive but are so interlocked that they behave like elastic bands: As the rod rotates, it wraps up these bands so they are pulled toward and then up the rod.

2.62 • Liquid rope coils

If you pour honey onto toast and adjust the pouring height, you can get the thin stream of honey to coil on the toast (Fig. 2-19). Other fluids also wrap on themselves if poured properly. For example, a wide stream of cake batter wraps back and forth to make folds like ribbon on a present. What causes this behavior?

Answer The fluids that exhibit *rope coiling* or *ribboning* are *viscoelastic*; that is, they are both viscous and elastic. When honey is poured from an appropriate height, it rope-coils or ribbons because of two factors. (1) When it reaches the pool of honey already on the toast, its high speed and high viscosity prevent it from flowing into the pool. So it is suddenly slowed by the collision with the pool, putting stress on the stream. (2) The stream thins as it falls and reaches the pool either as a thin cylindrical stream or as a thin, broad ribbon. If it is thin enough, the stresses on it cause it to buckle off to one side. A cylindrical stream continues to buckle so that it circles, forming a coil that may be hollow on the inside. A wider stream folds back and forth: As it buckles to one side, its cohesion pulls it back to center where it then buckles in the opposite direction, and so on. Generally, a higher fall means a higher frequency of coiling or folding, but the effect disappears if the fall is too high because then the fluid comes out of the container in spurts rather than as a smoothly flowing stream.

2.63 • Water waves

What causes water waves? That is, how are they generated?

Answer These two simple questions are not yet fully answered. However, a simple explanation is the following: A breeze or some disturbance in the air or water creates ripples. The ripples can then grow into larger waves as wind moves across them. In particular, the wind pushes on the windward side of a crest, comes over the crest, and then breaks up into vortexes on the leeward side. The air pressure is reduced in the vortexes, and so the pressure difference between the windward and leeward sides of the crest can push the crest downwind and also make it taller. In other words, the wind can feed energy into the crest. If the wind becomes stronger, the waves grow larger (and their wavelengths also change).

2.64 • Extreme and rogue waves

Most ocean waves have heights within a certain range of values, which can be correlated with wind and storm conditions. However, larger waves sometimes occur. If an *extreme wave* is described as having a frightening height, then a *rogue wave* would be described as having a terrifying height. It is preceded by a low point that is often characterized as a "hole in the water." Large ships that were strong enough to withstand violent storms have been ripped apart as they slid bow downward into such a hole only to be wrenched upward by a wave standing some 30 meters above it. The height of the rogue wave that hit the U.S. Navy steamship *Ramapo* in 1933 was 34 meters, as measured by the officer on watch by triangulation of the crow's nest against his view of the wave. (Doing physics in the face of death takes great physics courage.)

Both extreme and rogue waves have been spotted around the world, but the waters off the southeast coast of Africa produce more than their share of rogue waves, as verified by the many ships lost in the area. What causes extreme and rogue waves?

Answer For an ocean wave, you probably picture a sinusoidal wave (in the shape of a sine curve, with hills and valleys) moving over the ocean's surface. If two waves traveling in the same direction were to overlap, you can imagine that the *resultant wave* (what you would see) is simply the addition of the two waves. If the waves were exactly aligned (in phase), the hills and valleys of the resultant wave would be higher and deeper than those on the individual waves. And if many waves, moving in different directions, overlapped, the resultant wave might be confusing to figure out, but simple addition of the individual waves should still give the height and depth of the resultant wave.

Such simple addition of waves is said to be a *linear combination* of the waves. Extreme waves seem to be a *nonlinear combination*; that is, the combination of individual waves somehow generates hills and valleys that are too high and too low. Perhaps as the hills grow, the wind over them enhances their growth so that the final hill height is greater than would be expected. Or perhaps in certain situations the buildup of a resultant wave past a critical point modifies the individual waves and creates an even larger resultant wave. In short, some feature enhances the resultant wave. The odds are against an extreme wave occurring, but occasionally such a wave slams into a cruise liner or some other ship, surprising captains who tend to think in terms of linear combinations.

Rogue waves (also called *giant waves* or *freak waves*) are even more difficult to explain but must also be due to a nonlinear combination of waves. However, their occurrence off the African southeast coast is surely due to the opposition of the Agulhas current and the wind-driven waves in the region. The strong Agulhas current flows toward the southwest in a meandering path; the wind-driven waves are typically toward the northeast. As the waves force the current to meander, they can be focused much like light waves can be focused by a lens. With the correct conditions, this focusing generates the hole in the water that is followed by a huge wave leaning toward the hole.

2.65 • Waves turning to approach a beach

Waves can approach a beach from many directions, depending on the wind and the locations of distant storms. Why do the waves generally turn so as to be parallel to the beach (Fig. 2-20)?

Answer This tendency to turn, a form of *refraction* that is more commonly a topic in optics, is due to the decrease in speed of the waves as water depth decreases. As a particular wave crest crosses from deeper water to shallower water, the portion of the crest to cross first slows and then lags behind the rest of the crest. This lagging puts a kink in the crest: The slower portion in the shallower water is now traveling more directly toward shore than the portion still in the deeper water. Eventually all of the crest will have passed into the shallower water and be headed more directly toward shore.

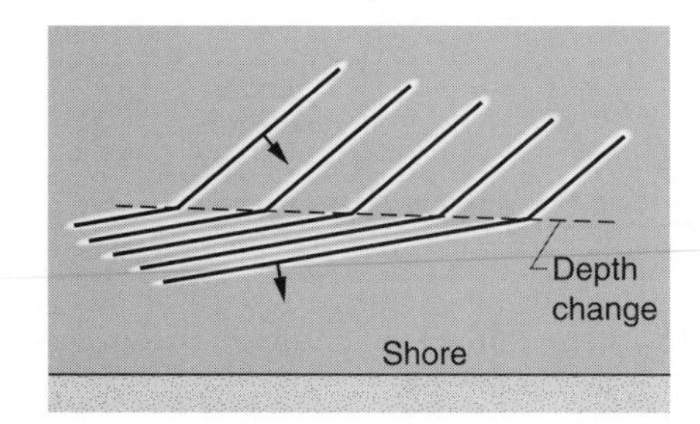

Figure 2-20 / *Item 2.65* Overhead view of ocean waves changing directions when water depth decreases.

The shape of the waves also changes because what we see traveling over the water is actually a sum of many individual waves with different wavelengths. The extent of the slowing, and thus also the turning, depends on wavelength, and so the individual waves slow and turn by different amounts.

2.66 • Waves pass through a narrow opening

When ocean waves pass through an opening that is somewhat wider than the wavelength of the waves, why do they flare (spread out) from the opening instead of traveling in their original direction (Fig. 2-21)?

Answer This tendency to flare, a form of *diffraction*, is due to the interference of a wave as it passes through the opening. Normally a straight wave traveling in a straight line is modeled as a line of many small wave-makers, each sending out a semicircular wave. The overlap and interference of all these waves continuously generates an advancing straight wave. However, when the wave enters a narrow opening, only the wave-makers within the opening survive. Their semicircular waves overlap and interfere but they are insufficient to produce an advancing straight wave. Instead they produce a wave that flares as it travels away from the opening. Moreover, the amplitude of the vertical motion of the water

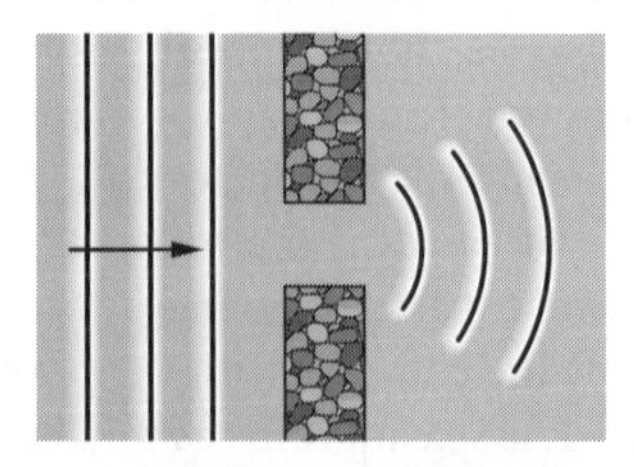

Figure 2-21 / *Item 2.66* Overhead view of ocean waves undergoing diffraction through an opening in a rock wall.

varies along this new wave. In some points, the vertical water movement is significant, but at intermediate points it is zero. So, if the opening is between breakers and the flaring wave reaches the beach, some points on the beach undergo significant wave action and others undergo no wave action.

Diffraction can also occur when a wave passes the end of a barrier: The portion near the barrier flares into the *shadow region*—that is, the region behind the barrier that seemingly is protected from the waves.

2.67 • Seiches and sloshes

When you carry an open container of liquid while walking, such as a bowl of water, why does the liquid slosh? What determines the sloshing frequency—that is, the number of times per second the water surface oscillates? Can a bathtub of water or a swimming pool be made to slosh? How about a pond, harbor, or lake?

Answer When you walk, your gait and your grip on the liquid container cause the liquid to move horizontally and vertically; that is, waves are set up on the surface of the liquid. Most waves interfere with one another in a random way but certain waves set up *standing waves* in which the pattern of vertical oscillations is repeated: Some points oscillate the most and some do not oscillate at all. The standing wave with the lowest frequency is called the *fundamental mode* and is nearly always set up. The frequency of the fundamental depends (at least approximately) on the horizontal dimensions of the container and on the depth of the water. When the frequency of your walk is approximately equal to the frequency of the fundamental, the sloshing can be so vigorous that the liquid spills. You can decrease that possibility by walking more slowly or changing your gait.

You can set up sloshing in a bathtub of water if you move a broad paddle back and forth in the water. Experiment with the frequency until you find the frequency for the fundamental. Then you can get such strong sloshing that you can easily flood the bathroom floor.

Sloshing can also occur in larger containers of liquid, such as a gasoline truck or a railroad tanker car. There, of course, uncontrolled sloshing can make the vehicle unstable and lead to an accident, and so, baffles are often installed inside the vehicle to reduce sloshing.

A swimming pool can be made to slosh at a fairly low frequency if people jump into the water in a coordinated and repeated manner, to build up the fundamental mode. A large mechanical plunger that oscillates at one end of the pool can do the same thing, which is a lot less fun.

Larger bodies of water, such as ponds, harbors, and lakes, will slosh in their fundamental mode if the water is made to oscillate by earthquake waves or air pressure variations (such as wind). Such larger sloshing events in natural bodies of water are termed *seiches*. In one famous event in March 1964, an earthquake in Alaska set up seiches all the way to the Gulf of Mexico. Most of these oscillations were too small to be noticed, but one measured 2 meters from the high-water point to the low-water point.

Harbors and tidal basins can also slosh if they are *pumped* by the tides or by a disturbance like a storm or a tsunami. They are then somewhat like a bottle or organ pipe that is being pumped by an oscillating source of air, except that the result is an oscillation of the water level instead of a sound wave.

The extent of oscillation in the harbor (and thus the possibility of damage) is generally larger for a smaller width of the harbor mouth (the opening to the ocean). One reason for this result, called the *harbor paradox*, is that a wide mouth allows the incoming wave energy to escape back out into the ocean whereas a narrow mouth more effectively traps that incoming energy. A similar result occurs with sound waves: If you blow across the narrow mouth of a partially filled soda bottle, you can produce a loud resonant sound in the bottle's air space. But if you blow across a wider-mouth bottle, the resonant sound is harder or even impossible to produce.

2.68 • Wakes of ducks and aircraft carriers

Why does a V-shaped wake form behind objects, such as ducks and aircraft carriers, moving across water (Fig. 2-22)? Does the shape or angular size depend on the speed of the object?

Answer The wake left by an object moving across water is approximately the same for any object at any practical speed, as long as the motion produces *gravity waves* (where the oscillations are controlled by gravity) rather than capillary waves (where the oscillations are controlled by surface tension). So, a duck and an aircraft carrier leave a wake with the same angle, about 39°. However, details of the wave structure inside the wake can be different for different objects, especially when viewed from overhead by radar (which is a subject of interest in military surveillance).

The pattern is due primarily to *phase waves* sent along the water surface by the disturbance of, say, a moving boat. A phase wave, which has the shape of a sine wave, travels by making the water surface oscillate. However, you cannot actually see the phase wave on water because the boat produces a great many of them, which overlap (or *interfere*) with one another. You can see only the *group waves*, which is the

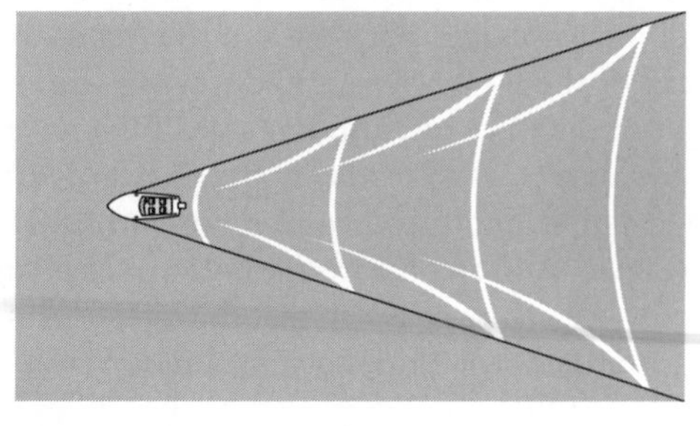

Figure 2-22 */ Item 2.68* Overhead view of wake behind an object moving across a water surface.

result of the overlap. The group waves seem to travel over the water, but they are actually continuously being recreated by the interference of the phase waves, which travel twice as fast as the group waves.

Water waves are further complicated by the fact that waves with longer wavelengths travel faster than those with shorter wavelengths. So, the long-wavelength phase waves tend to outrun the short-wavelength phase waves.

When a boat disturbs the water at point *A* while moving forward, phase waves move outward from *A* at twice the speed of the group waves their interference creates. Because the waves have wavelengths in a wide range of values, the phase waves and their group waves have a wide range of speeds. So, the pattern of waves sent out from *A* and from all other points along the boat's path is largely a mess. However, the waves produce prominent group waves along the boundaries of a V-shaped wake, with the boat at the apex and an angle of 39°. So, that is the wake shape that catches our attention.

If you look closely at a photograph of the wake, the interior of the V has many curved lines that make the wake resemble the structure of a feather. These interior lines are due to interference of group waves originating from many points along the boat's path.

If you are near a boat and its wake during sunlight, you might notice that the wake is calmer than the water outside the wake. In spite of all the waves generated by the disturbance of the boat, one result is that the wake probably contains fewer short-wavelength group waves than the water outside the wake. Sometimes this condition means that the reflection of sunlight to you is more mirror-like and thus brighter from inside the wake than from outside the wake.

2.69 • Surfing

What causes a surfer (on a surfboard) to move toward the beach or along a wave? Can you surf on top of the wave or on the backside?

Answer In open water, far from shore, waves travel at identical speeds. However, near the shore, the speed of a wave decreases as the water depth decreases. Thus, when an ocean wave travels through progressively shallower water as it approaches a beach, the bottom of the wave tends to slow. The top of the wave does not slow and so it tends to outrun the bottom of the wave, causing the wave to lean forward. The height of the wave can also increase. If the wave simply *collapses* or *surges*, it spreads in the forward direction, becomes less high, and thus is useless for surfing. However, if the wave *spills* (the top outruns the bottom) or *plunges* (the top outruns the bottom so much that the top plunges over to hit the base of the wave front, forming a tube of water), then a surfer can ride the wave.

The ride involves an interplay of three forces on the surfer. (1) Buoyancy, which is perpendicular to the water surface, occurs because the surfboard is partially submerged. (2) Gravity, which is downward, attempts to slide the surfer along the wave face. (3) Drag, which is along the water surface, opposes the motion of the board through the water and is due to the water pressure in front of the board and the friction between the board and the water as they slide past each other.

By paddling to get up to speed, a kneeling surfer can move from the back face, over the crest, and to the front face. Once positioned, the surfer stands and waits for a free ride (no more paddling). By adjusting the orientation of the board in the water, the rider can adjust the drag and the board's position on the front face. The three forces can cancel out (the surfer is in *equilibrium*) somewhere along the lower part of the front face. There, the buoyancy force is tilted in the wave's direction of travel and thus tends to propel the surfer. Gravity tends to pull the surfer down the slope but the water drag tends to oppose that motion, so the surfer rides the wave. To move around on the wave face or to move along the length of the wave, the surfer changes the board's orientation and thus the water drag. Generally, shifting the stance backward causes the rear of the board to dig more into the water, increasing drag and slowing the board, so that the rider climbs the front face. Shifting the stance forward causes the rider to speed up and move down the face.

A surfer cannot ride a wave if the wave spills or plunges enough to create turbulence. If turbulence appears simultaneously along the full length of a wave, the surfer will just wait for the next wave. However, if the wave comes in at an angle to the shore, the spilling and turbulence begin at one end of the wave and move along the length of the wave (the wave is said to *peel*). A surfer attempts to ride the wave in the spilling region just ahead of the turbulence. If the wave peels too quickly, the point of turbulence catches up with the surfer and then the ride is over.

Probably the most impressive ride is the *pipeline*, in which the top of a very tall wave plunges to hit the base of the wave, leaving a tunnel between the falling top and the rest of the wave. If a surfer can ride onto the front face of the wave before the pipeline begins, a ride through the pipeline is possible.

2.70 • Porpoise and dolphin motion

Porpoises and dolphins often accompany boats and ships, moving stealthily alongside the vessel about a meter below water. They may be upright, rolled over on a side, or even showing off by revolving around the body axis. But they do not appear to swim—they just move along as if attached to the ship, perhaps for hours. What propels them?

Answer The primary propulsion is due to the waves shed by the bow (or sometimes the stern). The porpoise or dolphin will position itself within the front of the wave, not too deep below the (slanted) surface. As the bow pushes on the water, forcing it forward, upward, and outward, the water pushes on the animal, propelling it forward. If the animal simply wants to ride the wave instead of play, it finds the depth where this forward force balances the drag from the water. Sometimes the animal can catch a free ride even if the bow wave is small, perhaps imperceptible to someone on the boat making the wave.

2.71 • Edge waves

If a paddle is oscillated vertically or horizontally in water, a beautiful pattern of waves can appear, with stationary crests that are *perpendicular* to the paddle, looking very much like teeth on a comb (Fig. 2-23). Thus, they differ from the usual waves made by a paddle, which have crests parallel to the paddle and move over the water surface. These peculiar waves, called *edge waves* or *cross waves*, were discovered by Michael Faraday on July 1, 1831, as he noted in his meticulous diary of scientific observations. The oscillations need to be uniform, the depth of the paddle insertion must be just right, and about a minute may be required for the waves to appear.

You can also see edge waves in a glass of wine that is full or almost full. Rub a clean, dry finger around the rim. If done correctly, the rubbing produces edge waves perpendicular to the rim. They can be so vigorous that they throw drops of wine into the air. Why does rubbing produce the waves?

Answer The normal waves produced by a paddle oscillating in water are a type of *capillary wave*, which are waves whose oscillations are controlled by surface tension instead of gravity. You might refer to them as ripples on the water surface to distinguish them from the larger *gravity waves.* Under the proper conditions, the paddle oscillations also set up a stationary pattern of oscillating water along the horizontal width of the paddle face. One surprising feature is that the frequency of water oscillations is half the frequency of paddle oscillations.

If you were to take a slow-motion video of the paddle as it begins to oscillate horizontally, you would find this sequence: Whenever the paddle moves forward, it pushes up a hill along its horizontal width; as it retreats, it leaves a valley along that width. As they are created, the hills and valleys move away from the paddle as normal capillary waves.

After the paddle has operated for a minute or so, you would find that each time the paddle moves forward, it would also push up a series of ridges perpendicular to the paddle and superimposed on the normal capillary hill. However, there are two sets of ridges. The forward motion of the paddle pushes up one set, then the next forward motion pushes up the next set, and so on. The ridges in one set occur midway between the ridges of the other set. Thus, the frequency at which either set appears is half the frequency of the paddle's oscillations.

If you rub the rim of a wineglass properly, you make the rim oscillate like a bell and, in fact, you might hear the glass ring. The rim is a circle, and the oscillations are either toward or away from the center of the circle, forming a pattern. Figure 3-4 (in Chapter 3) gives a possible snapshot at the instant your finger passes through the 12:00 clock-position on the rim. Maximum displacements occur at the 3:00, 6:00, and 9:00 positions, and the points of no displacement occur midway between those positions. As your finger moves, the pattern follows. The rim functions like an oscillating paddle, and a pattern of edge waves builds up.

Edge waves can also be produced in a Chinese brass water-spouting bowl. However, the handles are rubbed instead of the rim. If edge waves are set up properly, they can throw water up into the air by half a meter.

Edge-wave patterns and far more complex patterns (including beautiful patterns of stripes, hexagons, and circles) can be produced in a shallow layer of water and glycerol if the liquid is oscillated vertically.

2.72 • Beach cusps

What causes the pattern of cusps that adorns many beaches?

Answer The water pattern responsible for the cusps is mathematically similar to the edge-wave patterns in the preceding item. However, because the mathematics is challenging, we shall stick to a simple description: When a wave runs up onto a beach, its reach is roughly in the shape of a sine curve (Fig. 2-24). Where the wave advances most, it carries sand up onto the beach, forming a wet-sand cusp. The water retreats by flowing to the center of the cusp and then downward along a trough-like depression carved out of the sand. This *backwash* prevents a similar *run-up* from the next wave. Instead, the run-up of the next wave occurs off to both sides

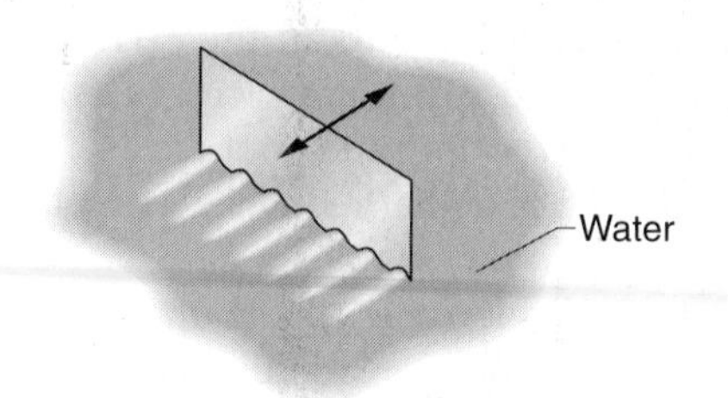

Figure 2-23 / *Item 2.71* Edge waves are perpendicular to a horizontally oscillating paddle.

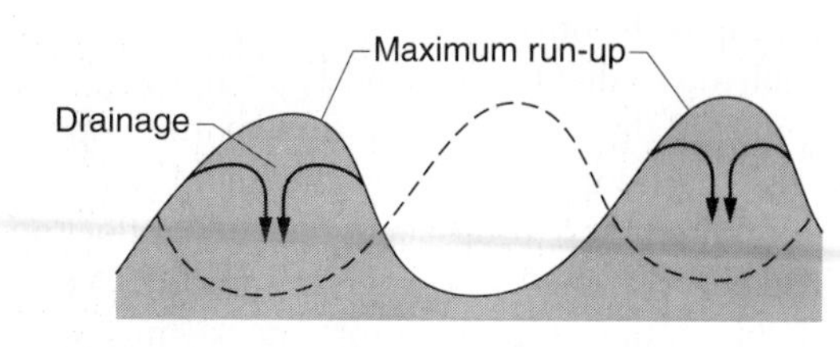

Figure 2-24 / *Item 2.72* Overhead view of wave run-up on a beach. The next run-up is shown in dashes.

of the cusp, where the previous wave advanced the least. The water deposits sand and again retreats along a central trough-like depression, and this backwash prevents the next wave from advancing up into this cusp. Thus, the pattern we see on the beach is due to the alternating run-ups in adjacent cusps.

2.73 • Oil and waves

Since ancient times, people have known that a layer of oil (such as olive oil) thrown onto open water can decrease or eliminate waves, even when wind is strong enough to generate waves. The effect was well known to Benjamin Franklin, who would carry a small amount of oil just in case he could demonstrate the effect. Once, while attending a party near a stream where a breeze was kicking up small waves, he walked somewhat upstream and made several "magic" shakes of his cane without allowing the partygoers to see its spray of oil. To their amazement, the waves almost immediately disappeared and the stream became flat.

When flying over open water, you can detect contamination slicks. They may reflect better than water, but more important, because a slick decreases or eliminates waves, its flat surface is brighter in sunlight than an adjacent uncontaminated water surface would be.

Why does oil calm open water?

Answer A layer of oil or other contamination calms the water for three reasons: (1) It is a viscous layer (its internal friction resists one layer sliding past another layer). So, as a wave is initiated, its energy is quickly removed. This effect is especially important for short-wavelength waves (waves with short distances between successive crests). (2) Normally waves grow from ripples because of wind blowing against and past the ripples. If the ripples are eliminated, then larger waves cannot appear. (3) The layer can interfere with the transfer of energy between long-wavelength waves and short-wavelength waves, with an overall decrease in wave activity.

2.74 • Floating drops

In some coffeemakers, individual drops of coffee fall onto a pool of coffee. Those drops should splash and quickly merge with the pool, but instead they race *over* the pool, perhaps running back and forth across the container several times.

Floating drops can also be produced when a common Styrofoam cup with coffee (or any other beverage) is rubbed across a tabletop so that the cup undergoes repeated sticking and slipping. If the staggered motion of the cup is fast enough, the ripples it produces on the liquid throw drops into the air. When those drops land, they might float on the liquid instead of immediately merging into it. When the stick-and-slip motion stops, the drops quickly merge.

When a steady stream of water from a faucet hits a flat basin, it can form a circular pattern around the impact point. Inside the circle, the water flow is fast and shallow; outside the circle, it is slower and less shallow. So, the circle is actually a wall at which a transition in the water occurs. If you release a drop from an eyedropper just upstream from the wall, the drop may float while anchored against the wall.

In all these cases, why do water drops float?

Answer A drop may be suspended above a pool because of electrical repulsion between molecules on the drop and those on the pool surface. If detergent is included in the drop and pool, detergent molecules tend to collect on the water surface with their hydrophilic (water-attracting) ends buried in the water and their hydrophobic (water-repelling) ends sticking out of the water. The hydrophobic ends on the bottom of the drop and the top of the pool can then repel each other, levitating the drop.

However, in the examples mentioned a more convincing argument for levitation is that a drop is supported by an air layer between it and the pool. Let's first consider the common situation of a drop released just above a stationary pool because this situation also involves an air layer. As the drop descends, the underlying air flows outward until the drop first touches the pool. Then a wave races across the drop, pinching off the lower half, which immediately merges into the pool. As the upper half descends, the underlying air partially supports it but flows outward until that remaining drop first touches the pool. Again, a wave races across the drop, pinching off the bottom half, which immediately merges into the pool. The cycle is then repeated, perhaps several times, before the last of the original drop merges into the pool instead of pinching in half.

Consider next a series of drops falling in a coffeemaker. In a splashing process discussed in the next item, a drop can land just as the crater left by the preceding drop is being filled by inrushing liquid. The inrush causes the new drop to bounce slightly off to one side. As the drop then descends, it tends to squeeze out the underlying air. However, because it moves, new air is always being forced under it, and so it is always supported by an underlying air layer. The sink example is similar, except that the drop is anchored at the wall and the water flow continuously drags new air beneath the drop to maintain the support.

If the pool and drop are made to oscillate, the motion can pump air under the drop fast enough to support the drop. This pumping action occurs in the Styrofoam cup that is dragged in stick-and-slip fashion across a tabletop. Any other means of making the drop and pool oscillate vertically would probably also work if the frequency of oscillation is about the same as that in the Styrofoam cup.

A drop can also levitate if its temperature or that of the pool is high. Then the vaporization of water can provide the gas that supports the drop. This mechanism is commonly called the *Leidenfrost effect* and is discussed in the thermal chapter (Chapter 4).

2.75 • Splashing drops

What becomes of a water drop, such as a raindrop, when it collides with a solid horizontal surface or a pool of water? Why do some drops splash (that is, throw bits of themselves upward or outward) and others do not?

When a blood drop falls or is thrown from a victim in a crime scene, a detective needs to determine the speed and size of the drop from the stain it leaves on a surface. An inherent difficulty is that the size of the stain depends on *both* the speed and size of the drop. That is, a larger stain could be due to a small drop with a higher speed or a large drop with a lower speed. Is there any way to analyze a bloodstain to unravel this information? (Imagine what Sherlock Holmes would do.)

Answer *Solid surface:* Depending on circumstances, a drop hitting a solid surface can either splash, spread out to coat (wet) the surface without splashing, or first bounce and then splash or spread out. In splashing, the drop forms a shallow layer with a *crown* (a raised rim) that probably throws off droplets as it ascends. The droplets form because the rim becomes unstable as it slows its outward motion. One of the waves that tend to develop along the rim during that stage becomes dominant, and its high points form ridges or fingers that can pinch off to form the droplets. Often the fingers can be seen in the pattern left by a splash. The wavelength that dominates the instability is approximately the circumference of the pattern divided by the number of fingers left in the pattern.

Water surface: A drop hitting a water surface can splash, merge with the surface, or float just above it. The floating, which is discussed in the preceding item, can occur only if the drop falls a short distance. For a longer fall, the drop normally pushes out a hemispherical crater in the water surface and then forms a crown around the perimeter of the crater. As the crown subsides and rushes back into the crater, the rapid inflow throws water upward in a central jet. The jet may pinch off one or more drops as it reaches maximum height. Finally, the jet subsides and the splashing ends.

In some cases, a crown will close over on itself to form a dome, and any central jet will be trapped unseen inside. If a drop hits the water without producing a crown, it produces a downward-moving vortex shaped like a horizontal doughnut: Water swirls downward on the doughnut interior and upward on the exterior. Food coloring in a drop can make a vortex visible. A drop thrown off by a central jet can also produce a vortex. A central jet is more apparent for a thin layer of water because the underlying solid boundary causes the crown to transform more vigorously into the central jet. The highest jet occurs when the depth of the water layer equals the radius of the crater produced by the drop. (Surprisingly, if the air above the water layer is either removed or replaced with a lighter gas, splashing can disappear.)

Drops falling onto water in a quick series can bounce from the water because, after the first few drops, the later drops hit the cavity just as the water is beginning to move upward to refill the cavity. Thus, any of these later drops can be simply thrown upward.

Drops of molten wax falling onto a smooth metal surface can undergo the same type of splashing as water drops but will solidify in the final stages. The final patterns are attractive when the perimeter's ridge or crown pinches off droplets because then the solidified droplets are left surrounding the main splash region. Drops of molten metal falling onto a flat, metal surface also solidify into interesting patterns. However, these drops seem to undergo more of *splat* than a *splash* because they flatten out across the flat surface, with fingers appearing along the perimeter. Some of the fingers pinch off into isolated bits of metal. Increasing the surface roughness generally reduces the number of fingers and increases their width.

To determine the speed *and* size of a blood drop from a crime victim, a forensic scientist needs to examine not just the size of the stain but also the number of fingers that are left around the perimeter. A higher speed produces a greater number of fingers. However, the analysis is complicated by the nature of the surface hit by the drop. A rougher surface tends to overlap fingers and reduce the stain size. Thus, extensive experimental work is needed to catalog the bloodstain properties on common types of surfaces, from concrete to paper to glass. Instead, when bloodstains are found at a crime scene, a sample of the solid surface with the stains should be taken back to a laboratory where experiments using blood drops of known size and fall height can be investigated with that surface.

2.76 • Bubbles in soda, beer, and champagne

Why do bubbles form in soda, beer, champagne, and other carbonated drinks after the container is opened? Why do they form only on the interior surface of the container and not within the liquid itself? Why do they grow larger as they ascend and why, especially in champagne, do they form *bubble trains*, where one bubble follows another? Why do bubbles generally rise faster in champagne than in beer? If you clean a drinking glass with detergent and then air-dry it before you pour in a carbonated drink, why is the bubble production almost eliminated?

If ice or salt is added to a freshly poured glass of beer, why does the bubble production increase vigorously, perhaps even to the point that the liquid overflows the glass? Why do the contents of a carbonated drink spray outward if the container is shaken before it is opened?

A shandy is a mixture of beer and a soft drink, usually lemonade or ginger beer. If the beer is poured onto the soft drink, nothing noteworthy happens. But if the soft drink is poured onto the beer, the eruption of bubbles may overflow the container. Why the difference?

If small lime pieces are placed in a glass of beer, why can they repeatedly travel up and down? When Guinness stout is poured rapidly into a glass, why do bubbles form into layers along the side of the glass, and why do those layers move *downward*?

Answer A carbonated drink is a liquid with much carbon dioxide in solution, held there under pressure. In champagne, the pressure might be six times normal atmospheric pressure. That is, both the liquid and the gas pocket above the liquid are under that much pressure. (If the cork is not released carefully from a champagne bottle, the pressure under the cork can shoot it out at 50 kilometers per hour, more than enough to badly injure an eye.) When the container is opened, the outflow of gas reduces the internal pressure, and then it is too small to maintain the full amount of carbon dioxide in solution. So, the carbon dioxide begins to come out of solution, either by crossing through the upper surface of the liquid (if possible) or by forming bubbles.

In general, a bubble can form and then grow only if its size exceeds a certain critical value. The trouble with a smaller bubble is that it has a highly curved surface. The mutual attraction of the water molecules along the surface tends to collapse the bubble in spite of the outward push of the bubble's gas. The surface of a larger bubble is less curved, and the inward force from the water molecules along the surface is not enough to collapse the bubble. However, bubbles larger than the critical size are unlikely to form in the liquid itself—they can neither grow to nor suddenly appear at that size. Thus, bubble *nucleation* (production) occurs only on a surface, primarily along the walls and the bottom (but also on any solid matter in the liquid). Eventually the amount of carbon dioxide remaining in the liquid is too small to grow more bubbles.

The most common explanation for bubble nucleation involves a scratch with a pre-existing bubble. If the scratch width is appropriate, the bubble surface may not be very curved and thus the bubble does not collapse. Moreover, carbon dioxide can slowly pass from the liquid into the bubble, inflating the bubble and increasing its buoyancy. Finally the bubble is large enough for most of it to pinch off from the scratch and float upward. The bubble production begins again with the small amount of gas left in the scratch.

Current evidence indicates that *most* of the bubbles in a drinking glass do not begin in scratches. Rather, they begin in cellulose fibers left stuck on the glass surface when the glass was last washed or dried with a paper or cloth towel. These hollow fibers contain trapped air that initiates bubble production. When the container is opened, carbon dioxide seeps into a trapped air bubble through the fiber's open ends. When the bubble grows large enough, part of it pinches off from one end, and the process is repeated. If the glass is not dried with paper or cloth or if it is washed with detergent and dried by clean air in a machine, the glass lacks these cellulose fibers and thus cannot produce bubbles. The only bubbles are from turbulence when the glass is first filled.

If a can of a carbonated drink is opened via a pull tab, almost the entire bubble production occurs on the submerged top surface of the tab. (Use a flashlight to look inside the rest of the can.) Presumably, the bubble production occurs where debris clings to the tab. If the can top was cleaned with paper or cloth for sanitary reasons, part of the debris is cellulose fibers.

Ice comes with air trapped at various sites on its surface and thus can serve like the fibers. Salt, however, is different. When someone salts a beer, the salt goes into solution and reduces the amount of carbon dioxide that can be dissolved in the liquid. The liquid already had too much carbon dioxide, and so carbon dioxide rapidly comes out of solution.

Once bubbles are released, buoyancy tends to accelerate them upward because they are lighter than the surrounding liquid. However, molecules such as proteins quickly begin to stick to the bubble, increasing the drag and slowing the ascent. Such slowing occurs much more in beer than in champagne because of beer's abundant proteins, and thus champagne bubbles outrace beer bubbles.

If a container of a carbonated drink is shaken just before being opened, the gas normally above the liquid is mixed into it as small bubbles. When the pressure is decreased, carbon dioxide can suddenly come out of solution by moving into these bubbles, and the bubble growth can be so vigorous that bubbles push liquid from the container, perhaps even shooting it out. To avoid the mess, a shaken container should sit for a while before being opened, to allow the bubbles to reach the top and pop. (I cannot see why tapping the container would speed up this process unless it somehow dislodges bubbles that happen to adhere to the container wall.)

The production of bubbles in a *shandy* depends on where the beer lies next to a wall, which is where bubbles can be produced. If beer is poured onto, say, lemonade, most of it initially forms a layer over the lemonade. So the bubbles are produced primarily in the top layer and can easily escape to the top surface in the usual manner. However, if lemonade is poured over beer, the bubbles are produced in the bottom layer and must climb up through the lemonade to reach the top surface. In addition, as some of the lemonade flows down into the beer, particles in it (such as pulp) can provide sites where bubbles can form and break free. In short, the rate of bubble production increases and all those bubbles must shoot up through the lemonade to reach the top surface. The result is rapid foam production and a good chance of overflow.

A small piece of lime, a peanut (not the dry-roasted type), and other items can collect enough bubbles near the bottom of the liquid to be floated to the top. Many of the bubbles pop there, and so the object sinks back to the bottom. As long as bubbles can still form, the process is repeated.

When Guinness stout is poured into a glass, the beer is initially filled with bubbles. When the bubbles run into each other, they can adhere to one another to form *bubble rafts*, which causes the bubbles to slow in their ascent. The rafts are separated vertically from one another by a distance that depends on the different rising rates of a free bubble and a bubble raft. (Bubble rafts can form in basaltic magma and lava flows if the layer is thick enough to allow the rising bubbles to form rafts before they reach the top of the layer.)

Some people argue that the downward motion of the rafts is an illusion, but the motion looks real and could be based on either one of the following effects, or both: (1) The bubbles in the middle of the glass will rise more quickly than those along the wall, which are hampered by friction from the wall and drag on the rafts. So, the liquid in the middle of the mug is pulled upward by the rising bubbles, causing replacement liquid to move downward near the wall. Thus, the rafts near the wall move downward. (2) Bubbles breaking free from the top of a raft are accelerated upward to the bottom of the next higher raft. Thus, each raft is losing bubbles from its top surface while gaining bubbles on its bottom surface, and the center of the raft is therefore moving downward.

2.77 • Soap bubbles and beer foams

What holds together a soap bubble, the kind you blew from a ring as a child? Why is soap or detergent required—can you blow a water bubble? Why can a soap bubble last longer if it also contains glycerin? (Mix the detergent, water, and glycerin in the ratio of roughly 1:3:3.) Why doesn't the liquid in the wall of the bubble simply drain to the bottom of the bubble, causing the top to pop open?

Why does the foam (the *head*) on poured beer last so much longer than the foam on poured soda? Why does beer foam eventually disappear?

Answer The wall of a soap bubble is a very thin layer of water with detergent molecules clustered along the inner and outer surfaces. One end of a detergent molecule binds to water (it is *hydrophilic*) and thus burrows into the water surface. The other end does not bind to water (it is *hydrophobic*) and sticks out of the surface. The forces that hold a bubble together are due to the *surface tension* of the water—that is, the mutual attraction of the water molecules. However, the surface tension of pure water is too strong to allow an inflated thin film of water to form a bubble. The detergent molecules clustered along the surface reduce the surface tension enough that the bubble does not simply collapse.

The water in the bubble tends to drain to the bottom due to gravity. However, as the top thins, the detergent molecules along the outer surface begin to be repelled by detergent molecules along the inner surface, and the thinning slows or stops. Still, the wall is thin enough that the film can rip due to evaporation, chance disturbances, or the diffusion (passage) of air through the film.

Glycerin stabilizes a soap bubble because its high viscosity (internal friction) slows the draining of water to the bottom of the bubble. It also decreases the evaporation from the bubble.

In beer foam, the liquid in the bubble walls drains slowly, and thus the walls thin until the bubbles pop. However, the draining is slowed by certain molecules that attract one another and also the liquid. Such stabilization is not used in other carbonated drinks, where a lingering head is not desired. Beer foam can be almost immediately eliminated if oil is added, as can happen if someone is drinking beer while also eating fried foods or wearing lipstick. The oil decreases the surface tension where it touches a bubble, and the surrounding liquid pulls the bubble apart.

Parts of the bubble walls also thin because their liquid tends to be sucked into the curved junctions where several bubbles meet, regions known as *Plateau borders* after the 19th century Belgian scientist Joseph Antoine Ferdinand Plateau. The pressure in the liquid is determined by the curvature of the surface due to the surface tension: Greater curvature means less pressure in the liquid walls. Thus the liquid pressure is less in the curved Plateau borders than in the flatter walls. So, liquid is pulled from the nearby portions of the flat walls into the Plateau borders, converting a *wet foam* to a *dry foam.*

The proteins that gather at the bubble walls help stabilize beer foam for at least two reasons: (1) They increase the viscosity and thus slow the drainage. (2) They also tend to prevent the two sides of a wall from getting close enough for the wall to rupture due to some chance disturbance, allowing the bubbles to *coalesce* (merge).

Even if the walls are stable, beer foam gradually changes because the carbon dioxide in a bubble *diffuses* (passes) through the walls. As a result, the bubbles along the top of the foam lose their gas and shrink. The smaller bubbles shrink fastest because, being highly curved, the gas is more tightly squeezed by the surface tension at their walls than in larger bubbles. The smaller bubbles also shrink because the squeezing causes their carbon dioxide to diffuse into adjacent larger bubbles, where the squeezing is less because the curvature is less. Thus, larger bubbles tend to grow at the expense of adjacent smaller bubbles.

One way to slow the diffusion and maintain the foam is to use nitrogen gas in the beer instead of just carbon dioxide gas. The nitrogen gas diffuses much more slowly through the liquid walls. However, because the foam of a nitrogen-laced beer is more stable, pouring the beer without the plentiful foam overflowing the glass requires patience. Guinness stout, for example, is notorious for the long *pull* required at a spout due to its nitrogen content.

Another way to slow diffusion is to chill the drinking glass. Then, as bubbles are generated along the glass wall, they contain colder gas when they join the foam at the top. The decreased temperature slows the rate of diffusion of gas molecules through the liquid walls.

Sometimes the top of the head on a glass of beer seems to undergo a sudden loss of bubbles, a process known as *cascade bursting.* The bubbles in the top have probably dried out and thus are fragile. When one of them ruptures, oscillations through the foam or the air trigger the others to rupture.

Someone skilled with Guinness stout knows the correct way of pouring it from a bottle: The bottle is suddenly inverted inside a glass. The beer is unstable, of course, and begins to flow out. However, the flow oscillates in a process known as *glug-glug*, with beer coming out in gulps on one side of the bottle opening as air flows inward in gulps on the opposite side. If the bottle opening is positioned in the foam forming above the poured liquid, the foam is caught up in the air inflow and pulled into the bottle. In the end, the glass is filled with beer and the bottle is filled with foam.

2.78 • Bursting bubbles

When a bubble on the surface of a liquid, such as water, bursts, why does it throw tiny water drops up into the air? When a bubble in a layer of champagne bubbles bursts, why do the adjacent bubbles form a flower-like arrangement, as though they were petals?

A soap bubble blown from a plastic ring can float in the air for a few seconds before it bursts. Does it disappear instantaneously, and where do all the soap and water molecules go?

Answer A bubble on a liquid surface bursts because the liquid in the thin layer forming its top surface drains until it ruptures. As the rupture opens up the full top of the bubble, the sides of the bubble are pulled down to its bottom by surface tension—that is, by the mutual attraction of the molecules along the bubble wall. Descending liquid from opposite sides of the bubble collide at the bottom and shoot upward, forming a *jet* (column of water). The jet is unstable and surface tension quickly pinches it off into drops, which are the drops thrown up into the air by a bursting bubble.

If the bubble bursts along the top surface of champagne as it is being drunk, the jets and droplets release aromas that reach the inside of the nose, enhancing the pleasure of drinking the champagne.

If a bursting bubble is surrounded by other bubbles, the downflow during its collapse sucks the surrounding bubbles toward it, stretching them into petal-like shapes that seem to grow outward from the central point of the burst.

A soap bubble floating in air will burst when it ruptures at some point on its surface. The rupture spreads as a circle, with the circle's rim gathering up the liquid in its way as it moves through the rest of the film at about 10 meters per second, too fast for you to see. The rim continuously throws off drops (many thousands in all) until it converges just opposite the bubble's original rupture point.

2.79 • Whales and bubble nets

Why do several types of whales release air to form bubbles when foraging for food such as krill?

Answer Whales can apparently trap their prey in nets (or curtains) of bubbles. The prey can, of course, simply swim through a bubble net but they are reluctant to do so when they are in a large group. Thus, by constructing a bubble net around or under a group of fish, a whale can herd the fish into a small region where it can then eat them efficiently. The fish do not appear to respond visually to a bubble net because the entrapment and herding can occur at night. Rather, the fish seem to respond to the noise produced as the bubbles in the net oscillate.

2.80 • Water striders

How does a water strider manage to rest on water or move over the water surface? Why does the motion generate waves in front of and behind the insect? A water strider does not make noise and lies immediately above the water, so how can it signal to other water striders that it wants companionship or that competitors for its companion should keep away?

Answer When a water strider stands, its weight is primarily on its middle and hind legs, which indent or dimple the water surface. They do not break the surface because of the water's surface tension—that is, the mutual attraction of the water molecules that makes the water surface act like an elastic membrane. A water strider can even bear down on the surface in the early stage of a hop and still not break the surface. If the water is shallow and the insect is brightly lit in sunlight, the indentations produce oval shadows on the bottom of the water. These regions are dim because as the light rays pass through the curved surfaces of the indentations, they are bent off to one side.

Of course, if water striders were too large, they would just sink in the water and would be called water sinkers. The ability of the water surface to support a normal (lightweight) water strider is due to the water resistance of the portion (tarsus) of each leg touching the water—the tarsus is *nonwetting*. If the tarsus were easily wetted, water could climb up onto a leg and the insect would sink. One nonwetting feature is a secreted wax covering the tarsus, making it *hydrophobic*. However, the primary reason the insect does not sink is the microscopic structure of a tarsus: It is covered with tiny hairs (*microsetae*) along which run tiny grooves. This microscopically rough, hydrophobic surface is very effective in not allowing the water to cover the legs. Without it, a water strider would still be able to stand on water but could never run or hop and would easily be some animal's snack.

A water strider runs by rowing its middle and hind legs. The propulsion comes primarily from the middle legs, which act like oars. As a leg sweeps backwards, it generates a U-shaped vortex tube in the water. The tops of the U on the water surface are two closely spaced vortexes that turn in opposite directions; these two tops are connected by the body of the U below the water surface. Because part of the water motion in the vortex tube is toward the rear, the insect is propelled forward. The research group that discovered this rowing-by-vortex-production built a mechanical water strider (dubbed robostrider) that sported steel wire legs and an aluminum body and was driven by elastic thread over a pulley. As the robostrider rowed its legs, a vortex pair was produced on each side during each backstroke. Nearby water striders were not amused.

The vortexes produced by the middle legs are usually difficult to see. Somewhat behind the insect, the vortex motion might change into waves, but because the waves are shallow and have a fairly long wavelength, they are also difficult to see. More visible are the short-wavelength waves that the insect's motion sends in the forward direction. The insect uses these forward-going waves, which might be visible as far as six or seven body lengths ahead of the insect, to detect prey, obstacles, or one another as they dash and zig-zag about the water surface. (Watch them for a while. In spite of their crazy maneuvers, they never collide.)

Water striders communicate with one another by pumping the water surface to send out waves at fairly high amplitude and at about 20 times per second. If an ant falls into the water and begins to thrash about, generating waves, the water striders in the region intercept the waves and then race directly to the ant at an amazing speed, for lunch.

Water striders avoid portions of the water surface covered with a thin film of contamination, such as oily material, because they cannot glide through the material or send their signal waves through it. If they accidentally get into one of these regions, they can escape only by hopping.

2.81 • Beading on rods and saliva threads

When no one is watching, reach inside your mouth and, with your thumb and first finger held together, pull off some of the saliva on the inside of your check. Then, with the saliva close enough to see, gradually separate your thumb and first finger so that a string of saliva stretches between them. During the separation, why does the string suddenly develop beads of saliva (Fig. 2-25)?

Dip a thin rod (or a fiber) into a cup of oil or honey and then pull it up vertically. As the fluid moves down the rod, why does it form beads, why does one bead seem to dominate and gobble up smaller ones in its path, and why do even more beads form after the big, gobbling-up bead passes?

Answer The surface tension of the saliva string (that is, the mutual attraction of the molecules) attempts to minimize the surface of the string. When the string has a moderate diameter during the early part of the stretching, the minimum surface area is that of a cylinder, and so the string is cylindrical. Waves from chance disturbances, such as the unavoidable slight shaking of your hand, run along the string, distorting its cylindrical shape, but surface tension quickly restores the shape.

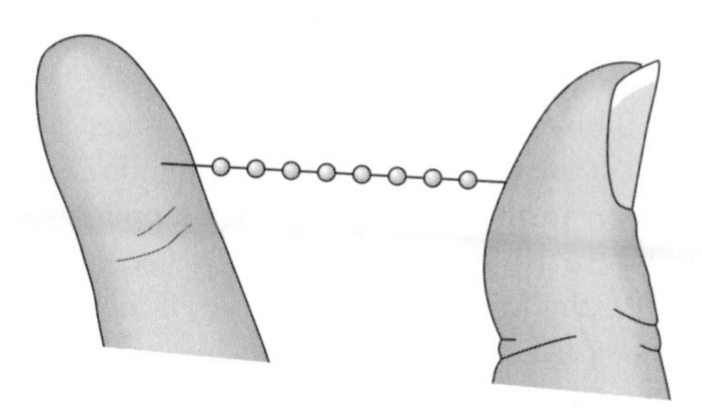

***Figure 2-25** / Item 2.81* Beads appear on a saliva thread stretched between thumb and first finger.

However, the string diameter decreases as your thumb and finger separate, and the string eventually becomes unstable to waves with a wavelength greater than the circumference around the thread. The reason is that the distortion due to such a wave actually decreases the overall surface area and so, once the distortion occurs, surface tension enhances it instead of eliminating it. The portions that become wider are pulled by surface tension into beads, and the portions that become narrow are thinned to become the narrow string between the beads. The spacing between the beads approximates the wavelength of the wave causing the transition. (If larger beads are separated by smaller beads, the beading process probably occurred more than once, with a different wavelength involved each time.) The final thread between the beads may be too small to see.

A thin layer of fluid on a rod (or fiber) is similarly unstable; chance disturbances and surface tension will reshape the layer into beads. If the rod is vertical, the beads can flow downward, especially a large one. Smaller beads in its path will merge with it, but after the large bead passes, the remaining thin film might also break up into beads. However, if the film is too thin, the downward flow prevents the beads from forming.

Certain spiders use the beading tendency when they construct their webs. After the basic web is built, the *capture threads* intended to catch flies are coated with a liquid that immediately pulls into beads on the threads. These sticky beads can ensnare a fly long enough for the spider to reach it after detecting the fly's thrashing via oscillations of the web.

Beading also shows up in the welding of steel. If the heat source moves along the steel within a certain range of speeds, a trailing series of *humps* (or bumps) are left as the trailing pool of liquid steel solidifies. Once the source moves past a point of melted steel, the surface tension of that liquid can form a bead before it solidifies. If the heat source moves too slowly or too quickly, the humps do not form.

2.82 • Rain harvesting by desert lizards

Some desert lizards are extremely good at acquiring drinking water on those rare occasions when dew might form or rain might fall (hence the name *rain harvesting*). For example, the Australian thorny devil (*Moloch horridus*) acquires water from dew by sitting down in it and from even a light rain by standing spread-eagle in it. How can either procedure allow a lizard to get a drink of water?

Answer The lizard soaks up water like a kitchen sponge. The spaces (channels) between the scales of the skin pull in the water by capillary action (that is, by the attractive forces among water molecules and between water molecules and the molecules along a pore). This process is so efficient that water is pulled into the skin as high as the top of the lizard's head.

To drink the water, the lizard makes small, repeated motions of its lower jaw, removing water from the channels near the mouth. As a lizard drinks, the water in those channels is replenished by water pulled from the rest of the skin. Gravity can also help move the water toward the mouth if the lizard stands with its head lowered and its rear end held high.

2.83 • Prey harvesting by shorebirds

Sprinkle a few tiny bits of Styrofoam onto water in a wide bowl and then try to pick up one of the bits by catching it between a thumb and a finger. The closing of the thumb and finger in the water probably just squirts water and the Styrofoam bit away from you. Now consider a shorebird that must catch planktons (which are small) in the water with its bill. Doesn't it have the same trouble as you do in catching the Styrofoam bits?

Answer Some shorebirds use surface tension to catch their plankton prey. One of these birds dips its bill into the water with its jaw almost closed and then separates its jaw a bit more. When it pulls its bill free of the water, a water drop bridges the two sides of its bill via surface tension—that is, by attractive forces between the water and the interior of the bill. Prey is trapped in the drop. To move the drop up to the throat so that the prey can be ingested, the bird gradually spreads the two sides of its bill. The drop continues to cling to the top and bottom sides of the bill but the separation of the sides tends to stretch the drop. To fight against this, surface tension pulls the drop up higher where the two sides are closer together. This process continues until the drop reaches the pharynx, where the captive prey can be swallowed.

2.84 • Drops and liquid films on solid surfaces

Why do some liquids spread over a solid surface, such as a glass tabletop, while other liquids bead up? Why can some drops cling to a surface even if the surface is tilted or if the drop is hanging from it?

When a liquid film drains down a surface of moderate slope, why does the lower edge usually become crooked or advance in spurts and at spots? You can usually see this result when using a rubber blade to clear soapy water or glass-cleaning fluid from a tilted car windshield. Scrape downward and then stop prematurely. Why does the liquid break up into fingers that run down the windshield?

When rainwater spreads down a vertical concrete wall as on common buildings, why is the advance usually uneven? When water gradually drains into a cave, why does it typically tend to form conical stalactites?

Answer The extent to which a liquid spreads on a horizontal solid surface depends on the attraction between the molecules in the liquid and the molecules in the solid. If there is a strong attraction, the liquid spreads and is said to *wet* the surface; if there is little attraction, the liquid tends to bead up. The extent of wetting (or beading) is often described in terms of the angle (the *contact angle*) that the liquid makes with the solid surface—a small angle corresponds to wetting and a large angle corresponds to beading. However, the contact angle is ambiguous because in any real situation (with a real liquid on a real, microscopically complex surface) it can be in a fairly wide range of values.

The details of how a liquid spreads over a solid are still not fully understood because they involve the atomic interactions along the edge of the liquid. In many cases, the edge moves only because a very thin *precursor film* spreads out slightly in front of the edge. The molecules in the liquid film then attract the molecules on the edge, moving the edge forward. Sometimes the edge becomes *pinned* (stuck) due to an imperfection in the surface or a point of strong attraction. If a drop begins to evaporate, the leading edge tends to contract but can again be pinned at points, making the contraction (called *dewetting*) uneven.

Some viscous liquids, such as oils and glycerin, have a peculiar way of spreading down a tilted plane: The advancing line quickly breaks up into uniformly spaced fingers, and the fingers then drain down the plane faster than the regions between them. The fingers form because the advancing line is unstable and chance disturbances create waves along it. One of those waves dominates the advancing line, creating strong downflows at regularly spaced intervals along the line.

If a liquid film spreads down a slope, a pinning point breaks up the uniform advance of the film, leaving a dry region below that point. You can see the results of pinning and instability when using a blade to clear soapy water from a car windshield. When rain spreads down a vertical concrete wall, the leading edge usually does not advance uniformly toward the ground. Some regions allow the water to descend faster than other regions, and that extra drainage in one spot can send a wide "finger" of water down the wall and to either side.

A stalactite consists of calcium carbonate that precipitates from water draining into the cave. If precipitation begins at a point on the cave ceiling, water tends to drain to the bottom of the bump that forms. Because the water layer tends to be thicker at the lowest point, the precipitation tends to be greatest there, causing the length of the bump to grow faster than the width and thus producing the general shape we associate with an ideal stalactite. However, if the rate of water drainage is slow relative to the precipitation rate, other shapes can form, such as rods and the beautiful twisted structures called *helictites.*

2.85 • Breakfast cereal pulling together

If two floating ovals of the breakfast cereal, under the trademark of Cheerios happen to float near each other in a bowl of milk, why are they pulled together? If you leave many ovals floating in the milk in random locations, why do they tend to cluster within a few minutes? Why do the ovals also flock to the sides of the bowl? These various effects are collectively known as the *Cheerios effect.*

Answer Next to an oval, the milk surface curves upward because of surface tension (Fig. 2-26); that is, the attraction of the water in the milk to the sides of the oval are enough to pull the water up along the oval in spite of the downward pull by the gravitational force. If two ovals come near each other, the surface between them becomes very curved, which puts a force on each oval that pulls them together. The attraction can also be explained in terms of energy: A curved surface requires more energy, and so the ovals come together in order to flatten the surface between them, thereby reducing the energy.

The liquid surface near the bowl wall is also curved up by surface tension, and so when an oval nears the wall, the intermediate surface becomes very curved. So, a force pulls the oval to the wall. If you fill the bowl completely with milk and then add a bit more so that the milk surface is slightly higher than the rim of the bowl, the liquid surface near the rim is curved downward. Now an oval that happens to come near the rim is forced away from it. This physics is the basis of a common pub challenge about something floating in a glass of water: Can you prevent the floating object from ending up on the wall of the glass?

A flat, double-edge razor blade can float on water if it is carefully lowered onto the water. Unlike the breakfast cereal oval, the razor blade floats slightly below the water level, and so the water surface curves downward to meet the blade. Still, if two razor blades float near each other, surface tension causes them to pull together so as to flatten the intermediate surface and reduce the energy.

In general, a material is called *hydrophilic* (roughly, "water loving") if water is attracted to it and *hydrophobic* (roughly "water hating") if water is not attracted to it. Two floating hydrophilic objects will attract each other, even from a considerable distance, and two hydrophobic objects will also. However, a hydrophobic and a hydrophilic object will repeal each other because if they were to come near each other, the curvature of the water surface would increase, which requires energy.

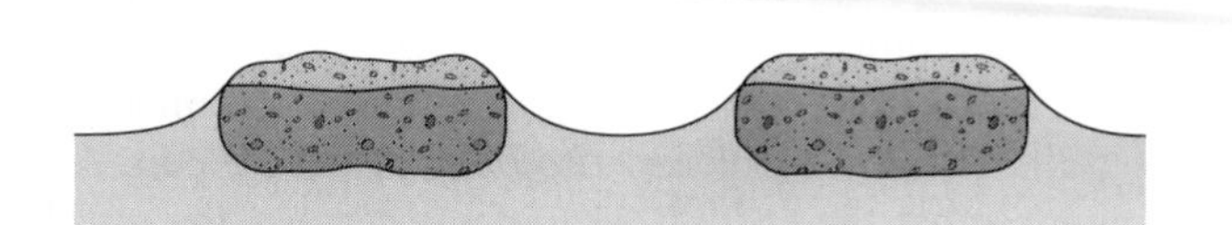

Figure 2-26 / *Item 2.85* Two ovals floating in milk.

2.86 • Sandcastles

What holds a sandcastle together? A pile of sand in a playground sandbox cannot be very steep and the sand can be fashioned into nothing more than a hill, and yet the wall in a sandcastle can be vertical and the features, such as a turret, can have sharp corners. Also, many naturally occurring sand formations, called *scarps,* occur with nearly vertical walls. What allows vertical walls?

Answer Dry sand is not cohesive because there is no force between sand grains to hold them together. Submerged sand is not cohesive because water can easily move between the grains, making the sand fluid. However, partially wetted sand can be quite cohesive. Water is attracted to sand grains and is said to *wet* the grain. When two adjacent grains have a small amount of water between them, the water clings to each grain by forming a *liquid bridge* between them. A bridge is roughly shaped like an hourglass: Water is splayed on each grain and the bridge has a narrow waist between the grains. The water is relatively immobile, and so it does not flow away from the grains or simply drain due to the gravitational force. It provides a cohesive force on the grains for two reasons. (1) Water molecules pull on the grains and on each other, an effect called surface tension. (2) Because the bridge has curved surfaces that are concave outward, the water pressure inside the bridge is less than the air pressure outside it, and so the grains tend to be sucked toward the lower pressure.

If sand becomes saturated with water, the grains are no longer joined by individual, immobilized water bridges and are instead lubricated by the water so that they slump. Practiced sandcastle builders spray their constructions so that water will be drawn into the surfaces to set up individual water bridges. If a sandcastle is left to dry, the outer surfaces lose their bridges to evaporation and soon slump off.

Wet beach sand is more cohesive than pure sand (silica) because it contains particles of clay and organic matter that can set up electrical bonds with the grains. In addition, a layer of sand can be coated with a salt crust that further adds binding forces between the grains. In the swash zone, where seawater periodically bathes the sand, the water stirs air bubbles into the sand, giving the sand a softer texture. Thus, the *hardness* of the sand can vary remarkably as you move from the dry sand at the top of a beach, down along the aerated and partially wet sand, then along the soaked sand, and finally onto the saturated sand submerged in the water.

2.87 • Appearance of bad coffee

If you examine a cup of bad coffee (the type kept hot for hours in second-rate restaurants), why can the appearance of the surface change as you insert and remove a spoon? With the spoon out, the surface lacks any luster, which is unappealing. With the spoon in, the surface has small shiny circles, which is even less appealing.

Answer Usually, bad coffee is bad because it has an oily layer on the top, which gives a dull, uninviting reflection. So, you can often tell from the appearance of the coffee whether it is oily. If you insert a spoon, the spoon can carry material that will spread out on the top surface, squeezing the oil into drops. Those drops, with their curved surfaces, can be small reflectors of any overhead lamps, thus giving many bright circles. When you pull the spoon out, the oil layer reforms and most of the reflections disappear.

2.88 • Tears of wine and other liquid surface play

In a glass with a strong alcoholic drink, such as wine or low-proof vodka, why do drops (said to be *tears of strong wine*) form, grow, and slide down on the glass wall above the liquid surface (Fig. 2-27)?

Answer Normally, a water surface will slightly climb the wall of a drinking glass because (1) glass molecules attract water molecules (there is *adhesion* between the two materials) and (2) water molecules attract one another (there is *cohesion* within the water). So, the water immediately next to the wall is pulled slightly up the wall by adhesion to form a film, and that water pulls up more water by cohesion, forming a curved surface near the wall.

The liquid film of wine climbs much higher because of an additional feature due to the difference in the surface tension in the climbing film and the bulk liquid. The molecules along the surface of a liquid attract one another and pull together, putting the surface in a state of tension so that we attribute *surface tension* to the surface. The surface tension in water is fairly large but that in an alcohol–water mixture is smaller. When an alcohol–water layer begins to climb the glass wall, the alcohol quickly evaporates, leaving primarily a water film on the wall. Because that water has a greater surface tension than the alcohol–water mixture in the bulk liquid, bulk liquid is strongly pulled up into the layer on the wall. Because the layer thickens, the upper edge can be pulled higher by the adhesion with the glass, so the film climbs higher than water alone would climb.

The downward pull of the gravitational force on the climbing film limits the climbing height. As alcohol evaporates from the film, the surface tension of the remaining water tends to pull the water into drops. The drops initially cling to the wall because of adhesion but eventually they grow too large and suddenly break free to run down the wall into the bulk. They can form provided the drink is not too diluted or too strong—the drink must have a mixture of alcohol and water to get the interplay of different values of surface tension in the bulk liquid and in the climbing film.

Figure 2-27 / Item 2.88 Tears form above the surface of strong wine.

When a fluid moves because its surface tension in one region differs from that in another region, the motion is said to be a *Marangoni* effect, named after one of its early investigators. The Marangoni effect can explain why some drops spread widely over a solid surface. The visible spreading can be preceded by a very thin layer in which evaporation is faster than in the rest of the drop. As with the tears of strong wine, if the evaporation from the thin layer increases the surface tension of the remaining liquid in the layer, then replacement liquid is drawn into the layer from the rest of the drop, tending to cause the drop to spread over the surface.

2.89 • Tia Maria worm-like patterns

The liqueur Tia Maria is often served with several millimeters of cream on the top and sipped through a straw. If the drink is left to stand for several minutes, why does vigorous movement appear, with the surface forming cells or worm-like tubular patterns?

Answer In one or more regions, alcohol diffuses (slowly passes) up through the cream, decreasing the surface tension in the cream due to mutual attraction of the molecules on the surface. The alcohol–cream liquid (with weak surface tension) is then pulled along the surface into the remaining regions of cream (still having strong surface tension). More alcohol rises to replace the liquid that is removed, and so on. For complex reasons, the presence of the cream (specifically, its resistance to motion) triggers circulation patterns of rising and descending liquid that can form into either isolated cells when the cream layer is thick or worm-like tubular rolls when the cream layer is thin.

2.90 • Patterns in hot coffee and other fluids

If you place a cup of hot coffee in glancing sunlight, you might be able to see patterns on the coffee surface: whitish regions outlined with dark lines that constantly form and reform (Fig. 2-28). Such patterns are called *Benard cells* after an early investigator.

If a layer of oil is heated in a skillet over a low flame, the oil displays little or no motion. But if the flame is gradually increased, the oil begins to move, forming Benard cells in the shape of polygons. (Glancing light is usually needed to make the pattern visible.) With a somewhat greater flame, the polygons may settle into a pattern of hexagons that resembles a honeycomb.

In a transparent cup of hot tea, gradually add milk near the wall. The milk sinks to the bottom of the cup. Add enough so that the bottom 75% of the cup has the readily apparent white milk. After a few minutes, why can horizontal bands appear in the milk portion of the cup?

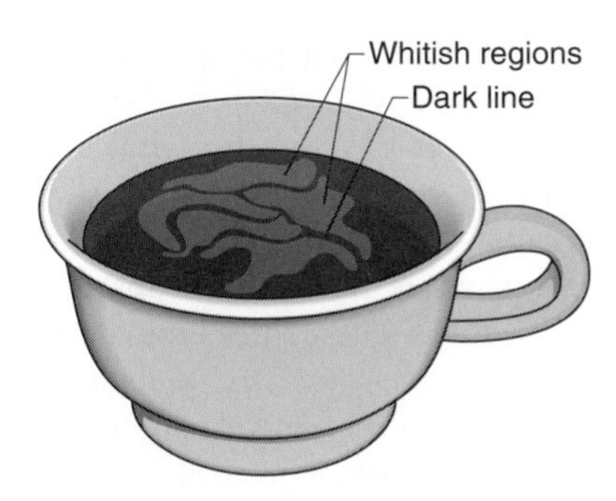

Figure 2-28 / *Item 2.90* Whitish regions and dark lines form over black coffee.

Answer As water evaporates from the top surface of the coffee, the surface cools, becoming somewhat denser. The resulting temperature difference (and density difference) between the top and bottom layers of coffee tends to cause circulation within the coffee. Imagine a parcel of liquid at the bottom. It is immersed in a layer of water that has the same temperature and density as itself and thus it tends to remain stationary. However, a chance disturbance to the cup can send the parcel upward into slightly cooler and denser liquid. There it is buoyant. As it accelerates upward into progressively cooler and denser liquid, its upward acceleration increases. Thus, the motion due to the chance disturbance is amplified.

A similar sequence can apply to a parcel of liquid on the top surface. If it happens to move down into the slightly warmer and less dense liquid, it is accelerated downward and the motion is amplified.

Because the coffee has an open surface, the motion across the surface is also affected by surface tension, which is due to the attraction of water molecules for one another. As the water on the surface cools, its surface tension increases slightly. So the surface tension in the region where (cooler) water happens to be descending is larger than the surface tension in the region where (warmer) water happens to be ascending. This difference in surface tension tends to pull the water across the surface from the ascending region to the descending region. Because greater surface tension tends to pucker water, the descending region forms a ridge that is slightly higher than the ascending region, something like a line of short hills fencing in a valley. The coffee surface is then covered with cells of broad regions (valleys) of ascending liquid and narrow ridges of descending liquid.

As the warmer water reaches the surface, some of it evaporates but, depending on humidity, the vapor can quickly condense to form water drops in the air just above the ascending region. The larger drops fall back to the liquid surface and the very small drops are carried away by the air currents above the hot coffee, but moderate drops can hover, being suspended by the upward flow of air and moisture from the coffee. When white room light or sunlight scatters from this thin cloud, the cloud is visible and whitish. The ridges marking the descending liquid lack the hovering drops and thus have the normal dark appearance of coffee. If you bring a charged object (such as a plastic comb charged by running it through hair) near the coffee surface, the hovering drops are removed electrically and the whitish appearance disappear.

Similar circulation patterns occur in an oil layer heated in a skillet. Whereas the coffee cools from above, the oil heats from below, but the important feature is that a temperature difference is set up between the top and bottom surfaces of the liquid. If the temperature difference exceeds a certain critical value, the convection becomes unstable to chance disturbances. The disturbances move portions of the liquid in various ways, and buoyancy and surface tension can overcome viscosity to set up cells of ascending and descending liquid. In some liquids, the motion sets up long roll structures, with liquid ascending on one side and descending on the opposite side. The polygons seen in the oil consist of broad regions of ascending hot oil and narrow lines of descending cooler oil. As in the hot coffee, the cooler oil has greater surface tension than the hot oil, and thus oil is pulled along the surface from an ascending region to a descending region.

The bands that can appear after milk is added to hot tea are due to horizontal rolls that run around the wall of the cup, a circulation pattern that may be caused by the cooling through the wall. The effect, first described to me in 1987 by Christian Roos of Germany, may develop into as many as eight bands, but you may need to experiment to get the right conditions for them to appear.

Benard cells can also occur in the melted wax of a large candle. The surface tension of hotter wax is less than that of cooler wax and so the variation in surface tension from the wick to the outer edge of the candle can drive the convection cells. If the candle is carefully extinguished, the cells can leave ridges in the wax as it cools and solidifies.

2.91 • Patterns in coffee stains

When coffee is spilled onto a horizontal surface and left to evaporate, why is the puddle's original position marked by a distinct brown ring? When pools of salty water are left to evaporate on, say, a sidewalk, why is the edge of the pool marked with a white ring?

Middle Eastern coffee is a strong mixture of water, sugar, and finely ground coffee grains that is brewed in an *ibrik* and then poured into a small cup, along with the coffee silt. As the coffee cools to a drinking temperature, the silt gradually sinks to the bottom of the cup. The drinker sips the coffee down to that lowest layer and then puts the cup aside. If the remaining liquid–silt mixture is left to evaporate for a few hours, the silt forms a surprising pattern of dark and clear thin lines around the edge of the liquid. The lines, each a few millimeters long and perpendicular to the edge, have a uniform spacing, as if drawn by an artist. What causes this pattern?

Answer When a puddle of coffee is left to evaporate on a solid surface, the puddle tends to shrink as it loses water. However, the perimeter (said to be the *contact line* because it marks the contact of air, liquid, and solid) can be snagged on some imperfection on the solid surface, a spot that is rough or chemically different. That is, the contact line becomes *pinned* and is unable to retreat from the spot.

The evaporation, which can be fairly rapid in the thin layer at the edge of the pool, leaves a residue of whatever is dissolved in the water, the *solute*. Because the contact line is pinned, coffee flows toward the edge from the middle of the puddle to replace the water lost to evaporation. Thus, more and more of the solute is deposited at the edge, building up the brown ring that is eventually visible. Once the ring begins, the contact line becomes more firmly pinned. However, as the liquid in the drop diminishes, the contact line can overcome the pinning and suddenly retreat inward. Then it is again pinned, and a new, smaller ring forms. Similar flows can leave a white salt ring around an evaporating puddle of salt water.

Similar flows can also occur at the edge of evaporating Middle Eastern coffee if the last of the liquid is in a cup with a sloped wall so that the edge of the liquid is shallow. In addition, the flow develops into a regular series of cells that brings the dark, fine coffee silt out to the edge and that brings liquid back from the edge. The outflow deposits solute at the edge; the inflow sweeps out any solute. The result is a regular pattern of short lines, alternating dark and clear, around the wall. Even if the coffee is stirred briefly, the cells are soon reestablished. If sugar is left out of the brew, the cells do not occur.

A simple explanation is that as water evaporates from the shallow edge, replacement liquid flows toward the edge, dragging some of the silt outward along the wall of the cup. That silt forms one of the dark lines in the pattern that builds up. Once the replacement liquid reaches the edge and begins to evaporate, it becomes more concentrated and thus denser, and so it begins to sink, sliding away from the edge along the curved wall of the cup. That inward flow drags silt away from the edge, clearing a narrow lane and forming one of the clear lines in the pattern. This silt distribution can occur only if the wall of the cup has a moderate slope. Neither a vertical wall (which does not offer any shallow edge) nor a nearly horizontal wall (which offers a wide shallow edge) will do.

2.92 • Breath figures

When you breathe on a surface such as a mirror or a lens in eyeglasses, why does the surface mist up? Why does a mirror in a steamy bathroom mist up?

Answer When your fairly warm breath meets a cooler surface, such as a mirror, the moisture in your breath begins to condense on the surface. The surface is likely to be covered with dust and the oil from fingerprints. Because the water molecules that condense on the surface are more strongly attracted to one another than to these contaminants, they tend to form small water drops between the contaminants. Thus, the water does not coat the surface uniformly but instead forms drops.

Initially the drops are tiny, but they grow and eventually begin to *coalesce* (merge), a process that is still not well understood. Because the spacing between the resulting larger drops increases, more fresh tiny drops begin to form between the larger drops. These three types of drop patterns (only tiny ones, then larger ones, and then larger ones with intermediate tiny ones) are collectively dubbed *breath figures* (because they can result from a breath).

Once breath figures form on a mirror, images become indistinct and the mirror seems to be covered with a white substance because the drops scatter the room's white light. If you rub your finger over the surface and repeat the breath-misting process, drops do not form in the rubbed regions because oil from your finger lowers the surface tension of the water too much for it to pull the water into drops. Instead, the water spreads out in a thin layer, a process known as *wetting*.

If drops continue to form, such as during a long hot shower in an otherwise cold shower room, the drops tend to grow and coalesce until some of them are too heavy to remain stationary. The gravitational force makes these heavy drops slide down the mirror. Because they run into other drops, an avalanche of drops is soon sliding down the mirror.

Breath figures can be dangerous if they form on eyeglasses or on a car's front windshield when viewing is essential. (That is, if you are the driver, you need to see the road clearly and not just vaguely or you will soon not see the road at all.) Some windshields are notorious for holding onto water drops while others are specially designed to shed water quickly. Some people use home remedies or commercial products in which they coat a windshield with a substance that causes the water to wet the windshield instead of forming drops.

2.93 • The lotus effect

Sprinkle water on the leaves of a lotus plant and the drops immediately bead and roll off. On the way, they can collect bits of dirt or dust and thus clean the leaf—the leaf is said to be self-cleaning. Water beads up on other surfaces, such as waxy leaves, but the beading on a lotus leaf seems remarkably different. What causes the beading on a lotus leaf?

Answer The ability of a water drop to spread on a solid surface is often called the *wettability* of the surface. If you could take a close-up view of a drop, you could see the angle that the drop takes at its perimeter on the surface. If the drop can easily wet the surface, it flattens out like a pancake and has a shallow angle at its perimeter. If the drop cannot easily wet the surface, it beads and has a large angle at its perimeter. On a lotus leaf, a water drop beads so much that it is almost a complete sphere.

One reason is that the material on the leaf does not attract water molecules (the surface is said to be *hydrophobic*). So, the surface tension (due to the mutual attraction of water

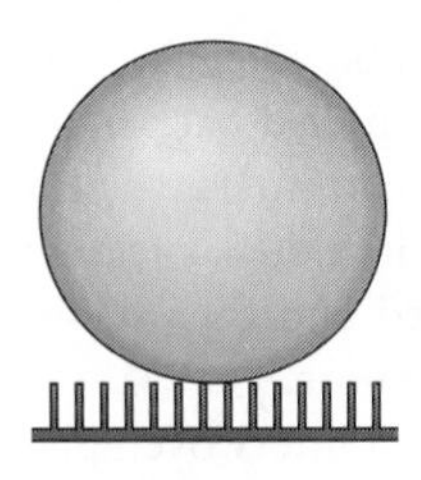

Figure 2-29 / *Item 2.93* A water drop forms a nearly perfect sphere on the spiked microstructure of a lotus leaf.

molecules) tends to pull the water surface into the tight curve of a sphere. Many other solid surfaces, such as some common leaves, are also hydrophobic and cause water drops to bead somewhat.

The extreme beading on a lotus leaf is due to the microscopic structure on which a drop initially sits. The surface has a fairly regular structure of spikes, much like those on a bed of nails (Fig. 2-29). The drop cannot slump into the spaces between the spikes because the material is hydrophobic and the spaces are much smaller than the drop. So, the spaces are filled with air and the drop sits on the tips of the spikes. The resulting minimum contact with the leaf allows the surface tension of the drop to pull the drop into a nearly perfect sphere. Even a gradual slope then allows the drop to roll off (it does not slide off). Dirt and dust along its path adhere to the water by various forces and are thus removed from the leaf.

Some self-cleaning household items that are based on the same principle are available. For example, a self-cleaning windowpane, with an appropriate array of microscopic projections, need never be cleaned because mist or light rain can form drops that slide down the pane, collecting all the dust, dirt, and grime. That is a very nice feature when the window is high off the ground, as on a skyscraper. Also, cars can be coated with a material that cleans itself in a light rain.

2.94 • Aphids and liquid marbles

An aphid enclosed in a plant gall must get rid of its excrement, honeydew, or the liquid will coat, entrap, and drown the aphid. The insect's solution to the problem is to roll the liquid out of the gall. How can a liquid be rolled?

Answer Special epidermal cells on an aphid secrete a wax material that fuses together and then disintegrates into a powder, coating the gall's inner surface. The honeydew becomes coated with this powder as soon as it is excreted by the insect. The powder also provides a microscopic rough surface on which the honeydew sits. Thus the honeydew balls up, becoming a nearly perfect sphere like a water drop on a lotus leaf as described in the preceding item. Because the ball is spherical and does not wet or stick to the gall, the insect can roll the ball out of the gall.

You can make a similar ball by mixing a droplet of water with soot or lycopodium powder. Because either material is hydrophobic, the powder grains stay on the droplet's surface. Once coated, the drop will be nearly spherical when sitting on many common surfaces, such as a horizontal glass pane. Normally water coats glass, but now it beads up because it is actually sitting on the microscopic projections that coat its surface.

2.95 • Paintbrushes, wet hair, and dunking cookies

Why do paintbrushes take up paint, and why do kitchen sponges and paper towels take up water and other spilled liquids? Why does long wet hair clump together?

Many people enjoy dunking a cookie (or biscuit) into hot tea or coffee because the temperature increase releases flavors and aromas. Why does the cookie soften and fall apart if it is submerged more than a few seconds? How can it be dunked so as to release the flavor and aroma and still remain firm, to be eaten instead of drunk?

Answer The bristles on a paintbrush attract the paint molecules, pulling the paint up into the space between the bristles. Because this motion is like liquid being drawn up into a narrow (capillary) tube, the force on the paint is said to be a *capillary force.* When the paintbrush is pulled back from the paint can, most of the paint stays between the bristles because of this attraction. When the bristles are sheared against a surface, such as a wall or canvas, some of the paint is scraped off but much of it can flow off because the bristles are momentarily splayed on the surface. The splaying increases the space separating adjacent bristles, weakening the capillary force and allowing the paint to flow.

A paper towel and kitchen sponge have many pores into which water can be pulled by a capillary force.

The strands of hair are held together by curved *liquid bridges* that span adjacent hairs. If the ends of the hairs are dipped into a pool of water, the water will climb up adjacent hairs while also bending them to pull them closer together.

The cookie consists of dry starch grains held together by a framework of sugar. When the cookie is submerged, the liquid is quickly pulled into its pores by capillary forces. The hot liquid quickly melts the sugar, ruining the framework, and the starch grains fall apart. If you want tea or coffee filled with starch grains, dunk the cookie vertically. If you would rather eat the cookie, slide it into the liquid at a slant so that the top of the cookie is not submerged. It can then remain firm enough to hold onto the wetted bottom of the cookie, provided the dunk is brief.

2.96 • Deep-fat frying of potatoes

When food like sliced potatoes or tortillas is fried in oil, the surface develops a flavorful crust while the interior remains tender. Why does the food take up oil, and why is most of the oil taken up *after* the food is removed from the frying apparatus?

Answer When a potato slice first touches the oil, the energy transferred from the oil to the potato raises the temperature of the potato surface. As the surface temperature approaches water's boiling point, water in the surface pores begins to vaporize and bubbles of escaping water vapor form at the openings of the pores, making the oil turbulent next to the potato. (You can hear the action.) As the surface loses its water, it hardens to form the crust associated with fried foods. The elevated temperature also causes certain chemical reactions in the surface that give it a characteristic fried-food flavor.

As the cooking proceeds, energy is transferred to the interior of the potato slice, cooking the interior. However, because the interior contains trapped water, the temperature there cannot exceed the boiling point of water by very much. Thus the interior can cook without becoming dehydrated or forming a crust-like texture.

Near the surface, however, water continues to evaporate from the pores, to a depth of a millimeter or two. When the slice is removed from the frying apparatus, it comes away with a coating of oil, which traps the water vapor remaining in the pores. As that vapor cools, it condenses to liquid water, which occupies much less volume than did the vapor. Because the gas pressure inside the pore decreases, oil is sucked from the external coating into the pore. This uptake can be enhanced by the attractive forces between the molecules in the oil and those in the pore walls, in an effect known as capillary action. In fact, this is the dominant effect when a thin food, such a thin potato chip, is fried until almost no water remains.

A cook wanting to decrease the oil uptake by fried foods should shake the oil from the food (or clean it off with paper towels) as soon as the food is removed from the oil bath.

2.97 • Ducks stay dry

In moderate climates, ducks (and other aquatic birds) need to stay dry because if they become wet, they lose the thermal insulation of the air layer between the feathers and the skin. Then they can lose thermal energy to the water faster than their metabolism can produce energy. Yet the feather layer is not waterproof because the feathers are obviously porous. So, how do the ducks stay dry when they float or swim?

Answer The feathers, which consist of keratin with wax and ester coatings, are *hydrophobic*. That is, they repel water, and thus water drops tend to run off them rather than wet them, such as when rain falls on a duck. However, this feature is not the primary reason a duck stays dry because when a duck floats, water should be pushed up through and between the feathers, pushing out the vital insulating layer of air and rapidly cooling the duck's skin.

Fortunately for the duck, the pores (open spaces) between and within the feathers are too small for water to enter, not even when the water pressure below the duck attempts to push water into the pores or to widen them. The reason has to do with the convex shape of the water surface as water attempts to enter an opening in hydrophobic material. With that shape, the water surface tends to be pulled back out of the opening by surface tension (due to the mutual attraction of the water molecules). Because a pore in a duck's feathers is tiny, the water surface is highly curved and thus surface tension prevents water from entering the pore.

Some types of fruit baskets consisting of interwoven plastic bands may look unfloatable because the bands do not give a continuous hull, but they float nonetheless. The water is unable to penetrate the open spaces along the bands.

2.98 • Cut potatoes, bird droppings, and a car

If the windshield wiper on your car breaks down, why can you still keep the windshield reasonably clear during a light rain if you rub it with the cut portion of a potato? (Granted, having a spare potato in the car may not be too likely.) When bird droppings adorn a car and it becomes wet with rain, why does the region adjacent to a dropping become drier sooner than the rest of the car?

Answer Visibility through a windshield decreases when water beads up on the glass, distorting your view through it. If you rub the exterior of the glass with a cut potato, the starch coating it leaves strongly attracts the water molecules and spreads the liquid out in a smooth layer. Visibility can then be reasonably clear.

When a bird dropping is partially dissolved in rain, the solution spreads over a small region around the dropping. Water landing elsewhere on the car beads up, especially if the surface has a waxy coating. When the rain ceases, the thin layers surrounding the droppings are the first to dry, well before the water beads completely evaporate.

Not all birds produce this wetting effect because of different diets. The ones that cause water to spread dine on fish and thus leave oily droppings. Such wetting can be a serious problem to electrical energy companies because the birds can soil the grounded supports of a power line. If the excretion is fluid, it can drain onto the power line just under the support, shorting out the line and causing *flashover*, which can shut down the power supply and extensively damage the power line. A buildup of bird droppings is also dangerous even if it is not fluid: During rain or the melting of snow, water can absorb charged particles from the droppings and become more conducting. As this water drains onto the power line, flashover can occur.

2.99 • Catapulting mushroom spores

Fungi such as mushrooms spread their spores in a variety of different ways. However, the most intriguing are the ballistospore fungi, which discharge their spores faster than the eye can follow. Each spore is attached to a stalk called the sterigma. Prior to discharge of the spore, a water drop forms at the base of the spore, near the spore's attachment to the sterigma. Within about 30 seconds, the drop grows to a diameter of about 10 micrometers, and then, very suddenly, the spore and the drop shoot off into the air. What propels them?

Answer When a ballistospore mushroom is ready to discharge a spore, it secretes certain compounds on the spore's surface to promote the condensation of water from the air. The condensation occurs most rapidly at the site where the drop forms, but it also occurs elsewhere on the spore where it forms a clinging film of water. As the drop grows in diameter and the film spreads over the spore, the film soon makes contact with the drop. At that instant, the surface tension of the water in the film yanks the water out of the drop and into the film. This yank gives so much momentum and kinetic energy to the water shooting into the film that the spore is pulled free of the sterigma and launched into the air. The launch acceleration has been calculated to be as much as 25 000 gs (or 25 000 times the acceleration of gravity), but the spore quickly slows because of air drag and so it does not travel far. Because the energy and momentum for the launch is derived from surface tension, the launch has been called a *surface-tension catapult.*

2.100 • Waves on a falling stream

Adjust the height of a finger held in a thin stream (a few millimeters in diameter) falling from a faucet. In a certain height range, ripples form on the portion of the stream just above the finger (Fig. 2-30). What causes these ripples? If you first coat the finger with liquid detergent (such as Tide or Ivory), why do waves form higher on the stream and not at the bottom?

Answer The ripples are due to waves sent *up* the stream by the impact of the stream on the finger. The waves are said to be a type of *capillary wave* because their oscillations are controlled by surface tension due to the mutual attraction of water molecules. In this case, the capillary waves move up the stream as fast as the water moves downward, and thus the waves are stationary relative to you. If you replace the faucet with a container leaking through a hole in the bottom face, the water speed in the stream decreases as the water level in the container decreases. That speed decrease causes the wavelength of the waves (the spacing between successive ripples) to increase until the stream becomes unstable, breaking up into drops.

The existence of the waves depends on the fairly high surface tension of the water. Adding liquid detergent lowers the surface tension. If the finger is coated with the detergent, some of it mixes up through the lower portion of the stream, decreasing the surface tension enough to eliminate the waves there. Then the flow through the lower section is smooth as if in a pipe and the waves form at the top of the pipe-like flow.

2.101 • Water bells, sheets, and chains

Hold a spoon, or some other rounded object, with its convex (outwardly curved) surface upward in the steady stream of a faucet. The water is deflected as a thin sheet that folds downward. Flat surfaces will work also, and one of the best is the plastic screw-off cap from a soda bottle. Insert two fingers into the cap and, with your fingers upward, hold the cap in the falling stream. If you are careful, the sheet folds almost back on itself, forming a *water bell* (Fig. 2-31). Many fountains have such curved sheets as a sort of water sculpture. What causes these *water sheets*?

You can produce water bells and relatively flat sheets by directing two thin turbulence-free streams toward each other. If the streams are vertical and at approximately the same rate of flow, they collide and spread out into a symmetric sheet. The sheet may disintegrate into drops, or it may bend down to form a water bell.

If the streams are directed downward and angled toward each other, they can form a liquid chain consisting of a series of loops with relatively thick rims. Successive loops are perpendicular to each other, making the arrangement resemble the links on a common type of chain.

Answer The water is held together by mutual attraction of the water molecules in an effect called surface tension. A sheet is curved downward by the gravitation force on it. If the sheet leaves the deflecting object with too much turbulence, it is unstable and quickly breaks up. With less turbulence, a closed water bell is possible—and it is beautiful.

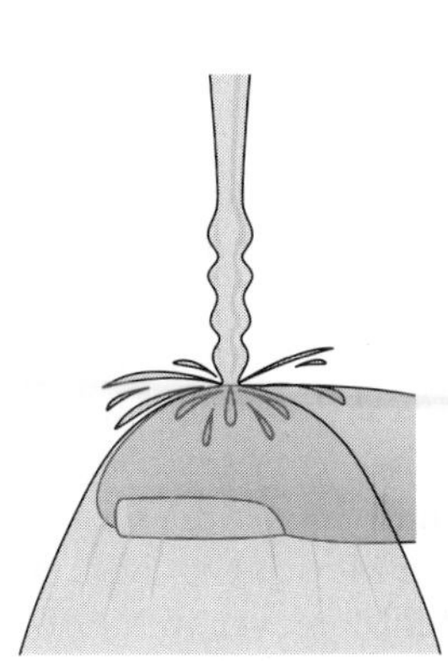

Figure 2-30 / *Item 2.100* Stationary waves form on a thin stream of falling water.

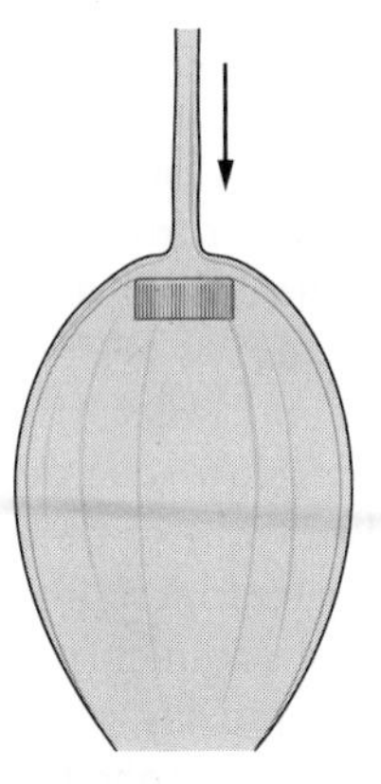

Figure 2-31 / *Item 2.101* Water bell formed by sheet of water deflected by a solid surface.

Some water sculptures use a broad, thin stream instead of a cylindrical water stream to create a water curtain that shoots off an edge and then folds downward. If the water flow is too low, however, a regularly spaced series of water columns will form instead of a single water sheet. The spacing is set by the surface tension of the water, which pulls the water into the columns.

Thin sheets of flowing water can also be made if two cylindrical streams are directed toward each other with approximately the same speed, with the outlets separated by a short distance. The impact of the streams shoots the water out to the side as a sheet. If the streams are vertical and directed toward each other, the sheet tends to be circular, breaking up into drops along its perimeter. If the streams are angled to each other, the sheet tends to be leaf-shaped.

If the streams are angled to produce liquid chains, the streams bounce from their impact and head away from each other with a thin sheet linking them. The surface tension pulls the streams back until they collide again. This time they head off in different directions but in a plane that is perpendicular to the plane of the first link. The bounce and the width of the chains decrease as the water continues to fall, until the links disappear and the falling water simply forms a cylinder.

2.102 • Stepping on a wet beach and into quicksand

If you step onto wet sand (not so wet that the grains swirl) and then lift your foot, why is the sand within your footprint relatively dry and why does it become wet again within a few minutes?

What causes quicksand, and how can you escape from it?

Answer Before you step on the sand, the grains are about as closely packed as they can get, and water fills the intermediate spaces. The sand looks wet because you see reflections off the water along the sand surface. When you step on the sand, you shear the sand by making parts move over or across other parts. This motion must increase the spacing between the grains. (The sand is said to be *dilatant* because shearing increases its volume from the initially closely packed state.) Water soon drains from the sand surface into the increased space between grains, leaving the sand surface relatively dry. Within a few minutes, either the grains slide back into closer packing or additional water comes in from the surrounding or underlying sand, and then the sand surface looks wet again.

If you had a squeezable bottle of sand and water, you could collapse the bottle a bit with a gentle, slow squeeze, which would allow the grains to move slowly out of their closely packed arrangement and also allow water to seep into the new spaces to lubricate the grains. However, an abrupt squeeze would attempt to move the grains too quickly, without the necessary water lubrication. The friction among the grains would be so great that you could not collapse the bottle at all.

Quicksand is a sand bed with a water influx, such as from a natural spring. The influx moves the grains apart somewhat and lubricates them so that they can slide over one another. If you step onto this arrangement, you can sink into the lubricated sand. If you struggle by trying to move your leg upward quickly, the quicksand suddenly becomes rigid and you cannot move the leg at all. The trouble is that sudden motion tends to increase the spacing between grains, but the sliding of grains against grains produces a lot of friction, preventing movement.

Quicksand is a dense fluid and, in principle, you may not sink in enough to drown in it. In an idealized situation, you might be able to lie down on it by bending at the waist, and then you could crawl by thrusting your hands along the surface, slowly extracting your legs. However, people experienced with quicksand point out that quicksand in wild country is far more dangerous than this idealized quicksand. It is likely to be hidden from view under standing or running water, and so even if you don't sink far into the quicksand, you could easily find your head below water. Also, because you fall into quicksand, you overshoot your floating level but, unlike in a swimming pool, you won't bob back up. Worse, you might divert the water flow that makes the sand *quick*, and then the sand around you becomes firm.

Experts advise that the only sure way to escape quicksand is to be prepared for an escape. When quicksand is a possibility, a person should have a bowline tied under the arms and around the chest, and someone at the other end of the rope should be ready to pull hard should the first person fall into quicksand.

2.103 • Collapse of buildings and a freeway

Just as the third game of the 1989 World Series was about to begin in Oakland, California, seismic waves from a magnitude 7.1 earthquake near Loma Prieta, 100 kilometers distant, hit the area, causing extensive damage and killing 67 people. Photographs relayed around the world showed a long stretch of the Nimitz Freeway, where an upper deck collapsed onto a lower deck, trapping motorists and causing dozens of deaths. Obviously, the collapse was due to violent shaking by the seismic waves. But why was that particular stretch of the freeway so severely damaged when the rest of the freeway, almost identical in construction, escaped collapse?

On September 19, 1995, seismic waves from an earthquake that originated along the west coast of Mexico caused terrible and widespread damage in Mexico City, about 400 kilometers from the origin. Why did the seismic waves cause such extensive damage in Mexico City but relatively little damage on the way there? Moreover, in Mexico City, why were buildings of intermediate height knocked down by the seismic waves but taller and shorter buildings largely left unharmed?

Answer The Nimitz Freeway collapse was confined to the stretch built on a loosely structured mudfill, which underwent *liquefaction* (or *fluidization*) during the shaking. That is, when the particles in the mudfill were shaken, they moved away from one another and became more fluid (they could flow) than solid. With the mudfill in a fluid state, the seismic waves had a much greater effect than in the surrounding regions where the freeway was anchored in rock deposits. One measure of the severity of seismic waves is the maximum speed they give particles as the particles are made to oscillate by the waves. In the mudfill regions, the maximum speed was at least five times what it was in the rock-deposit regions, and so the freeway was literally shaken until the decks collapsed on one another.

In some examples of liquefaction, houses have slid into the ground, as if caught up in quicksand. Also, geysers can develop where water and sand shoot up out of the ground.

The Mexico earthquake was a major earthquake (8.1 on the Richter scale), but the seismic waves from it should have been too weak to cause extensive damage when they finally reached Mexico City. However, Mexico City is largely built on an ancient lake bed, where the soil is still soft with water. Although the amplitude of the seismic waves was weak in the firmer ground en route to Mexico City, the amplitude substantially increased in the loose soil of the city. Also, when seismic waves entered the loose soil, some of them reflected between the top of the soil and the underlying firm material (*basement*). The waves with certain wavelengths reinforced one another, which increased and prolonged the motion of the ground. Accelerations due to the waves were as much as 0.20 *g*s (0.20 times the acceleration of gravity), and the frequency was (surprisingly) concentrated around 0.5 hertz. Not only was the ground severely oscillated for a surprisingly long time, but also many of the buildings with intermediate height had natural oscillating frequencies (*resonant frequencies*) of about 0.5 hertz. Many of these intermediate-height buildings collapsed during the shaking, while shorter buildings (with higher resonant frequencies) and taller buildings (with lower resonant frequencies) remained standing.

SHORT STORY

2.104 • Quicksand effect with grain

Falling into a large container of grain, such as a commercial grain bin or holding tank, is dangerous and can be lethal. In one case, a worker accidentally fell into a grain bin filled to a depth of several meters. He quickly sank to his armpits and was unable to free even his arms. Because he had a heart ailment, the pressure on his chest from the grain was threatening, and so rescuers immediately tried to pull him out, but they could not overcome the friction on his body. They then tried to dig him out, but grain continued to fill in around him and the airborne dust choked both the victim and his rescuers. Finally they lowered a cylinder around him, pushed it into the grain, and used an industrial vacuum cleaner to pull grain out of the cylinder, freeing the victim.

2.105 • Pedestrian flow and escape panic

When the density of pedestrians on a walkway increases, what do the pedestrians do to avoid a chaotic and disruptive flow? When crowds attempt to escape an enclosure (such as a room, building, or stadium) during an emergency or try to gain entrance to an enclosure during great anticipation and excitement, why does the motion become clogged and potentially lethal?

Answer Pedestrian movement can be a form of granular flow or even fluid flow. If you want to study pedestrian flow, find an elevated point so that you can watch a large portion of the flow.

When the density of pedestrians is low, each person or group (say, a family) will usually choose the most direct route to their goal although the route will not necessarily be straight if it is restricted to sidewalks and crosswalks. For example, if people can walk across a field at a county fair to reach a cotton-candy stand, they probably will choose a straight route to the stand. As the density of people increases, the route becomes zig-zag and the cotton-candy lover must occasionally pause to avoid colliding with other people. As the density increases further, people begin to organize flow lanes that resemble traffic lanes on a two-way street. As such *banding* or *lane-organizing* begins, people move in a coordinated way at a certain speed and with a certain body separation to avoid colliding. Now the cotton-candy lover must move in one or more flow lanes to get near the stand, and the overall route may be much longer than the original straight route.

When many clear-thinking people attempt to escape an enclosure through a fairly narrow exit path, they usually allow for one another to move into the path in a slow but continuous fashion. If those people are in a panic because they are trying to escape a danger (such as fire), they become jammed, forming arches around the exit. The pressure from people behind the arches can be so great that people within the arches cannot move in any direction, cannot even lift their arms, and may not be able to breathe properly, which leads to fainting while still standing up. In extreme situations, people in the arches can be lethally crushed against walls or barriers, or the walls and barriers can give way, allowing the people to fall to their death. The slow leak of people into the exit path relaxes the arches, but the escape time is greater than if the people exited smoothly.

If the onrush of people toward and into an exit causes some of them to fall, the bodies act as a trip obstacle for the oncoming people. The stack of bodies can become high enough to act as a barrier, obviously a dangerous situation.

To reduce the danger to exiting crowds, additional exits can be built. However, when a panicked crowd gets jammed up at an exit, the people may not realize that other exits are still open. Some stadiums are now designed with special exit corridors to reduce the possibility of jamming: These corridors flare toward the outside and are laid out as zig-zags so that no one can be trapped against a wall.

2.106 • Sandpiles and self-organizing flow

On a horizontal surface, slowly pour sand in a narrow stream to make a sandpile. Reasonably, the sandpile gets taller and wider. Why can't the side of the pile exceed a certain angle?

Slowly stir Tang (the orange grains used in an orange-flavored breakfast drink) and Nestea (the black grains used in making instant tea) and then slowly pour the mixture in a narrow stream. The grains form a pile just as sand does. However, the Nestea grains tend to gather at the bottom of the pile. Why the separation?

Slowly pour the stirred Tang–Nestea mixture into one end of a narrow transparent container as you watch through one of the transparent walls. Why do the grains form alternating bands of Tang and Nestea as the pile builds up (Fig. 2-32)?

Answer When the poured sand begins to form a cone, the grains on the slope lock together by their mutual friction and the slope angle gradually increases until it reaches a critical value. Then a few grains begin to slide down the slope, taking others with them until an avalanche occurs along part of the cone. Afterwards, the slope angle has a smaller value, said to be the *angle of repose*. Thus, the sandpile is said to organize itself—that is, adjust itself—to be at the angle of repose. Different powders and other granular materials (for example, glass beads, various seeds, dry couscous, and peas) have different angles of repose, depending on their average size and shape.

If two powders are stirred and then poured, they will probably separate somewhat as they run down the side of the pile and get locked into place. If the mixture is poured into a narrow container as described, you can see a cross section through the pile. Once avalanching begins, the larger grains tend to collect at the base of the slope and the smaller grains tend to get caught along the side, forming a layer. Then the larger grains begin to back up along the side until there is a layer of the larger grains. So, a series of avalanches alternates between forming a layer of the smaller grains (such as Tang in a Tang–Nestea mixture) and a layer of the larger grains.

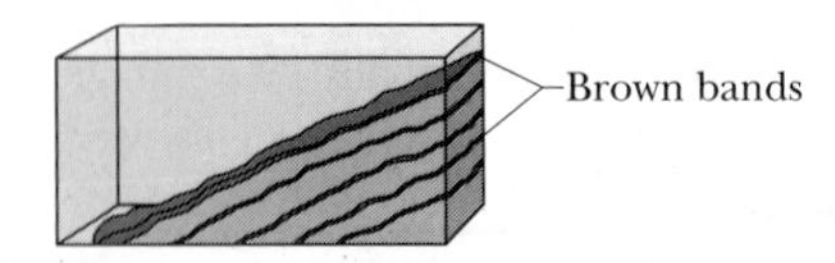

Figure 2-32 */ Item 2.106* Brown and orange bands form when a powder mixture is poured into a pile.

A vertical wall of granular material will both avalanche and slump. To see this, stand an open-ended tube upright on a table and fill it with a granular material. Then suddenly pull the tube upward to clear it from the grains. You'll have to look quickly or use a slow-motion camera to record the action. The column collapses in about half a second, but the way in which it collapses depends on the column's ratio of height to width. For a large ratio, the entire upper surface immediately moves outward, leaving a mound with a rounded point. For a small ratio, the outer part of the pile slumps and then the inner portion follows, leaving a mound with a sharper point.

2.107 • Flows in hourglasses and silos

If you place an hourglass on a sensitive balance to measure its weight, does the reading depend on whether the sand is flowing? Sand flows much like water. Why, then, don't we have water hourglasses?

When granular material, such as sand grains or glass beads, is put on a steep-enough ramp, it flows down the ramp. If the ramp is relatively rough and the granular material consists of particles of different sizes, why can the *front* (the bottom edge of the flow) break up into *fingers* that extend down the ramp?

Suppose that granular material flows down a ramp or chute to a barrier that stops it. If the supply of material is steady, why does the flow periodically turn on and off?

Answer Here is the traditional hourglass answer. The weight of the hourglass changes slightly when the sand flow begins and when it ends but is normal during the rest of the flow. (1) When the flow begins and before grains hit the bottom, the weight is less than normal because the grains in free fall do not contribute to the weight. (2) Once the grains begin to hit, their impact balances the weight loss of the freely falling grains, and thus for most of the time the weight is normal. (3) When the flow is ending and grains still hit but few are left in free fall, the weight is greater than normal.

Here are a few complicating details. (1) The grains actually begin to move before they reach the neck, and they have some initial speed when they begin to fall freely. So, their speed just before collision is larger than assumed in the traditional answer. (2) The sand piles up, so the point of collision shifts upward. The shift changes the severity of the collisions and also decreases the amount of sand falling at any given instant. (3) The flow may not be smooth for several reasons. A very narrow neck may allow grains to jam together to form momentary arches in or just above the neck. The arches may drag air into the lower chamber, increasing the air pressure there until air escapes back up through the neck, *slightly* disrupting the sand flow. This disruption can be regular enough that the hourglass is said to "tick."

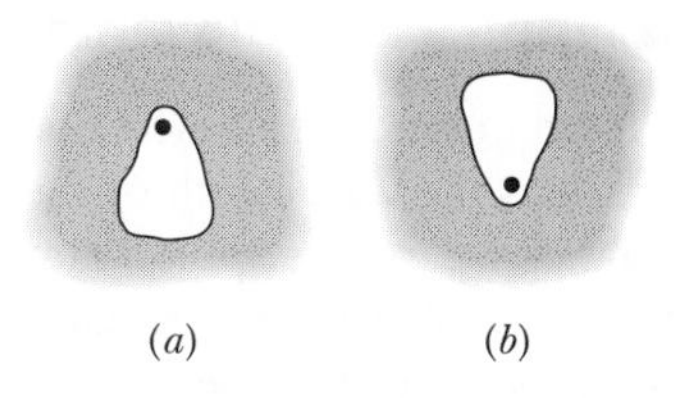

***Figure 2-33** / Item 2.107* Avalanche pattern in (*a*) a thin layer and (*b*) a thicker layer of granular material.

Arches can also form in the flow of grain in a silo. In some cases, the intermittent flow causes a silo to oscillate in what is called *silo quaking*. If large enough, the oscillation can produce sound known as *silo honking* and can rupture the silo, causing its collapse.

In a water hourglass, the rate at which water flows from the upper chamber depends on the height of the water in that chamber—greater height means a greater flow rate. In a sand hourglass, with the sand grains forming momentary arches, the flow from the upper chamber is independent of the sand height in that chamber—as long as sand is flowing, the flow rate is the same.

When a mixture of particles flows down a ramp, the courser particles tend to circulate in the flow in two ways. In a vertical cross section, they tend to circulate to the surface and to the front and down the ramp. As the flow moves past them, they are caught up and then return to the surface to repeat the cycle. In the meantime, these circulating paths vary across the width of the flow such that some are straight down and straight back up the ramp and some are bent left or right as we look at them. The former form the extended fingers down the ramp, while the latter form the space between the fingers.

If granular flow on a ramp hits a barrier and stops, the material begins to stack up and the stopping moves up the ramp until all the material on the ramp is stationary. Then, since the supply is steady, new material begins to flow down this "ramp" of stationary material, to repeat the cycle.

If a ramp is rough enough that granular flow leaves residual material on the ramp, you can increase the angle of the ramp a bit and cause avalanches wherever you disturb the remaining layer. If the layer is shallow, the disturbance (say, a poke from a pencil) produces a tear-shaped avalanche region that extends down the slope (Fig. 2-33*a*). If, instead, the layer is thicker, the disturbance produces a tear-shaped avalanche that extends *up* the slope (Fig. 2-33*b*).

2.108 • Brazil-nut effect and oscillating powders

Place a Brazil nut (or any other large nut) in a container and add enough peanuts (or any other small nut) to fill the container about halfway. If you shake the container vertically for a while, why does the Brazil nut come to the top surface of the peanuts?

In a large container of dried beans, bury a Ping-Pong ball and put a lead ball on top of the beans. If you swirl the beans by rotating the container horizontally, the lead ball drops out of sight while the Ping-Pong ball pops up. Why the difference?

Similar action can be seen when flour and fat are mixed during the preparation of pastry. In order to reveal the remaining lumps of fat, the cook might shake the mixing bowl—the lumps rise to the surface of the flour. Native Australian women once used a similar procedure (*yandying*) to separate edible grass seeds from the initial mixture with dust. A woman would tap or gently shake a flat container holding the mixture until the seeds formed one pile and the dust formed another.

Pour some Tang (the orange grains used in an orange-flavored breakfast drink) and some Nestea (the black grains used in making instant tea) into a transparent container, close the container, and shake the mixture. However hard you try to evenly and randomly disperse the Tang within the Nestea, you will always find orange islands of Tang in the Nestea. Why won't the Tang disperse?

Answer Two factors can cause a Brazil nut to climb up through a container of peanuts. One factor is that when the nuts are tossed vertically, the peanuts, being smaller than the Brazil nut, are likely to topple into the space below the Brazil nut and leave it more elevated with each toss. For the Brazil nut to descend, many of the smaller nuts would have to move out of the way, an improbable event. So, probability promotes the climb of the Brazil nut to the top surface. The Brazil nut might make the climb even if it is in a collection of objects with a somewhat smaller density.

The second factor for the climb is a circulation of peanuts caused by the vertical oscillations. The peanuts in the center of the container tend to move upward and those near the container wall, whose motion is retarded by the wall, tend to be forced downward by the central upflow. The Brazil nut can be caught in the circulation and brought to the top near the center.

The situation with the bean demonstration is different in two ways: The lead ball is *much* denser than the beans, and the friction between the beans is less than that between the peanuts due to a slick surface. If you swirl or oscillate the container, the lead ball burrows into the beans, readily sliding them out of the way. If you could somehow reduce the friction between the beans even more, they would behave more like a liquid and there would be no surprise that the lead ball sinks and the Ping-Pong ball bobs to the surface.

If you have a mixture of two materials, one material much larger than the other, you might find that the larger material gradually rises above the smaller material when the mixture is occasionally disturbed but not vibrated. This separation of materials commonly happens in packages of food products with two (or more) materials of different average size. And it drives producers of those products nutty because they want an even distribution of the materials when the package is opened. Each chance disturbance during the production, shipping, and purchase of the package allows the smaller material to shift into the spaces below the larger materials.

If you shake a mixture of Tang and Nestea, the smaller Tang grains tend to remain above the larger Nestea grains. Two mechanisms might be responsible for the Tang's inability to disperse. One mechanism is the same as that responsible for the climb of a Brazil nut to the top of some peanuts. The other comes from circulation cells that are set up in the mixture by the vibrations and the way the resulting forces on the grains are conveyed into the bulk of the mixture. The first mechanism seems to dominate if the vibrations have large amplitudes (vigorously shake the container up and down). The second mechanism seems to dominate if the vibrations have small amplitudes (strike the container with a fingernail hard enough to move the grains).

If a layer of Tang is spread over a vertically vibrating plate, the grains tend to collect into mounds. Once started, smaller mounds tend to move toward larger mounds, which they join. The migration appears to be due to the relative oscillations of the plate under a large mound (smaller oscillations) and a small mound (larger oscillations). The larger oscillations tend to throw more grains up into the air. The grains landing closer to the larger mound tend to stay in place more than the grains landing farther from the larger mound. Thus, there is a net motion of the grains toward a larger mound.

Here's a remaining puzzle: Half fill a fairly small cylindrical jar with salt and add a metal hex nut and a pushpin (the bulletin-board pin that comes with a plastic head). If you hold the cylinder upright while shaking it vertically, the nut goes to the top of the salt while the pushpin burrows out of sight. If, instead, you hold the cylinder horizontally while also shaking it horizontally, the pushpin goes to the top of the salt while the nut burrows out of sight.

2.109 • Avalanche balloon

An *avalanche balloon* is used by some skiers caught in a snow avalanche. The balloon, which is carried in a backpack, is initially uninflated. If an avalanche approaches, the skier pulls on a rip-cord device to inflate the balloon with nitrogen gas from a cylinder; the nitrogen flow draws in air from outside the balloon. As the skier and balloon are caught up in the avalanche, they tend to move to the top of the flow rather than be buried down within the flow. Thus the skier has a better chance of survival. Why does the skier move to the top of the flow?

Answer With an avalanche balloon attached to the back, a skier caught in flowing snow is effectively a Brazil nut caught in a shaken container of peanuts. The balloon provides an upward buoyancy force, because the gas it holds is less dense than the flowing snow. However, that force is insufficient to lift the skier to the top of the snow. Instead, the skier is lifted by the increased volume due to the balloon: The Brazil nut (skier and inflated balloon) is much larger than the surrounding peanuts (clumps of snow).

2.110 • Sand ripples and movement

Why can ripples form in sand on a desert floor (or a stream bed)? What determines the wavelength of the ripples—that is, the average distance separating ripples? How does a plant, such as a clump of grass, alter the ripple pattern? Why don't ripples commonly appear on snow beds?

Answer When the speed is high enough, the wind across an initially flat bed of (dry) sand can move the sand grains. A grain can *creep* along the bed or it can undergo *saltation*, in which it jumps and possibly bounces. If a grain lands on a flat portion of the bed, it jumps again, but if it lands on a raised portion of the bed (some bump that occurs by chance), the grain may get stuck (Fig. 2-34). As the bump grows taller, it collects more grains and it also shields the grains on the leeward side. Grains somewhat farther downwind, however, can still be made to jump and can thus get stuck on another bump. The bump can grow with a steep leeward side and a more gently sloped windward side. Wind deflecting over the crest tends to break into a vortex on the leeward side, which sends air *up* the leeward side, carving out sand to maintain the steep front. As the bump forms, it also moves downwind because grains on the windward side can hop over the crest. Some bumps move faster than others, and so many of the bumps merge or at least get close enough to affect one another.

Now move ahead by days, weeks, or perhaps years. This activity gradually results in the rippled patterns we see on the sand. Once established, the ripples are maintained by the wind and the saltation. Of course, if the wind changed dramatically, the pattern would be replaced by a new one.

The pattern formation can go much faster if the sand is submerged in a stream with a fairly regular flow. You might then be able to actually see the ripples form within a few minutes.

As wind whips around vegetation, with vortexes (or small whirlwinds) being formed, the orientation and spacing of the ripples vary from that existing on the windward side of the vegetation.

Saltation can also occur with snowflakes in a field of snow. However, ripples don't appear (or at least, are not prominent or common) for two reasons. (1) A flake tends to stick to any point it hits, regardless of whether it is a raised portion of the snow. (2) The snow field tends to form a frozen crust, especially after a sunny day, and then saltation cannot occur. However, before the crust forms, patterns can be left in the snow by strong winds, especially if they form whirlwinds on the leeward side of an obstacle.

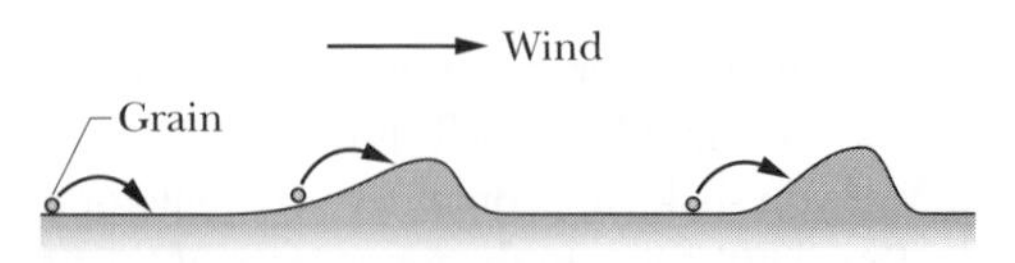

Figure 2-34 / *Item 2.110* Saltation of sand grains produces sand ripples.

2.111 • Sand dunes

Why do sand dunes form? Why can they move? When one sand dune runs into another, how can they merge and then separate? How can one sand dune cause another to split in order to provide a passageway for the first one? Why are the dunes on some deserts, such as in Libya, aligned in roughly parallel lines, as seen in satellite photographs?

Answer *Longitudinal dunes* are long mounds aligned with the prevailing wind. *Barchan dunes* are crescent shaped and perpendicular to the wind, with *horns* (tips) that point in the direction of the wind. All dunes result from humble sand ripples and are due to the ability of wind to toss sand grains. Over many years, perhaps centuries, the grains are caught up in ever growing mounds. Once a mound is established, it is so stingy with the grains that competing mounds die out or never really get started.

When you think of a sand dune, you probably imagine a barchan (moving, crescent-shaped) dune, although that type is fairly rare. A barchan dune gradually marches over a desert (or across roads and through small villages) because sand grains on the windward side are kicked up by the wind and thrown onto the leeward side. The leeward side eventually becomes too steep, and then avalanches bring sand down the slope to the base, decreasing the slope angle to stabilize the slope. On a time scale of years, the dune moves in the general direction of the wind.

Aerial photographs of some barchan dune fields show many of the dunes in a peculiar alignment, with the tip of one dune positioned in front of the midsection of another dune (Fig. 2-35*a*). Such an arrangement is due to how the wind is modified by the upwind dune. Given enough time, that wind modification can reshape a downwind dune to give this tip-midsection arrangement.

Watching a moving sand dune overtake another is almost impossible within a human lifetime, but the process can occur within minutes if miniature dunes are created in a flowing water stream. The interactions between two dunes are due to the flow alteration caused by the upstream dune.

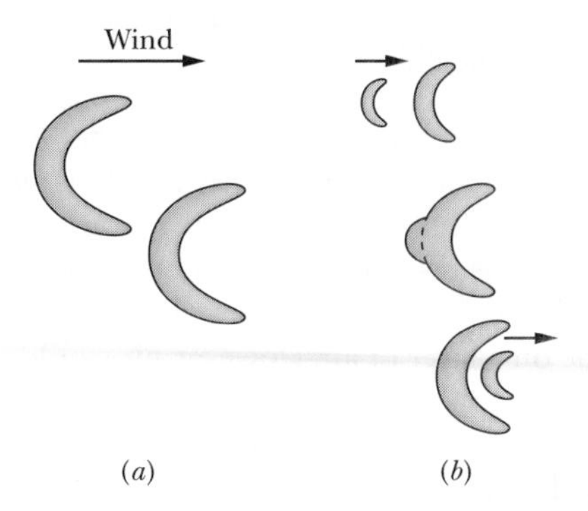

Figure 2-35 / *Item 2.111* (*a*) Overhead view of two barchan dunes. (*b*) Illusion of small dune passing through larger dune.

The flow can then erode the center of the downstream dune to (seemingly) provide a passageway. Instead, a small upstream dune might blend into a larger downstream dune, and then another small dune might emerge from the rear of the dune (Fig. 2-35*b*). Overall, the process gives the illusion that a small dune has passed through a larger dune.

The roughly parallel arrangement of longitudinal dunes is due to a vortex formation called *Langmuir circulation*. As wind moves over a flat expanse, it tends to break up into horizontal vortex tubes. If you looked along one of the tubes in the direction of the wind (with the air somehow marked with a smoke tracer), the air would gradually circulate either clockwise or counterclockwise as it moved away from you. Adjacent tubes have opposite circulation directions. Suppose we look along a clockwise tube; the airflow across the ground is leftward. The adjacent tube at the left has a counterclockwise flow and thus a ground flow that is rightward. Because these two ground flows converge, they tend to transport sand to the convergence point—that is where a longitudinal dune tends to form. On the other side of our tube, the ground flows do not converge and thus no dune tends to form. Because the tubes are roughly straight, the longitudinal dunes form along roughly straight lines that are separated by twice a tube's width.

2.112 • Yardangs and other sand cuttings

On a sandy beach, why are most of the stones leaning into the prevailing wind, and why does a sand ridge extend from many of the stones? On some beaches you can see turret-like sand structures: columns of moist sand extending from the general surface level, usually with different layers of sand evident. They look as though castle-like turrets have been cut out of a multilayered cake. The surrounding sand is usually dry.

Some of the most beautiful and eerie formations in the world can be found in sand deserts: *Yardangs* are rock formations that jut up from the sand, resembling inverted ship hulls. Some are hand size; others are hundreds of meters long. They occur on Earth and can also be found on Mars. How do yardangs form?

Answer When wind blows past a beach stone that stands at least somewhat upward on dry sand, the wind digs out a pit in front of the stone and drops some of that sand to make a ridge behind the stone. Eventually, the stone leans toward the pit and thus into the wind. If the stone is flat on top, the wind digs out the sand around the stone, leaving the stone on a pedestal. Then the wind digs out sand at the front of the pedestal until the stone leans forward into the wind. None of this digging and tilting by the wind happens if the sand is wet because the *liquid bridges* connecting the grains with one another hold the grains too tightly for them to be dislodged easily. However, wind can gradually dry the outer sand, and the grains can then be whisked away or sandblasted by airborne grains. Of course, water currents, such as wave swash, can erode the sand around stones.

The turrets occur in sand that is cohesive because of water from either precipitation, spray, or seepage from somewhat lower sand. The supply from the latter two sources might occur only in isolated regions. Those regions get wet, and the sand becomes cohesive. The regions may be covered with dry sand for a while, but eventually they are exposed by wind and sandblasted by airborne sand. The results are individual turrets or mounds.

Yardangs are rock products of wind and sand erosion. The wind carries off sand grains, exposing the rock, and then sandblasts the rock with streams of sand grains. The rock gradually is whittled into a narrow structure aligned with the prevailing wind. Many yardangs resemble a reclining cat. Indeed, the Sphinx may have been inspired by the yardangs that ancient Egyptians found in the deserts flanking the Nile River.

2.113 • Snow fences and wind deposits

A snow fence is a fence or line of vegetation placed to keep roads or train tracks relatively free of snow drifts. Where should such a barrier be placed? Won't a solid wall be more effective than a fence with open spaces? Or is a fence chosen because it is less expensive? How does snow accumulate around an obstruction such as a boulder or a tree trunk? In particular, what determines the holes or circular snow-free regions that can be found around tree trunks?

Answer The purpose of a snow fence is to force wind to dump its snow before it reaches, say, a road, so it should be erected well upwind of the road. A solid wall is not as effective because it causes the wind to swirl up into the air, which keeps the snow airborne. A fence with about half its area open to the wind and at least a few centimeters above the ground is far better.

Initially, a snow fence causes small swirls on its front and back sides, which keeps the two sides relatively snow-free. You can see this effect early in a snow season. Mounds of snow form on both sides of the fence, but the regions next to the fence have much less snow. The gap below a fence allows some of wind to contribute to the swirling on the back side of the fence, which clears that side.

As the snow accumulation increases (*matures*), the top of the snow pile on either side of the fence is extended toward the fence until it touches it. Then you see a continuous mound of snow with its high point at the location of the fence, but what you don't see is the hollow that is left on each side of the fence. Once the continuous mound forms, the fence is no longer useful.

As wind is deflected around a boulder (or even smaller rocks and other objects), it tends to dump snow at the rear while scooping out snow at the front and along the sides.

The circular depressions or snow-free rings around a tree trunk have two causes: The tree is an obstacle to the wind, which tends to swirl around the trunk. That motion carves out snow that falls next to the trunk. During the day, the tree tends to warm by absorbing infrared radiation in the sunlight. The limbs and trunk then reradiate some of that energy toward the ground. Because snow is an excellent infrared absorber, it absorbs nearly all of what it intercepts and tends to slump as it melts.

2.114 • Snow avalanches

Once a snow avalanche is triggered, how does the snow travel down a slope and how can it be stopped to prevent damage to a village near the bottom of the slope?

Answer Once an avalanche begins, the snow, especially powder snow, becomes airborne and advances as a turbulent cloud of particles. It quickly begins to trap air (*entrain* air) in the flow, diluting the concentration of snow particles; it also picks up more snow from the slope. The speed of the particles is greatest somewhat above the slope level, well below the top of the avalanche, but the particles move along complicated and ever changing paths instead of coming directly down the slope. The front of an avalanche might move as quickly as 100 meters per second; the height of an avalanche might be as much as 100 meters.

Tall walls, with sturdy back-supports, are positioned along the slope to stop an avalanche, but energy must be drained out of an avalanche before it gets to the walls. To do this, mounds are constructed uphill from the walls. The purpose of a mound is to deflect the snow up into the air, like a ski-jump ramp deflects a jumper. When the snow crashes into the slope somewhat below the mound, it loses much of its energy.

2.115 • Long-runout landslides

When a rock slope gives way, generating a large landslide, the debris can move down even a moderate slope and then spread for several kilometers along a flat valley, in a motion called *runout* or *sturzstrom*. In fact, it can move across a valley and up the slope on the opposite side. Why doesn't the friction between the debris and the valley floor or slope walls quickly stop the landslide?

Answer Most experts agree that landslides exceeding a certain volume of material have surprisingly large runouts because they ride on some type of a lubricating layer. However, there have been many suggestions about what comprises that lubricating layer. One popular idea was that the material rode on a layer of air. However, air would seemingly be driven out of the material fairly rapidly, and this mechanism would not explain the long-runout landslides that have occurred on the Moon.

Another idea was that pressure waves within the material would lift the material off the ground, thereby reducing the friction between the material and the ground. However, experimental evidence has not supported the idea.

Perhaps the most promising explanation is that the material moves over a thin layer of small oscillating debris, including material that is scraped off the ground. This oscillating debris acts somewhat like balls in a bearing, explaining two observed features of landslides: (1) Much of the material comes down the slope fairly well intact, with its original layering. (2) The material scraped off the ground can include water, which can then lubricate the sliding and allow a farther runout.

2.116 • Rockfalls

A rockfall is where a rock or a collection of rocks fall from the side of a mountain, usually the face of a cliff. Why do they finally fall, and what determines where the rock ends up? When a rockfall involves many rocks coming down a slope, why do the rocks tend to be sorted, with the larger ones at the bottom of the slope and the smaller ones higher up?

In July 1996, two successive rockfalls of huge granite chunks occurred near the Happy Isles Nature Center in Yosemite National Park, California. Each chunk slid down a steep slope and then went into projectile motion, hitting the ground after falling about 550 meters. The impacts produced seismic waves that were recorded as far away as 200 kilometers. More surprising, however, was the damage the chunks produced farther down the valley, up to 300 meters from where they landed: Over 1000 trees were knocked down or broken, a bridge and a snack bar were demolished, one person was killed, and several others were hurt. How could the granite chunks do so much damage 300 meters away?

Answer Most rockfalls are due to weathering of bedrock: (1) Fissures that collect water can be widened and extended when the water freezes, because the water expands. (2) The rock is weakened by chemical weathering, especially when moisture is available. Although any rock can undergo weathering, rockfalls occur only when the bedrock face is steep and when a section continues to be supported while fissures are driven into the face. At some point, the section's attachment or support is overwhelmed, and the rock breaks free.

Depending on circumstances, the freed rock can fall through the air, bounce down a fairly steep slope, tumble down a moderate slope, or slide down a shallow slope. It can also shatter into smaller rocks. With any of these outcomes, it loses much of its energy in the collision. It can also lose energy if it collides with trees, and thus a stand of trees is often grown as a rockslide barrier.

When a rockfall involves many rocks of various sizes, the rocks can become segregated along a slope because the larger rocks reach the bottom of the slope whereas smaller rocks tend to catch in the low points (the crannies) along the slope. In general, material along a slope tends to vary from fine grade at the top to coarse grade at the bottom. The foremost rock in some rockfalls ends up at a surprisingly large distance from the other rocks, presumably because it gains energy as other rocks hit it from the rear during the sliding and rolling portion of the fall.

In the rockfalls at Happy Isles Nature Center, the impact of each granite chunk produced an *air blast*, which is a pressure wave moving through the air from the impact point. The air blast from the second chunk, with about three times the mass of the first chunk, was especially destructive, creating winds up to 120 meters per second through the trees. In fact, the air blast from the second chunk was supersonic (it was a shock wave) because dust stirred up by the first impact reduced the speed of sound in the air from its normal value of 340 meters per second to about 220 meters per second. Near the impact, the air blast moved faster than 220 meters per second, thus faster than sound.

2.117 • Fluttering flags and ribbons

What causes a flag on a flagpole to flap even in a moderate breeze? Why does a sheet of paper flutter when held in front of a fan?

If you throw a roll of partially unrolled toilet paper into the air, why does the trailing (unrolled) section quickly form a wave-like appearance?

Answer Imagine that the plane of the flag is at an angle to the direction of the passing air so that the air pushes against one side of the flag. That push can simply straighten out the flag, making it extend in the direction of the airflow. Instead, the push can bend the flag. If the airspeed is above a certain critical value, this bending can become unstable and the flag will then flutter.

The fluttering has often been attributed to vortex formation by the flag. Indeed, whether the wind merely extends the flag or causes it to flutter, the free end of the flag *sheds vortexes*; that is, a series of vortexes, alternating on the left and right side of the flag, form and then move downstream (Fig. 2-36). The vortexes are larger if the flag flutters, but they are a product of, not the cause of, the fluttering and they can be present even if the flag is not fluttering.

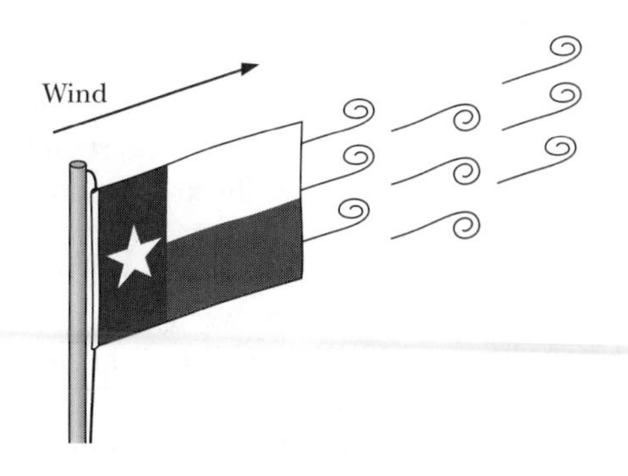

Figure 2-36 / *Item 2.117* Vortex formations alternate between left and right sides of a flag.

Other flexible sheets, such as paper, can also be made to flutter in a steady breeze or wind if the airspeed exceeds a critical value that depends on the type of sheet and its flexibility.

Any flexible ribbon that is weighted at one end and dropped will probably develop a wave-like appearance. The wave travels downward along the ribbon at half the speed at which the ribbon is falling, and usually the distance between bulges on the wave is greater for a longer ribbon. The wave appearance is likely due to the instability of the airstream that is forced to travel with the ribbon as it falls through otherwise stationary air. In simple terms, the captured airstream and the ribbon itself buckle as they fall.

2.118 • Fluttering fountains and pounding waterfalls

Many fountains have edges over which flowing water falls as a sheet into a pool. In some of these fountains, why does the sheet flutter at a frequency of a few times per second? (The frequency is a bit too low for you to hear sound waves produced by the fluttering.)

Larger and taller flows occur when overflowing water is allowed to pour through the exit conduits in a dam or over the top of a dam, to fall freely into a pool or stream. Why can they produce sound that is so intense that it is likened to some 20 high-speed trains?

A tall waterfall causes an appreciable oscillation of the surrounding ground, something that you feel with your body when standing near the waterfall. If you analyzed the shaking, you would find that the greatest amount of shaking occurs at a frequency related to the height at which water freely falls in the waterfall—greater height means smaller frequency. How does a waterfall cause the ground to oscillate, and why is the frequency related to the fall height?

Answer The key to a fluttering fountain is a layer of air trapped behind the sheet of falling water. When a chance disturbance causes a small oscillation of the water sheet at the top, the oscillation can grow as the water falls—that is, there is *gain.* At the bottom, the oscillation changes the air pressure in the layer of trapped air, which then pushes and pulls on the top of the water sheet—that is, there is *feedback* that reinforces the oscillation of the sheet. Although the effects are initially modest and perhaps even imperceptible, this gain–feedback loop builds them up until they are appreciable. However, the loop can occur only for certain flow rates, initial water speeds, and air-layer thicknesses, and so not all falling water sheets flutter.

The overflow from a dam sets up a similar oscillation in the water column and in the air between the column and the dam. The resulting motion of the water is known as *nappe oscillation* to hydraulic engineers. Dams are often constructed with *nappe interruptors* that spoil the smooth flow of the water, to prevent the oscillations by creating thicker curtains of falling water.

In a waterfall, the collision and turbulence of the water at the bottom of the fall produce sound waves inside the column of falling water. At certain frequencies, these sound waves set up *resonance* (that is, the waves reinforce one another) and then the oscillation can grow to be quite strong. The situation resembles the way sound waves set up resonance inside a tube of air having only one open end. The oscillation of the air molecules is zero at the closed end and maximum at the open end. The lowest resonant frequency, called the *fundamental frequency*, usually dominates what we hear leaking from the tube.

In the waterfall's column of water, the oscillation of the water molecules is zero at the top of the fall and maximum at the bottom, and the oscillations are greatest at the fundamental frequency. Sound waves at that frequency travel away from the falls through the air and through the ground. The higher the waterfall is, the lower the frequency is. For most eye-catching waterfalls, the frequency is too low for you to hear the sound waves sent through the air. However, you can feel those sound waves with your body and can feel the ground oscillate with your feet.

2.119 • Pulsating fountains

Many decorative water displays, both indoor and outdoor, have fountains that send jets of water straight up into the air. Why does such a vertical water jet pulsate even when the water flow is constant? You can both see and hear the pulsation.

Answer When the jet is first turned on, the water shoots to the maximum height allowed by gravity and its initial speed. From that maximum height, the top blob of water falls back onto the upcoming jet, flattening its top. Once this blob disintegrates into drops, the jet reappears and climbs back to the maximum height. The cycle of climbing, falling, and flattening then reoccurs. You can eliminate this periodic behavior if you tilt the jet so that its topmost water does not fall back on the jet or if you place an obstacle in the jet to prevent the water from reaching the maximum height.

2.120 • Pouring: inverted glass, yard-of-ale

Partially fill a drinking container (such as a normal drinking glass) that has straight sides and a wide opening. Cut a paper square somewhat larger than the opening and position the paper over the opening with one hand, spreading your fingers widely to press the paper against as much of the rim as possible. With your other hand, quickly invert the container; then remove your first hand from the paper. The gravitational force is pulling down on both paper and water; so, why don't they fall?

If you remove the paper or make it magically disappear, the water will pour out of the container. Why is this situation different from when the paper is in place? If the container is narrow (like a thin test tube), why can the water stay in place?

Figure 2-37 */ Item 2.120* A yard-of-ale.

A *yard-of-ale* is a quaint drinking glass (Fig. 2-37) that holds ale about a yard high. (Shorter versions with the same shape exist.) Normally you drink from a glass by holding it on your lips and tilting it upward to pour some of the liquid into your mouth in a controlled way. Why would using this procedure with a yard-of-ale leave you drenched? What is the proper way of drinking from a yard-of-ale?

Answer With the drinking container inverted and the paper in place, the column of water slumps, which decreases the pressure of the air trapped in the container. So, the pressure on the top of the column is less than the atmospheric pressure on the bottom of the column, and the pressure difference is enough to support the water. You can tell that the water slumps because of the visible downward bulge on the paper. The contact of water, paper, and rim may also provide some adhesion, but it is insufficient to support the water. If you repeat this demonstration with a rigid sheet, such as a glass plate instead of the paper, the column cannot slump and the water will just pour out.

If the paper were to disappear, the air pressure should still be able to support the water in the glass. That is, nothing in the support argument changes. However, the water is unstable to any chance disturbance sending waves along the air–water interface. The low points (the valleys) of any such wave tend to allow water to slump downward, and the high points (the crests) tend to allow air to move upward. One of the waves quickly grows, and air moves up one side of the glass while water moves down the other side.

The upward-moving air tends to pinch off at the bottom to form a bubble, and the downward-moving water tends to pinch off at the top to form a blob. As the bubble ascends and the blob descends, the air–water surface attempts to re-establish its initial flat state but it is just too unstable. So, a series of bubbles rise into the container as a series of blobs fall from it, in a process known by the onomatopoetic term *glug-glug.*

The water surface may not be sensitive to chance disturbances if the container is narrow because it is stabilized by surface tension. That is, the cling of the water at the surface to the sides of the container can overcome waves produced by a disturbance.

The glug-glug process is the reason why drinking from a yard-of-ale can be messy. The narrow neck on the glass prevents any easy passage of air up into the glass so that liquid can come out of the glass. Instead, instability along the liquid–air surface suddenly allows a large bubble of air to rise up through the neck and a large blob of liquid to pour out, too much liquid to be held in the mouth or quickly swallowed. Masters of the yard-of-ale know to tilt the glass while *rolling* the narrow neck between their two vertical hands. The rolling action swirls the liquid up against the neck walls and allows an air passage along the central axis of the container. Air can now easily move into the glass, and the drinker can control how much liquid comes out.

2.121 • Dripping

When water drips from a faucet, how does it break contact with the water remaining in the faucet? Is contact lost with a sudden snap or break, or with a long, drawn-out goodbye?

Answer Surface tension gathers a drip into a curved surface at the open end of the faucet. As the amount of water increases, the water sags into more of a spherical shape. When some critical point is reached, the gravitational force on the water causes it to descend rapidly. If you could watch this descent in slow motion, you would see that the falling drop is still connected to the water in the faucet by a cylindrical cord that rapidly thins. Suddenly the bottom of the cylinder forms an even thinner cord, which quickly breaks. Just after the break, the falling drop oscillates in shape and the remaining cylinder develops waves that rapidly grow into very small drops. These drops may pinch off from the cylinder and fall, or they may be pulled back up to the faucet.

A fluid with greater viscosity than water may develop many ever thinner cylinders, one after another, somewhat like the infinite series of cat-in-the-hat figures in a Dr. Seuss book, until the lowest and thinnest one breaks. Chance disturbance, such as air movement, faucet vibration, or even sound, probably causes the break.

2.122 • Soap bubble shapes

When you blow a bubble from a plastic loop, you inflate a soap film, forming a curved surface that moves away from the interior of the loop as it expands. The interior of the loop is then empty. Because the soap film is not a closed surface, how can a closed bubble form as the film breaks free of the loop?

Why is a freely floating bubble spherical? If a cylindrical film is formed between two nearby circular supports, what shape does the film take as the supports are gradually separated?

Answer As you inflate the soap film on a loop and a bubble forms, the portion of the bubble near the loop will suddenly develop a narrow waist and *pinch off* from the loop. As the waist collapses, the portion of the film on the loop flattens to fill the loop's interior and the portion that breaks free quickly becomes spherical. The bubble is spherical because the pressure must be the same everywhere inside the bubble. That pressure exceeds atmospheric pressure because surface tension tends to collapse the bubble. With the interior pressure the same everywhere, the curvature of the surfaces must be the same everywhere, and so the bubble is spherical.

If a bubble spans a short gap between two solid circular supports (paddles), the bubble forms a cylinder. The shape is stable because any chance disturbance automatically increases the surface area, and so surface tension pulls the film back into a cylinder. However, if the separation between the paddles exceeds their circumference, the bubble becomes unstable and then a chance disturbance will cause it to collapse into two smaller bubbles, one on each paddle.

You get the same results if you use two circular rings instead of paddles, provided that a film spans the interior of each ring to close off the ends. If one of those end films is ruptured, the pressure inside the bubble drops to atmospheric pressure, equalizing the pressures on the two sides of the bubble wall. This equalization means that the bubble's curvature must be zero, a condition that is met in a neat way: The bubble develops a waist and resembles an hourglass. Along a line extending from ring to ring, the curvature is concave (inward); along a line around the bubble, say at the waist, the curvature is convex (outward). Thus the overall curvature is zero, as required. This bubble is stable only if the ring–ring separation is much less than the circumference of the rings. If the separation is too large, the waist immediately narrows and pinches off, and then the remaining film on each ring shrinks to become flat.

2.123 • Bubble paths

A bubble released at the bottom of a tall cylinder of water should rise directly up the cylinder. In fact, small and large bubbles do exactly that. Why do bubbles of intermediate size take either a zig-zag or a helical path?

A bubble rising in water should be spherical, as small bubbles are. Why is a larger bubble flat on the bottom?

Answer The ascent of bubbles of intermediate size is still a subject of research. The zig-zag and helical motion appear to be due to vortexes that form on the underside of a bubble as the bubble shoves its way upward through the water. If the vortexes alternate left and right, the bubble can be deflected left and right. The vortex formations and the deflections appear to be associated with oscillations of the bubble.

If two rising bubbles get close enough, their motion can be altered by the flow of water past each other. They might tumble, be thrown apart, or be drawn together in a *kiss*.

The shape of any bubble is set by the surface tension of the bubble's surface and the fluid drag on the bubble as the bubble moves. For a small bubble, which has a tight curvature, the surface tension wins and the bubble is pulled into a sphere to minimize the surface area. In other words, the bubble has the least energy when it is spherical; any distortions increase the surface area and require extra energy. For a large bubble, which has less curvature, the fluid drag is important and the wake left by the rising bubble can flatten the bottom surface.

When a bubble is released in a very viscous fluid, such as syrup or various hair products, the bubble may take a long time to rise. Its shape can depend on how it is released. For example, if it is squeezed into the fluid, it may have a tail when it breaks free from its initial attachment and may retain that tail throughout its ascent.

2.124 • Antibubbles

Make a mixture of water and mild detergent in a container, and then suction some of it up into a squeezable bottle or a drop dispenser (such as the kind used to put a liquid medicine onto an eye or into a baby's mouth). Position the tip of the bottle or dispenser just above the liquid surface in the container, and squirt a small amount of the liquid through the surface. You produce normal bubbles and also some bubbles, dubbed *antibubbles*, that look and behave differently. What is an antibubble and why do they form?

Answer A normal bubble is a sphere of air with water and soap molecules along its surface. The soap molecules help stabilize the surface as they do in the soap bubbles blown in air by a child. Antibubbles consist of a thin spherical shell, with a sphere of water inside the shell and with water and soap molecules along both surfaces of the shell. Because an antibubble consists almost entirely of water, it lacks the buoyancy of a normal bubble and tends to hang in the water rather than ascend.

Here is another way of producing antibubbles. First make a cluster of three normal bubbles that touch one another as they float on water. Then allow a drop of water and detergent to fall into the dimple where the bubbles touch. An antibubble appears under the dimple.

In a normal soap bubble, fluid drains from top to bottom, thinning the top until it ruptures. In an antibubble, air in the shell rises to the top, thinning the bottom of the shell until the shell is breached by water. All this happens quickly, explaining why antibubbles last only briefly.

2.125 • Lifting rice with a rod

Firmly push a rod into a container of uncooked rice or some other grain. Why does the required force rapidly increase the deeper the rod moves, and why does it become very large as the rod nears the bottom of the container, where you may have to pound on the rod to move it even a little?

Once the rod is in place, tap or gently shake the container for a few minutes. If you then pull upward on the exposed end of the rod, why can you pick up the entire container of rice?

Answer These effects were first noticed long ago when merchants and customers would thrust poles into bags of grain to check the contents. As a rod is pushed into rice, for example, the friction on the rod increases for two reasons. (1) More grains press against the rod. (2) The deeper grains, which support the weight of higher grains, press harder. Thus, the total friction resisting the rod's motion increases with the rod's depth.

The friction increases even faster as the rod nears the bottom of the container. The effect is not well understood, but a reasonable guess is that great resistance to rearrangement of the grains comes from grain arches: Grains become locked into arches that resist the rod's motion. (Just as architectural arches can be strong, so can grain arches.)

When the container is tapped or shaken after the rod is inserted, the rice becomes closely packed. In particular, it is closely packed against the rod. When you pull up on the rod, the friction between the grains and the rod holds the rod tightly. Also, grains surrounding the rod are closely packed and hold tightly onto one another. And the grains next to the container wall hold tightly onto the wall. So, everything—rod, grains, and container—is locked together. However, if the rod or container is very slippery or if the rice is not tapped down in place, you will have a lot of rice to pick up from the floor.

If you were to pull hard enough to extract the rod in a slow, controlled way, you would probably find that the force on it from the rice varies in a periodic way. Although this variation is not well understood, it is probably due to the formation and collapse of arches near the rod.

2.126 • Throwing a discus

When a discus is thrown in a moderate wind, will it go farther if it is thrown against the wind or with it? At what angle should it be launched, and how should it be tilted? Why should it be spinning?

Answer The angle of the launch path is usually 35°. Some athletes argue that the plane of the discus should be tilted at the same angle, while others claim that the flight is longer if the tilt is smaller by about 10°. According to this claim, the smaller tilt means the discus has more lift during the descent stage and thus stays aloft longer. If the discus is launched without spin, it flutters and easily loses the desired orientation. A spinning discus is more stable, acting like a gyroscope in the sense that the axis around which it spins tends to point in approximately the same direction throughout the flight.

However, during flight, the orientation is not exactly constant because the drag force on the discus from the passing air is not uniformly distributed over the discus. Instead it is concentrated on the front and left side (assuming that the discus is thrown by a right-handed athlete with standard technique). Torques from the nonuniform drag rotate the front of the discus somewhat upward and the left side somewhat downward.

When a discus is thrown into a moderate wind, the pressure of the wind on the underside adds lift and gives a longer throw than in windless conditions. The advantage holds for wind speeds up to 15 or 20 meters per second; stronger winds upset the orientation and bring the discus down early. When the discus is thrown with the wind, the wind pushes on its top side, reducing lift and giving a shorter throw.

2.127 • Javelin throw

Throwing a javelin involves two important angles relative to the horizontal. One is the angle of the launch path; the other is the angle of the javelin shaft. To maximize the range of a throw, what values should the angles have? In particular, should they have the same value?

Answer Traditionally, the angles have both been 35°, a value wrung from countless trials that were plagued with many variables. Still, the traditional launch seemed reasonable. The initial alignment of the shaft with the path streamlined the launch. If the shaft were angled higher or lower, it would experience more air drag, and its flight would be reduced. However, some theoretical studies claim that the range lengthens if the launch-path angle is increased to 42° while the shaft angle is kept at 35°. Another study counters this claim with a practical point: When the launch angle is increased, the athlete probably cannot impart as large a launch speed because the launch is more awkward, and so the advantage is lost. Other studies find that the launch angle should be 32° and the shaft angle about 17°. The nose-down orientation of the shaft would certainly increase air drag, but it might also increase the lift (due to air pressure on the javelin's lower side) during the latter part of the flight. With more lift, the javelin would stay aloft longer.

A javelin normally noses over during flight and sticks in the ground. The rotation is due to forces on the shaft. The weight of the javelin is a force assigned to the *center of mass* (the center of the mass distribution). The lift is assigned to the *center of pressure* (the center of the pressure distribution), which is usually located behind the center of mass. During the flight, the lift rotates the javelin around its center of mass so that it sticks in the ground. Once rotated, the javelin is more streamlined with the passing air and no longer receives much lift. The range might be increased if the shaft is reshaped to bring the center of pressure closer to the center of mass. The relocation would reduce the rotation of the shaft and maintain lift during the descent.

2.128 • Two boats drawn together

When water flows past two adjacent boats facing upstream, why do the boats tend to pull together?

Answer When the water is forced into the confining space between the two boats, it increases in speed. The only

way it can get the energy required for the increased speed is to take it from the internal store of energy associated with pressure. As a result, the water pressure between the boats decreases. With normal pressure on the exterior sides of the boats and this decreased pressure on the interior sides, the boats are drawn together.

2.129 • Aerodynamics of cables and lines

A strong gust of wind can push any cable or electrical transmission line in the direction of the wind. Why do some cables and transmission lines *gallop* in a wind—that is, oscillate perpendicular to their length and to the direction of the wind? In some cases, this oscillation can short out adjacent lines, pull free a line, or pull down a support tower. The latter two results are more probable when a line also supports ice.

Cable oscillations plagued Le Point de Normandie, which was the world's longest cable-stayed bridge when it was opened in 1995. Although the cable galloping would not in itself cause the bridge to collapse, the motion would have prematurely aged the cables so that they would have required early replacement.

What causes the galloping of cables and lines?

Answer When wind blows past a cable, it can break up into vortexes on the leeward side. For a horizontal cable, the vortex formations will alternate between the top and bottom of the cable. Although these vortexes move downstream from the cable, their formation at or just behind the cable creates variations in the air pressure pushing against the cable. Air pressure is reduced at the site of a vortex, and so periodic changes in air pressure occur on the top and bottom of the cable. These vortexes and the air-pressure changes come at a certain frequency that depends on the wind speed and the cable diameter. If that frequency happens to match a frequency at which the cable can oscillate (a *resonant frequency*), the cable begins to oscillate in what is called *resonance*. That is, it gallops. Lines of different lengths will oscillate at different frequencies, but a wind with gusts could set a number of them oscillating at their different resonant frequencies.

To fix the problem at Le Point de Normandie, the bridge engineers hired mountain climbers to scale the cables and tie them together with ropes. Because adjacent cables had different lengths, they had different resonant frequencies. So, when two cables with different resonant frequencies were tied together at the appropriate points, one cable's oscillation tended to fight against the other cable's oscillation.

2.130 • Surf skimmer

To ride a surf skimmer, throw the circular board onto a layer of shallow water (as found at the water's edge on a beach) so that the board begins to glide over the water. Then step onto the board. If you ride the board correctly, you might glide as much as 10 meters. Why don't you just grind to a stop as soon as you put your weight on the board?

Answer The water does not act as a lubricant as it does for tires on a wet street when a car is being braked. Instead, the surf skimmer relies on the relative motion of the water as it moves over the water.

To ride a surf skimmer, the rider stands to tilt the board's front edge upward. Then the passing water collides with the board's underside, giving the board lift and thus allowing it to stay off the underlying sand. However, getting the correct stance takes practice. If the front edge noses up too much, too little of the underside of the board undergoes the collision and the lift is too small. And if the front edge noses up too little, the collision is too glancing to give enough lift. Of course, nosing down will immediately end the ride.

The air drag on a rider can be significant and is greater than the water drag on the surf skimmer. However, a rider can often lengthen a ride by crouching to reduce the cross-sectional area presented to the air.

2.131 • Buoyancy while turning a corner

If a helium-filled balloon floats in a car with the windows up, why does the balloon move relative to the ceiling when the car takes a sharp turn, and does it move outward from the turn or inward? If the turn is taken in cold weather when the car's heater is operating, why does the distribution of warm air shift during the turn, and which way does it shift?

Answer If the car makes a sharp left turn, you feel as though you are thrown outward from the turn—that is, to your right. The reason is that your upper torso tends to continue moving in the original direction while the lower part of your body is pulled, by a frictional force from the car seat, into the turn to the left. Thus, you lean outward from the turn. The air in the car also tends to move in the original direction, but the right-hand wall forces it to take the turn. This action increases the density of the air on the right side of the car. Helium, which is lighter than air, tends to float away from the denser air toward the less-dense air, and so it moves leftward, opposite the leaning of your body.

Warmer air is less dense than cooler air and tends to shift leftward during the turn. If you are the driver, you might feel it move across your face if the blower is not already directing air on your face.

2.132 • Wave reflection by sandbars

Why can a (submerged) sandbar near a beach reflect incoming ocean waves? Why can certain arrangements of sandbars (or submerged artificial barriers) strongly reflect ocean waves?

Answer We may see an ocean wave traveling across the top of the water, but the motion extends below the surface: Bits of water move in oval-shaped, vertical orbits as the wave passes, with the plane of the oval parallel to the direction of the wave's travel. A sandbar can block the orbital motion if the bar is not too far below the surface. Most of the wave still passes, but some of the wave reflects back into the ocean.

Much stronger reflection, said to be *resonant reflection* or *Bragg reflection*, can be produced by a series of long sandbars that reinforce one another's reflection. If the waves have a certain wavelength and their travel direction is perpendicular to the sandbar length, the reinforced reflection occurs if the spacing between sandbars is half the wavelength of the waves. Imagine a continuous wave reflecting from two successive sandbars. The portion that travels past the first sandbar, reflects from the second sandbar, and then passes the first sandbar again (but now moving outward) obviously has gone an extra distance. That distance is twice the distance between sandbars. If that extra distance is equal to the wavelength of the wave, then when this portion arrives at the first sandbar, it is in phase (or in step) with the wave just then reflecting from the first sandbar. So, these two reflections travel back into the ocean in phase, which gives a strong (high amplitude) resultant wave.

In short, when waves reflected by the sandbars travel back into the ocean in phase, the reflection is strong, so relatively little of the original wave remains traveling toward the beach. Such wave reflection could help protect beach and near-beach regions. If only one or two sandbars initially exist, waves could help move sand around by erosion and deposition, building up additional sandbars on the beach side of the original sandbars. Such wave action would build the additional sandbars with the proper half-wavelength spacing to give resonant reflections.

The problem with this argument is that waves arrive at a beach with a wide range of wavelengths and from a fairly wide range of directions. Resonant reflection would not occur for many of the waves.

2.133 • Rain and waves

Is there any truth to the old seafaring adage that rain calms ocean waves?

Answer The adage has some truth, provided that the wind accompanying the rain is not very strong. When a raindrop hits the water, it can send a vortex down into the water or cause the water surface to oscillate and even throw off water droplets. All this activity makes the top layer turbulent, which disrupts and diminishes waves with shorter wavelengths. However, if the rain is driven hard and almost horizontally by a strong wind, the rain and wind can produce and enhance short-wavelength water waves.

2.134 • A salt oscillator

Partially fill a transparent drinking mug with water. Then punch a hole in the bottom of a paper cup and lower the cup partially into the water, securing it in place with a clamp or by taping it to two dinner knives laid across the mug. Make a mixture of moderately salty water and food coloring in a separate vessel and slowly pour the mixture into the paper cup until the level is somewhat below the level of fresh water in the mug. A stream of dyed salty water will flow down through the hole. Then a stream of fresh water will flow up through the hole. This cycle of alternating flows should repeat itself every few minutes, perhaps for several hours. What drives this *salt oscillator*, as it is called?

Answer First, imagine that the hole is actually a short, narrow tube that is initially filled with the dyed salt water (Fig. 2-38*a*). An interface at the bottom of the tube separates the fresh water and salt water. Let's assume that the interface is initially in equilibrium; that is, the pressure just below the interface due to the fresh water equals the pressure just above it due to the salt water. Because salt water is denser than water, the equilibrium condition means that the water height above the interface must be shorter for the salt water than for the fresh water.

Although this arrangement is in equilibrium, it is unstable to chance, unavoidable disturbances. Suppose a disturbance sends a *small* amount of fresh water into the tube. Because the tube is so narrow, the height of the liquid in the paper cup does not change appreciably. However, the pressure changes because part of the tube now contains lightweight fresh water. So, the pressure just above the interface is now smaller than previously. The result is that more fresh water is pushed up into the tube. In fact, this push continues until the water height in the paper cup finally increases enough to re-establish equilibrium at the interface. At that point, the tube is filled with fresh water (Fig. 2-38*b*).

Again the equilibrium at the bottom of the tube is unstable. When a chance disturbance sends a small amount of salt water down into the tube, the weight increase in the tube pushes water out the bottom of the tube, causing even more salt water to be pushed into the tube. Eventually the arrangement returns to the original one with the tube filled with salt water. The cycle is then repeated many times.

If the paper cup has a hole instead of a narrow tube, we can treat the hole as a short tube. However, the intrusion of one liquid into the other is no longer gradual but is now

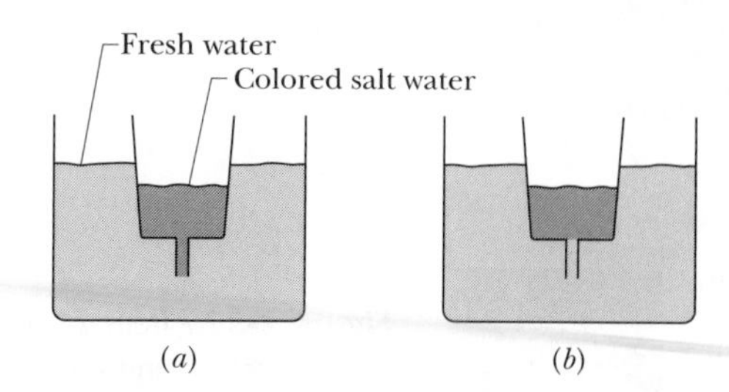

Figure 2-38 / *Item 2.134* (*a*) Fresh water will be pushed up into the narrow tube. (*b*) Colored salt water will now be pushed down into the tube.

quick enough that it is harder to turn off. (The intrusions carry momentum.)

Galileo described a similar demonstration: A globe with a narrow opening is filled with water and inverted in a glass of red wine. Wine then intrudes into the water until the globe is filled with wine and the glass is filled with water. Although Galileo did not describe any oscillations, we can assume they occurred.

2.135 • Salt fingers and a salt fountain

To see a formation known as *salt fingers*, partially fill a container with cool fresh water. Then carefully pour in warm, slightly salty water to which you have added food coloring to make it more visible. Pour with as little disturbance as possible by using a small fall height or pouring onto a floating object. The water on top is lighter than the water on the bottom: Although the water on top has salt, its higher temperature decreases the density to less than the fresh water. So, with lighter water on top, the arrangement should be stable. Why then, after a few minutes, do *fingers* of the dyed water reach down into the fresh water and those of the fresh water reach up into the dyed water (Fig. 2-39*a*)?

To make a *salt fountain*, partially fill a container with cold fresh water. Then punch a hole in the bottom of a paper cup, invert the cup, and lower it into the water (Fig. 2-39*b*). Now add a layer of warm water to the container until water begins to emerge from the hole in the cup. Then add a layer of hot salty water to the container. Finally, release a few drops of food coloring near the hole to make any flow visible. Why does water continuously emerge from the hole? In theory, a "perpetual" salt fountain could be erected in an ocean. Once started, water would continuously flow through a long pipe extending from the colder, less salty ocean bottom to the warmer, more salty top.

Answer The arrangement of warm, salty water lying over cooler, fresh water is actually unstable for two reasons: (1) Thermal energy is transferred fairly rapidly from the warmer water to the cooler water across their interface. (2) Chance disturbances send small waves along that interface, and one of those waves quickly builds up in size, producing the fingers.

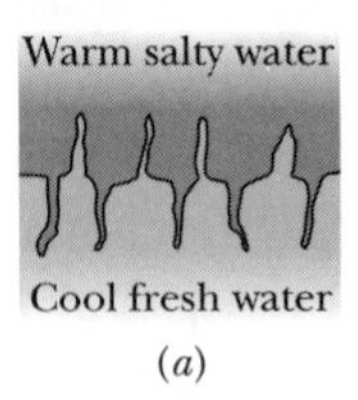

(*a*)

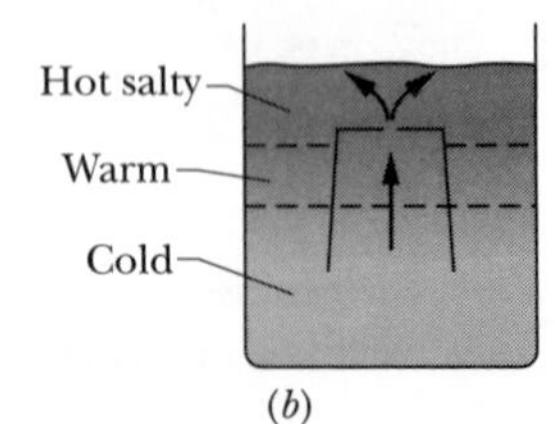

(*b*)

Figure 2-39 */ Item 2.135 (a)* Thin fingers extend vertically from interface between warm salty water and cool fresh water. (*b*) Water flow in a salt fountain.

To see how the instability works, consider one of those small waves. A high point (a crest) is a protrusion of cooler fresh water up into the warmer salt water, and a low point (a valley) is a protrusion of warmer salty water down into the cooler fresh water. Those protrusions should simply flatten out, except that the thermal energy moves into the upward protrusion and out of the downward protrusion. As it warms, the upper protrusion becomes lighter and thus pushes its way even higher. And as it cools, the downward protrusion becomes heavier and thus pushes its way even lower. So, the protrusions grow because of the transfer of thermal energy, and the initial chance wave forms into protruding fingers.

Similar fingering occurs if a (less dense) dyed sugar solution overlies a (more dense) salt solution. Both salt and sugar tend to spread (diffuse) across the interface between the two layers but the salt spreads more quickly. The protrusions of a chance disturbance should just flatten out, but the spread of salt out of the upward protrusions and the spread of salt into the downward protrusions cause the protrusions to grow into fingers.

In the salt-fountain arrangement, as cold water moves up through the cup, it warms because of the warmer water just outside the cup's wall. Thus, the upward-moving water becomes lighter and continues to move upward. As it reaches the hole, it finds itself much lighter than the surrounding hot salty water and so it emerges. A similar process would occur in the hypothetical ocean salt fountain: Once the flow begins, water moving up the tube would be warmed by the continuously warmer water outside the tube. Thus, the interior water would become lighter, and because it cannot gain salt through the tube wall, it would also be lighter than the outside water. So, it continuously moves up through the tube.

2.136 • Lifting water through tall trees

How does a tree, especially a tall tree such as a giant redwood, lift water to the leaves at its crown?

Answer The answer to this deceptively simple question is still very controversial. The generally accepted answer, known as the *cohesion-tension model*, is that evaporation of water from a leaf surface lowers the pressure in a continuous column of water running from the roots to the leaf. The column is said to be under *negative pressure* because water is pulled upward under tension. Water can, of course, be put under pressure, but the idea that it can be put under tension has long been challenged because the cohesion of water (the mutual attraction of water molecules) was believed to be unable to withstand tension. However, tension and negative pressure seem to exist in tree capillaries. In simple terms, as a water molecule evaporates from a leaf, a water molecule is pulled in at a root.

However, challenges to the cohesion-tension model continue. In some plants, water might be lifted in stages, somewhat like ships are lifted when moving through locks in a water canal. Also, environmental conditions such as drought can affect the way water is lifted.

2.137 • Windrows on water

When moderate winds blow over bodies of water, why do bubbles, seaweed, leaves, and other small floating objects form parallel lines called *windrows*?

Answer When the wind speed is in a certain range, it creates long horizontal circulation cells in the top layer of water. This circulation is called *Langmuir circulation* after Irving Langmuir who discovered it after seeing lines of Sargassum seaweed during a voyage across the Atlantic Ocean. The circulation causes a corkscrew motion of the water in the general direction of the wind. The directions of circulation of any two adjacent cells are opposite. Suppose that you were to look along a cell where water circulates clockwise. Then water circulates counterclockwise in the cell on the left and in the cell on the right (Fig. 2-40). That means that, on the surface, the circulation of your clockwise cell and the counterclockwise cell at your right converge, but no convergence occurs on the left side of your cell. Floating objects collected by the convergence form a line along the right side of your cell. Other lines form at the sides of other cells, and the spacing between the lines is twice the width of a cell. If objects are not floating on water, you might still be able to spot windrows if the circulations collect floating thin films (*monolayers*) into lines or lanes. The films decrease wave activity in the water, which makes the water reflect light differently than in film-free regions.

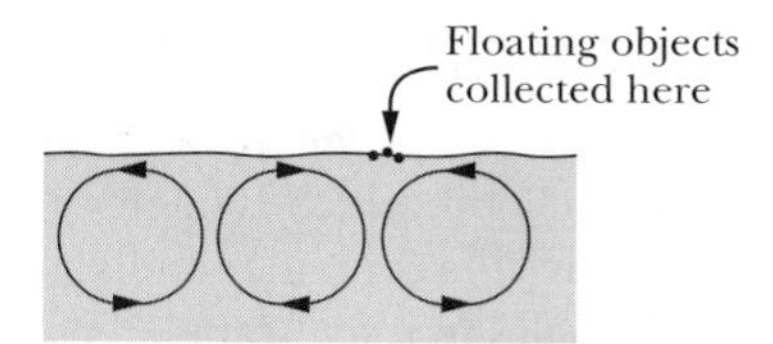

Figure 2-40 / *Item 2.137* Circulation cells in the water collect floating objects.

2.138 • Cloud streets and forest-fire strips

Why are clouds sometimes arranged in long thin lines called *cloud streets*? The arrangement is often difficult to spot from the ground, but when photographed from a satellite the lines can be so regular as to look artificial.

Answer The large-scale airflow in the lower portion of Earth's atmosphere often tends to form long, parallel vortex tubes extending in the general direction of the wind. Looking downwind, you would find that the air circulates in a corkscrew path along a tube: Adjacent tubes have opposite directions of clockwise and counterclockwise circulation. (like the cells in Fig. 2-40). Clouds tend to form where the flow of adjacent vortexes is upward. Thus, long streaks of clouds can form along the boundaries between adjacent vortex tubes, with a spacing of twice a tube's width.

Wind blowing across forest fires can also develop horizontal vortex tubes, and circulation within adjacent tubes alters the burning pattern. Where adjacent tubes have downflow, the trees are less likely to burn, because the circulation pushes the fire off toward the upflow regions. Thus, in a persistent wind the fire can march across a forest in parallel strips, leaving intermediate strips of unburned timber.

2.139 • Packing M&Ms

If you were to fill a jar with either candy spheres or M&M candies (the ellipsoid candies sold by Mars, Incorporated), which collection would weigh more if individual pieces of the two types of candies have identical densities?

Answer The answer may be obvious but nevertheless the answer (or better, the question itself) came as a surprise. Although the spheres are closely packed in the jar, there is still significant vacant space between spheres. Because the M&Ms are squat ovals, they closely pack with less vacant space. Thus, the M&M collection weighs more.

2.140 • A pile of apples

If you build a pyramid of apples or a mound of sand, where along the bottom of the structure is the force on the floor the greatest?

Answer If you construct a pyramid in which each object (take them to be blocks) are arranged in neat vertical columns, with no object straddling two columns, the answer is easy: The greatest force on the floor is beneath the tallest column, which is at the center of the pyramid, and the force becomes progressively weaker as you consider shorter columns toward the perimeter of the pyramid.

However, when you stack objects such as apples, sand grains, or other irregular objects, there are no neat vertical columns without straddling. Instead, each object within the pile rests on underlying objects that are displaced to the side. That arrangement can shift the force of support to the side of the pile. According to experiments, the maximum force on the floor beneath the pile usually lies in a ring around the base, between the center and the perimeter.

2.141 • Powder patterns

Patterns called *Chladni figures* are formed by sand on a horizontal metal plate when it is made to oscillate more or less continuously. The oscillations might be set up by a bow drawn across one edge, or the plate might rest on an upright speaker cone driven by an oscillator. What causes these designs? If the sand is replaced with fine dust (perhaps chalk dust), why can different designs appear? If both a mixture of sand and dust is used, why do the two components separate out?

Spread a fine powder in a fairly uniform layer over a horizontal glass plate, and tap the side of the plate about once a second with a plastic rod. Why does the powder form small, conical mounds after about 20 taps?

Answer At some points (called *antinodes*), the metal plate oscillates most, and at some points (called *nodes*) the plate does not oscillate at all. The antinodes can be adjacent and form lines across the plate; the nodes also can be adjacent and form their own lines. Any sand grains initially located at an antinode are thrown into the air, away from the lines, and tend to collect in the node lines. Once collected, the sand grains reveal the node lines and form one of the Chladni figures. Which of the figures actually appears depends on the shape of the disk and where it is held in place (where, for example, a clamp eliminates the possibility of oscillation).

Dust, being lighter than sand, is affected by air currents that are set up just above the plate when it oscillates. Next to the plate, air tends to move from a node to an adjacent antinode and then upward away from the plate. Thus, this airflow tends to carry dust from a node to an adjacent antinode, depositing it there as the airflow turns upward.

Chladni figures have been used in forensic analysis of smoke detectors. In certain types of sooty fires, the smoke particles tend to collect on the node lines of the pieces that oscillate when the detector sounds its warning. Later you can tell if an alarm sounded during a sooty fire by seeing whether a Chladni figure of soot accumulated on those pieces. If not, the detector may have failed to function.

When a plate holding fine dust is tapped, the tapping causes brief vertical oscillations of the plate, which tends to throw dust up into the air and also to cause the air to move. Suppose that dust happens to collect a bit more at spot A than in the surrounding region. The extra dust at A could modify the plate oscillations and the airflow such that dust in the surrounding region tends to move toward A when it is thrown up into the air. When a dust grain lands on a dusty spot, it tends to stick, whereas if it lands on a clear spot, it does not stick. Thus, the dust at A increases until A fully acquires the surrounding dust. This chance start and subsequent acquisition occurs over the plate surface with a roughly identical spacing between the mounds.

2.142 • A hydraulic oscillator

Figure 2-41 shows a U tube with water and two wide openings. The bottom center is heated while the upper ends are cooled, and everything is symmetric. Once the heating and cooling begin, why does the water oscillate left and right?

Answer The heating decreases the water density, and so the warmed water tends to rise; the cooling increases the water density, and so the cooled water tends to sink. Although the situation is initially symmetric, some slight disturbance from the environment sends more water up one arm than the other. Suppose the disturbance sends warm water up on the right side. The motion allows cold water to descend on the left side. The column on the right is then less dense than the column on the left, and the density difference drives more warm water up the right column, allowing more cold water to descend on the left side.

Eventually, the column on the right is sufficiently higher than the column on the left that the motion is slowed, stopped, and then reversed. In the meantime, the water along the bottom has been warmed. So, as the motion is reversed, warm water is driven up into the left column. The cycle is then repeated.

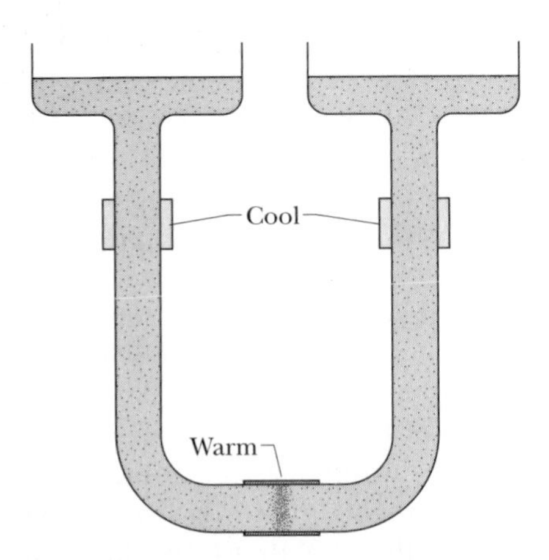

Figure 2-41 / *Item 2.142* Water will oscillate from side to side.

2.143 • Oil blobs moving through glycerin

Fill a container almost completely with glycerin, and fill the rest of it with a silicone oil, which is lighter and less viscous than the glycerin. Leave the container undisturbed overnight (to allow any air bubbles in it to escape), then close the container and invert it. Why do blobs of oil form in a fairly uniform pattern across the (now) bottom of the container and then rise in streams? This arrangement is the basis of several types of fluid novelty toys in which one fluid forms blobs that pass through a second fluid without mixing.

Answer The original arrangement is stable; that is, the oil floats on the glycerin. The inverted arrangement is unstable (an example of what is called *Rayleigh–Taylor instability*) and chance disturbances, such as the disturbances in the inversion process itself, send waves along the oil–glycerin interface as you might send waves along an air–water interface in a bathtub. One of the waves grows faster than the others and dominates. Where it has high points (crests), oil protrudes up into glycerin, and where it has low points (valleys), glycerin protrudes down into oil. The upward protrusions form into rising blobs that are fed by oil flowing underneath the downward protrusions, and the fairly regular spacing of the blobs over the bottom of the container roughly indicates the wavelength of the dominant wave.

Here's a similar demonstration: Allow a container of corn syrup to stand overnight. Arrange for a small tube to release a mixture of corn syrup and water near the bottom of the corn syrup in the container. The mixture forms a blob as it emerges from the tube. Because the water–syrup blob is lighter than syrup, the blob rises. As it rises, it leaves a trail behind it. That trail can act as a conduit for fresh blobs released by the tube.

2.144 • Ball in an airstream

To catch the attention of a customer, some department stores use a demonstration where a ball is suspended in an airstream. If the airstream is directly upward, the suspension may not be surprising because the impact of the air on the underside of the ball could balance the gravitational force on the ball. However, the attention-catching feature is that the stream is angled by about 45° from the vertical. How does it suspend the ball? If you give the ball a good knock so that it partially leaves the stream, why does it return?

Answer The key to the ball's suspension and stability is that it deflects the airstream. If the ball bobs downward in an apparent attempt to fall out of the stream, the stream flows over the top of the ball and down part of the back, where it breaks free, headed downward at an angle. Because the stream is forced downward, the ball is forced upward and back into the stream. No matter how the ball tries to leave the stream, it ends up deflecting the stream in that direction and thus is pushed back into the stream.

A ball can also be suspended in a vertical water stream. Although it bobs and attempts to leave the stream, it returns. The only difference is that the stream is water and not air, but the stream deflection still explains the ball's stability.

I once had a toy consisting of a plastic U-shaped tube, with a short narrow tube at one end. I blew into the narrow tube to lift a lightweight ball on a stream of air and to pull air through the rest of the U. As the ball rose, it passed the other open end of the U and was sucked in by the air circulation. The goal was to loop the ball around the U as many times as possible on a single hard blow into the narrow tube.

2.145 • Flettner's ship

In 1925, a ship designed by engineer Anton Flettner crossed the Atlantic driven not by the standard propeller jutting down in the water but by two large rotating cylinders jutting up into the air. How could rotating cylinders propel the ship through the water?

Answer The cylinders were driven by the wind but not in the way sails are driven. If the cylinder is stationary, the air flows symmetrically on the two sides of the cylinder, where somewhere near the rear it detaches from the cylinder and breaks up into vortexes. This arrangement provides some push on the cylinder because the windward side of the cylinder has greater air pressure than the leeward side (vortexes have low pressure).

However, the cylinder is pushed more substantially if it is rotating. On the side turning with the wind, the air tends to cling to the cylinder longer than previously; on the side turning against the wind, the air breaks free of the cylinder earlier than previously. The net result is that the airstream is deflected by the rotating cylinder, flung off to the side in the direction of spin. Because the stream is forced in one direction, the cylinder (and thus the ship) is forced in the opposite direction.

So, in principle, the ship can be pushed through the water by the deflection of its cylinders in the wind. In practice, the journey across the Atlantic must have been dreadful, requiring patience in adjusting the orientation of the ship in the wind and in taking a zig-zag course. (Well, maybe the ship's propeller was used more than reported.)

2.146 • Strait of Gibraltar, Strait of Messina, Strait of Sicily

When a ship pulls into a particular channel through the Strait of Gibraltar, why can it spontaneously rotate around the vertical or heave to the side? The Strait of Messina, which separates Italy and Sicily, has long been known for its treacherous waters, dubbed the *mad sea.* Homer, for example, attributed the water behavior to the monsters Scylla and Charybdis. On the other side of Sicily, in the Strait of Sicily that separates the island from Tunisia, high water sometimes attacks the large Sicilian fishing port Mazara del Vallo, sending a wall of water up the old estuary. What accounts for the odd behavior in these three straits?

Answer The odd behavior at the Strait of Gibraltar is caused by *internal waves,* which are waves generated within the tidal flow of water through the strait. Waves occur within this flow because the water from the Mediterranean Sea is saltier than the water from the Atlantic Ocean. (The Mediterranean water is saltier due to evaporation of water from the sea.) Thus the Mediterranean water is denser than the Atlantic water. When this denser water flows through the strait into the Atlantic, it must pass over a *sill* (a high point), which sends the water upward and over the incoming (lighter) Atlantic water. This is an unstable situation, with denser water moving over lighter water, and waves are generated. The waves are visible on the surface of the water only as bands of choppy water, but the movement of the waves can catch and turn ships.

Internal waves are also responsible for the mad sea in the Strait of Messina. There, a sill separates the denser, saltier water of the Ionian Sea south of the sill from the lighter, less salty water of the Tyrrhenian Sea north of the sill. Tidal oscillations in the Mediterranean Sea are generally quite small (only a few centimeters), but the oscillations in these two seas facing each other across the sill are out of phase (out of step). So, as high water occurs in one of them and low water occurs in the other, water flows over the sill. Because of the density difference, the flow generates internal waves. On the surface, the waves show up as bands in which the water is very choppy, as if kicked up by strong gusty winds. Thus, internal waves, and not Homer's monsters, are what have plagued fishing vessels in this region.

The strange waters in the Strait of Sicily are due to sloshing in the strait, an effect said to be a *seiche.* Resonance oscillations occur in the strait; that is, natural sloshing occurs much like you can set up sloshing in a pan of water as you carry it across the room. The sloshing in the strait can sometimes be large enough to send a wall of water, called a *hydraulic jump* or a *bore*, up the estuary.

2.147 • Granular splashing

If a heavy rigid ball is allowed to fall into a bed of much smaller, rigid balls, why can the impact send up a narrow, dramatic jet of the smaller balls?

Answer When the heavy ball burrows into the bed of small balls, it pushes out a cylindrical cavity. The balls that are pushed out of the way form a splash around the perimeter of the cavity. As the cavity collapses with balls flooding back into it, they collide and are deflected upward to form the jet.

2.148 • Slight ridge on moving water

When the light is right, why can you see a hair-thin line stretching across slowly moving water in a stream or brook? (You usually need the Sun to be low so that the light is glancing, but even then you need to try several viewing angles.)

Answer Most streams and brooks have layers of contamination on the water surface. The layers may be pollution, or they may be naturally occurring substances such as material from vegetation. The layers are usually too thin to see and may be only one molecule thick, in which case they are called *monolayers.*

When the slowly moving water meets a monolayer, the oncoming water piles up to make a very slight ridge before it manages to move under the layer. If the light is right, you can see the ridge because of its contrast with the moving water on one side and the stagnant layer on the other side. The ridge is commonly called the *Reynolds ridge* because Osborne Reynolds (1900) was one of its first investigators, although Benjamin Franklin (1774) and Henry David Thoreau (1854 and later) noted the effect earlier. You might also see a Reynolds ridge on a pond or on open water if a moderate wind pushes the uncontaminated water up against a layer of contamination.

2.149 • Meandering thin streams

If a thin stream of water runs down a smooth glass plane with a tilt of less than 30°, the stream is straight. If the tilt is greater than 30°, the stream might still be straight but it can instead meander left and right, forming either a constant meander pattern or one that continuously changes (Fig. 2-42*a*). What causes the meandering?

Answer When the *volume flow rate* (the amount of fluid flowing past a given point per second) is low, the gravitational force on the water tends to bring the water directly down the tilted plane. The surface tension (due to mutual attraction of the water molecules) tends to decrease the surface area and acts like an elastic membrane, keeping the stream straight. Along the first part of the stream, the gravitational force accelerates the stream. As the water speed increases, the cross-sectional area of the stream decreases because, with the water moving faster, less area is needed to move the same amount of water per second. However, as the speed increases, the drag on the water from the tilted plane increases until the drag matches the gravitational force. Thereafter, the speed and the cross-sectional area no longer change.

When the volume flow rate is somewhat greater, the water movement can make the stream unstable by setting up different speeds across the stream. This speed difference means that the shape of the stream is no longer symmetric: The surface tension on a more tightly curved side pulls inward harder than the surface tension on a less tightly curved side.

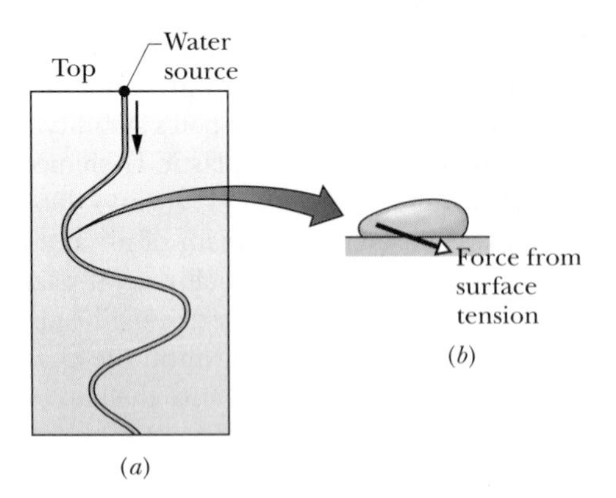

Figure 2-42 / *Item 2.149* (*a*) Looking straight down on a tilted plane; water stream meanders. (*b*) Cross section in a bend. The tight curvature on the left side produces a strong rightward force.

Suppose, by chance, a bend begins to develop in the stream. The bend becomes pronounced only if the surface tension in the stream produces enough force to send the stream diagonally across the tilted plane. Figure 2-42*b* shows an example: We see a cross section through part of a bend in the stream. The left side of the stream is more tightly curved than the right side, so the surface tension produces a greater force on the left side. That force is partially directed to the right, and the stream leaves the turn traveling along a diagonal to the right, which makes the bend pronounced.

When the volume flow rate is even greater, the onrush of water can defeat the surface tension tendencies. The stream might overshoot a bend, thus moving the bend. Or it might break apart along a diagonal, abandoning a bend as a new route is taken by the water; abandoned sections then slide down the tilted plane.

2.150 • Shaver clippings and camphor boats on water

If small hair clippings are emptied from an electric shaver onto water, such is commonly done when the shaver is cleaned over a toilet, why do the clippings race away from one another as soon as they hit the water?

An old amusement, now almost forgotten, is a camphor-driven boat. A lightweight "boat" is fashioned out of aluminum foil, with a wedge cut at the rear. The boat is carefully floated on water, and a small lump of camphor (which is a nonprescription chemical sold in pharmacies) is placed at the wedge, with a small amount submerged. The boat immediately begins to move in a generally forward direction. How does the camphor-in-water arrangement cause motion?

If a short length of certain types of cement (such as Devcon Duco Cement) is dropped into a shallow layer of water, why does the cement length wiggle and spin?

Answer When shaver cuttings are dropped into water, the oil on the hairs forms a thin layer on the water, which quickly reduces the surface tension. This oil layer and the cuttings are then pulled outward by the greater surface tension of the unaltered water surrounding them.

The camphor in a camphor-driven boat decreases the surface tension in the water at the rear of the boat because camphor molecules replace some of the water molecules, weakening the tension between the water molecules along the surface. The surface tension of the water in front of the boat is unchanged. The front and back of the boat are each pulled by the adjacent water, but the pull in the front is stronger, and so the boat moves forward. As the boat moves around on the water, the camphor left in the water gradually sublimes into the air or diffuses into the bulk of the water, and thus the water surface does not merely end up covered with camphor, which would cause the boat to stop.

If a sliver of camphor is cut with a curved arm and dropped into water, it will rotate. The concentration of camphor molecules in the water adjacent to the concave portion is greater than that adjacent to the convex portion. The reason is that the camphor molecules can move more easily away from the convex portion. The unequal concentrations produce unequal pulls from the surface tension. An arm is pulled more on the convex portion, and thus the sliver rotates with the convex portion leading the convex portion in the rotation.

If a floating ring surrounds the sliver (and is, say, 10 times the size of the sliver), a rotating sliver causes the ring to rotate in the opposite direction. If a camphor boat is placed in a figure-eight route (two circular routes joined at the middle), the boat can travel in a number of ways: It can remain in one of the circles, cross over at the middle while changing the turning direction between clockwise and counterclockwise, or cross over at the middle without changing the turning direction. The path taken at the cross-over point depends partly on random events, but it also depends on how much camphor lies ahead of the boat in the possible routes. If one route still has appreciable camphor and another route does not, the stronger surface tension in the second route will probably pull the boat into that direction.

If two camphor boats are put into a circular route or some other geometric route, the camphor left by one boat will affect the speed of the other boat and the two boats will soon fall into synchronized motion around the route, with the trailing boat keeping an approximately constant distance from the leading boat.

When short lengths of certain types of cement are dropped into water, they move by altering the surface tension in the water around them, much like camphor.

2.151 • Oil stains on a road

Why are oil stains on a road usually oval, with the long axis parallel to the traffic flow, and often annular (ring-like)?

Answer When an oil drop leaks from a moving vehicle, its speed through the air is initially the vehicle's speed. If the speed exceeds a certain critical value, the drop is blown into a bubble resembling a soap bubble on a circular hoop before the bubble breaks free. The inflated portion of the bubble is quickly blown apart, while the rim breaks up into droplets, which form an oval ring when they hit the road. If the pattern is examined soon after forming, the individual droplet stains can be distinguished.

Raindrops are limited in size by a similar process. If a falling drop becomes too large, the air blows it into a bubble and then pops the interior.

2.152 • Patterns of water drops falling onto glycerin

Why does a flower-like pattern gradually form when a water drop falls onto a thin layer of glycerin?

Answer The pattern is more striking when the water is dyed with food coloring. The splashing serves to split the drop into parts: First a central blob rises from the center of the impact and then a circular ridge (resembling a teacup) rises around the perimeter. Soon after the water is thrown outward from the impact, the water and glycerin begin to mix, and the fluid along the surface with the air begins to move. The motion is due to the differences in surface tension along the surface. Because water has a larger surface tension than glycerin, the water–glycerin mixture is pulled along the top surface radially inward toward the water. Because the lower fluid is retarded by the underlying glycerin, this radial motion sets up circulation cells within the water–glycerin layer. Around the perimeter of the splash region, chance variations in the shape of the boundary between water and glycerin are reinforced by the variations in surface tension. Within about 15 seconds, a flower-like pattern emerges.

2.153 • Olive-oil fingers on talc-covered water

In a clean shallow container, pour a shallow layer of water and spread a small amount of talc powder (such as "baby powder") on the water. Blow gently to spread the powder evenly (the surface should look dusty with the grains too small to perceive). Then touch the tip of a straightened-out paper clip to olive oil and then very briefly to the middle of the talc-covered water.

If the amount of talc is small, with well-separated grains, the oil merely moves the talc outward, clearing an approximately circular region. If the amount of talc is too much, the oil cannot move the talc at all and just sits on the water as a drop. However, if the amount of talc is moderate, the surface quickly shows a pattern radiating outward from the point touched by the paper clip. What causes the pattern?

Answer The oil attempts to spread to form a thin layer, perhaps a *monolayer* with a thickness of one molecule. If little or no talc is in its way, the oil easily pushes the talc outward. With too much talc in its way, the grains are in a logjam and cannot move, and so the oil sits as a drop. However, in the intermediate situation, the talc is not in a logjam but there is enough grain–grain contact to make the water effectively very viscous. Thus, the less viscous oil attempts to push its way into the more viscous talc–water mixture.

An interface between a viscous liquid and a less viscous liquid is unstable to chance disturbances, which tend to send waves along the interface. Each such wave tends to alternate between very slight intrusions of the oil into the talc–water and very slight intrusions of the talc–water into the oil. One of the waves dominates and its intrusions quickly grow to form narrow *fingers*. As the oil fingers grow, they clear out talc, leaving visibly cleared paths in the talc regions.

Such finger instabilities on an interface between two fluids are often studied in a Hele–Shaw cell, which consists of two plates of clear plastic separated by a narrow rubber gasket. The cell is filled with a fluid before being clamped shut. Then a second fluid is injected into the center via a syringe inserted into a small hole in one of the plates. The second fluid advances on the first fluid in a pattern of finger intrusions. Some patterns resemble ferns, others resemble flower petals, and others defy an easy description.

2.154 • Chicken-fat oscillator

In the middle of a shallow dish holding ammonia and dish soap, place a drop of liquid fat from cooked chicken. Why does the drop pulsate?

Answer This effect was discovered in the mid-1970s by Jeffrey May, a chemistry instructor at Cambridge School in Weston, Massachusetts, who was attempting to soak a pan that had been used to broil chicken. He ran hot water into the pan and then added dish soap and ammonia. Once *lenses* (islands) of oil formed on the top surface of the water, they began to pulsate. When he covered the pan to eliminate the evaporation of the ammonia, the pulsations ceased. He saw evidence that a "membrane-like coating" surrounded each lens, suggesting some interaction between the soap molecules and the oil.

This is similar to other systems that display oscillations in surface tension. The ammonia slowly breaks off oil from the chicken fat, allowing the oil to diffuse (gradually spread) over the water. The presence of the oil decreases the surface tension around the drop. Because the surface tension farther from the drop is stronger, the liquid around the drop is pulled radially outward, causing the drop itself to expand outward.

However, as the oil spreads, it encounters soap molecules, which lie along the water surface. Bits of oil become surrounded by soap molecules in an arrangement called an *aphron*: The hydrophobic end (does not attract water) of each soap molecule sticks into the oil and the hydrophilic end (attracts water) juts into the water. As the oil becomes trapped, the surface tension of the water increases, allowing the drop to relax and pull inward.

The cycle of expansion and contraction is then repeated because the outward motion of oil along the water surface causes an inward motion of water and ammonia farther down in the water. Thus, fresh ammonia arrives at the fat drop to start the next cycle.

Hiding under the Covers, Listening for the Monsters

Figure 3-1 / Item 3.1

3.1 • Howling of the wind

What causes the howl of a strong wind, which can conjure up images of werewolves baying outside your house on a dark and stormy night?

Answer If air blows past an obstacle, especially a projection such as a roof edge on a building or even the vertical edge of a building, the air breaks up into vortexes (swirls) that are carried downstream by the airflow. A vortex causes variations in air pressure, which travel away from the vortex as a sound wave—the howl. The sound can reach you directly if you are outdoors, but it can also be transmitted through window glass, doors, walls, and even your bedcovers to seek you out.

3.2 • Singing of telephone wires and pine needles

When a strong breeze blows across telephone lines or power lines or through a pine forest, why can you hear the lines or needles sing? That sound, coming and going with the random play of a strong breeze, is one of the soothing aspects of being in a pine forest on an autumn day.

Answer When the breeze flows past a slender cylinder such as a wire or a pine needle, the airflow tends to form vortexes downstream of the cylinder. The vortexes are said to be shed by the cylinder, first on one side, then on the other side, then back to the first side, and so on. The formation of a vortex changes the air pressure, and so a train of air pressure changes moves downstream from the cylinder and sends out a sound wave, said to be an *aeolian tone*. You hear the air pressure changes due to the vortex shedding when you intercept some of the sound wave. The faster the airflows past the cylinder, the more frequently the changes occur, and so the frequency of the sound is higher.

The cylinder can oscillate like a guitar string at certain frequencies, said to be resonant frequencies. If the frequency of the pressure changes happens to match one of these resonant frequencies, the cylinder will oscillate at that frequency. Now the cylinder's motion also sends out sound waves, and it may lock in (maintain) the frequency of the vortex shedding even if the speed of the airflow changes somewhat. When telephone or power lines oscillate, they are said to *gallop*. This can be of concern because galloping can rip out the rigs supporting a line on a pole or tower, especially if the rigs are also supporting ice that has formed on the line.

The whine from telephone wires may be the loudest and most shrill on very cold days because the cold temperatures cause the wires to contract and thus tighten between their supports. If the wires gallop, they can transfer motion to the supports, making them oscillate and thereby increasing the noise level.

3.3 • Whistles and whistling

How does a person whistle; that is, how is the sound produced? How does a whistling teapot whistle when the water is being heated? Countless whistling objects have been invented, but probably the most prominent are the English police whistle (single tube), the American police whistle, and the many musical instruments that produce sound via a whistling action.

Answer Any whistle depends on three features: (1) An airstream encounters an obstacle and breaks into vortexes. (2) The vortexes cause the air pressure to vary in a periodic way, so that a sound wave is emitted for you to hear. Either the vortexes themselves or the pressure variations of the sound wave feed back into the upstream portion of the airstream. (3) If the airstream is unstable (easily deflected or changed), this feedback increases the stream's instability, which enhances the production of vortexes at the obstacle. Once this vortex-production and feedback procedure is established, you hear a continuous sound—the whistle.

If you whistle by blowing through puckered lips, producing what is known as an *orifice tone*, vortexes are produced as the air is forced through the narrow opening of your lips. (Vortexes develop there because the flow through the center of the opening is faster than the flow closer to the lips.) Part of the sound waves generated by the vortexes travel back through the lips into the mouth (into the *vocal tract*). The frequency of this returned sound depends on the speed at which vortexes pass outward through the lips and the speed at which sound passes inward through the lips to the mouth. Sound can resonate in the vocal tract at certain frequencies called *formats*. If the frequency of the returned sound is close to the second format (the one with the second lowest frequency), the returned sound sets up resonance in the vocal tract. That is, the sound sets up waves that reinforce one another instead of canceling out one another. If resonance occurs at the second format, then the corresponding frequency is the one you hear.

You can change the format, and thus the whistling frequency, by changing the shape of the vocal tract, primarily by pushing your tongue forward or bringing it back. You can also change the whistling frequency by blowing harder, so that the frequency of the returned sound is higher and thus closer to a higher format (a higher resonant frequency).

The whistle of a teakettle at full boil is known as a *hole tone*. The whistling part of the kettle consists of a short cylinder with a hole at each end. As water boils (turns to vapor) inside the kettle, the air and water vapor that are pushed through the lower hole form an airstream that strikes the second hole, setting up vortexes within the cylinder. The resulting pressure variations of the airflowing through the second hole generate sound. Outside the cylinder, the sound waves are the teakettle's whistling. Inside the cylinder, the sound waves travel back to the first hole as a feedback that controls the airstream entering the cylinder. Thus, with this feedback disturbing the incoming airstream, the disturbance travels downstream to continue the vortex generation at the second hole.

A police whistle produces an *edge tone*: An airstream is blown onto or across the edge of a hole. The airstream breaks up into vortexes that then send out sound waves. The frequency of the sound is set by the resonant frequencies of the police-whistle cavity, much like resonance occurs in your mouth when you give a puckered whistle. However, in the American police whistle, a small ball bounces around inside the chamber, changing the shape of the interior (and thus the resonant frequency) and also disrupting the airflow, especially when it momentarily blocks the hole you blow across. The result is *warbling*, a coming and going of the whistle loudness and frequency.

A flute works in the same way: You blow an airstream across an opening and past an edge. The vortexes produce sound inside the flute chamber, primarily at resonant frequencies of the chamber. This resonance inside the chamber then feeds energy back into the vortex formation, so that the vortexes, the resonance, and the sound you hear are all continued.

One of the more curious edge-tone whistles is a Peruvian whistling bottle, which is a ceramic water bottle that was produced by the original Peruvian people up until the Spanish conquest in 1532. A few of the surviving bottles are now in collections. They consist of one or two chambers with a tube extending from one of the chambers and a whistle arrangement near the top. When someone blows into the tube, air escapes from the chamber through a hole near the whistle. The whistle consists of a small opening to a small air chamber. When the escaping airstream hits one side of the small opening, the resulting vortexes set up resonance inside the small air chamber.

The noisiest types of whistle are emergency sirens, which are still used on some emergency vehicles. (Modern police cars typically use electronic devices to emit the warning sound, but many fire trucks still employ true mechanical sirens because they can be very loud and arresting.) During World War II and the Cold War, huge sirens were constructed to warn civilians of a military attack. (Standing next to one of these sirens would shake you so badly that the ground would seem to be fluid.) Although emergency sirens have come in many varieties, most produced hole tones by sending compressed air through two grids of holes, with one grid rotating relative to the other. When the holes of the two grids rotated through alignment, hole tones were produced. The frequency of the sound was controlled by the rotation rate.

3.4 • Speaking and singing

How do you speak or sing? How do you whisper? Why are the words sung by a soprano so difficult to understand?

Answer Sound is produced by the muscles known as the vocal folds, which lie in the larynx. The folds on opposite sides of the throat are held closed while air pressure is increased in the lungs. The folds then suddenly move apart and air rushes past them, creating turbulence that causes the vocal folds to oscillate. These oscillations alter the air pressure, sending sound waves up into the vocal tract, which includes the upper throat, mouth, and nasal cavity. These sound waves have frequencies that match the oscillation frequencies of the vocal folds. The lowest frequency is the *fundamental oscillation* of the vocal folds. The other frequencies are whole-number multiples of that lowest frequency. For example, if the lowest frequency is 70 hertz, then the other frequencies are $2(70) = 140$ hertz, $3(70) = 210$ hertz, and so on.

The vocal tract is effectively a tube with one end closed (at the larynx) and the other end open (at the mouth and nostrils). Sound can set up resonance in this tube if the sound waves have the proper frequency, said to be a *format frequency*, or just *format*. These formats are not precise but are each spread over a range of frequencies. They are centered at odd-number multiples of the lowest format. For example, if the lowest format is centered at 500 hertz, then the other formats are centered at $3(500) = 1500$ hertz, $5(500) = 2500$ hertz, and so on.

When sound is fed into the vocal tract from the vocal folds, the vocal-fold frequencies can excite some of the formats. That is, sound waves at those formats build up inside the vocal tract so that the portion that leaves the vocal tract is loud enough to be heard. In effect, the vocal tract acts as a filter on all the frequencies being emitted by the vocal folds. You can change that filter by changing the frequencies of the formats, which is done by shifting the tongue's position, changing the opening of the mouth (or by pinching the nose closed), and by shifting the height of the larynx in the throat. (Classically trained singers usually do not want such larynx motion, because it interferes with their control of the vocal fold tension. So, they practice using muscles to hold the larynx in place.) You can change the frequencies of the sound waves being fed into the vocal tract by changing the tension in the vocal folds: Greater tension produces higher frequencies. Although this sounds complicated, most people learn to do it without thought by the time they are two years old.

Many animals can produce sound from their larynx. Some can control the frequency and amplitude of the sound from their mouth by controlling the larynx muscles or the size of the vocal tract. Some, such as the mynah bird, have enough control to mimic a human voice, but only humans can produce a rich tapestry of sounds (well, a rich tapestry if we leave out punk rock).

In a whisper, the vocal folds are relaxed so that the airstreaming past them does not cause them to oscillate. The (mild) turbulence in the airstream produces the sound that excites some of the formats in the vocal tract. Whispered speech is made by controlling and changing the size and shape of the vocal tract, largely with the tongue and lips.

To be heard in a large concert hall with an accompanying orchestra, a soprano must sing loudly at frequencies well above the normal lowest format of her vocal tract. (She needs to sing in a frequency range where an orchestra is generally quieter and where a person's hearing is best.) Although she can stress her vocal folds to produce a high frequency and then match that frequency to a higher-frequency format of her vocal tract, the resonance does not produce a strong sound.

Both the audibility and quality of the singing are better if, instead, she shifts her lowest format up to higher frequencies and then excites that format. To shift the format, she lowers her jaw to open her mouth and then shapes her lips into a smile. These actions effectively decrease the length of the vocal tract, which shifts the formats upward. Now, higher frequencies from the larynx can excite the first format of the vocal tract, which means that the singer can sing loudly at the higher frequencies. However, this procedure comes with a cost: The singer can no longer articulate certain sounds and words and thus cannot always be clearly understood by the audience.

3.5 • Speaking with helium

I'll describe this stunt, but do not do it yourself because it is dangerous and can even be fatal. If someone inhales helium and then speaks, why does the person sound like Donald Duck with his crazy high-frequency voice?

Answer As explained in the preceding item, the sound you hear from a person depends on the excitation of various formats in the vocal tract by the sound waves produced by the oscillating vocal folds. When a frequency of the vocal folds falls within the frequency range of a particular format, then sound at that format frequency is included in the person's voice. The frequency (either the center frequency or the range of frequencies) of each format depends on two factors. One is the shape and length of the vocal tract, something you control when you shift your tongue or change the opening of your mouth. The other factor is the speed of sound in the vocal tract.

Normally, of course, air is in the vocal tract and the speed of sound has a certain value (about 340 meters per second). However, if the air is replaced with an air–helium mixture, the speed of sound is much faster (perhaps 900 meters per second). This speed increase shifts up all the format frequencies. The oscillations of the vocal folds are approximately the same as in air, but now the higher frequencies of those oscillations excite the shifted-up formats of the vocal tract. The relative strengths of the formats may also change. The result is that the voice now consists of higher frequencies and is no longer familiar.

The danger here should be obvious: You can live only if you breathe air (or rather, the oxygen in the air), but if you foolishly fill your lungs with helium, you no longer breathe air. You are then in a race against suffocation. As the oxygen level in your blood drops, can you get the helium back out and the air back in fast enough to avoid suffocating or your brain undergoing oxygen starvation? We all have to die someday, but this is a really stupid way to do it.

3.6 • Throat singing

In Tuva, which is in southern Siberia, some vocalists can sing two tones simultaneously, a practice known as *throat singing* or *overtone singing*. One tone is a low-frequency drone; the other is a high-frequency sound that resembles a flute. How can someone sing two tones together?

Answer In normal voice, both spoken and sung, the harmonic frequencies provided by the vocal folds excite primarily the first format of the vocal tract (see the preceding two items). Some of the higher formats of the vocal tract are also excited, but they are not perceived separately by a listener; rather they are noted almost subconsciously as merely being characteristic of the voice (they are said to contribute to the *timbre*, which is a vaguely defined term).

In throat singing, the low-frequency droning produced by oscillating the vocal folds and upper throat is not especially tricky. The tricky part is producing the higher-frequency sound that seems to be unrelated to (or "floating on top of") the low-frequency droning. The idea is to make one of the high-frequency harmonics of the vocal folds match (almost exactly) one of the high formats of the vocal tract. If the match is made, resonance at that format is strong, and so is the sound emitted by the singer. However, making the match requires the singer to adjust both the vocal-fold action (how long the folds snap open and then close) and the vocal tract shape (the tongue is shifted). Almost anyone can learn to do all this, but it does take a lot of practice so that the sounds are musically pleasing.

3.7 • Snoring

A great many people snore while sleeping, often to the dismay of family members, and sometimes to the severe detriment of their own sleep and health. What causes snoring?

Answer Snoring occurs primarily when air is drawn into the lungs, either through the nose (with the mouth closed) or through both nose and mouth. The airflows past the soft pallet that forms the back portion of the roof of the mouth. (You can see it by looking inside someone's mouth.) If the air comes through only the nose, it enters the throat along the top of the soft pallet. If the airspeed exceeds a critical value, it pulls the soft pallet toward the back of the throat, partially obstructing the flow. The pallet flops first onto the tongue and then back to its original position. If, instead, the flow is through both nose and mouth, it is across the top and bottom of the soft pallet. In this case, the soft pallet flaps between the back of the throat and the tongue, alternating between obstructing the flow through the nose and the flow through the mouth. I can make myself snore (a theatrical snore, as if I were performing it in a play) by sharply inhaling through my mouth and nose. The obstruction, due to the flapping of the soft pallet, makes the airflow pulsate, which causes my nostrils to pulsate.

The motion of the soft pallet and the turbulence it creates produce sound waves in the throat. If the waves set up resonance in the throat (or the combined throat-mouth-nasal region), then the sound can be loud enough to wake family members.

A third cause of snoring is the periodical collapse of the pharynx (the flexible, collapsible upper portion of the throat, above the larynx). The collapse and subsequent reopening of the pharynx disrupts the airflow, setting up turbulence that produces sound waves.

3.8 • Purring and roaring

How does a cat purr and a lion roar?

Answer A cat purrs very much like you speak, as explained in earlier items, except that when air escapes through the vocal folds, the vocal folds oscillate to produce a buzzing sound in the vocal tract (throat-mouth-nose cavity). The buzz is at a frequency of about 25 hertz, which is probably too low for you to hear. However, it excites higher harmonics in the vocal tract, which you *can* hear radiated from the mouth and nose of the cat. The sound, that of a rolled "r," is usually a sign of contentment.

Some researchers believe that the ability to purr and the ability to roar are determined by the state of the hyoid, which is a structure that lies at the root of the tongue and is connected to the larynx. If this bone is fully ossified (a fairly rigid bone), purring can occur. If, instead, it is not fully ossified (not as rigid a bone), then roaring can occur. This probably means that the animal with a more flexible hyoid, as in the lion, can move the larynx down the throat to significantly increase the length of the vocal tract. This lengthening lowers the frequency of the sound the lion makes. The lion's larynx is also different from most other animals in that the vocal folds are thick and consist of elastic tissue, which can oscillate at low frequencies with fairly large amplitudes. Thus, the lion can roar!

SHORT STORY

3.9 • Sound from a *Parasaurolophus* dinosaur

The crest on a *Parasaurolophus* dinosaur skull contained a nasal passage in the shape of a long, bent tube open at both ends. The dinosaur may have used the passage to produce sound by vocalizing, in order to resonate the passage at its lowest frequency (at its fundamental), much like we vocalize by resonating our throat-mouth-nose cavity. Fossil skulls that contain shorter nasal passages are thought to be those of the female *Parasaurolophus*, which would vocalize at a higher frequency.

3.10 • Sounds of tigers and elephants

Part of a tiger's roar is below the audible range of humans, in the *infrasound* range. Does a tiger gain any advantage in emitting such low-frequency sound?

Although elephants hear best at a frequency of about 1000 hertz, when they call to one another, especially over large distances, they put much of the energy into sound between 14 and 35 hertz, which extends into infrasound. If you are near an elephant when it calls, you might feel the wave more than hear it. Does the high-energy, low-frequency call have any advantage over a high-frequency call? Elephants on a savanna call out nearly twice as much at night as during the day, either to seek a mate or to warn competing elephants to keep their distance. Is there any advantage to making these calls at night?

Answer The distance sound carries in a forest, the habitat of tigers, depends on the wavelength: Sound with longer wavelengths is absorbed and scattered less by the trees, brush, leaves, and grass than sound with shorter wavelengths. So, to call out for a mate or as a warning to other tigers, a tiger can send the signal farther with low-frequency (long wavelength) roaring than with higher-frequency roaring. (Besides, it is just scarier.) Other forest or jungle animals also depend on low-frequency communications. For example, cassowaries, which are the largest forest birds in the world, vocalize with deep booming sounds at frequencies as low as 20 or 30 hertz, which is as low as humans can hear. And the some of the sounds of the Sumatran rhino ("whistle blows") are in the infrasound.

On a savanna, the atmosphere often develops a nighttime inversion in which warmer air lies over cooler air. During an inversion, a low-frequency call can be effectively trapped below the warmer air. Thus, instead of the call spreading upward and being lost, most of it is trapped into traveling over the savanna to much greater distances (perhaps up to 10 kilometers) than during the day, when there is no inversion (perhaps only 2 kilometers). Higher-frequency sounds are less likely to be trapped below a warm layer and are also absorbed more by the air, and so a higher-frequency call by an elephant would not travel as far.

The best time for an elephant to make its call, so that the call spreads over the maximum possible area, is one or two hours after sunset, when the wind is low and an inversion has had time to develop. Later in the night, the wind might pick up; although a call might carry even better in the direction of the wind, it carries less well in any other direction and thus the total area in which the call can be heard is diminished.

3.11 • Bullfrog croaking

A male bullfrog croaks either to attract a companion or to warn away other males. How can such a small animal, with a small mouth, make such a deep, booming noise?

Answer A bullfrog emits most of its croak through its eardrums, not through its mouth. Reportedly, a researcher discovered this by (gently) pressing his fingers over the ears of a bullfrog and noticing that the bullfrog's sound level was greatly reduced. Later the demonstration was repeated with "frog earmuffs," which were pieces of foam held in place over the eardrums by a spring!

The sound originates in the vocal folds of the frog, just as it does in mammals. But it is then sent to the eardrums, which resonate at certain frequencies like drumheads. This resonance greatly increases the sound level at those frequencies and broadcasts the sound through the frog's environment. Before this role of the eardrums was discovered, many people believed that the resonance occurred in the *vocal sac*, the region of the throat that the frog inflates when croaking. Although gibbons, and some types of frogs and toads, might set up resonance in their vocal sac in order to be louder, the bullfrog does not.

3.12 • Crickets and spiny lobsters

How do crickets chirp and spiny lobsters rasp?

Answer A male cricket chirps for a mate by closing its right forewing over its left forewing, after having opened them. As the wings close on each other, a hard pick (a *plectrum*) on the top of the left wing rubs over a series of small ridges (a *file* of hooked teeth) on the bottom of the right wing. The pick hits ridge after ridge, causing the plectrum and file to oscillate, which causes much of the rest of the two wings to oscillate. In turn, the oscillations of the wings, especially in a fairly large region known as the *harp*, produce pressure variations in the air, which travel away from the wings as sound waves—the cricket's chirp. The frequency of the sound depends on the rate at which the plectrum catches and releases the ridges. That rate appears to be controlled by the oscillation frequency set up in the harp in each wing: As each harp moves, it twists either the plectrum or a ridge so that they disengage.

The male cricket's mating call comes with a price, however, because it attracts flies that can acoustically home in on the cricket, to lay eggs in the cricket. The eggs eventually hatch into parasitic larvae that burrow into and kill the cricket. (For man and cricket alike, mating calls often lead to trouble.)

A spiny lobster also rubs a pick (part of an antenna) over a file (a microscopically rough plate located below the eyes), but the situation is different because the pick is soft tissue and does not produce sound by striking the file's ridges. Instead, as the pick is rubbed along the file and over each ridge, the pick sticks on a ridge and is stretched before finally being released toward the next ridge. Upon release, the pick and file oscillate, producing sound—the rasp of the lobster. The rasping is used to startle predators and can function even when the lobster's exoskeleton softens during molting.

3.13 • Frog playing a tree; cricket playing a burrow

Why does a male of the tree-hole frog of Borneo (*Metaphrynella sundana*) chirp for female companionship while sitting in a tree cavity? Why does the chirping of a mole cricket increase in intensity and purity as the cricket digs and builds its burrow?

Answer A male tree-hole frog typically sits comfortably in a puddle of water inside a tree cavity, such as a hollowed-out tree stump. When it advertises for companionship, it experiments with its call by increasing and decreasing the frequency until it matches the lowest *resonant frequency* of the cavity. This is the lowest frequency at which sound waves reinforce one another within the cavity, allowing them to build up a strong (loud) resulting sound wave. Once a match is made, the sound that leaks from the cavity into the environment is loud and carries far to tell of the frog's loneliness.

The burrow-building cricket is similar, in that it digs a cavity in which its chirps cause *resonance*; that is, the chirp frequency matches the resonant frequency of the cavity. The cricket builds its burrow in stages, generally making a bulb and a horn-like section connecting the bulb to the outside, much like the open end of some musical instruments flare to allow sound to escape and be heard. The cricket pauses at the end of each stage to chirp in order to test whether the burrow will resonate. Eventually resonance is met, and the horn efficiently conveys the strong sound waves on the inside to the environment on the outside.

3.14 • Attack of the Australian cicadas

If a male of the Australian cicada *Cyclochila australasiae* calls out nearby while you sleep, you will be startled because the call is very loud (100 decibels at a distance of 1 meter). How can this insect, which is the loudest known insect but only 60 millimeters long, make such a racket?

Answer Each side of the cicada contains a drum-like structure with four vertical ribs that are flexed outward. A muscle acts to pull this structure inward so that the ribs, one by one in quick succession, suddenly flex inward. As it flexes, each rib emits a sound pulse—a click. The series of clicks sets up resonance in an air sac within the cicada's abdomen; that is, the sound waves reinforce one another to result in a large wave. This reinforced sound has a frequency of 4300 hertz and a sound level of more than 150 decibels (more than you would be subjected to at even a heavy metal concert). Sound is then emitted by the abdomen through an eardrum on each side of the cicada. Why the cicada does not go deaf is not well understood.

3.15 • Penguin voices

After diving into the water and eating, an emperor penguin must crawl back onto its home ice floe and return to its mate. In the winter, however, that mate can be anywhere among thousands of penguins huddled together to avoid freezing in the harsh weather of the Antarctic, where temperatures can be −40°C and wind speeds 300 kilometers per hour. Besides, penguins all look about the same, even to other penguins, and so a penguin cannot visually recognize its mate. How, then, does a penguin find its mate among thousands?

Answer Most birds vocalize by using only one side of their two-sided vocal organ, called the *syrinx*. Emperor penguins, however, vocalize by using both sides simultaneously. Each side sets up resonance in the bird's throat and mouth, much like in a pipe with two open ends; that is, sound waves that reinforce one another are set up to produce a strong net sound wave. The frequency of the net sound wave set up by one side of the syrinx differs from that set up by the other side. A listener perceives the average of the two frequencies but also perceives that the average sound warbles. That is, its intensity varies between loud and soft with a certain *beat frequency* equal to the difference in the actual two frequencies. Penguins can perceive these beat frequencies. Thus, a penguin's cry can be rich with different resonance frequencies and different beat frequencies, allowing the voice to be recognized even among a thousand other penguin voices.

3.16 • Whale clicks

A sperm whale vocalizes by producing a series of clicks. Actually, the whale makes only a single sound near the front of its head to start the series. What produces the rest of the series? How can researchers determine the length of the whale from such a series of clicks?

Answer Part of the sound produced by the whale at the front of its head emerges into the water to become the first detected click of the series. The rest of the sound travels backward through the spermaceti sac (a body of fat) in the head, reflects from the frontal sac (an air layer) at the back of the head, and then travels forward through the spermaceti sac. When it reaches the distal sac (another air layer) at the front of the head, some of the sound escapes into the water to form the second click, and the rest is sent back through the spermaceti sac. This cycle is repeated several times, producing several more clicks. The time interval between the successive clicks is related to the distance between the frontal and distal sacs, which is proportional to the size of the whale. Thus, by measuring that time interval, researchers can estimate the length of the whale.

3.17 • Reflection tone

When an airplane flies somewhat overhead and close enough to be heard, lower your head by stooping to the ground. Why does the frequency of the airplane noise increase as you lower your head?

Answer The sound you hear consists of the sound coming directly to you from the airplane and the sound that reflects to you from the ground. The two sets of sound waves undergo interference at your ears and you hear primarily the waves that constructively interfere (they reinforce one another rather than cancel one another). The height above the ground at which constructive interference occurs depends on the wavelength—greater (longer) wavelength requires greater height. As you lower your head, you move down to the heights at which shorter wavelengths (higher frequencies) undergo constructive interference. So, as you stoop, the sound you hear increases in frequency.

You can hear a similar effect if you walk away from a waterfall toward a vertical wall that reflects the waterfall sound to you, so that the reflected sound interferes with the sound reaching you directly from the waterfall. As you approach the wall, the sound you hear increases in frequency.

3.18 • Long-distance sound

My house is located in Cleveland Heights, Ohio, which lies above a flat region joining Lake Erie. A rail line runs along the flat region, at the foot of Cleveland Heights. I certainly cannot see the rail line because not only is it too far away, but also my line of vision is blocked by the ridge leading up to Cleveland Heights and by the thousands of trees and houses. Why then, on certain nights, is the rattle of trains along the rail line easily heard at my home?

When loud, repeated explosions occur at one location, as in repeated artillery fire, they might be heard only in zones surrounding the explosion site. If you were to drive away from the site of the explosions, the sound would decrease in intensity in the first (central zone), disappear in a second zone, and then reappear in a third zone. Why do the zones occur?

When the volcano of Mount St. Helens in the U.S. state of Washington exploded in 1980, the energy released was equivalent to several megatons of TNT. Why wasn't the blast heard any closer than 100 kilometers?

In World War I, near Messines, south of Ypres, Belgium, British servicemen spent a year digging 21 tunnels that extended underneath the German line, at a depth of about 30 meters. Once the tunnels were complete, about a million pounds of explosives were stashed in them, and in the middle of the night on June 7, 1917, the British set off 19 of these 21 stashes (two failed to go off), creating the largest man-made explosion to date. The explosion was heard in London and even Dublin, hundreds of kilometers away. How could the sound from the blast reach that far? (One of the remaining stashes went off unexpectedly during a lightning storm in 1955, fortunately killing only a cow. The remaining stash, at a location that is only vaguely known, has yet to go off, which is a concern for people living in the general area.)

Answer When a sound wave is sent up at an angle to the vertical, it changes its direction of travel if it encounters a change in air temperatures. The sound is said to be *refracted* or to undergo *refraction.* If the temperature decreases, the waves begin to travel at less of an angle to the vertical. If the temperature increases, they travel at more of an angle and can even be "turned over," so as to travel back to the ground. I can hear the distant rail noises on nights when the air well above the area is warmer than the air near the ground, a situation known as an *inversion.* Then some of the sound waves sent up from the rail line are bent back down into the Cleveland Heights region for all to hear (Fig. 3-2).

In times past, this longer transmission of sound during inversions was fairly well known. For example, Zulu men knew that they could hear one another across a valley as wide as 2 kilometers if they waited until evening when the air in the valley was cooler than the air above the valley.

When sound waves from an explosion travel far up into the air, they can be bent back down to the ground by the

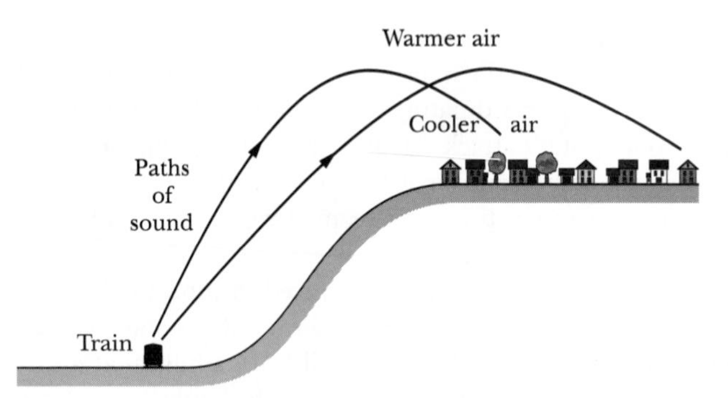

Figure 3-2 / Item 3.18 The paths of sound are bent back down to the ground by the increase in air temperature with height during an inversion.

increase in temperature found at the bottom of the stratosphere (below the stratopause, which is at an altitude of 42 kilometers) and at the bottom of the thermosphere (above the mesopause, at an altitude of 85 kilometers). The sound waves might then return to the ground at a surprising distance from their source, a much greater distance than sound waves traveling along the ground, which are blocked by trees, houses, and other obstacles. Thus, the sound can be heard in a zone farther than the first (central) zone. If this returned sound happens to reflect from the ground, it can take another "bounce" to come back to the ground in another, more distant zone.

When Mount St. Helens exploded, the compression waves (air is compressed by the outward-moving material from the explosion) formed too slowly for the human ear to respond. Thus, the compression wave was not heard (nor did it do any damage to windows and other fragile objects) in Toledo, Washington, 54 kilometers away. However, when the compression waves reached the stratosphere, they bunched up and were redirected back to the ground. When they reached the ground at distances greater than 100 kilometers, their pressure variations were rapid enough to be heard.

The sounds of the explosions near Messines similarly traveled up to the stratosphere and then back down to the ground. However, unlike the Mount St. Helens explosion, the compression waves at the explosion site were rapidly formed and produced tremendously intense sound near the soldiers.

The travel direction of a sound wave is also affected by wind. If the wave travels upward in the downwind direction, the wave's path is bent over so that the wave comes back to ground somewhere downwind. Sometimes this return of the sound is at a surprising distance.

3.19 • Acoustic shadows

In the U.S. Civil War of 1862 through 1865, field commanders on both the Union and the Confederate sides depended heavily on sound to determine when battles began and where they were located. On several occasions, a commander would split his troops to attack the enemy from two directions, but the only way of coordinating the assaults was when the noise from one group's attack signaled the other group to attack. Because the two groups may have been within a few kilometers of each other, this plan seemed reasonable and yet it sometimes failed in decisive battles.

A similar strange effect was noticed in June 1862 by the Confederate Secretary of War and one of his staff as they observed the battle of Gaines's Mill from a hilltop at a distance of no more than two kilometers. The battle down in the valley involved at least 50 000 men and 100 pieces of field artillery, and it created a horrible amount of noise and must have been deafening to the fighting troops. Yet the two observers heard nothing of the battle during their two-hour observation. How could such a battle be inaudible only a few kilometers away?

Answer There are three primary reasons why these tremendously loud battles could not be heard even a few kilometers away. (1) An intervening dense forest could have muffled the sounds by absorbing the sound waves. (2) The sound waves emitted at ground level could have been sent along paths that curved upward instead of paths tending to be horizontal. (3) The path taken by a sound wave could have been curved if either air temperature or wind speed changed with height.

Regarding the third reason, if the air temperature decreases with height, the sound waves sent out along paths that are somewhat angled upward end up traveling along even steeper paths and thus will not reach observers on the ground several kilometers away (Fig. 3-3*a*). The wind speed usually increases with height. In such a normal situation, if sound is sent in the direction of the wind, it tends to nosedive to the ground and thus can be heard (Fig. 3-3*b*). However, if it is sent opposite the wind, it tends to follow a path that bends up into the air and cannot be heard.

In some Civil War battles, the commander was upwind of the battle when there was a significant increase in wind speed

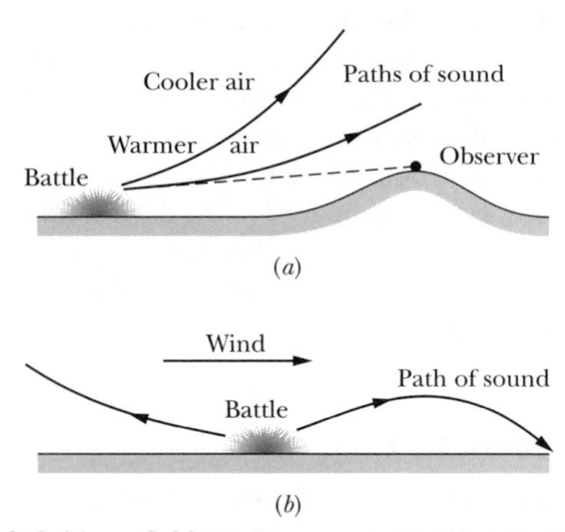

***Figure 3-3** / Item 3.19* (*a*) The sound of battle is bent away from an observer by the decrease in air temperature with height. (*b*) Wind bends the paths of sound.

with height. The commander is said to have been in an *acoustic shadow*. Even stranger were situations where sound waves were redirected upward by the temperature effect but then were redirected downward by the wind effect, ending up back on the ground far from the battle. So, distant soldiers heard the battle but not soldiers who were reasonably close to it.

3.20 • Hearing the Soviet subs

During the Cold War, the United States monitored the submarines of the Soviet Union by listening via an underwater network of acoustic antennas that could detect noises emitted by the submarines, such as the propellers. The curious feature was that the antennas were at midlatitudes and the submarines were 1000 kilometers away in the polar latitudes. How could even a noisy propeller be heard at that distance? The United States enjoyed this eavesdropping capability until a spy tipped off the Soviets.

Answer Some of the sound emitted by, for example, a noisy propeller becomes trapped in what is effectively a pipe, dubbed the *deep sound channel* (DSC), that runs between the polar latitudes and midlatitudes. The channel lies at the depth at which the speed of sound in the water is lowest. The speed of sound depends on both depth and water temperature. If we measure the speed as we descend, the temperature effect initially dominates and the speed decreases as the temperature falls. But eventually, the depth effect dominates and thereafter the speed increases.

So, there is some depth range at which the speed is lowest. If sound travels into this range, it can be trapped much like light can be in an optical fiber. If the sound is directed, for example, somewhat upward into the greater-speed region above the channel, the change in speed bends the path of the sound (sound is refracted) back downward. Similarly, if it is directed somewhat downward into the greater-speed region below the channel, the change in speed bends the path back upward. Noise from the Soviet subs operating in the polar latitudes became trapped in this acoustic channel and then traveled all the way to the antennas in the midlatitudes.

3.21 • Cheerleader's horn, foghorns

If a cheerleader yells directly at a large, noisy crowd, no one in the crowd will hear the yell. But if the cheerleader yells through a megaphone, the yell might easily be heard. How does a megaphone make a yell louder?

Why is the opening of a traditional foghorn wider vertically than horizontally? Doesn't that mean sound is wasted by being sent upward?

Answer When sound emerges from an opening with a size comparable to the wavelength of the sound, the sound waves *diffract*, or flare, into new directions. The smaller the opening is, the greater the flare is. When a cheerleader yells at the crowd, sound emerging from the mouth flares considerably, being sent not only forward but also rightward, leftward, upward, and downward. This flaring greatly reduces the intensity of the sound (the loudness) in any given direction. When the cheerleader yells through a megaphone, with its tapered sides, the sound emerges from a much larger opening (at the far end of the megaphone) and the flaring is considerably less. Thus, the sound is directed primarily in the forward direction and the intensity is much greater. So, a megaphone is used to reduce diffraction of the cheerleader's cheers.

A foghorn's intent is to spread a warning horizontally as widely as possible, so that someone on a ship approaching from any direction might hear the warning. Because horizontal flaring of the sound is desired and vertical flaring is not, the foghorn's opening has a narrow width and a large height.

3.22 • Direction of a whisper

In an open area (where there is little to reflect sounds to you), have someone speak at a fairly constant, low level while the speaker turns from facing you to facing away from you. You will probably be able to hear the words through much of the turn. Next, have the person repeat the turn while whispering at the *same* level (perhaps in a *stage whisper*, as if an audience is to hear it). Why does the whisper become inaudible faster than a normal voice as the person turns away from you?

Answer Here are two explanations, the easier one first: As explained in the preceding item, sound diffracts (spreads out in its direction of travel) when it passes through an opening with a size comparable to the sound's wavelength. The diffraction is less for shorter wavelengths. A whisper consists of shorter wavelengths (higher frequencies) than most of the sound in a normal voice. So, because a whisper involves less spread, a whispering person must be facing you for you to hear the sound.

Here is a more vague answer: Calculation of the spread of the sound from the mouth, first done by Lord Rayleigh in 1896, is difficult. Rayleigh, assuming a small source of sound was positioned on the surface of a sphere, found that the sound waves wrap around the sphere, the waves with shorter wavelengths wrapping around less than waves with longer wavelengths. So, a whisper, with its short wavelengths, does not wrap around the head as much as the sounds of a normal voice.

You can hear a similar effect while in an audience at an outdoor play that lacks adequate surfaces to reflect voices of the performers to you. A male performer might easily be heard even when he faces away from you, but a female performer, speaking just as loudly but at higher frequencies, may have to face toward you to be heard.

3.23 • Doppler shift

If you are stopped at a train-track crossing while a train roars through with its whistle blowing, why does the frequency of the whistle change, and does it change from high frequency to low frequency or vice versa?

Answer The motion of a sound source relative to a listener (you or any other sound detector) changes the frequency of the sound, an effect called *Doppler shift*. The primary reason for the shift is that sound is a wave. If the source is stationary relative to you, the high-pressure phases of the wave sweep past your ears at the same rate (frequency) as they are produced by the source. Thus, you hear the same frequency as the source produces, and there is no shift. If, instead, the source moves toward you, it is chasing the waves that it emits in your direction. The rate at which the high-pressure phases pass you is now greater than the source is producing, and so you hear a higher frequency. If the source moves away from you, the effect is just the opposite: you hear a lower frequency than what is produced. So, "toward you" means a shift up in frequency, "away from you" means a shift down, and the amount of shifting depends on the speed of the source. If the source's motion is misaligned with your direction, the effect (the amount of the shift) is less, and if the motion is perpendicular to your direction, there is no shift.

If you placed a sound detector between the train rails, you could measure a Doppler shift of the train whistle. The whistle would have a certain high frequency during its *entire* motion toward the detector (until the whistle is almost overhead) and then it would have a certain low frequency during its entire motion away from the detector.

If, instead, you placed the detector a safe distance off to the side of the track—say, 20 meters—the measurements would be different for a geometrical reason. As the train nears the detector, its velocity *toward* the detector is progressively less, and so the amount of the Doppler shift decreases. Thus, for much of the train's approach, the whistle has a certain high frequency as measured by the rail-mounted detector. However, as the train gets closer, the misalignment kicks in and the frequency quickly drops until, as the train's motion is perpendicular to the detector's direction, the frequency remains constant. Then the frequency quickly drops again until it reaches a certain low frequency, which is maintained as long as the whistle can be heard.

Suppose you are the detector and you are close enough to the track that the geometric effect can be neglected. You should perceive a *certain* high frequency as the train moves toward you and a *certain* low frequency as the train moves away from you. But surprisingly, you don't. Instead, you perceive a *continual increase* in frequency as the train moves toward you and a *continual decrease* in frequency as it moves away from you. Such a perceived frequency is often called *pitch*. In this situation, the pitch that is brought to consciousness is influenced by how loud the sound is. Because the whistle continually becomes louder as it approaches, you are fooled into perceiving that the frequency continually increases. And because the whistle continually becomes less loud as it moves away, you are fooled into perceiving that the frequency continually decreases. This effect of changing pitch is called the *Doppler effect illusion.*

3.24 • Bat finding an insect

When a bat walks over ground searching for insects (its prey), it is usually guided more by hearing an insect than by seeing it because a bat's vision is poor and they search at night. However, some types of bats can detect and capture an insect, such as a moth, when both are in flight. Although the bat flies quickly, it can home in on an insect and nab it. How does the bat detect not only the presence of the insect but also its direction and speed of flight?

Why do bats have a better time nabbing flying moths near a mercury vapor lamp (commonly used as a street lamp) than over an open field? Why does the advantage disappear if the lamp is a sodium vapor lamp?

When I explored caves in west Texas, I would spend an entire weekend underground. Twice a night thousands of bats would fly past me, first as they flew toward the cave entrance and then out of the cave to search for food and then another time as they flew back in to their roost deep in the cave. Never once, even in the complete darkness of a twisting cave passageway, did a bat fly into me or into a wall. How do they manage to avoid such a collision?

Answer A bat emits a burst of sound waves at frequencies that are too high for you to hear, in a region called *ultrasound.* The sound, which is probably emitted through the nostrils of the bat, reflects from objects in the bat's path, such as walls, spelunkers, and flying insects. The echoes from the objects alert the bat that the objects are in the way. However, this technique presents a problem for the bats in a swarm flying quickly through a cave passageway: How does any one bat distinguish an echo of its own sound from echoes of the sounds of other bats? The solution is that each bat's signal is distinguishable by its particular frequencies, changes in frequencies, and changes in amplitude. Still, the ability to pick out a signal among dozens or even hundreds of other signals, while quickly flying toward a wall, is remarkable.

The bat receives more information than just an echo because it is sensitive to the echo's shift in frequency due to its own motion. Suppose the bat emits sound at a certain frequency while flying toward a wall. The echo that returns to the bat has a higher frequency and is said to be *Doppler shifted.* The faster the bat flies toward the wall, the more the frequency of the echo is shifted. A bat uses the Doppler shift to determine its speed.

Some bats emit sound at a constant frequency and use the Doppler shift to detect not only obstacles but also insects. Other bats emit sound that sweeps through a frequency range. By analyzing the Doppler shift at different frequencies, a bat can determine surface features of a target and thus distinguish the echo returned by an insect from the echoes returned by leaves, for example. The task can be easier if the insect is flapping its wings in the bat's ultrasonic signal, because the variation in the wing tilt causes a variation in the echo return to the bat (at some tilts the echo is strong, and at other tilts it is not). This variation is a definite sign that the echo is from a flying insect.

Some bats prefer to hunt by flying low over water (*trawling*) because the flat surface of the water produces far less *echo clutter* for the bat to sort. Much of the reflection of the bat's signal off the water is away from the bat, but an insect will reflect sound directly back to the bat and will be prominent.

Certain types of insects are sensitive to the ultrasound used by bats. When one of these insects detects such a frequency, especially if the sound is loud, the insect immediately flies erratically and generally in a direction that decreases the intensity of the sound. Some of these insects have an even better defense: They emit a clicking sound that effectively jams the echo detection a bat needs to home in on the insect. The clicking is produced as a cuticle structure undergoes buckling motions as it is flexed. Each sudden buckle produces a sudden change in air pressure, and this repeated variation in pressure travels away from the insect in the form of sound in the ultrasound range. To confuse a bat, these clicks must arrive either simultaneously with the echo off the insect or immediately before it, so that the bat can no longer distinguish the echo.

At night, a mercury vapor street lamp attracts moths and other flying insects; thus a bat can find a tasty collection if it flies near the lamp. Curiously, some of these insects would flee when detecting ultrasound or they would send out a jamming signal, but near the lamp, they do neither. One conjecture is that because they do not fear bats during the daytime (bats sleep rather than hunt then), the bright, white light from the lamp fools them into thinking they are in daylight and thus are safe. A sodium vapor lamp emits a distinctly yellow light, which presumably a moth does not confuse for daylight.

3.25 • Bat finding a flower

How do the bats that feed on the nectar of flowers find the flowers? The pollination of many flowers, especially in the tropics, depends on such bat visitations. When a bat lands on a flower and juts its snout into a slit between petals to reach the nectar, it causes two other petals to catapult pollen up onto the bat's rump, so that the bat carries it to the next flower. Not only must a bat be able to locate a flower, but it must also locate the slit for its snout. How can it do all this with poor eyesight and in darkness? How does a flower prevent a second bat from landing until its supply of pollen has been replenished?

Answer A bat apparently can recognize a flower by the type of echo it receives when it sends its burst of ultrasound toward the flower (see the preceding item). In fact, the petals on some flowers are bell-shaped so that they can better return a recognizable echo to a bat. For example, flower petals on the plant *Mucuna holtonii* form a bell shape that strongly returns an echo to the bat even when it approaches at an angle to the face of the flower (the flower is an acoustic version of an optical *retroreflector* that joggers wear to be visible in headlights at night). The bell's top petal is raised when pollen is available. After a bat leaves the flower with the pollen on its rump, the top petal sags, ruining the bell shape. Thus, a second bat would not receive a strong echo from the flower. Later in the night, as the pollen supply is replenished, the top petal rises, reforming the bell shape and once again giving a strong echo so that another bat might land.

3.26 • Hearing underwater

When your head is underwater, why does sound emitted by someone off to your right seem to come more from the forward direction?

Answer One clue used by your brain to determine the direction of a sound source is the time delay between the arrival of the sound at the ear closer to the source and the arrival at the farther ear. For example, if the source is directly to your right, the time delay of 0.00058 second and your past experience properly tell you that the source is to your right, at an angle of 90° from the forward direction. However, if you and the source are submerged, the time delay will be only one-fourth as great (0.00014 second) because the speed of sound in water is 4 times the speed of sound in air. (The sound moves from the closer ear to the other ear much faster.) This shorter time delay and your experience improperly signal you that the source is at an angle of only 13° from the forward direction.

However, you probably cannot determine the angle very well because the ear-to-ear time delay is muddled by an additional effect. Sound can be more easily transferred from water into your head than from air into your head. So, with your head submerged, sound reaches the farther ear, not only by coming past your head through water, but also by coming through your head. The time delays for these two paths are somewhat different, which gives you conflicting clues as to the direction of the source.

3.27 • Cocktail party effect

In a small party with people standing and talking in pairs, each member of a pair stands at a "socially acceptable" distance from the other, and the two can hear each other without any trouble. However, as the density of people in the room increases, why does hearing become more difficult, and what does each member of a pair do in response? Why can a voice still be distinguished? You might notice these same effects in many other loud environments, such as a noisy restaurant or subway car.

Answer As the density of people increases, the background noise of their conversations (sounds coming directly to you from them plus sounds reflecting from the walls and ceiling, as well as from other people) also increases. When the background noise becomes about as loud as the conservation you are having, you and your partner automatically raise your voices, an effect known as the *Lombard effect*, after Etienne Lombard who studied the phenomenon in 1911. Because all the other pairs of people have the same problem of hearing, they also raise their voices, and so you still have a problem hearing your partner. At some point, to avoid screaming, you and your partner step closer than normal (inside the "personal space" of each other). If someone quiets the party, say for an announcement, and then the conversations resume, the voice levels quickly (exponentially with time) return to their former values. The Lombard effect has been studied in some animals, such as birds that automatically raise the sound level of their calls when confronted with increased background sounds from other birds.

If someone were to record your conversation with your partner, using a single microphone, and then play it back later (in a quiet room), you probably would not be able to distinguish your partner's words as well as you did when you heard them "live." The difference is that, in person, you hear the partner with two ears—the slight delay between what the ears hear and the slight difference in sound intensities at the ears help you to distinguish your partner's voice from the onslaught of other voices. The effect is dubbed the *cocktail party effect*. Being able to see your partner's mouth motion and "body language" can also help you fill in for words or even for entire sentences that you did not hear clearly. None of these clues are available if you listen to a single-microphone recording of the conversation. Then you must use other clues, such as searching for intelligent thoughts or recognizable tones buried in the background noise. Being able to pull a conversation out of such background noise is sometimes too easy, such as when you can easily hear the voice of an audience member who was sitting near the recorder when a bootleg recording of a concert was made. Being able to recognize familiar sounds in a noisy background is used by some animals, such as a chick king penguin hearing its parents in the din produced by thousands of other nearby king penguins.

3.28 • Sound emitted by the ears

About 60% of people emit sound from the ears, an effect known as *otoacoustic emissions* (OAE). Most emissions require a microphone and amplification to be heard, but some emissions are audible if you stand near the person in a fairly quiet room. Why do ears emit sound?

Answer When sound activates the eardrum, the oscillations are relayed into the inner ear (the *cochlea*), which consists of two fairly long compartments of fluid separated by the *basilar membrane*. The sound-sensing organ is the *organ of Corti*, which lies on this membrane. When a sound signal is relayed into the organ of Corti by oscillations of the basilar membrane, hair-like rods in the organ begin to sway, which trigger electrical impulses to be sent to the brain as information about the sound. This detection is highly frequency sensitive—sound with a particular frequency stimulates the hair-like rods in a particular region. This selectivity is due to a controlling system that feeds part of the signal back into the detection region. That feedback of energy can set the basilar membrane oscillating without an external stimulus, sending oscillations back to the eardrum, which then produces sound waves that travel out of the ear canal. Such sound waves are very minor in most people, but if someone ever says that their ears are buzzing, you might actually be able to hear the buzzing.

3.29 • Music in your head

Heavy-rock classics, such as songs by Iron Butterfly or Led Zeppelin, carry heavy bass lines. However, small loudspeakers, such as those in a car, cannot produce bass notes—the sound waves require long wavelengths, which cannot be produced on a speaker cone with a small diameter and depth. Still, the music sounds acceptable on those small speakers. So, how can you hear the bass line?

Answer The low-frequency sounds are produced in your head because of two effects. One has been dubbed the *missing fundamental effect*, which has to do with your perception of a harmonic series. Such a series consists of the lowest frequency (the fundamental) and higher frequencies (overtones) that are integer multiples of the lowest frequency. For example, if the fundamental is 500 hertz, then the series consists of 2(500) = 1000 hertz, 3(500) = 1500 hertz, 4(500) = 2000 hertz, and so on. Suppose that car speakers can reproduce any frequency above 800 hertz but cannot reproduce any lower frequency. If this harmonic series is sent to the speaker, the fundamental of 500 hertz cannot be reproduced, but the higher harmonics can be. However, the neurological system responsible for your perceiving those higher harmonics, and recognizing that they are part of a harmonic series, also creates a perception of the fundamental, even though the fundamental is not coming into your ears. So, when you listen to heavy rock on the speakers, the higher harmonic frequencies in a note are enough for your neurological system to discern the fundamental of the note. Exactly why neurological systems do this is not understood.

The second effect responsible for perceiving a bass line is that the ear's hearing mechanism is *nonlinear*. That is, it has a distorted response to the variations of any sound wave entering the ear. It presumably is nonlinear in order to respond to sound levels (loudness) over a very wide range and also to be able to sort sounds according to their frequencies. One by-product of this distorted response occurs when an ear deals with two frequencies—say, $f_1 = 1000$ hertz and $f_2 = 1500$ hertz. If the sound waves are fairly loud, a wave with a frequency equal to the difference in those two frequencies ($f_2 - f_1 = 500$ hertz) ends up being produced in the inner ear. Because these two frequencies are successive frequencies in the harmonic series listed above, the difference is equal to the fundamental frequency. Thus, even though the fundamental is not entering the ear, it is being produced in the inner ear by the ear's nonlinear response.

You can hear the *difference tone* $f_2 - f_1$ in other situations where two sound sources emit reasonably close frequencies *loudly*. They must be loud to cause the inner ear to produce distortion. For example, if one flute emits f_1 while another flute emits f_2, then you can hear a "ghost flute" at the frequency $f_2 - f_1$.

The difference tone can also be heard by blowing hard on a two-hole British police whistle. Covering the near hole with a finger, you hear the frequency due to the far hole being open. Covering the far hole, you hear the frequency due to the near hole being open. Leaving both holes open, you hear those two individual frequencies as well as a third frequency, the difference tone.

The difference tone is also employed in pipe organs: A pipe producing a low C at 16 hertz would require a length of about 10 meters, which is very long, heavy, and expensive. However, a pipe producing C at 32 hertz and a pipe producing G at 48 hertz, if played together loudly, produce the low C at 16 hertz as a distortion product in the inner ear. So, the two shorter, more affordable pipes make extra music in your head.

3.30 • Noise-induced hearing loss

When rock and roll began, countless parents scolded their teenagers that the music would ruin their hearing. Early studies showed that the prediction was just not true. However, as rock evolved and especially as it became highly amplified in concerts and nightclubs (and then highly concentrated in headphones), hearing damage began to show up. Indeed, after years of exposure to loud music either during performances on stage or playback on studio headphones, some veteran rockers developed severe problems. For example, Ted Nugent lost all hearing in one ear, and Peter Townshend (of the Who) and Lars Ulrich (of Metallica) developed a continuous ringing (*tinnitus*) that is pronounced enough to interfere with concentration and sleep.

Many DJs working in loud concerts and nightclubs have either a temporary loss of hearing after leaving work, a permanent loss of hearing, or tinnitus. Headphone damage is also beginning to show up in those people who cranked up the music on portable playing devices. Of course, many other sources of loud or abrupt sound can cause hearing loss—from leaf blowers, power mowers, nearby fireworks, shotgun blasts, jack hammers, jet engines, motorcycles, and race cars. Many people are now taking precautions. Some wear *passive* ear plugs (foam cylinders) that block up the ear opening. Others use *active* (noise-reducing) ear plugs and headphones that monitor any continuous background noise (such as from an airplane engine) and then cancel it out so that it is not heard.

Why do loud sounds cause these several types of hearing problems? How do noise-reducing ear plugs and headphones cancel noise?

Answer The details of temporary or permanent loss of hearing due to loud noise are not understood. The temporary loss may result in a decrease in blood supply to the inner ear because of constriction of blood vessels. The permanent loss may be due to the bending of the cochlea hairs that are responsible for transforming sound frequencies into neurological signals to the brain. If the hairs are bent and the signals change from the norm, the brain may think that the change indicates sound entering the ear, thus bringing that sensation of sound up to consciousness, producing the sensation of ringing.

A miniature device in active ear plugs and headphones monitors environmental sound and produces its own sound. If the environmental sound is fairly constant, then the miniature device produces a sound wave of the same amplitude and frequency. This would seem to make matters worse. However, the produced wave is exactly out of step (out of phase) with the wave from the environment, and thus the two waves cancel each other within your ear by destructive interference. The effect can be impressive: While wearing noise-reducing headphones with its switch off, the drone from airplane engines, for example, can be almost oppressive. But when the switch is flipped to activate the headphones, the drone suddenly drops to a faint whisper.

3.31 • Sound enhanced by noise

Normally, noise tends to *mask* (hide) a signal, as noise might mask a friend's voice at a very noisy party. (The *signal-to-noise ratio* is less than 1.0, which means the signal is lost in the babble.) However, in some situations, noise can actually increase the audibility of a signal. For example, if you listen to music while turning down the volume, the music is eventually too faint to be heard. If you then turn on a noise source that produces a fairly uniform sound (like a drone) and adjust the sound level of the source, you may find that you can again hear the music. How can noise make the inaudible music audible?

Answer The music consists of sounds of various sound levels, but if you turn down the volume until the sound is inaudible, you cannot hear even the loudest of the sounds. When you then turn on a fairly constant background noise, the sound level of the noise is added to the sound level of the music. Whenever the louder parts of the music occur, the added noise boosts those parts into the audible range. Now, you can start to pick out the rhythm of the music and maybe even some of the details. You are certainly not hearing high-quality music, because you miss all the low-volume parts. However, you hear enough to recognize the music.

3.32 • Stethoscopes and respiratory sounds

The sounds produced inside a patient, through the chest, back, or throat regions, can signal to a doctor that something is wrong. The doctor obviously cannot hear those sounds by simply standing near the patient. So, a stethoscope is used. Could the doctor better hear the sounds by pressing an ear to the patient (granted this could be very awkward)? What causes the sounds?

Answer The sounds are generated primarily by the blood flow through the heart and by the airflow through the lungs and throat. The sounds from the airflow are still not well understood but are usually attributed to turbulence that causes pressure changes in the air, which then send sound waves through the chest, back, and throat. Either unusually high turbulence or unusually low turbulence (*silent chest*) can indicate airflow problems and damaged lungs. Crackling and wheezing (which lasts longer than crackling) can indicate obstruction in the airway, which can be symptomatic of asthma.

The various sounds produced inside a patient are transmitted to the chest wall, the lower-frequency sounds being transmitted more strongly. However, the sounds are transmitted poorly across the chest–air interface. A doctor might be able to hear some of the sounds (certainly the heartbeat can be heard) with an ear pressed directly to the chest wall, because the sounds can set up resonance inside the ear canal. Indeed, pressing an ear to the patient was an early way of listening to sounds from the body. However, a stethoscope is less awkward and embarrassing. In addition, because low-frequency sounds can set up resonance inside the stethoscope tubes, the intensity of the sounds can be amplified by the stethoscope.

Traditional stethoscopes come in two main types: Either a metal diaphragm or a rubber bell is pressed against the chest. The chest sounds cause the diaphragm or the air within the bell to oscillate, which then causes the air in the stethoscope tubes to oscillate, so that the doctor can hear the oscillations. Both diaphragm and bell are wider than the tubes so that they collect sound over a fairly wide area on the chest but not so wide that the doctor cannot distinguish the location of a sound source within the chest. Experiments generally show the diaphragm type to be better than the bell type at transmitting sound to the listener, but many doctors still prefer the bell type.

3.33 • Tightening guitar strings and rubber bands

Why does tightening a guitar string increase the frequency you hear when you pluck the string at a certain point? If you tighten a rubber band by stretching it between thumb and finger while also plucking it, why does the frequency you hear either remain the same or decrease somewhat? Why should a guitar be played for a while offstage before it is played onstage?

Answer You hear sound from a guitar string when you pluck it because some of the waves you released along the string reinforce one another, a situation called *resonance*. The reinforcement means that the string's motion in the air is fairly large, which causes audible air-pressure variations. Most waves on a string do not result in significant string motion, but waves with certain wavelengths can produce resonance and thus sound.

For example, one of the waves sets up the resonance called the *fundamental*, which results in sound with the lowest frequency of all the frequencies the string can produce. The value of this frequency depends on the string length and also on the speed at which waves can run along the string. That speed depends on the string tension and density. So, three factors (length, tension, and density) determine the frequency of the sound emitted by the string.

If you tighten a guitar string, you increase the tension without significantly changing either the density or the length of the string. The result is faster waves on the string and higher frequency of the emitted sound. If you tighten a rubber band, you increase the tension but you also decrease the density and increase the length. The result is that the wave speed does not change appreciably. Thus, the frequency of oscillation and the frequency of the sound do not change appreciably.

As a guitar string is played, the motion increases its temperature and the string expands. Thus, the tension decreases and the frequencies that the string can produce also decrease. A performer does not want this to happen onstage, because then the string must be tightened to put it back in tune. So, a guitar is played offstage until the strings warm, and the tension is then adjusted to put them back in tune.

3.34 • Bowing a violin

How does bowing a violin string produce sound? Why do you hear almost no sound if you bow at the string's midpoint? (If the string does make a sound, you'll probably find it to be quite unpleasant.) Why is rosin applied to a bow?

A certain string, with a certain length, tension, and mass, can produce sounds with a series of frequencies that are said to be part of a harmonic series. For example, the lowest frequency, called the fundamental frequency, might be 500 hertz. The next higher frequency, called the second harmonic or first overtone is $2(500) = 1000$ hertz. You can find the rest of the series by multiplying 500 hertz by other integers (3, 4, 5, etc.). Any particular bowing location and finger stop (which determines the length of string that can actually oscillate) determines which of the series is actually produced by a string. Surprisingly to many violinists, a string can also be made to produce *subharmonics*—that is, frequencies that are even lower than the fundamental frequency, perhaps half as much. How can such subharmonics be played by a violinist?

Answer If you pluck a guitar string, the string oscillates in several of its resonant modes, where waves traveling through each other in opposite directions reinforce one another to produce an interference pattern. For each interference pattern, the string has regions of strong oscillation where it produces pressure variations in the air, and those variations travel away from the string as sound waves.

A bowed string is quite different because waves are not created on the string with a single pluck. Instead, as the bow is brought, say, upward across the string, the bow and string undergo a repetition of catch and release, or, as it is known, *stick and slip*. The string may initially catch on the bow and be brought upward, but eventually it slips off the bow. At that release, two triangular waves leave the release point in opposite directions on the string. One heads toward the near end of the string (the end nearest the violin player) and the other heads toward the far end of the string. Each wave reflects (and is inverted by the reflection) at the end of the string and then travels along the string to the opposite end. All the while, the bow is still moving upward across the string, but it is slipping over the string. At times, when the triangular waves return to the location of the bow, the string catches on the bow and is brought upward until it again is released. The triangular waves then continue.

The violinist must develop an intuitive feel and ear to synchronize the bowing with travel of the triangular waves without, of course, being able to see those waves. If the triangular waves are properly established, their oscillation produces variations in the air pressure, which travel away from the violin as sound waves. Their oscillation can set up resonance in the violin's wood structure and air cavity, and these two sets of oscillations produce additional sound waves at various frequencies, adding to the *richness* or *timbre* of the sound you hear.

A bow consists of horsehairs, which have a surface of tough scales and an interior of softer material. As a bow is used repeatedly, the hair gradually develops a groove along the side that rubs against the string, exposing the softer interior. When rosin is applied to the hair, rosin particles are partially embedded in the softer material; the exposed parts of the particles provide places on which the string can catch when it is being bowed. The bowing gradually removes the particles, which means that more rosin must be applied to the hair in order for the bow and string to have the proper stick–slip action.

I do not understand the published explanations of why bowing at the string's midpoint produces either no sound or a terrible sound. However, I believe it is due to the symmetry of the two triangular waves released by midpoint bowing. Normally, the waves travel different distances to reach the end of a string, but with midpoint bowing, they travel equal distances and arrive back at the bow simultaneously. In this way, they add, giving a larger than normal deflection of the string at the midpoint, which ruins the proper stick–slip action with the bow. The string then emits no sound or only a rude sound.

To play a subharmonic frequency, the violinist presses down hard on the bridge or even twists it while bowing a catgut string (not a synthetic string). Why this procedure decreases the frequency below the fundamental frequency has not been fully explained, but it appears to involve *torsion waves* that produce not just a sideward displacement of the string but also a twisting motion. Such waves could travel slower along the string than the normal triangular waves mentioned above, and that lower speed would result in lower frequencies of string oscillation and thus lower frequencies of sound.

3.35 • Flashing brilliance of a violin

For frequencies below about 1000 hertz, the sound from a violin comes from the direction of the violin. However, for higher frequencies, the sound can appear to come from different directions for different frequencies. Thus, as the sound changes frequency at the higher frequencies, the direction of the sound source seems to change, an effect sometimes called "flashing brilliance." The sound might then seem to be disconnected from the physical location of the violin. What causes this perception?

Answer When a sound source is small compared to the sound's wavelength, the sound seems to originate, reasonably, at the source even as the frequency is changed. However, when the sound source is large compared to the wavelength, different parts of the source can then act as separate sources, each beaming out their sound. At any given frequency, these sound beams overlap, producing an interference pattern with different intensities. The pattern shifts if the frequency is changed. If you hear a range of high frequencies, this difference in the interference pattern for the different frequencies can give the illusion that the different frequencies originate from different locations.

When a violin emits any frequency above the lower end of the audible range, the sound comes primarily from the wood. When the frequencies are above 1000 hertz, the wavelengths are small enough that parts of the wood emit as though they are separate sources, and then the illusion of flashing brilliance can occur.

3.36 • Conch shells

In times past, conch shells were blown to warn ships of the danger of rocks in dense fog. Today, they are blown largely for ceremony. To blow a conch shell, you first press your lips onto a narrow opening that is made by breaking or grounding off the tip of the shell. Why do the shells produce such a loud sound?

If you pick up a fairly large seashell on a beach and hold it near one ear, why do you hear sound from the shell that resembles the sound of waves breaking on a shore?

Answer Blowing a conch shell involves two sets of oscillations: Your lips are supposed to oscillate (buzz) somewhat like a guitar string and, if a match in frequencies occurs, your lip oscillations can set up oscillating sound waves inside the shell. You make your lips oscillate by blowing through them while they are up against the hole in the shell. If you do this carefully, the lips oscillate at several frequencies simultaneously in a series of values called a *harmonic series.* In one set of measurements, the lowest frequency, called the *fundamental frequency*, was 47.5 hertz and the higher frequencies were integer multiples of that value: $2(47.5) = 95.0$ hertz, $3(47.5) = 142.5$ hertz, and so on.

The oscillations of the lips create sound waves inside the shell, with the same frequencies as the lip-oscillation frequencies. Most of the sound waves simply cancel one another out, but some of them, at one of the *resonant frequencies* of the shell, can reinforce one another to build up a strong (loud) sound wave. In the experiment, the lowest such resonant frequency of the shell was 332.5 hertz.

The shell blower was able to make the shell sound off, because the seventh frequency in the harmonic series of lip oscillations was 33.5 hertz. Thus, the lip oscillations at that frequency produced *resonance* inside the shell at that same frequency, and the shell could easily be heard by someone else.

Normal environmental noises can also set up resonance within a shell. If you hold a shell near one ear, you can hear some of the resonant frequencies. The sound you hear from the shell probably fluctuates, because the environmental noise that produces it fluctuates. If you want to make sense of the fluctuating sound, especially because you are holding a seashell, you can easily imagine that the sound is of breaking waves.

The Stromboli Volcano in Italy is similar to the conch shell, in that fluctuating wind through its ducts can set up resonance in the ducts, making the volcano emit sounds with a fluctuating intensity.

3.37 • Didgeridoo

The didgeridoo (or didjeridu), the traditional musical instrument of the Australian Aborigines, produces a fairly continuous drone when played. It is simply a long tree branch with an end-to-end tunnel that has been carved out by termites. You play the branch by pressing your lips onto and into one end of the tunnel and then blowing in a certain way. However, producing a loud, fairly steady sound is notoriously difficult, even for someone skilled at brass instruments. How is the didgeridoo sound produced?

Answer A primary difference between playing the didgeridoo and a brass instrument is that with the former you must set up a strong resonance in your vocal tract (the combined cavities of your mouth, nose, and upper throat). That is, you set up sound waves in the vocal tract that reinforce one another to produce a strong sound wave. Then you allow some of that sound to leak into the didgeridoo in either of two ways: One way is to continuously allow the vocal-tract resonance to make your lips oscillate (buzz), and the portion of the lips inside the end of the tunnel sets up oscillations of air in that area. The second way is that you periodically open your lips to allow a burst of sound to escape from your vocal tract into the tunnel. You can change the sound emitted by the didgeridoo by changing the resonance in your vocal tract (by changing the oscillations of your vocal folds in the larynx or by shifting your tongue to change the shape of the vocal tract). In a sense, you are singing, humming, or droning into the didgeridoo, which you would not do while playing a brass instrument.

3.38 • Silo quaking and honking

If grain is allowed to pour from a silo (a construction for storing grain), why can the silo oscillate (dubbed *silo quaking*) and emit sound (dubbed *silo honking* or *silo singing*) that resembles the repeated blasts from a truck horn? (Some silos quake but don't honk, some honk and don't quake, some do neither, and some do both.) Silo honking is just annoying, but silo quaking can ruin the silo.

Answer Although grain flow from the bottom of the silo may be continuous, the descent of the grain within the silo is jerky. This intermittent motion may have several causes, one being the periodic formation and collapse of arches in the grain. However, the main cause appears to be intermittent sticking and slipping of the grain along the interior wall of the silo. That variation in the descent of the grain causes oscillations in the grain, which then causes the wall to oscillate. The walls act as giant sounding boards, broadcasting sound waves into the air. Sometimes the oscillations are strong enough to cause a silo to collapse.

In some silos the oscillations might also cause acoustic resonance in the column of air above the grain. That is, sound waves that have the correct wavelength to reinforce one another produce a large net sound wave in that column of air, somewhat like that produced in an organ pipe.

3.39 • Singing corrugated tubes

A corrugated plastic tube is sold as a musical toy. The tube, which may be about a meter long, is usually held at one end and moved in a small circle, causing the other, far end to move in a larger circle. For slow circling, nothing is heard, but for somewhat faster circling, the tube emits sound at a certain frequency. For even faster circling, the tube emits sound with a higher frequency. You might be able to produce four or five different frequencies by adjusting the speed. Why does the tube "sing"?

Answer As the far end of the tube moves in a large circle, the internal air is effectively slung outward along the tube. (The tube wall makes the air rotate but there is nothing holding the air in a circle, and so it flows outward along the tube.) As air leaves the far end of the tube, new airflows in the near end, and thus a continuous stream of airflows through the tube.

If the flow is fast enough, the internal ridges of the corrugated tube prevent a smooth flow. Instead, the airstream develops turbulence—that is, variations in air pressure. These air-pressure variations occur at frequencies within a certain range set by the speed of the airflow and by the spacing of the ridges. If a frequency within that range matches a *resonant frequency* of the tube, *resonance* is set up in the tube. That is, sound waves at that frequency reinforce one another and build up a strong sound wave. Part of this sound wave leaks from the open end and can be heard. Faster rotation and airflow through the tube shift the frequency range of the turbulence to higher values. One frequency in this new, higher range will match a higher resonant frequency of the tube, and you hear sound with a higher frequency.

The toy may not give you the lowest possible frequency, because the airflow through the tube for slow circling is too slow to result in turbulence. The lowest resonant frequency heard is the second harmonic (or first overtone).

You can get a corrugated tube to "sing" if you hold it out of a car while the car is moving. (Do not do this while *you* are driving!) Hold the tube end into the oncoming air so that air is forced through the tube.

3.40 • Coffee mug acoustics

Pour hot water into a coffee mug and then either tap the bottom of the exterior surface with your knuckle or tap the interior surface with a spoon while stirring. Note the frequency. Then add a powder, such as powdered instant coffee, and again tap. The frequency is now much lower but rises back to its original value within a few minutes. Why is the frequency lowered, and why does it then increase?

Answer When you tap on the mug with a spoon, you make the mug walls oscillate at a certain frequency and you set up momentary sound waves in the column of water. Here we are concerned with the second effect, and you minimize the first effect by tapping on the bottom of the mug with your knuckle or some other soft instrument. Of all the sound waves you produce, some have the proper wavelength to fit neatly in the column of water between the closed bottom and open top, in what is called *resonance.* Those sound waves reinforce one another, and thus make a large net wave. Some of the sound leaks from water and is heard, and the frequency is said to be a *resonant frequency* of the cup of water.

That resonant frequency depends on the height of the water column and the speed of sound in the water. The speed of sound in any material depends on the density of the material and how compressible it is. Greater density tends to yield greater speed, but greater compressibility tends to yield lower speed. In water, the speed is about 1470 meters per second.

When a powder is added to the water, air bubbles form on the powder grains. (The air is already dissolved in the water or it adheres to the grains as the powder enters the water.) The bubbles do not take up much volume (you don't see the water surface rise along the side of the mug), so they don't change the density of the water by much. However, they do significantly increase the compressibility, thereby decreasing the speed of sound in the water, which decreases the resonant frequency. Thus, the frequency you hear decreases when the powder is added.

Because most of the bubbles rise, they gradually reach the top surface of the water and pop open. As their number decreases, the frequency increases until it is back at its original value, before the powder was added. Because the best paper on this subject deals with hot chocolate instead of coffee, this frequency change is commonly known as the *hot-chocolate effect.* Pouring or stirring salt into a glass of beer can also give the hot-chocolate effect, but unless you get rid of the foamy head, you may not be able to hear it. (Besides, putting salt into beer is just *wrong.*)

3.41 • Bottle resonance

If you blow across the narrow open end of a bottle (soft drink or alcoholic drink), you can get the bottle to sound off. Indeed, by using a variety of bottles with different levels of liquid inside, you can play a tune. What produces the sound?

Answer When you blow across the lip of the bottle's open end, the moving air breaks up into turbulence, which contains pressure variations within a wide range of frequencies. You want one of those frequencies to match a *resonant frequency* in the bottle. That is, you want pressure variations to set up oscillations of the air inside the bottle that will then reinforce one another to build a strong sound wave. If you get a match in frequency, and a strong internal wave, part of the sound leaks out of the bottle and can be heard.

However, bottle oscillations are not like those we can set up in a simple tube. The difference is that the bottle has a neck, and the air in the neck (air plug) and the air within the rest of the bottle form what is called a *Helmholtz resonator.* The oscillation of this resonator is mathematically like the oscillation of a block on the end of a spring. Here the air plug acts like the mass of the block, and the rest of the air in the bottle acts like the spring. In a normal spring–block system, the block repeatedly compresses and extends the spring, always overshooting the midpoint, thus continuing to oscillate. In the bottle, the air-plug mass causes the air in the rest of the bottle to undergo repeated compression and expansion, always overshooting equilibrium and thus continuing to oscillate.

A given bottle (with a given air-plug mass and a given air spring) will oscillate strongly at a certain frequency, and if that frequency is available in the turbulence created at the bottle's lip, a strong sound wave will be built up inside the bottle. If you decrease the air volume within the bottle by partially filling the bottle with liquid, you increase the frequency at which the bottle will oscillate.

Some caves are well known for the strong wind through them, especially near the entrance. Because the wind reverses direction during the day, the cave is said to be *breathing.* Such a cave is another example of a Helmholtz resonator. External variations in wind and air pressure produce turbulence. The air in the narrow cave opening acts as an air-plug mass, and the air in the rest of the cave acts as a spring. The frequency of the oscillation is too low (0.001 to 1 hertz) to be heard, but you can feel the airflow (the wind) from the oscillations.

3.42 • Fingers on a chalkboard

Why does a squeaky door squeak? Why do fingernails drawn quickly across a chalkboard squeal? Why do tires screech when a car drags out from rest?

Answer These are three examples of a great many examples of what are called *stick-and-slip effects* (or, simply, *stick–slip effects*). Two surfaces move past each other while being forced against each other. In some cases, they might move smoothly, especially if they are lubricated. However, in some cases they first catch on each other, bonding together, then they stretch each other, and finally they release each other. Just after the release, as the stretching is undone, portions of the surfaces can oscillate, producing a sound wave that can be heard. Their motion might also set up oscillations in a larger region, which then acts as a soundboard, making the sound louder.

For example, when a fingernail is drawn across a chalkboard, the fingernail first catches and bends, and then it suddenly releases and slides over the board, oscillating and beating against the board. You hear the sound from both the tapping of the fingernail and from the oscillations the tapping sets up in the board, which acts like a sounding board. The motion of the fingernail is greatest at the outer end and least (zero) at the opposite end, much like the motion of a tree oscillating in a strong wind. Also like a tree, the frequency of oscillation depends inversely on the length of the fingernail. Because the fingernail is short, the frequency is high, which is one reason that the sound is so unnerving.

Rusty hinges on a door can squeak if portions rubbing together undergo repeated stick–slip. If you rotate the door faster, you might eliminate any chance of sticking and thus also eliminate stick–slip and the squeaky sound.

Tires slipping over dry pavement undergo stick–slip, which causes the tires to oscillate, producing sound. It is, in fact, the screech that some people prize in street drag racing. The tires will also screech if they lock up in a hard-braking stop (without any automatic braking system to intervene), but then the sound is not prized.

If you listen, you can find hundreds of other examples of sound being produced by stick–slip.

3.43 • Rubbing wineglasses

If you rub a wet finger around the rim of a wineglass or any of many other types of drinking glasses, you can make the glass sing. What produces the sound?

Answer As your finger rubs against the rim, the finger and rim are continually undergoing sticking and slipping. During the sticking phase, the rim is pulled very slightly in the direction of your finger's motion, distorting the shape of the rim. During the slipping phase, the rim breaks free of your finger and attempts to regain its original shape, but it ends up oscillating. The strongest oscillation is said to be *resonance*, in which the rim oscillates as shown in the overhead view of Fig. 3-4. This pattern of oscillation follows your finger around the rim, producing a pulsation to the sound (it comes and goes with a frequency of a few hertz, depending on the speed of the finger on the rim). The frequency at which the rim pushes on the air and the frequency that you hear are roughly proportional to the rim thickness and inversely proportional to the square of the glass's radius at the open end. Thus, generally the frequency is higher for a thicker rim and smaller radius. If you add liquid to the glass, you lower the resonant frequency because the liquid's mass decreases the rate at which the glass wall can oscillate.

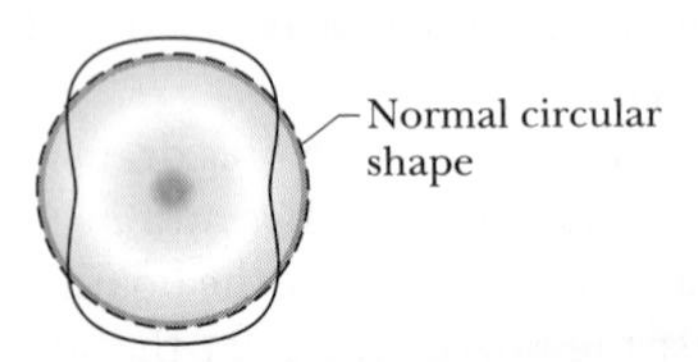

Figure 3-4 / Items 3.43 and 3.44 Exaggerated overhead snapshot of oscillating rim of a wineglass.

Some musicians are skilled at playing music on an array of glasses containing various levels of liquid (adjusting the level allows you to tune the glass). Benjamin Franklin, the famous inventor and statesman of early America, extended the idea of singing wineglasses by building the glass harmonica. This musical device, which became quite popular, consisted of glass rims that were mounted on a horizontal spindle. The rims were graduated in diameter, largest at the left, so that notes could be played on the glasses as they were turned via a foot treadle. A musician would press wet fingers against the rims to rub them as they turned.

Other quaint instruments can be made to sound off when they are rubbed and made to oscillate. A Chinese brass water-spouting bowl is one of the most charming. When it is partially filled with water and its handles are rubbed with dry hands, the bowl oscillates so vigorously that it can throw large water drops upward by half a meter.

3.44 • Shattering wineglasses with voice

Can a trained singer shatter a wineglass or some other glass container simply by singing, as portrayed in cartoons, comedy sketches, and advertisements?

Answer A wineglass can be shattered if subjected to very intense sound at a frequency matching its lowest *resonant frequency*—that is, the lowest frequency at which it would oscillate if thumped. At that frequency, the rim oscillates in the pattern of Fig. 3-4. As the oscillations build, a crack can develop, either at a microscopic defect in the glass or near a place of maximum oscillation. The repeated motion at a defect widens a crack and allows it to branch as it moves through the rim. The wineglass then quickly shatters. To do all this, the intense sound must be applied for a few seconds.

However, doing this with an unamplified human voice seems impossible because a steady frequency cannot be maintained for several seconds. In fact, experiments with unamplified singers have failed to break wineglasses.

3.45 • Murmuring brooks and rain noise

What produces the murmur of a murmuring brook and the plop of raindrops hitting a pond?

Answer The sound of water hitting water, either in a brook, waterfall, or rainfall, is due primarily to two mechanisms: The impact itself causes pressure variations in the air, which travel away from the impact as sound waves; we hear this sound as a noisy (brief) pulse. The impact also often traps air in the water as bubbles, which then generate sound as they oscillate in volume. That is, they increase and decrease in volume, sending pressure variations through the water and out into the air. Eventually, the air bubbles collapse or break open on the water surface, adding a faint splashing sound.

If a water drop in rainfall or a waterfall strikes a solid surface, such as sidewalk or rock, you hear only the impact noise because no air bubbles are produced or trapped. The next time you are near a sidewalk when rain begins to fall, see whether you can distinguish the change in sound from the first impacts (on the dry sidewalk) to the later impacts (on puddles that have collected on the sidewalk).

3.46 • Jar and beaker resonance

As you pour water into a straight-walled container such as a drinking glass, jar, or beaker, why does the frequency of the noise you hear increase?

Answer The air column inside the container (from the open mouth to either the liquid surface or the container bottom) acts like a tube with one open end. The noise from the splashing water (see the preceding item) produces sound over a fairly broad range of frequencies. One of those frequencies matches the lowest resonant frequency of the air column in the container. That is, the pressure variations at that frequency set up sounds in the air column that reinforce one another to produce a strong net sound wave. You hear part of the sound that leaves the air column, and that sound is primarily at the resonant frequency. (You also hear fainter noise coming directly from the splashing.)

The resonant frequency of the air column depends inversely on the length of the air column. So, as the container fills and the air-column length decreases, the resonant frequency increases. You can tell from the sound alone when the container is almost full.

3.47 • Rumbling from plumbing

What causes the rumbling you sometimes hear from pipes when you turn on a faucet?

Answer The sound from plumbing pipes is usually due to turbulent flow of water through the pipes, especially at turns and fittings where the water flow must change direction or get around obstacles. Turbulence in the water consists of vortexes that produce pressure variations. During those variations, the water pressure can decrease enough that air bubbles form from the air dissolved in the water, a process known as *cavitation.* The sudden presence of a bubble, the oscillation of the bubble, and the eventual collapse of the bubble send waves through the water. Those waves shake the pipes and you hear a rattle. Often you can eliminate the noise by decreasing the water flow, so that less turbulence is produced.

3.48 • Knuckle cracking

What causes the sound you hear when you pull on a finger to "crack" a knuckle? Why must you wait a while before you can crack the knuckle again?

Answer When you pull on a finger to crack a knuckle, you widen the space between the bones that form the knuckle and also decrease the width of the knuckle cavity. That cavity contains an initially thin layer of *synovial fluid* separating the bones. If the pull on the finger is done with enough force, the sides of the cavity can snap outward, increasing the width of the cavity and decreasing the pressure within the synovial fluid. This sudden pressure decrease allows one or more gas bubbles to form from the gas, primarily carbon dioxide, that is dissolved in the fluid. The sudden appearance of bubbles, called *cavitation,* sends a pressure pulse through the fluid, the knuckle cavity, and out into the air. When the pulse reaches your ear, you hear a cracking sound. To repeat this performance, you must allow 15 to 30 minutes for the cavity to recover its initial shape, for the synovial fluid to be squeezed back to a thin layer between the bones, and for the gas to be redissolved into the fluid. Until then, you need some other bad habit to annoy people around you.

3.49 • Korotkoff sounds

The traditional way to measure blood pressure is to inflate a cuff wrapped around the arm and then listen through a stethoscope as the cuff pressure is gradually reduced and the blood flow is resumed. The medical person records the blood pressure when certain sounds, called Korotkoff sounds, are heard through the stethoscope. When the first sound is heard, the blood pressure is recorded as the high number (systolic pressure) and when the last sound is heard, the blood pressure is recorded as the low number (diastolic pressure). What causes those sounds?

Answer Although Korotkoff sounds have been studied for about 100 years, their source is still debated. Here are two explanations:

Artery snapping: When the pressure in the inflated cuff decreases to the systolic level of the blood, blood begins to squirt under the cuff and into the lower arm, snapping apart the artery that had collapsed when the inflated cuff cut off the blood flow. That snapping sends a sound wave through the arm, which is heard in the stethoscope as a tap. As the pressure in the cuff continues to drop, the sound produced with each squirt dulls, and then it disappears once the cuff pressure reaches the diastolic level of the blood pressure. Thus, a medical worker records the cuff pressure at the first tap as the systolic blood pressure and the cuff pressure at the last dull thud as the diastolic blood pressure.

Cavitation: When blood begins to squirt under the cuff and into the lower arm, snapping apart the collapsed artery, the sudden decrease in pressure at the head of the blood flow causes gas (primarily oxygen, nitrogen, and carbon dioxide) to come out of solution to form bubbles. When a bubble collapses shortly later, with blood suddenly flowing to fill the space it had occupied, the sudden movement of the blood sends a sound wave through the blood and arm. That sound, or more likely the collective sound of collapsing bubbles that occurs just after each squirt of blood into the lower arm, is a Korotkoff sound. Such sound production continues until the cuff pressure is at the diastolic level, and then blood no longer squirts into the lower arm.

3.50 • Attack of the killer shrimp

The oceans are alive with sound, and in some places so much sound is produced by shrimp colonies that a submarine can avoid sonar detection by hiding next to a colony. The sound made by the snapping shrimp *Alpheus heterochaelis* occurs when it attacks its prey by using its oversized claw (it is much larger than the other claw). Surprisingly, it snaps the claw near, rather than on, the prey. Nevertheless, this action stuns or kills the prey, and then the shrimp can pick it up with its smaller claw, to be eaten. The peacock mantis shrimp (*Odontodactylus scyllarus*) also attacks its prey without striking it; instead, the shrimp abruptly flicks a feeding appendage toward it. For both types of attacks, what stuns or kills the prey?

Answer As the two sides of a claw snap toward each other, they expel a jet of water so quickly that the water undergoes cavitation. That is, dissolved air comes out of solution and forms bubbles. Those bubbles almost immediately collapse, sending sound pulses through the water that are loud enough for their pressure variations to stun or kill prey. The sound heard from a snapping shrimp is the collective sound from the collapsing bubbles, not the sound of the two sides of the claw striking each other.

The pressure variations due to bubble collapse are so severe that the collapse can produce a flash of light in what is generally called *sonoluminescence* (light produced by sound). However, for the shrimp the light production has been dubbed *shrimpoluminescence* (light produced by shrimp). The light appears because the collapse of a bubble very rapidly heats the air inside the bubble, causing the air molecules to ionize into electrons and positive ions. Almost immediately the electrons drop back into the molecules, losing their energy (they must lose energy to reenter) by emitting light. The light from a bubble collapse (or even a great many bubble collapses) is much too brief and dim for you to see and is simply a by-product of the cavitation process.

The peacock shrimp also produces sound pulses via cavitation, but the bubbles are due to the snapping motion of the feeding appendage.

3.51 • Sounds of boiling water

Put a pan of water on a kitchen stove, pull up a chair, and then listen to the sound from the water as it is heated while you also watch what is happening inside the pan. That sound is so common that you probably don't notice it, but you might subconsciously use it as a signal of when the water is in full boil. What causes the sound?

Answer The sound you hear from the water begins as an occasional hissing and grows into a continuous hissing. This sound is produced by air bubbles forming in the crevices (scratches) along the bottom of the pan. In a crevice, the increased temperature drives dissolved air out of solution and into the air bubbles. As each bubble grows, it oscillates, sending out sound waves through the water and pan walls—the collective sounds from these bubbles is the hissing. Once the bubbles grow large enough that buoyancy pulls them free of the bottom, they quickly rise to the top surface of the water, where they burst open with a faint splash.

As the water temperature continues to increase, most of the air is driven out of solution, and the formation of the air bubbles and the associated hissing die out. Then a sharper sound begins. Water begins to vaporize along the bottom of the pan, forming vapor bubbles in the crevices. However, the primary source of the sound you hear is not the formation of a bubble or even its oscillations. Rather, it is the bubble's collapse, in which the vapor suddenly redissolves and water rushes in to fill the space occupied by the bubble. That inrush sends out a *click* through the water and pan and into the air, which we hear.

When the water heats somewhat more, the vapor bubbles are large enough to break free of the bottom. However, they don't reach the top surface because as they rise from the very hot bottom through somewhat cooler water, they collapse, sending out harsh click sounds. These sound waves can set up resonance in the air above the water (between the water surface and roughly the top of the pan), in the water, and in the pan walls. That is, sound waves that reinforce one another give large net waves in these three regions, generating sound waves in the air that you can hear. The combination of these clicks and resonance sounds is the harsh sound you associate with water near the boiling point.

As the water continues to heat and the temperatures of the higher layers of water increase, the vapor bubbles eventually reach the top surface of the water without collapsing. There they burst open with a faint splash. Your audible clue that the water has come to full boil (boiling throughout its volume) is when the previous harsh sounds give way to the gentle splashing. Then it is time to pour the water for tea.

3.52 • Food-crushing sounds

You may not like to hear people bite into and chew their food, but can you tell anything about the food from the sound? For example, can you tell by the sound alone if an apple is ripe or if a tortilla chip is fresh? Food manufacturers often go to great lengths to see that their products have the "proper" sounds when eaten.

Answer You *can* tell from the eating sounds whether an apple is ripe or a tortilla chip is fresh. Indeed, I think that much of the snack-food industry is driven by the crunchy sounds made by their foods. (Throw in salt and fat, and what more would anyone want!) When a brittle material such as a tortilla chip is suddenly crushed, the chip breaks apart as many fractures race through its air-filled cells. These cells bend and break, and then the pieces momentarily oscillate, which produces pressure variations in the air and thus generates sound waves—you can hear someone eating a chip. The oscillations against the teeth send sound waves through the teeth and jaw to the ear. Thus, when you bite a tortilla chip, you can hear the waves not only through the air but also through this second path to the ear. A fresh tortilla chip is very brittle and produces high-frequency sounds (above 5000 hertz) as it fractures. An old tortilla chip has absorbed water from the air and may no longer be very brittle. When you bite into it, you might simply mash the chip without any rapid fracturing or oscillation of the pieces.

Crunchy apples can be distinguished from mealy (old) apples by the chewing sounds, especially in the first bite, which usually produces sounds with frequencies less than 2000 hertz. The difference has to do with the cells in the apple. A crunchy apple contains cells with water under pressure. When bitten, the cells burst open, producing sound in the frequency range of 100 to 1500 hertz. In contrast, the cells in a mealy apple are under less pressure (they have less turgor) because the carbohydrates (the pectins) in the walls have joined the solution (they have been *solubilized*). When bitten, the cells readily collapse with no explosion of material and little sound.

3.53 • Snap, crackle, and pop

One popular breakfast cereal in North America consists of toasted, puffed rice grains. When these grains are placed in milk, they make a crackling sound, hence the "snap, crackle, and pop" slogan long used by the manufacturer to market the cereal. Why does the cereal make the sound?

Answer Each grain is brittle and under stress; that is, the various parts of the grain pull tightly on one another. When a portion becomes wet, its rigidity decreases and the portions pulling on it rip it apart. This sudden motion causes momentary oscillations, which produce a faint pulse of sound that is more of a crackle than a snap or pop. If you eat this type of breakfast cereal, keep in mind that the sounds you hear are the dying shrieks of puffed rice grains.

3.54 • Sonic booms from aircraft and bullets

Airplanes were flown at speeds less than the speed of sound (they were *subsonic*) until 1947, when Chuck Yeager broke the so-called sound barrier by flying faster than the speed of sound (the airplane was *supersonic*). As supersonic flight became common, an irritating and sometimes destructive sound also became common—the *sonic boom* or *sonic bang.* Why can a supersonic aircraft produce a sonic boom that can startle people, scare cats, and even break windows? Can two people in a supersonic aircraft hear each other talk?

If a bomb blast damages a building (say, its walls), the blast pressure on the building must exceed a certain threshold value. Yet, a sonic boom might damage the building with a pressure that is 100 times smaller than that threshold. How is that possible?

Some bullets are supersonic. Do they also produce sonic booms?

When V-1 rockets flew into England during World War II, an observer would hear first the rocket's flight noise (a characteristic buzzing that acted as an unintended warning) and then the explosion of the rocket when it hit a target. Later, when V-2 rockets were used, an observer would sometimes hear those two sounds in reverse order: first the explosion (with no warning sound) and then, slightly later, the flight noise. Why was the sequence reversed?

On August 13, 1989, the space shuttle *Columbia* was headed toward Edwards Air Force Base in California when it flew over Los Angeles and then over Pasadena. The space shuttle was supersonic (moving at about 4600 kilometers per hour) and thus generated a loud shock wave (or sonic boom) that was heard in both cities. Curiously, the seismograph station at Pasadena received a prominent seismic wave from Los Angeles 12.5 seconds *before* it received the shock wave from the shuttle. How could the shock wave generate a seismic wave in Los Angeles?

Answer When an airplane travels through air, it pushes air molecules out of the way, which causes a variation in the air pressure. This pressure variation travels away from the airplane as a sound wave. Sound produced by the engines also travels away from the airplane. If the airplane is subsonic, you hear the sound from the engines and do not notice the pressure wave caused by the airplane's push through the air.

The situation is reversed when the airplane is supersonic. The pressure variations due to the airplane's push through the air still travel away from the airplane as sound waves, but now those waves are slower than the airplane itself, and they bunch to form a cone with its tip (apex) at the airplane. This cone travels along with the airplane as long as the airplane is supersonic. In a sonic boom, you hear primarily these bunched-up sound waves (said to be a *shock wave*) and not the sound from the engines.

As the airplane moves, say, horizontally, the lower part of the cone can sweep across the ground. If it sweeps past you, the air pressure on your eardrum first increases from its normal value, then decreases below its normal value, and then increases back to its normal value. (A graph of these changes resembles the letter N, and thus the shock wave from an airplane is often called an *N wave.*) These abrupt changes in the air pressure cause your eardrum to oscillate and thus for you to perceive a sound—the sonic boom.

The shock wave from an airplane actually consists of a number of individual shock waves from the nose, the engines, the wing–fuselage juncture, and the tail section. However, by the time these shock waves reach you on the ground, they may merge into a single shock wave, giving a single sonic boom. Sometimes, however, you can distinguish the individual booms.

The sound waves that form the shock-wave cone may not reach the ground because, as they travel downward, their paths are curved by changes in the air temperature along the way—the sound is said to be *refracted* or to undergo *refraction.* If they encounter progressively warmer air as they descend, the paths can be curved away from the ground enough that they miss the ground. The waves can also be channeled great distances (perhaps a hundred kilometers) if they are caught between layers with higher temperatures. Thus they can sometimes be heard by someone when no airplane is in sight (a spooky boom from the sky is apt to make you worry).

When supersonic aircraft, especially military aircraft, accelerate in the forward direction or make sharp turns, the shock waves can be emitted into various directions and some of them can intercept the ground at the same point. The combination of two or more shock waves means that the pressure variations are more extreme, producing a *superboom* that can be quite frightening to people on the ground. It is probably also the situation that damages buildings, especially if the rate at which the pressure varies happens to match the rate at which, say, a wall can oscillate. Then the oscillations caused by the shock waves can be large enough to break the wall.

Two people in a supersonic aircraft can certainly talk to each other. They are in air that is forced to move with the aircraft, and nothing special happens to the sound waves of their voices traveling through that air.

Part of the "bang" you hear from a rifle is the sonic boom of the fired bullet if that bullet is supersonic. The V-1 rockets sent into England were subsonic and thus their flight noise would reach an observer before the rockets did. The V-2 rockets, however, were supersonic and the rockets would reach an observer before the flight noise.

When the *Columbia*'s shock wave hit Los Angeles, it caused many of the tall buildings in the downtown area to oscillate, much like an earthquake would cause them to oscillate. The periods of these building oscillations (a period is the time for a complete oscillation) ranged from 1 to 6 seconds. As the buildings swayed, they produced seismic waves: A building oscillating with a period of, say, 2 seconds produced a seismic wave with a period of 2 seconds. The seismic waves traveled through the ground faster than the shock wave traveled through air, and so they reached Pasadena before the shock wave. The first-to-arrive waves from all the buildings arrived roughly in step with one another to produce a prominent disturbance at the seismology station. Thereafter, because the waves had a variety of periods, their overlap was a mishmash and many of them tended to cancel out one another, and so the disturbance at the station was diminished.

3.55 • Sonic booms from train tunnels

When the speed of fast trains in Japan was increased from 220 kilometers per hour to 270 kilometers per hour, the train tunnels began to boom whenever one of the trains traveled through it. These booms were as loud and startling as a sonic boom from a supersonic aircraft. Why did the speed increase produce the boom?

Answer A train, especially a fast train, must push its way through air, so it sends out compression waves in the forward direction. In the open, these waves quickly die out (*dissipate*) but in a tunnel they last longer. In fact, they can last long enough to bunch as a shock wave. When the shock wave reaches the end of the tunnel and escapes, it is powerful enough to produce a sonic boom. Although technology can increase the speed of the trains, this boom production acts as a limit on what speeds may be acceptable for any given train and tunnel design.

3.56 • Thunder

What causes thunder, and why can the sounds of thunder be anything from a nerve-racking clap to a drawn-out roll?

Answer The primary source of sound in thunder is the shock wave produced by the lightning, which is an electric discharge. The lightning's huge current runs between cloud and ground (or between cloud and cloud) in a narrow channel with a radius of only a few centimeters. Within the channel, electrons are removed from air molecules by the huge electric field set up by charges on the ground and in the cloud. The freed electrons are then accelerated by the electric field and collide with air molecules, transferring their acquired energy to the molecules. Because the gas of these molecules is then quite hot (the temperature may be 30 000 K), the gas expands. This process occurs so quickly that the channel of hot gas initially expands much faster than the speed of sound, which sends a shock wave of abrupt pressure variations into the surrounding air, producing the sound of thunder.

If you are standing close to the lightning, you hear a terrifyingly loud crack as the shock wave sweeps past your ears. If you are farther away, you first hear sound from the nearest part of the strike and then you hear sound from portions of the strike that are higher or farther from you. However, because the sound waves have spread out, this delayed sound may not be loud enough to terrify you. You probably will also hear reflections of the sound off hills, buildings, ground, and even clouds. These effects draw out the thunder into a roll.

You might hear a musical note if you are near lightning that consists of several rapid discharges. If the sound pulses from the discharges are closely spaced, you don't perceive them individually but instead perceive them as being part of a note. For example, if the time between successive pulses is

0.001 second, then you perceive sound with a frequency of 1/0.001 = 1000 hertz.

If the lighting is more than about 20 kilometers from you, you may not hear any thunder. As the sound travels through the air, it is refracted (its path is bent) by changes in air temperature (hotter air is less dense than cooler air, and such density changes can bend the path taken by the sound). Because the air is typically cooler at cloud level than at ground level, the sound traveling toward you from distant lightning is bent up and away from you. However, in some thunderstorms, the air near the ground happens to be cooler than the higher air, a situation called a *temperature inversion*. During an inversion, sound from a lightning strike that is initially headed upward can be bent down. Worse, sometimes the sounds emitted from various parts of a lightning strike can be focused (concentrated) in your direction. When that happens at night, forget about sleeping because you are going to be awake (and under the covers, hiding from the monsters) until the storm passes.

Some of the sound produced by lightning can be *infrasound*—that is, it can have frequencies too low for you to hear. The source of the infrasound appears to be the collapse of the electric field and charge distribution in a cloud when the cloud suddenly loses much of its charge in a lightning strike. The water drops in the cloud had been charged and were electrically repelling one another. However, once the lightning discharges these drops, removing that mutual repulsion, the drops suddenly move inward to make a tighter distribution. That sudden move produces pressure variations in the air that travel from the cloud to the ground as infrasound. The frequency (about 1 hertz) is too low to hear, but you might be able to feel it, especially if you are directly below the cloud. However, being there probably means you are going to pay a lot more attention to the nearby lightning strike than to any faint sensation due to the infrasound.

3.57 • Brontides—mysterious booms from the sky

Mysterious booms from the sky, even clear skies, have long been reported, and their sources have long been debated. For example, when the legendary explorers Lewis and Clark traveled through the American West, they initially dismissed stories from the native Americans about such booming, but they changed their minds when they too heard it near Montana's Rocky Mountains. *Brontides*, *mistpoeffers* (fog belchers), *Barisal guns*, and *brontidi* are some of many names that the booming has been given around the world. In modern times, reports of brontides sometimes reach newspapers, especially when large numbers of people hear them or if the brontides occur on a somewhat regular schedule. Can you suggest what causes them?

Answer In modern times, most of the brontides are probably due to sonic booms produced by supersonic aircraft. The aircraft can be quite far away from where the brontides are heard, thus making their origin seem mysterious. Although laws prevent aircraft from producing sonic booms over U.S. and Canadian cities, they can still produce them over thinly populated regions or over the ocean. For example, some widely heard (and widely reported) brontides on the East Coast of the United States were eventually connected to the sonic booms laid down by Concorde aircraft. Although a trans-Atlantic Concorde was required to drop below the speed of sound well before it reached, say, Kennedy Airport in New York, it was supersonic farther out over the ocean. The path taken by the sound waves in the sonic boom depended on how the air temperature and wind speed changed with altitude. In some situations, the sound waves could be channeled by temperature and wind-speed changes so that the sound waves were heard along the East Coast near Kennedy.

However, sonic booms cannot explain all modern brontides or *any* of the brontides heard before the first supersonic flight in 1947. Those other brontides may have several types of sources. They might be due to distant seismic disturbances if the ground shake is too weak to be noticed but the sound wave through the air is audible. Distant thunder could also be a source if the storm is too far away to be seen—the sound of the thunder can be heard if the atmosphere has the proper distribution of air temperatures. Then the sound might be repeatedly refracted wherever the air temperature changes with height, and thus the sound might travel a considerable distance. Such explanations are difficult to prove.

3.58 • Rockfall and tree downing

On July 10, 1996, two successive and dramatic rockfalls occurred near the Happy Isles Nature Center in Yosemite National Park, California. In each fall, a huge chunk of granite first slid down a steep slope and then went into projectile motion, hitting the ground about 550 meters farther down. The impacts produced seismic waves that were recorded on seismograph machines as far away as 200 kilometers. More surprising, however, was the damage the chunks produced even farther down in the valley, over 300 meters from where they landed: Over 1000 trees were downed, a bridge and a snack bar were demolished, one person was killed, and several other people were hurt. How could the granite chunks cause so much destruction in places they did not reach?

Answer The impact of each granite chunk at the bottom of its fall produced pressure variations in the air, which traveled away from the impact point as a sound wave. The wave, said to be an *air blast* from the impact, consists of a compression of air followed by an expansion of air. If you were standing in that air blast, you would be violently shoved, first in one direction and then in the opposite direction by the pressure variations, which were effectively a source of strong winds. The air blast from the second chunk, which had about three times the mass of the first chunk, was especially destructive, creating winds up to 430 kilometers per hour through the trees (this is comparable to winds near a tornado). In fact, the air blast from the second chunk was supersonic (it was a shock wave), because dust stirred up by the first impact reduced the speed of sound in the air from its normal value of 340 meters per second to about 220 meters per second, and near the impact point the speed of the air blast was greater than that.

3.59 • Popping bullwhips and wet towels

How do you pop (or crack) a bullwhip? Somehow you rapidly move the handle a short distance and the tip ends up with a high speed. How do you pop a towel, and why does it pop better when wet?

Answer To pop a bullwhip (or anything resembling a whip), the hand rapidly moves the handle to send a wave along the whip's length. A novice might use a simple wave, but an expert uses a loop wave (Fig. 3-5). As the wave reaches the end, the tip is rapidly accelerated (by as much as 50 000 times the acceleration of gravity) and its speed quickly exceeds the speed of sound. As with other supersonic objects, such as supersonic bullets and airplanes, the tip generates a sonic boom (or shock wave), which is the *crack* of the whip.

A wet towel serves better than a dry one because of the additional mass. You must work harder to initiate the motion but the tip ends up with more energy, enough to provide a painful slap on the skin of a locker-room victim.

Some paleontologists speculate that the sauropod *Apatosaurus louisae*, which was a dinosaur with a long flexible tail, could flick its tail much like a bullwhip, with the tip possibly exceeding the speed of sound and sending out a sonic boom.

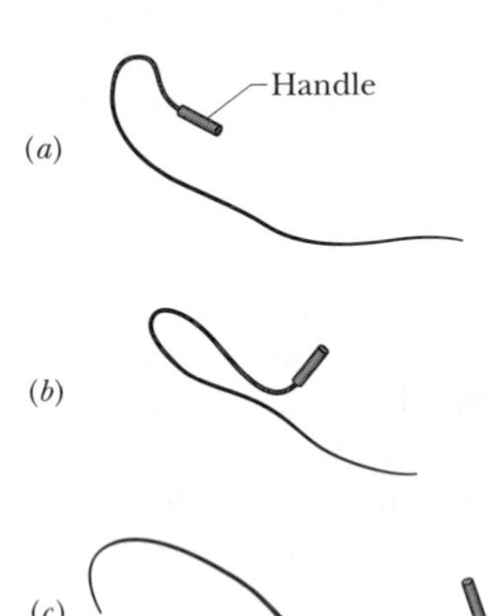

Figure 3-5 / *Item 3.59*
Three snapshots of loop wave traveling along whip as handle is brought quickly to the rear.

3.60 • Coughing and sneezing

What produces the sound of a cough or sneeze, and why is the cough or sneeze of some people so loud (and annoying)?

Answer When you cough, you expel air at high speed through the trachea and upper bronchi so that the air will remove excess mucus lining the pathway. You produce the high speed by breathing in a large amount of air, trapping it by closing the glottis (the narrow opening in the larynx), increasing the air pressure by contracting the lungs, partially collapsing the trachea and upper bronchi to narrow the pathway, and then expelling the air through the pathway by suddenly reopening the glottis. The airflow quickly becomes turbulent, which produces sound waves in both the air and the lung tissue. The vocal folds in the larynx do not produce sound at this stage, because they are held wide apart so as not to obstruct the airflow. However, toward the end of a cough, as the vocal folds come back together, the airflow can cause them to oscillate and produce sound.

During the explosive rush of air, the airspeed increases as the air travels through the contracted trachea and upper bronchi. In some people, I calculate that the airspeed reaches, or even exceeds, the speed of sound, and thus a mild shock wave (or sonic boom) comes up from the throat, making the cough especially loud. A hard sneeze can also create a shock wave. (People with supersonic coughs or sneezes are rarely appreciated in elevators.)

3.61 • Acoustics of rooms and concert halls

Some rooms and concert enclosures are acoustically terrible in that the audience cannot properly hear what is being spoken, sung, or played. For example, a concert played in a basketball arena is certain to be a mishmash of sounds. (In rock concerts, the volume is often cranked up until it is so high that no one in the audience cares about acoustic quality.) What determines whether the acoustics are proper?

Answer The sound that arrives at a listener can be grouped in three ways: the *direct sound* that comes directly from the source to the listener, the *early reflections* that arrive shortly later (within 0.050 second) from reflections off the walls or ceiling, and the *late reflections* that arrive after that. The early reflections should be loud. Because they arrive so soon after the direct sound, the listener mentally merges them with the direct sound. These early reflections should largely come from the left or right of the listener to reinforce the impression of being in a room, as well as to mentally expand the size of the source. Without these side reflections, the listener would think that the room is "dead" or that the source is small.

The late reflections should not be too loud, or they will overlap and *mask* whatever direct sounds arrive at the same time, and then the listener cannot distinguish the direct sound. (You may have noticed such masking if you yell out a sentence in some enclosure that gives strong echoes.) However, the late reflections should not be eliminated, because they give the listener a sense of being "immersed" in the sound.

Identifying the psychological effects of early and late reflections is an ongoing study, as is the design of concert halls. Those designs depend in part on whether the hall will be used by someone giving a speech, singing a song, or playing an instrument. In general, the walls of a concert hall consist of many projections or variations that can reflect sound into the audience. To decrease the chance of loud echoes, the back wall might be covered with a curtain, or part of the stage might be covered with carpet.

The acoustics in an orchestra pit are crucial to the musicians playing there. A musician hears not only the direct sounds from the other instruments but also reflections of those sounds. However, the more disturbing aspect for a musician sitting under the stage occurs when resonance is set up by sound waves bouncing between the pit floor and the covering (ceiling). (In resonance, the sound waves reinforce one another to give a big net sound wave.) If the resonance has a strong pressure variation near the height of a musician's ear, the source of that sound appears to be at about head level, not elsewhere in the pit as it actually is. This annoying distraction can be eliminated, or reduced, by properly designing the height of the pit and properly locating sound absorbers in the pit.

Old churches, with their hard walls, floors, and ceilings, typically have loud echoes lasting for several seconds. Organ music played in these churches sends a surfeit of sounds ricocheting between the hard surfaces. For example, in St. Paul's Cathedral in London echoes can last 13.5 seconds. Modern churches have shorter-lasting echoes so that the words spoken from the pulpit can be heard clearly. The echoes are shorter lived because the walls absorb sound better than the old stone walls. However, in at least one church the walls were coated to make them reflect better, making the church organ sound more gothic and thus more like an organ in a cathedral.

3.62 • Whispering galleries in various enclosures

Some enclosures have rooms that allow a whisper at one point to be heard by a listener at a second point, even at a surprising distance. According to legend, the "Ear of Dionysius" in Syracuse had this property—words from the prisoners in the dungeons were somehow channeled to the ear of the tyrant. Similarly, the dome covering the old Hall of Representatives in the Capitol Building in Washington, D.C., would reflect even a whisper spoken in confidence on one side of the chamber so that it was audible on the opposite side, perhaps to a member of the opposite political party.

More embarrassing, however, must have been the situation said to have existed in the Cathedral of Girgenti in Sicily. One of the church members discovered that when he stood at a certain point in the cathedral, he could hear the confessions whispered to a priest in the confessional near the far side of the cathedral. The confessions were fun to hear by that church member and his buddies—that is, up until the day his wife happened to be in the confessional.

Answer The whispering galleries found in various enclosures are unlikely to have been designed into the construction. They usually are in enclosures with elliptical cross sections, so that sound waves can be focused. An elliptical cross section will have two such foci. If someone speaks at one focus, the sound waves reflect off the ceiling and tend to converge at the other focus, assuming that the ceiling reflections are not blocked by ornaments or lamps.

3.63 • Whispering gallery in St. Paul's Cathedral

The dome in St. Paul's Cathedral in London contains a walkway around its perimeter, where spectators can see both the dome's interior and some of the main floor of the church far below. The walkway is circular, with a radius of about 32 meters. If you and a friend stand at opposite sides of the walkway and your friend speaks to you across the walkway's diameter, your friend will almost have to yell for you to hear the message. However, if your friend faces into the wall and speaks in a whisper, you can easily hear the message if you also stand near the wall. In fact, you can hear the message just as well if you stand at any other point along the walkway. That is, you don't have to be opposite your friend, so this is not a focusing effect as in the preceding item. How does the message reach you, why must your friend face the wall and be near it to be heard, and why is the message heard more clearly if it is whispered?

Answer Some of the sound waves leaving your friend's mouth cling to the wall by reflecting over and over again as they travel away from your friend. (The other sound waves might reflect once or twice and then be "lost" as they travel across the dome's interior.) The waves that cling to the wall are called *surface waves* or *Rayleigh waves.*

In 1904, Lord Rayleigh demonstrated this clinging effect: He bent a long metal strip into a horizontal semicircular wall and then placed a bird whistle at one end and a flame at the other end. The flame served as a sound detector because it noticeably wavered when sound disturbed it. When the bird whistle was blown, the flame on the opposite end of the semicircle wavered. However, when Rayleigh inserted a narrow barrier along the inside perimeter of the strip, the bird whistle could no longer make the flame waver. The sound that had been disturbing the flame did not come directly from the whistle to the flame; rather, it traveled along the semicircular wall with repeated reflections, forming a fairly narrow band of sound along the wall. When the barrier was placed next to the wall, it blocked the waves traveling along the wall. In simple explanation, we can say that the waves cling to the wall because they undergo repeated reflections that take them from one point on the wall to the next. Actually, the wave travel is far more complicated than that. Indeed, under some circumstances, these surface waves can actually travel along a flat surface where the explanation of repeated reflections would not make any sense.

The ability of sound waves to hug a curved surface as they travel along it depends on wavelength—shorter wavelengths work better because the points where a wave undergoes successive reflections are closer together. Thus, sound waves can travel around the walkway in St. Paul's Cathedral more successfully if they consist of short wavelengths. Such wavelengths correspond to higher frequencies, which occur in a whisper.

3.64 • Echoes from walls, corners, and forest groves

A standard echo is merely the reflection of sound waves back toward the origin of the sound. You probably have heard echoes in hallways or in many other enclosures with solid walls that reflect sound well. Some structures are known for multiple echoes, in which a sound reflects from many surfaces back to the origin, or repeatedly between two surfaces and back to the origin. A corner between three perpendicular surfaces, such as two walls and a ceiling, can form a *retroreflector*, which can efficiently bounce sound from two or three of the surfaces and back toward the sound source.

Some structures can produce echoes that are very difficult to explain. For example, if you stand beneath a masonry bridge that arches over water, a handclap can produce a host of echoes that probably depend on more than simple reflection of the sound waves. In fact, the echoes can be close enough that you hear them as a musical note.

If you ever find echoes from a grove of trees, notice the following: A high-frequency sound, such as a woman's scream, gives a fine echo, but a low-frequency sound, such as a deep baritone's roar, may not give any echo. If you sing a musical note (or even clap your hands), the echo is an octave higher (that is, has a frequency twice what you send out). Why does a grove of trees give such peculiar echoes?

Answer The ability of sound to reflect from objects, such as trees, depends on the wavelength of the sound. When the wavelengths are all larger than the object, a smaller wavelength reflects better than a longer wavelength. Thus, when you send out sound with a range of wavelengths, the smaller wavelengths reflect better from a grove of trees, for example. Because a smaller wavelength corresponds to a higher frequency, this means that the echo you hear will consist more of higher frequencies than lower frequencies, as opposed to what you sent out. This also means that low-frequency sounds will give poor or no echoes, while high-frequency sounds can give noticeable echoes.

If you send out a musical note, the sound has at least two components: a lower frequency (said to be the *fundamental*) and a frequency that is twice as much (said to be the *second harmonic*). Because it is a higher frequency, the second harmonic is far better reflected by the grove than the fundamental. So, even though what you projected is dominated by the fundamental, what you get in the echo is dominated by the second harmonic. Thus, the echo is (primarily) at twice the original frequency.

3.65 • Musical echoes from stairs and fences

If you clap your hands near a long flight of stairs or a long picket fence, why is the echo drawn out instead of being a single clap? Why does the frequency of the echo decrease with time? One of the more dramatic examples of this *musical echo* (or *chirped echo*), as it is called, can be heard after a handclap in front of the flight of stairs along the side of the Temple of Kukulkan, located in the Mayan ruins at Chichen Itza, Mexico. This steep flight consists of 92 stone steps, which some tourists climb.

Answer The frequency shift during the echo from a long flight of stairs is due to the angle at which the sound waves reach successive steps, not the frequencies in the sound pulse (such as a handclap). Take a side view of the steps, and assume that for the lower steps the paths of the incoming sound and the reflected sound are horizontal (Fig. 3-6). You get a return pulse from the lowest step (the one nearest you), a second return pulse from the next step (the next one nearest you), and then a third return pulse slightly after that, and so on. Each returning pulse is delayed from the preceding pulse because its path to and from the corresponding step is farther. You don't perceive these pulses individually; instead, you perceive the frequency at which they arrive. For example, if the time between pulses is 0.002 second, you perceive a frequency of about 500 hertz.

Now consider higher steps: The sound in the echo from them must take an oblique path, both in reaching a step and in returning to you. Thus the difference in the to-and-from path from one step to the next is greater for the higher steps than for the lower steps. So, the time between pulses is greater. For example, the pulse-to-pulse time might now be 0.003 second, which you would perceive as a frequency of about 333 hertz, which is less than the echo you receive from the lower steps.

The argument for the musical echo from a picket fence is similar, except that the reflecting objects (the pickets) are separated horizontally instead of vertically.

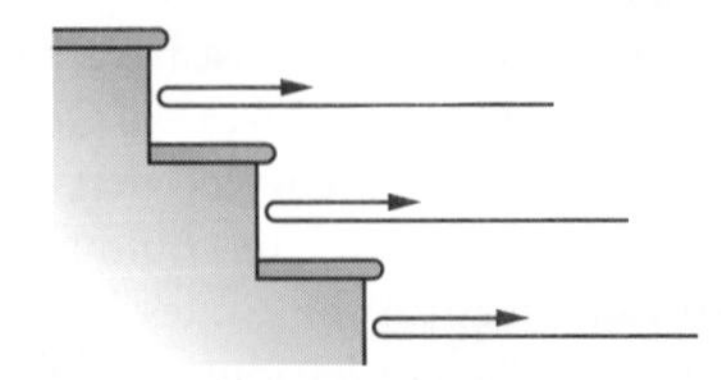

Figure 3-6 / *Item 3.65* Reflections of sound from steps on a staircase.

If sound waves are sent through a regularly spaced array of cylinders, the waves can undergo constructive interference (in which the waves reinforce one another) and destructive interference (in which they cancel one another). For some sound frequencies, the array strongly transmits the sound from one side to the other. For other frequencies, the transmission is greatly reduced. This experiment was done by sending sound waves through a minimalist sculpture erected as an artwork. The sculpture consisted of thin vertical cylinders laid out in a square array.

SHORT STORY

3.66 • Acoustics of ancient structures

Echoes probably played a role in the superstitious beliefs of ancient people. For example, some of the ancient rock petroglyphs in Australia were painted where echoes are still prominent. At some of those places you hear the best echoes when you are about 30 meters from the pictures, and so the echo seems to be sound emitted by the pictures themselves.

Some of the ancient rock art found in European caves occurs at points in the caves where echoes are strong. Perhaps those sites served in the superstitious rites of the people who made the paintings, and who may have chanted, sung, or drummed there.

The megalithic tombs found in Britain and Ireland have resonant frequencies at the lower end of the audible range. That is, sound waves with certain low frequencies reinforced one another to make large net sound waves. Such a tomb surely was not built for its acoustic properties. However, once the tomb was built, people may have discovered that by drumming at a certain frequency, they could set up resonance in the tomb. In modern times, resonance has been set up in the long entrance passageway that leads to the center chamber in the megalithic Irish monument Newgrange. A sound source placed in the central chamber was tuned until its frequency matched a resonant frequency of the passageway.

3.67 • Singing in the shower

Why does your singing sound better (or at least, *seem* to sound better) in a shower than elsewhere? The shower-stall singer often perceives the voice to be of better quality and so sings less guardedly.

Answer The answer here is far more complicated than I realized when I wrote about shower singing years ago. At that time I argued that the main reason an amateur singer sounds better is because the singer can set up resonance in a typical shower stall and thus build up the volume of the sound at the resonant frequencies. In setting up resonance, you fit an integer number of half wavelengths between two parallel walls or between floor and ceiling. When that condition occurs, the waves constructively interfere (that is, they reinforce one another) and build in sound level. So, the sound is loud, bold, and maybe even robust.

People who read my article suggested additional reasons. For one, the walls and floor (and perhaps the ceiling) in a shower stall are usually hard tile and thus reflect sound well. If you tried singing in an empty closet with the same dimensions as a shower stall, the reflections would not be as good and resonance would be harder to set up. (Besides, if you sing while closed up in a closet, you will only worry your family.)

Another reason is that because the reflections of your voice are returned so quickly by the nearby walls, you are immersed in your sound. So, you can hear a reflection of a note while still singing that note, and thus you can adjust your voice if it is somewhat off key. Of course, if the water is running loudly, its noise may simply mask all your off-key errors.

3.68 • Noisy upstairs neighbor

Living below a noisy neighbor can be trying, but what generally is the cause of the noise? Is it due to high heels clicking on a hardwood floor? Can the noise be decreased or eliminated if carpet is installed in the upstairs apartment? (That would eliminate any high-heel clicking.)

Answer Most of the irritating noise is described as "thuds" or "thumps" and, perhaps surprisingly, is not due to something like high heels clicking on the floor. Rather it is due to low-frequency noise generated by someone walking across the floor. The repeated footfalls cause the floor to oscillate like a drumhead, typically at a frequency between 15 and 35 hertz, which is at the low-frequency end of the audible range for most people. Such noise can be heard, and even felt, by the downstairs neighbor.

The high-frequency sound of heels clicking on a floor might be heard, but far more energy is transferred to the drum-like, low-frequency floor oscillations. Installing a carpet might actually worsen the situation because, with its softer surface, the footfalls can then transfer even more energy into the floor oscillations. The only solution is to switch to an apartment built when floors and supports were made with reinforced concrete.

3.69 • Booming sand and squeaking sand

On some beaches, the sand squeaks or whistles if you walk on it or push your hand or a plate down on it at angle of about 45°. In some deserts, sand dunes boom at a low frequency (100 hertz), sometimes with an intensity that makes conversation difficult. Some observers liken the sound to the drone of the Australian didgeridoo. How can sand make sound, and why doesn't all sand (on all beaches and all sand dunes) make sound?

Answer A sand dune slowly shifts across a desert because wind on its windward side carries sand grains up that side, and deposits them either on the top or high on the back side. This gradual transport eventually makes the slope on the back side too large to be stable, and then a layer of sand on the back side slides downward, reducing the slope. Thus, the upward transport of sand from the windward side and the eventual downward slide on the backside shift the dune across the desert.

On some dunes, the downward slide of sand can produce a booming sound, provided that the sand is fairly uniform in size and surface structure. The sand can slide down in more than one layer, each about 0.5 centimeter thick. As the layers descend, they oscillate perpendicular to the underlying sand surface and act much like an oscillating drumhead. When the sliding stops, so does the sound production. As they slide in a layer, the moving grains climb over one another and collide at a rate of about 100 times per second. This climb-collide rate and the layer-oscillation rate become coordinated (they are said to be in a feedback loop). So, the frequency of the sound produced by the layer oscillations is about 100 cycles per second—that is, 100 hertz.

If beach sand emits sound when walked upon, the footsteps must cause layers of the sand to slide over one another and oscillate, producing the sound waves.

The reason why only some sand is noisy is not fully understood. Apparently the grains of the noisy sand have acquired some special features that figure into the ability of the sand to move in fairly thin layers, with the motion causing the layers to oscillate. The most interesting possibility is that the grains have some special crust. Indeed, experiments with squeaky beach sand have revealed that the squeaking capability gradually disappears if the sand is rinsed in fresh water, and the capability cannot be restored even if the sand is reimmersed in salt water.

3.70 • Cracking ice and bergy seltzer

When you plop ice cubes into a room-temperature drink, what causes the cracking sounds you hear? When an iceberg begins to melt, it emits a different sort of sound, a "crackling, frying sound" known as *bergy seltzer* to people who hear it from submarines and ships. What causes bergy seltzer?

Answer The cracking sounds of an ice cube in room-temperature liquid are due to stresses inside the cube caused by the sudden increase in temperature on the cube's surface. The temperature increase tends to expand the ice; this puts the surface under tension, and a crack can suddenly rip through the surface. As the ice surfaces on either side of the crack move over each other or move apart, they produce changes in pressure in the liquid or the air, and those pressure changes travel away from the crack as sound waves.

Bergy seltzer is a different type of emission and occurs only with cloudy ice—that is, ice with trapped air pockets. As the ice surface melts, air can suddenly break out of an air pocket, shoving the water if the surface is submerged or shoving the air if the surface is not submerged. Either way, the sudden change in pressure travels away from the site as a sound wave, perhaps causing other sections of ice to oscillate. The collective emission, with its random changes in intensity, is the bergy-seltzer sound.

3.71 • Audibility through snow

Why can an avalanche victim buried in snow hear rescuers but the rescuers cannot hear the victim? The victim might scream or (as has been reported) shoot a revolver as a signal and still not be heard.

Answer The transmission of sound through layered snow is generally poor. However, the transmission up from the victim should be roughly the same as the transmission down to the victim. The primary reason that the transmission down is heard by the victim but the transmission up is not heard by the rescuers is that the victim is in a very quiet environment and the rescuers are not. Indeed, the rescuers may be making considerable noise as they search through the snow.

3.72 • Sounds of walking in snow

Why does snow squeak or squeal when you walk on it, and why is it more likely to sound like that when it is very cold?

Answer If the snow temperature is below approximately -10°C, the downward pressure of your footstep can cause some of the bonds between snow grains to snap or some of the layers of snow to suddenly yield and then slide over one another. Either action causes brief oscillations of the snow, which produce sound. If the snow is not as cold, the snow grains give too easily to snap or suddenly yield, because the bonds may be fewer or weaker than when the snow is very cold. The weakness could be due to partial melting, which lubricates any sliding. The melting might also be due to absorbed sunlight, especially on the top surface. Or perhaps at some points the pressure you apply is enough to melt the snow.

3.73 • "Can you hear the shape of a drum?"

This title and idea were published in 1966 by mathematician Mark Kac. His question can be rephrased: For a given flat drumhead, can you tell its shape from the frequencies the drumhead can emit? That is, after hearing many of those frequencies, can you tell from any one of them what the drumhead would look like in a photograph—what parts are oscillating and what parts are not?

Answer For a given string stretched between two supports, you *can* hear the shape of the string in that a certain frequency corresponds to a certain pattern of string oscillation. For example, the lowest frequency at which the string can oscillate corresponds to a certain pattern: The ends are

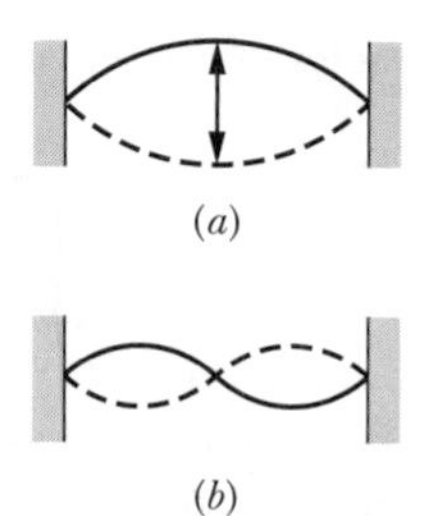

***Figure 3-7** / Item 3.73* Shapes of an oscillating string. Two snapshots of the string in (*a*) the simplest pattern and (*b*) the next more complicated pattern.

stationary (because they are tied in place), the center oscillates maximally, and intermediate points oscillate at intermediate amounts (Fig. 3-7*a*). The next higher frequency corresponds to the next more complicated pattern (Fig. 3-7*b*), and so on. These frequencies are said to be *harmonic frequencies* of the string, and the corresponding string shapes are said to be the *resonant modes.* Once you hear some of these frequencies, you can tell from any one of them the corresponding mode. Moreover, if you happen to already know the string density and tension, you can also tell the string length from the lowest frequency.

A flat drumhead has similar resonant modes and harmonic frequencies. However, the modes are complicated by the fact that the drumhead is two-dimensional. Circular drumheads are easy, but for any other drumhead, relating the oscillation mode (the parts that oscillate and the parts that don't) to the drumhead shape can be challenging. For most simple drumheads, it can be done. However, for more complicated drumheads, we cannot always work out the shape because at least two very different shapes can give the same set of harmonic frequencies. Nevertheless, even in those difficult situations, we *can* work out the area of the drumhead, and so you *can* hear the area of a drum even if you cannot always hear the shape of the drum.

3.74 • Infrasound

If you stand directly in front of the speaker system at a heavy rock concert, you are obviously going to be in discomfort. But can there be a quiet situation in which an acoustic effect from your environment makes you feel uneasy or nauseated?

Answer You may be in many situations where you are subjected to *infrasound* (sound with frequencies lower than you hear—lower than about 30 hertz) that is fairly intense. Your body may not shake in response to it, but your sense of equilibrium can be disturbed enough to make you feel ill, or you may be irritated by the rattle it produces in objects around you. A common example is the infrasound that exists in the passenger compartment of many types of vehicles. A quick exposure may have no effect, but a long car ride may leave you feeling "car sick." The unpleasant feeling is especially bad if the infrasound happens to set up *resonance* (where waves reinforce one another to produce a large net wave inside the car). Resonance can happen if a window is open and infrasound is part of the turbulence created as the trailing edge of the car window moves through the air. Infrasound is also created by the engine and the rolling of tires on the road, and its intensity increases if the car speed is increased. However, the improved aerodynamics of modern cars and their insulation against road noise have reduced the problem of infrasound.

Infrasound can also be generated when a strong wind blows past the corners or edges of buildings, breaking up into vortexes. The variation in air pressure can produce an infrasound wave that can cause tension in the building occupants. (The wind can also produce an audible sound—for example, the howl of a winter wind whipping around the building.) Here again the effect can be more severe if the infrasound sets up resonance in a room in the building, as might happen when a window is open.

Infrasound affecting a much greater area can occur when a strong airflow over a mountain range breaks into turbulence. In fact, some researchers propose a link between such infrasound production and widespread depression and increased suicide rates, although the link has not been established.

You are also subjected to infrasound from machinery (elevators, for example), ocean waves, explosions, and large storms. Even the infrasonic emissions of thunder of distant storms might affect you. The huge 1883 volcanic explosion of Krakatoa (near Java in the southeast Pacific) sent powerful infrasonic waves through the atmosphere. The waves were channeled between Earth's surface and the higher-temperature air in the stratosphere. As the sound waves penetrated the lower stratosphere, their paths were bent (they were refracted) back toward Earth's surface. At the surface they reflected back toward the stratosphere, and so on. No one could hear the audible portion of the explosion at any significant distance from the island, but the passage of infrasound was recorded on barometers worldwide.

Although you may be subjected to infrasound for much of the day, it probably has no effect because the intensity is usually small. However, if you need a handy excuse for why you did not complete your homework, why a relationship faltered, or why your favorite football team lost a game, blame it on infrasound.

3.75 • Sounds of corn growing

What makes the noise emitted by a cornfield even on relatively still nights? (Farmers dub the noise "the sounds of corn growing.")

Answer The sounds from a cornfield are made by leaves slapping against leaves when an occasional breeze moves through the field. The noise becomes more pronounced if the breeze increases. It also is more pronounced for more mature stalks, because then the leaves are larger, less constrained, and more brittle (drier), and the stalks sway more.

3.76 • Snapping cloth sounds

Hold a strip of cloth that is about 30 centimeters long with one hand on each end and the cloth hanging loosely. Then suddenly draw your hands apart to snap the cloth taut. Why does the cloth emit sound? Why is the frequency higher if the cloth is shorter?

Answer When the cloth strip snaps taut, it momentarily oscillates like a plucked guitar string, causing pressure variations in the surrounding air. Those variations travel away from the strip as a sound wave—the "snap" you hear. As with a guitar string, the frequency of the sound depends on the length of material that is oscillating. Shortening the length gives a higher frequency.

3.77 • Culvert whistlers

If you clap your hands at one end of a culvert, you will hear an echo that "zings." That is, it starts at the highest frequency and quickly descends to the lowest frequency. These special echoes have been dubbed *whistlers*. You can also hear a whistler if a friend claps hands at the far end of the culvert. A similar whistler sometimes occurs in a racquetball court, except that the frequency *increases* instead of *decreases* with time. Why does a whistler occur? That is, why would an echo's frequency change?

Answer The zing echo is dubbed a *culvert whistler* and can be heard in many types of pipes, including ones that might be short enough to use in a classroom. It is due to resonance, in which sound waves reinforce one another. However, here we can get by with a simple explanation. Let's assume that your friend's handclap, which emits a pulse of sound, is near the center of one end of the pipe and that the pipe has a length L. The sound can reflect from the sides of the pipe in many ways. For example, it might reflect at distance $L/2$ (halfway down the pipe) and thus make only one reflection.

More reflections require that the sound take a more zigzag path down the pipe, and thus it travels the length of the pipe more slowly. So, you first hear the single-reflection echo, then the double-reflection echo, and so on. The frequency you perceive is the frequency at which these echoes arrive at your ear. The first sequence of echoes (which requires only a few reflections) is rapid, and so you hear a high frequency (the *z* of *zing*). The later sequence of echoes (more reflections) is less rapid, and so you hear a lower frequency (the *ing* of *zing*). The limit is set by the echoes that reflect almost directly across the diameter of the pipe and thus hardly travel down the pipe.

The story is about the same if you listen for the echoes of your own handclap. However, this time the sound must reverse its direction of travel at the far end. That occurs for both a closed end (the culvert has a wall or cap there) or an open end. The latter may be surprising—when sound reaches the open end of a pipe, the abrupt transition to the open air causes some of the sound to travel back through the pipe. We say that some of the sound *reflects* at the open end. (Some musical instruments have a flared end to decrease this reflection and thus allow more of the sound to escape to an audience or microphone.)

When a racquetball slips across a wall or floor in a racquetball court, the sliding of the ball can occur in jerks, which sets the ball oscillating. The oscillations cause pressure variations in the air, and those variations travel away from the ball as sound waves, a squeal or squeak. Because the hard surfaces in the court reflect sound well, you hear not only the sound coming to you directly from the ball but also strong reflections from the surfaces. In fact, the sound reflects many times around the room (provided that the back wall does not have an appreciable open space for observers), and so you can hear echoes for up to 1 or 2 seconds. The frequency at which you intercept the echoes increases, and thus the frequency you perceive also increases.

3.78 • Slinky whistlers

Tie one end of a wire Slinky (a toy consisting of a coil, manufactured under trademark by Poof-Slinky, Inc.) to a wall and then stretch the Slinky by walking the free end away from the wall. Once it is stretched, tap on the Slinky with a pencil while listening to the free end. You will hear a *whistler echo*, in which the echo from the free end begins at a high frequency and quickly descends to a low frequency. What causes the whistler?

Answer Your tap on the Slinky sends *transverse waves* along the wire. Such waves cause the wire to oscillate perpendicular to its length, rather than along the length as *longitudinal waves* do. The speed of a transverse wave along the wire depends on the frequency of the wave: Higher-frequency waves travel faster than lower-frequency waves. When you tap on the wire, you send out a wide range of frequencies. When the waves reach the far end of the Slinky, they reflect and then return to you, the highest-frequency wave arriving first and the lowest-frequency wave last. The fact that the wire is wound in a helix does not seem to matter.

3.79 • Rifle-shot noises in permafrost regions

Historic descriptions of exploration in the permafrost regions of North America and Russia contain accounts of mysterious rifle-shot noises. In fact, one description accounts for the caribou's lack of concern for a real rifle shot by suggesting the caribou were accustomed to that type of noise. What could cause the rifle-shot noises?

Answer The permafrost regions are spotted with wedges of ice that extend down into the permafrost; these wedges are under tension and have numerous flaws, such as embedded bubbles. When the temperature radically drops, a vertical crack can appear in an ice wedge and then travel horizontally along the wedge. If the velocity of the crack is high, the sudden rupture at the front point of the crack can set up pressure variations in the ice and the surrounding air. Those pressure variations travel away from the advancing crack as sound waves that sound like shots.

3.80 • Hearing auroras and fireballs

Is it possible to hear auroras, those magnificent displays of light that occur at high altitudes in high latitudes? Some observers report hearing a crackling, swishing, or whistling that appears to be correlated with overhead auroras. Can you hear a meteor as it burns up in the sky? Some observers claim that they can hear a meteor either before or just as it is visible. This seems strange because meteors burn up at high altitudes. (A sonic boom is sometimes heard but that is not mysterious because it reaches the observer *after* the meteor passes. And if a meteor lasts long enough to hit something near you, or it hits you, then any sound from the meteor is quite clear.)

Answer Although the production of infrasound by an aurora has been recorded for a long time, there are no substantiated recordings of audible sounds. It is very difficult to think that a sound wave in the audible region can travel down through at least 100 kilometers of atmosphere and still be heard. Still, a number of people have reported sound that is correlated with auroras. Some of these events may be illusions (misinterpretation of environmental noises while watching a display, seeing a correlation when there is no correlation). Some of them may be due to the observer's breathing in very cold temperatures (−40°C or lower), because the water vapor in the breath can freeze and then fall to the ground, making a very slight swishing or tinkling sound. Some, however, might be real if there is a way to correlate an aurora and an electric field at ground level. Then (perhaps) such an electric field can produce electric discharges at sharp points, such as on the tips of a bush or the end of a metal rod.

If you hear a sonic boom from a meteor, that sound comes to you at the speed of sound, which means you hear the sound after the meteor has disappeared from view. How could you possibly hear sound at the same time you see the meteor, or even before you see it? The only way that can happen is if the meteor somehow produces an electromagnetic wave that reaches you at the speed of light. Such a wave could possibly cause objects around you to oscillate. If those oscillations were at a frequency in the audible range, you could hear the oscillations, and that would be the sound associated with the meteor. There is evidence that low-frequency electromagnetic waves can be produced when a meteor passes through the upper atmosphere.

3.81 • Australian bullroarer

A bullroarer is a blade-like piece of wood with a cord tied at one end. With your hand on the other end of the cord, you rapidly rotate the wood blade around your head to make a buzzing or roaring sound. (It is used in the movie *Crocodile Dundee II.*) What causes the sound?

Answer The wood blade rotates as it moves through the air, twisting up the cord in one direction, then untwisting and twisting up the cord in the opposite direction. This chaotic motion through the air sheds vortexes much like telephone wires do in a preceding item. The pressure variations within those vortexes cause the blade to oscillate, and the sound you hear is due to both the vortexes and the blade oscillations.

C • H • A • P • T • E • R 4

Striking at the Heat in the Night

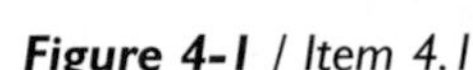

Figure 4-1 */ Item 4.1*

4.1 • Dead rattlesnakes

Because of their highly poisonous venom, rattlesnakes pose a danger to humans. When the snakes are discovered in residential areas, they are usually killed. However, the death of a rattlesnake does not immediately decrease its danger. Indeed, numerous people have made the error of reaching for a dead snake to remove it. Although the snake may have been dead for as long as 30 minutes, it can still strike a person, burying its fangs into a reaching hand and delivering its venom. How can a dead rattlesnake strike a reaching hand?

Answer Pits between each eye and nostril of a rattlesnake serve as sensors of thermal radiation. When, say, a mouse moves close to a rattlesnake's head, the thermal radiation from the mouse triggers these sensors, causing a reflex action in which the snake strikes the mouse with its fangs and injects its venom. Thus a rattlesnake can detect and kill the mouse even on a moonless night because the process does not require visible light.

The thermal radiation from a reaching hand can cause the same reflex action even if the snake has been dead for a while because the snake's nervous system continues to function. As one snake expert advised, if you must remove a recently killed rattlesnake, use a long stick rather than your hand.

4.2 • Fire-detecting beetles

The fairly small *Melanophila* beetles are known for a bizarre behavior: They fly toward forest fires and copulate near them, and then the females fly into the still smoldering ruins to lay their eggs under burnt bark. This is the ideal environment for the larvae that hatch from the eggs, because the tree can no longer protect itself from the larvae by chemical means or rosin. If a beetle were at the periphery of a fire, detecting the fire would be easy, of course. However, these beetles can detect a fairly large fire from a distance of 12 kilometers. How do they do that? This much is for sure: The beetles don't see or smell the fire from such a distance.

Answer The beetle has a pair of infrared-detecting organs along each side of the body, and each organ contains about 70 small knob-like sensors. A sphere in a sensor expands very slightly as it absorbs infrared light from the fire, and the expansion presses down on a sensory cell. This, then, is a mechanism that transfers energy from the infrared light to energy of a mechanical device. The beetle can locate the fire by orienting itself so that all four infrared-detecting organs are affected, and then it flies toward the fire so that the response of the organs increases.

4.3 • Bees kill hornet

The giant hornet *Vespa mandarinia japonica* preys on Japanese bees. However, if one of the hornets attempts to invade a bee hive, several hundred of the bees quickly form a compact ball around the hornet to stop it (they are said to *ball* the hornet). The hornet dies within about 20 minutes, although the bees do not sting, bite, crush, or suffocate it. Why, then, does the hornet die?

Answer After hundreds of Japanese bees form a compact ball around a giant hornet that attempts to invade their hive, they quickly raise their body temperature from the normal 35°C to 47°C or 48°C. If only a few bees did this, the energy transfer to the hornet would be insignificant because much of the increased thermal energy of the bees would be radiated away. But with the hornet trapped in a ball of hundreds of bees, the ball itself increases in temperature and significant thermal energy is transferred to the hornet. The higher temperature is lethal to the hornet but not to the bees.

4.4 • Huddling animals

Why do armadillos (perhaps a dozen of them) huddle at night? Why do emperor penguins (perhaps thousands of them) huddle during the Antarctic winter?

Answer Armadillos, emperor penguins, and many other warm-blooded animals huddle during cold weather in order to stay warm. If, say, an emperor penguin stands alone, it can lose a significant amount of thermal energy by conduction (to the ground), convection (to the air, especially if the air is moving), and thermal radiation (to the cold environment, including the sky). In the harsh environment of the Antarctic, where temperatures can be −40°C and wind speeds 300 kilometers per hour, individual penguins could perish due to the energy loss. The huddling is most important when the penguins breed during the winter. The resulting egg is incubated almost exclusively by the father, who keeps the egg from freezing by balancing it on his feet for months. During this incubation time, the father must fast because his food is in the water. So, with no incoming energy from food, he must be part of a huddle or his thermal energy losses will drive him into abandoning the egg to search for food.

By huddling (up to 10 penguins per square meter), the penguins significantly decrease the average thermal energy loss due to convection and radiation—only the penguins on the perimeter still suffer large losses, but they still benefit from their nearest neighbors. Here is another way of saying all this: If you place many individual "warm cylinders" in a cold environment, the total thermal energy loss can be very large because the total surface area through which the energy is lost is large. However, if you bundle those cylinders to make one very wide cylinder, the total surface area is less, and thus the energy lost through the surface is less.

4.5 • Space walking without a spacesuit

Some researchers speculate that a person could briefly walk in space without a spacesuit (as an astronaut does in the movie *2001: A Space Odyssey*) without dying. If the walk is far from the Sun, would the astronaut feel cold? Is there more danger to the astronaut than simply a lack of oxygen?

Answer One reason why the temperature of a room feels comfortable is that the infrared radiation sent to you from the walls and the infrared radiation sent to the walls from you are approximately equal. So, you gain energy at about the same rate as you lose energy. If the radiation to you drops significantly, you feel cool or cold. If you were to walk in deep space, away from your spacecraft, there are no walls and so you would feel very cold very fast. The rate at which you would lose thermal energy is about 800 watts. However, the lack of oxygen would be of much greater concern.

Your exposure to the vacuum would also be of concern. When water is exposed to a vacuum, it first boils (some of it vaporizes) and then it freezes. You have a lot of water in your body, and . . . well, maybe we should think about something more pleasant.

4.6 • Drops on a hot skillet, fingers in molten lead

If a metal skillet is heated to a temperature somewhat above the boiling point of water, and then a few drops of water are sprinkled into it, the drops spread out and last only seconds before they sizzle away. Surprisingly, if the demonstration is repeated with the skillet much hotter (with a temperature above 200°C), the drops bead up and may last as long as several minutes. How can they last longer when the skillet is much hotter? This effect is now known as the Leidenfrost effect, after Johann Gottlieb Leidenfrost, who studied it in 1756. (Earlier work by Hermann Boerhaave is not as well known.)

As the beaded drops dance on the skillet's surface, you might see small ones vibrate in roughly geometric shapes. A flash photograph will more easily reveal the shapes. If the vibrations are steady enough, you might "freeze" them in place with a continuously flashing strobe. Larger drops move around like sluggish amoeba. Occasionally either type of drop will emit a loud sizzle and a puff of steam. What accounts for these behaviors?

Does the Leidenfrost effect have anything to do with the old practice of someone preparing to iron clothing by testing the iron with a saliva-wetted finger? When the finger briefly touches the hot metal, why isn't the finger burned?

Since 1974, I have amused students with a stunt in which I briefly dip my fingers first into water and then into molten lead at a temperature of about 400°C. How does the water save my fingers? (I once foolishly left out the water, but realized the error immediately upon touching the liquid metal. The pain was excruciating.)

Before continuing, I must caution about the severe danger of this demonstration. Obviously, if I touch molten lead, my finger can be burned, and if the pot is tipped over, the lead could spill onto me and burn my body. However, there are two, more subtle dangers. If the lead is near its melting point, the sudden presence of much cooler water and a finger could solidify the lead around my finger. The lead would then be about 328°C and, soon afterwards, so would my finger. The other danger is that if there is too much water on the finger, some of the water will evaporate so forcefully that its expansion will hurl molten lead onto me. I have been badly burned on face and arms by such explosions.

A similar example of the Leidenfrost effect appears in Robert Ruark's best-selling novel *Something of Value.* To determine which of two men was telling the truth, a tribe forced the men to lick a very hot knife. The idea was that the man who was lying would have a dry tongue out of fear, and so he would be burned, while the man who told the truth would have a moist tongue and be spared any injury. The tribe did not know the Leidenfrost effect by name but realized its operating principle.

If liquid nitrogen is poured onto a flat surface, drops of the liquid will play about the surface like water drops do on a very hot skillet. Although the liquid nitrogen is at a temperature of about −200°C and should evaporate immediately, it lingers. How does the Leidenfrost effect allow it to stay?

You may have seen liquid nitrogen used in a common science demonstration where a flower is submerged in it. After the flower reaches the temperature of the liquid, it is withdrawn and then slammed against a tabletop. The frozen flower is so brittle that it shatters on impact and is left in small pieces scattered across the table.

For years I presented the flower demonstration in class and then, with a slight theatrical pause, placed the vat of liquid nitrogen at my lips and poured a large portion of the liquid into my mouth. Being careful not to swallow, I next breathed outward over the liquid, and a dragon-like plume emerged from my mouth. The plume was formed when the moisture-rich air from my lungs passed over the liquid nitrogen and some of the moisture condensed in the cold, forming water droplets that made the plume visible. Why didn't my tongue fracture just like the flower?

Here again there are several dangers. When I had the vat at my lips, the lips would sometimes freeze onto the vat's metal edge. Later I would have blisters at those spots. A more serious danger involves the natural tendency to swallow when something is in the mouth. Had I swallowed the liquid nitrogen, my throat and stomach could have been severely damaged because of their prolonged exposure, first to the cold liquid and then to the cold nitrogen gas that evaporates from the liquid.

Another danger caught me by surprise. In my last performance, the cold of the liquid or gas apparently thermally contracted two teeth sufficiently that their enamel fractured. I noticed nothing at the time, but during my next dental examination, my dentist said that, with close inspection, those teeth resembled a road map. She convinced me to drop the stunt.

Answer When a water drop approaches a metal surface that is hot but less than 200°C, the water spreads over the metal and quickly sizzles away. But when the metal is hotter than 200°C, the drop does not spread. As a drop approaches the metal, a fraction of the water on the bottom surface flashes to vapor and sets up a narrow cushion of water vapor on which the remaining drop comes to rest. The cushion is continuously replenished as more of the liquid on the bottom of the drop evaporates. Since the drop is kept from touching the metal, it is heated slowly by convection through the vapor and by thermal radiation from the metal, instead of rapidly by direct contact. Thus such a floating water drop can last quite a long time.

When I dip wet fingers into molten lead, some or all of the water flashes to vapor, and my fingers are then momentarily protected by a glove of vapor. Again, the vapor slows the transfer of heat. If the lead touched my skin, the heat transfer would be so rapid that even the briefest touch would result in a burn. If I were to touch a wet finger against a solid piece of very hot metal, the touch would be frictionless because of the vapor layer that is created. One blacksmith told me that the lack of friction is what prompts him into dropping a very hot piece of metal when he accidentally picks it up with his sweaty bare hand. If he reacted only to the sensation of pain, the signal would come too late and he would be badly burned.

When liquid nitrogen is poured onto a surface, such as in my mouth, a fraction of the liquid along the bottom side flashes to vapor and supports the remaining liquid, usually preventing any direct contact with the surface. Thermal energy is then slowly drained from the surface to the liquid by convection and radiation, but not as rapidly as would happen by conduction during direct contact.

An *inverse Leidenfrost effect* occurs when a very hot piece of metal is dropped into water. The water that first touches the metal flashes to vapor, which then covers the metal and delays its cooling. When the temperature of the metal surface drops below 200°C, the water begins to touch the surface and boil away.

SHORT STORY

4.7 • A rather dreadful swallow

In 1755, a summer storm shook the lighthouse at Eddystone near Plymouth, England. Henry Hall was the night watchman in charge of monitoring the lantern's candles, which sent light out over the waters. When he climbed up the narrow stairs at 2 A.M. to check the candles, he discovered that a stray cinder had set fire to the soot and tallow grease that had collected on the roof of the lantern. The roof was made from lead sheets that were supported by wood beams.

Although Hall feverishly tried to douse the fire with water, it rapidly grew into an inferno, consuming the beams and melting the lead. And then, just as Hall was flinging more water up into the fire, the roof support collapsed and molten lead poured down onto him. He was burned on his face and arms, and he also felt a searing pain down his throat and into his stomach. Apparently he had his mouth open during that last toss of water into the fire.

The fire spread to the rest of the lighthouse, driving Hall and two other workers out into the storm. When they were finally rescued and brought to shore, Hall managed to explain that he had swallowed molten lead, but he was thought to be merely in shock from the ordeal since he was 94 years old. The local doctor comforted Hall but remained skeptical about his story. How could anyone swallow molten lead and survive?

In fact, Hall did not survive for long—after 12 days he began to undergo convulsions and then, a few hours later, he died. When the doctor conducted an autopsy, he found an oval piece of lead that weighed almost seven ounces in Hall's stomach.

4.8 • Walking over hot coals

I first walked over hot coals as part of my physics lectures well before the fad hit the United States. I prepared a bonfire of common fireplace logs, allowed them to burn down into coals that were bright red, shoveled the coals into a wood trough that was lined with sheet metal and covered with sand, and then (with an assistant) proudly carried the trough into my classroom. I had been lecturing on the Leidenfrost effect (see item 4.6), and as I stripped off my shoes and socks, I briefly explained how the effect might help protect my feet during a stroll over the coals. The stroll took three footsteps, and although I felt some heat and my feet were dirtied from the ash, I was unharmed.

I repeated the stunt for two years, growing ever more confident. The confidence was unwarranted, because when I next walked over the coals, I suffered extensive burns. The pain was so intense that my brain turned off the information so that I could complete my 50 minute lecture. When I then stumbled to the infirmary, the pain returned in a flood.

In some of the popular "workshops" where one might learn how to walk over coals (often after paying a large fee), emphasis is placed on thinking "the right thoughts." Can any thought reduce the transfer of thermal energy to the feet? And if not, what then allows for a safe walk over the coals? Why does that feature sometimes fail, forcing the victim to contend with not only severe burns but also the chance of infection?

Answer Although I once argued that the Leidenfrost effect is the primary source of safety during the walk, I was finally convinced by physicist Bernie Leikind that something else is even more important. When I place a foot on the coals, the temperature of their surface is high, but the amount of their thermal energy is not. If my footfall is brief, only a little thermal energy is conducted into my skin, and so I might not be burned. Of course, if I linger, thermal energy can be conducted from the interior of the coals, and I can be burned badly.

Running is foolish for a practical reason—I might kick coals up onto the top of a foot where they would be in contact long enough to burn me. So, I walk at a moderate, but eager, pace.

The Leidenfrost effect is a secondary safety feature. When I walked over hot coals, I arranged to have sweaty feet. The sweat helped in three ways: It somewhat quenched the surface of the coals. It served to take up some of the thermal energy I encountered. And in places, it might have been vaporized to provide a brief Leidenfrost layer. Any of these factors helps if my footfall is a bit too long or if the coals are especially hot. I was usually so nervous about the stunt that my feet were naturally sweaty . . . except on the day when I was so self-confident that I took success for granted. Then, apparently, I needed the extra protection that my dry feet could not provide. In some of the fire-walking workshops, the participants are worked up emotionally, which might make their feet sweaty, and often they are led over grass wet from a garden hose or the dew before they step onto the coals.

(I have long suggested that when physics degrees are granted, they should not be based on some written final exam, but rather they should require that the degree recipients walk over red-hot coals. If the physics majors think the "right thoughts," that is, if they truly believe in physics, then they will be unharmed and can be handed their diplomas. To make things easier, the exam could be "open book"—they could be allowed to carry along a standard physics text for inspiration. I always carried my favorite, the original *Physics* by David Halliday and Robert Resnick, except on that dreadful day when in haste I forgot it and consequently had to learn to walk on the sides of my feet for two weeks, worrying about infection on the burned portions.)

SHORT STORY

4.9 • Fire-walking accounts

In 1984, a reporter for a San Francisco radio station attended a weekend fire-walking workshop at the invitation of a "psychic". The psychic claimed that no one had ever been hurt at the workshops, but when the reporter walked across the nearly three-meter-long bed of coals, she suffered first- and second-degree burns. Her tape recording, including her screams during the walk, was run on the station's news program the next Monday morning.

Also in 1984, a reporter for *Rolling Stone* magazine published an account of workshops given by a California "guru" who taught how "mind-control" can eliminate the burns from walking over hot coals if the participants strongly willed themselves to the task. Indeed, most participants escaped burns when they were emotionally revved up and then tested with the coals. Afterward, one claimed that if he had full control over his mind, he could even "survive a direct nuclear blast."

Fortunately, he was not around two nights later when a young woman with brain and spine injuries hobbled over the coals with two canes. She apparently believed the guru's spiel about how "thinking the right thoughts" can prevent burns from the coals. The *Rolling Stone* reporter noted that the average time of the participants on the coals was 1.5 seconds but that the young woman stalled there for 7 seconds when she collapsed from the pain. Before she fell onto the coals, she was grabbed and whisked away from the scene. She spent 12 days in a hospital with severe burns to her feet.

4.10 • Freezing and supercooling water

How does water freeze? Why can its temperature drop below the freezing point, perhaps by many degrees, without it freezing? Such cooled liquid water is said to be *supercooled*.

Answer Water needs a *nucleating agent* in order to freeze; that is, it needs a dust mote, some dissolved air, or some other material to begin the deposit of water molecules in the arrangement of an ice crystal. The reason has to do with the energy required for an initial ice speck of a given radius to grow to a larger radius. If the initial radius is smaller than a critical radius, growth requires a lot of energy, which normally means that growth is unlikely.

If the initial ice forms on a nucleating agent, growth might be easy because the radius may already be greater than the critical radius. However, if the ice lacks a nucleating agent, then its growth depends on the chance meeting of water molecules in certain orientations. The probability of this chance meeting increases if the water molecules are cooled below the freezing point, so that they become less mobile and thus more prone to get locked up as a solid. Therefore, water with few nucleating agents can be supercooled. Purified water in a clean container has been cooled to −20°C before it freezes, and water drops in clouds probably reach −40°C before they freeze. However, even common tap water, with its abundant nucleating agents, might not freeze until it is a few degrees below the freezing point.

In order for water to freeze at a water–ice surface, as in a freezing ice cube, it must lose thermal energy into the water or through the already-formed ice layer. If the loss is by conduction into supercooled water, the surface tends to form *dendritic ice*, which consists of beautiful fern-like extensions into the supercooled water. If the loss is by conduction through already-formed ice, the water–ice surface tends to be flat. If one point on the surface were to freeze faster than the rest of the surface, it would form a bump that would add to the distance through the ice. So, the freezing at the bump would then slow until the rest of the surface caught up with it, and then the surface would be flat.

4.11 • Eating sea ice

People living near the sea in the far north know that newly frozen sea ice is much too salty to eat or to melt for drinking, but sea ice several years old is fine. They also know that they can speed up the transition from unpalatable ice to palatable ice if newly frozen ice is pulled up out of the water, especially in warm spring or summer months. Why would the ice become less salty in warmer weather when evaporation seemingly decreases the amount of water in a given ice block, making the remaining ice even saltier?

Answer When seawater freezes, the salt (and other impurities) are squeezed into cells rather than being included in the ice crystal structure. Because these *brine cells* can be interconnected, the brine slowly drains out of the ice. If the ice temperature increases, as it might if the ice is hauled on shore and left in sunshine, the brine cells enlarge and the rate of drainage increases.

In addition to the drainage through interconnected cells, the cells migrate in the direction of higher temperatures. If the ice is part of an ice covering on the water, this direction is downward, because the bottom of the covering (at the water–ice surface) is at the freezing point and the external air (at the water–air surface) can be well below the freezing point.

In the brine cell of Fig. 4-2, the top of the ice layer is at the air temperature of −5°C, the bottom of the layer is at the freezing point 0°C, and the salty water inside the cell is at −2°C. That interior water is not frozen, because salt depresses the freezing point of water. (The salt molecules interfere with the ability of water molecules to lock themselves into the crystal structure of ice.) Ice at the bottom of the cell gradually melts, because thermal energy is conducted upward from the bottom of the ice layer. Water at the top of the cell gradually freezes, because thermal energy is conducted upward to the top of the ice layer. In this way, the cell migrates downward; when it reaches the bottom of the layer,

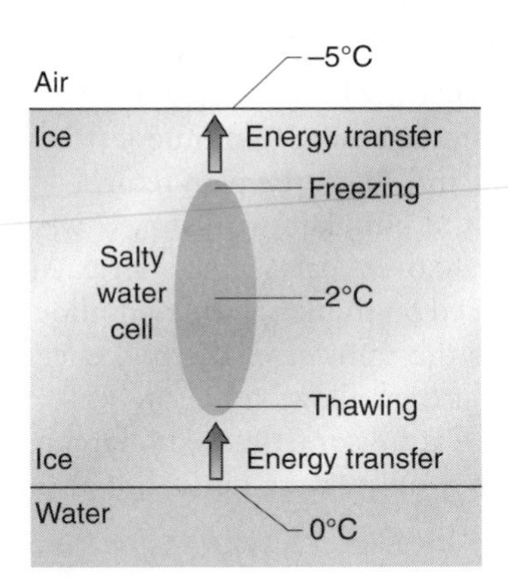

Figure 4-2 */ Item 4.11* A brine cell moves downward through an ice layer to underlying water.

it releases its salt into the water below the layer. Thus, the ice layer gradually gets rid of its salt.

The transfer at the bottom of a cell is related to the following surprising result: If an ice cube (pure water) at −1°C is placed in very salty water at −1°C, the ice cube dissolves even though the temperatures match. To explain the result, let's assume that the salt depresses the freezing point of the salty water to −2°C. Now consider the water molecules at the cube's surface. Those that are part of the cube are slightly warmer than the depressed freezing point in the salty water next to them, and so they tend to leave the cube. The molecules that are in the salty water next to the cube are slightly cooler than the normal freezing point in the pure ice of the cube, and so they tend to join the cube. However, the salt molecules prevent their joining by binding up the water molecules. (In water, a salt molecule will ionize and then water molecules will cluster around the ions like children around an ice-cream truck.) With water molecules leaving the cube and none joining it, the cube dissolves.

4.12 • Cooling rates of initially hot and warm water

The most controversial article I wrote for *Scientific American* concerned an old question: If equal amounts of water in identical open containers begin at different temperatures, one very hot and the other cooler, which will form ice first when they are placed in the same cold environment? Surprisingly, in some circumstances the initially hot water freezes first.

The result was known to Aristotle and to people in cold climates. However, scientists largely scoffed at the validity of the result until the 1960s, when a Tanzanian student, E. B. Mpemba, asked his high school teacher why an ice-cream mixture will freeze more rapidly in a freezer if it is inserted when hot rather than after it cools to room temperature. The teacher believed the claim only after Mpemba demonstrated the effect with water, and the result is now known as the Mpemba effect.

Why might initially hot water cool more rapidly and freeze sooner than an equal amount of initially warm (or sometimes even cool) water?

Answer One objection to the validity of the effect is based on common sense. If water sample *A* starts out hotter than water sample *B* and yet wins the race to the formation of ice, then at some time *A* must reach the temperature of *B*. If the samples are identical, won't they cool at the same rate from that matching temperature? One error in this argument is that we cannot assign a single temperature to either sample because they each have a range of temperatures within the water. So, the validation or rejection of the Mpemba effect requires a much more thorough investigation.

In fact, a convincing validation or rejection has yet to be found, largely because of the many variables that are involved. For example, variations of airflow and temperature in a common freezer might change the cooling rate from trial to trial, to give untrustworthy data where the Mpemba effect might erroneously be interpreted. So, many trials under controlled conditions are needed. A number of researchers who attempted to meet that requirement have seemingly demonstrated the Mpemba effect in controlled situations but without agreeing on the cause. Here are some of their points:

(1) There are greater losses of mass and energy in the evaporation from the initially hotter water. If the evaporation is eliminated by covering the containers, the Mpemba effect seems to disappear. (However, in special circumstances it may still exist even without evaporation.)

(2) Water undergoes a curious change in density as it cools from 4°C to the freezing point: Unlike most liquids, water expands during this last-stage drop in temperature. So, when a water sample drops below 4°C, the colder portions are lighter and thus rise, while the slightly warmer portions are denser and thus sink. This mixing brings the slightly warmer water up along the container's walls and to the uncovered top surface, allowing it to lose thermal energy. Experiments suggest that this mixing is more pronounced when water begins at a higher temperature. Thus, initially hot water may reach the freezing point first, largely because of this last-stage spurt in mixing and cooling.

(3) Water *supercools* (cools below the freezing point) before it suddenly begins to form ice. Initially cooler water supercools to a lower temperature than initially hotter water and thus takes longer to form ice than the latter.

4.13 • Water frozen by the sky

In some regions where freezers are uncommon, ice is made by leaving a shallow bowl of water outdoors overnight. The bowl is propped up or somehow insulated from the ground. The water will freeze, of course, if the air temperature drops below the freezing point. But on clear nights the water might freeze even if the air temperature stays somewhat above the freezing point. On such nights, what causes the freezing?

Answer On a clear night, you can regard the sky as being a single surface with a temperature that is below the freezing point of water. During the night, there is an exchange of infrared radiation between that surface and the water. The water, which begins at a temperature greater than the freezing point, initially emits more radiation than it absorbs from the sky, and so it cools. If the temperature of the air around the water is not too far above the freezing point, the water may lose enough thermal energy by this radiating process to freeze. The bowl must be insulated from the ground so that the ground cannot provide thermal energy through conduction to prevent the freezing.

4.14 • Saving the stored vegetables with a tub of water

When a hard winter's night affected my grandmother's home in Texas, she would worry about the jars of fruits and vegetables that were stored in the storm cellar in the backyard. To protect her jars, she would pull a large tub into the cellar and then fill it with water. How would such a measure help protect the jars from freezing and bursting?

Answer The large amount of water would forestall the temperature of the cellar from dropping below 0°C, the normal freezing point of water. As the water began to freeze, it released a great deal of energy, which maintained the temperature of the cellar at about 0°C. The watery solutions in the jars had somewhat lower freezing points because they were mixtures of various fluids, and so they did not freeze. The temperature in the cellar could fall below 0°C and endanger the jars only if the full tub of water froze, which was unlikely to happen overnight. A similar measure has been taken by motorists who realize that their car radiator lacks enough antifreeze to keep from freezing when a sudden cold spell hits: A tub of water is left near the radiator when the garage is shut for the night, and so the radiator does not freeze.

4.15 • Spraying an orchard to prevent freezing

When the fruit orchards of Florida are threatened by a hard freeze (temperatures lower than about −2°C), the plants are sprayed with water that forms a thin layer of ice on them. How could this procedure save the plants?

Answer The protection is *not* due to the ice layer that forms on the plants—it does not insulate the plants from the cold air. The protection comes from what happens to the water after it lands on the plants. There the water cools to the freezing point and then freezes; both processes require that the water release thermal energy to the plants so the water molecules can first slow in their thermal motion and then become locked up in the crystal arrangement of ice. The energy transferred to the plants and then to the air can keep the temperature of the orchard between −2°C and 0°C, which allows the plants to survive.

Spraying an orchard is tricky because if there is an appreciable breeze or if the air humidity is low, spraying can quickly ruin the plants. The reason is that water tends to evaporate from the drops as they fly through the air. Because evaporation requires a lot of energy, the temperature of the drops falls to the freezing point (or even lower if the drops supercool) before the drops reach the plants. The drops might partially freeze in flight or freeze immediately upon hitting a plant. Either way, far less energy is transferred to the plants, and the orchard temperature can fall below −2°C, killing the plants.

An orchard grower can tell from the ice on the plants whether spraying is helping or hurting the plants. If the process is working properly, the water drops spread on the plants before they freeze, forming a layer of clear ice. If the process is not working, the partially frozen drops freeze individually on the plants, forming a white and unclear ice because of the light scattering at all the boundaries between the frozen drops. Needless to say, when an overnight hard freeze threatens an orchard, a grower spends the night watching the temperature gauge and the transparency of the ice, rather than sleeping.

4.16 • Throwing hot water into very cold air

A novel amusement of people stationed at the Antarctic is to throw boiling water into the air when the air temperature is −40°C or lower. Why does the water make a "ripping" sound, as if protesting the low temperature? Why can breathing the cold air produce a tinkling sound?

Answer When water is flung into the air, it breaks up into drops. If the air is very cold, then as they fly through the air, the drops freeze and fracture. The composite noise of the fracturing is the ripping sound. Breathing can make a tinkling sound because the water moisture in the exhaled breath can freeze in the air. However, I don't know whether the sound occurs because of fracturing or because of the ice hitting the ground.

4.17 • Icicles

Why are icicles conical with a tip only a few millimeters wide? Why does a narrow column of liquid extend up through the center on an active (still-growing) icicle (Fig. 4-3)? Under what circumstances does that water freeze, and how does it manage to do so, considering its isolation in the center of the icicle? Why does a white line extend along an icicle's central axis? Why do horizontal ribs form along the side? Why are some portions of an icicle firm, while others are spongy enough to be easily penetrated by a penknife? Why are some icicles bent or twisted?

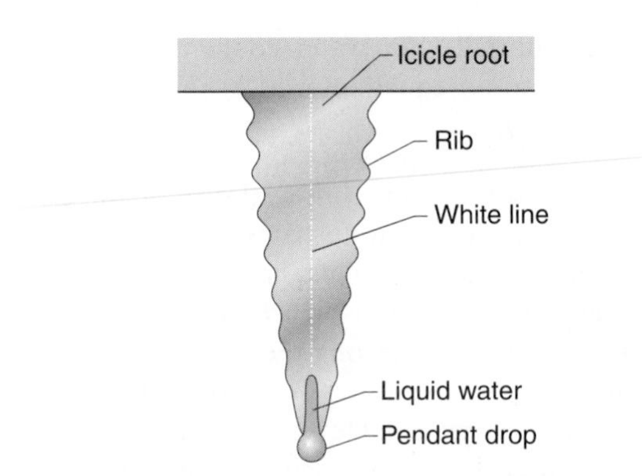

Figure 4-3 / *Item 4.17* Structure of an icicle.

Answer There are a lot of "whys" about icicles. Surprisingly, not all of them have been fully answered, but here is what I understand.

An icicle begins when water seeps off an elevated support, such as a roof gutter, and forms a pendant drop. The drop might entirely freeze, or maybe only the surface will freeze, forming a thin shell around the remaining liquid. As more water seeps onto the structure and freezes, the structure grows downward and outward.

Liquid can be held in an ice shell by surface tension, which is due to the attractive forces between water molecules. This liquid can freeze only if thermal energy is conducted upward from it and through the length of the icicle to the *root* (the icicle top). Thermal energy cannot be lost horizontally through the shell, because both sides of the shell (at the liquid surface and at the air surface) have the same temperature—namely, the freezing point of water. With no temperature difference in the shell's thickness, thermal energy cannot be conducted through the shell.

As the interior water freezes, dissolved air comes out of solution and forms bubbles that are then trapped in the ice along the central axis, where the liquid freezes last. These orphaned bubbles can scatter sunlight, and the composite of the brightly scattered light from them produces the white line seen along the icicle's axis.

The ribs along the side of an icicle probably begin by chance irregularities in the water flow down the side. Once established, they grow radially outward faster than the intermediate hollows for two reasons. They are coated with a thinner layer of water than the hollows, and they protrude into the cold air more and so are more exposed. For both reasons, the water on a rib is likely to freeze faster than water in a hollow. A hollow often freezes into a spongy water–ice network in which you might be able to insert a knife blade.

If the liquid sheath on an icicle begins to freeze (the air is cold and the water supply wanes), the external surface of the layer freezes first, momentarily trapping the remaining liquid beneath a skin of ice. When water freezes, it must expand. On an icicle, the expansion forces liquid through the ice skin in various spots. As the liquid emerges in those spots and freezes, it forms short spikes on the icicle.

If an icicle is buffeted by wind during its growth, it will become crooked and twisted. If it grows on a tree branch that progressively sags under the weight of the icicle, then it may end up being curved and appreciably nonvertical. Driven snow and uneven melting by sunlight also distort an icicle from its ideal shape.

If icicles form on a clothesline, telephone line, or power line during a freezing rain, they might have a regular spacing of a few centimeters. This periodic pattern is probably due to the tendency of the initial water sheath on the line to decrease its surface area by reforming into beads. A chance wave sets off this reforming, and surface tension then pulls the water into the beads, with a spacing that approximates the wavelength of that chance wave. The beads then grow into icicles.

4.18 • Ice dams at eaves

In cold climates, an ice dam can form along an eave, blocking the downward flow of water and pooling the water. Why does an ice dam form, and why can the pooling do extensive damage inside the building? Why do huge icicles tend to form on these buildings?

Answer Ice dams form on homes with sloped roofs lying over attics that are warmed by thermal losses from underlying rooms. The attic can melt snow and ice on the roof. If the meltwater trickles down to a cold eave, it can freeze at the eave instead of draining off the roof. The ice then grows up along the roof. A roof is waterproof as long as the water can flow down and off it. But when water is trapped by an ice dam, the water can move up through the shingles and then down along the underlying wood portion of the roof, which is not waterproof. Water can then drain down onto a room's ceiling or along the wall and ruin the plasterboard and paint.

A roof that tends to have ice dams can also have huge icicles. Instead of damming, the water happens to trickle off the eave along initially small icicles, increasing their length and weight as the water freezes on them.

Sunshine does not appear to play any significant role in the formation of ice dams and huge icicles. To eliminate them, ventilation openings are cut into attics to allow in cold air. Because the attic is then cold, it does not melt snow and ice on the roof, and thus the water supply to the eave is cut off.

4.19 • Rime ice and glaze ice on cables

When snow and ice collect on power-line cables, the additional weight can collapse the cables and their support towers. Such an event occurred in January 1998 in southern Quebec when *icing* brought down 1300 main towers and 35 000 secondary towers and eliminated electric power for over two million customers, some for weeks. Under what conditions does snow and ice collect on power lines? In particular, is the problem worse in colder air?

Answer Airborne water drops and snow can form two types of ice on a cable: *Rime ice* is a dry formation, with no liquid water on the cable (as water drops collect, they freeze). *Glaze ice* consists of an inner layer of both ice and liquid water and an outer layer of liquid water. The freezing line advances outward through the liquid, as *dendritic ice* (fern-like structures) extends into the liquid. The thermal energy released by the freezing water is conducted through the outer layer to the external cold air.

When glaze ice forms, some of the liquid water can drip off the cable, lessening the weight and the danger, but it can also form icicles. The icicles are spaced about 2 centimeters apart and grow both downward and horizontally as more water flows down their sides and freezes. They can eventually merge to form a curtain of ice. Not only does their weight threaten the cable and its support towers, but the icicles also increase the interception of water drops and snow. If a strong wind is blowing, the aerodynamic drag on the icicles can greatly increase the force on the cable.

Icing on a cable is probably most dangerous when the air temperature is only a few degrees below the freezing point, because then glaze ice can form. Airborne water drops and snow that hit the layer are sure to stick to it rather than just bounce off, as might happen with rime ice. Also, the dripping water can form icicles and make the situation worse because of the larger surface area and increased aerodynamic drag. Thus, if the air temperature increases, as from night to day, an initial formation of rime ice can become more dangerous if it transforms to glaze ice while water drops and snow are still being collected.

4.20 • Ice spikes and other ice formations

Why do most ice cubes that are frozen in a tray have a central upward bulge? Why do some ice cubes develop an upward spike? (More dramatic ice spikes can sometimes be found rising from birdbaths and other small outdoor pools of water during cold weather.)

Why does an ice covering on a puddle sometimes have a series of ridges forming rings in the puddle, and why are those ridges on the *underside* of the ice? How can the ice covering over some rivers develop a large, rotating disk that is separated from the rest of the ice by a narrow gap? (These disks have a diameter of about 50 meters, take about 1.5 hours to rotate full circle, and last for months.) Why do some ice coverings on a river develop a long crack that is shaped like a sine curve? Why does the top surface of ice on lakes, ponds, and even ice cubes sometimes contain ridges instead of being flat, even if the water was calm during the entire freezing process?

Answer When water molecules form ice, the water must expand. If it is in an ice cube tray, it can expand only upward. The center of one of the ice cube slots is the last to freeze, and so the expanding periphery pushes inward and upward.

That process can form a spike if the top of the water freezes into a thin layer of ice and then the underlying expanding water fractures the layer, forcing the remaining liquid water from the center of the cube out through the fracture. Usually this water merely trickles off onto the top of the ice, where it freezes, but sometimes it freezes into a hollow shell (Fig. 4-4). If the freezing rate is slow enough, more water can be squeezed up through the shell, trickling out the top, freezing, and extending the shell upward. When all the water has frozen, the shell forms a solid upward spike. These spikes are fairly rare, because their formation depends on a balance between the rate at which liquid is squeezed up through the fracture in the shell and the rate at which the remaining water in the cube freezes.

The ring-like ridges seen on the underside of a puddle's ice covering are due to the periodic freezing that occurs as water drains from beneath the covering. Initially, with the puddle full of water, a thin ice covering forms over the water, extending from one side of the puddle to the other. As water drains from beneath this covering, air seeps under its outer edge. At some stage, the drainage slows or stops. Under the ice, where air meets the water, some of the water freezes to form a downward ridge. Later, increased drainage removes more water, isolating that ridge. If the drainage again slows or stops, another downward ridge forms, closer to the center of the puddle. Thus, a number of ridges can form before all the water drains from the puddle.

Many of the curious patterns that can be seen in the ice on ponds and lakes occur because snow falls while the ice sheet is floating rather than held rigidly in place. The snow's weight forces the sheet downward, which pushes liquid water up through any existing opening in the sheet, or the water pressure punches out an opening at a weak spot. The upwelling water can then spread out over the sheet and through the snow, marking its path by melting some of the snow. If the opening is fairly large, the upwelling water might just freeze into a small circular disk that floats in the opening. If all this occurs on a river, the river's general flow across the uneven bottom surface of the disk can make the disk rotate.

The larger ice disks that form on some rivers in some winters are due to a whirlpool in the river. When ice floes from upstream collect in the whirlpool, they gradually meld into a single sheet that is rotated by the swirling water. As ice forms over the rest of the river, the rotation prevents the water between the sheet and the rest of the ice from freezing, and abrasion between the sheet and the rest of the ice gradually smooths the sheet into a circular disk.

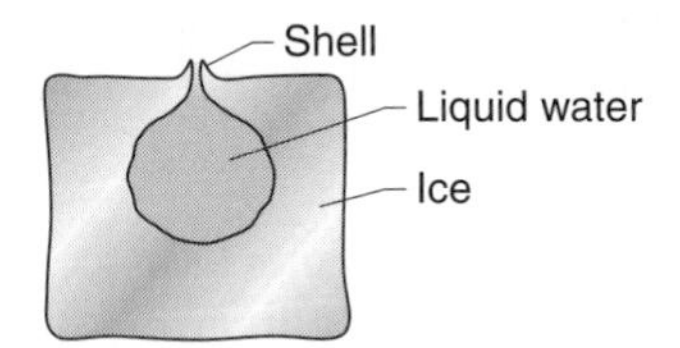

***Figure 4-4** / Item 4.20* Early stage in an ice spike formation.

If an ice covering is being pulled in opposite directions, as might occur if the covering is anchored on rocks while water flows under it, an initially straight crack can develop into a sine-curve crack as the crack travels across the covering. Such wavy cracking has been seen in glass plates that are pulled from a water bath and then past heating elements, which puts the glass under tension. Depending on the speed at which one of these glass plates moves, the crack can be straight (slow speed) or wavy (moderate speed), or it can split up into two or four cracks (high speed).

As water freezes, it forms hexagonal ice crystals. The axis through the center of the crystal, perpendicular to the two hexagonal faces, is called the *c* axis. The ice tends to grow most rapidly parallel to the hexagonal faces, in what is called the *basal plane*. Suppose a crystal begins to grow with its *c* axis vertical and thus its basal plane horizontal. Then it tends to grow horizontally and leaves a flat surface on the ice. If, instead, it begins with its *c* axis horizontal and its basal plane vertical, it cannot rotate because it butts up against adjacent crystals and buoyancy lifts it slightly so that its basal plane sticks slightly above the general level of the ice. Thus, it forms a ridge. If several adjacent crystals have this orientation, they form a series of ridges on the ice surface.

4.21 • Cloudy ice cubes

Why are ice cubes cloudy, and is there any way to make clear (transparent) ice cubes?

Answer Ice is cloudy because light scatters from structures and materials inside the ice. Some of the material can be impurities that are concentrated by the freezing process. For example, as it advances into the water, the freezing process drives impurities into the liquid next to the interface of the liquid and the ice and forces dissolved air to form bubbles there. As the freezing advances and drives more air into these bubbles, the bubbles become longer and surrounded by ice. Thus, long *wormholes* (hollows) extend toward the center of the cube.

The liquid–ice interface can advance only by conducting thermal energy from the interface to the surface of the cube, where it can be removed by cold air. The distance required of this conduction grows longer as the interface advances, and so the advance slows. Thus, wormholes are typically wider near the center of a cube (where the advance is slow) than near the surface (where the advance is faster). Some wormholes vary in radius because the freezer cycles between being on and off. Ice made from a brine (salty) solution can show more complex air bubbles than that made from tap water: Under ideal conditions, they can form tiny spirals or zigzag patterns.

To make clear ice, distilled water can be used to avoid impurities and the water can be boiled for about 15 minutes to eliminate most of the dissolved air.

4.22 • Figures inside melting ice

If ice, such as a common ice cube or naturally occurring ice, is placed in bright sunlight or in light from an infrared lamp, why do tiny figures appear inside the ice? They were called *liquid flowers* by John Tyndall, who first reported them in 1858, but now they are known as *Tyndall figures.* You might be able to see them with the unaided eye, but a jeweler's loupe or some other simple magnifying lens can bring out their details. Some Tyndall figures are hexagonal; others resemble the leaves of a fern; and some are simple ovals.

Answer The figures appear because the infrared light in sunlight, or from a lamp, penetrates the ice and is absorbed mostly at defects in the crystal structure. Some of the defects are situated where the atoms are not quite in the right order; others are where one ice crystal butts up against another; and still others are at locations of impurities. As the infrared light is absorbed, some of the ice melts or vaporizes, forming a cavity, which is seen as a Tyndall figure because of its contrast with the rest of the ice. Some of the cavities are filled with only water vapor; others contain both liquid water and water vapor. If the light is intense and the melting and vaporization processes are fast, fern-like cavities appear. If the light is not as intense, hexagonal cavities appear. Ovals are filled with water and appear where a fracture in the ice is compressed by ice on opposite sides. When the infrared light first shines on the ice, many Tyndall figures can appear simultaneously, probably because they are triggered by the release of stress in the ice as they form.

4.23 • Freezing of ponds and lakes

Why do ponds, lakes, and other bodies of water in regions with cold winters freeze from the top down? If, instead, they froze from the bottom up, they would probably not have any aquatic life.

What causes the striking radial patterns of ice that appear on some frozen ponds and lakes? They resemble crooked spokes on a wheel or petals arranged on a flower.

Answer As the water in a lake is cooled from, say, 10°C toward the freezing point, it becomes denser than the lower water and sinks to the bottom. Below 4°C, however, further cooling makes the water on the surface *less* dense than the lower water, so it stays on the surface until it freezes. Then the freezing progresses down into the body of the water. However, it progresses only by transferring thermal energy from the liquid water up through the ice layer via conduction. As the layer thickens, the process slows and can effectively stop. (The freezing of a lake is said to be a *self-limiting process* in that it stops itself.) Thus, a lake is unlikely to completely freeze during a winter, and so aquatic life can survive.

If lakes froze from the bottom up, the ice would tend not to melt completely during the summer, being insulated by the overlying water. After a few years, many bodies of open water in the temperate zones would be frozen solid all year round.

Liquid water at any temperature consists of clusters of water molecules that repeatedly form and break. As water is cooled below 4°C to the freezing point, however, these clusters last longer and are more extensive, and their presence requires, on average, more volume than when the water temperature is higher. This effect does not apply to seawater: Its density steadily increases as it is cooled to the freezing point, and only as ice forms does the density decrease.

If sunshine falls on a lake with a thin layer of ice, the light can warm the water just below the ice to 4°C, thus causing that water to sink and lighter, warmer water to rise to take its place. This wintertime mixing is necessary for some aquatic life.

If snow falls on a pond or lake with a layer of floating ice, the weight of the snow can submerge the ice layer, causing water to come up through holes in the layer or over the edge. Because the temperature of the water is above the freezing point, the water melts channels through the snow, moving radially outward from a hole, probably along crooked paths. If the air temperature drops and the sunshine disappears, these watery paths can freeze, producing an eye-catching pattern of crooked spokes or flower petals.

4.24 • Freezing carbonated drinks

If a bottle of soda or beer is cooled too long in a freezer, why will it burst? If the bottle is not left that long, why can it suddenly freeze when it is opened? If the liquid is cold (but not near freezing), why does opening the bottle produce a thin fog at the mouth, in addition to a spray of drops?

Answer A carbonated drink such as soda or beer consists primarily of water, and when water freezes, it expands to accommodate the ordered arrangement of water molecules in forming a rigid structure. Thus, when a carbonated drink becomes so cold that the water begins to freeze, there is a large outward pressure that tends to break the bottle. The freezing point of the drink is *depressed.* That is, it is lower than the normal freezing point of water, because the liquid is under pressure and because additives (especially alcohol) interfere with the ability of water molecules to form ice. However, most freezer compartments are cold enough to burst a bottle of the liquid.

When a bottle is opened, the pressure in it suddenly drops to atmospheric pressure and much of the carbon dioxide comes out of solution, forming bubbles that rise to the top surface. Suppose the temperature of the liquid is just above the depressed freezing point. As soon as the pressure drops, the freezing point rises and then the liquid is *below* the (new) freezing point, so the liquid tends to freeze. However, it needs *nucleating sites* for the initial ice crystals to form. The bubbles can initiate that process. If the bottle is transparent, you might see that the freezing begins at or near the top surface where the bubbles bunch, and then it moves downward, perhaps very rapidly.

When a cold carbonated drink is opened, the expansion of the gas above the liquid, out through the opening, requires energy. The expansion is so rapid that the only possible energy transfer is from the thermal energy of the gas. Thus, the gas loses thermal energy and becomes colder, causing the water vapor in the expanding gas to condense into water drops. Those airborne drops comprise the slight fog seen at the opening of the bottle.

4.25 • Bursting pipes

In houses where water pipes are exposed to very cold winters, why are the pipes subject to bursting, and why is the hot-water pipe more likely to burst than the cold-water pipe?

Answer A pipe with stationary water can burst due to the freezing of water if the ice forms a plug that traps water in a section closed by a valve. As more of the water freezes onto the plug, that water expands, putting tremendous pressure on the trapped water. Eventually the pressure can be high enough to crack the pipe. This possibility is greater in a pipe carrying water from a heating tank, because of the way water can freeze in that pipe. Ideally, water freezes at the freezing point of 0°C, but in practice water must cool a few degrees below the freezing point before it can freeze. Such liquid water is said to be *supercooled.*

Tap water that has not been heated in a heating tank has many impurities that can serve as nucleating sites. Once the water is supercooled by a few degrees, further removal of thermal energy from the water initiates the formation of ice crystals. The water first develops dendritic ice (which resembles ferns) that grow into the still-liquid water. Once the ice is established, a ring of ice forms along the pipe wall and gradually grows radially inward, until it is a plug that completely seals the pipe. The water between this ice growth and a closed valve has plenty of time to adjust to the expansion of the water during the ring's growth.

Water that has been heated in a heating tank freezes in a similar way, but the formation of the dendritic ice can be greatly delayed, perhaps by days. The reason is that when the water is heated, it loses much of the impurities that can nucleate the ice formation. Thus, this water can be supercooled to a lower temperature than the water in the cold-water pipe. When ice finally begins to form, it grows radially inward as dendritic ice and can rapidly seal off the pipe. If the ice then grows along the pipe, the expansion of the water, due to the freezing, can put tremendous pressure on the water trapped between the plug and a closed valve. Eventually the pipe, or a pipe connection somewhere between the plug and valve, can burst, even if the portion that bursts happens to be warm.

The burst will allow the trapped water to escape, which may not cause much damage. However, if the burst is not repaired before the ice plug melts, the consequent flooding can do lots of damage. To avoid this possibility, people living in very cold climates often leave outdoor water taps slightly open during the winter, so that the growth of an ice plug would merely push water out through the open tap rather than build up pressure.

Ice plugs can be a benefit in some situations. For example, an ice plug can act like a temporary valve on a section of pipe that must be repaired when the full system of water pipes cannot be turned off, such as in a hospital or housing complex.

In some situations, water must be sent through a cold pipe that has been drained, perhaps due to repair work. As it travels through the pipe, the leading slug of water can lose thermal energy to the pipe wall and thus drop to the freezing point. This more likely occurs for a buried pipe than for an above-ground pipe (at the same temperature) because thermal energy is more quickly drained from pipe walls by the surrounding ground than by surrounding air. If a pipe is only a few degrees below the freezing point, the water can supercool and then suddenly form dendritic ice that plugs the pipe. However, if the pipe is somewhat colder (below the supercooling temperature of the water), an ice ring forms on the pipe wall without dendritic growth, and the water keeps flowing.

4.26 • Touching or licking a cold pipe

If you touch a cold wood beam and a cold metal pipe that are at the same temperature, why does the pipe feel colder? Why can your hand stick to the pipe? In the movie *A Christmas Story*, one of the children takes up a dare and licks a cold pipe, only to find that his tongue sticks to it. Here is one of life's many rules: Don't lick a cold metal pipe.

Answer The sensation of how cold an object is to the touch greatly depends on the rate at which thermal energy is conducted away from your fingers by the object. Metal conducts thermal energy much better than wood, and so metal feels colder when it is at the same temperature as wood.

Fingers can stick to a cold metal surface, because the moisture on the skin can freeze into the tiny indentations in the surface. (A tongue would freeze even more so, because it has more moisture than fingers.) Pouring warm water over the skin–metal connection will usually melt the ice and free the fingers.

4.27 • Bumps in winter, pingos in permafrost

During a cold winter, why does the soil heave upward to form bumps, an effect known as *frost heave*? Not only can such bumps damage roadways, but they also endanger any motorist at high speed on them. You may be tempted to attribute frost heave to simple expansion of water when it freezes, but such expansion increases the volume by only about 10%, much less than would account for the large heaves.

How can cold weather break a rock? Must the rock go through a succession of freezing and thawing (*freeze–thaw cycles*) to be broken?

What causes *pingos*, which are conical hills that have formed in regions of permafrost? Some pingos are huge, with a height of 40 meters and a diameter of 200 meters.

Answer When the temperature of the top layer of soil (or the layer just below a roadway) drops below the freezing point, some of the water within the pores between the soil particles freezes and expands, causing a *primary frost heave*. As the amount of liquid water decreases, the pressure within the remaining liquid decreases also. Liquid water that is somewhat deeper is at a greater pressure, and so some of it is drawn up toward the frozen layer. As the liquid arrives, it too freezes and expands, producing a *secondary frost heave* that can greatly increase the height of the heave and so also its destruction of a roadway.

Much of the damage to a roadway occurs when ice within and below it thaws. Then the gravel below the road surface becomes saturated with water. When a car (or worse, a heavy truck) bears down on the road surface, the pressure in the water increases dramatically and the water can push back up onto the underside of the pavement. This water pressure can be enough to break the pavement. Further traffic then rips apart the pavement, forming a chuckhole, which is then gradually dug out by more traffic. In roadways with thin pavement or weak support, the pavement may give way under this pressure until a groove or depression forms in the roadway. When it hits the low point, a vehicle oscillates as it travels along the road. Each time it bears down on the pavement, it tends to create another low point. Given enough time, the roadway resembles an old-fashioned ribbed washboard against which laundry was rubbed.

Road damage is usually more extensive during mild winters than harsh winters, because mild weather gives water time to be pulled up under a roadway, whereas a hard freeze locks the water in place for a long time. Road damage is also more extensive if the roadway undergoes multiple freeze–thaw cycles rather than one severe freeze.

When water freezes in a crack of a rock, its expansion pries open the tip of the crack, both lengthening and widening the crack. The length and width increase even more if additional water is pushed into the crack, as in frost heave. However, such pushing in of water is diminished if the temperature is too low, and so the rupture of the rock proceeds fastest if the temperature is just below the freezing point. A succession of freeze–thaw cycles is not necessary. Sedimentary rocks are more susceptible to cracking than other types. Breaking apart crystalline rock is considerably more difficult, requiring long periods of subfreezing temperatures and a frequent or continuous supply of water.

Pingos come in at least two types, with at least two causes. The so-called *hydrostatic pingo* is usually an isolated mount that has risen up from a lake that has either been drained or filled in. Below the lake lies a layer of semifluid material that is surrounded by permafrost except on its top surface. As the lake loses its water, permafrost begins to form at the top surface. As the material freezes inward from the sides or upward from the bottom, the water is driven out of the material and forced up at the center of the former lake region. This upward push raises the pingo mound, and as the water freezes, it forms the ice core of the mound.

Hydraulic pingos are often clustered in groups and appear to be fed by ground water draining from nearby hills or mountains. The underground water somehow climbs up underneath a mound and then expands when it freezes, lifting the mound somewhat higher. How the water makes this climb is not fully understood. If the material underlying a mound consists of small pores, capillary forces can pull the water up through the material. That is, the molecular forces that cause water to cling to the material and to itself can pull the liquid up through the spaces between small material objects. (However, some hydraulic pingos occur over material that does not consist of small pores.)

4.28 • Arctic ice polygons

Why is ground in some of the Arctic lowlands and the subarctic tundra regions covered with large-scale polygons formed by wedges of ice? Such a wedge extends into the ground by several meters and along the ground by as much as a few hundred meters.

Answer When the ground temperature drops below the freezing point, the ground tends to contract inward (downward), which sets up stresses along the surface. At points, the stress can be large enough that the ground fractures, much like a pool of mud can fracture when drying. With time, a ground fracture can grow both vertically and (even more) horizontally. Where a developing fracture happens to extend toward an existing fracture, the developing fracture is steered by ground stresses to make a perpendicular intersection. Those stresses are already set up in the ground by the existing fracture and tend to pull the ground apart parallel to that fracture. If several fractures intersect at 90°, a polygonal pattern emerges.

After a fracture forms, it can be filled in by snow or frost or by water that melts from snow and then freezes. This, then, is the wedge of ice that forms the sides of the polygon. Similar polygons are seen on Mars, where sand now fills the fractures instead of snow or ice. (Mars also has giantic polygons that are, presumably, due to other causes.)

4.29 • Growing stones in a garden, patterned ground

In cold climates why are stones ejected from the ground during the winter? In some places, such as New England, the crop of stones is so abundant that they are gathered to build stone fences.

Why in some places do the emerging stones form circles, polygons, or stripes? Sometimes the pattern is so ordered and so extensive that it looks artificial. For example, on the island Spitsbergen, which is north of Norway, there are stunning arrays of stone circles within which the soil is relatively free of stones. What arranges these various stone arrays?

Answer Stones are brought to the surface by frost heaves, as explained in a preceding item. The *freezing line*, at which the temperature is 0°C, descends through buried stones faster than through the adjacent soil. The delay in the soil is due to the release of thermal energy when the water in the soil freezes. As the freezing line drops below a stone, it pulls in additional water from lower in the ground and also from the adjacent (still unfrozen) soil, and the expansion of that extra water as it freezes drives the stone upward. Whenever the stone moves, soil can partially fill in beneath it, wedging it so that it cannot descend later when the ice melts. After enough freeze–thaw cycles, the stone reaches the ground surface, perhaps to be gathered for a stone fence.

Upward-migrating stones have been found when ancient sandbeds have been excavated with vertical cuts. Stones are embedded at various heights within the wall of such a cut. Above a stone, the sand appears to be compressed; below the stone, a trail seems to mark where material filled in space left by a stone as it moved upward.

The various geometric stone patterns are due to decades or even centuries of frost heave. As stones emerge from the ground, they affect how fast the freeze line descends under them during subsequent freezing periods, which in turn affects how other stones are brought upward. During frost heave, some of the surface stones avalanche down the slopes formed by the heave and thus collect. Other stones can be pushed along already formed ridges of stones, thus elongating the ridges. Where stones are plentiful, they form circles, polygons, or labyrinths. Where they are not, they form islands. And where they are on slanted ground, they form stripes. Patterns like these have been spotted on Mars, suggesting freeze–thaw cycling of ground water there.

4.30 • Ploughing boulders

Why does a large boulder perched on a slope in a cold climate gradually move down the slope? Such a boulder is dubbed a *ploughing boulder* (using the British spelling) because a mound of soil typically lies in front of the boulder and a depression lies behind the boulder, as if the soil is on a field being plowed.

Answer The freezing and thawing of the soil around the boulder loosens the soil's grip on the boulder. When the soil freezes, the boulder is heaved upward slightly by the expansion of the water in the soil (water that is initially there and water that is drawn into the region from adjacent soil).

When the soil thaws, its water and any meltwater from snow piled around the boulder soften the ground. The gravitational force on the boulder can then pull it somewhat down the slope, and as it moves, it pushes up a mound and leaves a trailing depression. The motion is only slight, however, because the ground somewhat lower on the slope is not as water saturated and pliable.

SHORT STORY

4.31 • Dead-cat bomb and a frozen disappearance

When I reached MIT as a freshman, one of the stories circulated by the older students concerned one student, call him Fred, who was exceptionally angry with another student, call him Harry. One night when Harry was away, Fred snuck into Harry's dorm room with a dead cat and a large vat of liquid nitrogen, both taken from one of the campus labs.

Holding the cat by the tail, Fred dipped it into the liquid and waited until the body was as cold as the liquid. Then he lifted the cat and hurled it at one wall, where it shattered into many small pieces that littered the bed and other parts of the room. Within minutes the pieces thawed, leaving a horrible mess. (MIT students have been known for their clever pranks but not always for their compassion. In this case, I hope the story was fictional—an *urban myth* invented by older students.)

A similar situation lies at the core of a locked-room mystery by L. T. Meade and Robert Eustace. In their 1901 story, "The Man Who Disappeared" in *The Strand Magazine*, the authors cleverly set up the reader regarding the danger to the central character, Oscar Digby: He is charmed into dining with people who are obviously after his knowledge of a vast treasure. The police know of the risk and surround the house after Digby enters; not even a mouse could have crept through the ring of policemen. At midnight, the police march into the house and search it thoroughly, even ripping walls apart, but there is no trace of Digby. During their wait outside, they had heard a muffled banging but nothing more. What had become of Digby?

The answer was revealed when one of the men found slight traces of blood on a stone-crushing machine, which was near a large vat of liquid air. Apparently, Digby, once dipped and frozen in the cold liquid, had been pulverized by the machine and then scattered before the police closed in. Without a body, the police could do nothing but cringe at their discovery.

4.32 • Snowflake formation

What is responsible for the general shapes of snow crystals?

Answer The question here is deceivingly simple because its answer has yet to be worked out. The crystal *nucleates* (begins) on a dust speck. Water molecules gradually collect and bond to one another so as to give a hexagonal arrangement, said to have *sixfold symmetry* because the crystal consists of six almost-identical sections clustered around a center (like sections of a pie). The crystal grows as more molecules *diffuse* (drift) across the surface until they too bond. Most of the growth occurs at edges and corners because there the existing crystal is most exposed to the water vapor in the air.

The number of ways that these additional molecules can fit into the crystal is enormous and depends on the temperature and the density of the water vapor next to the crystal. For some values, a crystal grows to be a flat plate; for other values, it grows to be a sheath, hollow prism, column, or star. The fact that symmetry is preserved is surprising, and various mechanisms have been proposed to explain how the design and growth on opposite sides of a crystal manage to be almost identical. However, the crystal is small enough that the environmental conditions on all sides are probably the same, and they apparently determine the design and growth.

Although most snowflakes have sixfold symmetry, snowflakes with higher-order symmetry (12-fold, 18-fold, and 24-fold) have been photographed, but they are probably aggregates of two or more snowflakes, each with a common sixfold symmetry.

4.33 • Skiing

Why can a ski slide over snow?

Answer A ski can travel smoothly over snow because the friction between it and the snow melts some of the snow to provide a thin layer of lubrication. Faster skiing, with its greater production of thermal energy, tends to give better gliding, whereas slow skiing, with its poor energy production, can be difficult.

A ski that poorly conducts thermal energy works better because it keeps more of the thermal energy at the ski–snow contact area instead of just conducting it to the top of the ski. Skis with darker colors may be warmer in sunlight if they absorb more infrared sunlight than do lighter colors. If so, then they will be better lubricated even in the diffuse light on a cloudy day.

If the snow is very cold, the ski–snow friction cannot provide sufficient lubrication, and so skiing becomes difficult. When pulling sledges on ski runners over such very cold snow, some Arctic explorers liken the effort to pulling a sledge over sand.

A ski can also travel smoothly over snow, especially a quickly moving ski, because air trapped between the snow and ski helps support the ski, decreasing the friction. The ski is then a bit like a hovercraft.

4.34 • Ice-skating and making a snowball

Why can ice skates glide over ice? Can ice be too cold for skating? Does skating become more difficult if the ice is just below the freezing point? Why can you make a snowball out of snow, and why can't you do so when the snow is very cold?

Answer An ice skate can glide over ice only because of lubrication due to liquid water that is either already present on the ice surface or melted out of the ice by friction between the ice and the moving skate. That friction also warms the ice blade, the ice trail, and the bits of water and ice that are slung off to the sides by the moving blade. If too much water is on the ice surface, it produces a drag force on the blade, making gliding more difficult; this can be a problem when the ice surface is just below freezing, especially in bright sunlight.

In the past, the lubricating water layer was attributed to *pressure melting*, in which the pressure on the ice from the relatively thin skate blade decreases the freezing point from its normal value of 0°C. Thus, if the ice is not very cold, ice just below the moving blade would suddenly be *above* the freezing point and would melt. However, experiments and calculations show that pressure melting has little or no effect on ice-skating.

Although water molecules in ice are firmly locked into place by their bonds with one another, water molecules on the surface of the ice might be more loosely held and thus be in somewhat of a liquid state called *premelting* or *surface melting*. Those rather loosely held molecules, which exist in very thin patches rather than forming a uniform layer, offer less friction to a skate than the rigid ice itself.

To make a snowball, you squeeze and pat a handful of loose snow into a compact ball. As snow slides over snow during this process, portions melt due to their mutual friction and then refreeze as ice, which binds the snow together. Water in premelting patches on the snow may also freeze to form ice that binds the snow. If the snow is "wet" (easily melted by you or in the process of being melted by warm air or sunshine), the ball might end up with so much ice that it is more an iceball than a snowball. Of course, the traditional rules of snowball combat forbid the use of an iceball because it can be as hard as a rock.

You cannot make a snowball if the snow is too cold. The friction between the sliding portions of snow cannot melt the snow, and thus the snow cannot reform into ice. Without the ice bindings, the snow just falls apart.

4.35 • Ice walking

Why is walking on ice much easier if the ice is very cold? During which part of the walking process are you most likely to slip and fall? Why are some types of footwear better for walking on ice (less likely to slip) than others?

Answer Contrary to earlier thought, the pressure of your footfall probably produces little or no melting of the ice. So, the factor that can save you from slipping is a large amount of friction on your footwear. On wet ice (ice with a layer of slush or water), hard materials may work better, but on dry ice, soft materials may be better. In either case, cleats help for the obvious reason that your footfall can drive them into the ice like short nails, provided the ice is not terribly cold.

The most dangerous phase of walking is the *heel strike*, which is when you bring a foot forward and down, striking with the heel. You depend on friction to stop that foot, but on ice you have a good chance of overwhelming the friction between the heel and ice. Then that foot slides forward with no control, and so, without adequate support, you fall. If you have fallen like this, you now know to take small steps on ice so that a smaller amount of friction is needed to stop the motion of the forward-going foot. You also know to never run or jump forward on ice unless you happen to enjoy falling!

4.36 • Igloos

Can an igloo (that conical structure built of blocks of snow or ice) keep an occupant warm when the external temperature is below the freezing point?

Answer An igloo provides more protection against the cold than merely blocking the wind and thus decreasing the danger of windchill. The main point is that the walls provide thermal insulation, so that thermal energy radiated by a person's body or by a flame (even a candle) is only slowly lost through the walls. A well-made igloo is squat with an elevated "sleeping bench" across about two-thirds of the interior floor. Entrance is via a tunnel leading to the other, lower one-third of the floor. Once inside, a person climbs up onto the sleeping bench. Because warmer (lighter) air rises and colder (denser) air sinks, the air just above the sleeping bench is considerably warmer than the air in the low-floor portion, allowing someone to sleep in relative warmth. (Building an igloo with a high point would defeat this result, because the warmer air would rise above the sleeping bench.)

Thicker walls or walls with loosely packed snow (with plenty of air pockets) decrease the thermal conduction through the walls and thus keep the interior warmer. To further insulate the interior and to decrease any draft, the spaces between the blocks can be sealed with hand-packed snow both on the outside and the inside. The sealing snow and the interior surface of the blocks tend to melt and then refreeze into a protective ice layer.

4.37 • Snowrollers

On rare occasions, a gusty snowstorm can form large balls and cylinders of snow on open fields. Some of these snowrollers are tens of centimeters in diameter and have a mass of about 6 kilograms. The cylinders resemble a rolled-up sleeping bag or turf that has been rolled up prior to planting, except that the cylinders sometimes have a hollow interior. What produces these oddities?

Answer Snowrollers are believed to be created when fresh snow falls on an existing covering of old, crusty snow. If the temperature is near the freezing point, the fresh snow adheres to the crusty snow. If the wind is strong, it can catch a projecting patch of the composite layer and roll the patch over the field. As the patch rolls, it pulls up more snow and grows in diameter. When the wind is gusty, the snowroller might tumble in many directions and ends up looking like a sphere or American football, but when the wind is generally in one direction, the snowroller forms a cylinder. If the initial uplifted patch is long, the cylinder will have a hollow interior.

Snowrollers can also be started by individual snowflakes that are sent skipping over fresh snow by the wind, or by stones that are loosened by the wind and sent rolling down a snowy slope. You can tell if that happened by opening up the roller and finding the stone.

4.38 • Snow avalanche

What causes a snow avalanche?

Answer A complete explanation of what causes a snow avalanche is not yet available, largely because of the many variables involved. Thus, reliable predictions as to when and where an avalanche will occur are not available. However, much is known about avalanches. A *loose snow avalanche* begins at a point in either dry or wet snow that is not very cohesive; the slide is much like sand sliding down a sandpile. A *snow slab avalanche* is the motion of a slab that is fairly cohesive. The sliding can begin in a number of ways, such as when the load on it is increased by a skier or rain or when the slab is warmed.

Many skiers are killed every year by avalanches they initiate, but the details of the trigger mechanism are still not well understood. The mechanism involves the presence of a weak layer of snow buried below the snow slab. Here is one way such a weak layer can form. Early season snow falls on ground that is still above freezing, and then the temperature of the snow's top surface drops below the freezing point. The temperature distribution (warmer at the bottom of the snow, colder at the top surface) drives water vapor upward, where it condenses onto the snowflakes. The snowflakes transform into what is called *depth hoar,* a loosely bound collection of ice grains. This, then, is the weak layer. When fresh snow falls on it, the stage is set for an avalanche because the weak layer can be sheared; that is, it has little resistance to motion parallel to the layer, much like butter being spread over toast by a knife.

A skier can initiate the motion if the skier's downward force on the snow surface puts pressure on the weak layer. If the snow surface is relatively hard, the skier has no effect on the weak layer and does not cause the slab to move. If the snow layer is less hard, as it might be if warmed in sunshine, the pressure from the skier can rupture the weak layer, and then motion begins. Thus, a snow section that might be safe early in the morning when it is still cold can be dangerous later in the day after it has warmed.

4.39 • Patterns formed by melting snow

When snow melts on ground or paved surface, why are the longest-lasting snow lumps sometimes arranged in geometric patterns such as hexagons or rows?

Answer The snow and meltwater form a thin layer on a thin layer of ground in which the temperature can vary horizontally. Below these two layers is ground with a temperature that does not vary horizontally. By chance, some spots in the snow are closer to melting than the rest of the snow. Let's consider one of those spots. If the snow there is to melt, it needs thermal energy to free its water molecules from the rigid structure of an ice crystal. It can get this thermal energy from the surrounding snow out to some limiting distance. Because the surrounding snow loses energy, its melting is delayed and thus it forms the longest-lasting snow lumps. The spacing between these lumps is set by the limiting distance over which energy was transferred. So, we end up with snow lumps that are roughly spaced by a certain distance.

4.40 • Salting icy sidewalks

When sidewalks and roadways ice over during the winter, why do they clear when sprinkled with salt? Why is calcium chloride (rock salt) sometimes preferred over sodium chloride (table salt), other than reasons of cost?

Answer Suppose, first, that there is a layer of water on an ice surface and that both liquid and ice are at 0°C, the normal freezing point of water. The interface between the two states of water is a beehive of activity at the molecular level because molecules are continuously leaving the ice to join the liquid, and vice versa. However, if the number leaving the ice matches the number joining it, the amount of ice does not change.

If you sprinkle salt into the liquid, the salt molecules *dissassociate* (break up) into positive and negative ions. Water molecules are eager to cluster around each type of ion, and the ions are said to be *hydrated.* Because the water molecules are bound up like this, they cannot join the ice. Since there are fewer water molecules joining the ice but the same number of water molecules leaving the ice, the total amount of ice begins to decrease. That is, the ice begins to melt. If the meltwater dilutes the salty mixture enough, the situation return to the matching conditions of leaving and joining, and then the melting stops. Adding more salt restarts the melting.

The molecules in the liquid state are more energetic than those locked in the crystalline structure of the ice. When a molecule joins the ice, it must give up some of its energy; when a molecule leaves the ice, it must receive just as much energy. If the numbers of molecules leaving and joining the ice are matched, the energy given up by one process supplies the energy needed by the other process. But when salt decreases the number of molecules joining the ice, what supplies the energy for the molecules that continue to leave the ice? If the ice–water is outdoors, the supply comes from the sidewalk, roadway, and air. Although the ice melts, the temperature of the ice–water does not change and continues at the environmental temperature.

However, if insufficient energy comes in from the ice–water's environment, the liquid molecules must supply the energy for melting. That loss decreases the temperature of first the liquid and then the ice. In such a situation, the freezing point of water is said to be *depressed* by the presence of salt. The temperature decreases until there is again a match of molecules leaving and joining the ice.

The depression of the freezing point can also be seen if salty water is gradually cooled by an extremely cold freezer. However, there is a limit on just how far the freezing point can be depressed. For sodium chloride it is about −21°C, while for calcium chloride it is about −55°C. The lower limit for calcium chloride is one reason why rock salt is used on roads—it can clear the roads at much lower temperatures than table salt.

4.41 • Homemade ice cream

A home ice-cream maker consists of a central metal can that is surrounded by layers of salt and chopped ice. The exterior is a wood bucket. After the cream mixture is cooled in a refrigerator, it is poured into the can, and a stirring device is inserted. When I was young, I churned the cream mixture with the stirring device by turning a crank. Now I plug the ice-cream maker into an outlet and let a motor do the work.

Why is wood on the outside of the ice-cream maker? Why is the can metallic? Why are chopped ice and salt employed? Why must the cream mixture be stirred? Can't the mixture just be frozen in a freezer? What happens if the temperature of the cream mixture is not dropped much below the normal freezing point of water? What happens if the temperature is dropped too much?

Answer The freezing point of the cream mixture (the temperature at which ice begins to form within it) is lower than 0°C because the ingredients interfere with the formation of ice. To obtain such a low temperature, you salt the ice packed around the metal canister (see the preceding item). The ice and its melted water are then colder than 0°C and thus drain thermal energy from the cream mixture. However, you should not use too much salt or the ice–water will be so cold that it drains thermal energy too quickly. In that case, the cream mixture next to the container wall freezes quickly and retards the stirring. You want a more gradual drain of thermal energy so that the entire cream contents are in the same state. The bucket holding the ice should be wood or some other poor conductor of thermal energy so that the room does not melt the ice.

You use chopped ice because larger ice cubes make too few contact points with the container and so they cool the cream mixture too slowly. If you churn the mixture without sufficiently cooling it, you will separate out butter rather than make ice cream.

Churning has two purposes. (1) It disrupts the growth of ice crystals in the cream mixture by moving them around and by coating them with the cream. If the crystals were allowed to grow large, as they do if the cream mixture is merely frozen in a common freezer, the final product would be unpleasantly grainy. When the cream mixture is churned, the ice crystals remain small and the final product is smooth when eaten. (2) The churning's other purpose is to whip air bubbles into the cream mixture so that the ice cream ends up being a frozen foam, not a dense chunk of ice. The bubbles are stabilized by the fat globules in the cream mixture. The increase in volume due to the air bubbles is called the *overrun.* In a light and airy ice cream, half the volume might be air, giving an overrun of 100%.

After the mixture is churned, it is allowed to sit in the cold environment provided by the ice–salt mixture. During this period it is said to *harden* because the remaining liquid water is allowed to freeze. If all goes well, the final product is uniform with small crystals and thus smooth to the taste. However, if the ice cream melts and is refrozen, it will be full of grainy ice crystals.

An ice slurry will become grainy if stored in a freezer for too long (even if the temperature is always below the freezing point) because small ice crystals touching one another will merge to become larger ice crystals. (The process is driven by an energy savings due to the decrease in the total surface area because of the merging.) Ice cream in long-term storage can suffer the same result except the process is slowed by the coatings on the ice crystals.

Reportedly, U.S. fliers stationed in England during World War II made ice cream by putting a can of cream mixture in the rear gunner's compartment of a Flying Fortress aircraft, so that it endured the same cold and shaking as the rear gunner. When the fliers returned to their base, the ice cream was ready for eating.

4.42 • Drinking hot coffee, eating hot pizza

Why can coffee that is hot enough to burn someone be drunk (perhaps sipped) without harm? Why is eating hot pizza more likely to burn the mouth than eating hot soup at the same temperature?

Answer The danger of a burn obviously depends on the temperature of the food put into the mouth, but it also depends on the amount of food, how well the food can transfer thermal energy to the mouth, and how long the food is in contact with the mouth. Coffee can be sipped safely even if it is so hot that it would burn skin if spilled, as can happen when someone holds hot coffee while driving. In a spill, a fairly large amount of hot liquid will be held by clothing, thus maintaining the contact between liquid and skin long enough for an appreciable amount of thermal energy to be transferred to the skin.

In contrast, a sip puts only a small amount of liquid into the mouth where it stays in contact with any given portion of the mouth for only a short time. Sipping also helps in two other ways: (1) It mixes air into the liquid, cooling the liquid. (2) It breaks the liquid up into drops, which can individually transfer only a small amount of thermal energy wherever they touch inside the mouth.

Any food with hot cheese, especially if the cheese has been heated in a microwave oven, should be eaten with care for two reasons: (1) The surface of the cheese may not look especially hot, while the bulk carries a lot of thermal energy. (2) Worse, the cheese can cling to the upper surface of the mouth, allowing a large transfer of thermal energy from the cheese to the mouth surface. In fact, you can burn the roof of your mouth within seconds and then suffer for days.

4.43 • Boiling water

If you boil water in a pan, why do bubbles form well before the water reaches the boiling point, and why does the noise from the pan reach a maximum before the water is in full boil? In short, how does water boil when heated on a stove?

Answer The first bubbles to form are filled with air as it is driven out of solution by the increase in temperature. These air bubbles first appear at the bottom of the pan where thermal energy enters the water from the underlying stove.

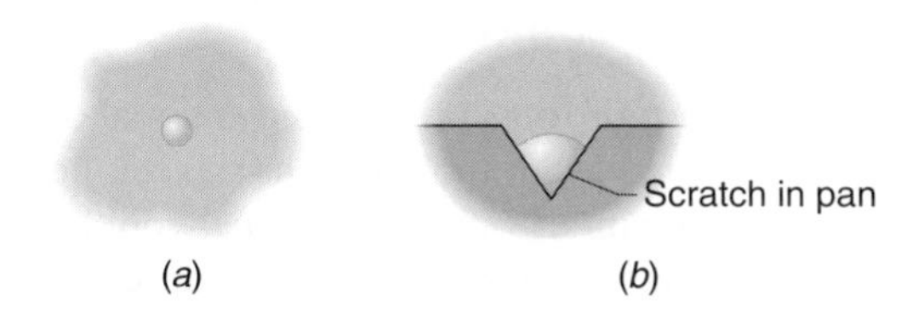

***Figure 4-5** / Item 4.43* (*a*) Small air bubble in water has a highly curved surface and large inward force. (*b*) Air bubble in crevice has a less curved surface and smaller inward force.

When an air bubble forms, the air pressure tends to push its surface outward but the surface tension (due to the mutual attraction of water molecules) tends to pull the surface inward. A small bubble with a very curved surface has a large surface tension, and so it collapses (Fig. 4-5*a*). Thus, air bubbles do not form in the bulk water, not even just above the pan bottom.

However, they *can* form in crevices (or scratches) along the pan bottom where they begin with a less curved surface (Fig. 4-5*b*) and thus a relatively small inward pull from surface tension. The outward pressure steadily increases as more air comes out of solution to enter a bubble. So, the bubble expands until it is large enough to pinch off from the crevice. Because it is lighter than the water, the bubble then rises through the water. This activity dies out once most of the air is driven out of solution.

Shortly later, the bottom of the pan becomes hot enough to vaporize water. Then vapor bubbles form in the crevices. At first, they collapse almost immediately, sending a *click* through the water and pan and out into the air. As more water vaporizes into the bubbles, they eventually become large enough to pinch off from the crevices and begin to ascend. However, they collapse as soon as they rise into somewhat cooler water, where the water vapor condenses back to liquid. Here again, every collapse sends out a click.

As the water temperature continues to increase, the bubbles rise progressively higher before they collapse and then, finally, bubbles reach the top surface. There they splash open with a much gentler sound. The water is then in full boil.

4.44 • Boiling an egg

What determines how much time is required to fully cook (hard-boil) an egg in water? Why is the required time increased when the cooking is moved to higher elevations? Why does an eggshell usually crack during the cooking, and how can the cracking be avoided?

Answer To hard-boil an egg, the yolk must reach a temperature of about 70°C. In boiling the egg, the only way to increase the yolk temperature is by conduction of thermal energy from the water to the yolk. The conduction rate depends on the temperature difference between the water and the interior of the egg. If the egg is placed in tap water that is then gradually heated to the boiling point, the cooking time is about 10 to 15 minutes. The cooking time is reduced if the egg is placed in boiling water, but then the eggshell cracks open, allowing egg material to ooze into the water.

The effect of altitude has to do with the ability of water molecules to successfully leave the water surface and thus for the water to vaporize. At room temperature, the water molecules on the surface are held loosely by attractive forces among them. If the water temperature is increased, those molecules have more thermal energy (they move around with more energy) and some of them can break free of the attractive forces, leaving the water surface. Some of these freed molecules collide with air molecules and simply ricochet back into the water. However, at the boiling point of water, the loss of molecules greatly exceeds the return of molecules by ricochet.

If we move the water to a higher altitude, we decrease the density of air molecules above the water and thus the chance of ricochet. Now the loss can greatly exceed the ricochet return, even at a lower temperature than required at the lower altitude. In short, the boiling point of water is lower at the higher altitude. Thus, the conduction rate of thermal energy into an egg is slower and the egg takes longer to become hard-boiled.

The shell of an egg will crack if the egg is cold when placed in the boiling water. The crack is due to the outward pressure of gas produced within the egg. If that pressure increases quickly, it can rupture the shell, and then you see a string of gas bubbles rising from the crack. Slower cooking might prevent an egg from cracking. Adding salt can cause coagulation of the egg white as it escapes, which seals up the crack. Piercing the egg with a needle to provide an escape hole for the gas might prevent the buildup of gas pressure.

4.45 • Cooking in a stove or over flames

Why can meat be browned if broiled over (or under) a flame, fried in oil in a skillet, or baked in an oven but not if it is boiled in water or cooked in a microwave? When browned, why doesn't the meat brown all the way through instead of only on the outside? If instructions tell you to cook a roast with a certain weight for a certain time and you have a roast with *twice* that weight, should you cook your roast for *twice* as long?

Answer The browning of meat (said to be a Maillard reaction) occurs when a carbohydrate unit reacts with an amino acid. Because the reaction requires high temperatures (higher than the boiling point of water), the meat must be heated by flames, hot oven walls, or hot oil. The first two methods transfer thermal energy to the meat surface primarily by infrared radiation; the last method transfers it by conduction and convection (from the skillet through the hot oil). The interior of the meat is gradually heated as thermal energy is conducted from the surface into the interior, but the interior does not become hotter than the boiling point of water and thus never browns. If the meat is cooked in boiling water or in a microwave oven, not even the surface exceeds the boiling point of water, and so it cannot brown.

Some meats, such as steak, can be seared over very hot flames and then cooked over medium flames as desired. The searing browns the meat but does not seal the meat as some cooks believe. The evidence is that the meat still loses fluids, so the searing is primarily for the taste.

Meat can be described as a protein matrix holding a fair amount of water. Initially the water is held fast because neither gravity nor a forceful push by a kitchen tool can force the water out of the meat. However, as the meat temperature increases during cooking, the water is set free and both gravity and a kitchen tool can force it out. Most of the water in any portion of the meat is lost by the time the temperature reaches 60°C. As the water leaves, the meat shrinks. Hence, a well-done steak can be surprisingly small.

Some people prefer to cook a roast over several hours by putting the meat in a stove set at the desired final temperature of the interior, which is less than the boiling point of water. In that way, the roast loses little of its water content during the cooking because the water on the surface never fully evaporates.

Cooking a roast in a hotter oven, or a steak over coals, can be tricky because the desired temperature at the interior can easily be overshot, and then the meat loses too much moisture to vaporization and becomes dried out. As the interior of the meat approaches the desired temperature, the meat must be checked frequently, either by a commercial temperature gauge (a thermocouple) or by the cook making a small cut to examine the color of the interior. The color is a loose guide to the temperature, because as the myoglobin in the meat denatures due to the temperature increase, it changes from red to gray-brown. (In the living animal, the myoglobin is responsible for taking up the oxygen transported into the meat from the lungs via the hemoglobin in the blood.)

To time the cooking of a roast or turkey by its weight according to a recipe can be guesswork because different ovens cook at different rates (their temperature settings are not well calibrated) and different pieces of meat conduct energy at different rates. However, here is a general rule: If you know the proper cooking time T for a roast of a certain weight, then the cooking time is $2^{2/3}T$ for twice that weight and $3^{2/3}T$ for three times that weight. Can you see the pattern? The multiplying factor on the weight is raised to the $\frac{2}{3}$ power.

4.46 • Campfire cooking

Cooking over a campfire may involve conventional cookingware (such as a skillet) and techniques (the skillet is held over the flames), but it can also involve unusual cooking techniques, especially in a "survival situation." How can you use aluminum foil, a large can, a paper bag, rocks, or an orange to cook, say, eggs or meat?

Answer The aluminum foil can be set up to reflect some of the infrared radiation from a campfire onto the food to cook it. One of the best ways is to mount the foil as if it were a tilted roof of a lean-to, so that it reflects the radiation down onto the food that lies on the "floor" of the lean-to.

You can fashion a stove from the large can by inverting it, punching holes near the top and folding up a flap from the bottom. Coals pushed through the flap will heat the can's interior, causing hot air to flow out of the holes at the top and cooler air to be sucked in through the flap. You can cook directly on the flat surface of the inverted can, or place a second, smaller can there in which to cook the food.

Food can be wrapped in aluminum foil and buried in the coals (which might then be covered with dirt). However, food tends to burn in places when cooked this way. A better technique is to wrap the food in two layers of foil with paper inserted between the layers. The air left in the spaces between these three layers slows the transfer of thermal energy from the coils to the food, making hot spots and burning less likely. A similar principle lies behind cooking with an orange: Slice off the top third of the orange, scoop out the rest of the orange, put food into that scooped-out cavity, replace the top as a lid, and then put the upright orange directly in the coals. The moisture in the skin of the orange decreases the chance of hot spots.

The temperature of water cannot normally be raised above the boiling point, which is considerably below the ignition point of paper. So, you can cook food in a paper bag as long as there is water or water-filled material covering the inside bottom of the bag. For example, you can crack one or more eggs into a paper bag, fold over the top several times to prevent moisture from escaping during the cooking, spear the top with a stick, and then, holding the stick at the opposite end, position the bag over hot coals. The water in the eggs prevents the temperature of the bottom of the bag from exceeding the boiling point of water, but that is hot enough to cook the eggs.

Chicken (or other fowl) can be cooked with hot rocks. First wrap dry rocks with aluminum foil, and then heat them on hot coals. (Do not use wet rocks because when they get hot, the water inside them can vaporize suddenly and explode the rock.) When the rocks are hot, move them to the interior of the chicken, wrap the chicken in foil, and cover the chicken with many layers of newspaper or leaves. The rocks will transfer energy into the chicken to cook it. The newspaper or leaves will insulate the chicken, so that the thermal energy is not simply conducted to the chicken surface and then lost before the chicken temperature can be increased properly. (Chicken must be cooked thoroughly to eliminate the possibility of salmonella poisoning.)

4.47 • Cooking pizza

Why does pizza develop a nicely melted cheese surface, with lightly browned spots, if topped with real cheese but not with fat-free cheese?

Answer A pizza is cooked by conduction from the hot pan on which it sits, infrared radiation from the oven walls surrounding it, and convection of hot air across its top (especially if the air is being forced to move by a fan). As thermal energy is gradually transferred to the interior, largely to cook the dough, the cheese is supposed to melt uniformly over the top and then lightly brown. The browning occurs where bubbles form in the cheese—that is, where water vaporizes to form bubbles of steam inside the cheese. As the tops of these bubbles thin during bubble growth, the tops can absorb enough thermal energy to turn brown.

If the pizza is topped with fat-free cheese, the water very quickly evaporates from the cheese, and the dried-out individual strands of cheese never melt and fuse, but instead just burn. To remedy this, fat-free or low-fat cheese is sprayed with an oil film when the pizza is prepared. Then the oil film slows the evaporation of water from the cheese, so that melting, fusing, bubbling, and browning can all occur.

4.48 • Heating in a microwave oven

How do microwaves in a microwave oven heat food? Why is a rotating platform used in most microwave ovens? Do microwaves cook food from the inside out? Why doesn't a microwave oven brown food (and give it the characteristic flavors of browned food) as gas or electric ovens would?

Why should you never heat a cup of water in a microwave and then spoon in a powder such as instant coffee or cocoa? Why should you never use a microwave oven to cook an egg in its shell or cook an unshelled egg with its yolk still intact even if the shell has been removed? Why is eating a pastry warmed in a microwave oven possibly dangerous?

Answer Microwaves are a form of electromagnetic radiation just as visible light is, except that microwaves have a much longer wavelength than visible light. ("Radiation" here does not mean nuclear radiation—the word refers to the fact that something is radiated or emitted.) Microwaves can penetrate most foods and are absorbed by the water inside the food. A water molecule forms an *electric dipole*, which is the electric analog of a magnet, with one end positively charged and the other end negatively charged (two *poles*, hence *dipole*). An electric dipole located in an electric field tends to line up with that field. The electric field in a microwave beam oscillates in direction and strength. Thus, the water molecules are constantly flip-flopping, trying to maintain alignment.

Energy is transferred from the microwaves to the molecules in places where this flip-flopping happens to break bonds between the molecules. That transferred energy ends up as thermal energy in the water, thus heating the water, as well as the food containing the water. The food never gets hotter than the boiling point of water, unlike food cooked in a gas or electric stove, which can brown the surface of the food with much higher temperatures by denaturing oxymyoglobin in the meat. Thus, microwaved meat is often described as "mushy and tasteless" because it lacks the flavors and surface texture on the surface that a gas or electric oven can provide.

Because microwaves penetrate a food sample, they can heat the entire sample simultaneously if the sample is fairly small, or heat it in a thick outer layer if the sample is large. In the latter case, time would be required for the thermal energy to be conducted to the center of the sample.

Pizza slices and pastries with jelly (or jam) interiors can be dangerous if heated in a microwave oven and then immediately eaten. The sauce and the jelly interior warm much more rapidly than the crust because they have much more water. So, when the food is removed from the microwave, the crust may not feel hot but the sauce or jelly might be so hot that it can quickly burn the interior of the mouth.

The microwaves are beamed from a microwave device called a magnetron. In order to spread the microwaves around the sides of a food sample, the sample is usually rotated in the beam by a turntable. Earlier oven models used a metal fan with tilted blades to reflect the beam. As a blade moved through the beam, it would reflect the beam into a wide range of angles to "spray" the food sample. Without such a spray of reflections or a rotation of the food, there are *active spots* (where food will be heated quickly) and *inactive spots* (where food will be heated only slowly). You can record the pattern of active and inactive spots by placing uniform layers of cheese over the oven floor (with the turntable removed). The active spots are located where the cheese first melts and bubbles.

When water is heated on a conventional stove by a flame along the outside bottom surface of a pan, the energy transferred from the flame vaporizes water to form bubbles of water vapor in crevices along the inside bottom surface. Those crevices are required. In order for a bubble to be created and grow larger, it must fight against the surface tension of the water molecules (their tendency is to cling together because of attractive forces between them). A tiny bubble in the bulk of the water has little chance in fighting against the surface tension because its surface with the water is so sharply curved. That curvature means that the inward forces from the water are very large, and so the bubble will probably just collapse. However, if a tiny bubble forms in a crevice, its surface with the water is not so curved, the inward forces are not so large, and the bubble can not only survive but actually grow as more water is vaporized to join it. So, in conventional heating, crevices on the bottom surface of the pan begin the boiling process.

When water is heated to the boiling point over a flame, thermal energy is fed into the water at a surface where vapor bubbles can form. The water cannot exceed the boiling point because once that point is reached, the thermal energy begins to transform liquid into vapor to form the bubbles.

When water is heated in a microwave oven, the process is very different, because thermal energy is absorbed in the bulk water and not at a surface. In the bulk water, the surface tension of the water will collapse any vapor bubble that tries to form. So, the water does not change from liquid to vapor. As more thermal energy is absorbed by the water, the temperature rises above the normal boiling point, and the water is said to be *superheated.* Eventually, vapor bubbles begin to form in spite of the surface tension on their surfaces.

Suppose, however, that you remove a cup of superheated water from the microwave oven before the bubble production begins. Without the bubbles, you have no clue that the water is hot. If you dump a spoonful of powder, ice chips, or any other fine-grain material into this superheated water, water-vapor bubbles suddenly form in all the nooks and crannies on the surfaces of those items. The water can then flash into a boiling state with such vigor that the very hot water is thrown out of the cup. Occasionally someone is badly burned by this process.

An egg, intact yolk, or any other closed container of water will probably explode in a microwave oven. The water is heated until it flashes to vapor with a sudden outward push that bursts open the container. Sometimes the egg does not explode until it is jostled as it is being removed from the microwave oven. Then, not only does it make a big mess, but it can burn someone, perhaps on an eye surface. Smaller explosions, called *microwave bumps,* can occur with foods such as green beans and lima beans, which contain small amounts of water in what is an effectively closed container.

Of course, popcorn popped in a microwave oven is an explosion: The water inside each kernel suddenly vaporizes and blows the kernel outward into a fluffy shape that is just right for a late-night snack. However, just bathing the popcorn kernels with microwaves will not give a bagful of uniformly popped popcorn because each kernel contains so little water. To pop the popcorn more rapidly, the bottom of a popcorn bag has a card containing material that readily absorbs microwaves. The card, which grows hot very quickly, transfers energy to the kernels touching it, so that they explode, which causes other kernels to fall onto the card. This procedure continues until most of the kernels have been popped.

A leaky microwave oven is obviously a danger. (The most common leak is around an oven door that sags from long use.) Not only can the microwaves possibly heat portions of your body that should not be heated (such as the eyes), but long-term damage could also occur.

4.49 • Popping popcorn

Why does popcorn pop? That is, what produces the expansion and the sound?

Answer Popcorn is a special type of maize, grown for its ability to explode when heated by hot air or grease or when heated in a microwave oven. (In the microwave oven it heats by absorbing microwaves directly and by touching a special card that rapidly heats by absorbing microwaves.) The *pericarp* part of a popcorn kernel is a small, closed container of starch and liquid water. As a popcorn kernel is heated, that water partially vaporizes but largely remains liquid. Because the liquid is trapped in a container, the pressure increases and, as a consequence, so does the boiling point of the trapped water.

When the water reaches about 180°C at a pressure of about 8 times atmospheric pressure, the pericarp walls burst open, the pressure drops to atmospheric pressure, and the boiling point drops to its normal value. Thus, the water in the pericarp is suddenly well above the boiling point, and it vaporizes so rapidly that the evaporation explodes the hot, molten starch to many times its original volume. The sudden expansion against the air sends a sound wave through the air—the *pop* of the corn.

Because the volume expansion of popcorn gives the product its desired fluffy texture, manufacturers want to maximize that expansion. Popcorn with a greater amount of water in its pericarp would tend to give a greater explosion and thus a greater expansion. However, popcorn with pericarp water exceeding a certain amount explodes poorly, if at all, because the additional water reduces the rigidity of the pericarp walls.

4.50 • Cooking scrambled eggs

Why are eggs stirred when making scrambled eggs? Why should they be cooked over low heat?

Answer To cook scrambled eggs, you add the eggs (and milk, if you like) to a moderately hot skillet and then stir continuously. The purpose of the stirring is twofold: You want to unravel some of the raveled-up proteins in the egg to create a mesh, and you want to break any curds that form as the heat coagulates the egg. If you pause, the egg along the skillet surface will receive too much thermal energy and will cook to a crisp. If the skillet is too hot, then while the proteins unravel and heat, they lose the water molecules attached to them. The water forms drops and then pools. Some people like eggs with water pools. Others want the egg cooked so much that the pools dry out (which may leave the scrambled eggs about as tasteful as warm yellow cardboard).

If the transfer of thermal energy is slow enough to avoid water separation from the proteins, and if the stirring avoids overcooking and large curds, then the scrambled eggs should be pulled off the fire early so that the thermal energy remaining in the skillet completes the cooking. The eggs are then moist (but not wet), uniform in texture, and tasty. Salt can be added only when the eggs are ready to be eaten because any earlier use helps separate the water from the proteins. If vegetables or other products are to be added to scrambled eggs, eliminate their water content by draining or cooking them, and wait until almost the last minute to fold them into the eggs in the skillet.

An omelet differs from scrambled eggs in at least two respects. First, to produce air bubbles, the egg whites are separated from the yolk and are beaten to entrap air bubbles into the mesh of unraveled proteins. Then this aerated mixture can be rejoined with the yolks. Second, you want a crusty layer for a framework in which the rest of the egg mixture will cook. Thus, you leave the mixture a bit longer on the skillet, so that a crust forms along the bottom. Then the omelet can be folded over. The egg mixture between the crusty top and bottom surfaces then cooks, with air bubbles and vapor bubbles expanding.

4.51 • Geysers and coffee percolators

Why do water, steam, and other materials shoot from a geyser instead of just pouring out? Why are some geysers, such as Old Faithful in Yellowstone National Park, somewhat periodic?

A common type of coffee percolator consists of an inverted funnel that rests loosely on the bottom and supports an upper basket with the coffee grounds (Fig. 4-6*a*). How is the coffee brewed in the percolator?

Answer Old Faithful, probably the most studied geyser in the world, has been monitored with a video camera in situ as the geyser fills up with water and then explodes. It is actually a crack that reaches a depth of some 200 meters. Along the sides of the crack, cool water, hot water, and steam pour in from fissures in the adjacent rock. The heat source for all the hot water is magma at a depth of several kilometers.

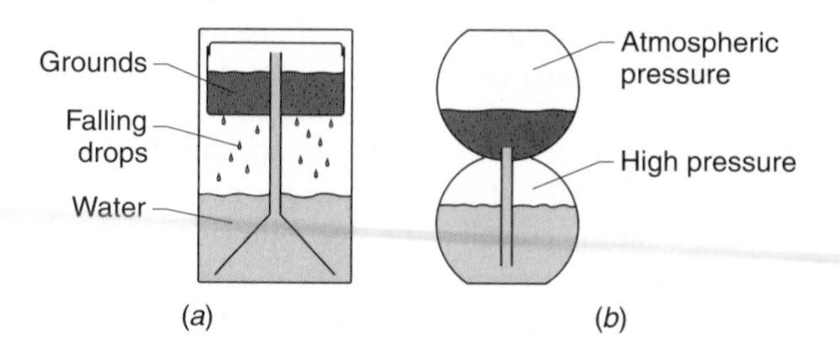

Figure 4-6 / *Item 4.51* (*a*) Inverted-funnel percolator. (*b*) Bulb percolator.

The action starts when some of the water 6 or 7 meters down in Old Faithful becomes hotter than the boiling point of water at that depth. (The boiling point increases with depth, because the water pressure also increases with depth.) Steam bubbles rise up from the level of boiling, transferring thermal energy into overlying water, where the temperature is less than the boiling point at that depth. However, that overlying water is soon brought to its boiling point, and then the sudden expansion of water flashing to steam ejects the column of water and steam out of the geyser.

The process repeats itself, but the time required depends on how much water was left in the geyser by the previous eruption. Old Faithful eruptions usually have two periods, a short interval and a long interval, one followed by the other.

When an inverted-funnel percolator is placed over a hot burner, the water inside the funnel is heated to the point that some of it vaporizes along the bottom surface. The sudden expansion of the vapor pushes liquid up the narrow stem of the funnel; the water spills out into the basket and drips downward through the coffee grounds. The process is repeated until the coffee has the proper strength.

Another popular percolator design consists of two bulbs that are snugly connected with a rubber gasket, one above the other (Fig. 4-6*b*). Grounds are placed in the higher bulb, which is open at its top; water is placed in the lower bulb. A small tube extends from below the water level in the lower bulb up into the higher bulb. When the water in the lower bulb is heated sufficiently, the expansion of air and water vapor in it pushes much of the water up the tube and into the higher bulb. Then the percolator is set aside. As the lower bulb cools and its water vapor condenses, its air pressure decreases and becomes lower than atmospheric pressure. Because the higher bulb is open, the pressure on its liquid is atmospheric pressure. Due to the pressure difference between the two bulbs, liquid is pushed back down the tube and into the lower bulb. This motion forces the liquid through the grounds, strengthening the brew. When all the water has returned to the lower bulb, the coffee is ready.

4.52 • Toy putt-putt boat

The toy boat shown in Fig. 4-7 is propelled by a discharge of water from the two pipes that run from a "boiler" to the rear of the boat. (The boiler could be merely several turns of tubing.) To prepare the boat, you fill the boiler and pipes with water, float the boat in a pool of water, and then place a lit candle beneath the boiler. As the water in the boiler heats and evaporates, the increased pressure shoves the water in the pipes out to the rear of the boat.

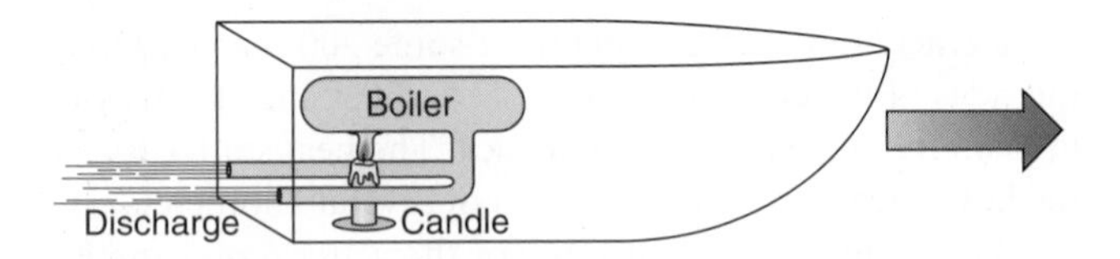

***Figure 4-7** / Item 4.52* Discharge of water propels the toy boat.

The curious feature is that once the water is discharged, the propulsion does not cease. Instead, the boat is pushed forward every few moments in a stutter-step fashion. What provides the continued propulsion?

Answer When water is expelled from the pipes, some of the steam generated in the boiler moves into the pipes, where it condenses because of the cooler environment. The movement and condensation both act to lower the pressure in the gas. As the pressure falls, water from behind the boat is pulled into the pipes, refilling them. Then the whole cycle of discharge and refilling repeats itself, with the boat being shoved forward again.

The rapid expulsion of water is in the form of a jet directed toward the rear, which requires that the boat move forward. The boat does not move backward during the refilling stage because the water intake is not a jet. Instead it is slower and from a wide range of angles (approximately a hemisphere). Thus, the force tending to pull the boat backward is weak and cannot overcome the water drag on the boat. So, the boat moves forward during each discharge stage but is stationary during each refill.

4.53 • Thermal effects on lengths

Why are most bridges constructed in sections with short gaps between the sections? Why were old rail lines constructed with short lengths of rail with intermediate gaps? Such rail construction produced a noisy, jolting ride—"clink, clink, clink"—because the train wheels would hit those gaps and then oscillate, shaking the passengers and generating sound. Why don't modern rail lines have such gaps?

Answer Most materials used in bridges expand when heated and contract when cooled. If such a material is subjected to significant variations in temperature during the year, then its change in length must be allowed in the construction. Otherwise, expansion would lead to buckling.

Rail lines were originally constructed with short lengths of rails with intermediate gaps to allow for such expansion. Modern rail lines have almost no gaps left for expansion—they are *continuous welded rails.* They are so securely fastened to the cross ties, which are firmly held in place, that buckling is rare. The rail is normally set when the temperature is about midway between its yearly extremes.

When hot oil begins to flow through an oil pipeline along an ocean floor, the pipeline might buckle due to thermal expansion because it is not secured in place. Unless the buckling is extreme, it probably will not be a problem.

When the Concorde aircraft was built, its design had to allow for the thermal expansion of the fuselage during supersonic flight because of frictional heating by passing air. The temperature increased to about 128°C at the aircraft nose and to about 90°C at the tail, cabin windows were noticeably warm to the touch, and the fuselage length grew by about 12.5 centimeters.

Dental materials used in filling cavities and fashioning crowns are carefully designed so that their thermal expansion and contraction match the surrounding tooth. Otherwise, eating ice cream and then drinking hot cocoa would be a never-forgotten experience.

Many factors led to the widespread power outage that left 50 million Americans and Canadians without power in August 2003, but one curious instance was reported about a transmission line in Ohio. The current through the line was high that day, which led to a higher than normal rate of heating inside the line. As the temperature of the line increased, the line lengths between any two support points also increased, and so the line segments began to sag. One of those segments sagged sufficiently close to a tree for its current to spark through the tree to the ground. This eliminated that transmission line's ability to carry current, and its loss apparently contributed to the instability of the power-grid system, leading to the system's shutdown.

4.54 • Collapse of railroad storage tank

Railroad tank cars are extremely durable and normally can be damaged only in a high-speed crash. However, they can also be ruined when certain principles of physics are ignored. Here is an actual case: The interior of a tank car was being cleaned with steam by a crew late one afternoon. Because the job was unfinished at the end of their work shift, they sealed the car and left for the night. When they returned the next morning, they discovered that something had crushed the car in spite of its extremely strong steel walls, as if some giant creature from a grade B science fiction movie had stepped on it during a rampage that night. What crushed the tank car?

Answer When the tank car was being cleaned, its interior was filled with very hot steam, which is a gas of water molecules. The cleaning crew left the steam inside the car when they closed all the valves on the car at the end of their work shift. At that point the pressure of the gas in the car was equal to atmospheric pressure because the valves had been opened to the atmosphere during the cleaning. As the car cooled during the night, the steam also cooled and much of it condensed, which means that the number of gas molecules and the temperature of the gas both decreased while the volume was constant. As a result, the gas pressure decreased. At some point that night, the gas pressure inside the car reached such a low value that the external atmospheric pressure was able to push the car's steel walls inward, crushing the car. The cleaning crew could have prevented this accident by leaving the valves open, so that air could enter the car to keep the internal and external atmospheric pressures equal.

4.55 • Drying of hanging laundry

Before laundry dryers were common, laundry was "hung out to dry" on clotheslines: Clothes were pinned with clothespins to the line, from which they hung, in sunshine or shade, until they were dry. Why does a shirt hung in this way dry? In particular, why does the drying start at the top and proceed downward?

Answer You might think that a hanging shirt dries from the top down because water drains downward and then drips from the bottom. In fact, you are right, at least for the first 30 minutes or so of the drying process. As water drains to the bottom, it forms drops that are eventually large enough to break free and fall. However, when this draining stops, the shirt is still wet. If it were merely heated in sunshine, the entire shirt would then dry at roughly the same rate and not from the top down. That top-down drying process is due to the convection of air along the shirt due to temperature changes.

After water drains from the shirt, the shirt is still wet because water is trapped by surface tension in pores (open spaces between fibers). Let's consider a slug of water in a vertical pore. Such a slug has a curved surface with air at the top and bottom. Because of surface tension (due to the mutual attraction of water molecules), the curvature at the two surfaces produces a net force on the slug. If the top of the pore is wider than the bottom of the pore, the net force is downward, in the same direction as the gravitational force on the slug. So, the water will move downward. However, if the top of the pore is narrower than the bottom of the pore, the net force from surface tension is upward and can hold the slug in place in spite of the gravitational force. Water slugs trapped in such pores remain in the shirt after drainage has stopped.

Water gradually evaporates from these trapped slugs. The evaporation requires energy to free water molecules from the surface. So, the evaporating water carries away some of the energy from the remaining slug, cooling the slug and also the surrounding fabric and air. Because air becomes denser when cooled, this cooled air descends along the shirt, picking up moisture from lower trapped slugs. Thus, the drying line that divides the dry and wet portions of the shirt begins at the top of the shirt and is then driven downward by the descending cool air.

If you want a physics excuse for relaxing in sunshine for hours, hang up a wet shirt and then say that you are inspecting the march of the drying line down the shirt due to convective drying.

4.56 • Warm coats

If you walk into a cold room while dressed for the beach, why do you become cold? How can a coat keep you warm?

Answer You feel cold if you lose more thermal energy to your environment than you receive from it, and there are four types of losses. (1) In conduction, you lose energy by direct contact with another object that is colder than you are, which happens when you sit on a cold bench. (2) In radiation, you lose energy by radiating infrared light to your environment. You also gain energy by absorbing infrared radiation emitted by the environment, but if the environment is colder than you are, that gain is less than your loss. (3) In convection, you lose energy as air flows past you. If the air is colder than you are, you lose thermal energy via the collisions of air molecules with you. (4) You can also lose energy through the evaporation of sweat from your skin, which is one reason why you sweat when you exercise. The change from liquid to vapor requires energy, which comes from your skin. If you sweat while in a breeze or wind, the evaporation rate increases, and so the rate at which you lose energy also increases.

The purpose of a coat (or clothing in general) is to decrease all these types of energy losses. For example, animal hide, such as leather, decreases the convection and evaporation losses due to wind by blocking it. A coat can also provide a layer of semitrapped air around part of your body. Because thermal energy is conducted rather poorly through air, the layer helps insulate you. Using several layers of clothing beneath a coat helps even more because you then have several layers of semitrapped air to insulate you.

Fur on a coat helps keep you warm because air is somewhat trapped among the hairs. However, if you are standing in a wind, that air is easily blown away. So, you would be warmer if you wore the coat inside out, to keep the hairs out of the wind.

If you expose bare skin, such as face or fingers, to a cold wind, your sensation of cold is approximated by the *windchill index*, which is the temperature in a windless situation that would yield the same sensation. Coming up with an accurate calculation of the windchill index is tricky, because it involves your ability to adapt to cold air; some people adapt very easily and others very poorly. Of course, the danger of a cold wind is the possibility of frostbite, which occurs when the skin begins to freeze. Generally, skin does not freeze at temperatures above $-10°C$ regardless of the wind speed, but the danger increases rapidly for lower temperatures and greater wind speeds.

4.57 • Warm plants

When a late-winter snow happens to fall on skunk cabbage (*Symplocarpus foetidus*) of North America and Asia, why does the snow around the cabbage soon melt? During heavy snow, a cavity can develop around the cabbage.

Answer The skunk cabbage is one of several plants that are able to raise their temperature well above the temperature of the environment. Thus, the skunk cabbage can melt the snow around it because it loses more energy by infrared emission due to its elevated temperature. Similar to birds and mammals, these plants are said to be *thermoregulating*, because they can maintain their temperature even as the temperature of the environment changes. Some of the plants, such as the inflorescence (cluster of small flowers, or florets) of *Philodendron selloum* might be warm to the touch (that is, warmer than a human) and generating thermal energy at a rate rivaling that of a small cat. (Roger S. Seymour, one of the principle researchers in this field, says that he sometimes imagines the plant as being a cat growing on a stalk.)

4.58 • Polar-bear hairs

Why are the hairs on polar bears hollow?

Answer The white hairs on a polar bear trap the visible and infrared portions of sunlight, because those portions are reflected and transmitted down into the pelt to reach the skin. There it is absorbed, increasing the thermal energy of the skin. (The ultraviolet portion of sunlight is also absorbed by the hairs, but ultraviolet light contributes little to the warming of a bear.) The thermal energy of the skin is maintained partially because the hairs are hollow and conduct thermal energy poorly. (The notion that the hollow hairs somehow function as optical fibers is just a myth.)

4.59 • Black robes and black sheep in the desert

It is commonly believed that white clothing is cooler than black clothing when one is in a hot, arid environment. Yet, the Bedouins, who live in the challenging high temperatures of the Sinai desert, sometimes choose a black robe over a white one. Why such a choice?

If you were lost in a desert, would your chance of survival improve if you stripped off your clothing and eliminated their absorption of light?

The sheep of the Bedouins are usually black, not because of selective breeding, but apparently because of natural adaptation to the environment. Why would black coats on the sheep aid their survival?

Answer A black robe may absorb more sunlight and heat to a higher temperature than a white robe, but the temperature of the air inside the robe and of the skin of the Bedouin does not primarily depend on the robe's color. A higher temperature of the black robe is probably offset by the greater convection of air through the robe: Air enters at the bottom and, because of heating, rises and escapes at the neck. The robe is somewhat like a chimney. If the Bedouin is in a gusty wind, air circulation is further aided by a bellowing of the robe.

If you are in a desert and must wear clothes that are tightly gathered and do not allow much air circulation, white clothing will heat your skin less than black clothing. You normally would not want to go without any clothes because of the danger of sunburn. If water is plentiful, the clothing should be porous so that the evaporation of perspiration from your skin might cool you. However, when water is scarce, you also need clothing to reduce the evaporative losses from your skin, or you will become dangerously dehydrated in a short time. In Herbert's science fiction classic *Dune*, the desert people live in so hostile an environment that they wear sealed suits to trap precious body moisture.

The black coat on a Bedouin sheep allows it to survive the harsh winters of the Sinai. The color of the coat does not matter until the sheep stands in direct sunlight. Then the greater absorption of sunlight by a black coat warms the sheep and reduces metabolism. Since food for sheep is scarce in the winter, lower metabolism is an advantage.

4.60 • Cooling rate of a cup of coffee

Suppose that you just made a hot cup of coffee but don't intend to drink it until later. Suppose also that you take your coffee with milk. To have the coffee as hot as possible when you begin to drink it, should you add the milk immediately or just before you begin to drink it? Should you stir it in the meantime? Should you place a spoon in it? Does a metal spoon have a different effect than a plastic spoon? Does the cooling rate of the coffee depend on whether the mug (or even the liquid) is black or white?

Answer There are three factors that must be juggled here: (1) The hotter the coffee is, the faster it will lose heat. (If this were the only important factor, then the milk should be added immediately to diminish the temperature and the loss.) (2) Adding a volume of cooler milk to the volume of hot coffee yields a mixture with an intermediate temperature; the temperature decrease of the coffee will be greater the hotter it is when the milk is added. (If this were the only important factor, waiting to add the milk would be better.) (3) The presence of milk will probably decrease the evaporation of water and also the associated loss of heat.

One set of researchers reported that black coffee cools about 20% faster than white coffee under normal circumstances, probably due to the third factor rather than any difference in the emission of infrared radiation. They also found that if the milk is colder than room temperature (perhaps it has just been taken from the refrigerator), then the coffee will be hottest if the milk is added immediately. However, if the milk is warmer than room temperature (which is uncommon), the correct time to add the milk depends on a number of factors, including how long you want to wait to drink the coffee. Thus, in general, add the milk immediately.

Stirring will hasten cooling because it brings hot liquid to the surface to evaporate. A metal spoon will conduct heat up its length if left in the coffee (it acts like a *heating fin*), but a plastic spoon will probably have little effect.

The question of color involves the rate at which a surface radiates energy. In visible light, a white surface radiates more energy than does a black surface, but the principal loss of heat by radiation from the surface of a mug (or the liquid itself) is in the infrared range. For that range, black and white surfaces radiate about the same, and so the color of the mug is unimportant.

A lid on a coffee cup, or a layer of whipped cream on coffee, will keep the coffee hot for a longer period because of the decreased evaporation and associated heat loss.

4.61 • Cool water from porous pottery

In hot, arid climates, why does water kept in porous (leaky) pottery and placed in a windy, shady spot become cool? When my parents took me on a car tour of the southwestern United States, they strapped a porous water bag onto the front bumper. When we stopped to drink the water, the air and car were very hot but the water was always cool. Why was the water cool?

Answer During evaporation, water molecules leave the bulk water to move through the adjacent air. Energy is expended to free those molecules from the attractive forces among the molecules on the water surface. If random motion happens to bring those molecules back to the surface (their collision with air molecules happens to bring them back), the energy is returned. But if the air is moving in a breeze, those freed molecules are removed and cannot return the energy. In that case, the water surface loses energy. If this energy loss is quick enough, the water temperature drops before there is a significant transfer of energy to the water from the warm environment. Thus, if a porous pot is kept in shade, a breeze can cool the water by removing the molecules evaporated from water that has seeped through the pot's wall. Similarly, if the water bag on my parents' car was shaded by the car, the passing air could cool the water by evaporating molecules that seeped through the bag's wall.

This cooling process is also used in other ways. On a picnic on a hot day, for example, food or wine can be kept fairly cool for hours if it is enclosed in porous pottery that has been soaked in water. In fact, you might keep yourself cool on a hot day by soaking your clothes with water and then standing in a shady, breezy spot.

The South American frog *Phyllamedusa sauvagei* reportedly uses evaporative cooling in a unique way. Normally it spends the day with its eyes closed, but when its body temperature increases above about 40°C on a hot day, the frog periodically opens and closes its eyes. When opened, the eyes protrude and cool by evaporation. When closed, the eyes retract and press against the brain. Thus, the frog apparently can cool the brain via the evaporation from its opened eyes.

4.62 • Dunking bird

The dunking bird (Fig. 4-8) is a popular toy, both in and out of the classroom. To initiate the bird's bobbing, you wet its head. The bird then gradually leans forward until it suddenly rotates into a nearly horizontal orientation. After a bounce or two, the bird rights itself. If you have provided a glass of water into which the bird can dip its beak, the action repeats itself indefinitely. What drives the bird into this motion? Are there other ways you can make the bird bob without wetting its head with water? If so, the bird will bob even with high humidity, which otherwise stops the bobbing when water is employed.

Answer The body (lower portion) of the bird is partially filled with a liquid that is easily vaporized, typically methylene chloride. The open spaces in the body and head contain the chemical's vapor. The head and beak are covered with felt. The neck consists of a tube that extends from the head down into the body. Somewhere along the neck, the bird is supported by an axle that is free to rotate around the legs and base of the toy.

After the head is sprinkled with water, the water begins to vaporize into the surrounding air. Because the transition of liquid to vapor requires thermal energy, the evaporation cools the felt, head, and the vapor inside the head. (Blowing on the felt aids the evaporation.) As the temperature of the vapor inside the head decreases, so does its pressure. Since the vapor pocket in the body is not directly connected to vapor in the head, it remains at a higher pressure, and the pressure difference between the two pockets gradually drives liquid up through the tube. The shift of liquid makes the bird top-heavy and rotates it around the axle supports in the legs. The rotation is initially gradual, and then the bird suddenly rotates forward to an almost horizontal orientation. It then

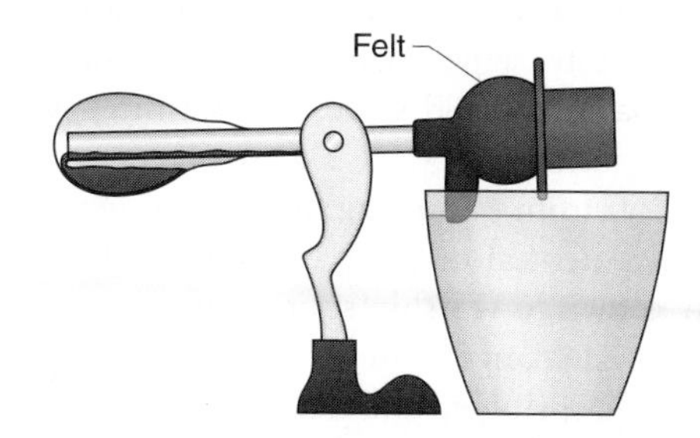

Figure 4-8 / *Item 4.62* Dunking bird that is about to bounce back to its upright orientation.

either bounces slightly upward from the edge of a water glass or off the portions of the leg supports that limit its rotation.

Just as the bird is almost level, the lower end of the tube rises above the liquid in the body, and the two vapor pockets are momentarily connected and thus equalize in pressure. During the upward bounce, the tilt of the tube allows liquid to flow from the head back into the body, which restores the original weight distribution, and the bird then fully rights itself. If the bird wet its beak during its plunge, the added water seeps through the felt on the head, and the whole dunking process is repeated.

There are several ways to promote the dunking without wetting the bird's head with water. For example, you might coat the felt with alcohol, which would vaporize even if the air humidity is high. (For example, strong whiskey will work but may ruin the bird's felt.) You might place the bird in bright sunlight and shade its head and the upper reaches of its neck but not its body. If you also paint the body black, you might drive the bird into frenzied activity.

SHORT STORY

4.63 • Large dunking birds

In the 1960s, a scaled-up dunking bird was suggested as being ideal for the irrigation demands of the arid Middle East. (Patents for similar engines date back to 1888.) A bird would be positioned alongside a water-filled canal. Once started with a sprinkling of water on the bird's felt-covered head, the bird would dip forward and force an attached scoop and tube to pick up water from the canal and pour it onto a higher level of ground (the field). Each dip forward would allow the bird to wet its beak in the canal water.

To increase the economy of the action, a second scoop tube could be attached to the opposite side of the bird and the bird placed between two parallel canals. Then with both the bob forward and rearward, water could be delivered from a lower canal to a higher canal, which would be at field height.

Here is a bolder scheme. Suppose that we erect a giant dunking bird in the shallower waters off the coast of California. Ropes attached to the bird would be connected to geared wheels on an apparatus on shore. As the bird continuously dunked and replenished the water on its beak, it would force the wheels to rotate, and the motion could be converted to electrical power. A flock of giant birds lining the California coastline could serve to offset the power-hungry needs of the state.

I backed off from proposing this scheme when an associate found a potential danger: Throughout history people have been known to worship sources of energy, such as fire and the Sun. The danger with my proposal is that dunking-bird cults could spring up and that their members would pay tribute to the birds by lining the beaches to bow in unison with them. Since we have enough cults, I dropped the idea.

4.64 • Heat pipes and potato stickers

A pork roast or large turkey can be cooked in reduced time if a *heat pipe* is inserted into the meat, at an upward angle. The device consists of a closed, hollow tube with a wick and a small amount of liquid (perhaps water). The lower end is either a large, solid metal cylinder or a narrow piece with several fins. The hollow nature of the device seems wrong. The idea is to convey thermal energy from the hot environment of the oven into the interior of the meat. So, wouldn't a solid metal rod do better at transporting thermal energy?

When a potato is baked, many cooks insert a nail (or nail-like rod) into the potato to conduct thermal energy into the interior of the potato. Since metal conducts thermal energy better than the potato itself, the nail should reduce the time needed to cook the potato. Why, then, do most of these "potato stickers" fail to reduce the normal cooking time of one hour by more than a minute or two?

Answer A solid metal rod transports thermal energy into a roast by conduction, but the process is slow. The transport is much faster with a heat pipe. In fact, when the interior of the meat is still cool, a heat pipe may transport thermal energy several thousand times faster than a solid rod with the same dimensions.

The rapid transport is due to the liquid in the pipe and the design of the exposed end. The end has a large surface area, so that it readily absorbs thermal energy, both from the hot air and by intercepting thermal radiation in the oven. Eventually, the liquid inside the pipe vaporizes, which uses up a large amount of thermal energy. The hot vapor then rises through the slanted pipe and into the cool interior of the meat. The cool environment causes the vapor to condense and release all the thermal energy used in vaporizing it. The released thermal energy is then conducted away from the pipe and through the meat. In the meantime, the condensation either runs back down the pipe or along the wick. As it reaches the lower part of the pipe, it is again vaporized and the cycle is repeated. Because so much thermal energy is involved in the vaporization and condensation processes, the thermal energy transport into the meat is much more rapid than if the energy is conducted along a solid metal rod.

A common potato sticker fails to reduce cooking time significantly because the exposed end is so small that it absorbs thermal energy from the oven very slowly. If the exposed end were more massive or came with fins, the plan would work.

4.65 • Foggy mirrors

If you take a hot shower when the air is cool, why does a mirror or a windowpane in the room *fog up*, and why does the fogging start at the top? Why can you eliminate the fogging if you put a thin coating of soap (or detergent) on the mirror before you shower? (You can, instead, use the tip of a bar of soap or a detergent-coated finger to write a message on a mirror. Before the next person showers, the message is almost invisible, but once the uncoated portions of the mirror fog up, the message becomes apparent because the coated portions are clear.)

Why does water sometimes appear on roadways when there has been no rain and all else is dry? On a cold winter day, why does condensation form on the interior of a glass window pane rather than on the exterior? Is it because the humidity is higher indoors? No, the humidity is usually lower. (That is why someone in a cold climate develops "dry skin" during the winter and also why electrostatic shocks are more common in winter than in summer.)

Answer The amount of water vapor contained in the air is often given as a *relative humidity*, in comparison to a *saturation limit*. For example, a relative humidity of 50% means that the amount of vapor is half the saturation limit. During a hot shower in a closed room, the relative humidity can reach 100%, and then as additional water vapor is forced into the air by the shower, some of the vapor condenses into drops onto various surfaces, including the mirror.

The other reason for the mirror drops is that the saturation limit is lower for cooler air. If the mirror is cool when you shower, it cools the very humid air that reaches it, reduces the saturation limit of the air, and causes some of the vapor to condense. Normally, the hot, humid air from the shower fills the top of the room. Thus, this process of condensation onto the mirror begins at the top of the mirror.

Although the mirror may appear to be clean, it is covered with dust and films (such as the oil in a fingerprint). The condensed water molecules are more strongly attracted to one another than they are to the contamination on the mirror, and so their mutual attraction pulls them into little beads that spot the mirror and reduce its ability to reflect clearly.

You can eliminate the beading if you smear a thin layer of soap over the glass. The soap reduces the water's surface tension (due to the mutual attraction of the water molecules), and so the water spreads out in a smooth layer over the soaped regions. The smoothness of the water restores clear reflections. You can get almost as good an effect if you use oil from your finger instead of soap.

A roadway can sometimes become wet while everything else remains dry if the surface radiates its thermal energy and becomes cool enough. It then cools the adjacent air and might also reduce the saturation limit of the air. If so, some of the water vapor in the air condenses onto the roadway.

4.66 • Condensation on eyeglasses

When someone who wears eyeglasses walks from the cold outdoors into a warm room, why do water drops form on the lenses? If the person refrains from wiping the lenses clean and instead waits for a while, why do they begin to clear, and where does the clearing appear first?

Why do the lenses *fog up* if the person walks from a room at normal temperature and into a sauna? Why do they eventually clear, and where does the clearing begin in this case? Does the clearing's starting point depend on the makeup of the frames or how the lenses are curved?

Answer In either case, a lens cools the air next to it, reduces the air's *saturation limit*, and causes some of the water vapor to condense on the lens. (See the preceding item.) The condensed liquid then beads up and fogs the lens. That is, the beads distort the transmission of light so much that images are no longer seen clearly.

When a person enters a warm room after being in cold, outdoor weather, the clearing first appears at the rim nearest the nose. The nose heats that part of the rim by radiation, by conduction through the support on the nose or through the intermediate air, and by convection as air warmed by the nose rises from it. The thermal energy is then conducted into and through the lens, warming it, vaporizing water drops, and thus clearing the lens.

When a person enters a sauna, the clearing process is different because the temperature difference between the air and the lens is greater. Now the nose is not very important in the clearing process; more important are the water beads forming on the lens. When water vapor condenses to liquid, the molecules must give up some of their energy. In the sauna, the released energy warms both the lens and its frame; however, the temperature increases faster near the center of the lens, because the frame usually has a greater *heat capacity* (it requires more thermal energy to increase in temperature). The other source of thermal energy is the convection of the hot sauna air past the lens.

If the exterior of the lens were flat, the clearing would appear first in the center of the lens and gradually spread to the rim. If the exterior of the lens is curved outward, the protrusion of the surface into the passing air enhances central clearing. If the exterior of the lens is curved inward, the center is somewhat removed from the passing air and is not warmed as rapidly. In an extreme case, the clearing might start near the rim in spite of the rim's large heat capacity.

4.67 • Water collection in arid regions

How can the tenebriorid beetle *Stnocara*, which lives in the arid Namib Desert of southern Africa, collect drinking water from the early morning fog that blows over the desert? How do people living in the Atacama Desert of northern Chile collect water from the air drifting in from the Pacific Ocean?

Answer The beetle stands in what is called a *fog-basking* position, with its head downward on the windward side of a sand dune and its rear raised into the wind so that it forms a ramp. The rear is covered with overwings that are covered with randomly spaced bumps. The bumps are *hydrophilic*; that is, water molecules can bind to them. The low regions between bumps are waxy and thus *hydrophobic*; that is, water molecules do not bind to them. Fine mist collects on a bump to form a drop. When the drop is small, the hydrophilic force holding it to the bump is greater than the weight of the drop, which tends to pull the drop down the ramp. Eventually the drop is large enough to break free of the bump. Then it slides down the ramp, partially guided by bumps on the way, until it reaches the beetle's mouth, where it is consumed.

People living in the Atacama Desert collect water from fog that drifts in from the ocean by erecting large nets onto which the water can condense, forming drops. Eventually, the drops are large enough to slide down the net to a collection region. Some 11 000 liters of water can be collected in one day, enough for a village.

Certain stone constructions now found on the Crimean peninsula appear to be condensation traps onto which dew from very humid nights would collect. After sunset, the stones radiated infrared light to the night sky. On a clear night, they probably radiated more energy to the sky than the sky radiated to them, and thus they cooled below the air temperature. Water vapor coming in contact with the relatively cool stones condensed to form drops. As the drops grew, they were eventually large enough to break free of their perch and flow into either a collecting depression in the stone or into pipes to be carried to a central container. The efficiency of collecting water in this way was probably quite low.

Some survival books illustrate that dew can be collected by a trap. First, a hole about a meter wide and a meter deep is dug in the soil or sand in a sunny location. An open-top can is centered at the bottom and a tube (for drinking) extends from the bottom of the can to the top of the hole. Then a plastic sheet is lowered into the hole so that it has a slope of about 45° and its lowest point is located directly over the can. The top of the plastic sheet is held in place by dirt or rocks around the rim of the hole. As this arrangement warms in sunlight (it is something like a greenhouse), water evaporates from the soil at the bottom of the hole. When the Sun goes down and the hole cools, water vapor condenses to form drops on the underside of the plastic sheet. Once the water drops are large enough, they slide down to the lowest point of the sheet and fall into the can. After enough water collects in the can, the water can be sucked up the tube and drunk. A cactus can be chopped up and slipped into the hole to provide more moisture. Seawater can also be used, because as the water is evaporated, the salt is left behind.

Stories about naturally forming dewponds that provide drinking water for people or animals are only myths, because dew provides too little water to form a sizable body of water.

4.68 • Mud cracks

Why does mud develop cracks that initially tend to be perpendicular to each other at points of intersection? Why can the cracks then form polygons (the mud resembles a tiled floor), and why might the edges of a polygon curl up, perhaps so much so that the top layer curls into a tube, breaks free of its base, and rolls away?

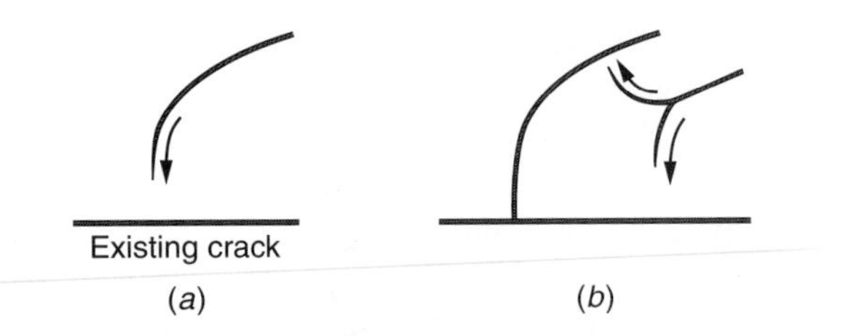

***Figure 4-9** / Item 4.68* (*a*) Developing mud crack makes a perpendicular intersection with existing crack. (*b*) Developing crack bifurcates and will complete a polygon.

Why do giant polygons sometimes develop in arid basins after rainfalls? These polygons may be 300 meters across and formed by fissures that may be 1 meter wide and 5 meters deep.

Answer As a flat layer of mud slowly dries out, starting at the surface, the top of the layer contracts and develops stresses (one portion is pulled by the surrounding portions) whereas the bottom of the layer is held in place by the underlying ground. At random points on the surface, the stress becomes large enough to crack the surface, which reduces the stress. Those cracks then grow in length horizontally and vertically (down to the underlying ground). Where a developing crack happens to extend toward an existing crack, the developing crack tends to be steered by the stress in the surface to make a perpendicular intersection (Fig. 4-9*a*). (The tendency for the surface to be pulled apart is greatest parallel to the existing crack.) After an initial stage of crack formation, a secondary system of cracks develops in the mud. These cracks may each start in a straight line, but as they extend into the bulk of the mud, they tend to *bifurcate* (split) as in Fig. 4-9*b*. Depending on the rate at which the mud dries out, the intersection of the secondary cracks with the first-stage cracks tends to split the mud into polygons.

If a thin, top layer of a polygon dries out quickly, it contracts. The contraction can cause the layer to curl up to be concave. Once the edges rise, the underside of the layer can dry out and the layer can curl up enough to form a tube. More rarely a mud polygon dries more slowly on its top surface than lower in the mud, probably due to quick drainage as occurs on a slope. Then the edges can curl *downward.*

The giant polygons appear to form for the same reasons as the small mud polygons.

Cracks and polygons in mud are examples of cracking behavior induced by *desiccation* (drying). You can find many other examples, such as in the desiccation of paint, a cornstarch–water mixture, and a water–coffee grounds mixture. If you experiment with the latter two examples, you will find that the typical size of the polygons depends on the depth of the cracking layer: Smaller-size polygons occur with a thinner layer, until the layer is so thin that a crazed, irregular pattern occurs instead of polygons. You can also alter the pattern by greasing the inside bottom of the container, to decrease the friction between the material and the container: Less friction (less stress) leads to fewer cracks.

4.69 • Inflating juice containers on airplanes

Even if the cabin pressure is not lost on an airplane, certain containers may be unsealed when they reach you. For example, a container of cream for coffee might be open along the edge of its flexible top. If it is not, notice that the top bulges outward. When you later descend for landing, the bulge gradually disappears. What causes the opener to open or bulge while in flight?

Similar physics can lead to comic relief on a tedious flight if someone shakes up a sealed container of juice and then hastily rips off the foil top on the near side. Some of the juice is likely to squirt all over the person. A seasoned traveler knows to open the seal slowly and with the opening facing away.

Why do your ears sometimes ache as you ascend in an airplane? Why do they sometimes seem "clogged up" after you land?

Answer The air pressure in an airplane is maintained mechanically, but it is less than the air pressure on the ground. So, as the airplane climbs and the air pressure decreases, air or other gas inside flexible containers of cream, juice, and salad dressing tends to expand. Sometimes, the seals on the containers are forced open. If a container of juice or salad dressing is still closed and you shake up the contents, you coat the underside of the seal. If you suddenly rip off the seal, the expansion of internal gas through the opening will blow the contents outward.

The discomfort that can be felt in the ears while flying is due to the air pressure in the middle ear, which is located behind the eardrum. Normally the air pressure there is adjusted to be equal to the ambient air pressure via the Eustachian tube, which runs to the back of the nose and throat. However, if the Eustachian tube does not open while the airplane ascends, the external air pressure on the eardrum is the reduced cabin pressure while the internal air pressure is the ground-level pressure. The pressure difference tends to push outward on the eardrum. The problem can be avoided by yawning or swallowing, which opens up the Eustachian tube so that the higher pressure in the inner ear can be reduced.

When the airplane descends, the external pressure increases to the ground-level value and then the pressure difference tends to push the eardrum inward. The problem may not be easy to solve, because the lower pressure in the inner ear tends to keep the Eustachian tube closed. Still, yawning and swallowing may open up the tube and allow air pressure in the inner ear to increase to the ground-level value.

The decreased pressure in an airplane can sometimes be annoying. Bottles or even seemingly strong beverage cans can burst open if they are packed in a suitcase, because the luggage area of an airplane is usually not pressurized. A similar problem can occur if a deceased person is shipped in the luggage area—the coffin must be tightly secured to keep it closed.

4.70 • Inflating bubbles and balloons

If you inflate a spherical rubber balloon, why is the inflation difficult at first and much easier when the balloon is partially inflated? If the balloon is cylindrical, why does the inflation begin in one place instead of throughout the balloon? As you continue to blow into the balloon, why does the bulging travel along the length of the balloon?

Suppose two soap bubbles of different radii are connected by a tube that has a valve that is closed (Fig. 4-10*a*). What happens to the bubbles if the valve is opened so that air can flow between the bubbles? If the bubbles are replaced with rubber balloons, what happens when the valve is opened?

Answer If you inflate a spherical soap bubble, you must provide air pressure that is larger than the existing air pressure inside the bubble. The internal air pressure depends on the curvature of the bubble's surface. To see the point, consider a patch on the surface (Fig. 4-10*b*). It is pulled along its sides by adjacent sections of the surface. The pull at the left side and the pull at the right side are partially inward toward the center of the bubble; it is that inward portion that determines the air pressure. When the bubble is small and highly curved, the inward pull on a patch is large and so the internal pressure is also. The bubble is then difficult to inflate. When the bubble is larger and its curvature is less, the inward pull is small and so the internal pressure is also. The bubble is then easier to inflate.

A rubber balloon is different in that the stretching of the membrane during inflation acts to increase the pressure. During the initial stages of inflation, the resistance to stretching drives the pressure up and requires a large pressure from you if you are to inflate the balloon further. However, once the balloon has reached a certain size, the subsequent decrease in curvature acts to lower the internal pressure, and then further inflation is easier. The ease is further aided by a

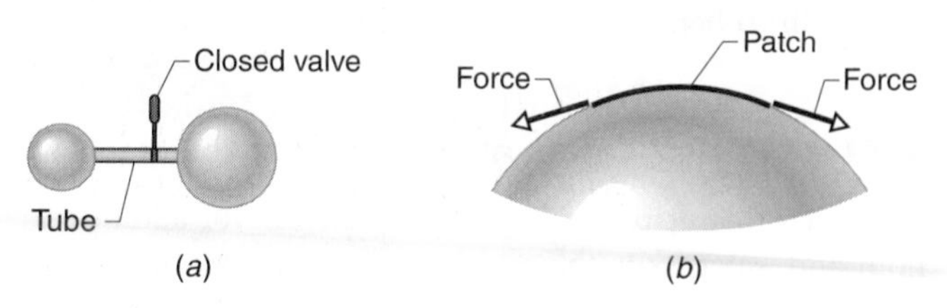

Figure 4-10 / *Item 4.70* (*a*) Two bubbles (or balloons) on a tube closed by a valve. (*b*) Forces on the left and right edges of a patch of the surface of a bubble.

lowered resistance to stretching of the rubber at about the same size of the balloon. (The resistance returns in full force when the balloon is much larger.)

Another factor is also important. When you blow into the balloon, you insert a certain volume of air from your lungs, the "puff volume." When the balloon is small, the added volume requires that the surface area increase significantly, which appreciably increases the resistance to further stretching. When the balloon is large, the added volume is small compared to the existing volume and does not require as large an increase in the surface area, nor as much increase in the stretching of the rubber.

A peculiarity of some rubber balloons is that even if they are ideally spherical in nature, they may be noticeably nonspherical when their inflation is in a certain range. Soon, after the ease of the inflation sets in, and before the inflation again becomes difficult when the rubber is highly stretched, the balloon may develop an appreciable bulge on one side. M. J. Sewell of The University of Reading noted this peculiarity: "Over the whole range of inflation, nature does not prefer a sphere even when offered one."

When you inflate a cylindrical balloon, it first bulges at its weakest point, usually the part nearest the opening. The section that joins the bulge to the uninflated portion is concave along the length of the balloon. When you blow in additional air, the tension in the concave section helps to expand the balloon there, and the front of the bulge moves along the length of the balloon.

When two bubbles are connected by an opened tube, higher pressure in the smaller bubble drives air through the tube to the larger bubble, which has smaller pressure. The small bubble collapses while the larger bubble expands. This commonly occurs unnoticed in the head on a beer that contains carbon dioxide bubbles. The bubbles are not connected with tubes, of course, but the carbon dioxide gas can *diffuse* (spread) from one bubble to another through the bubble walls. The smaller bubbles lose their gas to adjacent larger bubbles and eventually collapse, a process known as *Ostwald ripening.* However, the rate of diffusion for nitrogen is much less than for carbon dioxide, and so those beers (such as Guinness stout) that contain nitrogen in place of carbon dioxide tend to have much longer lasting bubbles.

If the bubbles are replaced with balloons, the results can be different. Depending on the total amount of air in them, they can end up with either the same radius or with one larger than the other.

4.71 • Making cakes at high altitudes

Why does the recipe for angel food cake call for additional flour and water when the cake is made at high altitude?

Answer The rise of the cake depends on the expansion of air bubbles trapped in the whipped batter and on the production of steam as the water (part of which comes from the eggs) vaporizes. Since the atmospheric pressure is lower for higher altitudes, the bubbles can expand so much that the strength of the cake's infrastructure is overwhelmed and the cake falls. One solution is to decrease the sugar, which tenderizes the infrastructure, but the cake becomes tougher and less sweet. Another solution is to increase the amount of flour to strengthen the cake. The amount of leavening might also be decreased to decrease bubble production.

Water boils at a lower temperature at high altitudes than it does near sea level. That fact means that the cake will lose more water to evaporation when prepared at high altitudes, because vaporization is easier to produce. To offset the loss, the recipe calls for more water.

4.72 • Champagne in a tunnel

In November 1827, when the tunnel beneath the Thames River in London was completed, dignitaries went down into the tunnel to celebrate. Because the tunnel was kept under pressure during the construction (to help hold back water in the ground below the river), the dignitaries first entered an air lock where the pressure was increased until it matched the tunnel pressure. While inside the tunnel, the dignitaries toasted the project with champagne, but the champagne was disappointingly flat (it did not bubble much when the bottles were opened). Still, celebrations continued until the dignitaries returned to the surface via the air lock.

When the air-lock pressure was reduced to the normal (external) pressure, the dignitaries were in much discomfort, and one had to be rushed back to the air lock, to be *repressurized* (put back under the higher pressure). What was wrong?

Answer The carbon dioxide dissolved in champagne comes out of solution to form bubbles (the froth) when a bottle is opened. Before then, the contents are under considerable pressure, and the carbon dioxide vapor at the top of the bottle is in equilibrium with the carbon dioxide in solution. That is, on average, a carbon dioxide molecule leaving the solution to be part of the vapor is matched by a carbon dioxide molecule leaving the vapor to be dissolved. However, as soon as the bottle is opened, the pressure of the carbon dioxide vapor is reduced and, for a while, the carbon dioxide is not in equilibrium. That is, much more carbon dioxide leaves the solution than reenters it. Of course, that gives the flow of bubbles—carbon dioxide vapor momentarily surrounded by liquid film.

Such is what normally happens, but in the pressurized tunnel, the air pressure was large enough that bubbles hardly formed. Thus, much of the carbon dioxide remained in solution. When the dignitaries drank the champagne, a lot of dissolved carbon dioxide was ingested. When they left the air lock and were exposed to lower air pressure, the dissolved carbon dioxide suddenly could come out of solution, which tended to inflate various internal organs, producing (at best) belches and (at worst) inflated stomachs and bladders.

Until fairly recently, construction workers had to work under pressure while building tunnels under rivers and bays. When the workers finished their shift and returned to the surface, they had to undergo a decompression schedule much like deep-sea divers. The problem is that when air is breathed under pressure, nitrogen molecules from the air are forced into the bloodstream. When the worker comes out of the pressurized part of the tunnel, the reduced pressure on the body and in the lungs allows the dissolved nitrogen to form bubbles. The bubbles tend to move with the blood flow, collecting into "slugs" if they move into larger veins (toward the heart) or becoming wedged and blocking the blood flow if they move into smaller veins (away from the heart). Either way, the victim can suffer terrible pain, long-term disability, or even death.

These days, tunnels are constructed with drilling machines that are guided remotely, and workers go into the pressurized section of the tunnel much less frequently. (In some cases, tunnels are no longer drilled but are constructed of prefabricated sections that are dropped down onto a river bed and then fitted together.)

SHORT STORY

4.73 • Stuck in a bottle

A young girl, determined to get the last of a chocolate drink clinging to the bottle's interior, forced her tongue into the bottle and then inhaled sharply to pull the liquid into her mouth. Her tongue became stuck because she reduced the air pressure inside the bottle by removing some of the air. Either she sucked some of the air out when she inhaled or she pushed some of it out past her tongue as she wedged her tongue into the bottle. Either way, the pressure difference between the interior air and the exterior air meant that she could not get her tongue out. Neither could an emergency-room crew until a professional glass cutter showed up with a glass-cutting device.

4.74 • Wintertime thunder

Why is a thunderstorm (and thus thunder and lightning) less likely in winter conditions than in summer conditions?

Answer Thunderstorms develop when the lower part of the atmosphere is unstable—that is, when parcels of warm air rise rapidly due to buoyancy (they are less dense than cooler air). Such is the case when the air temperature decreases sharply with height—a parcel of warm air near the ground is pushed upward into the cooler air.

However, if the air contains water vapor, the change in temperature with height need not be extreme. The reason is that as the parcel ascends, some of its water vapor can condense to form drops. That change from vapor to liquid releases a large amount of thermal energy, which makes the parcels warmer. Thus, the buoyancy on the parcel increases and the parcel accelerates upward. This sets the stage for instability and thunderstorms.

In winter conditions, the temperature decrease with height is usually gradual and the air parcels near the ground are too cool to contain much water vapor. With the upward acceleration less likely, the air is too stable for a thunderstorm to form. Still, on occasion, I have heard thunder during a snowstorm.

4.75 • Stack plumes

When the wind is gusty, it can whip the plume of smoke or condensation from a chimney stack into a chaotic, ever changing pattern. But how about when the wind is uniform, both in time and with height? Shouldn't the plume then climb along a slant to the vertical while widening? Strangely, it might not; instead, it might form one of the patterns shown in Fig. 4-11*a*. What accounts for the shapes? Why do some wind-bent plumes split into a pair of plumes (Fig. 4-11*b*)?

On a windless day, most plumes rise while spreading horizontally, forming a narrow, upright V. Why do some stacks emit a plume that first narrows and then widens?

Answer In some industrial stacks, the gas is propelled upward so that it and the smoke that it carries is less likely to settle back to the ground and create a pollution problem. Further rise of the gas will occur only if the gas is hotter than the surrounding air. The temperature difference creates buoyancy that accelerates the gas upward. In other stacks, including a home fireplace chimney, the gas is not mechanically propelled and its rise depends entirely on buoyancy.

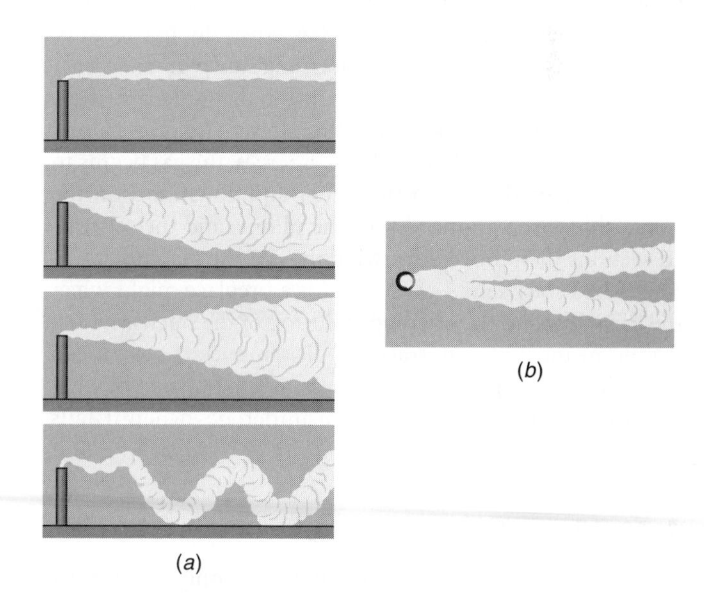

Figure 4-11 / *Item 4.75* (*a*) Smoke plumes from tall stacks in a steady horizontal wind. (*b*) Overhead view of plume that splits.

Even if the gas is initially hotter than the air, it may not be able to rise because ascent would cool the gas. The cooling is due to the decrease in air pressure with height. As a parcel of gas rises, it expands against the decrease in pressure. The energy from the expansion comes from the random motion of the gas molecules—as they slow and give up energy, the temperature of the gas decreases. If the gas temperature drops below the air temperature, the gas experiences a *negative buoyancy* and is forced downward.

Ascent is also influenced by the water content of the gas. If the gas cools to the point that the water vapor condenses to form drops, the transition of the vapor to liquid releases energy that helps warm the gas even though the gas is expanding. Similar considerations must be made for descent of the gas. In that case, the increased air pressure causes the gas to contract and warm. If the descent is to continue, the gas must remain cooler than the air.

The relative temperature of gas and air determines which direction the gas takes, but the shape of a plume that is bent over by wind depends on the air turbulence (swirling). If the turbulence consists of small swirls, the plume will gradually expand as it moves away from the stack. If the swirls are larger, the plume might *loop*. (The looping is partially illusionary, because any given parcel of gas does not move up or down but in an almost straight line. The appearance of loops is due to a succession of parcels that are forced to move in different straight lines by the turbulence.)

As an example, consider the case of *coning*, which occurs when any ascent of gas rapidly cools it below the air temperature while any descent rapidly warms it above the air temperature. The gas is then partially trapped at about the height of the stack, but mild turbulence can gradually mix it vertically. *Fanning* occurs when the air temperature actually increases with height, a condition known as *inversion*. Then the gas is more narrowly trapped and turbulence can mix it only horizontally. In either case, if the gas has especially hot regions, those regions can burst upward, forming side plumes.

If a plume splits, the division is due to the rapid rise of gas from the center of the stack and the slower rise of gas along the wall of the stack. As the plume emerges, the gas begins to circulate upward through the center and downward on the sides. This vortex motion splits the plume, and the separated sections then move downwind in separate directions.

When a stack emits gas slowly (as a fireplace might when the fire is low and the opening at the top of the stack is wide), the plume must first contract and pick up speed before it can widen into a V. The contraction is an inverted version of what a smoothly flowing water stream undergoes when it falls from a faucet.

4.76 • Smoke signals and mushroom clouds

Native Americans knew how to signal over long distances by controlling the smoke from a bonfire. They would place plentiful brush in the fire to increase the smoke and then would briefly cover the fire with a wet blanket. When they pulled back the blanket, a burst of smoke would climb into the air and could be seen by a distant observer. If the signaling was made in the early morning or late afternoon, why did the smoke spread out horizontally when it reached a certain height? The smoke then resembled the cap on a mushroom or an atomic bomb explosion.

Native Australians also used bonfires for signaling but chose to lift the burning materials with long poles instead of covering it with a blanket. As several people lifted, others threw on fresh brush. The lifting increased the flow of air through the fire and fanned the flames, and the added brush enhanced the smoke. If the signals developed into "mushroom caps" and if the natives carefully coordinated their efforts, they could send up six caps, each neatly stacked above another. How did the natives control the height at which each cap appeared?

Why do large explosions, including ground-level and above-ground nuclear explosions, produce mushroom clouds?

Answer The puffs of smoke spread horizontally at the height where the gas carrying the smoke has cooled to the temperature of the surrounding air. (See the preceding item.) The mushroom cap appearance is most noticeable when the air has an *inversion* (the air temperature increases with height), such as it might have when the ground is cool in early or late day. The Australians controlled the height of each cap by adjusting how fiercely the fire burned. When the temperature within the fire was increased, the burst of smoke and hot gas rose higher.

In World War II, smoke screens were used to hide targets, such as factories, from nighttime aerial attack, especially on nights that were well lit by moonlight. A thick, oily, black smoke was produced by burning diesel oil in the same devices used to blanket orchards with smoke on nights when the fruit was threatened by a deep chill. The smoke did not rise appreciably, because it was not very hot when it exited the tall chimney on a burner. Thus, barring any breeze, it would form a shallow layer that could hide or disguise a target.

A nuclear explosion produces an incandescent fireball that rapidly heats the air. That air then quickly rises, sucking ground-level air, dirt, debris, and water moisture up into its wake to form the mushroom stem. As with fires, the air eventually cools to the temperature of the surrounding air and thereafter spreads horizontally to form a mushroom cap.

4.77 • Fire in a fireplace

Why can you never get a good fire going if you stack kindling around a single log? Why, instead, should there be a stack of at least three logs? Why should the logs be occasionally stirred as they burn? How does a fire in a fireplace heat a room? Why does a poorly built fireplace smoke—that is, send smoke into the room—whereas a well-built fireplace doesn't?

What is the purpose of the glass fronts that are sold to cover the opening of a fireplace? Is there a more efficient way of stacking logs so that the fire might better heat a room?

Answer The surface of a log will burn when its temperature is raised above some certain value, the *combustion temperature* or *ignition temperature.* Kindling may initially produce burning on the side of a log, but after the kindling is consumed, infrared radiation by the log's surface and convection of the hot gases away from the log soon lower the surface's temperature below the combustion temperature. To keep a log burning, you need one or two more logs stacked next to it. The burning surfaces then heat one another by radiation, convection of hot gas, or conduction through the intermediate air, and the temperature of the logs remains above the combustion temperature.

This description is adequate for a campfire, but a fire in a fireplace differs because it is partially enclosed. Gas and heated bricks at the top of the hearth form a hot layer that emits infrared radiation to the burning surfaces, helping to keep them hot. The partial enclosure can also limit the required supply of oxygen, restricting the burning. In an idealized situation, equilibrium is reached between the thermal energy being produced by the fire and the thermal energy being lost by convection of hot gases up the chimney (hopefully not out into the room), conduction through the back wall of the fireplace, and radiation out into the room. The fire can flare up from the equilibrium state if more wood is added or if the airflow into the fireplace and up the chimney is increased.

The warmth in the room is largely due to the thermal radiation sent through open spaces between the logs. A log is occasionally stirred to move unburned regions into the region between the logs. Some of the burning surface then faces the room and briefly provides thermal radiation, while the fresh surface moved into the interior bursts into flames.

Hot air and combustion gases rise up into the chimney, passing through a narrow portion where the width is controlled by a *damper.* Their rise is prompted by buoyancy, because their high temperature makes them lighter than the air in the room and outdoors. A taller chimney gives a strong *draw* (pull up the chimney) because it allows the buoyancy to accelerate the air and gases to a greater speed before they reach the top. If a wind blows across the top, it snares the escaping gases and air and enhances the draw. A poorly drawing chimney will *puff* if discharges of hot gases alternate with cold external air sinking into the chimney.

A chimney might discharge smoke into the room for several reasons. Included is the possibility that when air is drawn into the fireplace, it hits the back wall and then circles forward back into the room over the fire, carrying along some of the smoke. A well-designed fireplace has a high interior so that the front wall blocks such swirling. If the damper or the associated section of the chimney is too open, downdrafts of outdoor air might send smoke into the room. When the fire is low or the fireplace is uncommonly large, such downdrafts are more of a problem.

Since a fireplace heats a room primarily by thermal radiation, the walls at the back and on the sides are often slanted so that they will scatter the radiation into the room. The radiation is more directed if the logs are stacked on a support so that they form a slot, with a large log at the back and smaller logs at the top and bottom of the slot. The fire is contained inside the slot, and the embers face the room rather than the brickwork of the fireplace.

Since the fire sucks air from the room, it drains warmth from the room. To alleviate the loss, an enclosure of heat-resistant glass can be mounted across the fireplace. The glass allows a view of the fire and radiates the thermal energy it receives into the room while also blocking the large loss of warm air from the room. The fire is still fed by air that is pulled through open ports at floor level, where the room air is the coolest.

For some fireplaces, reverse airflow can occur when there is no fire in the fireplace. Such fireplaces are normally shaded from sunshine, so they can contain relatively cool air while sunshine warms the external air. The cool air in the chimney is denser and heavier than the external air. So it sinks to the bottom of the chimney and out into the room, pushing air out through either open doors or windows or through cracks around them.

Similar airflows occur in some multiple-opening caves, where one opening is much higher than the other. Then the passages from the lower opening to the higher opening act as the chimney. When the chimney air is warmer than the external air, as in the winter, the air flows in through the lower opening and out through the upper opening. When the reverse is true, the air flows in the reverse direction. Such a cave is said to be *breathing.*

4.78 • A candle flame

How does a candle burn; that is, how does it consume its fuel? Why is the light from a candle flame largely yellow, and why do blue regions usually form along the side of the flame (Fig. 4-12)? Why is there a dark cone between the wick and the yellow part of the flame? Why do some candles smoke; why do some flicker? Why is soot from a flame black, and yet white vapor is emitted from a candle that has just been extinguished?

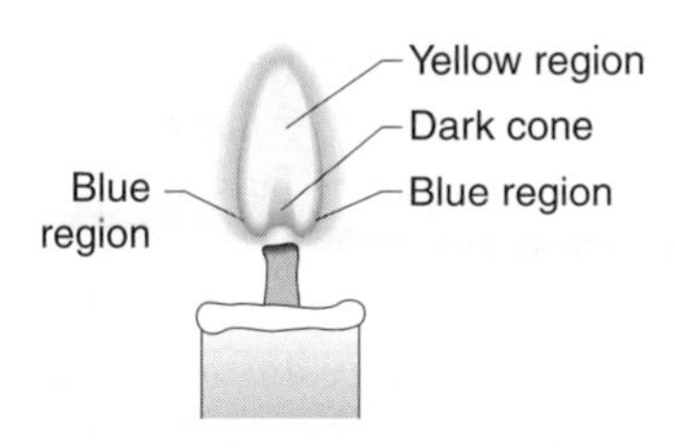

Figure 4-12 / *Item 4.78* Structure of a candle flame.

Answer The wax fuel for a candle is either paraffin or stearin (stearic acid) or a combination of the two. Thermal radiation from the flame is intercepted by the wax, melting and liquefying it. The liquid is drawn up into the wick by capillary action (that is, the molecules in the wick attract the molecules of liquid, which are held together by their mutual attraction). As the wax climbs the wick, it is vaporized by the hot environment of the flame and is then carried up and also outward by the flow of hot gases (convection).

Some of the hydrocarbons released by the vaporization are carried up into the dark cone just above the wick, but they are probably no hotter than about 600°C to 800°C. Oxygen is needed to burn the hydrocarbons, but oxygen must diffuse (spread) into the flame and not much oxygen reaches the dark-cone region. (A candle is said to be one type of *diffusion flame.*) So, with low temperature and little oxygen, the hydrocarbons in the dark cone don't emit much light.

Some of the hydrocarbons are carried out into the blue regions, which are said to be the *reaction zone.* There the plentiful oxygen reacts with the hydrocarbons, breaking them down into smaller molecules and producing the hottest region of the flame. Included in the smaller molecules are molecular carbon C_2 and the hydrocarbon CH. When those molecules are produced in the excited state, they quickly de-excite by emitting light in a series of certain wavelengths. Most of the bright emissions lie at the blue end of the visible spectrum, and so you see blue emitted from the side of the flame. (In an otherwise dark room, a faint glow might appear to surround a candle flame. However, that glow is due to the scattering of the flame's light within the eye. Artists often depict the glow by drawing short, wavy lines extending from a candle flame.)

The hydrocarbons that rise into the yellow portion of the flame from the dark-cone region or the reaction zone form tiny solid carbon particles that then burn in the oxygen. The particles become so hot that they are incandescent, giving off the yellow light that dominates the light from the candle.

If the rate at which hydrocarbons enter the yellow region is matched by the rate at which the solid particles are consumed, the flame is smokeless. However, if the supply rate is too large (as can happen with a wick that is too wide), the flame can smoke—that is, give off dark fumes, which are carbon particles incompletely burned in the flame. If you insert a paper clip into any candle flame, you can interrupt the burning and cause soot to be released, which ends up coating the paper clip.

If there is a mismatch between the fuel-supply rate and the fuel-burn rate, the flame can either just die out or it can flicker. Dying out occurs if the flame fails to liquefy enough fuel or the capillary action fails to pull enough fuel up the wick to keep the flame alive. Flickering occurs if there is a feedback mechanism between the flame and the liquid supply. For example, suppose the flame flares and thus increases the thermal radiation and the amount of liquefied wax. If the wick is slow to bring this extra liquid to the flame, the flare consumes the available fuel at the top of the wick and the flame dims. When the extra liquid finally arrives at the top of the wick, the flame flares again, and the cycle is repeated.

To blow out a candle, you must blow long enough to remove not only the incandescent carbon particles in the yellow region and the reacting hydrocarbons in the blue region but also the hydrocarbons that will vaporize from the hot wick as you blow. In fact, just after a candle is extinguished, hydrocarbons are still vaporized from the wick, but they no longer form carbon particles and burn. If they remain individual molecules, you probably don't notice them, but if they clump together, they increase their scatter of light enough for you to see a wispy, whitish stream rising from the wick.

4.79 • Spraying a fire

Why does water suppress a fire of burning wood? Why do firefighters normally adjust the nozzle on a fire hose so that the water emerges as a fine spray rather than a stream? When firefighters break into a burning room that has been closed, why do they normally spray the room's ceiling rather than the floor even if the fire is located only on the floor?

Answer Water can suppress or diminish a fire by several processes. (1) It absorbs thermal energy from the burning surfaces and the vapor products (which may be burning) and thus cools them enough that they no longer burn. (2) It absorbs some of the thermal radiation emitting by the burning substances and thus decreases the chance that surrounding materials will become hot enough to burn. (3) It takes up room in the air and thus decreases the amount of oxygen reaching the burning materials, which need oxygen to continue burning.

In the first process, the rate at which thermal energy is absorbed depends on the surface area of the water. Because the total surface area increases if a stream is converted to a spray, the nozzle is adjusted to give off a spray.

When a closed room is burning, the required oxygen is soon depleted, leaving unburned fuel in the air; in fact, the hot materials in the room may continue to release unburned fuel. Because that fuel is hot, it tends to collect near the ceiling. If a firefighter suddenly opens the room, air tends to flow in through the bottom of the opening, supplying oxygen that soon reaches the unburned fuel near the ceiling. That fuel suddenly bursts into flames, and because of the air inflow at the bottom of the opening, there is an outflow of this burning fuel through the top of the opening. Unburned fuel carried out of the room by this outflow meets the abundant air, producing a fireball that shoots from the room in what is called a *backdraft* or *flashover.* This whole process occurs so quickly that the fireball can envelop the firefighter who opened the room. So, as an opening is made, a firefighter immediately sprays water near the ceiling to cool the hot fuel floating there and thus decrease the chance of a backdraft. The occurrence of backdraft and the need of this preventative measure appear to have increased as rooms and buildings were made more airtight in order to decrease the cost of cooling or heating them in adverse weather.

4.80 • Cooking-oil fires

Common cooking oil can *self-ignite* (burst into flames) if its temperature exceeds the so-called *ignition point.* For example, canola oil will self-ignite at a temperature of about 330°C. Water can put out a fire, but should it be used to put out burning cooking oil, such as when a skillet of oil catches on fire on a kitchen stove?

Answer The standard advice about a cooking-oil fire is to smother it with a lid or some other solid piece of metal, in order to eliminate the oxygen supply and contain the hot oil. If water is thrown onto the hot oil, the drops enter as they would enter a pool of water, but then each drop separates into many droplets. These droplets heat so rapidly in the oil that they flash to steam. As liquid water expands to become water vapor, the sudden increase in volume shoots the overlying hot oil in many directions. This airborne oil does not cool appreciably during flight and can burn skin or countertop. If it hits a flame either directly or via splashing, then it can ignite and flare up along walls. Thus, throwing a fistful of water at a skillet of burning oil can be disastrous because of the water's explosive conversion of liquid to gas.

4.81 • Brush fires and forest fires

How does a grass fire, such as those seen in Texas rural areas, spread? How does a forest fire spread, and how does it ignite houses when it reaches an inhabited region?

Answer A brush fire spreads largely by the flames along the fire edge touching unburned material just beyond the fire edge. The spreading is accelerated if wind can blow the flames over the edge into that unburned material. It is also accelerated if the unburned material is somewhat higher, as is the case when the fire climbs up the side of a canyon. Burning embers, lofted by the convection of the hot gas from a fire, can then land in the unburned material, setting it on fire.

A forest fire in which the material near the ground is on fire spreads in much the same way, but a *crown fire,* in which even the overhead canopy is on fire, spreads largely in a different way. The transfer of energy from the fire to the unburned material is primarily by thermal radiation. The situation is similar if you stand in front of a roaring bonfire: You feel warm, perhaps even uncomfortably hot, because of the thermal radiation you intercept. In a crown fire, the unburned material becomes so hot that it ignites, and thus the fire spreads. The thermal radiation originates in two principal regions: in the burning solid materials (tree trunks) and in the burning gases (in the flames at the crown). Radiation from the solid materials is blocked by airborne particles and trees and does not penetrate into the unburned material very well. Radiation from the flames can penetrate better, especially if wind bends the flames in the direction of the unburned material so that the side (or full length) of the flame faces somewhat downward toward the unburned material. Much of the radiation from the side of the flame is then intercepted by the unburned material.

As a crown fire approaches a house, the external walls facing the fire tend to heat to the ignition point. However, overhanging eaves can help shield the walls from thermal radiation. Trees between the fire and walls can also shield the walls. The best situation is one in which the trees burn poorly and do not contribute to the crown fire. (Of course, a house can also be ignited by embers blowing onto the roof.)

4.82 • Firestorms

One night during World War II, the German city of Dresden suffered massive incendiary bombing by Allied airplanes. As the bombing began, the ambient winds in the region were weak, only about 4 meters per second. Why did the winds increase to 20 meters per second (over 40 miles per hour)? (Some firefighters estimated the wind speeds to be greater, perhaps 50 meters per second.)

Answer Within about 30 minutes after the bombing began, the fires across the city spread inward toward one another and merged into one huge fire, creating the conditions of a blast furnace in the city streets. The hot gases were strongly accelerated upward, because they were lighter than the surrounding cooler air. Those gases cooled as they rose by emitting infrared radiation and by mixing with the surrounding air, but the cooling process was slow enough that the column of hot gases rose about 7 kilometers. At the top of the column, the gasses spread horizontally because they had the same temperature as the surrounding air.

This huge upward flow of gas pulled (sucked) air along the ground and into the burning region, thus creating strong winds that whipped the fire into a frenzy. Normally, wind spreads a fire, but the radial inflow in Dresden kept the fire stationary, creating a situation now called a *firestorm.* The fire caused a tremendous loss of human life and an almost complete destruction of the buildings in the region.

In some firestorms, the upward flow of air begins to spiral, forming a vortex. Smaller (but still intense) fires also develop vortexes that resemble dust devils. These vortexes are dangerous because they help spread the burning material.

4.83 • Temperature regulation in mounds and buildings

The termite mounds of northern Australia are called magnetic not because the mounds or the termites are actually magnetic, but because these tall wedge-shaped mounds are oriented north–south, as if they were a needle on a compass. The proper name of the termites is *Amitermes meridionalis*, because the mound orientations are along meridians. Why do the termites prefer that orientation for their mounds?

In moderate climates where the summers are not terribly hot, some buildings can still be uncomfortably warm even with the windows open. Is there some way to better ventilate such a building so that the expense of cooling via air-conditioning can be avoided?

Answer The magnetic termite mounds are cleverly designed to maintain the internal temperature. The broad tall face on the east side of the wedge absorbs plenty of sunlight as the Sun rises. The opposite face absorbs sunlight as the Sun sets. When the Sun is high in the sky, the sunlight falls on a relatively small cross-sectional area. That is, the area seen from overhead is smaller than the area seen from the west or east side. Thus, the amount of sunlight absorbed during the hot day is less than during the cooler morning and evening hours. Overall, the temperature inside the mound is approximately constant all day.

Some buildings, such as some built in England, now include a *solar tower* of transparent glass on the side (or corner) that faces the Sun during the day. The top of the tower contains a vent that can be opened or closed as needed. The bottom of the tower is connected to each floor in the building. The Sun warms the air in the tower. Because warmer air is lighter than cooler air, that warm air moves up the tower and out through the vent, pulling air in through any open windows in the building. If the building and tower are designed correctly, there can be a steady flow of air through all open rooms in the building.

Traditional buildings in Iran show remarkable adaptation to an environment that is hot during the day and cool at night. The buildings are clustered to shade one another. Some have a *wind tower* that catches the wind, so that the wind can push air down the tower, through an underground tunnel (where the air is cooled by the ground), and then into the basement of a building. If water is available, either on damp tunnel walls or as a fountain in the basement, the air is further cooled by evaporation of the water. That is, thermal energy is taken from the air, the tunnel, or the fountain water to change the water from a liquid to a vapor.

Some buildings are domed with an open cap at the top. As wind passes by a cap, it entrains warm air at the inside top of the dome and carries it away. This allows cooler air to flow into the building at ground level or (better) along underground tunnels.

4.84 • Warmth of greenhouses and closed cars

Why is a greenhouse relatively warm? Does it have a special type of glass that somehow traps thermal radiation (infrared radiation)? Why does the interior of a closed car become hot if the car is left in direct sunshine on a hot day?

Answer The primary reason that a greenhouse is warm is that the enclosure cuts off or severely limits air circulation. Thus, warm air is not allowed to rise out of the greenhouse to be replaced by cooler air flowing along the ground; also breezes are not allowed to displace the internal warm air. (A common myth is that the glass or plastic roof of a greenhouse somehow traps thermal radiation. Unfortunately, because the term *greenhouse effect* is often applied to the trapping of thermal radiation by Earth's atmosphere, the idea of such trapping is erroneously carried over to a greenhouse.)

A closed car parked in direct sunlight on a hot day is like a greenhouse. Its interior can become very hot, because air circulation is eliminated. In fact, if the sunlight enters through the front windshield, the dashboard and steering wheel can become hot enough to burn skin. Establishing circulation by lowering the windows or opening the doors can (slowly) decrease the temperature. Because black paint absorbs visible light more readily than white paint, you might think that a black car would be hotter than a white car. However, heating a car is primarily due to the absorption of infrared radiation, not visible light, and the two paints probably absorb about the same in the infrared range.

4.85 • Heat islands

Why is the temperature of an urban area, especially the center of a city, usually warmer than the surrounding countryside? For example, during the summer the climate in a city might be hot and stagnant while the countryside is pleasant. Does such an urban *heat island* result primarily from the higher concentration of heat-generating machines in a city?

Because of a heat island, spring budding may start noticeably sooner in the city than in the countryside, and the autumn loss of leaves may come later. Another consequence is that dew forms more rarely in a city than in the countryside.

Answer Several factors contribute to a heat island in an urban area: The tall buildings block and channel the winds that would otherwise cool the area. There is less loss of thermal energy due to evaporation, since rain and snowmelt are quickly drained into the sewer system. Salting further aids the removal of snow from the roads. The paving and building material are more apt to absorb and store sunlight than are grassy or wooded areas.

If the buildings are approximately the same height and cool by radiating their heat from the roofs at night, a layer of cool air can form at roof level. That layer can then inhibit the rise of warm air from street level and thus trap thermal energy within the city. If the city is covered with a thick layer of airborne particles (pollution), the situation can be worse: The top surface of the layer can radiate to the sky and thus provide further cooling at roof level. Although the city will cool somewhat during the night, it does not cool as much as the countryside.

In hot regions, such as in the Southwest of the United States, the absorption of sunlight by surfaces can pose serious danger. For example, asphalt pavement can easily reach 70°C, which exceeds the temperature of 44°C at which a surface will burn skin when touched. Anyone touching the pavement, such as victims in a traffic accident, can be badly burned. Even standing in an empty parking lot covered with asphalt can be difficult because of the intense infrared radiation emitted by the asphalt.

4.86 • Rubber-band thermodynamics

Quickly stretch a rubber band while holding it against your upper lip. Why does the rubber band become warm enough for your lip to sense? With the rubber band still stretched, hold it away from your lip for a few minutes, put it back on your lip, and then let it quickly contract. Why does it become cool?

Answer The rubber in a rubber band consists of long-chained molecules that are coiled up in a spaghetti-like fashion, with a multitude of cross links. When you stretch the rubber band, you stretch out those molecules and part of your work goes into the thermal motion of the molecules. The warmth you feel on your lip is due to that increased thermal motion. If you allow the rubber band to contract, the molecules do work in coiling up; the energy needed for this work comes from the thermal energy of the molecules, and so the rubber band feels cool.

If a rubber band is warmed, the additional thermal energy of the molecules allows the molecules to coil up tighter, thus shortening the rubber band. If a rubber band is cooled, the loss of thermal energy means that the molecules cannot coil up as tightly, thus lengthening the rubber band.

The fact that a rubber band will contract when heated and expand when cooled can be put to use in a machine, albeit one that is only a novelty. A wheel is mounted so that it rotates about its central axis. A second axis is offset from the rotation axis and rubber bands stretch from this second axis to the perimeter of the wheel. This offset of the second axis means that the stretching of rubber bands is not symmetric around the wheel: Some are stretched more than others. The wheel is then submerged to the depth of its radius in a container of hot water. Thermal energy from the water causes submerged rubber bands to contract, and the asymmetry of the rubber bands then causes the wheel to turn slowly. As rubber bands emerge from the water, they cool and become less stretched. As they re-enter the water, they contract again.

4.87 • The foehn and the chinook

Foehn is a general name given to a warm dry wind blowing down mountain slopes. Originally this type of wind was noted in the Alps, where a sudden foehn could dramatically melt and evaporate snowbanks. In the United States, this type of wind is called a *chinook* (after the Chinook tribe), which blows down the eastern slopes of the Rockies. In one extreme example, a chinook blowing into Harve, Montana, raised the temperature from −12°C to 6°C (or from 11°F to 42°F) in about three minutes. What causes a chinook or foehn?

Answer Although many factors involved in these winds have not been fully sorted out, some of them have been identified. Let's consider the chinook. As air moves from the Pacific to the Rockies and then up the windward side of the Rockies, the air dries by condensing out much of its water vapor. This change from vapor to liquid releases energy, and so the air is warmer than it would be without the change. As the air then moves over the Rockies and down the leeward slopes, it warms even more because it moves into progressively greater pressures. (You may notice the same effect when pumping the tire of a bicycle.) So, as the air reaches the base of the Rockies, it is warm and relatively dry and thus can quickly melt and evaporate any snow.

One researcher described how he drove a car from a cold valley up into a chinook that had picked up moisture by evaporating snow. Within seconds of driving into the winds, his cold windshield was covered with frost as moisture condensed onto it from the wind. Had he been driving at freeway speeds, his sudden inability to see through the windshield could have been disastrous.

4.88 • The boiling-water ordeal

An example of "magic" is the boiling-water ordeal that members of the Japanese Shinto have demonstrated. In this ordeal, a performer dips two clumps of bamboo twigs into boiling water and flings the water into the air, showering himself and the fire below the pot that holds the boiling water. Great clouds of steam pour from the fire when the water hits it, but the performer is unharmed. Why isn't the performer burned by the boiling water?

Answer The water flung in the air consists of many small drops. These drops cool rapidly because they carry a small amount of thermal energy that can be transferred quickly to their surfaces and then to passing air. When the drops land on the performer, they might be warm but will not burn the skin. If the same amount of water were to fly through the air as a single glob, it would lose less energy to the air because its surface area would be less than the total surface of the individual drops. So, upon landing, the glob would be hotter than the individual drops and could burn the skin. (Of course, if the performer were to pour the boiling water onto his skin, the water would probably not cool at all before landing and would certainly burn the skin.)

4.89 • Energy in a heated room

Suppose that you return to your chilly dwelling after snowshoeing on a cold winter day. Your first thought would be to light a stove. But why, exactly, would you do that? Is it because the stove would increase the store of internal (thermal) energy of the air in the dwelling, until eventually the air would have enough of that internal energy to keep you comfortable? As logical as this reasoning seems, it is flawed, because the air's store of internal energy would not be changed by the stove. How can that be? And if it is so, why would you bother to light the stove?

Answer A dwelling is not airtight (in fact, an airtight dwelling would not be safe). As the air temperature is increased by the stove, air molecules leave through various openings so that the pressure inside the dwelling continues to match the atmospheric pressure outside the dwelling. Although the kinetic energies of the remaining molecules increase, the *total* kinetic energy does not increase because *fewer* molecules are in the dwelling.

So why does the dwelling feel more comfortable at the higher temperature? You have a tendency to cool because (1) you emit infrared radiation and (2) you exchange energy with air molecules that collide with your body. If you increase the room temperature by lighting the stove, (1) you increase the amount of infrared radiation you intercept from the surfaces in the dwelling (walls, ceiling, floor, furniture, etc.), replacing the energy you lose through infrared radiation, and (2) you increase the kinetic energy of the colliding air molecules and you gain more energy from them.

4.90 • Icehouse orientation

Before the refrigerator was invented, people in northern climates would store winter ice in icehouses for use in the summer. One of the features required of a good icehouse was proper orientation: Reportedly, its opening should face the east, so that sunlight would pour into the icehouse soon after sunrise. Doesn't that seem incorrect, because wouldn't the direct sunlight warm the interior and thus melt the ice?

Answer The purpose in the icehouse orientation was to eliminate (or at least diminish) the influx of moisture-laden air. If such air were to enter the icehouse, it would condense on the cold surfaces of the ice. In order for water to condense, it must lose a large amount of thermal energy so that the water molecules can settle into a liquid state. That release of thermal energy into the ice would hasten the melting of the ice.

So, the building plan was to have sunlight enter the icehouse during dawn to warm the internal air and decrease the humidity and the chance of condensation. The condensation problem was probably worse during the night, but the Sun does not shine at night. So, using the morning sunshine was the best possible solution.

4.91 • A radiometer toy and its reversal

A radiometer was a device invented in 1872 by William Crooke to measure the energy emitted by a light source, but today it is a novelty or toy sold in science shops. Inside a sealed, partially evacuated glass bulb, four vertical metal vanes are attached to a metal hub that can rotate around a vertical needle. The vanes have the same arrangement of colors: white on one side and black on the other side. When the device is mounted near a light source, the vanes and hub rotate around the vertical needle, rotating faster for brighter light. What causes the rotation, what is its direction (does, for example, the black side of a vane lead), and how can it be reversed?

Answer The motion is often attributed to the pressure of light, but that effect is far too small to observe with the toy and, besides, it would yield a rotation that is opposite what is seen. Here is the argument: Light can push on object, and the push is greater if the light reflects from the object. Thus, light shining on the vanes will push on the white sides more than the black sides, and the vanes should rotate with the black sides leading. Were the bulb almost fully evacuated, the vanes would indeed turn like this.

However, the pressure on the vanes due to the residual air gives a *much* larger effect. Because light (infrared radiation and visible light) is absorbed more on the black side of a vane than on the white side, the black side becomes slightly warmer than the white side. Because the residual air molecules run into a vane, they push on the vane. The faster the molecule is moving, the greater the push. The air molecules on the black side of a vane move faster than those on the white side because of the temperature difference. Thus, the push on the black side is greater than that on the white side, and the vanes rotate around the support pin with the white side leading. After a while, the two sides of each vane reach the same temperature (they reach *thermal equilibrium*), and the effect disappears and the vanes stop rotating.

To reverse the motion, put the toy in a refrigerator. The black side of each vane loses thermal energy slightly faster than the white side via infrared radiation, and so the white side then has the higher temperature and the greater push from the air. Again, the rotation continues until thermal equilibrium is reached.

4.92 • Water wells and storms

When my grandmother was young, her drinking water was hand-pumped from a water well. She claimed that during stormy weather, the water was easier to pump but had too much suspended matter to drink safely. The result seemed to be independent of whether rain fell. Artesian wells also seem to be sensitive to the weather, flowing more freely during stormy conditions, but again the result does not depend on the rainfall. What makes wells respond to a storm?

Answer Although the general water level in a well is governed by the local rainfall or snowmelt, changes in the barometric pressure can vary the water level by several centimeters. When the barometric pressure drops during a storm, the well level will rise. The resulting increased water flow through the ground may pick up enough sediment to make the water unfit to drink.

The air in a cave system can also respond to changes in the barometric pressure: When that pressure drops, air flows out of the cave, and when the pressure increases, air flows into the cave. The motion is most noticeable in the flow through a narrow passage, because the airspeed there is greater.

4.93 • Insect and shrimp plumes

Why do insects (such as mosquitoes and flying ants) sometimes form a plume over a tree? These *insect plumes* can be so dense that they resemble smoke, as if the tree contains a small fire. Sometimes the plumes form over groves and steeples. On one occasion, a fire department rushed out to fight a church fire, only to find that the plume over the steeple consisted of only insects.

Why do shallow-water brine shrimp sometimes form a plume over an underwater rock lying in sunlight? Why does the plume, which can be quite thick, rise up from the rock but bend away from the direction of the Sun?

Answer In the early evening, the trees may not cool off as quickly as the surrounding ground and thus warm air may rise from them. The insects are apparently attracted to that warm air and possibly also to the moisture that may condense out of it as the rising air cools.

The brine shrimp rise in a similar convection column of water that is warmed by the sunlight. Although they enjoy the warmth and possibly the nutrients that might be carried along by the warm water, they do not like the sunlight, and so they veer away from the direction of the Sun as they rise. Once they reach the water surface, they swim back to the bottom and re-enter the convection column, to ride it again.

C • H • A • P • T • E • R 5

Ducking First a Roar and Then a Flash

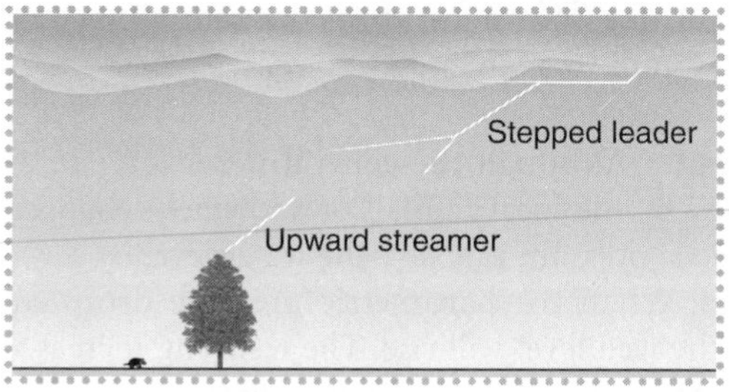

Stepped leader
Upward streamer

Figure 5-1 */ Item 5.1.*

5.1 • Lightning

What causes lightning, and why are sound and light produced? Since it can be seen from great distances, is lightning wide?

Answer Lightning is a very large electrical discharge (spark) between clouds and the ground. Although details of the discharge have been worked out and measured, the ultimate cause of the charges in the clouds and the triggering event for lightning are still not well understood. The standard explanation for the charges is that collisions between hail and smaller ice crystals transfer electrons to the hail, which then falls to the base of a cloud. Thus, because electrons are negatively charged, the base becomes negatively charged; because the top of the cloud lost those electrons, it becomes positively charged. A small amount of positive charge also lies somewhere near the base.

The ground normally has abundant electrons that can move about, but when a charged cloud is overhead, those electrons are driven away by the negative charge in the base of the cloud. Thus, because it loses electrons, the ground below the cloud becomes positively charged. That charge and the charge arrangement in the cloud produce a large electric field between ground and cloud. If the field exceeds a critical value, a discharge occurs, starting at the base of the cloud, where some of the electrons suddenly jump to the small amount of positive charge there.

A *stepped leader* then begins to snake its way toward the ground, *ionizing atoms* (removing the outer electrons) and bringing some of the negative charge down from the cloud. This snaking, which is too faint for you to see, is done 50 meters at a time (hence "stepped"), with many downward-forking routes. Although lightning usually appears to be vertical to an observer on the ground, the path is primarily horizontal. Only as the path nears the ground does the lightning seem to "notice" objects, such as a tall tree, on the ground.

Channels of ionized atoms develop upward from those objects. When one of these *upward streamers* happens to meet the descending stepped leader, completing a *conducting path* between ground and cloud, electrons near the ground are accelerated downward to the ground by the electric field along the path. This dumping of electrons to the ground, called the *return stroke*, proceeds rapidly up the conducting path, until the base of the cloud is reached. Because the electrons are accelerated, they strongly collide with air molecules along the path, knocking out electrons and greatly raising the temperature of the molecules. Because of the heating, the air expands so rapidly that it produces a shock wave, which is the "crack" of thunder. As freed electrons recombine with air molecules, the bright light of the lightning is produced. Although the lightning can be bright and powerful, the conducting path along which all this activity occurs is probably less than a centimeter in diameter.

Once a conducting path is established, a cloud can send down several pulses of electrons, as more electrons move from the rest of the cloud to the top point of the path. You might see these multiple pulses as a flickering lightning stroke. If a strong wind blows the path sideways during multiple pulses, you may see a bright "ribbon" of lightning rather than a single stroke.

Most lightning involves a downward-traveling stepped leader and a transfer of electrons from the cloud to the ground. However, a downward-traveling stepped leader can also start in the higher, positively charged region of a cloud, with electrons then being transferred from the ground to the cloud. And stepped leaders can also begin on the ground or, more likely, from tall structures like a skyscraper, and travel upward. Such a stepped leader traveling to the lower part of the cloud will transfer electrons to the ground, and one traveling to the upper part of the cloud will transfer electrons to the cloud. You can identify upward-traveling stepped leaders by the fact that they fork *upward. Spider lightning*, the beautiful, slow-moving, spread-out display of lightning that decorates the bottom of storm clouds, is usually due to cloud-to-cloud discharges during the last stages of a storm.

5.2 • Lightning: people, cows, and sheep

Why is a direct lightning hit usually fatal? Why are shoes and clothing sometimes ripped from a lightning victim? If a person is caught outdoors during a lightning storm, what can the person do to reduce the danger? For example, should the person hide beneath a tree or stand in an open field? Should the person stand still, crouch, or run? Why can a person's hair stand up, and is that a sign of danger?

How can groups of people, such as the players on the field during a baseball game, be in danger from lightning? After all, lightning could hit only one of the players, and yet sometimes all the players are knocked down.

In a lightning storm, why are cows, horses, and sheep usually in more danger than a person?

According to many stories, Benjamin Franklin, the famous American scientist and statesman, flew a kite during the approach of a thunderstorm in order to demonstrate the electrical properties of such storms. Why wasn't he killed by lightning?

Answer A person can be injured or killed by lightning in five basic ways.

(1) The most obvious way is through a direct hit by lightning, which can send a large amount of current (moving electrons) through the chest, stopping the heart, paralyzing the muscles required for breathing, and causing internal burns. If the victim is very wet, then much of the current might stay on the outside of the body and then the lightning hit might not be lethal.

(2) Injury can also occur if a person is touching an object, such as a car, that is struck by lightning. Some of the current is then diverted through the person.

(3) A person can also be hurt while standing near an object, such as a tree, that is struck by lightning. Part of the current can jump through the air to the victim in what is called a *side flash*. If the victim is lucky, the current may be too small to be lethal.

(4) A more subtle way of injury or death lies in the *ground current*, which is the lightning current along the ground. If a victim is standing with one foot closer to the strike point than the other, the ground current can detour up one leg, across the torso, and down the other leg (Fig. 5-2*a*). If the amount of current is small enough, the victim might be paralyzed only temporarily. Ground current can knock down a group of people, such as the players in a baseball game.

(5) The fifth way is even more subtle. As explained above, in common lightning a stepped leader snaking its way down from the cloud is met by a short upward streamer in which

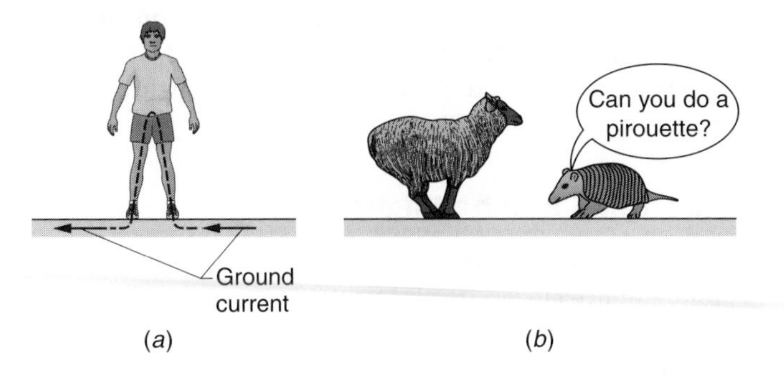

Figure 5-2 / *Item 5.2* (*a*) A ground current from a lightning strike can deviate up through the body because of separated feet. (*b*) A sheep guarding against a ground current.

air undergoes ionization. Once contact is made, the full discharge with its huge current occurs. Other upward streamers also occur without making contact with the stepped leader. Although the full lightning current will not be set up in those dead-end upward streamers, they nevertheless are channels in which electrons are ripped from air molecules. If such a streamer is set up through a person, the electron flow through the person can be lethal.

Burns on a person's skin are sometimes *dendritic* (fernlike in their forking, a pattern known as a *Lichtenberg pattern*), because the current spreads outward over the skin surface from some initial point. (Someone with a keen imagination might take the pattern to be a photographic image of a flower, a landscape, or their favorite religious leader, but lightning does not make photographs or draw religious pictures.) If the person is wearing metal, even a bra with wire supports, the metal temperature may increase enough to burn the victim. If the person's clothing and shoes are very wet, they can be ripped and blown off as the current heats the water, causing it to flash to vapor with an explosive increase in volume.

The best advice for a person caught outdoors during a storm is to move away from tall trees or other tall, conducting structures that lightning might seek, try to find a ravine or other low point, to lower the head so that lightning does not seek out the head, crouch, and pull the feet together to decrease the possibility of ground current through the torso. Running might be a good option, even though the head is kept high, because only one foot is on the ground at time. Cows, horses, and sheep are usually in more danger from ground currents than people, because their front legs and their hind legs are well separated, which can result in a greater amount of ground current directed across the body. People can stand with feet together, but sheep cannot (Fig. 5-2*b*).

When someone is knocked out by lightning, the heart often restarts automatically but the lungs do not. Thus, getting the victim breathing again with mouth-to-mouth resuscitation is extremely important. A defibrillating device is needed if the heart fails to restart or if it is in fibrillation.

Lightning can sometimes find a victim inside a house, entering via an external (and ungrounded) television aerial (but not "rabbit ears" on top of a television set), the telephone line (but not a cell phone or a telephone connected with fiberoptics), the plumbing, or the household wiring. Generally, during a lightning storm, play a quiet game of cards or talk to someone on your cell phone, and save the bath for after the storm.

Here is another safety point: Lightning may be most frequent as a storm first comes overhead, but it can still occur as the storm is leaving, and occasionally, less cautious people come out of hiding too early and are killed by lightning.

If a person's hair stands up, the electric field between ground and cloud is very strong and lightning could strike at any moment. Thus, such a person should hide immediately. (This most definitely is not a time to pose for a goofy-hair photograph. Run! Hide!) In the normal arrangement of charges, with the cloud base being negatively charged and the ground being positively charged, the hair strands all become strongly positively charged. Thus they repel each other and attempt to move as far apart from one another as possible, even if they must move upward against the gravitational force on them.

According to unverified stories from anglers, storm clouds might cause a fishing line to hover just above the water after being cast. If the stories are true, the line and the water surface must have the same sign of charge. The water could be charged because of the overhead clouds; the line could be charged for the same reason, or because it acquired charge when being pulled off the reel during the casting.

Benjamin Franklin was not killed during a kite-flying experiment because he never did the experiment. Only someone with very poor judgment would fly a kite with a storm approaching; Franklin was a very smart person. He did, however, give the impression that he had performed the kite-flying experiment.

5.3 • Lightning: vehicles

Why is a person in a car usually safe from lightning? Why is an airplane probably not safe from lightning?

Answer A car is a very good place to hide from lightning because the car body conducts electricity. So if the car is struck by lightning, the current will probably stay on the exterior. However, a convertible (with a nonconducting roof) offers little protection, and a car with a plastic body may offer no protection. A person in a car during a lightning storm should avoid touching the outside of the car or anything that is attached to an external antenna. Keeping the windows up so that they are covered with rain (which is conducting) might help. Normally a car has four poorly conducting tires, but those tires offer no insurance against the car being hit by lightning, which has jumped through several kilometers of poorly conducting air.

Because an airplane is made of metal, it too offers protection to the occupants. However, those airplanes made with nonconducting materials are somewhat like a convertible and offer less protection.

An airplane is, of course, more vulnerable than a car if it is flying, because sensitive electronic instruments required for the flight may be harmed or destroyed by the current directly or by the electromagnetic field momentarily set up by the lightning. If the current reaches the fuel tanks, either directly or through a stream of unburned fuel ejected from an engine, then the tanks may explode.

When an airplane is part of the conducting path of a lightning strike, the path taken by the current across the airplane usually depends on where the strike initially hits. If the hit is at the front of the airplane, the strike will probably run through the airplane and exit at the rear. If the hit is near the rear, then the exit will probably be nearby.

An airplane can also initiate a lightning discharge, even in clouds where there is no other lightning. For all these reasons, and because of the severe turbulence that occurs in a storm, pilots avoid electrical storms or any cloud system in which the airplane may trigger lightning. Still, most commercial airplanes are eventually hit by lightning.

5.4 • Lightning: trees, towers, and ground

Why can lightning blow apart a tree or set it on fire? How can tall buildings escape damage by lightning? How can lightning dig a trench or produce sand structures (almost sculptures) called *fulgurites*?

Answer When lightning strikes a tree, it can scar the tree, blow off chunks of bark, blow the tree apart, set the tree on fire, or do nothing to the tree. The damage depends on how wet the bark is and whether the lightning jumps to the sap. If sufficient current goes through the sap, the sap can be vaporized so rapidly that its expansion blows apart the tree. Rapid expansion of rainwater under a layer of bark can explode or crack the bark. Most lightning strikes to trees do not set the tree on fire, presumably because the current is so brief that the bark is not sufficiently heated to ignite. However, many strikes do cause fires (and thus forest fires), because they last as long as a second and thus significantly heat a tree.

Lightning can be initiated from a tall building if the building sends up a stepped leader to an overhanging cloud. The current in the strike is carried through lightning rods on the building or through the building's metal superstructure. When a building, such as a church with a prominent steeple, is struck by lightning without the protection of a lightning rod, the current can blow apart wet regions just as with a tree, and the wood can be set on fire if the current runs long enough.

When lightning strikes moist ground, it vaporizes the water so rapidly that the dirt is thrown off to one side, leaving a trench. The air blast from the sudden heating of the air in a lightning strike can also dig into the ground.

When lightning strikes quartz sand, the current can raise the temperature of the sand above the melting point. The sand soon cools, forming a thin fused cylinder along the tortuous path taken by the current through the sand. The resulting structure of fused sand is a fulgurite, which is highly prized when excavated intact from a beach.

5.5 • Bead and ball lightning

What produces the luminous spheres sometimes seen (and photographed) in electrical storms? *Bead lightning* is a string of bright spheres or somewhat elongated spots left in the sky by a lightning flash. *Ball lightning* is a more mysterious luminous ball, with a diameter of about 20 centimeters, that floats over the ground for several seconds. Some disappear quietly and others with an explosive pop. Ball lightning reportedly can travel through glass (from one side to the other) without harming the glass. It has been seen sliding over power lines and across indoor floors (from one wall socket to another). And it has been seen floating down aisles of airplanes, from one end to the other (which must make aisle seating a bit less popular). When a ball touches someone, the person can be stunned, knocked down, burned, or mentally impaired. If you see ball lightning, move away from it.

Answer There is no generally accepted explanation of bead lightning. Presumably, the beads are regions that remain hot, and thus luminous, after the rest of the lightning path has cooled too much to be luminous. Perhaps the residual hot spots are points where the path was kinked.

There is also no generally accepted explanation of ball lightning. Indeed, there are many theories, but none can predict the observed properties of ball lightning, especially its lifetime. A similar type of luminous ball, called a *plasma ball*, can be produced in a laboratory or power station when a powerful electrical discharge occurs. In the discharge, air molecules are ionized; that is, electrons are pulled out of the molecules, so that the region then has individual negative charges and positive charges. This state (*plasma*) will last for less than a second before the electrons and ionized molecules recombine.

Although that lifetime is considerably shorter than the several seconds or more attributed to ball lightning, the most enduring explanation of ball lightning is that it is a plasma ball produced by either a direct lightning strike or an upward streamer. Presumably, the discharge ionizes either air or the material (ground, lightning rod, etc.) at the lower end of the strike or streamer. However, if a plasma ball is produced by lightning, it must have a peculiar interior to last several seconds instead of quickly collapsing. In addition, it must not be terribly hot, because the ball does not rise as hot air would. It also cannot simply be St. Elmo's fire, the visible discharging that occurs at the points of conducting objects, because ball lightning moves and St. Elmo's fire does not. So far, we have no convincing model for the structure of ball lightning.

5.6 • Sprites

For decades, pilots flying near a thunderstorm at night occasionally reported seeing huge flashes far above the storm's clouds, just after seeing lightning below the clouds. However, the high-altitude flashes were so brief and dim that most pilots figured they were just illusions. Then in the 1990s, the flashes were captured on video and dubbed *sprites*. If sprites are associated with lightning that occurs between ground and clouds, why do they appear only high above the clouds and not immediately above them?

Answer Sprites are not well understood but are believed to be produced when especially powerful lightning occurs between storm clouds and the ground, particularly when the lightning transfers a huge amount of negative charge from the ground to the clouds. Just after such a transfer, the ground beneath the cloud has a complicated distribution of positive charge. The negative charge in the clouds and the positive charge on the ground set up an electric field above the clouds and between clouds and ground.

This field tends to ionize atoms and molecules in the air; that is, it pulls out electrons. However, ionization can occur only if the electric field exceeds a certain critical value, and that value depends on the air density. Immediately above the cloud, the electric field is strong, but the air density is too great for the air to become ionized. Much higher above the cloud, the field is somewhat weaker but the air density is much less, and ionization occurs. So, at those higher altitudes, the electric field not only frees electrons from molecules, but also accelerates them so that they then collide with other molecules, primarily nitrogen, causing the atoms to emit light. According to some researchers, a sprite is the collective emission of the molecules in such collisions. However, the complete sprite mechanism is likely to be more complicated than this collision model. In addition, researchers must explain the various shapes of sprites as well as *elves*, which are ring-like structures that expand away from sprites.

5.7 • Lightning rods

Does a lightning rod actually protect a building from a lightning strike and, if so, how does it give such protection? Does the rod increase the likelihood of a strike hitting the building? Should the upper end of the rod be sharp or blunt?

Answer The primary purpose of a lightning rod is to give lightning an easy path to the ground if the stepped leader of the strike comes close to the building. Thus, if the rod is to function, it must be connected to the moist, conducting region below the ground surface. The rod has no effect on where the stepped leader begins along the base of a cloud. In fact, it has no effect until the stepped leader comes down near the ground, and only then does an *upward streamer* (along which ionization occurs) travel up from the rod to meet the leader. The meeting completes an ionized and charged path between the ground and the cloud base. Hopefully, then, the current in the discharge does not enter the building or the building's walls, where it might electrocute the occupants or start a fire.

To function, a rod should extend above the highest point of the building. As a general rule, it then provides protection in a region resembling an inverted cone, with the apex at the tip of the rod. Any stepped leader coming into this imaginary cone will supposedly strike the rod instead of the building.

In the past, people argued that the upper end of a lightning rod should be sharp to attract lightning. The argument is based on the fact that a sharp end creates a stronger electric field than does a blunt end. Since a strong electric field might increase the chance of an upward streamer rising to meet a stepped leader, we might conclude that a sharp end is desired. However, a counterargument is that a sharp end increases the ionization of the air molecules around the rod, which decreases the chance of an upward streamer.

Experiments with lightning rods are difficult to conduct, because laboratory arrangements are never exactly like the natural arrangements, and natural arrangements depend on the chance occurrence of lightning. However, experiments suggest that slightly blunt ends are struck more often than sharp ends.

Because a lightning rod cannot influence the occurrence of lightning, it cannot promote the discharge of an electrified cloud. So, it does not bleed off charge from the cloud and make lightning less likely, as Benjamin Franklin, the inventor of lightning rods, had originally predicted.

5.8 • Sweaters, playground slides, and surgery rooms

If a person peels off a jacket or sweater while working at a computer, the computer can be destroyed. If a child slides down a plastic playground slide and then reaches for another person, the child can experience a painful surprise. If a surgeon does not wear the right type of shoes during surgery, the patient can be fatally injured. What is the danger in these situations, and why is the danger decreased if the humidity is high?

Answer When different materials come in contact, electrons can be pulled from one surface to the other, leaving the first surface positively charged and the second surface negatively charged. If the surfaces are rubbed together, more points come into contact and thus more charge is transferred. The abrasion caused by the rubbing can also increase the transfer.

Such transfers are called *triboelectricity* or *contact electrification*. If the air is humid, the surfaces are almost immediately neutralized by airborne water. But if the air is dry, the surfaces can be so highly charged that sparks will jump through the air from one charged surface to another. Often the sparking will be between a charged surface and conductor, such as another person or something metallic. If a person walks across certain types of carpet on a dry day and becomes, say, negatively charged, then a spark can jump between an outstretched finger and a metallic doorknob, or something electrically connected to the ground, such as a faucet or a computer keyboard.

The charging occurs when a shoe makes contact with the carpet and then moves up and away from the carpet as the person walks. The contact leaves the shoe with excess electrons. When the shoe moves away from the carpet, some of those excess electrons are driven by mutual repulsion onto the rest of the person's body. Thus, each footstep tends to increase the number of excess electrons on the body, putting the body at an electric potential of several thousand volts.

If the person touches another conducting object, at least some of the excess electrons move to that object. If the touch is with the broad part of the body, such as the back of a hand or the side of an arm, the electron transfer is over a large enough area that the person may be unaware of the transfer. However, if the person brings a finger near the conducting object, the person may be *quite* aware of the electron flow. Because a finger is a pointed object, the excess electrons on it can produce a strong electric field between the finger and the object. That field can even be strong enough to pull electrons out of the intermediate air molecules, producing a conducting path between the finger and the object. The excess electrons on the person can then easily move from the finger along the conducting path, in a spark that can be seen, heard, and felt. If you want to avoid sparking, use a broad portion of your body to make contact instead of a finger or reach out with a metal key so that the key takes the spark. (Sparking to a friend's ear lobe in a "sneak attack" is a sure way to end the friendship.)

Some types of fabric cause charge transfers when touching skin or other clothing. A sweater is notorious for producing sparks when pulled off in dry weather. A child sliding down a plastic slide can become so charged that the child may come off the slide at an electric potential of 10 000 volts. If the child reaches for another object, especially a grounded conductor, a very painful spark may jump between the child and object to neutralize the child.

Sparking can be a very serious danger in surgery. If any flammable vapor is present, a spark might ignite the vapor. However, the widespread use of flammable anesthetics was terminated in the 1950s, and so this danger has decreased. The sparking between two surfaces can also kill a person if one of the surfaces is in the interior of the body. Normally the skin offers a huge resistance to the flow of electrons and thus protects the heart. But if the electron flow is introduced directly into the conducting fluids inside a body, the flow across the heart might be enough to disturb the normal electrical regulation of the heartbeat. The possibility of such a *microshock* worries surgical teams so much that their clothing is chosen to reduce the chance of sparking, and their shoes are usually partially conducting so that charge bleeds off to the floor as quickly as it might be produced by the clothing. The floor is also partially conducting so that the charge can then move to a ground connection.

Office and industry workers who use computers or other sensitive electronic equipment often wear grounded wristbands to provide a conducting path between their body and the ground. The path is usually not highly conducting so that the charge on the person is reduced gradually rather than very quickly as in a spark.

5.9 • Cars, fuel pumps, and pit stops

When you exit certain cars and reach back to shut the door, why are you shocked? (Why aren't you shocked on all cars?) When you slow as you enter a toll-booth center, why is the toll-booth attendant reluctant to reach out for your money immediately?

Gasoline vapor is flammable, yet filling a car with gasoline at a gasoline station is relatively safe unless you are foolishly smoking or pouring the gasoline onto a very hot portion of the car. Nevertheless, some people have accidentally ignited fires as they pump gasoline into their car. In some videotaped examples, a person inserts the gas pump into the car gas port, sets the flow on automatic, and then climbs back into the car to get warm or to search for something. After a few minutes, the person climbs out of the car to manage the end of the pumping, but as the person reaches for the pump handle, the vapor ignites. Why?

Answer A moving car is charged by tire–pavement contact, which moves electrons from one surface onto the other, because the attractive electric forces on one surface overwhelm the attractive electric forces on the other surface. Let's assume that the electrons are pulled from the pavement onto the tire. Those electrons can then move through the tire and across the metal connections of the car, charging the car to a potential of 10 000 volts or even more.

If you stop the car, and thus also the charge transfer at the tire–pavement contacts, the charge on the car will bleed off through the tires. However, this discharging rate depends on how well the tires conduct. If they are made with conducting carbon black (held together by polymers), the bleeding is fairly rapid. If, instead, they are made with nonconducting silica (bonded with polymers), the bleeding may take a long time.

Suppose the tires on your car are fairly conducting, and so the car takes more than a minute to discharge significantly. Suppose also that you emerge from the car soon after it stops and touch only a plastic (nonconducting) handle to open the door. When you reach back to push the metal door closed, the abundant electrons on the car will jump across an air gap to reach your fingers, so that some of them can spread out over your body or move through your body to ground. Thus, a spark jumps between the car and you. This is a caffeine-free way to wake up after the morning drive to work or school. However, if you want to avoid the shock, either wait a few more minutes to allow the charge to bleed off or push the door shut with your foot or rear end. A blunt object, such as your rear end, decreases the chance for the ionization of air molecules that allows for a spark.

When you ride in a car, you might remain electrically neutral if you touch only plastic (nonconducting) parts of the interior. However, you are still charged by induction. That is, the movable electrons in your body tend to move away from the electrons that accumulate on the conducting parts of the car surrounding you. Suppose you drive up to a toll booth while charged like this. If you and the (grounded) toll-booth attendant immediately reach out to exchange money or a toll card, a spark can jump between you and the attendant as those electrons on your body attempt to get away from one another.

The spark is unlikely on humid days, because the airborne moisture quickly neutralizes the charge on both you and the car. So, on a dry day, the attendant will probably wait a few tens of seconds before reaching out to you, to allow the charge on the car to bleed off and the charge induced on you to decrease. However, if you have been waiting in a queue, the charges have probably disappeared by the time you reach the booth.

Some fires at gasoline stations have been caused by the gasoline flow into the car's gas tank because of a design feature of the gas tank entrance, but that design feature has now been fixed. The trouble was that the gasoline becomes charged when flowing through a pipe or tube. The gasoline immediately next to the wall, in what is called a *boundary layer*, does not move. As the rest of the gasoline flows past the boundary layer, electrons are transferred from the boundary layer to the moving gasoline. As a result, the boundary layer is left positively charged and the negatively charged gasoline flows into the gas tank.

If the tank is plastic and thus nonconducting, that negative charge collects on the interior surface and repels electrons in any nearby conducting part of the car. Those repelled electrons move away from the tank and some can end up near the gas hose. If they spark to the hose, the spark can ignite the gasoline vapor released during the gasoline flow. To fix the problem, the hose connects the car to ground so that charge cannot collect on the car near the hose.

A person can become charged in a car even if the car is stationary because the contact between the person's clothing and the car-seat material can result in a large charge transfer. Suppose a person begins fueling a car with the pump set on automatic and then climbs back into the car for some reason. After sliding on the car seat, the person may return to the pump with a lot of charge. If a spark jumps between the person and the pump, it can ignite the gasoline vapor and start a fire. A person can avoid the danger by not going back into the car or by neutralizing the charge by touching a metal pole before touching the fuel pump.

A race car makes a pit stop after traveling very fast and thus after accumulating a lot of charge due to the contact of tire and pavement. The pit crew often must immediately start fueling the car, either with a hose or an upturned fuel container. The process quickly produces fuel vapor at the fuel port on the car. To avoid sparking into that vapor, which could be disastrous, the car either is grounded as soon as it stops (a long conducting rod can be extended to the metal framework of the car) or is fitted with tires that conduct well, so that the charge rapidly bleeds off through the tires. The latter solution is not always desirable, however, because tires that conduct (recall that they contain carbon black) tend to wear out faster than tires that do not conduct (they contain silica).

SHORT STORY

5.10 • Shocking exchange of gum

Here is a classic story from a 1953 physics journal; the stunt it describes is dangerous—don't you dare try it. A professor was driving fairly slowly when two of his friends drove up alongside him, matching his speed. The friend in the passenger side of the second car reached out (while the car was moving) to hand a package of gum to the professor. When the separation between their hands narrowed to a few centimeters, a "terrific discharge" occurred between them, momentarily disabling them. Luckily, the professor did not veer into the second car before he could regain his senses and the control of his car.

The spark occurred because the motion of the two cars charged the professor and the other person by different amounts, perhaps even with charges of different signs. When their hands were close enough, electrons from one hand jumped through the air to reach the other hand, to reduce the charge difference.

5.11 • Danger of powder floating in the air

Why is an electrostatic spark dangerous in airborne powder, such as in a coal mine or flour mill?

Answer In bulk, the powder particles may not burn at all, but when they are airborne, each particle is surrounded by air and thus has an abundant air supply that allows it to burn very quickly. In fact, once burning begins at any point in the dust, the thermal energy is transferred through the dust (from one grain to the next) so rapidly that the dust explodes. That is, a lot of energy is released in an uncontrolled way, with a rapid increase in temperature and pressure. In a grain silo, the explosion may simply destroy the structure; in a coal mine, it can kill the miners. In spite of today's precautions, such explosions still occur frequently.

In some situations, a spark from an electrical discharge can trigger the explosion if it provides sufficient energy. The spark could be associated with malfunctioning electrical equipment, but more likely it is due to two charged objects discharging to each other, or to a charged object discharging to the ground at some grounded point.

For example, in the 1970s, an explosion occurred in chocolate crumb powder as it was blown through plastic pipe into a silo. As the powder grains were shaken from bags into the pipe system, and as they touched one another and the pipe walls while moving through the pipe system, the grains became charged. When they shot from the last pipe into the silo, a spark jumped between the grains and some grounded point in the silo. The spark might have started with the airborne grains as they fell to the mound that had already formed in the silo. Alternatively, it may have started at the tip of the cone formed on the mound, as grains slid down the slopes of the mound. (The electric field was strongest at or near that tip, so sparking, in which a strong electric field ionizes air molecules, could have started there.)

Actually, sparking probably occurred frequently in the silo but without enough energy to ignite the grains. The explosion came when, by chance, the energy of a spark (or perhaps the combination of several, nearly simultaneous sparks) was above the minimum necessary value for an explosion. Engineers cannot eliminate static charge and sparking in powder industries. Instead, they try to keep the sparking energy below that minimum necessary value.

5.12 • Danger of aerosol cans

If a spray of dry powder or liquid from an aerosol can hits an open flame, such as in the kitchen, why can the spray become a "flame thrower"? (Never spray near an open flame, because you can end up setting yourself and the room on fire.) Why do some aerosol cans emit flames even when the spray is not near an open flame?

Answer The particles in the spray might be flammable, and the high speed of their release from the can sets up the effective flame thrower. If the can shoots out a dry powder, the powder grains and the can itself can become charged. If the can is not touching something conducting, such as a person, the charge can build up until there is enough charge for the can and sprayed powder to spark to each other. If that spark provides enough energy, the power will be ignited. However, if a person is touching the can, as is likely, then much of the charge on the can spreads to the person, and not enough is left on the can to cause a spark.

5.13 • Danger of spraying water

When the shower in a typical bathroom is turned on and the door is closed, why can a strong electric field build up in the room's air? What produces the strong electric fields found near waterfalls? In times past, the cargo tanks of crude-oil ships were cleaned by spraying water into them under pressure. Why did the practice sometimes cause the tank to explode? To prevent such explosions, the amount of oxygen in the tank was decreased by pumping an inert gas into the tank. Why did tanks still explode?

Answer When water hits a solid surface and creates a spray, as it does at the base of a shower, the drops become charged: The larger drops usually become positively charged (they lose electrons), and the smaller drops usually become negatively charged (they gain those lost electrons). Because the larger drops fall out of the air rather quickly, only the smaller, negatively charged drops are left floating in the air. If the ventilation is low, the number of charged airborne water drops can increase dramatically, producing a large electric field, but this situation presents no danger in a bathroom or near a waterfall.

During the cleaning of an oil tanker, water drops become charged when they leave the hose as a spray and when they splash on the tank floor or walls, filling the tank with a charged water mist. Sparks can then jump between these airborne charged particles and either a large conductor, the spray nozzle, or a grounded point. If the tank still contains vapors from the crude-oil cargo, the sparks can ignite the vapor, causing it to explode.

One solution to this danger is to pump an inert gas into the tank before the cleaning, so that the oxygen supply might be too low for an explosion. However, the procedure was initially flawed, because the generator supplying the gas caused charge separation within the gas (the gas became charged). Until the problem was recognized and corrected, tanks were still exploding during cleaning.

5.14 • Ski glow

What causes the glow of skis that is sometimes noticed by nighttime skiers?

Answer When a ski rests on snow, charge is transferred between the ski and the snow. The actual mechanism of transfer is generally called *contact electrification* or *triboelectrification.* Although the mechanism is complicated, we can simply say that electrons (the ones free to move) can be pulled (transferred) from one surface to the other surface.

Let's assume that the ski is not metallic and does not contain metal staples. Then, as charge accumulates on the bottom of the ski because of the electron transfer, the ski material becomes electrically polarized. That is, the positive and negative charges within the molecules become slightly separated. The result is an electric field across the thickness of the ski, with the top and bottom of the ski having charges of opposite signs. For example, if the ski pulls electrons up from the snow, the bottom of the ski becomes negatively charged and the top of the ski becomes positively charged. (The ski is then a capacitor.)

When a ski slides over snow, this charging effect is much more pronounced, and many small sparks can jump between the snow and the top or bottom of a ski. At night, with dark-adapted eyes, a skier might see some of those sparks.

5.15 • *Hindenburg* disaster

The pride of Germany and a wonder of its time, the zeppelin *Hindenburg* was almost the length of three football fields—the largest flying machine that had ever been built. Although it was kept aloft by 16 cells of dangerously flammable hydrogen gas, it made many trans-Atlantic trips without incident. In fact, German zeppelins, which all depended on hydrogen, had never suffered an accident due to the hydrogen. However, on May 6, 1937, as the *Hindenburg* was ready to land at the U.S. Naval Air Station at Lakehurst, New Jersey, the ship burst into flames.

Its crew had been waiting until a rainstorm partially left the area, and handling ropes had just been dropped from the ship to a Navy ground crew, when ripples were sighted on the outer fabric of the ship about one-third of the way from the stern. Seconds later a flame erupted from that region, and a red glow illuminated the interior of the ship. In about 30 seconds the burning ship fell to the ground, killing 36 persons and burning many others. Why, after so many safe flights of hydrogen-floated zeppelins, did this zeppelin burst into flames?

Answer As the *Hindenburg* was ready to land, and after the handling ropes had been dropped to the ground crew, the ropes became wet (and thus able to conduct a current) in the rain. They *grounded* the metal framework of the zeppelin to which they were attached; that is, they formed a conducting path between the framework and the ground, making the electric potential of the framework the same as that of the ground. This should have also grounded the outer fabric of the zeppelin, except that the *Hindenburg* was the first zeppelin to have its fabric sealed with several different layers having large electrical resistance (they conducted current very poorly). Thus the fabric remained at the electric potential of the atmosphere at the zeppelin's altitude of about 43 meters. Because of the rainstorm, that potential was large relative to the ground.

This was a dangerous situation: The fabric was at a much different potential than the framework of the zeppelin. Apparently, charge flowed along the wet outer surface of the fabric and then sparked inward to the metal framework of the zeppelin. There are two basic arguments about how the sparking caused a fire. One argument is that the spark ignited the sealing layers. The other argument is that one of the handling ropes ruptured a hydrogen cell, which released hydrogen between that cell and the zeppelin's outer fabric. (The reported rippling of the fabric supports this argument.) The sparking then ignited that hydrogen. Either way, the burning rapidly ignited the hydrogen cells and brought the ship down. If the sealant on the outer fabric of the *Hindenburg* had been more conducting (like that of earlier and later zeppelins), the *Hindenburg* disaster probably would not have occurred.

5.16 • A gurney fire

Often, a burn victim is treated while lying on a gurney in an enclosed chamber filled with oxygen-enriched air. Once a treatment session is over, a hospital worker pulls the gurney and patient from the chamber onto a trolley, to be rolled away. On at least two occasions, the gurney caught fire at the end that was last to leave the chamber. A burning gurney holding a patient already suffering from burns is a dangerous situation, and obviously, fires burn easily in air rich in oxygen, but the question remains: What caused the gurneys to catch fire?

Answer Investigators soon realized that charge separation occurred between the patient and the gurney. Let's assume that the patient lost electrons to the gurney, which thus became negatively charged. Some of the electrons in the metal framework below the gurney were then pushed away, leaving the top of the framework positively charged. This arrangement of a negatively charged gurney and a positively charged framework then resembled a capacitor, the electrical device used to store charge in an electric circuit.

Could a spark between the gurney and the framework ignite the gurney? Apparently not, for two reasons: (1) The electric field between the two was insufficient to ionize atoms (to remove electrons from atoms so that there was a conducting path along which electrons from the gurney could move to the framework). (2) The energy associated with the charges was insufficient to cause a fire.

However, the situation changed as the gurney was pulled off the framework, because the charge on the gurney was then forced into a decreasing area to be near the charge on the metal framework beneath it. This concentration of charge increased the electric field and the associated energy until a spark erupted between gurney and framework, with enough energy to ignite the gurney.

5.17 • Glow in peeling adhesive tape

After dark-adapting your eyes for about 15 minutes, peel tape from its dispenser roll at a steady pace. What causes the faint glow along the line at which the tape separates from the roll? If you peel the tape near the antenna of a radio tuned to an unused (empty) channel, why does the peeling generate noise on the radio? Why can high humidity eliminate both the glow and the radio noise?

Answer As the adhesive on the tape fractures and then detaches from the tape still on the roll, charged particles (positive ions and electrons) collect into patches on the two surfaces. These patches tend to neutralize one another by sparking before the surfaces get too far apart, either by jumping between the surfaces or along one of them. Because air flows into the widening space between the two surfaces, the sparks jump through air. The sparks primarily consist of electrons, and the air primarily consists of nitrogen molecules. So, as they jump from a negative patch to a positive patch, the electrons tend to collide with nitrogen molecules, exciting them. The molecules almost immediately de-excite by emitting light in the blue end of the visible spectrum (in addition to ultraviolet light). The faint glow seen along the line of separation is the composite of light emitted by the de-exciting nitrogen molecules and by the sparks.

The sparks also emit in the radio-frequency spectrum. So, when tape is peeled near a radio antenna, the antenna picks up some of the radio-frequency emissions. The intensity of the radio noise is roughly proportional to the intensity of the visible light.

High humidity provides moisture that flows into the line of separation with the air. The moisture neutralizes charged patches on the tape, eliminating sparking.

In the days when all photographs were taken with film, sparking was a major nuisance in the film-processing business. When the film unrolled from rolls or passed over rollers, sparks occurred as the surfaces separated. The film was exposed wherever the sparks emitted light, and thus a pattern of sparks could be seen when the film was later developed (which is not what anyone wanted on a family photograph).

5.18 • Parsley, sage, rosemary, and thyme

If sticky tape is peeled from a plastic surface and a fine mixture of two powders is blown lightly over the region where the tape had been, why do the two types of powder separate, with one collecting in certain regions and the other collecting in other regions?

The lightly blown powder mixture can also reveal something about a spark, such as the spark that might be generated when a person walks on certain carpets and then brings a fingertip up to a large metal object or plumbing. A square of Mylar is first taped along its edges to, say, a metal cabinet. Then a person walks over the carpet to become charged (not all carpets will do this, and high humidity can ruin the experiment). If the charged person brings a fingertip (or metal key held in the hand) up to the Mylar, a spark suddenly jumps across the gap. Why is the direction of electron flow in that spark revealed if the Mylar is then blown lightly with a mixture of two fine powders?

Here is how the powders can be applied: They can be finely ground herbs or a toner powder from a copy machine. (Caution: The powders can be messy and can ruin clothes and computers.) Choosing powders with different colors makes the separation more apparent. The powders are put into a flexible container along with metal bolts, and then the container is shaken vigorously so that the bolts mix up the powders. If the container has a nozzle at its opening, a quick squeeze of the container causes the powder mixture to be blown over the plastic surface. Only a brief puff of powder is needed. If a squeezable container is not available, the powder mixture can be sprinkled (lightly) well above the surface to allow the powder to drift down onto the surface, and then the surface can be tilted and gently tapped to remove any excess powder.

Answer When certain powders are shaken together, the contact between the different types of grains causes a separation of charge. That is, one type will gain electrons at the expense of the other type. When the sticky tape is peeled from a nonconducting surface, it leaves patches of negative charge and patches of positive charge. The charge will soon decrease because of air humidity. (It also decreases if the "nonconducting surface" actually conducts somewhat.) However, if the powder mixture is dusted lightly over the region so that the grains drift down onto the charged regions, the negative grains will collect in the positive patches and the positive grains will collect in the negative patches. If the powders have different colors (such as black toner and brown cinnamon), the patches are visible. Some powder mixtures work better than others. For example, crushed paprika and photocopier toner work well together, but paprika and flour attract each other so strongly that they virtually ignore the charged patches on the dusted site and settle out uniformly.

When a spark jumps between a fingertip and Mylar that is stuck on a large conductor (such as a metal shelf), the charge flow leaves charged regions in the Mylar, at least until air humidity neutralizes those regions. If a mixture of cumin and toner powder is blown lightly over the Mylar, it will reveal one of two general types of patterns, called *Lichtenberg patterns* after George Christoph Lichtenberg, who discovered them in 1777.

If the person has become negatively charged by walking on the carpet and thus has an excess of electrons, electrons jump from the fingertip to the Mylar and produce a circular spot of negative charge on the Mylar, centered on the spark (fine radial lines might be visible). If, instead, the person has become positively charged and has too few electrons, electrons break away from atoms in the Mylar, flow along forked lines to the spark point, and then jump from that point to the fingertip. The forked lines are left positively charged. So, if dusting the Mylar reveals a negatively charged circular spot, the person was negatively charged, and if it reveals positively charged forked lines converging to a point, then the person was positively charged. Some lightning victims have the forked-lines version burnt into their skin by the lightning.

Some scientific supply stores sell beautiful forked-line patterns that have been produced in small cylinders or slabs of Plexiglas. To produce the pattern, the Plexiglas sample is moved through the beam of an *electron accelerator* (a device that electrically accelerates electrons to a fairly high speed); the electrons stop within the Plexiglas and are trapped there. The Plexiglas is then (immediately) put on a grounded plate while a grounded, pointed conductor is pressed against the opposite side of the Plexiglas. The high concentration of electrons within the Plexiglas produces a fairly large electric field, especially at the pointed conductor where a spark occurs. The high temperature produced by the spark carbonizes the Plexiglas along the spark path, creating a conducting path. The electric field then extends from this path out into the rest of the Plexiglas. Sparking occurs along these new lines of electric field, producing more carbonized paths, until the electrons left by the accelerator are drained to the pointed conductor. The collection of carbonized paths forms the tree-like, branching structure seen inside the Plexiglas.

5.19 • Wintergreen glow in the closet

You and a friend first adapt your eyes to darkness for about 15 minutes in a closet or outside on a moonless night. Then have your friend chew a wintergreen LifeSaver candy (a candy in the shape of a marine lifesaver and infused with the oil of wintergreen) with the mouth as open as possible so that you can see inside. Why does each bite initially produce a faint flash of blue light, and why do later bites fail to produce the light? (If you don't want to eat the candy, squeeze it with a pair of pliers until it fractures.)

Why does tonic water have a faint blue tint?

Answer Whenever a bite breaks one of the mint's sugar crystals into pieces, the pieces will probably end up with different charges. Suppose a crystal breaks into pieces *A* and *B*, with *A* negatively charged and *B* positively charged (Fig. 5-3). Some of the electrons on *A* will then jump across the gap to reach *B*. Because air has moved into the gap after the crystal breaks, these electrons jump through air. Those that collide with nitrogen molecules in the air transfer energy to the molecules, exciting them. When the molecules de-excite, they emit in the ultraviolet range, which you cannot see. However, the wintergreen molecules on the surface of the candy pieces absorb ultraviolet light and then emit blue light, which you *can* see—it is the blue light coming from your friend's mouth. This process of absorbing light in one wavelength range (here, in the ultraviolet) and then emitting light in a longer wavelength range (here, blue light) is called *fluorescence*.

The quinine in tonic water is like wintergreen oil in that it absorbs ultraviolet light and then emits blue light, giving tonic water its faint blue tint. You can see the tint better if the tonic water is near a fluorescent bulb in an otherwise dark room. The quinine will then convert some of the ultraviolet light from the bulb into blue light. The effect is decreased if the illumination is through a plastic or glass wall such as with a bottle of tonic water, because plastic and glass absorb ultraviolet light. The effect is increased if you illuminate the tonic water with a black light (ultraviolet) lamp.

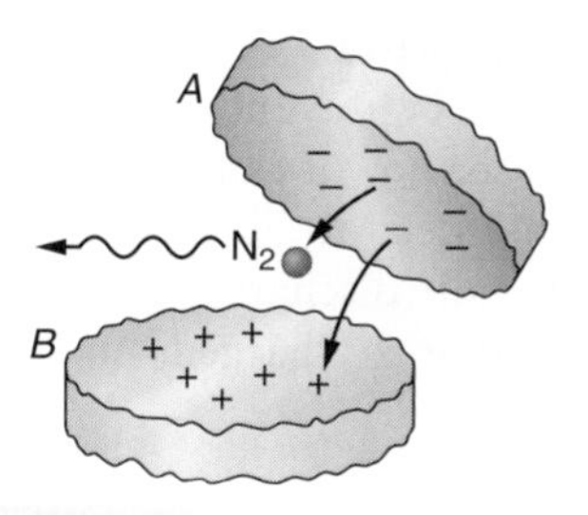

***Figure 5-3** / Item 5.19* Two pieces of a wintergreen LifeSaver candy as they fall away from each other. Electrons jumping from the negative surface of piece *A* to the positive surface of piece *B* collide with nitrogen molecules (N_2) in the air.

5.20 • Earthquake lights

In some regions, earthquakes have reportedly left the night sky red or been accompanied by luminous regions on the ground or luminous objects moving through the air. What could cause these lights, which are collectively called *earthquake lights*?

Answer The earthquake lights are still highly controversial, in spite of hundreds of reports and a number of reliable photographs of the lights. The lights may come in more than one type and could have more than one cause. Of the many explanations about the lights, here are two: (1) Light can be emitted when rock is put under enough stress to rupture; the rupture emits fine dust, gas, and free electrons. Presumably the electrons can then excite air molecules, causing them to emit light. (2) The earthquake motion could release flammable gases trapped underground, and the lights are the emissions as these gases are ignited, supposedly by sparking between charged surfaces or particles.

5.21 • St. Elmo's fire and Andes glow

What causes the electric sparks sometimes seen at the top of a ship's mast or at the tips of other slender objects? The sparks are called *St. Elmo's fire* or *corona*. What causes the very rare glow seen on the distant peaks of the Andes at night?

Answer St. Elmo's fire is due to electrical breakdown of the air adjacent to fairly sharp, conducting objects such as a ship's mast, an antenna, or an airplane wing. When the electric field in the air is stronger than normal, it can be very strong at the tip of a conducting object, where charges in the object can gather. If the field strength in the air next to the tip exceeds a critical value, the field can pull electrons out of the air molecules and accelerate them. When these electrons collide with air molecules, they excite the molecules and also make them move faster. The eventual de-excitation of the molecules produces light that can be seen. The increased motion of the molecules means that the temperature of the air has increased, which can lead to the sizzling or hissing sometimes heard with the discharge. St. Elmo's fire is not considered to be dangerous.

The Andes glow is not understood and sightings are very rare. I don't see how it can be St. Elmo's fire, because an observer would need to be fairly close to see the light from such small-scale discharges. More likely, it is larger-scale discharges due to charged, blowing snow on the mountain peaks.

5.22 • High-voltage lines

Why is electrical energy transmitted at high voltage and low current rather than high current and low voltage? (Because the power is the product of the voltage and current, the power could be the same in the two situations.) Why is the transmission done with alternating current (AC, in which the current alternates in size and direction) rather than direct current (DC, in which the current does not vary)?

When a high-voltage power transmission line requires repair, a utility company cannot just shut it down, perhaps blacking out an entire city. Repairs must be made while the lines are electrically "hot"—that is, still active. The repair technique involves a helicopter hovering next to the high-voltage line while a technician sits on one end of a platform attached to the landing frame beneath the helicopter. How does the technician avoid being electrocuted while reaching for the line and then holding it?

In some regions, high-voltage lines are a major threat to bird populations. Obviously, the birds can be hurt or killed if they fly directly into a line. But what is dangerous about perching on a line or the pole or tower supporting a line?

Answer When electricity is sent through a line, part of the electrical energy is lost to thermal energy, as the electrons (which make up the current) collide with atoms and molecules along the conducting route. The amount of electrical energy lost this way is equal to the product of the line's resistance and the square of the current. Thus, to keep the loss low, electrical energy is transmitted at low current. To meet a certain power demand, this means that the voltage must be high, 765 000 volts for example. At the distribution point where the energy is shuttled off to a home, a transformer changes the electricity to a lower voltage (which is safer) and a higher current (which can be limited by circuit breakers and fuses).

The original electrical supply in the United States was DC and came from the company of Thomas Edison. Later, an AC supply was offered by George Westinghouse. The competition between the men was rather fierce, each trying to demonstrate that his method of transmission was safer than the other method. Representatives of Edison had several public demonstrations in which they brazenly electrocuted dogs to show the danger of AC. However, Westinghouse eventually won the competition, largely for a practical reason. He could transmit at high voltage and then use transformers to switch to low voltage at the homes. Edison, on the other hand, could not transmit at high voltage and thus would have needed to build an electrical generating plant every four or five kilometers, which was clearly impractical.

When a technician approaches a "hot" high-voltage line to repair it, the electric field surrounding the line brings the technician's body to nearly the electric potential of the line. To match the two potentials, the technician then extends a conducting *wand* to the line; a spark jumps between the line and the outer end of the wand, which can numb the arm for a while. To avoid being electrocuted, the technician must be isolated from anything electrically connected to the ground. To ensure that the body is always at a single potential—that of the line being worked on—the technician wears a conducting suit, hood, and gloves, all of which are electrically connected to the line via the wand.

A bird may be able to perch safely on a high-voltage line because its resistance to current is greater than the resistance of the section of line between its feet. However, if a large bird lands near enough to a grounded part of the support pole or tower, it can "short out" the line in what is called a *flashover*—current flows from the line through the bird and to the ground, killing the bird.

Although that type of flashover is possible, a more likely type involves bird droppings (the urine–feces mixture excreted by a bird). If the bird perches on part of the grounded support pole or tower, such as a cross arm, under which the line is strung, then any fluid excretion can electrically connect the bird to the line, causing a flashover. Droppings can be a problem even if they are not particularly fluid because they can build up with time. Then, because of rain, freezing rain, or melting snow or ice, a water stream can electrically connect the droppings to a line. Such electrical connections are already a problem where there is plenty of snow and ice, but the bird droppings make the problem even worse because of the water's ability to conduct electricity increases when it absorbs ions from the droppings.

5.23 • Current, voltage, and people

Which of these can injure or kill a person: current or electric potential (voltage)? How is a person harmed? What accounts for the danger of working with electrical appliances on wet floors, something we are all cautioned against?

Answer Harm is done to a human body by the current (the flow of electrons) through the body. Electric potential is related to the possible flow of current and can be linked either to the energy that is available for the flow or the force causing the electrons to move.

For example, in the United States, if one hand touches a live (electrified) household wire while the other hand touches a connection that is *grounded* (electrically connected to the ground), the potential difference between the two hands is 110 volts, which can drive current between the two hands. However, the amount of current also depends on the body's electrical *resistance* to the current. Usually the resistance is primarily due to the skin, and dry skin has high resistance. So, when an electrician accidentally "grabs" 110 volts between the hands, the high resistance of the skin may keep the current below the lethal amount.

However, if the skin is wet, has open sores, or is covered with conducting gel, the current meets little resistance, and a dangerous amount of current can exist in the body. Similarly, if a person stands on a wet floor while touching a live wire (or an ungrounded electrical appliance), a dangerous amount of current can exist between the hand and the feet.

Although the response to currents through the body differs between people, genders, and applications of direct current (DC) and alternating current (AC), here are some general responses:

Less than 0.001 ampere of current: no perception

0.001 ampere: tingling or sensation of heating

0.001 to 0.010 ampere: involuntary muscle contraction, pain

0.10 to 0.50 ampere: ventricular fibrillation

0.50 to a few amperes: heart stops but can restart if current stops

More than a few amperes: heart stops, no breathing, burns

If the current causes only involuntary muscle contraction, the initial contraction might just be painful. However, if the victim is unable to let go of the current source, the body's resistance can gradually decrease, allowing progressively more current through the body and increasing both the pain and the danger. If a rescue is attempted by pulling the victim off the source, the rescuer could also end up with involuntary muscle contraction and be "frozen onto" the first victim, facing the same gradual increase in current.

If the heart has been put into ventricular fibrillation, its uncoordinated, random contraction and expansion does not pump blood, with dire consequences to the brain. A rescue team with a defibrillation device is needed very rapidly.

If current stops the heart, as is actually done with a defibrillation device, the heart can restart itself. However, the breathing, which is stopped by the contraction of the chest muscles, may not restart on its own. The victim will then need mouth-to-mouth resuscitation to restart the breathing before oxygen starvation damages the brain.

Burns are caused by collisions of the moving electrons (which comprise the current) with the atoms and molecules along the current's pathway through the body. If the burns are exterior, they might be repaired, but internal burns are difficult to treat.

SHORT STORY

5.24 • An act of indiscretion

Late one night, Dr. Milton Helpern, the Chief Medical Examiner for New York City, received a phone call about a distraught family: One of their members had died that night in a New York subway station, apparently by throwing himself from the waiting platform down onto the rails. In that system, the third rail is "hot," acting as the source of electricity that drives the subway trains. Apparently the victim was electrocuted by bridging the third rail with at least one of the other two rails, thus grounding the third rail and allowing a large current through his body.

At the family's request, Dr. Helpern conducted the autopsy of the victim but could find no physical evidence of stroke or heart attack that would suggest the victim collapsed accidentally onto the rails. He did, however, find curious burns on the victim's thumb and first finger of the right hand, as well as on a private part of the body.

Dr. Helpern then began to investigate the man's background, discovering that the victim would become belligerent when drunk. To display that belligerence, he would often urinate in public. Dr. Helpern concluded that the victim's last act of belligerence was to urinate from the waiting platform down onto the rails, including the third rail, not realizing that urine is a good conductor of current. The current through the victim left the burns found on the thumb, first finger, and the private part of his body.

5.25 • Use of current in surgery

Electrosurgery is a medical procedure in which a narrow conducting probe applies a high-frequency alternating current to a patient. It allows a surgeon to make an incision while also coagulating the exposed blood vessels (by heating) so that no unnecessary bleeding occurs.

The electrode (and thus the region of incision) must be part of a complete circuit in order for the current to exist. In one type of procedure, the circuit consists of the probe, the patient, and an electrode placed under the patient. In the early days of this procedure, patients ended up being badly burned. Can you see the (perhaps obvious) reason?

What is the danger when the electrode is applied to an organ that is attached to the body via a slender stalk, as might occur when electrosurgery is used in circumcision?

Answer The current is purposely concentrated at the point of an incision, but it should be allowed to spread over a much larger region where the electrode is placed under the patient. Otherwise, the current will burn the body at its contact with the underneath electrode. So, that underneath electrode is broad (to spread the current) and is fitted so as to make good contact with the body, not contact at only a few places, or contact only next to bony regions where the current might be concentrated. In the early days of this procedure, these safety precautions were not taken and patients were badly burned.

When the electrode is applied to an organ on a stalk, the current tends to be concentrated at the base of the stalk, a situation known as *current crowding*. Thus, the base of the stalk can be rapidly heated and destroyed, and this occurred in several tragic operations before the danger of current crowding was understood.

5.26 • Surgical fires and explosions

Surgical teams take extraordinary precautions to ensure that fires and explosions do not occur near or within a patient undergoing surgery. Prior to the 1950s, the flammable anesthesia that was used was a serious risk. Since then, the frequency of fires and explosions has decreased but is still not zero. Here are two fairly recent examples:

Tracheostomy: When a very overweight man underwent surgery for obstructive sleep apnea (airway obstruction associated with snoring), a tracheostomy was performed to provide oxygen to the patient via an incision through the trachea in the neck. In this procedure, an incision is made at the trachea, and a tube (the endotracheal tube) is inserted to carry 100% oxygen to the patient. However, the thick layer of fat on the man's neck made the tracheostomy difficult, and blood continued to ooze. At one point, one of the bleeding vessels near the tracheal incision was fused closed with electrocoagulation, in which a high-frequency alternating current was applied to heat the vessel. The region near the incision immediately ignited, producing a flame that shot upward from the neck by 0.5 meter. The flame was put out by covering it with surgical sheets and then dousing the remaining flame with a nearby saline solution. What caused the flame?

Polyp removal: During colonoscopy, a colonoscope is inserted through the rectum to search for and remove polyps in the colon. When one is found, it is snared and removed by heating it with current sent through the snare. The point of attachment to the colon wall is then cauterized with current to stop any bleeding. During the cauterizing phase of one such routine operation, there was a loud explosion, a blue flame shot out of the free end of the colonoscope for almost a meter, and the patient screamed and tried to leave the table. What caused the explosion?

Answer Tracheostomy: The current used in the electrocoagulation heated the abundant fat near the incision point, which was bathed in 100% oxygen. The fat quickly caught fire. In other cases where fires have occurred in tracheostomy or in surgical work involving the mouth, nose, or throat, an electrical heating device or a laser has ignited sections of plastic used in the procedure. (Plastic can easily burn when in 100% oxygen.)

Polyp removal: The human gastrointestinal system produces hydrogen and methane gas, which are flammable and explosive. For example, 40% of the gas in the large intestine may be hydrogen and methane. As many young men know, the gas expelled from the intestines is flammable, which leads to an amusing demonstration. If electrical cauterization is performed in hydrogen, methane, and oxygen, the heating (or sparking) can cause the gases to explode, burning and rupturing the intestines. Thus, any such surgical procedure requires that the intestines be empty, and so the patient fasts for up to a day. If there is still concern, the surgeon may flush the intestine with a nonflammable gas prior to the procedure.

Flammable gases can also be produced in the stomach when it does not empty properly (the pyloric valve is too narrow, a serious condition called pyloric stenosis). To relieve some of the pressure due to gas in the stomach, a person might belch. In one recorded case, a man lit a cigarette just as he belched uncontrollably: The cigarette shot from his mouth like a rocket and his lips and fingers were burned. In another case, a man was bending over a table to light a cigarette on someone else's lighter. The belch came up through the nose, and a flame shot out of each nostril, making him look like a medieval fire-breathing dragon. In another case, a surgeon opened up the stomach by using an electrical cutting device instead of a scalpel. The sparking from the cutting device reached the internal gas, which then ignited and burned with a bright blue flame for about 10 seconds.

5.27 • Lemon battery, tingling of teeth fillings

A crude but novel battery can be made by inserting a zinc probe (galvanized nail) into a lemon and then inserting a copper coin into a slit cut in the side of the lemon. The potential difference between the probe and the coin is about 1 volt. If several such lemon batteries are connected *in series* (one after the other) to a small lightbulb, the bulb will emit light, albeit weakly. If they are also connected to a capacitor, the charge they generate can be stored in the capacitor and then later dumped into a camera strobe to set off the strobe. How can a lemon battery produce current and a potential difference? Other food items can be substituted for the lemon.

You may have noticed a similar production of current and potential difference if you have metal fillings in your teeth and if, for some reason, you chewed on a metal strip such as aluminum foil. What causes the tingling you feel in the tooth and nearby gums?

Covering leftover food with aluminum foil is a common kitchen practice. However, if the food lies in a stainless-steel container and if the foil makes contact with the food, the foil may end up dissolved into the food at those points of contact. What causes that?

Answer The atoms of a given material have a certain tendency to gain or lose electrons from other, neighboring atoms of a different material. When a galvanized nail is inserted into a lemon, the zinc on the nail tends to lose electrons to become positive zinc ions, and a certain electric potential is associated with that tendency. Near the copper coin in the lemon, the hydrogen ions in the lemon juice tend to gain electrons to become neutral hydrogen atoms, and a certain electric potential is associated with this tendency. If the nail is electrically connected by wire to the coin, then the electrons lost by the zinc at the nail can move through the wire to the coin, to be gained by the hydrogen ions. Thus, this lemon battery can supply a current (electron flow) through the wire, and that current is driven by the electric potential difference between the nail and the coin (or actually, the lemon juice near the coin).

Similar electron loss and electron gain occur when aluminum foil touches a metal filling, with saliva between the two surfaces in several places. The foil–saliva–filling functions as a battery, sending current through points of direct contact between foil and filling, or through the surrounding gum.

A similar process occurs in the stainless steel–aluminum foil arrangement. The steel–food–foil can act as a battery, sending current through the points of direct contact between the foil and the steel bowl (probably all along the bowl's rim where the foil is usually pressed down). As the foil is oxidized, with aluminum atoms being converted into aluminum ions, the ions dissolve into the food, especially if the food is something like tomato sauce. Here is a food storage tip: Use plastic wrap instead of aluminum foil, or a plastic container instead of a stainless-steel container.

5.28 • Electric fish and eels

Fish, such as the giant electric ray *Torpedo nobiliana* of the North Atlantic and the electric eel *Electrophorus* of the Amazon, can produce sufficient current to kill or stun prey, or even to stun a person. (For example, the torpedo discharges in a pulse of 50 amperes at about 60 volts.) Long ago, electric fish were sometimes used for medicinal purposes, such as when an electric ray would be placed directly on the point of pain in a persistent headache (an early form of shock therapy). The electric properties of fish were known to early hunters, who quickly learned which fish should not be grabbed with bare hands or handled via a conducting spear.

Many other fish produce an electric field to navigate through dim or dark waters or to locate objects, including one another. In fact, these fish can alter their electric field in such a way as to be identified. How can an animal produce current, electric potential, and electric field?

Answer The source of the electric effects of fish can be traced to the cells known as *electroplaques*, which are similar to nerve and muscle cells. Normally the membrane of an electroplaque passes potassium ions but not sodium ions, and so the concentrations of sodium and potassium ions differ across the cell membrane. Because these ions are charged, this difference in concentration produces a difference in electric potential across the membrane.

When the fish wants to discharge, a nerve impulse changes the membrane so that it can then pass sodium ions, and then suddenly the difference in electric potential across the membrane changes and charged particles flow through the membrane (that is, there is a current through the membrane). The change in potential difference and the amount of current are both small. However, the fish may have several thousand electroplaques connected in series (one after the other, Fig. 5-4*a*) to build up the total potential and total current.

The total current is intended to leave the fish at one end (head or tail), traverse the water (thus possibly through prey or a person), and then reenter the fish at the opposite end. However, if the fish had only a single series of the electroplaques, the total current through the fish would stun or kill the fish itself. To avoid such a result, the fish has hundreds of series arrangements connected in parallel (Fig. 5-4*b*), so that the total current is split up evenly among those parallel paths.

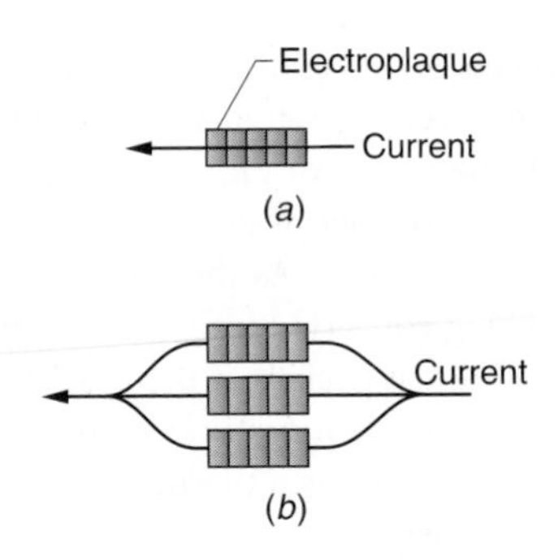

Figure 5-4 / *Item 5.28* (*a*) A series of five electroplaques within an electric eel. (*b*) A parallel arrangement of three series of electroplaques.

Thus, the current along any one path through the fish is insufficient to hurt it.

Electric fish living in salt water differ from those living in fresh water because the salt water offers far less resistance to the current. Thus, the salt-water fish needs fewer electroplaques in each series arrangement to get enough current through the surrounding water to stun or kill prey.

Weakly electric fish do not attempt to send a pulse of current through the surrounding water; instead their electroplaques simply generate a weak electric field in the water as a probe. Because they are extremely sensitive to the strength of that field, they can tell when other objects come into the field, changing it. In addition, they can change the field in a characteristic way so as to communicate to other fish of their kind.

5.29 • Charging by blown dust, sand, and snow

How can a wire fence be electrified by blowing snow? Sometimes a long wire fence collects so much charge that a person touching it can receive a large shock and might even be knocked down.

When dust or sand is blown hard by wind, such as in dust storms, dust devils, and tornadoes, why is the material highly charged? In the few cases where people have looked up into the interior of a tornado funnel and lived to talk about it, they described an interior illuminated with a shimmering light, crisscrossed by long discharges.

What causes these several examples of electrification?

Answer The process by which blown snow becomes charged is not well understood, but here are the general methods: If two neutral ice crystals of different temperatures collide, the warmer crystal becomes negatively charged and the cooler crystal becomes positively charged. If two ends of a neutral ice crystal are at different temperatures, the warmer end is positively charged and the cooler end is negatively charged. So, if the ice crystal is broken apart in a collision, the two pieces leave with different signs of charge. Once charged, the crystals can charge a fence by touching it.

Dust that is blown through the air by a storm, a dust devil (whirlwind), or a tornado generally becomes charged by contact with the ground and other dust grains. (The contact of two objects is enough to cause electrons to be transferred from one object to the other.) Whether the dust is charged negatively or positively depends on the nature of the dust and the ground. In some situations, the dust loses electrons when it touches the ground and thus becomes positively charged; in others, the dust gains electrons and thus becomes negatively charged. Once airborne, the dust grains can exchange charges during collisions.

Dust devils on Mars can be much larger than those on Earth, and so they might be much more charged. However, there is a limit to how much charge a dust devil can have. Here is the reasoning: As charge increases, the electric field at the surface of the dust devil also increases. Eventually the electric field is so strong that the surface begins to spark (that is, discharge), which drains electrons from the surface. Once the sparking stage is reached, any charge gained by the dust devil's motion is immediately lost to sparking.

A tornado is charged not only by the dust it kicks up, but also by the electric charges in the large thunderstorm producing it. Thus, the lights seen in tornado funnels are most likely due to discharges between charged pockets of dust and debris. In addition, some of the lights associated with a tornado could be regular lightning. However, the old idea that a tornado is caused by an approximately continuous flow of current between the clouds and ground has been discarded.

5.30 • Lightning-like discharges above a volcano

The plumes from some erupting volcanoes, such as the Sakurajima volcano in Japan, develop electrical discharges that flash over the crater, lighting up the sky and sending out sound waves that resemble thunder. What causes these sound-and-light shows?

Answer The discharges are due to charged particles carried aloft by the plumes rising from or shot from a volcano. The plume may be dominated by positive charge, but usually it also contains regions of negative charge. These regions can discharge to one another or to ground. The current in a discharge can heat the air so intensely that the air expands faster than the speed of sound; such expansion sends out a shock wave, which reaches an observer (who is at, hopefully, a safe distance) as a loud boom.

Several effects might account for the charged particles being in the plumes: (1) If water suddenly encounters molten lava, it can bead up in what is called the *Leidenfrost effect*, floating on a vapor layer. Any such large drop quickly splits into charged smaller drops, which are then carried into the atmosphere by the rising plume of hot air and water vapor. (2) Magma becomes charged when it fractures as it either hits water or crashes through the upper end of the volcano conduit and then is ejected in the plume.

Once the charged particles are aloft, collisions can transfer charge from one particle to another or even cause additional charging, as occurs in wind-blown dust.

5.31 • Bacterial contamination in surgery

Surgical teams take extraordinary measures to avoid bacterial infection of a patient. Masks are donned, hands are meticulously cleaned and gloved, and instruments are sanitized at high temperature and in alcohol baths. Recently a subtle source of bacteria was discovered in operating rooms, one that had been overlooked for years. Here is an example of the situation: In endoscopic surgery, a surgeon manipulates an optical-fiber system through an incision, the throat, or the colon. The optical-fiber system brings an image of the interior to a viewing screen on a video monitor. The surgeon can advance the optical-fiber system or employ surgical instruments attached to it. For example, a polyp can be snared and removed. One advantage of using the optical-fiber system is that the head surgeon can coordinate the efforts of the team members by pointing out objectives on the monitor, where everyone can easily see the progress.

Somewhere in this procedure lurks a hidden source of bacterial contamination. Can you find it?

Answer To create an image on the viewing screen, especially with traditional monitors, electrons are shot toward the screen from the back of the monitor. To attract these electrons, the screen is kept positively charged. The charged screen also attracts airborne particles floating around in the operating room, such as lint, dust, and skin cells. If an airborne particle is negatively charged, it is pulled onto the screen's exterior surface. If, instead, it is electrically neutral, some of its electrons can be pulled to the side of the particle nearest the screen, giving the particle an *induced charge*, where one side is negative and the other is positive (Fig. 5-5*a*). The negative side is pulled toward the positively charged screen while the positive side is pushed away from the screen. Because the negative side is closer to the screen, the pull toward the screen wins this tug-of-war.

Because many of the particles collected on the screen's exterior surface carry bacteria, the screen becomes contaminated with bacteria. Suppose that a surgeon's gloved fingers come within a few centimeters of the screen, pointing to a particular part of the image, say, in explaining a surgical concern to the rest of the team. The positively charged screen pulls electrons from inside the fingers to the fingertips (Fig. 5-5*b*). The negatively charged fingertips then cause particles (airborne or on the screen) to collect on the gloves at the tips. When the surgeon next touches the patient with the contaminated gloves, the bacteria ends up on or (worse) inside the patient's body. To avoid this risk, surgeons are now warned not to bring fingers near a video monitor.

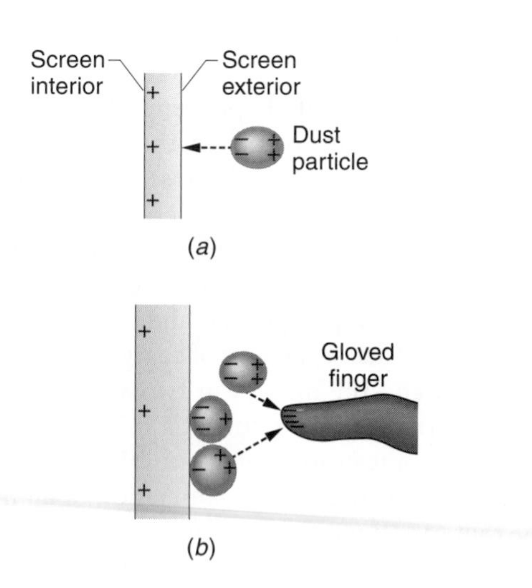

Figure 5-5 / Item 5.31 (*a*) A cross section of the viewing screen on a video monitor. The positively charged screen produces induced charge on a nearby neutral dust particle. (*b*) A gloved finger (not to scale) brought near the screen has an induced charge and can attract dust particles from the air and from the screen.

5.32 • Bees and pollination

Bees help pollinate flowers by collecting pollen at one flower and carrying it to another. This is not an accidental procedure; that is, the bee doesn't just accidentally brush up against the pollen. Instead, the pollen grains actually jump to the bee at the first flower and then jump away from it at the second flower. What causes pollen to jump?

Answer After a bee leaves its hive, it usually becomes positively charged during its flight through the air. When the bee hovers near a flower's anther (Fig. 5-6*a*), which is electrically neutral, the electric field due to the bee produces *induced charge* in some of the pollen grains on the flower. Such a grain is electrically neutral but the electric field of the bee redistributes the grain's charge: Some of its electrons move to the side facing the bee, to be as close as possible to the positively charged bee. This motion leaves the far side of the grain positively charged. The grain is still neutral but now has negative charge on one side and positive charge on the other.

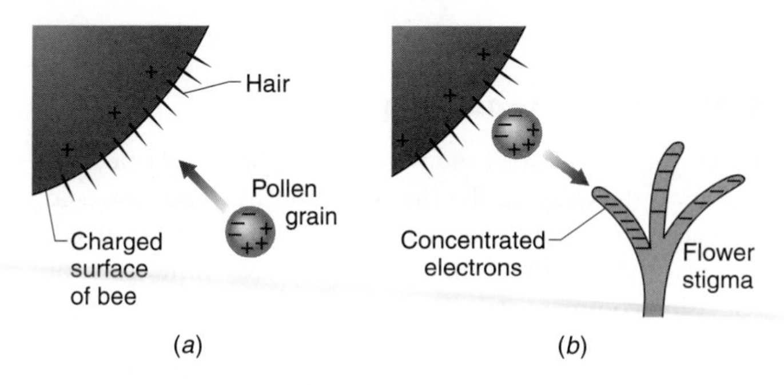

Figure 5-6 / Item 5.32 (*a*) Positively charged surface of a bee separates charges in pollen grain. Grain jumps to bee. (*b*) Grain jumps from bee to the concentration of electrons on flower stigma.

The negative side is pulled toward the bee; the positive side is pushed away by the bee. Because the negative side is closer to the bee, the pulling wins and the grain jumps through the air to land on the bee. (Actually, it lands on hairs of the bee. If it touched the bee's charged body, it would lose its electrons. Then the grain would be left with only its positive charge, and so it would be forced off the bee, never making the trip to the next flower.)

Delivery of the pollen to the next flower occurs when the bee comes near the flower's stigma, which is electrically connected to the ground. The bee's electric field pulls electrons in the stigma to be as near as possible to the bee, making the top of the stigma negatively charged (Fig. 5-6*b*). The pollen grain on the bee is still pulled toward the charge spread out on the bee but it is now pulled even more strongly toward the concentrated charge at the top of the stigma. So, the grain jumps from the bee to the stigma, pollinating the flower.

SHORT STORY

5.33 • Fire ants and electrical equipment

As the imported fire ant *Solenopsis invecta* Buren moved north into the United States from Central America, swarms of the ants would attack and destroy electrical equipment in the open (such as the switch boxes for traffic-control lights) and alongside buildings. One initial explanation was that the fire ant was attracted to the electric fields or the magnetic fields (or both) of the electric circuitry. That would have been very interesting—how would an ant detect either type of field?

However, as researchers probed the ant behavior, they discovered a far simpler explanation: As ants randomly crawl into electrical boxes, one of them happens to short out the circuit by spanning two bare wires or a bare wire and a grounded point. That ant is either killed or made very, very angry (fire ants already have mean dispositions). Dead or alive, the ant emits chemicals that agitate nearby ants, which then swarm to the first ant and also become electrocuted. Eventually there are so many dead ants shorting out the equipment that either a circuit breaker shuts off the current or the equipment is destroyed by too much current.

5.34 • Plastic food wrap

When you fold plastic food wrap over a glass or plastic bowl and press the wrap on the rim, why does the food wrap stay in place? You might even be able to turn the bowl upside down without spilling the contents, as sometimes seen on television commercials.

Answer The plastic food wrap probably contains charged patches when you pull a length of the wrap off the roll in the dispenser box. The charges were left there during the manufacturing process: Patches with excess electrons are negatively charged and patches with an absence of electrons are positively charged. Patches with charges of opposite signs attract one another, which is one reason why the plastic wrap will fold back onto itself (making it unusable) or back onto the roll.

When the food wrap is placed against the rim of a bowl, charge is transferred between the two surfaces in what is called *contact electrification*. For example, the food wrap might pull some of the electrons on the rim up onto the food wrap, leaving that portion of the rim positively charged. The negatively charged wrap and the positively charged rim would then attract each other.

In addition, a type of molecule-to-molecule attractive force, called the *van der Waals force*, can act between the rim and the wrap. This force is due to an electrical interaction in which a very slight separation of positive and negative charge in a molecule on one surface causes a similar separation of charge in the nearest molecule on the other surface. Each separation of charge is said to be an *electrical dipole*, and the dipoles on two surfaces attract each other. Although weak, this attraction can help hold the wrap to the rim or cause the wrap to cling to itself.

5.35 • Flies on ceilings, geckos on walls

A fly can adhere to a smooth surface because it secretes an oil that adheres to both the surface and the fly's foot. And some beetles can adhere to a smooth surface by a suction mechanism. But what about the gecko? Its feet are dry and cannot set up suction. Yet a gecko can even run up a smooth wall and crawl across some ceilings, moving forward or backward. How do they stick to the surface, and how do they rapidly unstick themselves so that they can run along the surface?

Answer The foot of a gecko has about half a million hairs known as *setae*. Each seta has hundreds of projections with triangular or leaf-shaped ends known as *spatulae* (because they are in the shape of a spatula). When the gecko presses a seta against a wall, all those spatulae adhere to the wall by a force known as the van der Waals force. This force is due to the electrical interaction in which a very slight separation of positive and negative charge on one surface causes a similar separation of positive and negative charge on the second surface. Each separation of charge is said to be an *electrical dipole*, and the dipoles in two surfaces attract each other. This arrangement occurs at a million or more points when a gecko puts its foot against a wall. Although each van der Waals force is weak, the combination of all those forces on the foot can support the gecko. Even if the wall is microscopically rough, a foot puts enough of the spatulae against the wall to support the gecko.

Each seta is angled to the wall, and adhesion occurs when the angle is kept relatively small. To unstick its foot while running, the gecko peels the seta from the wall by pulling away from it so as to increase the angle. The spatulae then come away one by one, freeing the seta.

5.36 • Meringue pie

In making a meringue pie, egg whites are first beaten until they are somewhat stiff. Then a small amount of sugar is beaten into them. They are then ready to be put onto a pie and baked.

Why does even a small amount of yolk ruin the meringue? Why are the egg whites beaten, and why do they stiffen when beaten? Why does excessive beating ruin the meringue?

Answer Egg whites consist of several types of proteins, which are huge molecules with complicated three-dimensional structures. One purpose of beating the whites is to partially unravel these molecules by breaking some of their weaker internal points of attraction. The weaker bonds include the ionic bond (where charges of opposite signs attract each other), the van der Waals force (where separated positive and negative charges in one place in the molecule attract separated positive and negative charges in an adjacent place), and hydrogen bonds (where hydrogen acts as a go-between to keep two atoms together). Once the proteins are unraveled, they attach to one another, making a mesh.

The other reason the egg whites are beaten is to trap air in that mesh. Here the yolk can ruin the dessert, because it is too heavy and too viscous to allow enough air to be added. A cook wants to put a meringue with lots of air bubbles into an oven where heating will expand the bubbles, making the dessert even lighter. If the egg whites have been beaten correctly, the air bubbles are entrapped by thin films of water that are bonded to the egg white mesh. These films will stretch with the expanding air bubbles, keeping the air trapped in the meringue. However, if the egg whites are overbeaten, the water separates out from the proteins and the mesh becomes too firm (too well bonded) to expand properly in the oven. Then the air bubbles merely pop open, causing the meringue to collapse, a cook's nightmare. To avoid overbeating, an experienced cook stops beating the egg whites when they lose their sparkle and are on the verge of forming water drops.

If egg whites are beaten in a copper bowl, some of the copper atoms are taken up by the whites, bonding with the sulfur portions. Those portions cannot then bond in the protein mesh, which prevents the mesh from becoming so firm that it squeezes out the water.

5.37 • Sauce béarnaise

Sauce béarnaise is a notoriously difficult sauce to prepare and can "go bad" even if the cook does everything correctly. It is a warm concoction, consisting primarily of dilute vinegar, wine, egg yolks, and butter, and it is served with broiled red meats, chicken, fish, and poached eggs. The sauce is supposed to be a smooth blend of the ingredients but it can be ruined if butter suddenly separates from the rest of the ingredients, forming unsightly pools. The questions here are: Why does the butter pool when the sauce fails? What normally keeps the butter from pooling?

Answer The sauce can be viewed in either of two general ways: It is a *colloidal suspension* of semisolid particles (butterfat) in a liquid of primarily water and acetic acid (the vinegar). It is also an *emulsion*—that is, a dispersion of two immiscible liquids (liquids that do not mix, here the butterfat and water), with the butterfat forming tiny drops in the water.

In the colloid model, the butterfat drops tend to attract one another via a weak force known as van der Waals force, which is due to the electric dipole (a separation of positive and negative charges) in the molecules. However, the drops also have negative charge on their surfaces; so, as they move toward one another, threatening to touch and coalesce, the repulsion between them tends to keep them apart. The danger comes when the sauce is warmed, because then the drops move about with more energy and can collide in spite of the electric repulsion. If they do begin to bind together (they are said to *flocculate*), the problem might be too little charge on the drop surfaces. Many chefs recommend that lemon juice be vigorously whisked into the sauce. The whisking breaks up the flocculated butter, and the lemon juice presumably provides additional charges to keep the resulting drops separated.

In the emulsion model of the sauce, the butter drops are *stabilized* (maintained as drops instead of being allowed to flocculate) by lecithin molecules distributed on their surfaces. Each lecithin molecule, which comes from the egg yolk, has a water-binding end (said to be its *polar end*) extending away from the surface and into the water. That end binds up water molecules so that each drop is surrounded by a layer of water bound to the polar ends of the lecithin molecules. This bound water interferes with the flocculation of the drops. If the sauce begins to flocculate, it may not have enough lecithin. Many chefs recommend that more yolk be vigorously whisked into the sauce, to break up the flocculated butter and to provide more lecithin.

In practice, vigorous beating and either the lemon-juice or the egg-yolk remedy will work (and thus the remedy gives little clue as to the correct model of the sauce). A skilled cook knows that any remedy should not be overdone or the taste and appearance of the sauce will be ruined even if the sauce has no flocculation. A skilled cook also knows that overheating is the bane of the process, because at elevated temperatures the thermal motion of the drops is greater (increasing the likelihood of them colliding with one another) and the yolk will curdle (becoming unsightly and no longer serving to stabilize the sauce).

5.38 • Lodestones

Naturally occurring magnetic rocks, called lodestones, were discovered long ago, notably by the Chinese who treated them first as novelties and then realized their value in navigational compasses. How do rocks become magnetized, and why don't all rocks become magnetized?

Answer A lodestone is a piece of iron ore that can retain its magnetization after being magnetized and thus is said to be a type of *permanent magnet*. That retention is due to a cooperative quantum effect in the interaction of the electrons in the iron atoms. In nature, iron ore can be magnetized by two processes: being heated and then cooled in the Earth's magnetic field, and being near the very large current in a lightning strike.

In the first of these processes, iron ore is heated in a lava flow and then cooled. When the iron ore is very hot, it loses its magnetic properties because the thermal energy of the atoms totally disrupts the cooperative interaction among the electrons. When the ore cools and that cooperative interaction is reinstated, the magnetic direction of the electrons tends to mimic the direction of Earth's magnetic field at the ore's location. The better this alignment is within the ore, the stronger the resulting lodestone will be. Lodestones produced in this way can be found in veins buried in iron mines or lying exposed on the surface due to weathering.

In the second process, a large current from lightning passes through, or very near, the rock. The magnetic field produced by the movement of electrons in that current aligns the magnetic direction of some of the iron compounds in the rock, and that alignment is retained by the rock after the lightning event. This type of lodestone usually occurs near the surface (because lightning typically does not penetrate the ground by more than a few meters) and is probably found in isolated patches.

5.39 • Earth's magnetic field and archaeology

Because Earth's magnetic field gradually changes, the direction of north indicated by a compass also changes. For many reasons, researchers want to know the direction of magnetic north at specific times in the past, but finding historic records of compass readings is rare. However, researchers find help in ancient clay-walled kilns that were used to bake pottery and in very old paintings, such as the murals in the *Bibliotheca Apostolica Vaticana*, a celebrated hallway at the Vatican. How can a kiln or a mural indicate the direction of magnetic north?

Answer The clay in the walls and floor of an ancient kiln contains the iron oxides magnetite and hematite. In general, such iron materials contain individual grains in which there are *domains*, which are regions in which the magnetic fields of the material are uniform. A grain of magnetite consists of multiple domains that are microscopic; a grain of hematite consists of a single domain that may be as much as a millimeter wide.

When the clay is heated to several hundred degrees Celsius (as the kiln is used), the domains of both grain types change. In magnetite the walls of the domains shift, so that domains more closely aligned with Earth's magnetic field grow, while others shrink. In hematite, the domains rotate to more closely align with Earth's field. For both processes, the result is that the clay has a magnetic field that is aligned with Earth's field. When the kiln cools after use, the arrangement of the domains, and thus the magnetic field of the clay, is retained, an effect known as *thermoremanent magnetism* (TRM).

To determine the orientation of Earth's field when the kiln was last heated and cooled, an archaeologist outlines a small area of the floor, carefully measures its orientation relative to the horizontal and to geographic north (the North Pole), and then removes that section of the floor. By next determining the direction of the section's magnetic field relative to the section's dimensions, hence to its position in the kiln, the archaeologist can determine the direction of Earth's field when the kiln was last used. If the age of the kiln is found by radiocarbon dating or some other technique, the archaeologist also knows when Earth's field had that direction.

Many old mural paintings contain hematite. Artists' pigments are a suspension of various solids in a liquid carrier. When a pigment is applied to a wall as a mural is being created, each hematite grain rotates in the liquid until it aligns with Earth's magnetic field. When the paint dries, the grains are locked in place and thus record the direction of Earth's magnetic field at the time of the painting.

A researcher can determine Earth's field direction at the time a mural was painted by determining the orientation of the hematite grains in the paint. A short section of sticky tape is applied to a portion of the mural, and the orientation of the tape is carefully measured relative to the horizontal and to today's direction of magnetic north. When the tape is peeled off the mural, it carries a thin layer of the paint. In a laboratory, the tape section is mounted in an apparatus to determine the orientation of the hematite grains in that layer of paint.

5.40 • MRI complications

MRI (for *magnetic resonance imaging*) is a technique to image the interior of a body (humans, animals, fossils, and many other objects). The technique was originally called NMR for *nuclear magnetic resonance*, but the name reportedly was changed to MRI when the Cleveland Clinic (in Cleveland, Ohio) bowed to public pressure after announcing that a *nuclear* facility was planned. (The public seemingly did not realize that "nuclear" refers to the central objects in all atoms, including those in their own bodies.)

MRI uses electromagnetic waves in a certain frequency range (said to be radio waves) to penetrate the body and to flip some of the protons in some of the nuclei within the body. Those protons are initially aligned by a large magnetic field, and after they are flipped by the electromagnetic waves, they rather quickly regain their alignment. Monitoring equipment records the progress of that realignment, and very sophisticated computer programs transform the recordings into images of the material containing those protons. This is entirely safe because the magnetic field and radio waves cause no harm. Indeed, radio waves from various nearby radio stations, telecommunication antennas, and even telecommunication satellites penetrate you all the time.

If the procedure is safe, why then, on *very rare* occasions, have patients been burned during the procedure? Why have some patients with tattoos undergone a "tingling" or "tugging" sensation in the vicinity of the tattoos, and why have some of them been badly burned? Why is the procedure either not allowed or not recommended for a patient with metallic implants (such as in the eye, eyelid, dental work, or certain breast enhancements, or as part of a heart valve)? Why is the procedure often not advised for someone who has worked in the welding or metal-grinding business?

Answer Before the danger was appreciated, some patients were burned when electrical leads for monitoring the patient were allowed to touch the patient at more than one point. In one case, a pulse oximeter was attached to a finger of a sedated patient. That single attachment between the finger and the monitoring equipment (which was outside the MRI apparatus) posed no danger. However, the wire extending from the finger happened to touch the patient's arm. The section of wire between finger and arm and the part of the arm between those two points then functioned effectively as a complete conducting loop. When the radio waves were turned on, their rapidly varying magnetic field created a current around this loop. The large resistance to the current at the skin–wire contact points led to the severe heating and burning there. However, because the patient was sedated, the injury was not discovered until after the patient was withdrawn from the MRI apparatus.

A second way of burning a patient occurred when a *long* electrical lead that extended to the patient acted like a receiving antenna to the radio waves. An appreciable electric field built up along the lead, and the field at the end of the lead was strong enough to cause sparking, which burned the patient.

Some of the black or blue-black pigments in either a tattoo or permanent eyeliner contain a *ferromagnetic material* (the iron oxide magnetite). When a patient with such a pigment is moved into or out of the magnetic field in a MRI apparatus, or when the strength of the magnetic field is varied, the ferromagnetic material tends to be reoriented, much like a compass needle is reoriented when moved into or out of a magnetic field. Some patients sense this as a tingling or tugging of the skin. In the few cases where skin was burned, the tattoo design included a complete (or nearly complete) loop of the ferromagnetic pigment. Presumably, the radio waves can create a current in such a loop, and the current is great enough to heat and burn skin.

Metallic implants complicate an MRI analysis because they tend to distort the image. If they consist of ferromagnetic material, they can also rotate like a compass needle when the patient is moved into or out of the magnetic field. Often the motion is not even noticed, but rotating a metallic heart valve or an embedded steel bullet might be dangerous. Motion of eye implants or eye debris from welding or metal grinding is also of concern. In the past, the ferromagnetic ports used in some breast implants tended to heat due to induced currents. However, MRI technicians now know about these possibilities and have set up guidelines for your protection.

SHORT STORY

5.41 • Magnetic search for the Garfield bullet

In 1881, James Garfield, then President of the United States, was shot twice by an assailant in a railway station in Washington, D.C. Although one bullet merely grazed an arm, the other was buried below his pancreas. However, the doctors could not locate the second bullet because they lacked any way of looking inside a body without cutting it open. Because the bullet had been deflected by a rib after it entered the president's body, not even the entrance wound gave enough clues about the bullet's location.

Alexander Graham Bell, who is linked with the invention of the telephone, offered to help by using what he called an *induction balance*. The apparatus consisted of (a) an electromagnet powered by batteries and (b) a small coil connected to a telephonic receiver (the listening part of the telephone at that time). Because the battery sent a steady current through the electromagnet, a steady magnetic field was produced by the electromagnet. Bell held the small coil in that field such that the field was perpendicular to the plane of the coil. If he somehow changed the field through the small coil, current would appear in the small coil and produce a clicking sound in the receiver.

Bell's plan was to pass the electromagnet and small coil over Garfield's body. If the apparatus passed over the bullet, the bullet would alter the amount and direction of the magnetic field through the small coil. Thus, the bullet would be detectable by the clicking it would produce.

Unfortunately, the bullet was made of lead and thus did not alter the field by very much (steel would have altered it much more). Also, the bullet was too far from the surface of Garfield's body to cause much change in the field. After many attempts, Bell gave up the search; Garfield died about a month later.

5.42 • Magnets, tattoos, and body jewelry

Why can a strong magnet cling to and pull on skin containing a tattoo with black or blue-black lines? Body jewelry has long been fashionable, but some people elect to hold the jewelry in place with magnets instead of undergoing piercing. What is the danger of wearing magnetically attached nose rings?

Answer The black and blue-black lines in tattoos are usually produced by injecting the skin with a pigment of iron oxide (magnetite), which is ferromagnetic and attracted to a strong magnet. Thus, a strong magnet can cling to the tattoo site. More curiously, a strong magnet can cause some of the magnetite to migrate through the dermis, to collect under the magnet at the interface between dermis and epidermis.

This migration tendency can be used in tattoo removal. A laser that emits light pulses in the near infrared (just barely out of the visible range) is first used on the tattoo to disrupt and disperse the ink pigments in the dermis and to open up the epidermis. Then a small but very strong magnet is taped over the tattoo to draw some of the ferromagnetic pigments up through the epidermis to the magnet, where it can be removed. Thus, the tattoo becomes less distinct.

One danger in wearing magnetic body jewelry was exhibited by a girl who wanted to wear magnetic earrings as nose rings. Each ring was to be held by a magnet placed inside a nostril, against its outward side. However, when she attempted to mount the second ring, the two magnets strongly attracted each other and jumped to the septum (the partition separating the two nostrils) fairly high up in the nose. There they stubbornly clung to each other across the narrow septum, requiring her to make a trip to a hospital emergency room to have them removed.

5.43 • Breakfast and cow magnetism

If you pass a strong magnet over a slurry of milk and certain types of breakfast cereals, why does the cereal collect on the magnet? Why are magnets commonly slipped down into the rumen of cattle?

Answer These types of breakfast cereals are advertised as being "iron fortified," which means that they contain iron filings, to provide iron in your diet. Similarly, some bank notes (paper money) contain iron compounds in the printer ink, making a bank note pull toward a strong magnet.

The cow magnets are intended to collect scrap iron that the cow accidentally swallows when eating grass or hay, so that the scrap iron cannot damage the rest of the cow's digestive system. The magnets are inexpensive and can be found at your favorite cow-supply store.

5.44 • Electric guitars

Soon after rock began in the mid-1950s, guitarists switched from acoustic guitars to electric guitars—but it was Jimi Hendrix who first understood the electric guitar as an electronic instrument. He exploded on the scene in the 1960s, ripping his pick along the strings, positioning himself and his guitar in front of a speaker to sustain feedback, and then laying down chords on top of the feedback. He shoved rock forward from the melodies of Buddy Holly into the psychedelia of the late 1960s and into the early heavy metal of Led Zeppelin and the raw energy of Joy Division in the 1970s, and his ideas continue to influence rock today. What is it about an electric guitar that distinguishes it from an acoustic guitar and enabled Hendrix to make so much broader use of this electronic instrument?

Answer Whereas the sound of an acoustic guitar depends on the acoustic resonance produced in the body of the instrument by the oscillations of the strings, an electric guitar is a solid instrument, so there is no body resonance. Instead, the oscillations of the metal strings are sensed by electric *pickups* that send signals to an amplifier and a set of speakers.

The wire connecting a pickup to the amplifier is coiled around a small magnet just below a string. The magnetic field of the magnet produces a north and south pole in the string section just above it. That section then has its own magnetic field. When the string is plucked and thus made to oscillate, its motion relative to the coil changes the amount of its magnetic field within the coil, creating a current in the coil. As the string oscillates toward and away from the coil, this current changes direction at the same frequency as the string's oscillations, thus relaying the frequency of oscillation to the amplifier and speaker.

On a Stratocaster electric guitar, there are three groups of pickups, placed at the near end of the strings (on the wide part of the body). The group closest to the near end better detects the high-frequency oscillations of the strings; the group farthest from the near end better detects the low-frequency oscillations. By throwing a toggle switch on the guitar, the musician can select which group, or which pair of groups, will send signals to the amplifier and speakers.

To gain further control over his music, Hendrix sometimes rewrapped the wire in the pickup coils of his guitar to change the number of turns. In this way, he altered the amount of current that would be created in a coil, and thus the coil's sensitivity to the string's oscillations.

5.45 • Electric-guitar amplifiers

Solid-state physics and solid-state electronics have radically changed modern life. For example, early computers relied on bulky vacuum tubes and took up the space of a large room. Today, far more powerful computers rely on tiny transistors in integrated circuits and take up only the space on your lap (or less). Vacuum tubes are seemingly a thing of the past; indeed, they are no longer taught to electrical engineering majors. However, many of today's hard-rock guitar players insist on amplifiers using vacuum tubes and shun those using transistors. Why do guitar rockers choose tube amplifiers instead of transistor amplifiers?

Answer The mechanical oscillations that a player sets up on a string of an electric guitar produce electrical oscillations in a pickup coil positioned just under the string. Those electrical oscillations must be amplified so they can drive a speaker system to produce sound for the audience. When the electric guitar became popular in rock in the early 1960s, the amplifiers used tubes because transistor amplifiers were not yet dependable. As rock moved into psychedelia and then heavy metal, guitarists cranked up their amplifiers in order to shake their audiences. Such high amplification by a tube amplifier introduces significant distortion in the final sound, and that distortion quickly became identified with the sound of rock.

Transistor amplifiers do not produce the same type of distortion when driven hard—they are said to produce a "clean" sound. Thus, rock guitarists shun them today, because they do not produce the "proper" sound of rock. Jimi Hendrix, who was the first person to understand the electric guitar and its amplifier as a combined musical instrument, once said, "I really like my old Marshall tube amps, because when . . . the volume is turned up all the way, there's nothing [that] can beat them . . . ."

5.46 • Auroras

If you are outside on a dark night in the middle to high latitudes, you might be able to see an aurora, a ghostly "curtain" of light that hangs down from the sky. This curtain is not only local; it may be several hundred kilometers high and several thousand kilometers long, stretching around Earth in an arc. However, it is only about 100 meters thick. What produces this huge display, and what makes it so thin?

Answer Auroras can sometimes be associated with flares on the Sun if the particles emitted in the flare affect the magnetic and electric fields in Earth's atmosphere.

Auroras occur when electrons are accelerated in the altitude range of 3000 to 12 000 kilometers and are then funneled along the lines of Earth's magnetic field into the high latitudes, toward the magnetic north and south poles. Because the field lines converge down toward a pole, the electrons move into lower altitudes where the air is denser, and where they collide with atoms and molecules, exciting them. The atoms and molecules then de-excite by emitting light; an aurora is the light (both visible and invisible) that is emitted. For example, green light is emitted by oxygen atoms and pink light is emitted by nitrogen molecules. However, the light may be so dim that only white light is perceived. Sometimes the display may seem to move across the sky as if blown by wind, but the motion is an illusion.

Because the electrons are funneled along converging field lines, they converge into a fairly narrow region. Thus, only the atoms and molecules in that narrow region contribute light to an aurora. However, this simple explanation predicts much thicker auroras than what is observed, and more complicated explanations are currently being worked out.

5.47 • Solar eruptions and power outages

At 2:45 A.M. on March 13, 1989, the entire power-grid system for the Canadian province of Quebec failed, leaving millions of people without power on that cold night. In fact, many power-grid systems in the Northern Hemisphere malfunctioned that night, creating a nightmare for the engineers who maintained the systems. The cause was not a sudden overtaxing demand for power or a failure of aging equipment. Rather, the cause was an explosion—a *solar flare*—that had occurred on the Sun's surface three days earlier. How can a solar explosion shut down a power-grid system?

Answer In a solar flare, a huge loop of electrons and protons extends outward from the surface of the Sun. Some solar flares explode, shooting those charged particles into space. On March 10, 1989, a gigantic solar flare exploded toward Earth. When the particles arrived three days later, they transferred their energy to Earth's *magnetosphere*, a high-altitude region in which the dynamics are controlled by magnetic and electric fields. In particular they fed energy into the *electrojet* there.

Because it is a current, the electrojet sets up a magnetic field around itself, including the ground and the long power lines of the power-grid system. It was the effect of that magnetic field on the power lines—or rather, how that field varied—that caused the problem in Quebec.

At one end of such a power line, a *step-up transformer* increases the electric potential so that the electrical energy can be transferred along the line at a very high potential. At the other end, a *step-down transformer* decreases the electric potential to the low value that is used in a home. A transformer consists of two coils (the *primary* and the *secondary*) with different numbers of turns wound around an iron core. The normal alternating current in the primary produces an alternating current in the secondary at either a higher or lower voltage, depending on whether the secondary has more or fewer turns than the primary.

Both transformers on a line are grounded; that is, they are electrically connected to the ground. When a magnetic field is set up by an electrojet, it can pass through the effective loop (Fig. 5-7) consisting of the power line (top part of loop), the ground lines on the transformers (the two sides of the loop), and the ground (the bottom of the loop). However, a constant magnetic field through this loop does not create a problem. The problem occurs if the field varies.

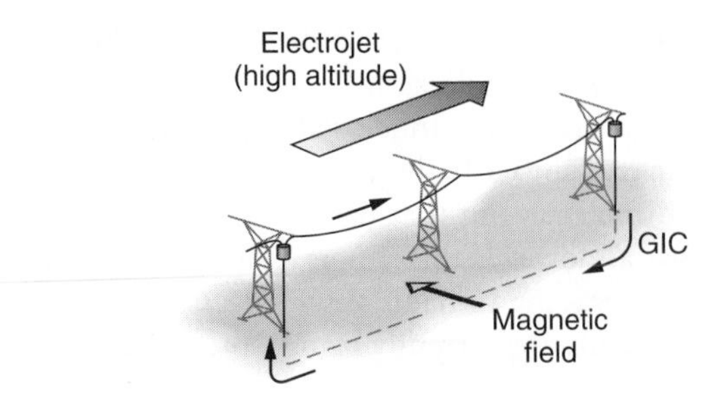

***Figure 5-7** / Item 5.47* An electrojet (current) produces a magnetic field through a vertical loop formed by a transmission line, the ground, and the wires grounding the transformers (located inside the cylinders at the ends of the transmission lines). Variations in the field generate a current (GIC) around the loop.

On the night of the blackout, the field was changing wildly and abruptly because the energy from the solar flare was causing the electrojet to change similarly. Whenever a magnetic field in a loop changes, it creates a current in the loop. When such a current occurs because of the electrojet, the current is called a *geomagnetically induced current* (GIC). So, on the night of the blackout, the transmission line carried a large and abruptly changing GIC in addition to its normal current.

Transmission of power by a power-grid system depends on the proper variations in current and voltage throughout the system. The presence of the GIC through the transformers in the Quebec system ruined the ability of the transformers to transfer the alternating current from the primary to the secondary. The result was that the current and voltage in the secondary were highly distorted and no longer varied in the proper way; this distortion disrupted the power transmission and burned out some of the transformers, shutting down the system. Today, whenever a solar flare explodes toward Earth, power-grid engineers are immediately warned about grid disruptions.

GIC can occur on any long conductor, such as telecommunication cables and the Trans-Alaska Pipeline. In fact, it was first noticed (but not understood) when long telegraph lines were in use about 150 years ago: On some days the lines already had current in them without being connected to their batteries as normally required. GIC can also occur along the ground itself, which can affect the corrosion in long pipelines electrically connected to the ground.

In an action similar to that causing GIC, a current can be set up in long underwater cables due to water motion, such as tidal motion. Because the water is conducting and moving through Earth's magnetic field, electric current is driven through the water (and thus along a cable in the water). The circuit is completed by the current coming back through the seabed.

5.48 • Levitating frogs

A frog (and other small animals) can be levitated by a magnetic field produced by a solenoid (a coil of wire carrying current). However, the frog is clearly not magnetic; otherwise, every time it hopped through a kitchen it would be slammed up against the metal door of the refrigerator. Given a sufficiently large magnetic field, you could also be levitated, but you certainly don't suffer the refrigerator-door effect. How can these biological materials be levitated?

Answer Several frogs have now become famous for being floated in the magnetic field of a solenoid. (The frogs are in no discomfort; the sensation is like floating in water, which frogs enjoy.) The solenoid was mounted vertically and the frog was placed near the upper end, where the magnetic field spreads from the solenoid. Although a frog is not normally magnetic, it does have magnetic properties when placed in a magnetic field. The frog (as well as people and many other materials) are said to be *diamagnetic.* In such a material, a magnetic field alters the electrons in the atoms, causing the material to become magnetic. So, when a frog is placed in the spreading magnetic field at the top of a solenoid, the frog is repelled upward by the field. The frog rises to the point where the upward push balances the downward pull by gravitation, and there the frog floats.

If the frog is replaced with a small magnet, the small magnet is unstable and will not float. The frog differs from the small magnet in that its magnetic properties depend on the strength of the magnetic field from the solenoid. For example, if the frog moved away from the solenoid to where the field is weaker, its magnetic properties would also weaken, whereas those of the small magnet would not change.

A small magnet *can* be made to float if it spins and precesses as a top. The delightful toy marketed under the trademark Levitron is based on this idea: A rapidly spinning magnetic top floats a few centimeters above a magnetized ceramic slab. However, as air drag slows the top's rotation, the spin rate is eventually too low for the top to be stable, and it finally falls.

5.49 • Fizzing sound from a magnet

Put an audio cassette player (music was once recorded on magnetic tape) in its play mode without a cassette in place (or with a blank cassette) and turn the volume control to maximum. Then bring a strong magnet up to the play head. Why does the motion produce a fizzing sound on the player?

Answer The play head in the cassette player is ferromagnetic and consists of many magnetic domains, or regions in which the magnetic properties are uniform and produce a magnetic field in a certain direction. However, the field direction differs from domain to domain. As you bring the magnet near the play head, the domains shift abruptly to bring their magnetic fields into alignment with the magnet's field. As these domain fields change, a varying current is produced in a coil wrapped around the play head. Those changes in current are amplified and fed to a speaker, where a fizzing sound is heard—the coming and going of current as the magnetic field of the domains shift.

5.50 • Currents in you at a train station

We all live in the Earth's electric field, which makes the electric potential at nose height different from that at the feet. Why then do we not feel an electric current through our bodies?

People waiting near the tracks of an electrically driven train have sometimes noticed a tingling if they touch a conducting object, such as a pipe, that is connected to ground. What causes the tingling?

Answer We don't have a current through us due to Earth's electric field because the density of charged particles in the surrounding air is too small to supply a perceptible current.

If the train is driven electrically by an overhead line, it probably carries alternating current (AC). Because such current continually changes direction and strength, the magnetic field it produces in the surrounding region also changes direction and strength. In conductors this variation in magnetic field produces currents, but the currents are probably too small for a person to notice. However, if the person touches a larger conducting object, such as a metallic sign, the currents can be larger and noticeable.

Splashing Colors Everywhere, Like a Rainbow

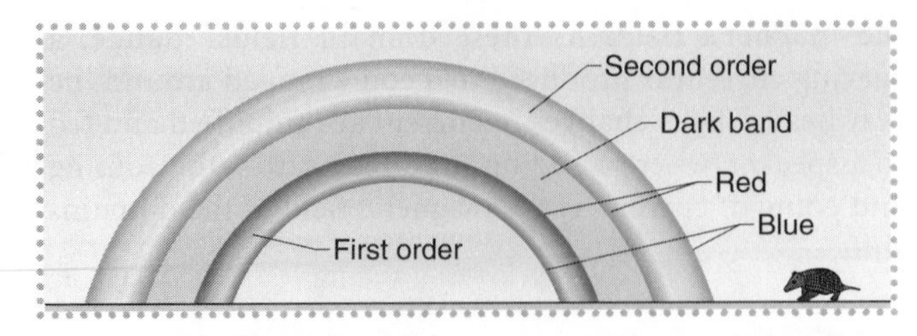

Figure 6-1 */ Item 6.1*

6.1 • Rainbows

Why do rainbows appear in some rain showers but not all? Why are they partial circles? Can a rainbow ever form a full circle? How far away is a rainbow—could you ever walk to one end of it? Why are rainbows usually visible only in the early morning or late afternoon?

Normally, you see only one rainbow, but sometimes you might find two, each a partial circle around a common point. What is that point? Why is the color sequence of the two rainbows reversed? Why is the region between the rainbows relatively dark? Why is the upper rainbow broader and dimmer than the lower one?

Why are the legs of a rainbow usually brighter and redder than the top? What produces the faint, thin bands that can occasionally be seen just below the lower rainbow?

Why are colors seen in only two bands and not throughout a rain-filled sky? If a third rainbow is possible, is it near the first two? Can thunder alter a rainbow?

Answer Rainbows result when falling water drops spread white sunlight into its various colors, concentrating the colors in a band, the rainbow band. Since bright sunlight must illuminate the drops, rainbows are not seen when the cloud cover is extensive. Light undergoes *refraction* (its path is bent) when it enters and leaves a drop. The extent of refraction depends on color. For example, because the path of blue light is bent more than the path of red light, blue light and red light leave a drop at slightly different angles.

The most frequently seen rainbow involves light rays that enter a drop, reflect once from the interior surface, and then exit toward you. This rainbow, called the *primary rainbow* or *first-order rainbow* because there is only one reflection, has red higher than the blue. The *second-order rainbow*, which requires two internal reflections, has the opposite sequence of colors because of the different geometry of the light paths involved. The additional reflection allows further spreading of the colors within each drop, which results in a wider and dimmer bow. The bow is also dimmed because some light is lost at each reflection point as it leaves the drop, leaving less light for the rainbow.

All the falling, illuminated raindrops refract light and separate colors, but only those drops at certain angles happen to send the colored rays toward you. The drops creating the first-order rainbow must be about 42° from the *antisolar point*, which is directly opposite the Sun's position relative to you. To find the rainbow drops, point your outstretched arm toward the antisolar point (at the shadow of your head) and then lift it upward or in some other direction by 42°. Your arm then points toward where drops would give you the first-order rainbow. The drops for the second-order rainbow will be about 51° from the antisolar point.

Since the drops must be at certain angles relative to the antisolar point, the rainbows form circular arcs around that point. From an elevated position, such as in an airplane, you might see full circles. Rainbows have no true distance from you—all drops along the proper angles (regardless of their distance from you) can contribute color. So, you cannot march to the end of a rainbow (to find a pot of gold). Also, the rainbow you see is personal; someone standing next to you sees colors from a different set of drops.

A rainbow is usually visible only in the early morning or late afternoon because during midday the antisolar point is too far below the horizon. Still, you might be able to see a rainbow if you look down on the drops from an elevated point.

The third- and fourth-order rainbows (requiring three and four internal reflections, respectively) lie in circular arcs around the Sun (rather than the antisolar point) but are too dim to be seen in the glare from that part of the sky. There are rare reports that third-order rainbows have been spotted, but the colors are more likely due to ice crystals. The fifth-order rainbow (five internal reflections) lies between the first and second rainbows but is much too dim to be seen, as are all other possible rainbows.

The intermediate region between the first- and second-order rainbows is dark compared to the regions just below and above the rainbows because drops in that intermediate region do not redirect light rays toward you, while the drops below and above the rainbows do.

The legs of a rainbow are often brighter and redder than the top of the rainbow because of several factors, one involving the size and shape of the drops. The colors of a rainbow should be more distinct with larger drops, because the additional light path within larger drops allows the colors to separate more. However, air drag flattens the larger drops as they fall. Along the legs, the light passes through a horizontal circular cross section of each drop; such a cross section is ideal for producing bright, distinct colors. At the top of the rainbow, the light passes through a noncircular cross section, which yields duller, less distinct colors.

The legs may also be brighter, because the drops in the legs are better illuminated by sunlight skirting beneath an overhanging cloud bank. They are redder if that light loses all but the red end of the spectrum as it travels a long distance through the air to reach the drops.

The faint bands that can be seen just below the first-order rainbow and (more rarely) just above the second-order rainbow are called *supernumeraries.* They reveal that the colors of a rainbow are not produced by the drops acting like simple prisms. Instead, a rainbow is actually an *interference pattern* created by light waves passing through each drop and then overlapping. The colors you normally see are the brightest portions of the interference pattern. For example, the bright red occurs where overlapping waves of red light are in step with one another and thus reinforce one another.

If the drops are approximately the same size, you might see the faint supernumeraries. When the drops are nonuniform in size, the supernumeraries overlap too much to be distinguishable, and you see just an overall dull white illumination.

Although simple models of the rainbows work well with drops larger than about 0.1 millimeter, much more complex models are required for smaller drops and are still being researched.

Thunder causes the water drops to oscillate, which smears or eliminates the colors because of distortion in the shape of the drops. Oscillations, due to the buffeting by the air as the drops fall, may also smear the colors, especially for larger drops.

6.2 • Strange rainbows

Why are some rainbows white and some red? What accounts for the rarity and lack of color of rainbows seen in moonlight? What are the shapes and colors of bows seen in fog, on a cloud, or on a dew-covered lawn? If you see a rainbow on a body of water while a rainbow appears over that water, is the former merely a reflection of the latter?

On rare occasions a seemingly vertical band of colors can be seen adjacent to the lower portion of a normal rainbow. What produces this extra band?

The normal rainbow is produced with visible light. Are rainbows also produced with infrared and ultraviolet light?

Answer The separation of colors in a rainbow is smaller with smaller drops. One reason is that for a smaller diameter, the colors have less chance to spread inside a drop. If the drops are quite small, the colors overlap to yield a white rainbow.

A red rainbow can appear when the Sun is low. Since the sunlight must then travel a long distance through the atmosphere, the scatter of the light from the air molecules along the path removes most of the blue end of the visible spectrum, and the drops are illuminated primarily with red light.

At night, rainbows can appear in light from the Moon, but they lack color because your eyes detect colors poorly in the dark. Moon-lit rainbows are rarely noticed because of their dimness and because people don't look for rainbows at night.

Rainbows can be seen in fog, on a cloud bank, or on a dew-covered lawn, but they are difficult to spot because the drops are often too small to give distinct colors, and the bows may be lost in the overall glare. They are white bands in the shape of hyperbolas or ellipses because of your perspective in seeing them on a horizontal surface. You can also see a bow on a body of water if the pond's surface is partially covered with floating drops.

If a normal rainbow forms over a body of water, you might see a *reflection rainbow* in the water. However, this reflection rainbow is not just a reflection of the rainbow you see above the water, because a different set of raindrops are involved. To form a reflection rainbow, light rays must enter raindrops, reflect once (or twice) inside them, leave them, and then reflect from the water surface to you. The drops doing all this are angled differently in your view than the drops giving you the normal rainbow. As a result, the reflection rainbow does not overlap the normal rainbow.

The seemingly vertical band that is sometimes seen near the foot of a normal rainbow also results from a reflection of light by a water surface. However, in this situation, the light first reflects from the water surface and then illuminates raindrops. Those drops that are angled properly in your view to send you colored rays produce the vertical band (Fig. 6-2). In some rare cases, you may see a complete extra rainbow arched over the normal rainbow. (This extra rainbow is sometimes misinterpreted as being the third-order rainbow.) The normal rainbow is centered around the antisolar point, but the extra rainbow is centered around a point that is shifted upward from the antisolar point because of the change in the geometry due to the reflection. If only the legs of the extra rainbow are visible, they may seem to be vertical but are actually curved.

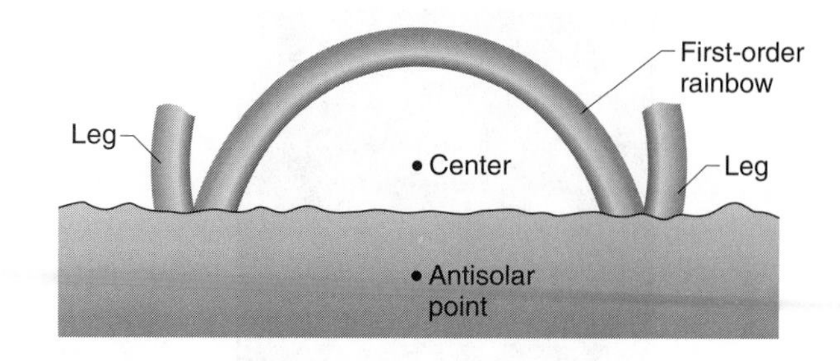

Figure 6-2 / *Item 6.2* Light reflected by water can form a rainbow that is centered on a point higher than the first-order rainbow. Here, only the legs of the extra rainbow are shown.

The ultraviolet light and infrared light in sunlight can also form rainbow bands. Although they are invisible to our eyes and do not have color in the normal sense, these bands can be detected with the appropriate equipment.

6.3 • Artificial rainbows

When water is sprayed into direct sunlight near you, why can two partially overlapping rainbows be seen? When a searchlight is angled up through light rain at night, why can two bright bands be seen in it (Fig. 6-3)?

In some locations, a rainbow-like display might be seen in the street even when the street is dry. There have been rare stories of rainbows being seen in mud and other unexpected places. What produces these colors?

You can see rainbow spots on a single drop of water that hangs from a paper clip if you direct a beam of light onto the drop in an otherwise dark room. With some care, you can see colored spots that correspond to the first dozen orders of rainbows (that is, up to a dozen internal reflections).

Answer When water drops are nearby, each eye sees the drops from a different perspective so that you perceive two rainbows that only partially overlap. When the drops are distant, your eyes have essentially the same perspective of the drops and the rainbows overlap completely.

Light from a searchlight is refracted and separated into colors by raindrops falling through the beam. Some of the drops are angled correctly in your view to send rainbow rays to you. The band farther from the searchlight corresponds to the first-order natural rainbow (the lower natural rainbow) while the other band corresponds to the second-order natural rainbow. As the beam rotates, the locations of the drops sending rainbow rays to you move up and down the beam; thus the bands move as well. The colors of the bands are dull largely because your eyes detect colors poorly at night.

Dry-street rainbows are due to tiny, transparent glass spheres sometimes added to the stripes on a street to reflect car light back to a driver, for the stripes to be more visible at night. If many of the spheres break free and spread over the street, they separate direct sunlight into colors just as water drops do. The other strange rainbows are more difficult to explain but are probably due to water drops, bits of glass, or other objects that break up white light into colors.

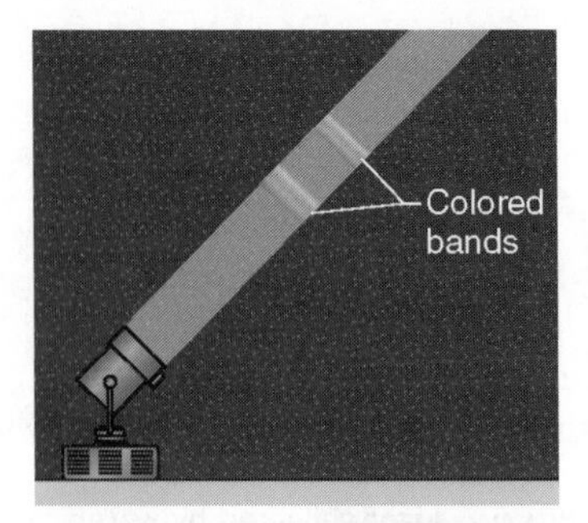

***Figure 6-3** / Item 6.3* Colored bands seen in a searchlight beam on a rainy night.

6.4 • The daytime sky is not dark

Why is the sky bright during the daytime? Apparently, the atmosphere somehow deflects the light toward you. However, if air is transparent, why doesn't the sunlight pass through it without deflection?

This question is often answered in terms of *Rayleigh scattering*, a model about how light scatters from air molecules. Albert Einstein pointed out that were this answer complete, the sky would be dark during the daytime.

To follow his argument, consider an overhead air molecule that scatters light to you. For simplicity, suppose that sunlight has only one wavelength. You also receive light scattered by other molecules that lie along the path extending from the first molecule to you. One of them should be positioned so that the light wave it sends to you arrives exactly out of step with the light wave from the first molecule. These two waves cancel to give darkness (Fig. 6-4). Since, on average, every molecule should have a partner molecule that cancels the light sent your way, you should receive no light, and the sky should be dark except directly toward the Sun. Right?

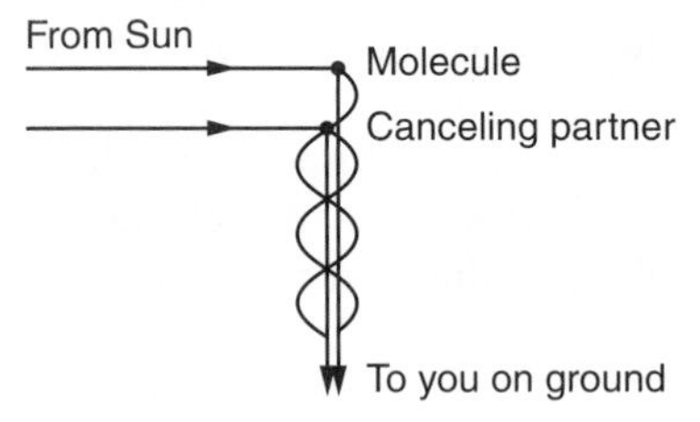

***Figure 6-4** / Item 6.4* Light waves cancel when scattered by two molecules half a wavelength apart.

Answer Light scatters from the air molecules according to Rayleigh's model, and Einstein's argument should apply. However, as Einstein noted, the sky is not dark because the atmosphere's density is not uniform. Moreover, the molecules continuously move and accumulate briefly, removing the possibility that at any given instant the light scattered from *every* molecule is eliminated by a partner molecule—the sky is bright because the density of air molecules is nonuniform and fluctuates in time.

6.5 • Colors of the sky

Why is the daytime sky blue? Is the blueness due to water or aerosols in the air, or is it due to the air molecules themselves? What would happen to the blueness if the atmosphere were much thicker or thinner? Why isn't the sky violet (deep blue)?

Why is the region near the horizon whiter than the overhead sky? Where is the bluest section of the sky? Why isn't the whole sky uniform in color? Is the sky blue on a moonlit night? (Although the sky is too dark for your eyes to detect any color, the sky may still be colored.)

Why are sunsets red? Shouldn't the last colors be red and yellow, giving a mixture that is orange? Why is there sometimes a sharp line of contrast in the red region of the sky?

Answer The daytime sky is mostly blue mainly because the air molecules more strongly scatter the blue end of the spectrum out of the sunlight than the red end. So, when you look at the sky away from the Sun, you intercept light that is dominated in intensity by the blue. It is not a pure blue because you also intercept other, fainter colors that are also scattered by air molecules.

Although violet light is scattered even more strongly than blue light, the sky does not appear to be violet because sunlight is dimmer in violet than in blue and also because your eyes are much less sensitive to violet than to blue.

This type of scattering by molecules is usually called *Rayleigh scattering*, after a model developed by Lord Rayleigh around the end of the 19th century. He was convinced initially that a pure gas (having no aerosols or dust) was transparent and that therefore the blue of the sky was due to the scattering of light from small particles, not molecules.

Although water and ozone molecules in the atmosphere absorb the red end of the spectrum and so leave the blue end, they do not appreciably color the sky blue. There is too little water in the atmosphere for its absorption to matter, and sunlight during the daytime travels too short a distance through the ozone layer for ozone absorption to matter.

The sky on the horizon is white because light from distant molecules undergoes multiple scattering before it reaches you. The nearer molecules send blue light in your direction, as do farther molecules. However, the long travel distance of the light from farther molecules means that the light undergoes additional scattering, with the blue component becoming weaker. That light ends up being red-dominated and combines with blue-dominated light from nearer molecules to create the whiteness of the horizon sky.

The sky at night is blue but the light is too dim for your visual system to perceive the color. However, the blue can appear in long-exposure photographs.

When you see a sunset, you intercept light that has traveled a long way through Earth's atmosphere rather than directly down through the atmosphere from an overhead Sun. Along this long route, the blue end of the spectrum has been scattered out by air molecules, leaving primarily the red and yellow end of the spectrum headed toward you. The peak in intensity should be at a wavelength of about 595 nanometers, and the net color should be orange. However, most sunsets are shifted more into the red because small particles in the atmosphere help scatter out all but red light.

A sunset may sometimes have a sharp division in the sky colors if you view it through a *thermal inversion* in which the air temperature increases with altitude. When light near the sunset passes through such a variation in air temperature, it is refracted and you do not intercept it. You still intercept light from somewhat lower and somewhat higher sections of the sky, but because of the missing section in your view, the color of the lower section does not blend smoothly with the color of the higher section.

6.6 • Blue mountains, white mountains, and red clouds

Suppose that you examine the color of a range of dark mountains stretching away from you. Why are the intermediate ones blue and the somewhat more distant ones even bluer, but the ones on the horizon are white? Why are somewhat distant, brightly lit snowfields sometimes yellow? Why are very distant clouds sometimes red? You might think that this coloration occurs only during sunsets when some of the sky is red, but it can also occur even when the Sun is high.

Answer A dark mountain at an intermediate distance is blue because, as explained in the preceding answer, the air between you and the mountain scatters more of the blue end of the spectrum to you than the red end of the spectrum, and so you see a blue tint overlaid on the dark image of the mountain. If a mountain is somewhat more distant, there is even more air between you and the mountain to scatter blue-dominated light, and thus the blue tint is more perceptible. If a mountain is on the horizon, it appears whitish for the same reason that the horizon sky itself appears whitish (see the preceding answer).

The color of a somewhat distant, brightly lit snowfield differs from the color of a mountain at the same distance, because the snowfield scatters bright white light toward you but the mountain does not. As the light scattered by the snowfield travels toward you, the blue end of the spectrum is weakened by scattering due to air molecules along the way, and the yellow–red end tends to dominate the light reaching you. However, the illuminated air molecules between you and the snowfield also send you light, which is dominated by the blue end of the spectrum. The combination of the blue-dominated light from the air molecules and the yellow–red-dominated light from the snowfield is a whitish light with a yellow tint.

Like a distant snowfield, a distant cloud also scatters bright white light toward you, and you might think the cloud would also have the same yellow tint. However, a very distant cloud can have a noticeable red tint. This difference in coloration is due to your ability to see clouds at greater distances than snowfields. The added distance means that the light from the clouds undergoes more scattering by air molecules, which reddens that light.

6.7 • Sailor's warning

Is there any truth to the saying, "Red skies at night, sailors' delight; red skies in the morning, sailors take warning"?

Answer The saying may have some truth where storms usually approach from the west and do so in what are called *storm systems*. If the western sky is red at sunset, then the area west is free of storm clouds that would block the sunlight skirting Earth's curve, and you would probably have good weather for a few days. However, if the eastern sky is red at sunrise, then the area east is free of storm clouds, and the next storm may be coming in soon from the west.

6.8 • Sunsets and volcanoes

Why do volcanic eruptions produce brilliant sunsets worldwide? Such unusual sunsets were apparently the inspiration behind several paintings by Edvard Munch, including his famous painting *The Scream*, which depicts someone in existential despair against a blood-streaked sky. Munch would have seen brilliantly colored sunsets in his homeland of Norway soon after the explosion of the island Krakatoa near Java in 1883. That explosion hurled ash into the upper atmosphere, and the ash spread around the world. When the ash reached the high northern latitudes of Norway, Munch must have found the sunsets to be particularly wrenching and threatening.

Answer The ash and other particles spewed upward by a volcanic eruption form a layer at an altitude of about 20 kilometers. Part of this material is sulfur dioxide, which reacts with the ozone at that altitude, forming sulfates that then precipitate to form an aerosol.

This layer of ash and aerosol curves around Earth. The colors you see in the sky around a sunset are a combination of sunlight scattered from that layer and from the air molecules below and above it. The light reaching you *from below the layer* tends to be red, because it passes through the dense portion of the atmosphere, losing much of its blue due to scattering by the plentiful air molecules along its path. The light reaching you from this layer also passes low through the atmosphere, but it may lose some of its red light to absorption by the ozone. The light reaching you from above the layer passes through thinner air and loses only a little of its blue. When you look at the sky around a sunset, you can intercept light with a variety of stunning colors that can vary across the sky and also from night to night. Because the aerosols and volcanic ash can last for months, so can the striking sunsets.

If distant clouds below the western horizon block some of the light reaching the underside of the layer, you will see horizontal variations in color that may produce a sunset with strikingly different color distributions on the left and right sides of the Sun.

6.9 • Bishop's ring

In August 1883, the island of Krakatoa near Java in the Southeast Pacific erupted in a huge volcanic explosion. In September, Reverend Sereno, Bishop in Honolulu, described a "peculiar corona or halo extending 20 to 30 degrees from the Sun, which has been visible every day with us, and all day, of whitish haze with pinkish tint, fading off into lilac or purple against the blue." What causes this halo, now called *Bishop's ring*, which often accompanies major volcanic eruptions? What determines its size and the color of its edge?

Answer Bishop's ring is due to the *diffraction* of light by the small particles spewed into the upper atmosphere by a volcano. (Diffraction is a type of scattering; here it means that small particles spread white light into its various colors.) The larger particles from the eruption settle out of the atmosphere, but the small particles remain aloft and are spread around the world by atmospheric circulation at high altitudes. When you look off to the side of the Sun, in addition to the normal sky light, you also intercept light that has been diffracted in your direction by these small particles. This additional light makes a circular region around the Sun of exceptional brightness, as if there is a ring around the Sun.

The size of the ring depends on the extent of the diffraction, which is larger for smaller particles. The edge of a ring can be red because red light, with the longest wavelength in the visible spectrum, is diffracted (spread) most. Thus, a particle on the edge of the ring off to, say, your right can diffract red light to you but not the less-diffracted blue light. However, the edge can be purple if you intercept both the diffracted red light and the normal blue light of the sky at the edge. The ring has a sharper and more colorful edge if the dust particles all happen to be about the same size and thus diffract light similarly. If it has a large range of sizes, the ring is diffuse and whitish.

6.10 • Cloud-contrast bow

During your next airplane flight, examine the texture of the clouds near the antisolar point (the point directly opposite the Sun). Chances are that you can see more texture in the clouds within the region about 42° from the antisolar point than outside it. Why should the texture change so?

Answer You can distinguish features in the clouds when there is good contrast between regions that brightly scatter sunlight and adjacent regions shadowed by the clouds. The brightest scatter of light by the drops in the clouds occurs within 40° of the antisolar point, and so you see the best contrast within that region.

6.11 • Sky colors during a solar eclipse

During a complete solar eclipse, why does the horizon turn red while the overhead sky turns bluer than it was before or after the eclipse?

Answer Normally, the light from the direction of the horizon is white. Air molecules that are not too distant from you scatter more blue light in your direction than red light. Much more distant air molecules scatter the same color distribution toward you, but since the light must travel a greater distance to reach you, it loses much of its blue because of scattering by the air molecules along the way. Thus, closer molecules contribute blue light while distant molecules contribute red light, and because the combination appears white, so is the horizon. However, when you are in the shadow cast by a full solar eclipse, the closer molecules are not illuminated, and you receive only the red light from the distant molecules, and so the horizon looks red.

The overhead sky is bluer during a full eclipse, because the shadow eliminates light scattered to you from the lower portion of the overhead air. That light is normally tinted red, because to reach the overhead air, the light travels through the denser part of the atmosphere. Along the way, it loses much of its blue because of scattering by the plentiful air molecules, and so when it finally scatters toward you from overhead, it is dominated by red light. The light that scatters to you from the higher portion of overhead air travels through the less dense part of the atmosphere, encounters fewer molecules, and thus loses less of its blue. So when it finally scatters toward you from overhead, it still has most of its blue. The color of the overhead sky is normally a combination of red-tinted light from the lower air and bluer light from the higher air. However, an eclipse eliminates the red-tinted light, and so during an eclipse the overhead sky is bluer than normal.

6.12 • When the sky turns green, head for the cellar

When I was young and living in Texas, my grandmother always routed us to the cellar whenever a severe storm turned the sky greenish, because she (and others) believed that the green color signaled the possibility of a tornado. Why would the sky turn that color instead of just becoming dark?

Answer Two features are required for the green light: (1) The incident light must have lost the blue end of the visible spectrum due to scattering by air molecules; thus it must be from a low Sun. (2) As this light passes through the water drops in a cloud, the water absorbs the red end of the visible spectrum. If portions of the cloud have the proper thickness (as might be the case in a severe storm), perceptible light emerges from the cloud that is depleted at both the blue and red ends of the visible spectrum, and so that light is tinted green or green-yellow, the remaining colors in the visible spectrum (Fig. 6-5).

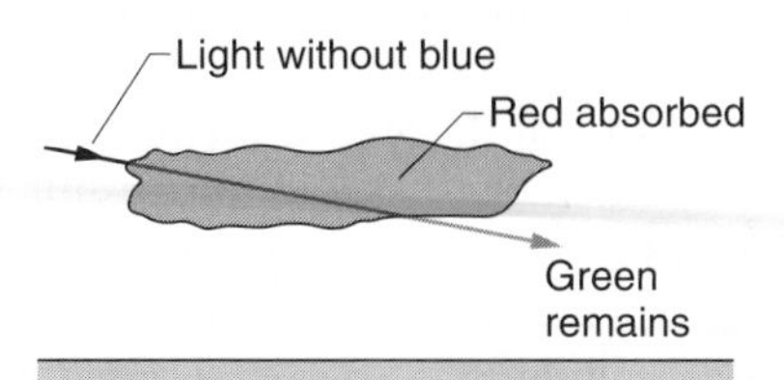

Figure 6-5 / *Item 6.12* Light from a low Sun loses the red end of the spectrum in the cloud, and only green light remains.

6.13 • Enhancement of overhead blue

Why does the overhead sky turn bluer during a sunset? Shouldn't it become red because sunsets are red?

Answer During a sunset, the light that scatters to you from overhead has traveled to that region along a long slanted path through the atmosphere and also through the ozone in the stratosphere (Fig. 6-6). Because ozone absorbs in the red end of the spectrum, and because the light travels so far through the ozone, this light becomes blue-dominated even before it scatters to you. This bluing makes the overhead sky extra blue during a sunset, especially about 20 minutes after the Sun has disappeared below the horizon.

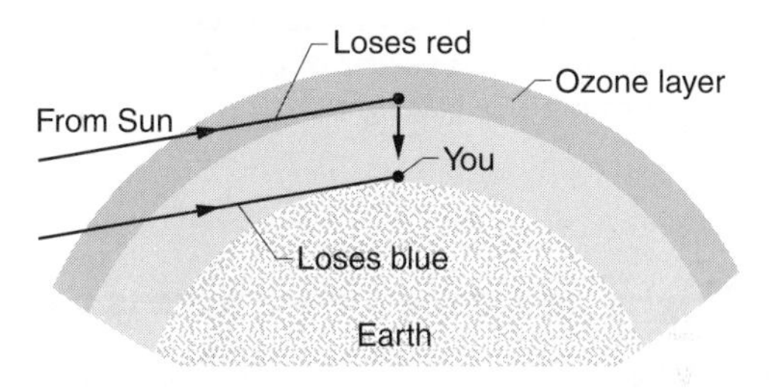

Figure 6-6 / *Item 6.13* Sunlight passing through the lower atmosphere becomes red. Sunlight passing through the ozone layer becomes blue.

6.14 • Dark patch and rosy border during sunset

During a sunset, why does a dark patch rise out of the eastern horizon (Fig. 6-7*a*)? Why is the patch's top border, which is called the *belt of Venus*, often red or orange? Why is the interior sometimes a faint blue?

Answer The dark patch is Earth's shadow on the atmosphere; the shadow rises in the east as the Sun descends in the west. The top of the shadow is illuminated with light that is reddened during its long passage through the atmosphere, as many air molecules scatter blue out of the sunlight (Fig. 6-7*b*). Some of the light reaching the top of the shadow then scatters back to you, and so the top appears to be red.

There are several reasons why the interior of the shadow may be blue. First, the light from inside the shadow cannot come directly from the Sun but must be scattered there by high-altitude air that is still directly illuminated. Because the high-altitude air is thin, that light does not lose as much blue as the light traveling through low-altitude air. If it travels appreciably through the ozone layer, it becomes even bluer, because ozone absorbs the red end of the spectrum. Some of this blue-dominated light scatters into the shadow and then scatters to you. Because the background of the shadow is dark, sometimes you can see this dim blue-dominated light.

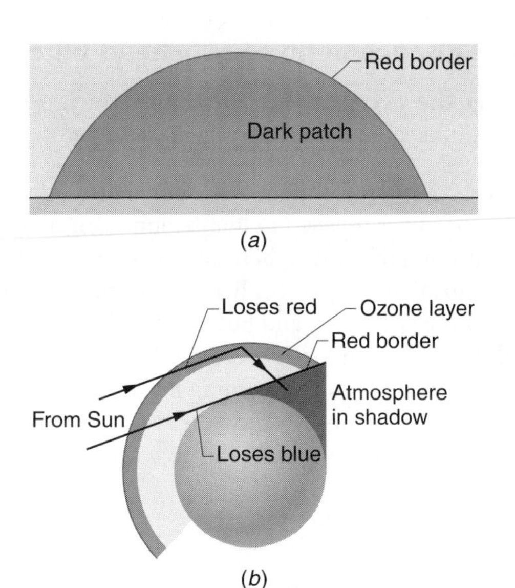

***Figure 6-7** / Item 6.14* (*a*) A dark patch rises in the east during sunset. (*b*) Red light forms border of Earth's shadow on sky. Interior is faintly colored by blue light.

6.15 • Bright and dark shafts across the sky

Sometimes when the Sun is low, bright or dark shafts fan across the sky, diverging from clouds near the Sun or converging toward clouds near the point just opposite the Sun. If you are very lucky, you might even see shafts over most of the sky. Why do they appear, and why aren't they parallel? After all, the Sun is so distant that sunrays are approximately parallel.

Answer The shafts have a variety of names, including *sunbeams*, *rays of Buddha*, and *Buddha's fingers*. They are almost parallel but appear to diverge or converge because of your perspective. (A similar illusion of convergence can be seen if you look along straight railroad tracks that extend a long distance away from you.) The shafts usually form when clouds near your view of the Sun throw their shadows across the sky. If there is only one small cloud, its shadow is a dark shaft. If the clouds are more extensive, you see bright shafts because of light sneaking through spaces in the clouds. (In some locations, shafts can form when light sneaks through mountain tops.) Some of the light then scatters to you from dust, rain, snow, aerosols, or air molecules along the light's path; you can distinguish the bright shafts because of their contrast with the intermediate regions of shadow.

The shafts are difficult to see overhead, because in that perspective you are looking through their narrow width and therefore intercepting only a little of the scattered light. The shafts are easier to see when you look toward the Sun or the point opposite it, because then you view somewhat along their lengths. That means you intercept more of the scattered light, improving the contrast with shadows.

Similar shafts of light are commonly seen when bright, direct sunlight passes through very dusty air in a room that is otherwise dimly lit. You can see the shafts because dust scatters the light to you and because that scattered light is not lost in the normally brighter light reflected from the furnishings behind the shafts.

6.16 • Blue haze, red haze, brown haze

Some vegetated mountain ranges, such as the Blue Ridge Mountains of Tennessee and the Blue Mountains of Australia, are known for their blue haze. The haze is not due to smoke, because the regions are relatively uninhabited. It is also not due to windswept dust because the blue disappears during strong winds. It is not a fog because it is most obvious during warm weather. What produces the blue haze?

Sometimes when the ground or ocean is covered with haze, that surface may be invisible from a plane flying at cruising altitude. Why is this haze often red?

Why is the haze in a city often brown? Is the color due to the haze particles absorbing certain other colors? Or does the scattering of light from the particles leave a brownish light? Does the color depend on the colors of the background against which you view the haze?

If you look toward the Sun through an extensive haze, why is the Sun surrounded by a bright white region? Why is the region surrounding the Sun sometimes red?

Answer The blue haze is due to aerosols that are produced when large molecules, such as terpenes, are released by vegetation. In addition, particles of wax, from the pointed ends of pine needles and other plant surfaces, may be released when strong electric fields are created by passing electrified clouds. In either situation, the particles are small enough that they preferentially scatter blue light to you from the sunlight illuminating them, leaving the region with a blue hazy look.

The red haze is likely due to dust and aerosols that are somewhat larger in size (about 0.1 micrometer in diameter). Particles of that size preferentially scatter red light from the sunlight.

The haze in a city consists of water droplets into which various compounds have dissolved. One of the compounds is nitrogen dioxide, which absorbs enough of the colors in sunlight to make the light from the haze brownish. The color of the background, such as brick buildings, can also play a role if a lot of the light illuminating the haze comes from the background.

When you look up at the Sun through an extensive haze, the suspended particles scatter additional sunlight in your direction. The scatter is primarily in the forward direction, and so the brightest area is seen near the Sun. When the Sun is high, the haze is illuminated by white sunlight, and so the area surrounding your view of the Sun will probably be white. When the Sun is low, the haze is illuminated by light that has been reddened by a long passage through the atmosphere, and so the surrounding area is red.

6.17 • Lights of a distant city

As you drive toward a distant city at night, why do you see a dirty orange glow over it? Why is the light from a distant Christmas tree primarily red even if the tree is actually covered with lights of many colors?

Answer Several factors color the region above a distant city, one of which is the color of the lights in the city. Even if the lights are white, however, the region will probably be orange or red if a haze hangs over the city. When light scatters from the haze and travels to you, the blue end of the spectrum is diminished by scattering from air molecules along the way, and only the red end of the spectrum reaches you. In addition, the initial scatter of light by the haze may selectively send the red end of the spectrum toward you if the particles are 0.1 micrometer or somewhat larger in size.

If you look at a distant Christmas tree, the blue light traveling toward you is similarly weakened by scattering from the intermediate air. If the distance is sufficient, you see only the red lights on the tree.

6.18 • How far is the horizon?

Is the horizon where Earth's curvature prevents you from seeing any more-distant point on Earth's surface? How does the distance to the horizon depend on your elevation above Earth's surface? Does the horizon remain distinct if you ascend?

Answer Because the air density decreases with altitude, the visible horizon may lie beyond Earth's curve. Consider rays of light originating somewhat beyond the curve and angled in your general direction. If the rays traveled in a straight line, they would pass well overhead and not be seen, but the decrease in air density encountered as they climb causes them to *refract* (bend) slightly downward toward you. You may intercept some of these rays, and thus you see a horizon beyond Earth's curve. Generally, the higher you are, the more distant the horizon is. However, for altitudes above a few kilometers, the extensive horizon disappears because of blurring due to the extensive scattering of light by particles in the atmosphere.

6.19 • Color of overcast sky

Why does the color of an overcast sky in the countryside change with the seasons, appearing to be greener in the summer than in the winter?

Answer The green tint during the summer is due to light that first scatters from vegetation and then scatters to you from drops in the clouds.

6.20 • Maps in the sky

In ice fields of the far north, large maps of the surrounding terrain sometimes appear on the base of overhanging clouds or haze. Such a map, called an *ice-blink* or a *cloud map*, enable a traveler to pick a route through the ice field if kayaking or over the ice if sledding. It may show the ice field as far as 30 kilometers. What accounts for ice-blinks? Can they be seen under other circumstances?

Answer The ice regions reflect more sunlight onto the base of the overhanging clouds than the open water. The variation of illumination on the base of the clouds thus mimics the ice field—dark regions correspond to the waterways while brighter regions correspond to the ice regions. Similar maps can appear on an overhanging haze.

6.21 • Brighter when it snows

Some observers have noticed that during a wintertime fog, visibility noticeably improves when snow begins to fall. And when there is no fog but the sky is overcast, the brightness seems to increase quickly if snow begins to fall. Why does snow alter visibility and brightness?

When the sky is overcast, why is snow at the horizon brighter than the adjacent sky?

Answer When snow falls through a fog, the snow crystals sweep out some of the suspended water drops. They also steal water molecules from those drops that remain in suspension, thereby reducing the size of the drops. Both factors can diminish the extent of the fog and improve visibility. On fogless days, a sudden snowfall increases the brightness because light is brightly reflected from the fresh snow on the ground.

During an overcast day, snow at the horizon is brighter than the adjacent sky for three reasons: (1) Drops in the cloud scatter sunlight primarily in a forward direction, and so you intercept more light from the overhead portion of the overcast sky than a portion near the horizon. Thus, the sky near the horizon is comparatively dark. (2) Snow scatters light strongly in all directions, and so you intercept significant light from snow near the horizon. Thus that snow is bright. (3) When you view a border separating regions differing in brightness, your visual system will enhance that difference in order to make the border more distinct.

6.22 • The end of a searchlight beam

Searchlights were used in World War II to locate enemy aircraft on dark nights; these days they are used to attract customers to movie premiers. Why does a searchlight beam abruptly end instead of fading out or simply extending "forever"?

Answer The beam dims along its length because it spreads and because air and airborne particles scatter the light. (If something did not scatter light in your direction, you would not see the beam.) The dimming due to the scattering is rapid, giving a rather sudden end to the beam.

SHORT STORY

6.23 • Newgrange winter-solstice sunbeam

Newgrange is a large mound built by Neolithic people in about 3150 B.C. in what is now Ireland. It consists of a stone lattice overlaid by dirt, with a single entrance facing south and a 20 meter passageway extending to the mound's center. A small opening, a *roof box*, lies over the stone slab, forming the top of the entrance. The purpose of the roof box was lost over the ages until 1969; upon sunrise at the winter solstice, the Sun sends a pencil beam of light through the roof box and along the passage way to illuminate the burial chamber at the mound's center. This is no fluke alignment but one that was engineered as the Neolithic people wrestled their stone lattice into place when constructing Newgrange.

Similar optics occurs at MIT: Near sunset on certain days, the Sun happens to send light from the main entrance (77 Massachusetts Avenue) along a main corridor, a distance of 251 meters. Although there is no official burial chamber at the end of the corridor, many students have met academic death in or near the corridor.

6.24 • The green flash

As the Sun sets on a clear horizon, you might see a brilliant flash of green just as the top of the Sun disappears. The color is not an afterimage, because it has been photographed and it can also be seen just as the Sun rises. In high latitudes, the flash may be visible for longer times, perhaps 30 minutes, as the Sun moves along the horizon during its rise at the end of a long winter night. Clear horizons such as over the ocean are normally necessary.

Although much rarer, a red flash may be seen when the Sun peeks out beneath a cloud while low in the sky. (If you search for either the green or red flash, take great care when looking directly at the Sun, because the light can easily damage your retina even if you feel no pain. Never look at a high Sun for more than a second and never with binoculars or a telescope. When the Sun is about to set, the light is dimmer because of the absorption during its long travel through the atmosphere to reach you, and the observation is somewhat safer.)

What produces the green flash and the red flash?

Answer Several factors are involved in the production of the green flash. The principal factor is the separation of colors in sunlight as the light path is bent by Earth's atmosphere. Here is the traditional argument: When the Sun is low, its image is spread slightly according to color. Lowest is the image formed by the red light, and progressively higher are images formed by yellow, green, and blue light. Until the Sun is setting, you see only a single, composite image. But just as the top edge of the Sun slips below the horizon, the images in the red and yellow end of the spectrum are eliminated and only images in the blue and green are left. However, the blue image is too dim to be perceived because of the scattering of light during its long travel through the atmosphere. So, the last, fleeting image is dominated by green light—the green flash.

However, there is a twist to this traditional argument. If you watch the Sun as it sets, the eye photopigments responsible for seeing the red end of the spectrum stop working—they are said to be bleached. If the last light you see from the Sun is primarily yellow, you will perceive it as green because of that bleaching. If you were to suddenly turn toward the setting Sun as the last light is visible, you might see the last color as it truly is—namely, yellow—which is what a photograph would also show. This physiological switching of color is not a problem if you view a rising Sun, because the photopigments are not bleached.

If this argument is complete, why don't you see a green flash with every sunset on a clear horizon, and how can a green flash be photographed? An additional factor accounts for the rarity. The flash can be enhanced by atmospheric layers where the temperature varies. Sometimes the layering can separate the top edge from the rest of the Sun's image and magnify its size. When the magnified top edge turns green, you have the best chance of seeing the green flash. Under these conditions, the green flash truly occurs and can be photographed.

The exceptionally rare red flash occurs when the low Sun protrudes below a cloud such that you see only its red image, the other colors being slightly too high to pass beneath the cloud. The rarity seems to be due to the need for the Sun to be quite low but still visible beneath a distant cloud bank.

6.25 • Distortions of the low Sun

If you examine the Sun while it is low on a clear horizon, you may find that it appears to be oval instead of its actual circular shape. It may also be broken up into horizontal layers, distorted into some other shape (such as an uppercase omega Ω) or composed of separated images. Why does the image change in these ways?

Answer When the Sun is low, its image can be shifted upward by the refraction (bending) of the path taken by light as the light passes through Earth's atmosphere. Moreover, since the density of the atmosphere decreases upward, the bottom of the Sun's image is shifted upward more than is the top of the image. This variation in the shift decreases the height of the image. With the horizontal width unaltered, the image is an oval with its short axis vertical. The shift in the image is great enough that just as the image reaches the horizon, the Sun may truly be below Earth's curve.

The more complicated distortions to the image result from either reflection of the Sun by calm water or refraction through the atmospheric layers where the temperature varies. In a simple model of the lower atmosphere, the air temperature decreases with altitude, but if there are layers where the temperature happens to increase, the change in refraction gives you a layered image of the Sun or one that consists of separated pieces. The layers can also produce a mirage of the Sun that seems to join the bottom of its normal image, yielding a net image that is distinctly noncircular.

6.26 • Red Moon during lunar eclipse

When the Moon is fully within Earth's shadow during a full lunar eclipse, why is the Moon briefly red? Why isn't it red when there is only a partial eclipse with, say, 50% coverage?

Answer During a full lunar eclipse, the Moon should not be illuminated at all. However, sunlight passing through Earth's atmosphere is refracted into the shadow and dimly illuminates the Moon. As the light passes through the atmosphere, the blue end of the spectrum is diminished by scattering from air molecules, and so the light that reaches the Moon is dominated by the red end of the spectrum (Fig. 6-8). Some of the light then reflects to you from the lunar surface, producing a red Moon.

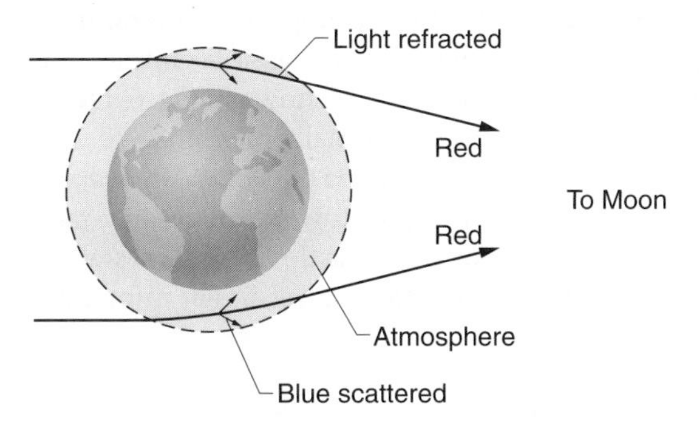

Figure 6-8 / *Item 6.26* Light rays lose their blue component while passing through Earth's atmosphere.

During a partial lunar eclipse, you intercept not only the red light that is reflected from the shadowed portion of the Moon but also the much brighter white light reflected from the rest of the Moon. Unless at least 70% of the Moon is covered, you cannot distinguish the dim red light next to the brighter white light, and so the shadowed portion merely looks dark. Similarly, you cannot distinguish the red light even during the early or late stages of a full lunar eclipse, because you intercept too much white light.

6.27 • Crown flash

When lightning flashes in the main body of a cloud, you might spot a brightening effect that ripples outward to the top of the cloud. Is this brightening, which is called *crown flash* or *flachenblitz*, an unusual electrical discharge or some peculiar reflection of the light from the lightning?

Answer Crown flash is a reflection of light from the lightning by ice crystals that are flat hexagonal plates. Normally, these plates fall with a flat side downward. However, after a lightning discharge, the plates flutter and some of them happen to briefly go through the orientation that reflects sunlight or the lightning's flash to you. The reason for the flutter is not known. It is due either to the sound wave from the lightning (it reaches you as thunder) or to the change in the electric field of the charged particles within the cloud.

6.28 • Oasis mirage

On warm days you may see a distant pool of water lying on the ground, but once you reach the spot it is dry. The water looked real, being blue and covered with small ripples. This classic *oasis mirage* can not only be seen but also photographed.

You can often see the mirage at night if you examine the light from an oncoming, distant car. Just below where you see a headlight, you may also see a streak of light on the road. If the streak is dull, you are seeing a weak reflection of the headlight by the road. However, if the streak is bright, you are probably seeing a mirage of the headlight. What produces this type of mirage? Can a bird flying over a road see the roadway mirage? That is, can a bird be fooled into thinking that an underlying road is a water stream?

Answer The mirage of water lying on the distant ground is actually an image of part of the sky just above the horizon in that direction. The ground (or any other surface) absorbs sunlight and heats the air next to it. If the air temperature markedly decreases with height, the oasis mirage can appear. As light from the low sky travels toward the ground, passing through air with a continuously increasing temperature, it is continuously refracted (bent) upward until it finally heads upward at a shallow angle to the ground (Fig. 6-9).

If you intercept this light, your brain automatically interprets its origin as being a bright spot on the ground, along a straight, backward extension of the intercepted light ray. That bright spot is, of course, an illusion but it seems real. Moreover, if the actual origin is from a blue sky, the bright spot can appear blue, as if it is water. If the air is turbulent, the refraction of light noticeably varies and the spot shimmies, as if ripples play on the water.

The oasis mirage can appear in cold environments, since it does not require hot air, only a decrease in air temperature

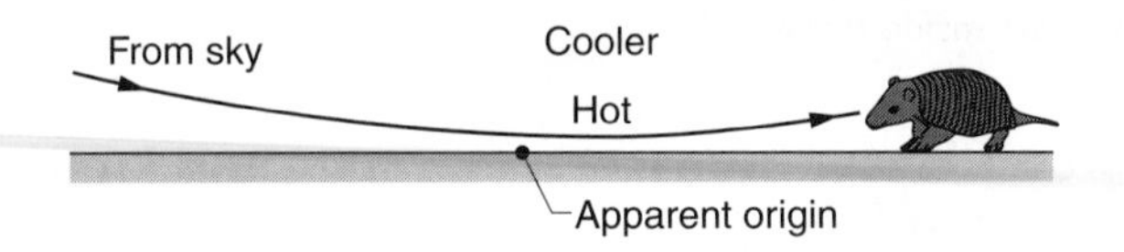

Figure 6-9 / *Item 6.28* Path of light from the low sky bent by a change in the air temperature. Observer perceives that the light originated on the ground.

with height. It is commonly seen on roadways because most pavements absorb sunlight and readily heat the adjacent air. It is usually more apparent if you view the scene from near the ground or look through a telescopic lens at distant ground.

Distant objects can also produce mirage images if light from them refracts through air near the ground. This type of mirage, as well as the oasis mirage, is said to be an *inferior mirage* because the image appears *below* the source of the light.

The night version of the mirage is due to a layer of heated air that lies on the roadway. The pavement may still be hot from daytime heating by the Sun, but it might also be warmed by the tires of passing cars and trucks.

Since the change in the direction of travel by light is slight, a flying bird cannot see a mirage of water on an underlying road. It might see a mirage off in the distance as you do, but the mirage continuously moves as the bird moves, similar to how the water mirage moves along a road as you drive.

6.29 • Wall mirage

A distinct mirage can sometimes be seen if you look along a long wall that faces the Sun. Stand at one end of the wall with your eyes near the surface while a friend stands at the other end. You may see a mirror-like image of your friend within the wall, seemingly connected at points to your friend. You may even be able to see two mirage images of your friend. I have photographed such mirage images with a telescopic lens mounted on a camera that I held right next to a wall. The technique worked best when I stood at a corner of the wall, so that the lens was aimed directly along the wall.

Answer The wall mirage has the same explanation as in the preceding answer, except that the layer of hot air lies vertically against the wall. Some of the rays that leave your friend skim through the hot air, refracting slightly away from the wall (Fig. 6-10). If your eyes, or the camera, are close enough to the wall's surface, you intercept some of those rays, and they appear to originate inside the wall.

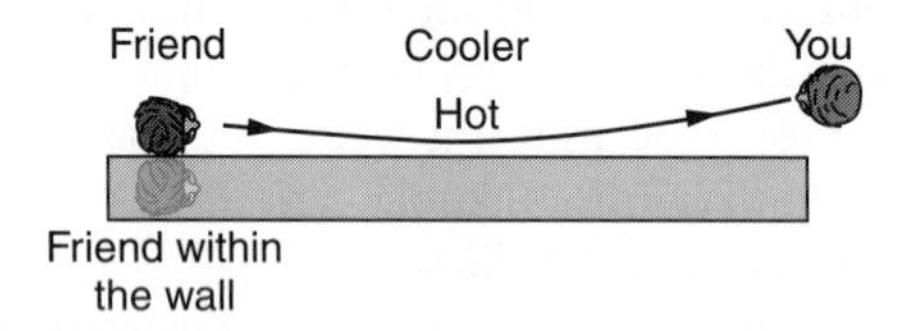

Figure 6-10 / *Item 6.29* Overhead view of wall heated in sunshine. Path of light ray bent by change in air temperature. Mirage appears inside the wall.

6.30 • Water monsters, mermen, and large-scale mirage

In some places in the late afternoon or early morning, a certain kind of mirage can allow you to see a mountain that lies beyond Earth's curve and thus should not be visible. The mirage may begin as a hazy patch above the horizon and sharpen until the mountain becomes distinguishable.

This type of mirage may account for how Erik the Red discovered Greenland. According to legend, when he was exiled from Iceland by the other Vikings, he headed directly toward the nearest part of Greenland, although not even the highest parts of Greenland should have been visible from the highest parts of Iceland under normal conditions. Perhaps on occasion he had spotted a mirage of the nearest part of Greenland and knew there was land.

Sightings of water monsters, such as the Loch Ness monster, may actually be examples of a similar but more local mirage. Under certain conditions and with the right perspective, a common log floating in water may seem to extend upward like the neck on a monster, no longer being recognizable as a log. The neck may even seem to oscillate, as if the monster is swimming.

Mermen, reported by medieval seamen, were huge monsters that rose out of the sea. They had shoulders like men but were handless and narrow at the waist. Since none were ever seen close up, no one knew if they had skin like a man or scales like a fish. Mermaids were similar but had breasts, heavy hair, webbed hands, and tails. How might such apparitions arise? Why did reports of them drop off once ships with high decks came into use?

Sometimes distant objects appear as several images in a paper-doll-like display. Some classic examples, involving images of a distant sailboat, are shown in Fig. 6-11*a* but without the distortion that is normally present.

The most beautiful mirage is the *Fata Morgana*, where distant objects appear to be towers of a fairy-tale castle. According to legend, the castle is the crystal home of Morgana the fairy. In some cases, a similar mirage can create

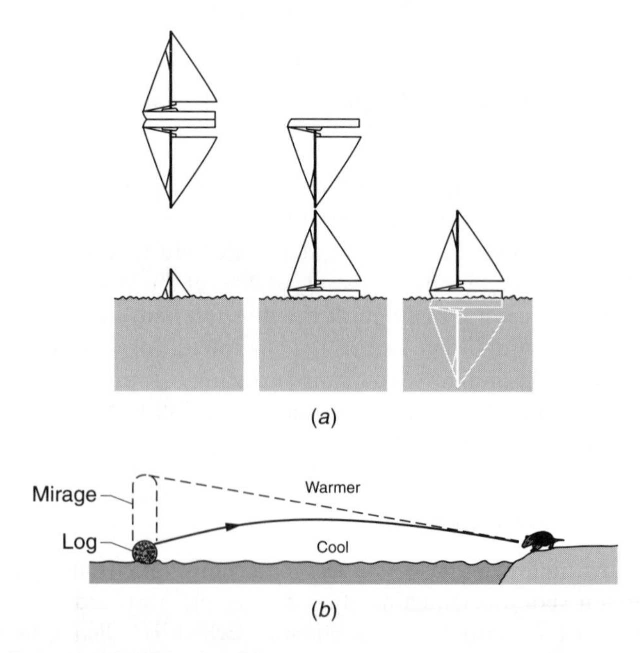

Figure 6-11 / *Item 6.30* (*a*) Paper-doll-like mirage. (*b*) Path of light from a floating log.

the illusion that distant people are walking on water that actually lies between you and them.

In 1597, while stranded on the Arctic islands of Novaya Zemlya during a search for the Northeast Passage, some of the crew of Captain Willem Barents saw the Sun's first appearance at the end of the long winter darkness. However, the Sun was still 4.9° below the horizon, so how was it seen? Such mirage images of the Sun are often highly distorted, perhaps taking on the appearance of a stack of pancakes. The images usually lie in a dark "window," with brighter sky above and below it.

The *hillingar effect* is a mirage that seems to lift the horizon (usually at sea), leaving it flat or shaped like a saucer. During the *hafgerdingar effect* (the word is Icelandic for "sea fences"), the horizon becomes irregular as if bounded by randomly placed vertical structures.

On U.S. Highway 90 near Marfa, Texas, observers have reported persistent displays of moving lights at night. The displays, known as the *Marfa lights*, wiggle across Mitchell Flat, just to the south of the highway.

How do these various types of mirages form?

Answer All of the examples of mirage are due to the refraction of light rays from their normal straight paths by layers of air in which the temperature varies with height. (See the item on the oasis mirage.) In the more complicated examples, several factors may be involved, each producing mirage images, sometimes inverted, to form a paper-doll-like display or strangely extended and distorted composite images.

Sightings of the Loch Ness monster, or the Manipogo monster in Lake Manitoba, most likely involve distorted images of logs or other objects floating in distant water. The conditions are ideal for such a mirage when cold water cools the air just above it while sunlight warms the higher air. Then some of the light from, say, a log is refracted downward from its initial upward travel so that you intercept it (Fig. 6-11*b*). The refraction is slight, so you must be near the water level to intercept the light.

Depending on the actual temperature variation with height, you may see several images or one extended image, none of which resembles a log. If the refraction varies because of rapid changes in the temperature distribution, the serpent-like object in the water may seem to swim. When a mirage, like this one, appears *above* the source of the light, it is called a *superior mirage.*

Statistics support the possibility that the Loch Ness monster is a mirage. About 77% of the monster sightings occurred from May through August, when the water warms more slowly than the air, resulting in an increase in air temperature with height. About 84% took place when the water was fairly calm, another condition enhancing the visibility of a superior mirage. Furthermore, many of the observations were made by people near the water level, where they would intercept refracted rays.

Mermen and mermaids were probably stretched and distorted images of walruses and whales as seen by seamen positioned near the water surface. The mirage disappears if the observers are too close to the animals or too high to intercept the light that refracts to create the mirage. So, after the height of ships increased, reports of the mythical beings decreased.

The Fata Morgana is due to refraction that vertically extends the image of small, distant objects, so that they appear to be towers or walls. If you view a distant person over an intermediate body of water, the image may be a superior mirage: You see most of the body as a lifted image, while the feet, which are not seen, seem to be submerged, as if the person were walking over the water.

Mountains that are too distant to be seen directly can sometimes be seen by means of a mirage if some of the light from them is refracted around Earth's curve. Images of the Sun can also be channeled around the curve. The most striking example is the Novaya Zemlya mirage, in which light rays are refracted downward at high altitude, pass by Earth's surface, return to high altitude, and then are refracted downward again, as if constrained by some large, curved pipe. Although the channeled image of the Sun remains bright in spite of the long travel through the atmosphere, the surrounding sections of sky grow dim because of scatter by air molecules and end up being a dark window around the Sun's image. The brighter regions above and below the dark window are parts of the nearby sky.

The hillingar and hafgerdingar effects also involve the downward refraction of light by air that increases in temperature with height. The hafgerdingar involves irregular patches of air temperature, whereas the Novaya Zemlya and hillingar mirages require the distribution of air temperature to be stable over a wide region.

Atmospheric refraction of light at night is responsible for a host of strange, transient lights such as the Marfa lights. The sources of the lights could be stars or planets near the horizon (in which the images are an inferior mirage) or the headlights on a distant car or train (in which the images are a superior mirage). The Marfa lights are an example of headlight mirage: They are produced by distant cars headed toward Marfa on a highway coming across Mitchell Flat. The teenagers of Marfa should be thankful for this lively example of optics, because it gives them a ready excuse for parking in the dark along U.S. 90.

6.31 • A ghost among the flowers

Can you explain the following account of a ghostly image? One hot afternoon, a woman was gathering flowers on steamy ground, with the Sun somewhat off to one side. Suddenly she noticed movement in front her, and then she gradually realized that she was seeing an image of herself, with detail and color. Needless to say, the apparition soon unnerved the woman, and she fled.

Answer The apparition may have been produced in this way: Sunlight reflected from the woman's body to the drops in a mist rising from wet, hot ground. There, some of the light scattered from the drops back to the woman. I suspect that the woman must have been brightly illuminated by sunlight, while the mist was in a shadowed area, making the scattered light perceptible. This arrangement would then be somewhat like the ghostly images at Disney World, where bright images are projected onto a curtain that is so thin and porous that it is almost invisible in dim lighting.

6.32 • Shimmy and twinkling stars

When you view a distant object through air over a fire or a heated surface such as a sunlit road, why does the image of the object wiggle, an effect called *optical shimmy*? Why is such distortion more difficult to see above roadway at your feet than more distant roadway?

Why do stars twinkle? Do they twinkle more in the summer or winter? Why do they sometimes fluctuate in color? Why don't the Moon and planets twinkle? If you were in space looking at the stars, would they still twinkle?

Answer Shimmy is due to air turbulence that is produced by a heated surface. As light rays from an object pass through the turbulence, the incessant variations in the air density refract the rays in randomly changing directions, and the image that is seen distorts and dances.

For shimmy to be perceptible, the light rays normally must follow a long path through the layer of heated and turbulent air. If you look almost perpendicularly at a hot surface such as pavement, the rays you receive travel too short a distance through the turbulence to display shimmy. If, instead, you take a slanted view of the surface, the rays travel much farther through the turbulence, and then you can see shimmy. Usually this requirement means that the hot surface must be distant, such as a distant section of roadway. Even then shimmy may not be noticeable if the road texture is uniform. Shimmy is much more noticeable if the road has a striped pattern that wiggles.

Sometimes you see an effect related to shimmy: Fleeting dull shadows can appear on a flat, white surface because turbulent air refracts sunlight coming through it. Sometimes, refraction focuses the sunlight into a relatively bright patch, and sometimes it flares the sunlight so that, being spread out, it forms a relatively dark patch. An unstable arrangement of warm air flowing under cold air can also lead to shimmy and dull shadows. Because warm air is less dense than cold air, the boundary between the two temperatures is unstable and tends to form wave-like shapes that can focus the light in some regions and flare it into other regions.

Similar variations in refraction of starlight by the atmosphere slightly and rapidly alter the apparent position of a star. The apparent motion is visible because the star is, in your view, a bright point on a dark background. In addition, the changes alter the phase of the light waves reaching you. When light waves arrive *in phase* (in step), their interference is constructive and the star is brightest; when they are *out of phase*, their interference is destructive and the star is at its dimmest.

Your visual system sums the star's image over a short interval, but the variations in position and intensity are still visible. Were you in space the stars would not twinkle, but they would still appear to have small points because the light scatters within your eyes.

The Moon and planets are too large in your view to twinkle. Although each point of, say, the Moon wavers just as a star does, it is not an isolated point of light on a dark background and the wavering is imperceptible.

Stars twinkle more in the summer because the atmosphere is more unstable due to additional heating by the Sun during the daytime.

If a star is near a clear horizon, it may also fluctuate in color. The long passage of its light through the atmosphere allows air molecules, dust, and aerosols to scatter out some of the light's component colors; the star is then no longer its original color, which may be white. Incessant variations in the scattering alter the color you see.

6.33 • Shadow bands

For several minutes before and after a total solar eclipse, the ground may be covered with faint, rippling bands, each a few centimeters wide. Some observers report that the bands move. What produces the bands?

In 1945, another type of dark-band pattern was reported by Ronald Ives, who spotted it on six occasions while looking down from an elevated spot onto flatlands during sunset. The bands were several kilometers wide and traveling at about 60 kilometers per hour. What produced this set of bands?

Answer The eclipse shadow bands are most likely due to the focusing (producing bright bands) and flaring (producing dark bands) of the sunlight as the sunlight passes through the turbulent pockets of air in Earth's atmosphere. Their visibility is best when the eclipse is nearly complete and the visible portion of the Sun forms only a narrow crescent. Then the light occupies a very small angle in our view, and redirection of the light by high-altitude turbulent pockets can produce visible bands that are parallel to the crescent. When the eclipse is less complete and more of the Sun is visible, the turbulent pockets producing the bands are lower and the bands are less visible (more "washed out"). At their best, the bands are difficult to see because the contrast between adjacent bright and dark bands is weak. Because turbulence rapidly varies, the bands and their contrast also vary. That variation in the bands can give an illusion of motion.

The rare and rarely studied sunset shadow bands are probably also due to the focusing and flaring of sunlight by turbulence in the atmosphere. The bands were visible to Ives because the sunset narrowed the visible portion of the Sun, as in an eclipse. He would not have noticed the bands if he had been at ground level.

6.34 • The 22° halo and sun dogs

Sometimes the Sun is surrounded by a bright circle that may be colored red along its inner edge and blue along its outer edge. The angle between the circle and its center is 22°, and so this display is called the *22° halo.* (You can easily measure this angle: Open your hand while extending your arm toward the Sun, palm outward. The part of the sky you see along a straight line between the tips of your thumb and little finger occupies about 22°.) Sometimes the Sun is flanked on one or both sides by a bright, brilliantly colored spot called a *sun dog* or *parhelion.* What produces the 22° halo and the sun dogs?

Answer The 22° halo is produced by sunrays refracted (bent) by high-altitude ice crystals. The crystals are hexagonal columns called *pencil crystals* that fall approximately broadside while tumbling. When sunlight passes through a pencil crystal, it refracts and travels in a new direction that is 22° or greater from its initial direction. The deflection is concentrated at 22° and so the light is brightest at that angle. When you look 22° away from the Sun in any direction toward these falling ice crystals, you intercept some of that bright light and see a portion of the halo.

The halo may be colored because the refraction by the crystals separates the colors in the initially white sunlight. Red light is refracted slightly less than blue light, and the brightest red ends up on the inside of the halo.

The sun dogs are also due to refraction through hexagonal crystals but the crystals are flat instead of columnar. These *plate crystals* fall face down while fluttering. They redirect light in your direction only if they lie approximately along a horizontal line through the Sun, so the sun dogs are to the left and right of the Sun. When the Sun is low, the sun dogs are about 22° from it. When the Sun is higher, the sun dogs are slightly farther away. They are usually colored because of the color separation by the crystals, and the red part is closest to the Sun.

The halo and sun dogs can appear even during the summer because ice crystals can still form in the high-altitude cold air.

6.35 • A sky full of halos, arcs, and spots

In addition to the 22° halo and the sun dogs of the preceding item, there are many other halos, arcs, pillars, and bright spots that might appear in the sky. Figure 6-12 shows some possibilities, but not all of them can be present simultaneously because they may require different elevations of the Sun. Their shapes can also change with the Sun's elevation. Some of them are so rare that only a few people have seen or photographed them.

One commonly seen display is a *pillar of light* that extends above or below the Sun or Moon. You might even be able to see such pillars stretching into the night sky from the lamps that line a street or parking lot.

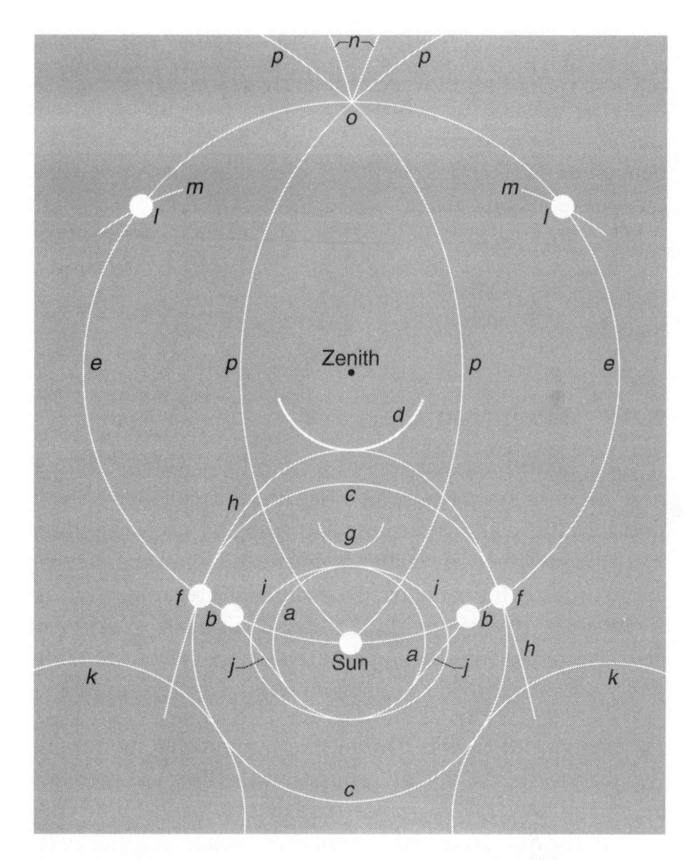

Figure 6-12 */ Item 6.35* Some possible bright sky displays. (*a*) 22° halo. (*b*) Sun dogs to 22° halo. (*c*) 46° halo. (*d*) Circumzenith arc. (*e*) Parhelic circle. (*f*) Sun dogs to 46° halo. (*g*) Parry arc. (*h*) Supralateral tangent arcs to 46° halo. (*i*) Tangent arcs to 22° halo. (*j*) Lowitz arcs. (*k*) Infralateral tangent arcs to 46° halo. (*l*) Paranthelia. (*m*) Paranthelic arcs. (*n*) Narrow-angle oblique arcs to anthelion. (*o*) Anthelion. (*p*) Wide-angle oblique arcs of anthelion.

When you fly, you might like to search for the *subsun,* a bright spot that lies below your view of the Sun and seemingly tags along with the airplane.

What is responsible for these various displays?

Answer The displays are due to falling ice crystals intercepting and redirecting sunlight toward you. Some of them are due to pencil crystals and others are due to crystal plates (see the preceding answer). Some of them, such as the *circumhorizontal arc,* involve light refracting through the crystals. Others involve light reflecting from the crystals. A few, such as the *Lowitz arcs,* require plate crystals that spin rapidly as they fall.

The pillar of light that extends above or below the Sun or Moon can be due to either pencil crystals or crystal plates, but the latter is responsible only when the Sun or Moon is low. In that situation, light reflects to you from the flat face of the plates as they fall approximately broadside. Slight fluctuations in the orientations of the crystals smear the region from which you receive light, thus forming the pillar. When the Sun or Moon is higher, the pillar is due to light reflecting from the sides of the pencil crystals that fall with their long axis horizontal. At each point along a column, some crystals happen to be tilted just right to reflect sunlight to you. The composite of these reflections forms the column.

The subsun is actually a version of a light pillar except that you look down on the plate crystals. When they are almost uniformly horizontal, they produce a mirror-like image of the Sun.

6.36 • Mountain shadows

If you examine the shadow of a mountain while standing near the mountain's summit when the Sun is low, you will find that the shadow is triangular, with the apex extending away from you (Fig. 6-13). Why do all mountains give approximately the same triangular shadow, regardless of the actual shape of the mountain or details along its sides? Why does the tip of the shadow have a distinct spike extending either left or right when you stand well below the summit?

Answer When you stand at the summit of a mountain, the shadow forms on the ground and within the aerosols below you. Most of the shadow is so distant that details of the sides of the mountains are too small for you to perceive in it. The triangular shape of the shadow is due to perspective: The sides of the shadow begin off to your left and right but appear to converge at a distant point on the ground and within the aerosols. The taper of the sides is like straight railway tracks that would seem to converge at the horizon if you were to look along them. The shadow itself is like your shadow on a lawn when the Sun is low—the shadow of your feet is normal size but the shadow of your head is tapered.

When you stand well below the summit and in the shadow of the mountain, you look out through the shadowed aerosols (rather than down on them). If you stand in the exact center of the shadow of the mountain, the sides of the shadow still taper directly away from you, but the end may also seem to extend upward. If you move from the center to, say, your left, you shift the summit and its shadow to your right. Moreover, because you are seeing the shadow through the aerosols, the shift of any part of the shadow increases with the distance of that part. Thus, the shadow of the summit, being the most distant, is shifted the most. As you follow the tapering of the shadow away from you, it curves upward but also toward the right, forming a spike at the end.

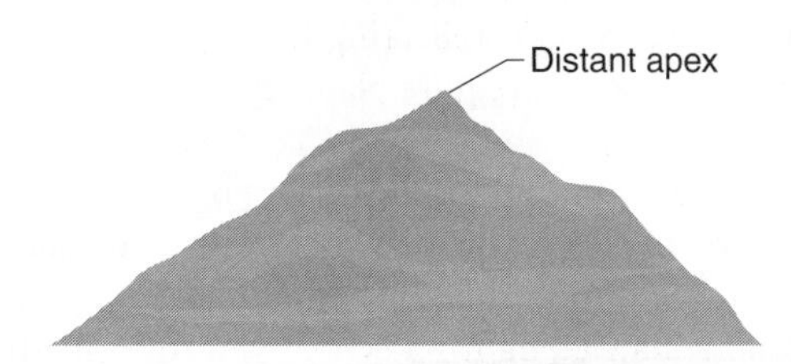

Figure 6-13 */ Item 6.36* Shadow of mountain on underlying aerosols as seen from near the summit.

6.37 • Disappearing cloud shadows

Suppose that you are flying over the ocean while isolated cumulus clouds dot the sky. Why can you see their shadows if you sit on the sunlit side of the airplane but not on the opposite side?

Answer When you look toward the ocean on the sunlit side of the airplane, most of the light you intercept is sunlight that has reflected or scattered off the water surface. This light is called *glitter*. If a cloud blocks part of the sunlight before the light reaches the water, you see a shadow on the water. The shadowed region lacks glitter and is thus darker than the surrounding water.

On the other side of the airplane, the reflection and scatter are primarily away from you. Although some of the intercepted light is a reflection of the sky, most is *upwelling*, which is light from the Sun and sky that penetrates the water surface and is then scattered back through the surface by suspended materials (or by the bottom). When a cloud casts a shadow on this side of the airplane, it does not appreciably reduce the upwelling in the shadowed region because the rest of the sky still illuminates that region. Thus, you cannot see the shadow. However, should the shadow drift from the water to land, the shadow is suddenly perceptible. Now there is no upwelling, and the only light you intercept from the ground is sunlight (not sky light) that is scattered from the ground. A cloud blocks the sunlight in the shadowed region, preventing such scattering, and thus you can see the shadow.

6.38 • Colors of the ocean

Ocean water does not have a single color. Instead, it can vary from the blue of a clear sky to the gray of an overcast sky. At times it may be white or reddish, or it may be bluish-green, green, or yellow. On occasion it may even be brown. What accounts for these colors? Why does the color often depend on your visual angle of the ocean surface?

While swimming deep in water, hold a flat, white object horizontally. Why is the object's top surface colored differently than its bottom surface?

Answer The colors of the ocean are varied because many factors are involved. Suppose that the water is pure, that the atmosphere is missing, and that the bottom of the water is too deep to scatter light to you. Then the water would be black with, perhaps, a faint blue tint due to the water molecules, which gradually absorb the red end of the spectrum and scatter the blue end.

In less ideal circumstances, suspended material can alter the water's color by selectively absorbing or scattering certain colors in the light. The bottom surface similarly alters the color by selective absorption, provided the water is shallow enough that light from the bottom is perceptible.

A small contribution to the water color is due to the overhead sky. If the sky is blue, the water may seem a bit bluer because you intercept some of the skylight reflecting from it. Similarly, if the sky is gray, then the water might be gray. However, if you face into the bright glare from the water on a sunny day, you will probably find that the water seems to be white because most of the light you intercept is sunlight that reflects from the water to form that glare.

The color of the light illuminating the top surface of a submerged white object has a certain tint because the light is modified by absorption and scattering as it travels down through the water. The color of the light illuminating the bottom surface is modified even more because the light travels farther through the water (down and back up). So, the top and bottom surfaces have different tints.

6.39 • Reflection glitter of Sun and Moon

What accounts for the shape of the bright area on open water when the Sun or Moon is low over the water? How does the shape change as the Sun or Moon changes elevation?

When the Moon is somewhat above the horizon, you might see a dark triangle just above the bright region on the water. What produces the dark triangle?

Answer If the water were perfectly flat, you would see a mirror-like image of the Sun or Moon in the water below the horizon. The image would be as far below the horizon as the light source is above it. However, when the water surface is covered with waves, the light reflects from a myriad of tilted surfaces and you see many, fleeting images of the Sun or Moon, whenever and wherever a surface has a tilt to send you a reflection. On average, these images lie in an oval or a path (either is called the *glitter path*) that extends away from you, with the left and right sides of the region of images converging toward the point of the horizon just below the Sun or Moon. The region of images is oval when the Moon or Sun is high; it is more of a path stretching to the horizon when the Moon or Sun is low.

The dark triangle above the lunar glitter path is probably an illusion due to contrast, because the bright glitter path just below the horizon lies next to the dark sky just above the horizon.

6.40 • Rings of light

Normally, the images reflected by waves on a water surface are distorted so rapidly that you cannot perceive them, but you can "freeze" them in place by photographing them with a camera that has a fast shutter. Suppose you photograph the reflection of a ship's mast. In the photograph, you might find that part of the image of the mast is a squiggly line, while other parts form isolated and closed loops. A squiggly image may not be difficult to explain, but what produces a loop? Can there be isolated, incomplete loops? Other objects also form curious reflections. Sections of land produce *landpools* (isolated patches of images of the land) while sections of sky produce *skypools.*

Answer The distortion seen in either a squiggly line or a loop is due to the curvature of the waves. When an image of a mast is a closed loop, the sky to one side of the mast ends up inside the loop while the sky on the other side ends up outside the loop. Loops from extended objects, such as a mast, are always complete. A point source of light, however, can yield a complex reflection with "loose ends" that correspond to the opening and closing of the shutter.

6.41 • Shadows and colors in water

Why can you see your shadow in a muddy pond or a clear puddle but not in a deep, clear pond? Why must the water be very muddy if you are to see shadows of other people?

While casting a shadow onto a water surface covered with small, randomly oriented waves, examine the surface around the shadow of your head. Bright spokes appear to extend from the shadow, as beautifully described by Walt Whitman in his "Crossing Brooklyn Ferry" in *Leaves of Grass.* What produces the spokes?

In clear water about a meter deep, examine the edges of the spots of light formed on the bottom by sunlight streaming through the leaves of an overhead tree. If the Sun is at your back, the spots are white. If you face the Sun, the spots have colored edges, red on the near side and blue on the far side. What accounts for these colors, and why does their appearance depend on your direction of view?

Answer You can see a shadow only if it is noticeably dimmer than its surroundings, like the situation when your shadow falls on a sidewalk. If you pour clear water over the sidewalk, your shadow is then less distinct because the top surface of the water reflects the sky and surrounding objects to you. Because some of the reflected images overlay your shadow, that region is not as dark as previously.

The contrast of your shadow with its surroundings is even weaker if the clear water is much deeper, because then the bottom surface in the surroundings no longer strongly scatters light to you. However, if the water is somewhat turbid, your shadow is more perceptible because light scatters to you from material suspended in the water around your shadow. Your shadow is then three-dimensional rather than flat as it is on a sidewalk. For this reason, you may not be able to see the shadow of someone else, because when you look toward it your view is along a slanted path through both illuminated and shadowed regions in the water. If you gradually increase the turbidity of the water, that other shadow gradually rises toward the water surface, and with extremely muddy water, it lies on the surface and is easily seen.

When waves move over moderately clear water, the waves focus light down into the water where suspended material then scatters light back out of the water. If you intercept this scattered light, you see bright lines where the light is focused and dark lines where it is not. The lines lie parallel to a line between the Sun and your eyes. Consequently, they seem to converge at (or radiate outward from) the point directly opposite the Sun—that is, within the shadow of your head. (Their appearance is like the apparent convergence of long, straight railroad tracks when you look along them.) The apparent rotation of the spokes is an illusion in which your brain forces order on a randomly changing pattern.

Spots of light in shallow, clear water are white when you have the Sun at your back even though the sunlight in each beam separates into colors upon entering the water. The red and blue rays from any given beam strike the bottom of the pool at different points and scatter in many directions. However, the rays that your eyes intercept must travel back along their original route, which is back in the direction of the Sun. Thus, the colors recombine as they emerge from the water, and so you intercept white light and see a white spot on bottom of the pool.

When you face the Sun, you intercept colored rays from different points of scatter on the bottom of the pool (Fig. 6-14). The spot you see on the pool bottom for a narrow beam is mostly white, because from points within the beam you intercept rays of all colors. But from points along the far and near edges of the spot, you can see colors. The far edge is formed by blue light as the initial beam of sunlight refracts into the water: The blue ray is bent more and ends up farther from you on the pool bottom than the other colors. Similarly, the near edge is formed by red light, because the red ray is bent less and ends up closer to you. Thus the far edge scatters blue light to you and the near edge scatters red light.

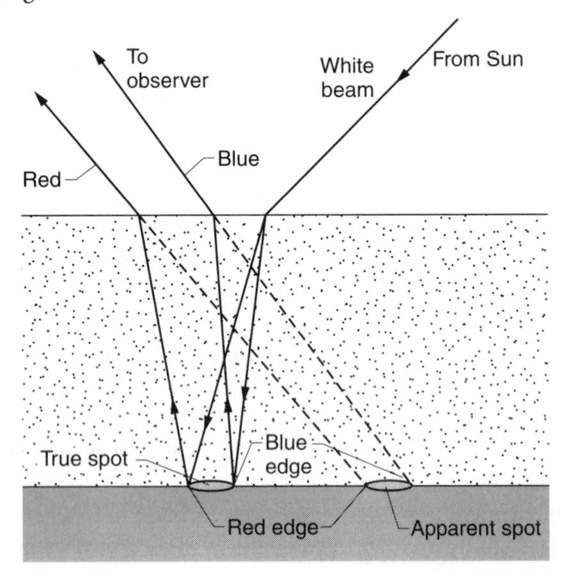

***Figure 6-14** / Item 6.41* Observer facing Sun sees colored edges at bottom of a pool.

6.42 • Color of your shadow

If you examine your shadow on fresh snow, you may find that it is tinted. What causes the coloring? Can the tint vary?

Answer Your shadow should be black, but if the sky is blue, skylight might illuminate the shadowed snow, giving it a faint blue tint. If you examine your shadow on another surface, the color of the surface may tint the shadow. If the region surrounding your shadow is brightly colored, you may instead see the complementary color in the shadow, an illusion produced by your visual system. You can distinctly see such coloring in the shadows thrown by performers on a stage when they are brightly lit by a colored spotlight.

6.43 • Seeing the dark part of the Moon

When 75% of the part of the Moon facing you is illuminated by sunlight, why can you still see the remaining 25% even though it is not in sunlight?

Answer The dark portion of the Moon is faintly illuminated with *earthshine*—that is, light that scatters from Earth. You are seeing some of the light that has been scattered back to Earth by the Moon. By examining that returned light, researchers can determine how light from Earth would look to an observer deep in space. The sky is mainly blue to us; it is also mainly blue to a space observer. Earthshine also reveals the infrared light scattered by vegetation. This work is being done with the hope that analysis of light from a planet orbiting another star might reveal whether the planet has atmosphere and vegetation.

6.44 • Heiligenschein and opposition brightening

On a morning when the grass is wet with dew, examine the shadow of your head on the grass. It may be surrounded by a bright light called the *heiligenschein* (Fig. 6-15*a*). Similar bright regions can be seen surrounding your shadow in many other environments: dry grass or other vegetation, shallow expanses of water covered with ripples, and a variety of dry, rough surfaces. You might even see a bright streak that rises from the shadow of a moving car when the shadow extends across a grassy field.

When you fly, look for heiligenschein on the ground at the aircraft's shadow point (the point directly opposite the Sun). As the shadow point sweeps over grass, trees, bare ground, pavement, and clouds, the heiligenschein appears and disappears. If you are near the ground, you might see sudden bright flashes of light within the heiligenschein. If you are well above the ground, you might see a thin dark line extending from the heiligenschein. What produces heiligenschein, bright flashes, and the dark line?

When certain dew-covered plants are illuminated at night, as with a flashlight or camera strobe, the plants glow in the

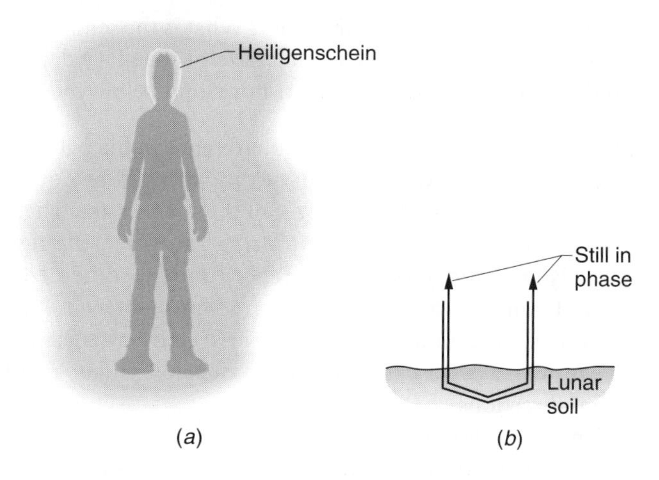

Figure 6-15 / *Item 6.44* (*a*) Shadow on dewy grass with low Sun. (*b*) Paths of two rays through lunar soil.

light as seen by the person holding the flashlight or strobe but not seen by someone off to one side. What causes this glow and why don't all plants glow in this fashion?

Just as the Moon becomes full, any given illuminated region on its surface suddenly peaks in brightness, an effect known as *opposition brightening*. In fact, any such region may be 25% brighter at full Moon than a day earlier or later. What causes this sudden increase in sunlight scattered by the lunar surface? (Before astronauts landed on the Moon, NASA was concerned that backscatter of sunlight from the soil might blind the lunar astronauts if they did not have proper viewing shields on their helmets.)

When a lawn or sports field is mowed in a certain way, the grass resembles a chessboard with bright and dark squares. What produces the variation in brightness? Does the old saying, "the grass is always greener on the other side of the fence," have any scientific merit?

Answer First consider dry grass. When you look at the region around the shadow of your head, you see only grass blades, not the shadows they each cast because the shadows are behind the blades. So, the region is bright because you receive only reflected sunlight from it. When you look away from the bright region, you begin to see some of the shadows cast by the blades, and the net brightness diminishes. In contrast, the region around the shadow of your head is bright.

Heiligenschein from rough terrain may be produced by tiny depressions that are roughly shaped like right-angle corners. Such regions reflect light back toward the source, and you intercept some of the returned light. Other shapes and materials on a surface may contribute additional return of light, especially when the surface is porous and has "tunnels." When you look around the shadow of your head, you see light scattered back to you by the tunnel interiors but you don't see it when you look off to the side.

When the grass is covered with dew, light can enter a dew drop, reflect from the back surface and grass blade, and then leave the drop, heading approximately back toward the Sun. If you look at grass around your head's shadow, you intercept some of this returned light and so that grass is bright. But if you look at grass off to the side, you don't intercept any of the returned light and so the grass there is not as bright.

A spherical drop focuses sunlight to a small spot slightly behind the back surface. The returned light is brightest if the grass blade lies right at this focal point, but usually the drop rests directly on the blade. However, certain types of leaves are covered with thin hairs that can hold a drop slightly above the leaf surface. These leaves give very bright heiligenschein.

When a flashlight beam illuminates drops on vegetation, part of the light is sent back in the general direction of the flashlight as heiligenschein. The returned light is much brighter if the drops are nearly spherical. On the waxy leaves of certain plants, drops tend to bead up into spheres. So, when a light beam is directed onto such a plant, the plant looks radiant in comparison to other plants.

The bright flashes you see within the heiligenschein while flying are due to *retroreflectors*, which are usually plastic or glass beads that send light back toward its source. For example, the paint on a traffic sign might have tiny embedded retroreflectors to send light from a pair of headlights back to a motorist so that the sign is easily seen at night. When the shadow of your airplane passes by such a sign, you intercept some of the light the paint sends back in the general direction of the Sun. The dark line that sometimes accompanies heiligenschein is a shadow of a *contrail* (condensation trail) left by your aircraft.

When your aircraft is low enough for you to see its shadow on the ground, you might find a bright band surrounding the shadow. That band can be due to light scattered back in the general direction of the Sun by water drops on vegetation (thus heiligenschein) or they might be airborne. However, you might see a bright band even if no water drops are present (such as over dry or barren land). Also, you might see a bright band around the shadow of *another* airplane, which certainly cannot be due to water drops. In these cases, the bright band is a perception illusion. When a dark region (such as an airplane shadow) lies next to a brighter region in your view, your visual system produces a bright band (known as a *Mach band*) along the border of the two regions.

When sunlight reaches the Moon's surface, it can scatter in almost any direction. However, the brightest scattering is back in the general direction of the Sun because of the way the light waves can reinforce one another in that direction. Two slightly separated light rays can take the same path through the lunar soil but in opposite directions and then emerge heading back toward the Sun (Fig. 6-15*b*). The waves of these two rays are approximately *in phase* (in step) and so they reinforce each other; that is, they combine to make bright light. During a full Moon, we move into this bright light scattered back toward the Sun, and so the Moon is especially bright. When we intercept light scattered on other nights, the rays have emerged in random ways from the soil and generally do not reinforce one another, so the Moon is less bright.

Some types of moss can also brightly scatter light back toward the Sun with waves reinforcing one another. You might see this bright return if your shadow point falls on fairly distant moss, but it is more commonly seen in aerial photographs over mossy regions.

The checkered design left when a lawn is mowed is due to the orientation of the grass blades left by the mowing machine. In some regions, the blades are oriented so as to reflect light to you, while in other regions they are not and you instead see some of their shadows.

The grass on the other side of the fence may look greener because your oblique view of it may not reveal the underlying brown soil that you more easily see when looking straight down. (Surely, there is a lesson about life and relationships here, but I leave that for you to work out.)

6.45 • Grain field waves

Sometimes a field of grain or tall grass appears to be undergoing wave motion, with bright and dark regions seemingly moving over the field as waves do over the ocean. The wavelike motion disappears if you are too close to the field. What accounts for the motion?

Answer The apparent waves are due to gusts of wind that force the grain or grass to oscillate. When the sides of the vegetation happen to face you, they reflect light to you, giving the region a bright appearance. When the sides do not face you as much (the stalk bends over toward or away from you), the reflection to you is less and the region appears darker. As the gusts come and go, so do the bright and dark appearances. To get the impression of waves, you need to be far enough from the vegetation that you cannot see their details.

6.46 • Glory

If you stand on a mountain with your back to the Sun and peer down into a thick mist that is illuminated with direct sunshine, you may see a pattern of colored rings around the shadow of your head. The pattern is called the *glory*, the *anticorona*, or the *Brocken bow*. You may feel saintly because you'll never see such a pattern around a companion's head.

The glory can most often be seen when you fly above extensive clouds. It forms around a point that is directly opposite the Sun. If the airplane is reasonably close to the clouds so that you see its shadow, the glory is centered on the point in the shadow that corresponds to your location in the airplane. Often the glory is circular, but sometimes it can be greatly distorted into an oval. If the point opposite the Sun passes intermittently over clouds and ground, the glory will come and go and at times may be replaced by heiligenschein, a brightening effect due to the tendency of terrain to scatter light back toward the Sun.

What is responsible for the glory? What is the sequence of colors in it? How does the angular size of the glory depend on the size of the drops in the mist or cloud below you?

Answer The glory is due to interference in the light that is scattered back toward the Sun by tiny water drops. You see the glory when you intercept some of this backscattered light. The scattering is a type of *diffraction* in which the light spreads and undergoes interference; that is, some of the waves reinforce one another (*constructive interference*) and some tend to cancel one another (*destructive interference*). The result is a pattern in which different colors end up in different directions.

The model for diffraction by the smaller drops involved in a glory is very complex. Here is a simplified explanation: The scattering consists of two light components. One component consists of light waves that enter the drop, reflect inside, and then exit back toward the Sun. The other component consists of light waves that skim along the back surface of the drop and then head back toward the Sun. These two components undergo reinforcement or cancellation, and the composite of this interference from many drops forms the pattern of bright, circular bands, with red on the outside of the pattern and blue on the inside.

The angular size of a glory depends on the size of the drops: The larger the drops, the smaller the glory. Usually the drop sizes vary over a wide range, and so the colors tend to overlap and become difficult to perceive. However, if the range happens to be narrow, the colors can be distinct and several complete spectra (from blue to red) can be seen.

The glory is distorted into an oval if the Sun is low and you see the glory spread out over a long cloud bank in which the drop size changes with distance from you.

6.47 • Corona

The corona is a bright region surrounding the Sun or Moon. Sometimes the region consists of colored rings. I once spotted a lunar corona with two complete sets of colored rings and portions of a third, outer set; the sight was magical. What produces the corona, and what is the color sequence in it? Why are the colors visible only sometimes? What determines the size of the corona?

Answer The solar and lunar coronas are due to the diffraction of light as it passes by water drops in the clouds on its way to you. Diffraction is a form of scattering in which the light is spread into an interference pattern. Some waves reinforce one another to produce bright light, and some waves tend to cancel one another to produce darkness. The diffraction spreads the colors of the initially white light, with red spreading through a greater angle than blue. As a result, the outer fringe of a corona is often red. The extent of spreading for the colors depends on the size of the drops. You see distinct rings of color when the drops are approximately identical in size, a few micrometers in diameter. When the sizes vary over a wide range, the colors overlap to yield a white corona (with, perhaps, a faintly red edge).

The rings around the Sun in Vincent van Gogh's *Red Vineyard at Arles* probably have nothing to do with corona. Rather he painted them to represent radiance. He commonly saw such rings around light sources because his vision was affected by his medical use of digitalis, which was at a toxic level.

6.48 • Frosty glass corona

When passing a frosty window on a cold night, you may find that the interior lights are surrounded by colored rings. Why do they appear? Why are they surrounded by a dark ring?

Answer The corona in a window is similar to the atmospheric corona of the preceding item: Both are due to the diffraction of light by water drops, but here the water drops have condensed on the glass. The outer dark ring is the part of the diffraction pattern where light waves arrive at your eye out of phase (out of step) with one another and thus cancel one another.

If the frosty window is moving, as on a train, you can see a corona on each passing source of light, which seem to sweep past you. However, if the light source flickers, such as a mercury street lamp does, you might also see vertical bright and dark bands superimposed on the corona.

6.49 • Iridescent clouds

Why are some clouds tinted with faint colors, often pink and green?

Answer When water drops or ice crystals are a few micrometers in diameter, they diffract (spread) the sunlight into an interference pattern in which waves reinforce one another or cancel one another. At some angles in the spreading, the light waves are in phase (in step), reinforce one another, and thus result in bright light. At other angles they are out of phase, cancel one another, and thus result in dim or no light. This results in different colors at different angles. However, this color spreading depends on the size of the drops or ice crystals. If they all have about the same size, the colors are vivid; if they have a range of sizes, the colors overlap to give a dull color or whiteness.

Because the light diffracts primarily in the forward direction, a cloud must be almost in line with your view of the Sun if you are to intercept some of the colors. Also, a cloud should be tenuous (thin); otherwise, the color spreading occurs so many times as the light passes through the cloud that the colors overlap too much to be distinguishable. In a thick cloud, only the tenuous edges show colors.

6.50 • Blue moon

My grandmother lived in a Texas town so small that, according to her, something exciting happened there only once "in a blue moon." But what is a blue moon?

Answer A blue moon is produced by atmospheric aerosols consisting of particles with radii of about 0.4 to 0.9 micrometer. The particles may have been spewed into the upper atmosphere by a volcano or a large forest fire. They may initially be of the proper size to produce a blue moon, or they may grow to the proper size as water condenses on them. When moonlight travels through the particles, the red and yellow portions of the spectrum are scattered off to the side of your view, and you receive primarily the blue and green portions of the spectrum. The Moon is then blue or blue-green. However, if the Moon is near the horizon, the light must take such a long journey through the atmosphere that the air scatters out much of the blue, and so you see a green moon. These same results apply to the color of the Sun, which can be blue or green if seen through the appropriate aerosol.

6.51 • Yellow fog lights

Can yellow lights on a car penetrate a fog better than white lights?

Answer If the fog drops are smaller than about 0.2 micrometer in radius, the blue and green end of the spectrum is scattered more than the red and yellow end. In fact, yellow light penetrates the fog deeper than the other colors and can be scattered or reflected back to you by the roadway. However, if the drops are somewhat larger, about 0.6 micrometer, the result can be the opposite, and then blue or green penetrate better. If the drops are even larger, all colors penetrate about the same. To complicate the matter even more, the drops might contain impurities that absorb certain colors. So, we have no definite answer here.

6.52 • Dark when wet

Why is wet sand much darker than dry sand? Why is wet hair darker than dry hair? When you drive along a road at night without any illumination by street lamps, why do the painted lane stripes almost disappear during rain?

Answer When sand is dry, much of the light scatters only once or twice before leaving the sand and little of the light is absorbed: The sand is bright, perhaps even dazzling. When sand is wet, the light scatters many times and much less of it leaves the sand: The sand is darker. Two models explain the additional scattering when the grains are coated with water: (1) Light can become trapped within a water layer and is gradually absorbed as it repeatedly reflects inside the layer. This explains why hair is darker when wet. (2) When the sand

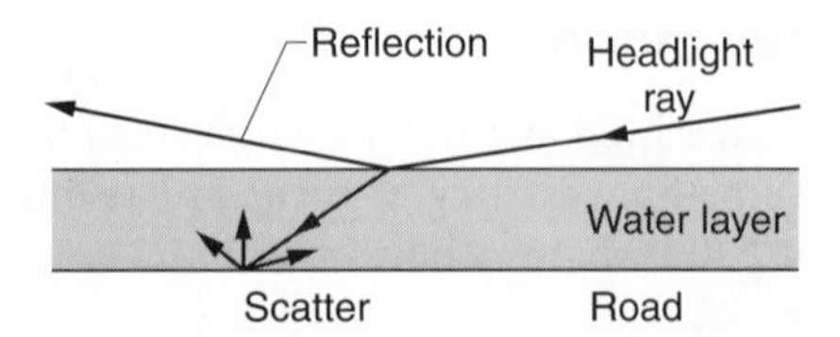

Figure 6-16 / Item 6.52 Water layer on road diminishes the amount of light that reaches the road surface.

grains are wet, light scatters more in the forward direction—that is, deeper into the sand bed—and thus light has less chance of coming back out.

When a road is dry, light from your car's headlights scatters in all directions from the road surface. Enough of the light returns to you that you can distinguish the stripes and some of the surface texture of the road. When the road is covered with a water layer, some of the light from the headlights undergoes a mirror-like reflection at the air–water surface, and less light now illuminates the road surface (Fig. 6-16). Once any light scatters from the road surface, it must again pass through the air–water surface to reach you. At that surface some of the light (perhaps even all of the light, depending on the angle) reflects back to the road surface. The result of these several factors is that the light returned to you may be too dim to distinguish the lane stripes and road texture.

6.53 • Colors of snow and ice

Why is fresh snow usually white, and why is a hole dug into the snow sometimes blue? What is the color of ice? Why are Antarctic icebergs sometimes green while Arctic icebergs never are?

Answer When you view fresh snow in white sunlight, you intercept light that reflects from the surfaces of the crystals and also light that travels through some of the crystals. The light that is reflected retains the white of the incident sunlight. The light that passes through some of the crystals undergoes a slight absorption at the red end of the spectrum and ends up slightly bluish, but this coloring is too faint for you to perceive and thus the snow appears white. If light reaches a hole by traveling through the overlying snow, it passes through a great many crystals. In this case, the blue tint may be perceptible.

Large sections of ice, such as in an iceberg, are blue if the light travels a meter or so through the ice, either from one side to a second side or simply back out the first side after scattering from internal imperfections. However, some of the icebergs that are calved from the Antarctic ice shelves are noticeably green. That green tint is due to the absorption of blue light by marine phytoplankton that were gradually added to the base of an ice shelf as seawater froze to the base. The absorption of blue light shifts the color of the light transmitted through the ice into the green. If the section of ice overturns after it is calved, the green-tinted ice is exposed on the resulting iceberg. Green icebergs do not appear in the Arctic, presumably because the ice shelves move and calve into the sea too rapidly for appreciable accumulation of phytoplankton.

6.54 • Firnspiegel and snow sparkles

Occasionally the reflection of sunlight from a snow field is dazzling and filled with color. What causes this spectacle, which is called *firnspiegel* (German for "ice mirror")? What causes the more common, subtly colored sparkle that can be seen in a fresh snow field? Why do the sparkles often appear to lie above or below the surface of the snow?

Answer Firnspiegel appears on bright, sunny days when the top layer of the snow melts and then refreezes, forming a thin layer of ice crystals. The thermal energy for the melting comes from sunlight penetrating the snow layer and reflecting many times until much of it is absorbed. The thermal energy is then transferred to the surface, where small sections of snow melt and subsequently are refrozen by cold air. The result is a layer of ice crystals that act like prisms in the sunlight, sending bright light and separated colors to you.

The more common snow sparkle is due to the snow crystals themselves acting like prisms, producing pinpoints of color. Since your eyes are separated, they each see a different array of sparkles in nearby snow. When a sparkle seen by one eye lies near a sparkle seen by the other eye, your brain automatically fuses the two sparkles and brings up to consciousness the illusion that a single sparkle lies either above or below the top surface of the snow. The single sparkle is below the surface when the left eye sees a sparkle just to the *left* of the sparkle seen by the right eye (Fig. 6-17*a*). It is above the surface when the left eye sees a sparkle just to the *right* of the one seen by the right eye (Fig. 6-17*b*).

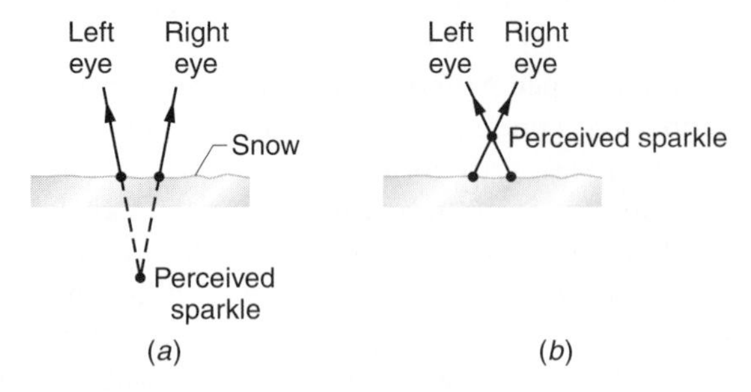

Figure 6-17 / Item 6.54 Two adjacent sparkles in snow can give the illusion of a single sparkle (*a*) below or (*b*) above the snow surface.

6.55 • Whiteouts and snowblindness

What conditions lead to a loss of visibility and orientation in a snow field, a condition known as *whiteout*? If the light is bright, why do shadows disappear? Sometimes a whiteout can hurt the eyes, perhaps even permanently blinding them (*snowblindness*). Are whiteouts more likely on a sunny or a cloudy day?

Answer Whiteout has two forms. When a ground blizzard whips up loose snow, limiting visibility to a few meters, you may become lost after walking only a short distance. A different whiteout appears when the ground is covered with snow and the sky is covered with white clouds. Since both brightly reflect light, the illumination becomes diffuse enough to eliminate shadows. With the ground as bright as the overhead sky and clouds, the horizon disappears and the sky and snow merge into a single white surface. You might then have the impression of being in a vast white emptiness. In his diary and stories of five years of polar expeditions, Vilhjalmur Stefansson recalled that whiteout was not likely on clear days or very overcast days. Rather it threatened on days that were overcast just enough to hide the Sun but still allow sufficient sunshine to come through. Then you might not be able to see an ice mound half your height, much less a small mound that could trip you.

If the visible light and ultraviolet light are bright, a whiteout can cause pain to the eyes and even permanent blindness. Until modern times, some of the native inhabitants of Canada and Alaska would diminish their exposure to the light by wearing goggles that consisted of a piece of wood or bone with a narrow slit over each eye.

6.56 • Yellow ski glasses

Some skiers claim that, on hazy days, subtle snow bumps are better seen through yellow-tinted glasses. The famous polar explorer Vilhjalmur Stefansson also recommended amber glasses for travel across snow and ice fields. Does vision actually improve with yellow glasses in these situations?

Answer Here is one reason why yellow sunglasses may help: The haze diminishes the visibility of bumps in the snow because it scatters sunlight into the shadows of the bumps, making those shadows less visible because then they have almost no contrast with the surrounding snow. If the haze consists of very small particles (with radii smaller than about 0.2 micrometer), they scatter the blue and green portion of the spectrum more than the red and yellow portion, and little of the yellow light is scattered into the shadow regions. If you wear yellow sunglasses so that you see only the yellow light scattered from the snow, you might be better able to distinguish the shadows and recognize the presence of bumps.

Here is another reason: Yellow sunglasses may enhance the apparent (not actual) brightness of a scene, whether it is snowy or not. The effect is due to the action of the rod photoreceptors in the retina. If they are stimulated by light at the yellow–red end of the spectrum but not by light at the blue–green end, their signal to the brain interacts with the signal from the cone photoreceptors in the retina, resulting in an increase in apparent brightness.

6.57 • When the ice grows dark

When a frozen pond thaws in the spring, why do portions of its ice turn dark?

Answer Here is one reason: As the ice melts, some of the top surface degenerates into a fragile structure of vertical, pencil-width crystals that are interspaced with water. Previously the ice was bright because sunlight reflected uniformly from it, but now sunlight is reflected many times between the crystals, growing dimmer with each reflection. Less light is reflected to you, and so the ice is darker.

Here is another reason: When the water freezes quickly, the dissolved air is forced into bubbles that are trapped in the ice. Because the freezing rate is typically fastest near the water surface, the top layer of ice can have many trapped bubbles. They each scatter sunlight and give the ice a bright white appearance. In the spring, sections of the top ice layer melt, exposing lower ice layers that have fewer trapped bubbles. Thus, the exposed lower-level sections are darker than the remaining top-layer sections.

6.58 • Bright clouds, dark clouds

Why are most clouds white and bright? Why are some clouds dark? Why are the edges of some dark clouds bright (such a cloud has a "silver lining")?

Answer Clouds are white for three reasons: (1) The water drops scatter different colors in the white sunlight about equally well. (2) They absorb little of the light, and thus there is no coloring due to absorption. (3) They scatter the sunlight many times before the light heads in your direction. Any collection of small scatterers having these three properties is white in sunlight. A cloud may appear to be black out of contrast with its surroundings or because it is thick enough to disallow much penetration by the light. A cloud that is dark, maybe even black, as seen from the ground will be white and bright if seen from above in an airplane. In that view, the only dark clouds are ones that are not illuminated by direct sunlight.

The water drops in a cloud strongly scatter light in the forward direction; the scattering in any other direction is much weaker. So, if a dark cloud lies near the Sun in your view, the drops along the edge of the cloud will strongly scatter light in your direction. Although the interior of the cloud in your view is dark, this strong scattering through the tenuous edge makes the edge relatively bright. If the cloud is not near the Sun, the edge lacks this strong scattering of light in your direction and you do not see a bright edge.

6.59 • Noctilucent clouds

In latitudes around 50°, ghostly, silver-blue clouds occasionally appear well after sunset, especially during summers in the British Isles and the Scandinavian countries. What causes these clouds, which are called *noctilucent clouds*, meaning "luminous night clouds"? Why are they seen only well after sunset, and why do they sometimes have a wave-like appearance? Why did they first appear in 1885, and why have the brightness and the number of appearances generally (but not always) increased since then?

Answer The clouds form at high altitudes (about 80 kilometers) in the part of the atmosphere called the *mesosphere*. Hence, they are often called *mesospheric clouds*. Being at high altitudes, they are still illuminated with sunlight even when you have been in the dark for a hour or so. They probably consist of tiny bits of ice that form on dust, which may come from comets, meteors, and (sometimes) volcanoes. The clouds are too thin to be seen during the daytime or even during sunset. Their wavy appearance is likely due to *density waves* (wave-like variations in pressure and temperature, often called *gravity waves*) traveling through the clouds.

The first appearance of noctilucent clouds was apparently due to the 1885 explosion of the Krakatoa volcano near Java. That fantastic explosion hurled both dust and water into the high altitudes. At about 80 kilometers, water collected on the volcanic dust (and perhaps also on comet and meteor dust) to form the small (submicrometer) particles that made up the first reported noctilucent clouds. The general increase in both the number and brightness of noctilucent clouds since 1885 is due to the increased production of methane by industry, rice paddies, landfills, and livestock flatulence. The methane works its way into the upper region of the atmosphere, undergoes changes, and results in an increase of water molecules and bits of ice for the noctilucent clouds.

6.60 • You in a looking glass

Here is a common question about images in a flat mirror: Why is your image reversed left and right but not up and down?

Suppose that the top of a flat mirror is level with the top of your head. How long must the mirror be if you are to see your feet in it? Does the answer depend on your distance from the mirror? If you move away from a flat mirror, will you then see more or less of yourself in it?

Answer Your image in a flat mirror is a front–rear inversion, not a left–right reversal. Notice, for example, that everything that is on your left remains on your left in the reflection. The confusion arises if you mentally rotate yourself around a vertical axis until you are aligned with the image in the mirror. What you then would call your right hand is actually the image of your left hand. However, the mirror does not perform such a rotation. To see this point, turn rightward so that your left arm is near the mirror and notice that the mental rotation no longer makes any sense.

If the top of the mirror is level with the top of your head, the mirror needs to be only half your height for you to see your feet in it. Then you can see your head via a reflection at the top of the mirror and your feet via a reflection at the bottom of the mirror. These reflections do not change if you change your distance from the mirror.

6.61 • Reflections off water and a stage mirror

If you look at a scene while also looking at its reflection from calm water, is the reflection a mirror image of the scene?

If you look at the reflection of the overhead ceiling or your face in a cup of tea, why is the image distorted near the cup wall? If you hold the tea in direct sunlight when the Sun is high, why can you see two tiny images of the Sun for some orientations of the cup? Why are the distortions near the wall different if the cup is overfilled so that the liquid surface is slightly above the cup's edge?

If you tilt a partially filled cup of milk so that the milk covers only part of the bottom of the cup, why is that bottom portion outlined with a clear band?

If an actor (stage, television, or film) looks into a mirror and you can see the actor's face centered in the mirror, what does the actor see in the mirror?

Answer A direct view and a water-reflected view generally differ because foreground objects block background objects in somewhat different ways. The reason is that to reach you, the direct rays are probably almost horizontal and the reflected rays are initially headed down toward the water (Fig. 6-18).

In a cup of tea, the liquid surface curves slightly near the cup wall because the cohesion of the water and the adhesion between the water and the wall pull the liquid up the wall. (The climb is usually said to be due to the *surface tension* of the water.) The liquid surface acts like a mirror by producing images of whatever is above it. While the flat portion gives undistorted images, the inwardly curved (*concave*) portion compresses the images. If you hold a cup of tea in sun-

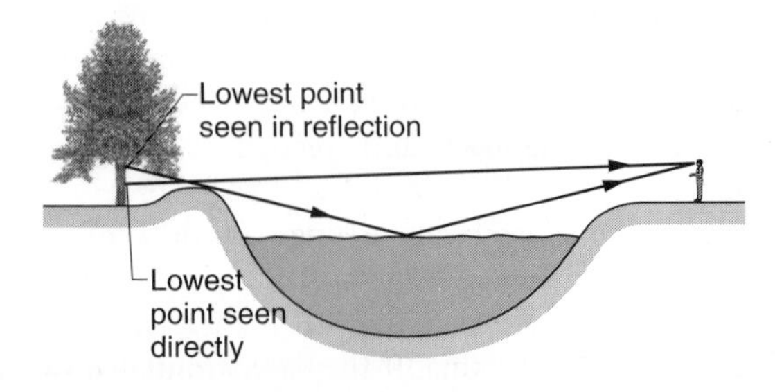

Figure 6-18 / *Item 6.61* A direct ray and a reflected ray reach an observer along different paths.

light when the high Sun is generally at your back, you can see two compressed images of the Sun: One is sent to your eyes by the curved surface at the wall, while the other is first sent from the curved surface to the flat surface and is then reflected to your eyes.

If the cup is overfilled, the liquid surface near the cup's rim bulges outward (it is *convex*). Again the images are compressed, somewhat like the images seen in store mirrors to discourage shoplifting.

A cup of milk is white because the light first penetrates the milk surface and then scatters to you from the milk particles (such as fat). This scattered light is bright enough to mask any mirror-like reflection from the milk surface. However, the region of the clear band around the milk in a tilted cup contains relatively few particles, and so the light scattered to you by the particles is faint. Also, the curved surface concentrates the light reflected by the surface. The result is a bright featureless band.

An actor is directed to hold a mirror to reflect a facial image to the camera or audience. However, what the actor sees in the mirror is not the face but the camera or audience. The same optical misdirection and intended error has also appeared in numerous paintings.

6.62 • Pepper's ghost and the bodiless head

In 1863, John Henry Pepper of the London Polytechnic Institution invented an illusion in which a moving and speaking apparition floated in midair on a stage. A related illusion has long been popular in sideshows. Entering a dim tent you might discover a person's head on a table. Although the person speaks to you, the space beneath the table is empty. How are these illusions accomplished?

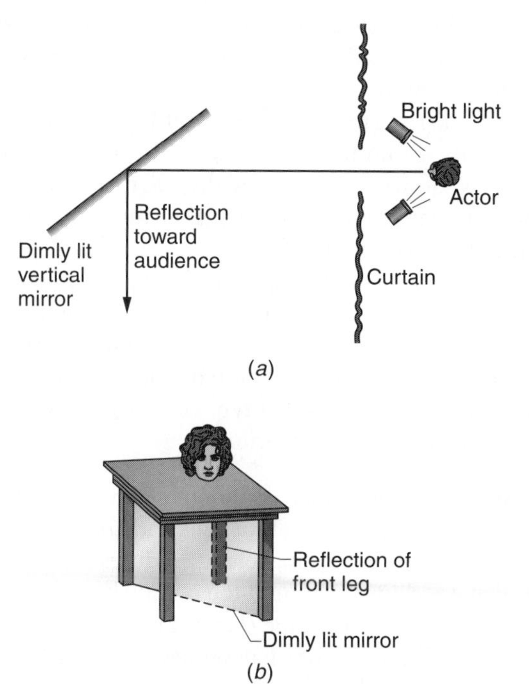

***Figure 6-19** / Item 6.62* Arrangements for (*a*) Pepper's ghost and (*b*) the bodiless head.

Answer Pepper's ghost is a reflection of an actor from a large mirror or sheet of glass on the stage (Fig. 6-19*a*). The actor, who stands in the wing out of the audience's direct sight, is well illuminated while the stage is dim. Although the audience can see the reflected image, they cannot see any evidence of the reflecting surface.

The body of the bodiless head is below the table but hidden from sight by a mirror. When you look below the table, you see a reflected image of a front leg, but that image appears to be the leg on the far side of the table (Fig. 6-19*b*), thereby giving the illusion that the space beneath the table is empty.

6.63 • Tilt of windows for air traffic controllers

Why are the windows surrounding the working area of an airport's control tower tilted with their top edges outward? The front windshield of a car is tilted in the opposite way in order to streamline the car. How might that design interfere with a driver's view of what lies ahead?

Answer If the windows of a control tower were vertical, the controllers would see reflections of themselves and their consoles in the windows. Since they need a clear view of air traffic at the airport, the windows are tilted outward to send unwanted reflections onto the ceiling, which is painted black to absorb the light.

Because of the inward tilt of a car's front windshield, a driver sees a reflection of the dashboard superimposed on the view of what lies ahead. If the dashboard or something on it is bright, the driver may have difficulty in perceiving oncoming dark vehicles.

6.64 • Images in two or three mirrors

How many images of yourself do you see while standing in front of two adjacent, flat mirrors that form an angle, as you might find in a clothing store (Fig. 6-20*a*)? How does the number of images depend on the angle and your position in front of the mirrors? How many images do you see if a third mirror is added so that you are enclosed in a triangle of mirrors? What do you see if you stand between two parallel (or almost parallel) mirrors while looking into one of them?

Answer To determine the number of images in the two-mirror arrangement, first draw an overhead view of the mirrors and then, going both clockwise and counterclockwise toward the rear of the mirrors, add imaginary mirrors that form the same angle as the real ones, as in Fig. 6-20*b*. Each new pie-slice section you add contains an additional image. The tricky part comes when you add the rearmost sections to the drawing because they may overlap. These sections can

(a)

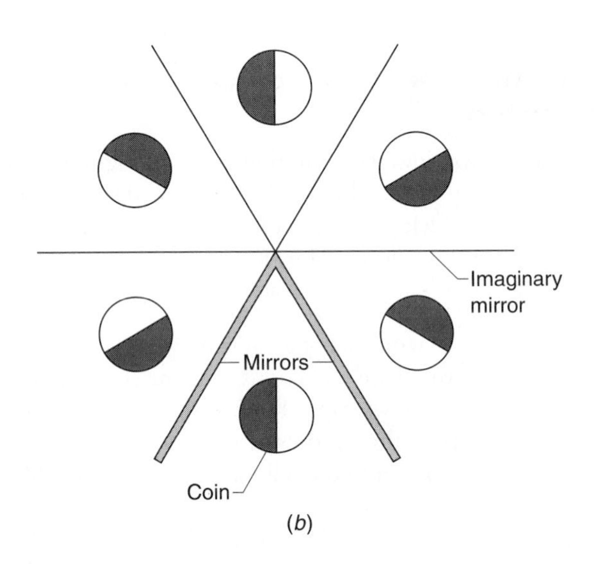

(b)

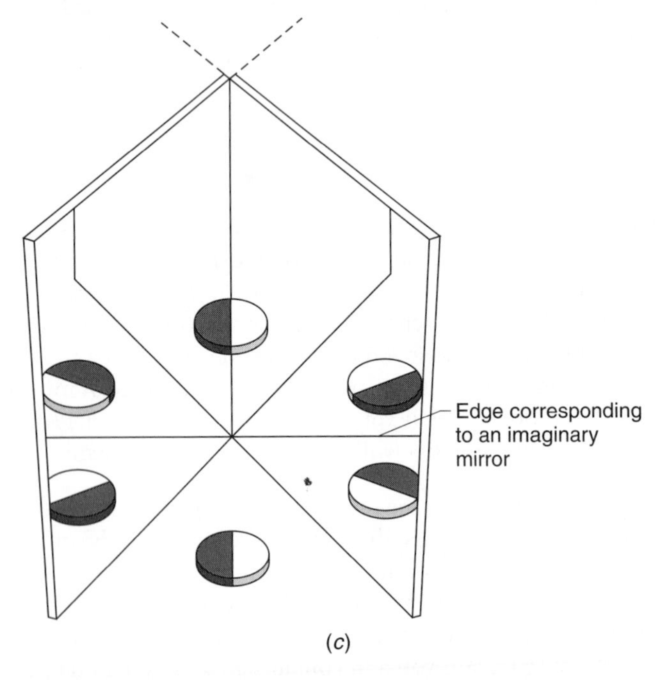

(c)

Figure 6-20 / *Item 6.64* (*a*) Standing between two mirrors. (*b*) Schematic of the images. (*c*) The images you would see.

contribute between one and four images, depending on how much they overlap and on how you position yourself between the mirrors.

Now count the number of images in the drawing. That is the number you see when you look into real mirrors (five images in Fig. 6-20*c*). In addition, you see the pie-slice sections bordered by images of the mirrors themselves.

When a third (real) mirror is added to enclose you in a triangle, the number of images becomes infinite, in principle, because light from you or other objects can be trapped inside the triangle, reflecting a great many times. In practice, the number of images is finite because absorption and imperfect reflections gradually weaken and blur the images, especially with inexpensive mirrors that have the reflecting surfaces on the back of the glass.

Something similar happens if you stand between two parallel mirrors while looking into one of them. In principle, you see an infinite number of images of, say, an arm stretched parallel to the mirrors. However, you can see multiple images of your head only if the mirrors are somewhat slanted. Can you explain why?

A novelty device employs a mirror that is parallel to and behind a half-coated (partially reflecting) mirror. Between the two mirrors are small lights. When you look into the device through the half-coated mirror, you see multiple reflections of the lights, creating the illusion that a series of lights recedes away from you to a great depth. Sometimes the rear mirror has a convex section so that the center of the array is also filled with images of the lights.

6.65 • Kaleidoscopes

In a typical inexpensive kaleidoscope, you can see a single cluster of images symmetrically arranged around a common point. How do more expensive kaleidoscopes produce many clusters of images? How many different types of symmetric arrangements within the clusters can exist in a single kaleidoscope? What arrangements of the mirrors yield images that do not change (shift) when you change your angle of view into the instrument?

If the mirrors are slanted such that the opening at one end is smaller than at the other end, what do you see inside the kaleidoscope? How do some types of kaleidoscopes create colors when there is nothing colored (such as colored glass or plastic) at the far end? What type of images do you see in a round tube with a shiny interior?

Answer Most inexpensive kaleidoscopes have two mirrors that extend along the length of the tube and are at an angle of 60° to each other. This arrangement produces five reflection images that are clustered around the point where the mirrors meet at the far end of the tube (Fig. 6-21*a*). Also included in the cluster is a direct view of whatever lies between the mirrors at the far end. Thus, there is *sixfold symmetry* in the display. If the angle is varied, the number of images and the type of symmetry change (see the preceding item).

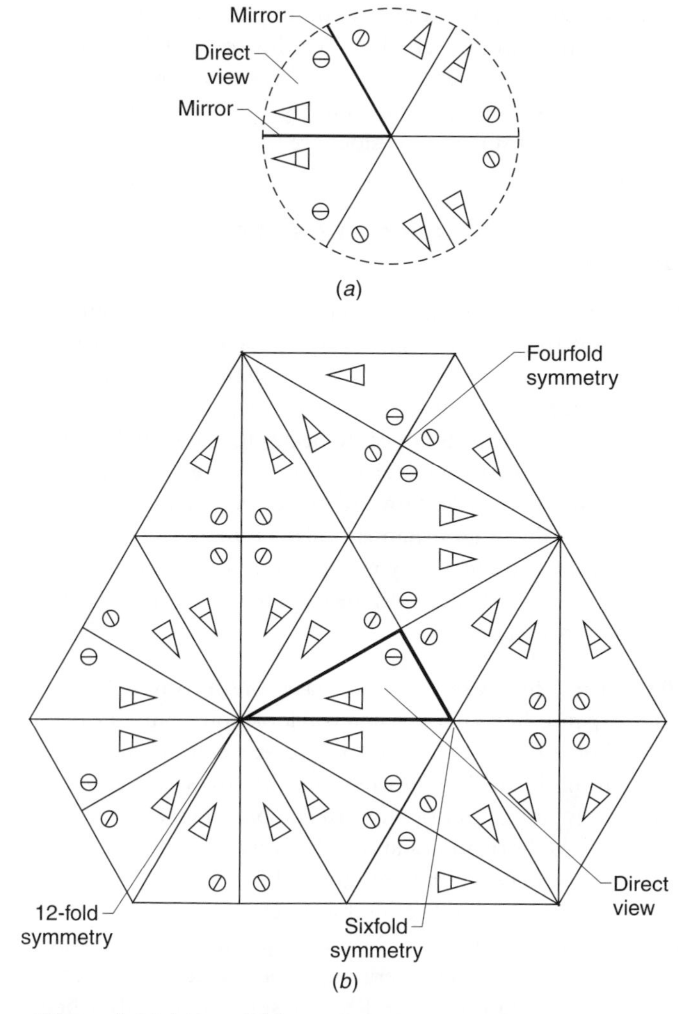

***Figure 6-21** / Item 6.65* (*a*) Images seen in a two-mirror kaleidoscope. (*b*) Part of the field of images seen in a three-mirror kaleidoscope with angles 90°, 60°, and 30°.

Better kaleidoscopes contain three or four mirrors. (The reflecting coating is probably on the front surface of each mirror because with a rear coating, light reflects from both the coating and the front of the glass. The resulting slightly displaced reflections give murky images.) With three or four mirrors, you see a vast array of images at the far end of the kaleidoscope. If three mirrors form an equilateral triangle, the images are arranged in clusters with sixfold symmetry. If the mirrors form some other type of triangle, there will be two or three different types of symmetry in the clusters. Figure 6-21*b* gives an example.

Except for the following four arrangements of mirrors, what you see shifts as you change your angle of view into a kaleidoscope: (1) If a kaleidoscope has four mirrors, they must form a rectangle or square at the opening. If it has three mirrors, they must form (2) an equilateral triangle, (3) a right triangle with angles of 60° and 30°, or (4) a right triangle with two 45° angles.

If you look into the wider end of a kaleidoscope with slanted mirrors, the reflection images produce a geodesic sphere that seems to float in empty space. If you look into the narrower end, you seem to be inside a geodesic sphere.

The colors you see within a kaleidoscope can be produced by colorless plastic pieces that are sandwiched between two polarizing filters.

If you view a point source of light through a circular tube with a shiny interior, you see a series of narrow rings.

6.66 • Mirror labyrinths

The Hall of Mirrors that once stood in Lucerne, Switzerland, was an elaborate maze of mirrors where I quickly became lost. The floor plan consisted of equilateral triangles. Full-length mirrors were positioned along some borders of some triangles. When I stood in any of the triangles, I saw six apparent hallways extending away from me, with a jumble of reflection images between the hallways. What produces the hallway illusion? What lies at the end of such a hallway? Could a person hide from me in a mirror maze, or is the entire interior visible from any given location in the maze?

Answer The hallways in the Hall of Mirrors were produced by light rays reflecting at angles of 60° from the mirrors. Figure 6-22*a* shows a simple version of such a maze. You

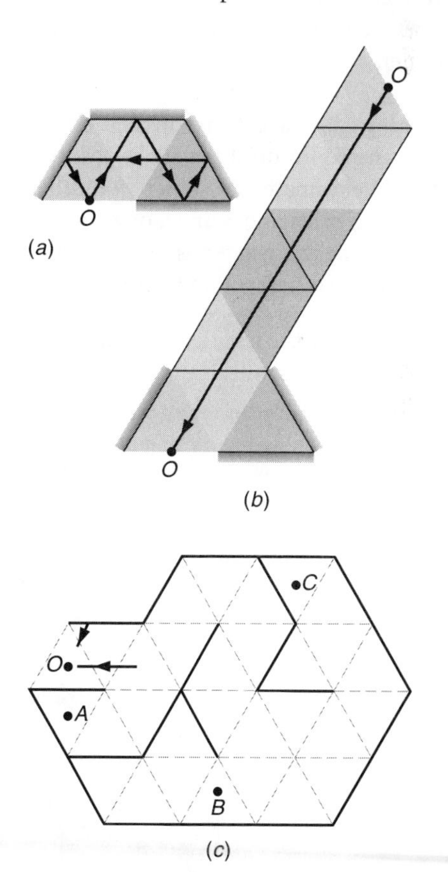

***Figure 6-22** / Item 6.66* (*a*) Overhead view of a simple hall of mirrors. A light ray leaves you at *O* and returns to you via reflections. (*b*) The hallway you see. (*c*) Larger hall of mirrors based on triangular floor sections. The solid lines represent mirror walls.

stand at point O; a light ray leaves you, reflects four times within the maze, and then returns to you. When you look in the direction of that returning ray, you see a hallway stretching away from you (Fig. 6-22*b*), with your image at the end of the hallway because the light ray originated from you. In such a simple version, no one could hide from you within the maze because every triangular floor section shows up at least once in the hallway. However, someone might be able to hide from you in more complicated mazes. For example, from point O in Fig. 6-22*c*, are points A, B, or C visible in hallways?

6.67 • A sideshow laser shoot

As you stroll through a carnival's sideshows, passing familiar games of skill and chance, you spot a new amusement, "The Laser-Blast Target Shoot." Intrigued, you enter, finding yourself in the corner of a rectangular room with walls covered with ideally reflecting mirrors (Fig. 6-23*a*). At your corner there is a powerful laser locked into place horizontally and at an angle of 45° to the walls. At each of the other corners, a clay armadillo target smirks at you.

The attendant behind you explains that you are to shoot the laser after first guessing whether a target will be hit and, if so, which one. He also points out that the room has been precisely constructed with a length of 7 units and a width of 4 units. He then abruptly leaves, as if your corner might be the real target.

Will you hit a clay target or yourself, or will the light reflect around the room until the slight absorption at each reflection finally eliminates it? What would happen if the dimensions were 7 units and 3 units, or 8 units and 3 units? Bravely you squeeze the trigger as you try to envision the multiple reflections you are about to create.

Answer As long as the length and width of the room are both integers, you will not shoot yourself and are sure to hit a clay target. To find out which one, you might trace the shot around a sketch of the room or you might use the following recipe. If the ratio of length to width can be reduced (for example, 8/4 can be reduced to 2/1), do so, and then consult the three possible results shown in Fig. 6-23*b*, where the odd and even dimensions of the sides are the key.

6.68 • Dark triangles among the decorations

Closely pack several shiny, spherical ornaments (as put on Christmas trees) in a single layer on black cloth or paper. When you look down onto the array with bright light behind you, you can see distorted reflections of yourself. Strangely enough, the ornaments appear to be hexagonal with dark triangles between each cluster of three, adjacent ornaments (Fig. 6-24*a*). If you point toward one of the ornaments, your image in all the other ornaments also points toward the one you have singled out. Can you explain these results? You can see the reflections better with the large reflecting spheres sold as garden ornaments.

Answer Suppose that you place a reflecting ball on a wide expanse, like a floor. When you look down onto the ball, you will see distorted images of yourself and nearly everything around you. The *horizon* is marked by whatever sends an initially horizontal ray toward the ball (Fig. 6-24*b*). The image of the horizon lies above the *equator* of the ball. You see the floor between the horizon and the equator.

If you place two reflecting balls next to each other, they each reflect rays as in Fig. 6-24*b*, but now the rays from below the horizon can reflect several times before being intercepted by an overhead observer. Because some light is absorbed at each point of reflection, these rays form dimmer images.

If you arrange a cluster of three balls, the array of images is even more complex, with many dim images due to multiple reflections below the horizon. In any one ball, the horizon lies along an approximately straight line, and the

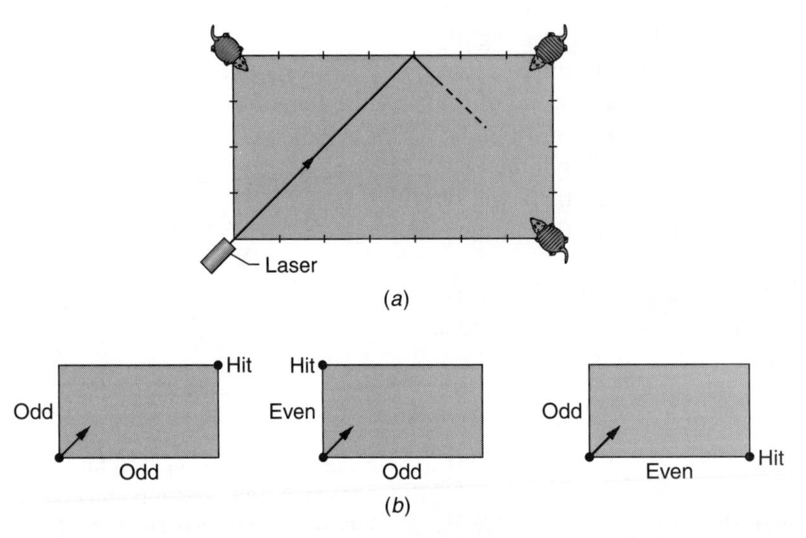

Figure 6-23 / *Item 6.67* (*a*) Overhead view of room with mirror walls. (*b*) Determining which corner gets shot.

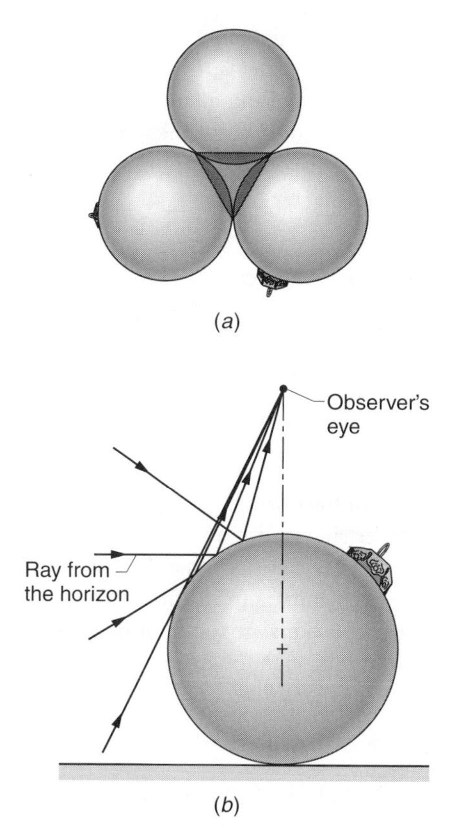

***Figure 6-24** / Item 6.68* (*a*) The dark triangle seen in a cluster of reflecting ornaments. (*b*) The reflection of light rays from a ball to an observer.

composite of the straight lines in the three balls forms a triangle. Images inside the triangle are dim due to multiple reflections. Because the balls probably block the ambient light from reaching the exposed floor between the balls, even the observer's direct view of the floor inside the triangle will be dark.

To understand why reflected images of your finger all point toward a single ball, first consider pointing off to the right in front of a single flat mirror. Your finger and its image point toward a common spot off to the right. You would find the same result if you replaced the mirror with a reflecting ball, except now the reflecting surface is curved. Any other reflecting ball, such as in the array of balls, also gives an image pointing in the direction of your finger.

6.69 • Shiny turns to black; blacker than black

A common double-edged razor blade is shiny. However, if you stack many of them together and then squeeze the stack, the side of the stack along which the sharp edges are aligned is dark. How can shiny surfaces be dark?

Black cardboard is obviously blacker than white cardboard. Can you devise a way in which white cardboard might turn out to be blacker than black cardboard when both are illuminated by the same lamp?

Answer When a light ray enters the space between the beveled edges of adjacent blades, it reflects several times before exiting (Fig. 6-25). About 45% of the incident light is absorbed at each reflection. So, the light that finally leaves the blades is reduced in intensity to only a few percent of the initial intensity, giving the dark appearance.

To make white cardboard blacker than black cardboard, construct a box with the white cardboard, darken the *exterior* with flat black ink, and make a hole in one side of the box, keeping the diameter of the hole less than 10% of the box's edge length. Then direct a lamp onto the side with the hole. The light that enters the hole scatters multiple times within the box. At each scattering point, only a little of the light is absorbed by the white cardboard, but because of multiple scattering, the light that finally emerges from the hole is quite dim.

In contrast, the light that illuminates the inked exterior is appreciably absorbed by the ink, but because this light scatters only once, you actually intercept brighter light from the exterior than from the hole. Thus, the hole (which gives light from the white interior) is darker (blacker) than the black cardboard.

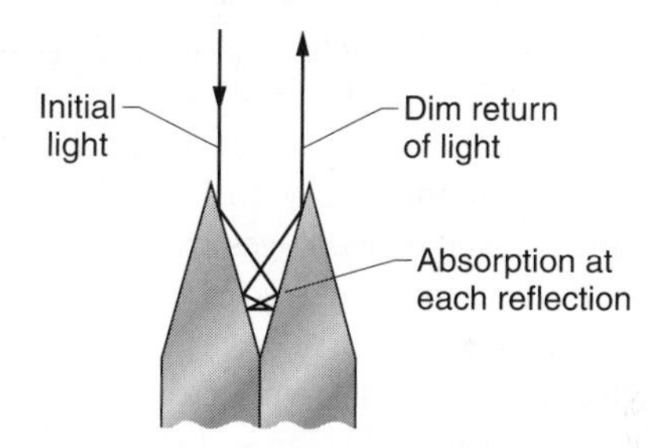

***Figure 6-25** / Item 6.69* A light ray reflecting between the beveled edges of two razor blades.

6.70 • Retroreflectors

A *retroreflector* is a device that reflects a beam of light back toward its source regardless of the angle at which the light enters the device. Sheets of tiny retroreflectors are commonly sewn on the clothing of joggers to make them more visible at night: When a car's headlights shine on the retroreflectors, light is sent back to the driver so that the jogger is noticed and avoided. Why then don't you see a mirror-like image (or even a distorted image) of your face if you look into a nearby sheet of retroreflectors?

Retroreflectors are sometimes fastened to roadways to make the lanes visible even when a roadway is dark (because it is either wet or poorly lit). Retroreflecting sheets are also added to many traffic signs so that the signs are more visible at night. You might see evidence of these retroreflectors while flying during the day: When you are at low altitude, look near the airplane's shadow. Sometimes you will see a brief flash of light when the sunlight is reflected back in the general direction of the Sun by the retroreflectors on the signs. (You will probably see other flashes of light that are not near the shadow point of the airplane; those flashes are due to mirror-like reflections of sunlight from metal or glass surfaces or bodies of water.)

Retroreflecting sheets have a remarkable ability to remove distortion from a beam of light. For example, a photographic slide can be projected through a ruffled layer of plastic (which distorts the image from the slide) and then onto one of the sheets. When the light returns through the plastic and onto a viewing screen, the distortion created during the first passage is largely removed and the image is almost complete.

How does a retroreflector work, and how does it remove distortion?

Answer Retroreflectors come in two main types: spheres and corners. As a light ray enters a sphere, it is bent and directed toward the back of the sphere, where it reflects; it then exits the sphere headed back toward the light source. When a ray enters a corner (for example, a cubical-corner indentation in a plastic sheet), it reflects from the internal faces two or three times and then leaves the corner headed back toward the source. If the retroreflectors of either type were perfect, the light would return directly to the source. A jogger's retroreflector is imperfect, slightly spreading the returned light so that some of it is intercepted by a driver.

If you hold a strip of these retroreflectors in front of your face, you will not see an image of your face. The reason is that the only rays intercepted by your eyes are ones that left them. The rays that left your nose are returned to your nose, but you don't see with your nose.

A retroreflecting array can remove distortion because it sends light back through the same distorting region. For example, if a ray is deflected leftward during the first pass, the deflection is reversed during the second pass and so the ray ends up parallel to its initial direction. Such reversal of distortion works even if the distortion is due to a turbulent flame because the light returns to the flame before the turbulence has time to change.

SHORT STORY

6.71 • Landing in the dark behind enemy lines

During World War II, the British Office of Strategic Services was faced with the difficult task of landing small planes behind enemy lines in darkness. To do this without tipping off the enemy, they cleared a stretch of ground to act as a short runway and outlined it with small retroreflectors consisting of three mirrors fitted together to form the corner of a cube. "The pilot wore a flashlight bulb at the middle of his forehead, and while the light it gave out was hardly visible from the ground, the reflections from the mirrors were enough to outline the landing area to him. When the plane had departed, the ground crew located the mirrors with flashlights, and evidence of the temporary runway was gone." (letter, H. B. Clay, 1986)

6.72 • One-way mirror

How can a mirror transmit light in only one direction?

Answer A one-way mirror is like a common mirror, except that it lacks a backing (such as cardboard) and it leaks a small amount of light from one side to the other. The illusion of light being passed in only one direction relies on the fact that the room on one side of the mirror is brightly illuminated while the room on the other side is dim. In the bright room, the reflections off the mirror are so bright that they mask the faint light coming through the mirror from the dim room. In the dim room, the room reflections are too faint to be seen against the bright light coming through the mirror from the bright room.

6.73 • Rearview mirror

How do you get bright images from a car's rearview mirror during the day but only dim (and thus nonblinding) images at night?

Answer The rearview mirror is a glass wedge having a reflecting coating on the side opposite you. When the mirror is positioned for daytime viewing, light rays originating

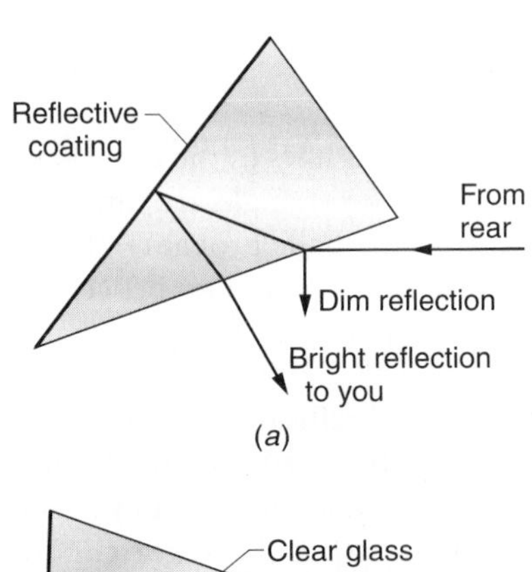

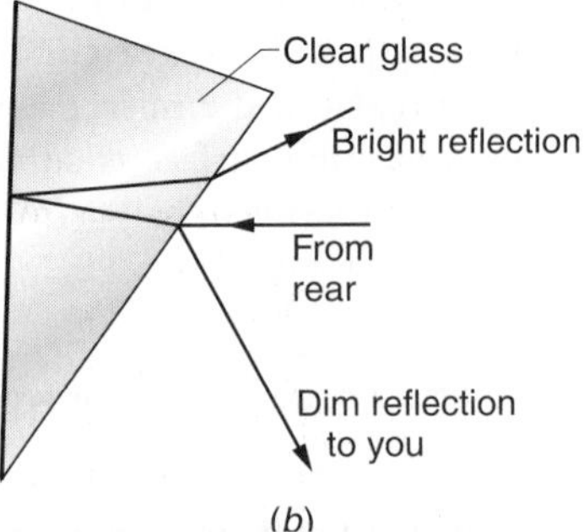

***Figure 6-26** / Item 6.73* Rearview mirror for (*a*) daytime viewing using the bright reflection off the rear reflecting coating and (*b*) nighttime viewing using the dim front-surface reflection.

behind the car enter the wedge and reflect to you from that coating, giving you a bright image of the scene behind the car (Fig. 6-26*a*). To reposition the mirror for nighttime viewing, you flip the bottom of the wedge toward the rear of the car to send those coating-reflected images onto the ceiling and to allow light that is reflected by the front surface of the wedge to reach your eyes (Fig. 6-26*b*). Although the front-surface images are dim, they are bright enough for your dark-adapted eyes to see.

6.74 • Sideview mirror

The purpose of a sideview mirror is to allow you to see a trailing car in an adjacent lane. However, many arrangements of a flat mirror leave a *blind spot*, which is a region in which the trailing car is too close to be visible in the mirror (Fig. 6-27). The danger is that you may switch lanes without being aware of a slowly passing car. To diminish the blind spot, should the mirror be near you or near the front of your car?

Answer The mirror should be located near the front of your car. Then an image of a passing car is always in the mirror until the front of the car becomes visible in your peripheral view. With the mirror in its normal location next to the driver, a small car that is slowly passing might disappear for many seconds.

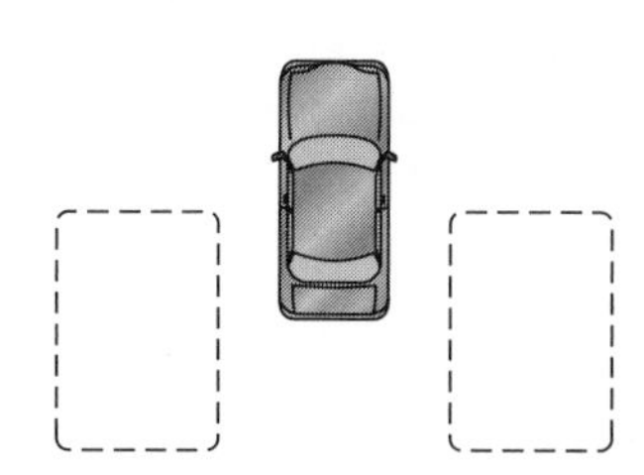

Figure 6-27 / *Item 6.74* Blind spot in a sideview mirror, roughly located. Generally, blind spots differ on the two sides of a car.

6.75 • *A Bar at the Folies-Bergère*

Edouard Manet's *A Bar at the Folies-Bergère* has enchanted viewers ever since it was painted in 1882. A rough outline is given in Fig. 6-28. In the foreground, the painting shows a bartender whose eyes betray her fatigue as she stands behind her bar. In a large mirror behind her, we see reflections of her, a customer, various bottles on the bar, and the crowd in the room. Part of the painting's appeal is a subtle distortion of reality that Manet hid in it—a distortion that gives an eerie feeling to the scene even before you recognize what is "wrong." Can you find it?

Figure 6-28 / *Item 6.75* A sketch of Edouard Manet's *A Bar at the Folies-Bergère*.

Answer The images reflected by the mirror have the correct shapes, but their locations are wrong. When you first look at the painting, you may sense this even before you realize why. The bottles at the left of the painting are actually close to the back of the bar, but in the reflection, they are close to the front of the bar. The image of the woman should be behind her, not well off to the right. And most troubling, the woman looks right at you, but in the reflection there is a man directly in front of her—thus, that man must be you. But if that is so, your image should not be well off to the right as painted; in fact, the woman's body should be blocking your view of your image.

6.76 • Renaissance art and optical projectors

Some modern painters and historians contend that European painters of the 15th and 16th centuries may have used a curved mirror to project an image onto the canvas. If so, those painters could have simply traced the outlines and details of an image onto the canvas and then filled it with colors in something like a paint-by-numbers procedure. How can we tell if such a technique was used?

Answer The photographic realism of certain Renaissance paintings, such as *Husband and Wife* by Lorenzo Lotto, suggests that a concave mirror was used to project a scene onto a canvas. An artist would mount such a mirror in front of the scene to be painted and, slightly off to one side, he would mount his easel facing the mirror. He would then orient the mirror so that its (inverted) image of the scene fell on the canvas. Normally, a painter must labor to put the painted images into proper perspective, to give the painting a feel of three-dimensional realism, as if the viewer is seeing the original objects and not just a flat representation of them. However, if a painter were to project the scene onto the canvas and trace the outlines of the objects, the perspective would automatically be incorporated.

Analyses of the paintings show that, in fact, they contain many errors in perspective, strongly arguing against the use of mirrors (or any other optical devices). For example, two parallel lines extending away from the painter should converge toward a single point called the *vanishing point.* An artist seeking realism in a painting would locate such a point and carefully draw all the lines of perspective relative to it. In the analyzed paintings, the lines of perspective in different parts of a painting were directed toward different vanishing points, suggesting that they were drawn freely instead of simply traced on a projected image.

6.77 • Anamorphic art

Some paintings and drawings produced in the 15th through 18th centuries were purposely distorted so that objects in them were not immediately recognizable, perhaps to disguise a political statement against a king. For some pieces you must view the art from an edge before you can recognize the objects depicted in them. For other pieces you must view a reflected image of the art to perceive the objects. For example, you might have to look at the reflection provided by a shiny cone or cylinder placed at the center of the art. Why are the objects recognizable in such strange viewing schemes and not in the normal fashion?

Answer If you view an anamorphic painting in the normal manner, its figures create images on the retina that are too distorted to be recognized. However, when you view the painting in the special way intended by the artist, the retinal images of the figures are near enough to their normal shapes that you can recognize them.

For example, suppose that a cat's face has been painted by an artist as he views his work by reflecting it from a shiny cone placed at the center of the canvas. When the cone is removed, the face is unrecognizable because of a distorted perspective. (The eyes are well separated; the chin is wide and curved around the painting; and the composite certainly is not cat-like.) When you put the cone back in place and look at the reflection of the painting from the cone, the distortion is removed and the image on your retina is easily recognized as being a cat's face.

6.78 • The bright and dark of street lamps

When two identical street lamps (the old-fashioned, unshielded, light-polluting type) are lit, where are the points of maximum and minimum brightness along the intermediate sidewalk? If a straight row of evenly spaced, identical lamps is lit, where are the points of maximum and minimum brightness along the row? Is there some other arrangement of the lamps (in a row) that will increase the brightness at the dimmest points?

Answer With only two lamps, the point of minimum intensity is midway between them, and the points of maximum intensity are at a distance from a lamp base that depends on the height and separation of the lamps. In a long row of evenly spaced lamps, the points of minimum intensity point are also midway between the lamps, but the points of maximum intensity are now at the base of the lamps. Let D be the distance between the lamps in that situation. The brightness at the dimmest points can be increased if the lamps are rearranged into pairs with a separation of $D/2$ within a pair and a separation of $2D$ between the centers of the pairs. In fact, you might be able to find a similar arrangement that gives almost uniform illumination along the row.

6.79 • Multiple images from double-pane windows

If you look at a light source at night through a window consisting of two panes of glass, you may see multiple images of the source. If such a window is used on an airport control tower, the extra images can create a hazard because they might be interpreted as being additional aircraft. What causes the extra images, and how does their spacing depend on the angle at which the light reaches the window? Does the spacing depend on weather conditions?

You can see similar multiple images at night when you look at airport runway lights through the window of an airplane. Once you are airborne, if you turn on the lamp over your seat and hold a shiny object in the light, you may see several reflections of the object in the window. What accounts for the extra images?

Answer When you look at a light source through a double-paned window, the primary image is due to light passing directly through both panes of glass (Fig. 6-29). The dimmer, secondary images are due to light reflected between the panes or even between the two surfaces of a single pane. The brightest of them is due to light reflected first from the inner pane and then from the outer pane before it reaches you. The other images involve even more reflections. The images are easier to see if they are well separated, which occurs when the light from the source is oblique to the window.

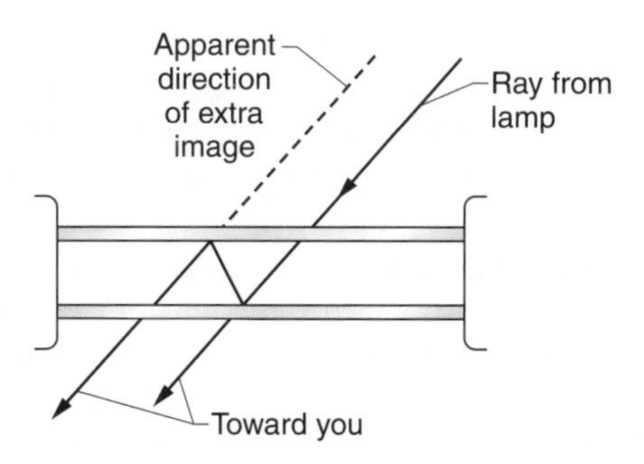

Figure 6-29 / *Item 6.79* Light paths through a double-pane window.

Sometimes the air pressure between the panes differs from the external air pressure, forcing the panes to bulge inward or outward. Their curvature increases the separation of the images. To remedy this problem in air-traffic control, such a window has some means of equalizing the air pressure between the panes with the external air pressure. For example, the window may have a small hole through both panes.

The extra images that can be seen in an airplane window are due to three panes in the window. When you have a slanted view into the window, you see multiple images due to multiple reflections of the light. When the airplane is at high altitude, the outermost pane bulges outward because of the low external air pressure, and its curvature can distort the reflections from it in surprising ways.

6.80 • World's most powerful searchlight

In a 1965 paper that chronicles how wishful thinking sometimes invades science, R. V. Jones described two British dock workers who suggested that the brightness of a searchlight could be greatly increased. Starting with the light emitted by a carbon arc, an ellipsoidal mirror was to focus an image of the arc. Then a second mirror was to use that image to focus another image. Then a third mirror was to use that image to focus yet another image. And so on, until a final mirror was to focus its image back onto the carbon arc, thereby making the arc much brighter than it was originally. This amplifying procedure was to be repeated many times until, when the searchlight was needed, one of the mirrors was to be removed to allow the beam to escape. That beam would have been exceptionally bright. What was wrong with this plan?

Answer According to Jones, when the workers submitted their amplification plan to the British authorities, they received a reply that the plan could not work because it violated the second law of thermodynamics (which involves a quantity called *entropy*). The dock workers quickly apologized, saying that they had not realized they were breaking any official regulation.

Of course, we cannot start with a certain amount of energy and then somewhat increase it without any additional input of energy. So, reflecting the light many times might concentrate the light but certainly won't increase its energy.

6.81 • Archimedes' death ray

Historians have long argued whether Archimedes defeated the Roman fleet during their siege of Syracuse in 212 B.C. with the use of a *burning mirror*. According to legend, Archimedes arranged for mirrors to focus sunlight onto the side of a Roman ship, igniting it. After enough of the ships were burned and sunk, the rest retreated. Is such a feat possible?

Answer Burning wood at a distance with a set of flat mirrors or a single curved and focused mirror is possible, but it is highly unlikely that Archimedes employed this technique. The more conventional weapons during his time would have served better because of several problems in using mirrors.

One problem has to do with focusing: To gain the necessary intensity on the wood, the mirror must focus the light onto a small spot. A single flat mirror will not create such intensity (thankfully, actually, because otherwise any careless use of a cosmetic mirror in sunlight could start fires). To focus the light to a greater intensity, a series of flat mirrors arranged roughly in a parabola would be needed. However, to set a ship on fire, the focal distance would have to be adjusted to match the distance to the ship, a requirement that would be daunting during battle.

A second problem has to do with the time required for the focused light to ignite wood. Since the ships would probably be moving (and rocking) during the assault, holding a focused beam of sunlight onto one spot on a ship long enough to ignite it would have been highly impracticable. Since the wood on the side of the ship was probably wet, the task would have been effectively impossible for Archimedes. In short, the legend about Archimedes is just an interesting myth.

In 1993, Russia put a plastic mirror with a 22 meter diameter into orbit to test the possibility that sunlight could be reflected into the Russian high latitudes during the dark hours of their long winter. The mirror produced a dim spot of light, several kilometers in width, that swept across Europe during the test. Several observers saw the light in spite of the cloud coverage that night.

SHORT STORY

6.82 • Illuminating a referee

In Arthur C. Clarke's short story, "A Slight Case of Sunstroke," a game of football (soccer) was played between two rival countries before an audience of 100 000. About half of the spectators were military personnel, who were admitted free and honored with large, handsome programs that had been bound with shiny metal resembling silver to commemorate the event.

The game was intensely anticipated, especially since the home team had lost the game the preceding year because the head referee had been bribed by the visiting team. Actually, the home team also bribed the referee, but apparently not enough.

Since, according to the rules, the visiting team had the right to choose the referees for a game, they again chose the same man. The audience was anxious to see how fairly he would call the game this time. Early on, most of his calls seemed fair enough but then, just after the visiting team scored a goal, the referee nullified a countergoal by a home player and gave the visiting team a free kick, which they made. So, with the home team down by two points, anguish swept through the home spectators.

However, their hopes were lifted when the gritty home team fought back with a goal that was so clean that not even the biased referee could question it. Soon afterward, the home spectators rose to their feet when one of their players managed to bring the ball through several defenders and kick it through the goal again, tying the game. Amid the cheers came the referee's whistle. He disallowed the goal with the astonishing claim that the player had handled the ball.

Some of the home audience became enraged and threatened to storm the field, but not the disciplined military personnel among them. After the teams retreated to the sidelines, leaving the referee isolated in the center of the field, there was a shrill bugle call, and in unison the military personnel lifted their shiny programs and angled them in the bright sunlight. With a brilliant flash, the referee was transformed into a smoldering heap from which rose a column of smoke.

In some countries, football is taken quite seriously.

6.83 • Spooky lights in a graveyard

In an old graveyard about a mile from the tiny Colorado town of Silver City, visitors sometimes gather at night to watch curious lights dance among the black marble gravestones. The site is dark because the town is distant and the area in the opposite direction is nearly deserted. The lights are usually white points, but sometimes they seem larger and take on a blue tint. Are ghosts coming out to play or is there a more rational explanation for the lights?

Answer The black marble of a gravestone can act like a mirror when light rays approach the stone surface obliquely, but they are absorbed when their angle to the surface is larger. If you walk through a dark graveyard, some surfaces of the stones will be oriented properly to reflect light to you from the town or stars. The reflections might be off the broad face of a stone or off the straight or curved sides. As you move, the reflected images will seem to move, or come and go in a perplexing way, giving them life-like behavior. They are also animated by temperature fluctuations that can incessantly vary the paths taken by the light rays; you probably have seen similar animation, called *shimmy*, in light that passes over fires or hot roadways.

6.84 • What a fisherman sees of a fish

If you are to shoot a submerged fish with an arrow, should you aim directly toward the fish? Any skilled hunter knows that generally you should aim low. Why? Does the apparent location of the fish change if you tilt your head so that your eyes lie on a vertical axis? (Granted, trying to shoot a fish with your head held horizontally is an invitation to shooting yourself in the foot.)

Answer When light rays from the fish pass through the water surface and into the air, they refract (bend) away from the vertical (Fig. 6-30). Once they reach your eye, you make sense of their origin by mentally extending lines back along the rays but without allowing for the bending. As a result, you perceive the fish to be higher than it truly is, and so you need to aim low—that is, below where you see the fish.

The apparent position of the fish is further complicated by the way you assign distance to an object and by how much your eyes must converge to see it. Because the ray reaching your left eye is bent toward your left and the ray reaching your right eye is bent toward your right, your eyes must converge more than if no bending occurred. Based on the convergence, you perceive the fish as being closer than it is. If you tilt your head to bring your eyes onto a vertical axis, while also looking along a slant into the water, the fish seems even closer and higher in the water.

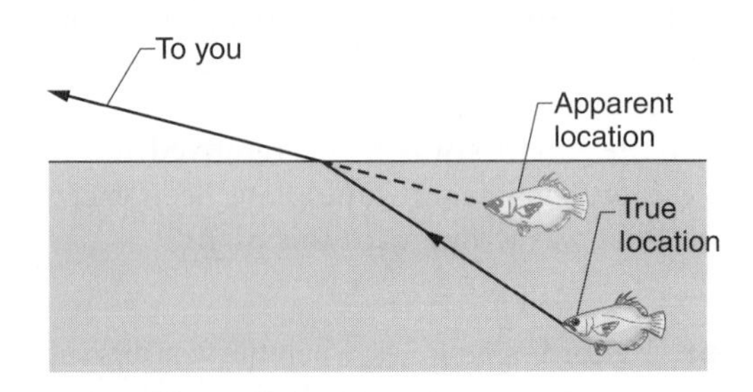

Figure 6-30 / *Item 6.84* Apparent and true location of a fish.

6.85 • What a fish sees of the fisherman

What of the external world can a fish see when looking upward? Is the image of the world distorted by the passage of the light rays through the water surface? Does the image depend on the depth of the fish?

In fly fishing, if you cast the fly toward a fish, must you position it precisely, or can the fish see a more distant floating fly?

Suppose you lie on your back on the bottom of a shallow pool. Is your view of the external world similar to that of a fish? Does your view change if you wear a mask having a flat plastic face and containing air? If you have eyesight that requires no correction, why is your underwater vision so poor when you do not wear a mask containing air? If you are nearsighted or farsighted, is your vision improved when underwater?

When an archer fish sees an insect in vegetation hanging over the water, the fish sticks its snout slightly out of the water and then squirts a water jet from its mouth to the insect, to knock the insect down into the water where it can be captured and consumed. When the fish aims its water jet, its eyes are underwater. Does it aim at the point where it sees the insect?

Answer A fish's view of the external world is distorted by refraction of light rays passing from air through the water surface to reach the fish's eyes. This refraction bends a light ray at the surface so that it becomes more vertical in the water. The amount of bending is zero for an initially vertical ray and is progressively greater for an initial ray angled more from the vertical.

The refraction means that the external world is seen within a circle on the water surface directly above the fish (Fig. 6-31). An image of the horizon of the external world lies on the perimeter of the circle, and images of external objects lie within the circle.

Outside the circle, the fish sees mainly a mirror-like reflection of the bottom of the pond if the water is shallow, or a murky surface if the water is deeper or dirty. If you want the fish to see all of a fly you cast, the fly should be near the center of the circle. If it is near the perimeter, the dry portion of the fly is highly distorted and probably lost in background clutter of images from trees and other objects on the horizon. If the fly is outside the circle, the fish can see only the submerged portion of it. The deeper the fish is, the smaller the circle is.

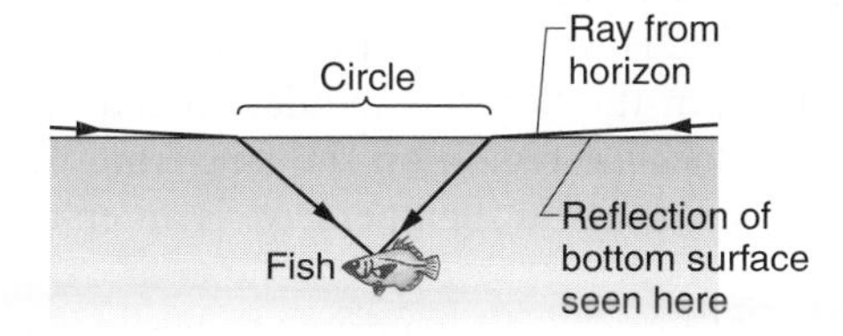

***Figure 6-31** / Item 6.85* External light reaches a fish through an overhead circle.

If you have normal vision, you will not see a clear image of the external world in such a circle while submerged. Most of the focusing of light by your eye results from refraction of the light as it passes from the air into the cornea. When water is next to the cornea, this bending is almost eliminated and so you probably see only poorly underwater.

Your vision improves if you wear a mask, because the air next to the cornea restores the refraction of light into the cornea and thus the focusing. However, the light must also refract from the water, through the transparent plastic face of the mask, and into the air trapped in the mask. This additional refraction eliminates formation of the overhead circle on the water surface. The external world then looks almost as it does when your eye is above water.

If you are nearsighted, your eyes focus an image of a distant object in front of the retina. If you diminish the refraction at the cornea by submerging the eyes, the image might fall on the retina and thus be seen clearly.

Sometimes an archer fish will squirt water at an insect directly over it, in which case the fish's line of sight is not affected by refraction. However, for other directions, the refraction does matter and the fish must allow for it either by trial and error (until the water jet finally hits the insect), by learning from prior experience, or by genetic encoding.

6.86 • Reading through a sealed envelope

Suppose I demonstrated my mind-reading skill with the following: I'll have you write a word on a sheet of paper and then place the paper in a standard white envelope where it fits without any need of folding. You seal the envelope. Before you hand me the envelope, you examine it. The word is not visible through the envelope, and the room lacks any bright light that could possibly penetrate the envelope to reveal the word as a shadow.

As I hold the envelope, I'll have you think of the word so that I can "see" it in my mind. Although I occasionally look at the envelope, I never attempt to hold it up against a light. After a few minutes of concentration, I tell you the word.

The truth is, of course, that I have no mind-reading powers. How do I manage to determine the word? Here's a hint: Similar physics makes clothing, especially white cotton as in a tee-shirt, semitransparent when wet.

Answer After I receive the envelope, I secretly apply a small amount of grease to it. Normally, light does not penetrate an envelope because it scatters from the fibers and fillers within the paper of the envelope. We can explain the amount of scattering in terms of the *index of refraction* of a material, which is a measure of how fast light travels in that material. The paper contains pockets of air with a low index and fibers and fillers with a higher index; the large difference in these indexes leads to a large amount of scattering at all the air-pocket surfaces.

I can reduce the amount of scattering by letting the paper absorb grease, which has an intermediate index of refraction. The index does not change very much at an interface between air and grease or an interface between grease and the fibers and fillers. Because the light does not scatter as much, it penetrates the envelope more and can illuminate the ink or pencil marks of the word and the paper on which they lie. The dark marks largely absorb the light, but the surrounding paper scatters some of the penetrating light back through the greasy envelope and finally to me. I can distinguish the marks (and thus the word) by their dark contrast with their surroundings.

SHORT STORY

6.87 • Sword swallowing and esophagoscopy

Today, *esophagoscopy* is a routine procedure in which a viewing device consisting of optical fibers is extended down the throat of a patient and into the stomach so that a doctor can examine that route. The viewing device is curved in order to take the turn from the mouth to the esophagus. Part of it carries light to the end of the device to illuminate the passageway or stomach. Another part carries an image of the illuminated interior back up to equipment that displays the image on a monitor. The doctor can manipulate the device to change which part of the interior is illuminated and might be looking for signs of cancer or an ulcer or maybe even for illicit packages of drugs that some people attempt to smuggle by swallowing them ("body packing").

Modern esophagoscopy is well understood but its beginnings were a bit strange. A crude *endoscope* consisting of a straight tube illuminated by a candle had already been used to examine the lower end of the colon. A straight tube was also used in the first esophagoscopy, but the tube was too short to reach the stomach. However, that pioneering doctor hit upon a way to use a longer tube: He employed a sword swallower, a person who can tilt the head back, relax certain muscles along the esophagus, and make a fairly straight passage from the pharynx to the stomach. When the doctor illuminated the free end of the inserted tube, he saw the stomach's interior and modern esophagoscopy began.

6.88 • Shower-door optics

Hold a strip of transparent plastic tape immediately above printed words and then move the tape toward you. Why can you initially see the words through the tape but not when the tape is more than half a centimeter from them? Why is the body of a bather easily seen through a frosted (textured) shower door only when the person is immediately next to the door (Fig. 6-32*a*)?

Some art museums use similar optics in their displays of paintings. The paintings are protected by a layer of glass or plastic, but if the layer were a normal pane, your view of the painting would be marred by the layer's stray reflections of the room. To avoid the distraction, the layer is slightly textured. Why does the texturing eliminate the room reflections without distorting your view of the painting?

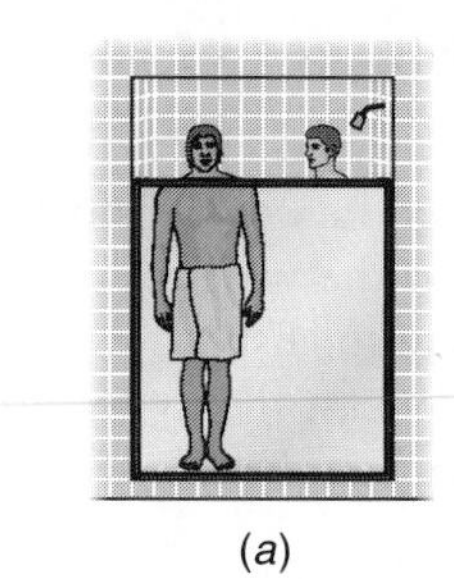

(*a*)

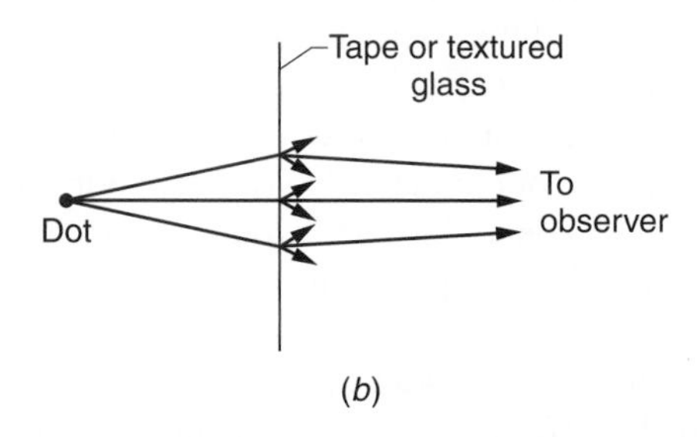

(*b*)

Figure 6-32 / *Item 6.88* (*a*) The bather is visible only when near the frosted door. (*b*) Scatter of light by irregularities.

Answer When you look at a dot through a normal pane of glass, your eyes intercept some of the light rays emitted by the dot and your brain makes sense of them by automatically extending them backward to create a perceived image of the dot. That image is clear and sharp because the rays that you intercept come from a tiny region on the glass along your line of sight of the dot.

When a strip of tape is substituted for the glass, each ray entering the tape is scattered by the tape's irregularities into a conical spray of rays (Fig. 6-32*b*). You now intercept additional dot rays from points on the tape that are not along your line of sight of the dot. Still, your brain extends any intercepted rays backward in an attempt to perceive its source. If the tape is close to the dot, the spread of rays is small, as is the region on the tape from which you intercept the rays, and you can still perceive a fairly clear and sharp image of the dot. However, when you increase the distance of the tape from the dot, the angular spread of the rays increases, as does the region on the tape from which you intercept rays. What you then perceive is an enlarged and vague dot.

If you look through tape at a printed page, your ability to read the print depends on your success at distinguishing the letters in the print. When the tape is more than half a centimeter from the print, the spread of the letter images causes too much overlap for you to distinguish them.

The situation is similar with the bather behind a textured shower door and with the covering over a painting. In the situation of the covering, the painting is close enough for you to distinguish details, but items in the room are sufficiently distant that their reflected details blur and cannot be recognized. In the situation of the shower door, you might wonder why the texturing is usually placed on the exterior of the door. If the texturing were drenched with water, some of the water would fill in the crevices of the irregularities, giving a smoother surface, thereby reducing the spreading of the light rays. The bather would then be clearly visible even if relatively far from the door. (Thoughtful optics can aid the modesty of an observed bather.)

6.89 • Magic of refraction

A magician folds a newspaper around a test tube, breaks it into small pieces with a hammer, and then pours the pieces into a transparent beaker of water. After a short chant of magic words urging the pieces to bring themselves back together to re-form the test tube, he reaches into the water, discovers that the pieces have done exactly that, and pulls out the test tube. How is the trick accomplished?

Add a clear glass marble to a shot glass, and place them over some printed material. When you look down through the marble, the print is unreadable. How might you bring the print into focus without moving the shot glass?

Answer To set up the test tube trick, first put a test tube in the water. You can see it in the water because light passing through the water is either reflected or refracted by the test tube. You can hide the test tube by dissolving sugar in the water. When the sugar–water solution and the test-tube glass have the same optical properties (when they have the same *index of refraction*), light passing through the water can pass through the glass without any telltale change in direction of travel. Thus the submerged test tube is invisible. So are the broken pieces of a (second) test tube when you pour them into the sugar water. When you reach in to pull out the first test tube, search for it with your fingers.

The marble focuses the light from the printed material so strongly that you cannot resolve the image. To reduce the focusing, pour water into the glass. The indexes of refraction of water and marble are close, and so light rays undergo little refraction when they pass from one to the other. Your eyes can then focus the emerging light rays enough for you to read the print.

6.90 • The invisible man and transparent animals

H. G. Wells wrote a novel about a man who became invisible (Fig. 6-33). Is such a thing optically possible? Would a man be invisible if he became transparent like fine glass? If a man is invisible, can he see? Why is your eye transparent but not your skin? Can any animal be largely transparent?

Figure 6-33 / Item 6.90 The invisible man relaxing in his favorite chair.

Answer An invisible man is, of course, impossible. Were he merely transparent (like fine glass), the curved portion of his body would act like a complicated lens, distorting your view of the background as he walked in front of you. He would also reflect light from his surface, just as an ice sculpture does. To eliminate distortion and reflection, the man would need to have the same optical properties as air, which means he would need to consist of air, an impossible requirement.

For the man to see, he would have to focus light and then absorb some of it. If the eye lens is to focus light, its optical properties must be different from that of air. If the retina is to absorb light, it must be at least partially opaque. Both factors would be apparent if you looked at his eyes. However, suppose that he focused light by means of tiny specks (see the material about pinspeck cameras in item 6.102) and then absorbed only a fraction of the light. He might then go undetected.

When visible light is sent into the human body, it scatters from the collagen, membranes, and various other components along its route, at points where the optical properties change. The scattering is significant because those variations in the optical properties occur over distances *larger* than the wavelength of the light. Any image sent through the skin is largely scrambled by this scattering, and so a human is not transparent in visible light. (However, there are ways in which the scrambling can be "undone" by computer analysis, and thus an image *can* be passed through human tissue.)

The cornea and lens of the human eye are transparent to visible light in spite of the collagen fibers in the cornea and the crystalline proteins in the lens. The reason is that the fibers and proteins are densely packed and have what is called a *short-range order*. That is, the fibers or proteins in a small region (a few fiber diameters) all have the same orientation. The dense packing means that the changes in optical properties occur in a distance *smaller* than the wavelength of light. As a result, light is primarily scattered in the forward direction—that is, toward the retina. Thus, light can carry information of an image through the eye to the retina, where the light is detected and recognition of the image begins.

Some aquatic animals minimize their visibility by reflecting light so that you see more of the ocean instead of the animal. Such reflections might hide an animal's eyes so that a predator does not recognize them, or they might hide the gut, which is opaque because of the food in it. The transparency achieved by some aquatic animals is not currently well understood but is surely due to minimizing the variations in the optical properties of the biological components so as to minimize the scattering of light. The variations that *are* present occur in distances that are smaller than a wavelength of light, and so light always scatters in a forward direction as if the variations were not present. A few animals are transparent for a simple reason—they can flatten themselves until the amount of scatter within them is almost imperceptible.

The Hawaiian bobtail squid hides itself by using unique proteins in stacks of platelets. Those platelets function as thin films that can reflect light, like a series of parallel soap films reflects light. What is curious about this squid is that the light is produced by bacteria in an organ on the squid's underside. When the squid is illuminated by, say, moonlight, it wants to avoid casting a shadow on the seafloor, which would reveal its presence. So, it alters the oxygen flow to the bacteria to provoke them into emitting light, and then the platelets reflect that light into the shadow region, eliminating the shadow.

6.91 • A road made crooked by refraction

If you sit in a window seat of a jet airplane, either over the wing or near the wing's trailing edge, watch as your view of a straight roadway slips past the leading edge. Often, the road section nearest the wing appears to become kinked (Fig. 6-34*a*). As more of the road seemingly slides beneath the wing, the kink travels along the road. What causes the kink?

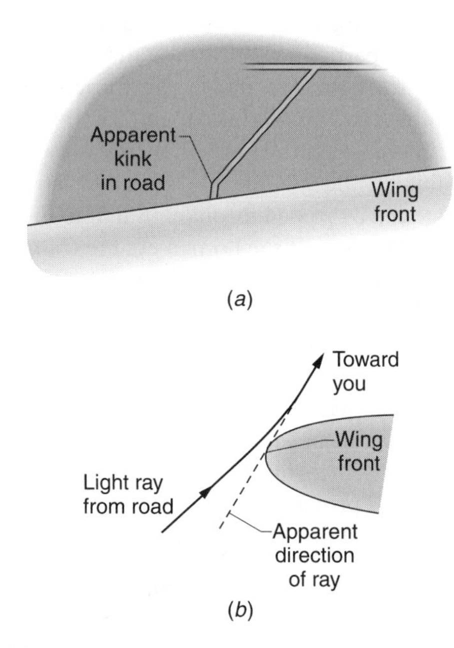

Figure 6-34 / *Item 6.91* (*a*) Your view of roadway. (*b*) Side view of a light ray passing the wing front.

Answer The light rays coming from most of your road view travel along a straight line, and you see the road's true shape. However, the light from the section that you see quite near the wing's front edge must pass through air that varies greatly in density because of the intrusion of the wing into the air. The variation in density refracts (bends) the light rays upward (Fig. 6-34*b*). When you intercept them, they appear to have originated lower in your view than they actually did. The kink is the point separating the affected rays from the unaffected ones.

6.92 • Watering plants in sunlight

Some gardeners claim that you should not water the lawn or shrubbery when sunlight is bright because the drops on the blades and leaves can focus the light enough to burn those surfaces. Is the claim valid?

Answer No researcher has reported seeing damage to leaves in such a situation. Indeed, one researcher reports that the presence of water can actually cool a leaf. Calculations suggest that focusing and subsequent heating are significant only if the drop beads up on a leaf. On most plants, the water tends to spread (the water is said to *wet* the leaves). However, some plants, such as the lotus plant, have leaves with special microscopic structures that cause water to bead into nearly perfect spheres. Although such plants might be in danger of overheating in direct sunlight, those near-perfect spheres tend to just roll off the leaves.

The advice of watering at night instead of during sunlight does have merit in arid regions. Once the sun goes down, the water has a better chance of soaking into the ground than just evaporating.

6.93 • Starting a fire with ice

In Jules Verne's story *The Desert of Ice*, Captain Hatteras and a few loyal men were abandoned near the Arctic by a mutinous crew during an attempt to reach the North Pole. Although the abandoned men possessed wood for a fire, they lacked any sparking materials or other means by which to ignite the wood. Faced with a long trek over the ice field to reach another ship, the abandoned men knew that they would soon freeze to death. However, the ship's doctor hit upon a scheme by which ice could be made to ignite kindling. Can you guess how? Will such a technique actually work?

Answer According to the story, the doctor fashioned a convex lens from a clear section of ice (it lacked air bubbles normally trapped in ice during the freezing process). With a hatchet he chopped out the section and roughly shaped it. Then he smoothed it with his knife and the warmth from his fingers. As he held the ice lens in the bright sunlight, he adjusted its height so that the point of concentrated sunlight (the *focus*) was positioned on the kindling. Within seconds, the kindling ignited.

The idea for this scheme may have originated with William Scoresby, a noted British scientist who is remembered for his pioneering Arctic work. He once described how his roughly formed lenses of transparent ice could ignite wood, melt lead, and light a sailor's pipe. More recently, Matthew Wheeler of McBride in British Columbia told me how he took photographs with a camera using an ice lens instead of its normal lens.

You might also be able to start a fire with a lens in common eyeglasses. If the lens is designed for someone who is farsighted, it has a focus that can be positioned on kindling. However, if the lens is designed for someone who is nearsighted, the lens does not focus the rays. Thus, the fire-starting story in *The Lord of the Flies* is flawed: Piggy is very nearsighted, and Ralph could not have ignited wood with the eyeglasses as described.

6.94 • Diamonds

Why do diamonds sparkle? What produces their colors, and why are the colors more brilliant in a larger diamond? Why is a diamond dark if you look through its bottom surface at a small source of light? Why does grime on the bottom surface decrease the sparkle seen through the top surface?

Answer If a diamond is to sparkle with color, light entering the top surface must separate into colors and return through the top surface. Thus, when the light reaches the bottom surface, it should reflect entirely and not escape through that surface. To avoid that loss, the bottom surface is angled sharply to the light's direction of travel, which causes all the light to be reflected by the surface. The light is said to undergo *total internal reflection.* Thus, if you look up through the bottom surface, the view is dark. However, if the bottom surface is coated with oil or some other grime, some of the light can escape into the coating, decreasing the diamond's sparkle. So, to keep a diamond brilliant, clean *both* top and bottom surfaces.

One measure of how well a material can separate colors when illuminated with white light is the *index of refraction* assigned to the material. Diamond, with a high value of the index, separates colors much better than glass, with a low value. Thus, fake diamonds made of glass may sparkle if cut with many facets, but they lack the play of colors you see from a diamond. In principle, a large diamond is much more colorful than a smaller one because the longer travel distance across the diamond increases the color separation in the light.

6.95 • Opals

What produces the striking colors of an opal? The color production must be different from that in a diamond because the size of an opal does not determine the color separation. Also, the colors are different. If you rotate a diamond below a bright white light, you see a variation in color across the full visible spectrum. If you rotate an opal in that light, you see a narrow range of colors. What determines the difference between the colorless potch opal and the prized black opal?

Answer An opal is not a crystal but an amorphous silica with a small amount of water. The silica is in the form of tiny spheres (with diameters of about 100 nanometers) that are closely packed somewhat like oranges in a container. The spaces between the spheres contain air, water vapor, or liquid water. This arrangement of spheres and spaces forms an array in which the optical properties vary. When white light passes into opal, it undergoes diffraction (a type of scattering) by the array such that different colors are sent back out of the opal at different angles. The angle at which any particular color is diffracted depends on the periodic spacing of the array (the diameters of the spheres) and the angle of the incident light. If an opal moves while you examine it, you see points of different colors flashing on and off—this display is called the *fire* of the opal.

If the colors are brilliant and distinguishable, as in a black opal, they must each consist of only a narrow range of wavelengths. To produce such narrow ranges, the spheres must be nearly uniform in size so that the diffraction from any particular part of the opal is uniform. However, the best displays, with brilliant colors seen from any angle, occur in opals where the silica stacking varies in orientation and arrangements from region to region; the stacking is said to have *faults.* The beauty of a black opal with faults is enhanced by the presence of small particles (carbon, iron oxide, or titanium oxide) that absorb the undiffracted light, providing a dark background to the colorful light you intercept and making that light more perceptible. In potch opal the range in the sizes of the spheres is wide and the colors are not brilliant. The range is smaller in white opals but still wide enough to yield a murky white *opalescence.*

6.96 • Alexandrite effect

The color of most gems is reasonably the same under sunlight and incandescent light, but certain gems, such as alexandrite and tanzanite, can undergo a remarkable shift in color—from blue-green in sunlight to yellow-red in incandescent light. When the first alexandrite gemstone was discovered in Russia's Ural Mountains in 1831, this color shift was named the *alexandrite effect* to honor Tsar Alexander II of Russia. What causes this color shift?

Answer A gem displaying the alexandrite effect transmits light well in the blue-green and red portions of the visible spectrum but not in the intermediate portion. Also, the transmission in the blue-green portion is better than in the red portion. When one of the gemstones is viewed in sunlight, which consists of the full visible spectrum, the transmission in the blue-green portion dominates what you see from the gemstone. Incandescent light is produced by a hot filament in a bulb and is generally dim at the blue end of the spectrum. Thus, when the gemstone is viewed in incandescent light, the blue-green transmission is much dimmer than the red transmission, and the stone should reasonably be redder than in sunlight. However, it is even redder than this explanation predicts. Apparently the extra red is a product of your visual system; that is, part of the stone's color is generated by your brain.

6.97 • Star sapphire

When you look down on a star sapphire that is illuminated by a small light source, why do you see a six-point star floating above the stone?

Answer The star is produced by light scattering from needles of titanium oxide grouped into one of three orientations that are separated by 120°. The light scattered by each group forms a line in your view. So, you see three lines that intersect at their midpoints and form the six rays of a star. If the gem is cut with a round or oval dome (instead of being cut with facets), the star appears to lie somewhat above the dome, giving the floating appearance. The image is *virtual*, meaning that it is something your visual system conjures up to make sense of the light. A luminous star would not appear on a card placed at the star's apparent position.

6.98 • Patterns from a glass of wine, a window, and a drop of water

Examine the sunlight shining through a glass of white wine and onto a table. The table is not uniformly lit by the glass but is covered with one or more bright lines called *caustics*. The curvature of the glass redirects the light to produce the lines. Windows reflect similar patterns onto their surroundings. However, the most common caustic is probably the one formed within a ceramic or plastic coffee cup (not Styrofoam). The caustic consists of two bright, curved lines that intersect (Fig. 6-35*a*).

A greater variety of caustics can be produced by sending a laser beam through a ruffled or rippled sheet of plastic such as found in the coverings around fluorescent lights in offices. If you move the plastic to music, you can create a miniature laser light show. You can also use a flat, clear sheet of plastic onto which a distorted mound of transparent glue has been added. Kinetic caustics can also be seen on the bottom of swimming pools as the sunlight travels through the ripples on the water surface.

The most intriguing of all caustic patterns can be seen if you look at a street lamp at night through a water drop. The drop might be on a lens of your eyeglasses or on a windowpane (in this case, bring your eye very close to the drop). If the drop is small and irregular, the pattern you see consists of outwardly concave bright lines that intersect (Fig. 6-35*b*). If the drop is larger and clings pendulously, the bottom of the pattern is similar to that of the smaller drop, but the top is a bright, outwardly convex line (Fig. 6-35*c*). Just within it are stars that can be made to dance if you gently wiggle the drop. If you can rotate the drop around your line of sight of the lamp, a cusp initially at the bottom of the pattern shrinks and then enters the interior to become a star.

What accounts for the patterns seen in these varied circumstances? Can the patterns be broken down into basic units?

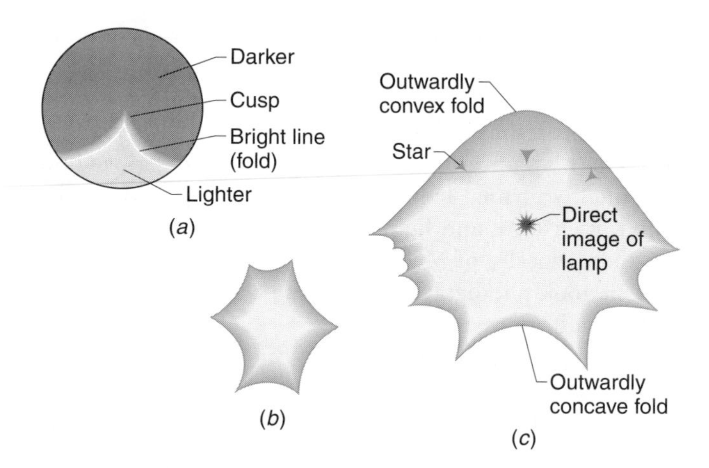

Figure 6-35 / *Item 6.98* Patterns of light produced by (*a*) a cup of coffee in sunlight, (*b*) a small water drop near the eye, and (*c*) a larger water drop near the eye.

Answer The caustics can be reduced to two basic shapes: *folds* (the curved lines) and *cusps* (the intersection of two folds). The basic units are optical examples of the mathematics called *catastrophe theory*. They appear because a surface (wine glass, rippling water, etc.) bunches the rays by refraction or reflection, thus concentrating the light into bright lines. More precisely, the light is concentrated into a three-dimensional structure in the air. When your eye or a viewing surface intercepts the light, it is taking a two-dimensional slice through that structure.

The three-dimensional structures come in three varieties, and each has a *singularity* in which the caustic in a slice is the most compact. If a slice is taken somewhere else through one of these three structures, the structure is said to be *unfolded*.

Sunlight reflected from a building window onto a wall can form an image in the general shape of the window but with curved edges, which are folds. The folds are either concave or convex depending on whether the windows bulge inward or outward. When the window is outwardly convex, the reflection creates an oval pattern; when it is outwardly concave it yields a bright cross. A double-paned window with both types of curvature can create both patterns.

6.99 • Shadows with bright borders and bands

When you take a bath, examine the shadow cast by a pencil as it is illuminated by a single, overhead lamp. (But do not position a lamp such that it might fall into the water and electrocute you!) If you hold a pencil above the water or fully submerge it, the shadow on the bottom of the tub resembles the pencil. However, if you hold it partially submerged and at an angle, it casts two sausage-shaped shadows that are separated by a white band (Fig. 6-36*a*). What creates this display, called the *sausage-shadow effect*?

Next, insert the pencil vertically through the water surface. Adjust the depth of the lower end. If the end is near the bottom of the tub, the pencil makes a small shadow. Why is the shadow replaced with a bright spot if you move the end up toward the water surface?

Float a flat, double-edge razor blade (not the kind with a reinforced side) by carefully lowering the blade onto the water. If the water is only a few centimeters deep, the edges of the blade's shadow are normal. Why do the edges have a bright border if the water is deeper (Fig. 6-36*b*)? Are the edges bright if you lift the blade so that it is slightly above the normal level of the water?

Why does a floating hair often produce a string of shadow sections, some with normal edges and some with bright edges?

Run a finger or pencil through bath water that is at least six centimeters deep. Why are dark circles with bright edges left playing over the bottom? You may see similar shadows on the bottom of a sunlit swimming pool when someone swims or climbs out of the water.

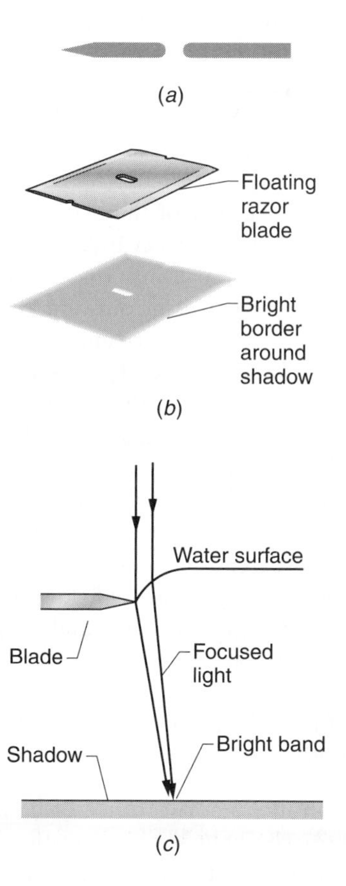

Figure 6-36 / *Item 6.99* Shadows of (*a*) a pencil and (*b*) a razor blade in shallow water. (*c*) How the water curvature near the object focuses light rays to form a bright band at shadow's edge.

Answer In the sausage-shadow effect, water is pulled up along the pencil by surface tension (due to the attraction of the water molecules to one another and to the pencil). This leaves the water surface concave. When light passes by the pencil and through the curved water surface, it spreads into part of the shadow, producing an illuminated gap between the shadows created by the dry and submerged sections of the pencil.

The weight of the razor blade depresses the water surface so that the surface is convex. The light passing through the curved surface tends to be focused, but if the water is shallow enough, the rays are intercepted by the bottom before they become focused and the shadow has a normal edge. If the water is deeper, the rays focus along the edge of the shadow, giving it a bright border (Fig. 6-36*c*). When you lift the blade, the water is pulled upward into a concave shape that tends to spread the light into the shadow region without any focusing, and so the shadow edges are normal.

A floating hair can create sausage shadows, normal edges, and bright edges, depending on how the water surface is curved along its length.

When you run an object through water, you leave dimple-like vortexes swirling along the surface. The innermost part of a vortex is concave and spreads the light, creating a dark circle on the bottom where the light is weak. Somewhat farther out in the vortex the surface is convex. Light passing through that section is focused onto the edge of the shadow, giving it a bright border. If the water is deep enough, the border expands into a band of light, with the inner and outer edges especially bright.

6.100 • Bright and dark bands over the wing

A 1983 report by A. Hewish describes a pair of dark and bright bands that he spotted along the length of a wing on a jet airliner during flight (Fig. 6-37*a*). The dark band was 1 to 2 centimeters wide and had low contrast with the rest of the sunlit wing. The Sun was at an elevation of about 25° above the wingtip, and the bands were visible for more than an hour. As the airplane descended, they shifted toward the leading edge of the wing, finally disappearing.

I have seen similar bands and have also seen a related effect when the Sun was on the opposite side of the aircraft:

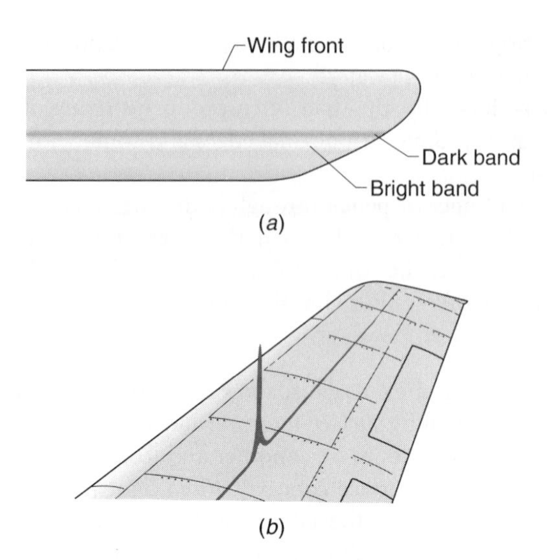

Figure 6-37 */ Item 6.100* (*a*) Bright and dark bands seen along an airplane wing. (*b*) A dark band seen standing up from a wing.

A dark band stood upright about halfway out on the wing, noticeably distorting the details of the wing's outer half (Fig. 6-37*b*). When I moved my head forward or backward along the fuselage, the band moved too. Sometimes I saw two bands.

What produces the bands?

Answer Let's take the viewpoint from an airplane, with air flowing past a stationary wing. The flow over the wing creates a *shock front* where the speed of the oncoming air abruptly decreases and the density of the air correspondingly increases. The shock front stands perpendicular to the wing and extends along its length. (If the shock front were visible, it would resemble a porous barrier along the wing.) When light rays skim through the shock front, they can be perceptibly refracted because of the change in air density. In Hewish's observation, light rays that were headed toward a particular place on the wing were redirected by refraction to a place somewhat farther back on the wing. So, a bright band forms there and a dark band forms where the light would have landed without the shock front.

The dark band I additionally found was also due to light rays skimming through a shock front, but they originated from details on the sunlit wing. When the rays from adjacent details on the wing passed through the shock front, they were bent by different amounts. The resulting spread in the rays left the region between the details relatively dark, and the composite of these dark spots formed the dark band I saw.

SHORT STORY

6.101 • Shock waves from the *Thrust SSC* car

When the jet-powered car *Thrust SSC* set the land-speed record in 1997 as it raced across the Black Rock Desert of Nevada, it was *supersonic* (faster than sound). Not only did observers hear a loud boom produced by the merging of the individual shock waves from various portions of the car, but also evidence of the individual shock waves was captured in photographs of the car. In those photographs we should be seeing an undistorted view of the hills beyond the car's path, but when the light rays from those hills came through the shock wave on their way to the camera, their directions of travel were changed slightly by the variation of air density within the shock waves. This redirection distorts a photograph's image of the background hills, thus revealing the shock waves. In one of the photographs, I can see four shock waves extending upward from the car.

6.102 • Pinhole and pinspeck cameras

A pinhole camera consists of a small, circular opening through which the light must pass to reach the film. How can such an arrangement form an image on the film? How large should the pinhole be? Some tiny reflecting surfaces, such as a glass fragment, can function in the same way to produce an image.

You can make a large version of a pinhole camera by allowing light to pass through a small opening in a window curtain and into an otherwise dark room. On the far wall you might see an inverted image of the scene outside the window. This arrangement was quite a novelty to people in the past.

An interesting ping-pong-ball version of this effect was published by Patrick A. Cabe of the University of North Carolina at Pembroke. Briefly you do this: (a) Blacken one hemisphere of a ping-pong ball. (b) Drill a 2 millimeter diameter hole at the center of the blackened side. (c) Roll black construction paper into a cylinder slightly smaller than the ball. (d) Squeeze the ball into one end of the cylinder, with the black hemisphere facing outward. (e) Tape the seam of the construction paper in place and then tape the ball in place without blocking the hole. (f) Look into the open end of the cylinder while you point the far end toward a brightly lit scene. You will see an inverted image of the scene on the wall of the ball inside the cylinder.

A pinspeck camera consists of a small, opaque dot positioned in front of the film (the dot might be drawn on an otherwise transparent plastic sheet). A large opening in front of the dot acts like a camera's field stop that limits the extent of light passing the dot. What kind of image is formed on the film from such an arrangement?

You can also set up a demonstration in which a pinspeck blocks some of the light from a fluorescent tube that illuminates a screen. What kind of image forms on the screen?

Answer Consider a small source of light in front of a pinhole camera. The light waves from the source diffract through the pinhole; that is, they spread from the pinhole and undergo interference where some waves reinforce one another and some waves tend to cancel one another. The result is simple: The light forms a small, bright spot on the film. The spot is an image of the source; other small sources of light also form such images. If the pinhole is too large, the images overlap, perhaps too much to be distinguished. The overlap decreases if the pinhole is made smaller, but then the intensity of the individual images also decreases. So, what is the best size for the pinhole?

The answer depends on the manner in which a light wave from an object spreads as it travels to the film. When the wave reaches the plane of the pinhole, it is said to be divided into circular zones, each centered on the pinhole. The light that passes through the central zone arrives at the film *in phase* (in step), so the waves tend to reinforce one another (they have *constructive interference*) and produce a bright image.

If the pinhole fits the central zone exactly, the image is at its brightest. If the pinhole is smaller, it blocks some of the light through the central zone, and the image is dimmer. If the pinhole is larger, it allows some of the light in the next zone to reach the image on the film. Because the rays of that light take longer, oblique paths to the film, they arrive somewhat *out of phase* (out of step) with the central-zone rays. The result is that some of the waves cancel one another (they have *destructive interference*), which dims the image. The optimum situation is when the pinhole is small enough to pass light only through the central zone. Then an image is bright and sufficiently sharp.

The opaque dot in a pinspeck camera casts a shadow spot on the film for each small source of light on an object in front of the camera. The composite of the shadow spots forms a *shadow image* (or *negative image*) of the object.

If the dot is placed between a fluorescent tube and a screen, a shadow image of the tube appears on the screen: The dot casts a shadow of each portion of the tube onto the screen, and the composite of the shadows is a dim image of the tube. The image is not totally dark because any part of it is still illuminated by most of the tube.

6.103 • Solar images beneath a tree

During a solar eclipse, what produces the many small images of the Sun in the shadow cast by a tree? Are solar images in the tree's shadow at other times? Why do shadow images of leaves, sometimes with a pair of edges, one inside the other, appear beneath a tall canopy of leaves? Why don't they appear beneath shorter trees?

Answer The eclipse images are produced by small holes in the leaves or between adjacent leaves. Each hole functions like a pinhole camera (see preceding item), throwing an image of the Sun into the tree's shadow on the ground. The images are produced even when there is no eclipse, but they may be more difficult to distinguish because the overall glare of sunlight from the sky and the landscape partially illuminates the shadow. During an eclipse, that glare is diminished by the overall darkening and the images are then more perceptible. In either situation, the images are much easier to see on a flat surface than on uneven or grassy ground.

A leaf shadow seen beneath a tall canopy of leaves is cast by a low-lying leaf that is illuminated by sunlight streaming through a hole higher in the canopy. If a low-lying leaf is illuminated by two higher holes, you might see two overlapping shadow images, with one leaf-edge image inside a second leaf-edge image.

6.104 • Lights through a screen, lines between fingers

If at night you view a distant bright lamp through a wire window screen that is at least several meters from you, the light from the lamp forms a pattern of dark and bright lines (Fig. 6-38*a*). What creates the pattern? A similar pattern can be seen if you look at a bright lamp through the fabric of most common umbrellas. In some situations you might see colors. If you sight through a thumb and finger that are almost touching, while in a brightly lit room, why do you see multiple dark lines between them (Fig. 6-38*b*)?

Answer These patterns of dark and bright lines are usually attributed to the diffraction of light, which is an effect that causes light to flare when it passes through a narrow

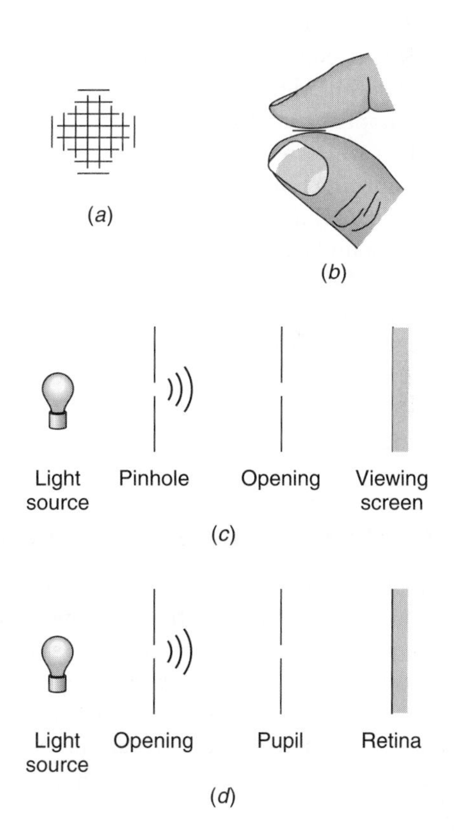

***Figure 6-38** / Item 6.104* Pattern seen (*a*) through a mesh and (*b*) between thumb and finger. (*c*) Light sent through a pinhole illuminates a narrowing opening, producing a diffraction pattern on the viewing screen. (*d*) Light sent through that same opening illuminates your pupil, producing a diffraction pattern on your retina.

opening or past a narrow object. Moreover, at some angles the waves in the flared light are *in phase* (in step) and reinforce one another (undergo constructive interference), producing a bright line. At other angles the waves are *out of phase* (out of step) and tend to cancel one another (undergo destructive interference), producing a dark line. If this flared light falls on some featureless viewing surface, the bright and dark lines form a pattern.

However, this pattern can appear only if the light going through the narrow opening is *coherent*; that is, the waves must be in step (or nearly so) before they undergo diffraction. Light from most common sources, such as any bulb, is *incoherent*; that is, the waves are generated randomly without any coordination. Coherent light can be produced by passing incoherent light through a pinhole (Fig. 6-38*c*). Because a pinhole is small, all light waves passing through it are almost identical and thus are almost in step. When the light reaches a narrow opening, such as in umbrella fabric or window screen, the light is diffracted by the opening, and a diffraction pattern is formed.

If you remove the pinhole so that the incoherent light from the light source (such as a lightbulb) falls directly on the narrow opening, the diffraction pattern disappears. Light still flares at the narrow opening, but the flaring changes from instant to instant as uncoordinated light waves randomly pass through the opening. What you see on the viewing surface is a featureless illumination.

However, suppose you replace the viewing screen with your eye so that you directly see the light from the narrow opening. The opening now acts like the pinhole in the traditional setup (Fig. 6-38*d*). Because the opening is small, the light waves spreading from it are almost in step (coherent) when they reach the pupil of your eye. Thus, they undergo diffraction when they pass through the pupil and produce a visible diffraction pattern on your retina. You perceive this pattern when you look at a distant lamp through window screen or umbrella fabric at night or between closely spaced thumb and finger.

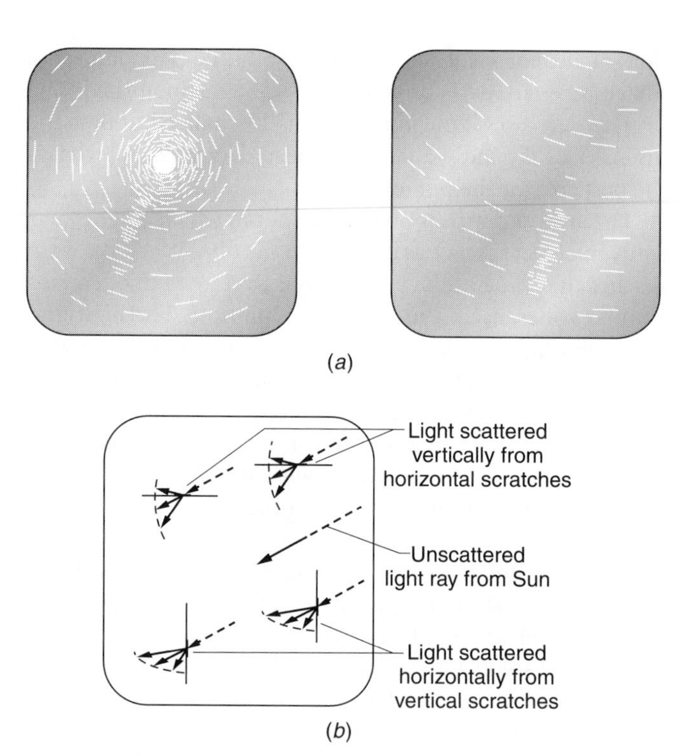

Figure 6-39 / *Item 6.105* (*a*) Pattern of concentric bright scratches in a window toward the Sun and slightly away from it. (*b*) Scatter of light by a scratch. Light approaches from outside the window and scatters into a room.

6.105 • Bright scratches and colorful webs

When you look toward the Sun through the window of an airplane or building, why does the window have bright scratches that are arranged in concentric circles (Fig. 6-39*a*)? Why do the scratches sometimes appear to give streaks of light that point toward the Sun? Why are the scratches usually invisible when you do not look toward the Sun?

If a spider web lies near your view of the Sun, why is it colorful, and why do the colors vary if you change your perspective of the web? (The colors may be faint; you can see them better if there is a dim background to the web, but the web must be in direct sunlight.)

Answer The arrangement of scratches in concentric circles is an illusion. The window is likely to have myriad short, randomly oriented scratches, but only some of them are bright in your view, and those create the illusion. When light scatters from a scratch, the scatter is confined to a flat fan in a plane that is perpendicular to the length of the scratch (Fig. 6-39*b*). To see the scratch, your eye must be in the fan.

If a scratch is directly to the left or right of the Sun, you will be in its fan of scattered light only if the scratch is vertical. If the scratch is above or below the Sun, you will be in its fan only if the scratch is horizontal. The requirement can be generalized. If a scratch is to be seen, it must be tangential to a circle around your view of the Sun. So, when you see many short scratches, you have the illusion that they actually lie in circles around the Sun. If instead the scratches are clustered, they might seem to form a radial streak extending from the Sun.

You can be in the fan of scattered light from a scratch only when the scratch is near your view of the Sun. Otherwise you miss the scattered light, and the scratches are then usually imperceptible.

Extreme situations of scratched airplane windows have occurred when an airplane has accidentally flown through the plume of an erupting volcano. For example, a jumbo jet was flown through the plume from Mount Galunggung in Indonesia, cutting off all four engines and sandblasting the cockpit windows. The pilot managed to restart three of the engines and headed for an emergency landing at Djakarta in the dark. However, he could not see through the windows. Instead, he had to rely on his copilot, who was squinting through a few millimeters of unscarred window, trying to line up the airplane with the runway lights.

When a web lies near your view of the Sun, light scatters (diffracts) from the silk threads and also from the sticky spheres that a spider leaves along some of the threads in order to snare prey. The scatter from a thread is much like that from a scratch on a window: The scattered light flares outward in a fan. The scatter from a sphere is a cone of light. With either type of scattering, the angular extent of the scatter differs for different wavelengths, which means that the scattering can spread the incident white sunlight into a pattern of distinguishable colors. As you change your view of the web, you can change which colors you intercept. The same spread of colors can make the bright scratches on a window colorful.

6.106 • Bright streaks in a car windshield

When you drive through rain at night, street lamps and other light sources produce streaks of light on your front windshield (Fig. 6-40*a*). The streaks are straight or curved, are directed toward a common point, and rotate about that point as the car moves. You may also be able to see a daytime streak if you view the Sun through the windshield.

Often, the streak from a lamp or the Sun appears to have depth—it seems to be a bright lane that leads from the windshield to the light source. When a streak is curved, it resembles a lane that runs along a valley and then up a hill. What produces a streak and what gives it depth?

A variety of other, dimmer streaks or patches can be seen in a windshield at night. The patches can be randomly placed but are sometimes organized in a straight or curved line. One type of streak seems to pass through the light source like the one just described, but it is not directed toward any particular point on the windshield and can sometimes be seen through other types of windows. You might also see them through eyeglasses.

Answer The wiper grinds circular grooves in the gummy material that adheres to the windshield. A groove scatters lamp light into a flat fan that is perpendicular to the length of the groove. (See the preceding item.) The composite of the scattered light from many adjacent grooves forms the visible streak. Depending on the curvature of the windshield, the streak can be straight or curved. Either way, the lower end of the streak points toward the center of the circles—that is, toward the wiper's pivot point.

A streak seems to have depth because each eye actually sees a different streak. The separation between the two streaks is smallest at the ends near your view of the light source and largest near the wiper's pivot point. If your brain fuses the two images into a single image, it interprets the sections with larger separations as being nearer than the sections with smaller separations. The appearance of depth may seem strange for a wide light source, because then the streak is widest near your view of the source, but that part of the streak seems to be the most distant. The result is exactly opposite the apparent distance and width that you would see in an actual path extending away from you.

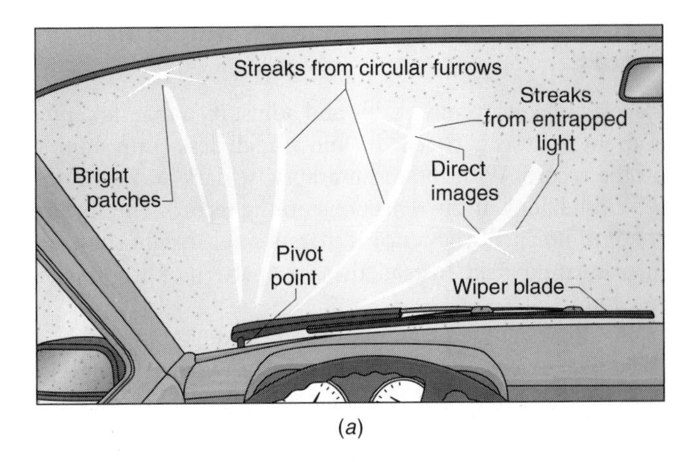

(*a*)

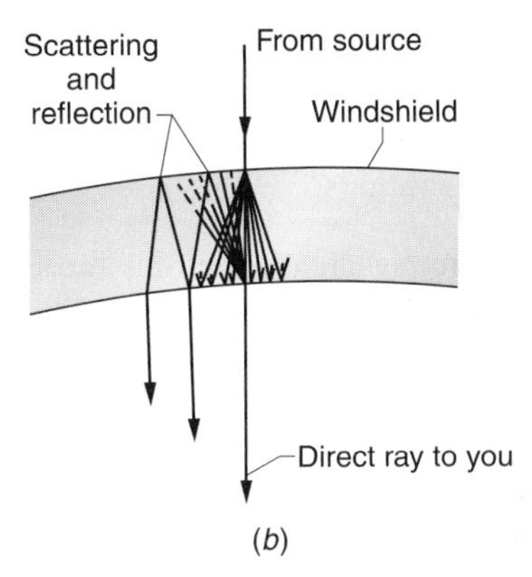

(*b*)

Figure 6-40 / *Item 6.106* (*a*) Patterns seen in a car windshield. (*b*) The internal reflection of light within the windshield.

The second type of streak that can be seen on a windshield and many other types of glass or plastic is due to light reflecting inside the layer (Fig. 6-40*b*). The light originates in the beam that runs directly from the source to your eye. When the beam passes through the layer, some of it is reflected internally. That light might reflect many times inside the layer. At each point of reflection, some of the light leaks out. If the layer is curved with its concave side facing you, you might intercept part of the light that is leaked at points to the side of your direct view of the source. What you see is a streak extending from the direct view. If you somehow block the direct beam from reaching the glass, say, with a finger by reaching out the car window to the front of the windshield, the streak disappears. (The car should be stopped, of course, because doing this experiment is not worth your losing your arm in a traffic accident.)

The bright patches are due to reflections or scattering from regions of gummy material. If the layer is curved, you might intercept some of the light and thus see the patch.

6.107 • Reflections from a phonograph record

Place a vinyl record on a table and adjust it so that its center is about midway between you and a small desk lamp shining on the record. With the room otherwise dark and with one eye closed, look at the reflections on the record. The surface is not uniformly bright, nor is there a single bright spot as a mirror would yield. Instead, there is a pattern of bright, narrow lines.

By adjusting the position of the record, you can make a cross-like pattern or one in which one or two hyperbola-like lines lie on the record. In both cases at least one of the lines passes through the center of the record. Sometimes you will see an especially bright spot on the record—it is always on the line that passes through the center. Similar patterns of bright spots can sometimes be seen on venetian blinds when you look through them at a street lamp at night. Can you explain these observations?

Answer You receive light from only those places on the record where the grooves have the proper tilt to reflect light toward your eye. All the grooves reflect light up to the height of your eye, but only some of them are properly tilted to reflect the light both up and horizontally so that the eye intercepts the light. Those grooves lie along the bright lines of the patterns you see. The especially bright spot is a mirror-like reflection of the lamp.

The reflection patterns on a record are similar to the light streak seen in a car's windshield (see preceding item), and if the light source is bright enough, you might perceive depth in the patterns as with the windshield light streak.

6.108 • Colors on finely grooved items

When white light illuminates a music or movie disk or certain types of patterned paper, the reflections are highly colorful. You can also see colors from a vinyl music record if you tilt it correctly in the light.

Although a street lamp may appear to emit either white or yellowish light, the several component colors in its emission can be seen and photographed if you attach an inexpensive *diffraction grating* (a plastic piece with many fine grooves or slits) to the lens of the camera. The direct image of a lamp as seen through the camera appears to be normal, but a spectrum of component colors is spread out on each side of that view.

In each case, why does the initial light end up in colors?

Answer The light scattered (diffracted) by the finely grooved structures undergoes *constructive* and *destructive interference*. Consider a spot that appears to be red. When the white light scatters from that spot, the red components that are sent toward your eyes undergo constructive interference; that is, the red waves end up in step and reinforce one another. The other color components that scatter to you from the spot undergo destructive interference; that is, the waves are out of step and tend to cancel one another. So, what you see from the spot is light that is dominated by red. Other spots, being angled differently in your view, send light to you that is dominated by other colors.

If there is to be perceptible color separation, the separation of grooves in your view (and the lamp's "view") must be small and about the size of the wavelength of visible light. If the light shines directly down on a vinyl record and you also look directly down, the grooves on the record are too widely spaced to yield colors, and you see only the black of the record's dyed plastic. To separate out the color components of white light, you must arrange for the initial beam of light and for your view of the record to be nearly along the plane of the record. Then in your view (and the lamp's view), the grooves on the near and far sides of the record are spaced sufficiently close to separate the colors.

6.109 • Anticounterfeiting: Optically variable devices

One security feature on credit cards, driver's licenses, and many other types of identification cards is a *variable display* that changes when the card is tilted. Such a variable display is often called a hologram, which is a type of photograph that gives depth and perspective to its images. Indeed, the earliest versions of these variable displays on credit cards were holograms. However, the use of holograms as a security measure was not very successful for two reasons. First, they were dim, murky, and difficult to see in the lighting of a typical store. Second, and far worse, they were easily duplicated by counterfeiters. Today the variable displays are bright, sharp, and easily seen in store lighting. Better, they are extremely difficult to counterfeit. How are variable displays made bright and counterfeit-proof?

Answer Most credit cards now carry *optically variable graphics* (OVG), which produce an image via the diffraction of diffuse light (such as store light) by finely textured regions called gratings. The gratings send out hundreds or even thousands of light waves. Someone viewing the card intercepts some of these waves, and the combined light creates a virtual (imaginary) image that is part of, say, a credit-card logo. For example, in Fig. 6-41*a*, gratings at point *a* produce a certain image when the viewer is at orientation *A*, and in Fig. 6-41*b*, gratings at point *b* produce a different image when the viewer is at orientation *B*. These images are bright and sharp because the gratings have been designed to be viewed in diffuse light.

An OVG is very difficult to design because optical engineers must work backwards from a graphic, such as a given logo. The engineers must determine the grating properties across the OVG if a certain image is to be seen from one set of viewing angles and a different image is to be seen from a different set of viewing angles. Such work requires sophisticated computer programming. Once designed, the OVG

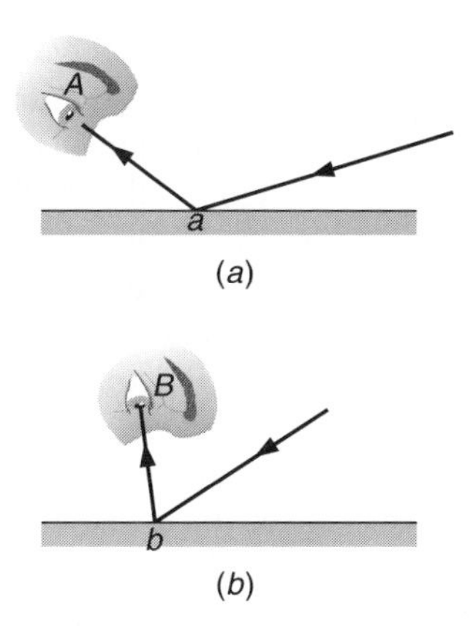

Figure 6-41 / Item 6.109 (*a*) Gratings at point *a* send light to a viewer at orientation *A*, creating a certain virtual image. (*b*) Gratings at point *b* send light to the viewer at orientation *B*, creating a different virtual image.

structure is so complicated that counterfeiting is extremely difficult.

6.110 • Colored rings from a misty or dusty mirror

After taking a hot shower and steaming up the bathroom mirror, turn off the lights, face the mirror, and burn a match somewhat off to one side of your direct view of the mirror. The flame's image in the misty mirror will be surrounded by colored rings. You will also see the rings if the mirror is covered with dust or powder instead of water droplets.

Next, arrange for the misty mirror to be in front of you in a dim room with a brightly lit window at your back. Adjust your distance from the mirror until you see rings superimposed on your reflection in it. At the proper distance, I find that the images of my eyes are each replaced with an array of colored rings, looking like psychedelic graphics. Again, the mirror can be dusty instead of misty.

What produces these rings?

Answer The two demonstrations involve different ways in which the light is scattered (diffracted) by the particles (either droplets or dust motes) on the mirror. There are two basic arrangements:

In one arrangement, the rings are centered on a small light source held slightly off to one side of your view of the mirror. In this arrangement, each particle scatters light back from the mirror and into a circular pattern in which the center is bright and the surrounding rings alternate between being bright and being dark. Since the angle of scattering depends on the wavelength of the light, the colors appear at different angles and thus are separated in your view. Specifically, red appears on the outer edge of a ring and blue on the inner edge, because red light scatters through a slightly larger angle than blue light. You actually see only a portion of the pattern from each particle, but the composite of the individual portions from many particles is itself a circular pattern that you perceive as centered on the reflected image of the light source.

In the second arrangement, the rings are centered on your eyes and the light source is large and behind you. In this arrangement, the rings are due to two sets of light scattered to you by the particles (Fig. 6-42). Consider one particle. (1) It can scatter light (ray *A* in the figure) to the back of the mirror where the shiny surface then reflects the light to you. (2) It can also scatter light to you that has already been reflected from the back of the mirror (ray *B*).

Those sets interfere with each other when they reach you. For some of the particles the two sets arrive at your eyes out of phase and destructively interfere—that is, cancel one another. For other particles, the two sets arrive in phase and constructively interfere—that is, reinforce one another. The composite pattern is one of concentric, bright and dark rings, with the bright rings colored because of the slight color separation involved in the scattering.

Which type of interference pattern you see depends on the orientations of you and the light source. You can overlap the two types if you place a small light source directly between your eyes and the mirror.

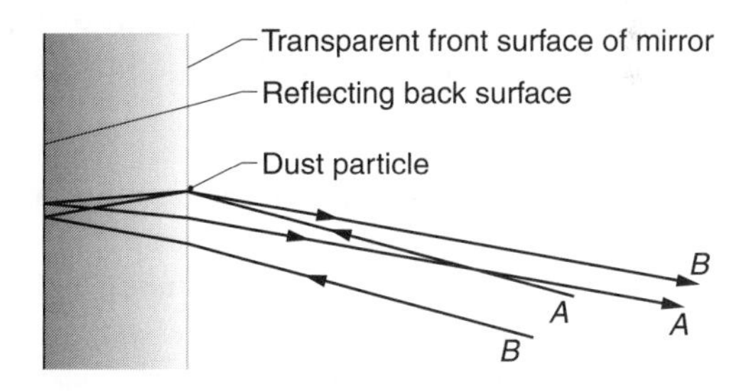

Figure 6-42 / Item 6.110 Two ways in which light scatters to an observer from a dust particle on the front of a mirror.

6.111 • Color of milk in water

Arrange for a narrow beam of white light to shine through a small aquarium (or any other container with clear, flat sides) filled with water. With the lights off, begin to add drops of whole milk (not low-fat milk). Initially, you see little of the beam in the water, but soon the beam becomes quite visible. As you continue to add drops, monitor the color of the beam in the water from the side and from a view almost directly toward the light source. Eventually you will find that the beam is bluish when seen from the side and reddish when seen head-on. Why the colors? Why is milk that you drink white instead of red or blue?

Answer Milk contains small fat globules that scatter light. The blue end of the spectrum is more strongly scattered off to one side than is the red end of the spectrum. So, when the concentration of globules is large enough that the colors are first perceived, the light scattered to the side is dominated by blue, while the light that is scattered in the initial direction of the beam is dominated by red.

Your ability to see the colors depends on the low concentration of the globules. When their concentration is as high as it normally is in milk, the light scatters many times before it leaves the container. Then, for any direction of view, you receive as much of the red end of the spectrum as the blue end, and the combination is white.

If you tip a glass holding a small amount of milk such that the edge of the milk lies on the bottom of the glass, you can see a bright clear band along the edge. The band lies in the region where the milk forms a curved surface because it climbs the glass surface by a short distance. The light reflected from that curved surface is concentrated and thus brighter than the light reflected by the rest of the milk surface. This concentrated light masks the scattering by the particles within the milk, so you don't see the white of the milk in this band across the bottom.

6.112 • Color of campfire smoke

When the smoke rising from a campfire is viewed against a dark background such as surrounding trees, it is tinted blue, but when it is seen against a bright background such as the sky, it is tinted yellow, red, or orange. Why do the colors differ in the two viewing situations?

Answer The smoke particles from the campfire are small enough that they scatter the blue end of the spectrum more strongly than the red end. So, the light continuing in the original direction weakens in blue and becomes yellow, red, or orange. If you view the smoke against a dark background, the light source must be behind you (it might be the Sun or a bright portion of the sky). Blue light scatters in your direction. If you view the smoke against a bright sky, that portion of the sky is the light source. You then intercept light that has become depleted in blue after passing through the smoke; thus the light is dominated by the red end of the spectrum.

6.113 • Ouzo effect

Certain aniseed-based alcoholic drinks, such as *ouzo* in Greece, *le pastis* in France, *raki* in Turkey, and *sambuca* in Italy, display a peculiar behavior: At one point as water is gradually added to the fairly clear drink, the drink suddenly turns milky white. What causes this change? That is, what does the water do, and what accounts for the change in the drink's appearance? The effect can be reversed if more alcohol is added.

Answer Each of these drinks is a solution with a uniform distribution of aniseed oil and ethanol alcohol. When a light beam (such as sunlight) is sent into the solution, it emerges on the opposite side as a beam. When water (a third liquid) is added, things can change because the aniseed oil cannot dissolve in water. Initially the drink is transparent (a light beam still travels through it as a beam). However, when the water content reaches a certain fraction of the liquid, said to be a *critical value*, the oil molecules spontaneously form drops that are suspended in the liquid. The drink is said to undergo a *phase transition*, switching from a homogeneous (uniform) liquid–liquid solution to a nonhomogeneous liquid–droplet dispersion (or emulsion). The drops scatter visible light. So any beam sent into the drink is scattered into many directions, which gives the drink a milky appearance. If more of the alcohol is poured into the drink, dropping the water content below the critical value, the phase transition is reversed and the drink becomes clear again.

6.114 • Colors of oil slicks, soap films, and metal cooking pots

Why do oil slicks on wet streets display colors? Why are the colors absent if the street is not wet? Why can the colors occur even if the Sun is hidden behind clouds?

Why are soap films and bubbles colorful? Why do the colors disappear when the film becomes especially thin or thick? Why don't you see similar colors from a microscope slide or a windowpane?

Suppose that a thin soap film is suspended vertically while illuminated from the front with a beam of white light. As gravitation drains the film, the horizontal colored bands on the film begin to migrate downward, and then the top portion turns black (provided the background behind the film is dark). How can the film be black, even though it is brightly illuminated from the front? If you examine the black region carefully, you might find spots that are especially black. Why do they appear? Why is the band just below the black region white instead of blue? (Blue light has the smallest visible wavelength, and so it should correspond to the very thin portion of the film below the black region.)

Sometimes cooking pots have areas that are colorful even though they have been carefully cleaned. What produces the colors? Similar colors appear if oil drains from a polished metal surface. Why don't such colors appear if the surface is not polished?

Answer A transparent layer with a thickness that approximately matches the wavelength of visible light can produce colors when illuminated with white light. Suppose that a ray of a single color is incident perpendicularly on such a film (Fig. 6-43). Some of the light reflects from the film's front surface while some of the rest travels through the film, reflects from the back surface, again travels through the thickness of the film, and then emerges. When your eye intercepts these two waves leav-

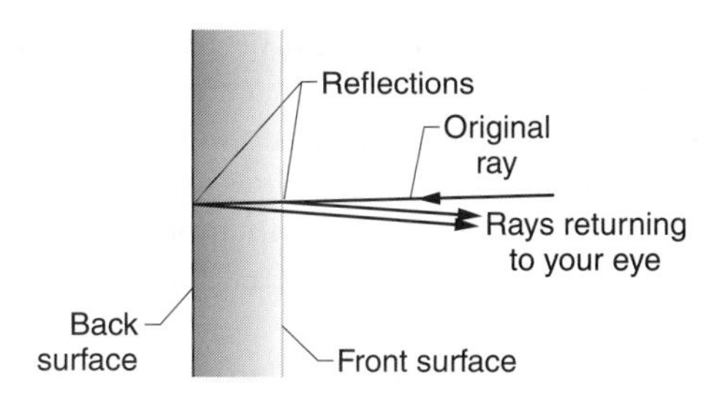

Figure 6-43 / *Item 6.114* Light reflecting to your eye from the front and back surfaces of a thin film.

ing the film, they undergo interference. If the waves happen to be in phase (in step), they constructively interfere (they reinforce one another) and you see a bright color on the film. If they happen to be exactly out of phase (out of step), they destructively interfere (they tend to cancel one another) and you see darkness on the film.

The film thickness is one factor that determines whether a particular color is bright. So, when gravitation drains a vertical film, making it thicker toward the bottom, different bright colors appear at different heights. If you change your perspective of the film, you change the distance that light travels through the film to reach you. Thus, you change which colors are bright. Colorful displays that change color as you change your perspective are said to be *iridescent.* (A blue shirt that is blue because of a dye is not iridescent.)

You can see iridescent colors on an oil slick if the oil forms a thin horizontal film on water. The colors may be visible even when the Sun is hidden, provided that part of the sky is brighter than the rest of the sky. If the oil lies on a dry street, the thickness varies across the film because of the rough texture of the underlying street. The colors from different thicknesses overlap, and the pattern is dull, perhaps even colorless.

When the top portion of a draining, vertical soap film becomes much thinner than the wavelength of light, all light waves reflecting from it almost completely cancel one another and the top portion is black. The thinning can be momentarily arrested at this stage for two reasons. (1) The layers of soap molecules on opposite sides of the film are now close enough to electrically repel each other. (2) The water molecules along each surface apparently become ordered (somewhat like the order in an ice crystal) and begin to overlap. Because overlapping requires energy, the thinning tends to stop at this point. In spite of these reasons, the thin film is unstable and likely to collapse suddenly and burst. However, if the film contains charged impurities, it may be able to thin even more without bursting. Such areas are especially black parts of the thin black region.

The band just below the black region is faintly blue because blue light undergoes *partial constructive interference* (the waves are somewhat in step and thus somewhat reinforce one another). However, the blue is difficult to see, and you may find that the first distinguishable band below the black region is white. In that part, the film has a thickness such that all colors in the visible spectrum undergo partial constructive interference, and their combination appears white. Below the white band we find a band of yellow-red (orange) and then a band of blue-red (purple). Only then do we find a band of approximately pure blue.

As a film drains and thickens at the bottom, the colored bands at the bottom begin to overlap. Eventually, the overlap is enough to eliminate any color, and that part of the film is white.

However, colored bands would not appear even if we switched the light source to a pure color. The problem is the random emissions of light waves from any common lamp. From moment to moment, short lengths of waves (said to be *wavetrains*) are emitted by a lamp. If a film is so thin that the waves reflecting from the front and back surfaces are always part of the same wavetrain, then a certain type of interference can occur and a persistent colored band can be produced. However, if the film is thicker, the two reflecting waves can be from *different* wavetrains, and then the interference can switch randomly from being constructive at one instant to being destructive at the next instant. Thus, no persistent band can be produced. So, interference patterns and iridescence do not occur for thick films, microscope slides, windowpanes, glass drinking glasses, and so on.

Colored bands appear fairly often on dry metal cooking pots because of thin layers of metal oxide. Also, if the metal has a thin layer of oil and is shiny, the oil layer can produce interference colors. However, if the surface is rough, the light reflecting from the back surface of the oil layer is scattered in random directions, thereby ruining the display of colors.

6.115 • Structural colors of insects, fish, birds, and monkey butts

A yellow canary is yellow because a pigment in its feathers absorbs all but the yellow component in white light. Most colors in the everyday world, including those of animals, are due to similar pigmentation. However, many animals are colored not by pigmentation, but by some curious optical feature of the structure of their external surface (wing, shell, feather, skin, etc.). What produces the coloration in the following examples?

Some butterflies and other insects have wings that are iridescent; that is, the colors change as you change your perspective. One beautiful example is the wing of a Morpho butterfly. Although the pigmentation on the wing is brown (as can be seen on the wing's bottom surface), the top surface is a brilliant, iridescent blue.

Herring use similar (but somewhat more complicated) optics to make themselves silver-white. One benefit of this coloration is that a predator has a more difficult time distinguishing a herring from water.

Some tropical gyrinid beetles (whirl-a-gig beetles) show strong reflection of (white) sunlight when viewed at a certain angle but attention-grabbing iridescent colors when viewed at other angles. Several other types of beetles display similar strong reflections and iridescent colors. Perhaps the most interesting structural colors are produced by the scarabeid beetles, whose surface acts like a type of liquid crystal to reflect bright light of various colors. In contrast, tiger beetles use optics to camouflage themselves by reflecting only the color that matches that of the local soil.

Some mammals have brightly colored skin. For example, the male mandrill monkey has blue skin on its face, rump, and scrotum. Although this coloration is striking, it is not iridescent.

A different type of optics is used to produce the blue areas and white areas of tent caterpillars and the blue on a blue jay.

The forewings of the Hercules beetle are either yellow or black, depending on the humidity. If the humidity suddenly changes, the beetle needs only a few minutes to change its color.

Answer Many of the colored areas on butterflies and moths are iridescent because of transparent, cuticle-like scales that produce interference of light. On the top surface of a Morpho wing, the scales are arranged in a terrace-like structure. The scales have a thickness and vertical separation such that when white light passes down through them, reflections of the blue component by successive scales (say, the top one and the next lower one) result in constructive interference. Reflections of the other color components of the white light result in either partial or fully destructive interference, and so you see blue from the wing. As you change your viewing angle, you slightly change the path of the light that reaches you; this variation alters the wavelength, or tint, of the light that undergoes constructive interference. Because the tint you see on the wing depends on your perspective, the wing is iridescent.

The coloration of herring is also due to optical interference by scales, but there are three overlapping arrangements of scales, each one producing constructive interference of reflected light in a different part of the visible spectrum. When you intercept these three types of bright reflections, you perceive their combination to be white. However, the whiteness is different from normal white paint because there is a subtle change in the coloration when you change your perspective of the fish. The herring is difficult to spot in water because the reflected light is almost the same as the ambient underwater light.

A tropical gyrinid beetle has a surface covered with narrow scales that are aligned to act as a diffraction grating, which is an optical device normally consisting of many parallel, thin grooves that produce an interference pattern. If you intercept the central bright portion of the beetle's diffraction pattern, you see bright white light. If you intercept the off-center bright portions, you see colors that have been spread enough by the diffraction to be distinguishable. When one of these beetles circles in water, the rapid changes in intensity and color can confuse a predator.

Reflections from a scarabeid beetle are due not to scales but to a peculiar arrangement of microfibrils. They are aligned in layers, with the alignment in one layer rotated from the alignment in the next lower layer. As sunlight penetrates the layers, reflections from appropriately spaced layers undergo constructive interference and emerge from the beetle as bright, colored light.

Tiger beetles might be brown or black to your eye (to match the soil in which they are found), but they are actually multicolored. For example, the wing covers of the tiger beetle *Cicindela oregona* have circular patches of blue-green surrounded by red. Both colors are produced by interference due to cuticle-like scales on the wing. When you view the beetle naturally, your visual system combines the intercepted colors, bringing a color of brown up to consciousness. Similar combination of colors from very small, individually imperceptible regions occurs when you view color monitors and pointillism paintings.

The blue skin seen on mandrill monkeys and some other mammals is due to the somewhat periodic (*quasiperiodic*) arrangement of collagen fibers in the dermis. In any given microscopic region, these fibers are parallel and have a width and spacing that scatters blue light back out of the skin. In some places, such as a face, the collagen fibers are so plentiful that blue dominates the light coming back out of the skin. In other places, such as in the rump, there are fewer collagen fibers, but the layer lies over melanin that absorbs the light passing through the collagen. With the dark background provided by the melanin, the blue light coming back out of the skin is arresting. However, it is not iridescent, which requires more widespread alignment of the scattering sections.

The blue region on a tent caterpillar is due to light scattering by transparent filaments of cuticle-like material covering the top surface. Below the filaments is a dark surface. The filaments are so small that they scatter primarily blue light back toward you; the rest of the light continues to the dark surface, where it is absorbed. Thus, you see primarily blue light from the region. A white region differs in that the underlying surface reflects or scatters light toward you and thus is not dark. Here, blue light from the filaments is lost in the brighter white light from the underlying surface.

The blue on a blue jay feather is due to the preferential scatter of the blue end of the spectrum by small alveolar cells in the barbs of the feathers. Colors on other birds can be due to a combination of scattering and pigmentation, or various ways in which the light undergoes interference.

The hard leathery forewing of the Hercules beetle consists of a thin transparent top surface that lies over a yellow spongy layer. Below the spongy layer is black cuticle-like material. When the spongy layer is filled with air, light scatters from it with a loss of most colors except yellow, so the wing appears to be yellow. When the beetle is in high humidity, the spongy layer fills with water, and more of the light passes through that layer to reach the black cuticle material where it is absorbed. The wing then appears to be black.

6.116 • Pearls

What causes the luster and iridescent colors seen on pearls and the interiors of their shells (mother of pearl)?

Answer Because the colors are iridescent (they change with a change in viewing angle), they must be due to interference of the light waves rather than a simple absorption by pigments. A pearl consists of *nacre*, which is a brick-and-mortar arrangement of calcium carbonate (calcite or aragonite) and a matrix of large biological molecules. Flat plates of aragonite (the bricks) are separated by very narrow gaps of matrix material (the mortar). The luster and colors seen in a pearl appear to be due to the interference of light waves scattering from the gaps rather than within the bricks. A wider gap gives bright interference (*constructive interference*) for longer wavelengths (toward the red end of the visible spectrum). In order for a particular color or wavelength to be *pure* (clearly distinguishable) in a pearl, the gaps should have approximately the same width and the separation between gaps should be approximately the same, so that the light waves scattered from the gaps reinforce one another. If the gaps have a range of widths and a range of separations, the colors are *washed out* or *dull*. Some of the black pearls have this coloration effect but also seem to have a pigment that is responsible for its dark background.

6.117 • Protuberances on insect eyes and stealth aircraft

An eye of an insect consists of many facets called *ommatidia* through which light passes to a receptor, where the visual processing begins; the insect sees a mosaic of the images produced through the ommatidia. Many insects have ommatidia with smooth external surfaces. Why do some insects have ommatidia with tiny, tapered protuberances on the external surfaces? The physics of those protuberances also shows up in the absorbing surfaces of stealth aircraft and in the coatings of double- and triple-paned windows.

Answer One measure of the optical properties of a material is the *index of refraction*, which is related to the speed of light in the material. When light encounters a boundary at which the index of refraction changes, some of the light reflects from the boundary and the rest travels through it. For example, if a light beam is sent into a layer of glass, some of the light is reflected at the air–glass boundary.

A similar reflection occurs at the air–ommatidia surface in an insect eye because the index of refraction of the ommatidia is greater than that of air. Thus, normally some of the light incident on the insect's eye is reflected and does not participate in the vision process. However, those ommatidia with protuberances have much less reflection; so more of the light enters the ommatidia. The advantage of a protuberance lies in its tapered structure: thin on the outer end and progressively wider toward the base (Fig. 6-44). So, as light enters the ommatidia, it does not find a single boundary at which the index of refraction sharply changes but instead a gradually increasing index of refraction as it moves along the protuberance. This gradual increase reduces the amount of reflected light, so more of the light enters the eye.

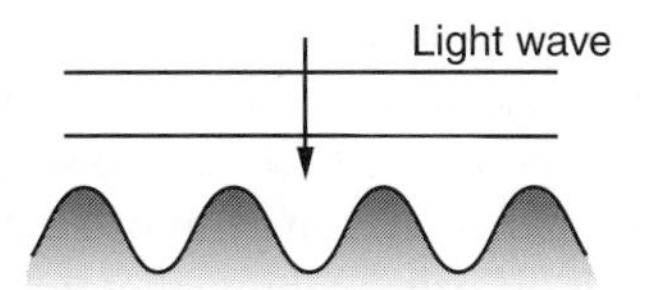

Figure 6-44 / *Item 6.117* Protuberances on insect eyes decrease the amount of reflection from the eyes.

Reflection in windows with multiple layers of glass can produce multiple images of whatever lies outside the windows. If the window is part of an air-traffic control tower or a cockpit, those extra images can be dangerously misleading. If the window is in a cold climate, any reflection at the window means less light to warm the interior. To decrease the reflections for these reasons, one or more of the panes can be coated with a plastic film consisting of protuberances.

One reason that a stealth aircraft is difficult to spot with radar is that the surfaces are coated with protuberances of radar-absorbing material. If the surface were flat, only part of the radar signal would be absorbed and the rest would be reflected. (The situation would be similar to light shining on black glass: Although much of the light is absorbed, enough reflects to you to reveal the presence of the glass.) The surface on a stealth aircraft has protuberances with separations smaller than the wavelength of the radar waves, which are a form of microwaves. As a radar wave moves along a protuberance, it is gradually absorbed and thus little is reflected back to a radar detector.

6.118 • Iridescent plants

Plants in the extreme shade of a tropical rain forest receive little light. This fact may account for the blue-green iridescence of some ferns and flowering plants growing there. The leaves of other plants have a velvety sheen that is produced by convex epidermal cells. What advantage would these features provide to the survival of plants in low illumination?

Answer The iridescent blue-green of the ferns is due to the optical interference of light reflecting from stacked layers having different optical properties. Specifically, layers alternate between having high and low values of the index of refraction, a measure of the speed of light in the material. The layers also alternate in thickness. The result is that the layers act as an array of thin films. Reflected waves in the blue end of the visible spectrum emerge roughly in phase (in step) and reinforce one another, and so we see blue-green light. Transmitted waves in the red end of the spectrum continue through the layers roughly in phase and reinforce one another, so red light is transmitted to the leaf interior (to the chloroplasts where photosynthesis takes place). This arrangement appears to increase the amount of light absorbed by a leaf versus what it absorbs without the optical interference taking place. The fruits of some plants also have an iridescent blue because of thin layers causing optical interference.

The convex epidermal cell in a velvety-sheen leaf is shaped like a lens to focus light onto the underlying chloroplasts. This focusing at least doubles the concentration of light on the chloroplasts, allowing the plant to live in low illumination. (The sheen, a side effect, is the mirror-like reflection of light from the sides of the cells.) The leaf may also have an iridescent film to reduce the amount of light reflected from the leaf.

In other types of leaves, the reflection of light is minimized by the shape of the cells. These cells do not focus light onto the chloroplasts particularly well, but the increased light within the cells is apparently still beneficial.

6.119 • Anticounterfeiting: Color-shifting inks

Governments worldwide scurry to stay ahead of counterfeiters who are quick to use the latest technology to duplicate paper currencies. Some of the security measures used to thwart counterfeiters are security threads and special watermarks (both of which can be seen if the currency is held up to the light) and microprinting (which consists of dots too small to be reproduced by a scanner). The feature that is the most difficult for a counterfeiter to duplicate is probably the variable tint that results from color-shifting inks. For example, the "20" in the lower right corner of the front face of a U.S. $20 bill contains color-shifting ink. If you look directly down on the number, it is red or red-yellow. If you then tilt the bill and look at it obliquely, the color shifts to green. A copy machine can duplicate color from only one perspective and therefore cannot duplicate this shift in color. How do color-shifting inks shift colors?

Answer The color-shifting inks used on paper currencies depend on the interference caused by thin transparent flakes suspended in regular ink. Light penetrating the regular ink above the flake travels through thin layers of chromium (Cr), magnesium fluoride (MgF_2), and aluminum (Al). The Cr layers function as weak mirrors, the Al layer functions as a better mirror, and the MgF_2 layers function like soap films. The result is that light reflected upward from each boundary between layers passes back through the regular ink and then undergoes interference at an observer's eye.

Which color undergoes constructive interference (where the light waves reinforce one another) depends on the thickness L of the MgF_2 layers. In U.S. currency printed with color-shifting inks, the value of L is designed to give fully constructive interference for red or red-yellow light when the observer looks directly down on the currency. When the observer tilts the currency and thus each flake, the light reaching the observer from the flakes undergoes constructive interference for green light. The shift to this other wavelength is due to the longer path taken by the light through tilted flakes. Thus, by changing the viewing angle, the observer can shift the color. Other countries use other designs of thin-film flakes to achieve different shifts in the currency colors. Color-shifting inks and paints are now available for commercial use.

6.120 • Color saturation in flower petals

The color of many flowers differs from petal to petal when the petals are on the flower, but if you pull the petals off and lay them out flat, side by side, they have the same color. What produces the variation in color in the natural arrangement?

Answer If the petals are flat and side by side, you intercept light that reflects from a petal surface only once. The reflection, which scatters light in many directions, tends to remove some colors from the light via absorption by molecules in the petal. For example, a red petal tends to remove the blue end of the visible spectrum so that you see primarily red. However, if light reflects only once, as with the laid-out petals, only a little of the blue end of the spectrum is removed and the light is *unsaturated* red (dull red).

The situation is similar if the petals are approximately flat (in the same plane) when they are on the flower. Again, you intercept light that reflects from a petal only once or that passes through a petal only once. The situation is different if the petals are closely packed and at a variety of angles in your view. Then some of the light you intercept has reflected more than once from the petals—in effect the petals form a "leaky light trap." Because each reflection tends to remove certain colors and leave others, the remaining colors become more *saturated* (purer). Moreover, if you look from petal to petal or maybe even across an individual petal, the saturation differs from point to point.

6.121 • Yellow brilliance of aspen trees

During the autumn, when the leaves have changed to yellow, why are the leaves on an aspen tree more brilliant when they are seen in the general direction of the Sun than when they are seen in the opposite direction?

Answer Once aspen leaves turn yellow, they tend to absorb the blue end of the visible spectrum, leaving the yellow-red end, regardless of whether you see them against the Sun or in the opposite direction. When you see them against the Sun, all the leaves remove the blue end of the spectrum equally well, and so the light transmitted through the leaves is a brilliant yellow. When the leaves are in the opposite direction, they do *not* remove blue light equally well. If the light reflects from the top face of a leaf, blue is absorbed well and thus is not reflected. But if the light reflects from the bottom face, blue is not absorbed well and thus remains in the reflected light. When the leaves are opposite the Sun, some of the leaves present their top face to you and some present their bottom face. Because you intercept blue from those bottom faces, the aspen leaves are not as brilliantly yellow as when seen in the direction of the Sun.

6.122 • Colors of eyes

What is responsible for the colors of human eyes: blue, green, and brown? Why do some people begin life with blue eyes, only to have the color soon change to brown?

Answer Blue eyes are due to the preferential backscatter of blue light by proteins, fats, and other particles in the aqueous material of the iris. The color is visible if the material is backed by a dark layer. If the backing is lighter or if there are pigments on the surface of the iris, blue is not perceived and the eyes may look brown. They are green if there is a pigment that reduces white to yellow, because the backscatter of a combination of blue and yellow is seen as green. The blue of the eyes may change with age if the particles within the iris grow large enough to scatter all colors equally well instead of preferentially blue.

6.123 • So cold I turned blue

Why does pale Caucasian skin turn blue when cold? Why does a close shave on such skin leave the skin tinted blue? Why are veins blue instead of red? After all, blood is red, not blue.

Answer Some of the particles in the surface of Caucasian skin scatter more blue light than other colors. However, the tint is faint enough to require a dark backing to be seen. If the skin of a person with a heavy growth of facial hair is examined after a close shave, the blue can be seen because of the dark background of the hair stubs beneath the skin's surface. When a pale Caucasian becomes cold, the flow of blood through the skin capillaries can decrease enough that the skin loses its normal pink coloring, and the scattered blue light can be distinguished. Similarly, the skin of a corpse turns blue.

Veins appear blue because red light penetrates skin deeper than blue light. To see why this matters, consider the light scattered back out of two adjacent regions of skin: As shown in Fig. 6-45, region *a* is skin that is *not* over a vein and region *b* is skin that *is* over a vein. In region *a*, we intercept a certain amount of blue light and a certain amount of red light, both due to scattering. In region *b*, we intercept the same amount of blue light—it is unaffected by the underlying vein because

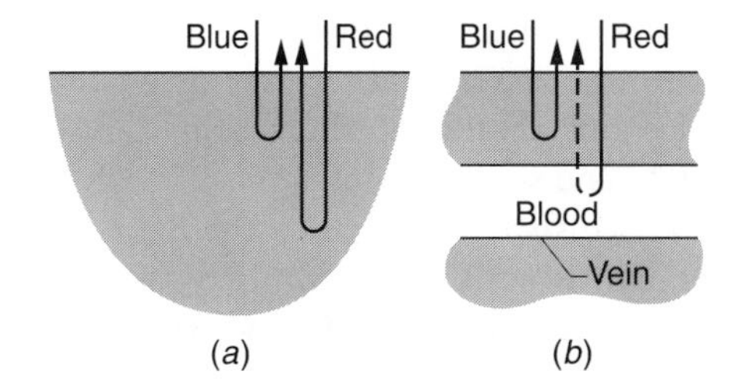

Figure 6-45 */ Item 6.123* (*a*) Skin scatters both blue light and red light back out into the air. Red light penetrates deeper. (*b*) A vein absorbs much of the red light; only weak red scatters back out into the air.

blue light does not penetrate deeply enough to reach the vein. However, red light does reach the vein and is partially absorbed by the blood. Thus, *less* red light is scattered back out of region *b* than region *a*.

Because the two regions are adjacent, we subconsciously compare their colors. Objectively, we have equal amounts of blue light in the two regions and less red light in region *b*. Because of that red depletion, we perceive region *b* to have more blue than region *a*. Thus, the vein below region *b* appears blue—that is, your brain is coloring the vein.

6.124 • Speckle patterns

On many surfaces illuminated with bright sunlight you can see grainy patterns (speckle patterns) of bright and dark points, often with bright colors. First try a flat black surface. Once you know how to recognize the pattern, you can see it on other surfaces such as shiny metal or even a fingernail. The patterns are considerably more pronounced if the light source is a laser.

If you examine a speckle pattern produced by a laser while moving your head sideways, you may find that the pattern appears to move in the same or opposite direction, or it may seem to boil with motion lacking any net direction. What determines the type of motion you perceive?

In some cases speckle can be kinetic even if you are stationary. Partially fill a spoon with milk (avoid skim milk) and place it in bright sunlight. At the shallow edge of the milk, bright points of color dance. If a red apple or red tomato is illuminated with red light from a helium–neon laser, the speckle on the surface of the fruit fluctuates. What accounts for these displays?

Answer A speckle pattern results from interference of the light waves reflecting from the surface. The waves approach the surface approximately in phase with one another, but that condition can change because the surface is microscopically rough. So, light waves reflecting to you from, say, a low point on the surface travel slightly farther than those reflecting to you from an adjacent high point. Thus, depending on circumstances, the waves arriving at your eye can be in or out of phase and can reinforce one another or tend to cancel one another. In short, some spots in the pattern are bright and some are not.

You don't see such a pattern in the light from a lamp because its light is emitted randomly by atoms. Thus, the waves illuminating a surface may be in phase at one instant but not in the next instant. The speckle pattern changes from instant to instant, too fast for you to perceive, and you see just an illuminated surface free of any apparent pattern. If you want to see a speckle pattern, you must use coherent light (waves with an approximately fixed phase relationship). Practically, this means you must use either sunlight (which is partially coherent) or laser light. Although the light emission from the Sun is certainly random, the Sun is so distant that it acts like a point source of light, for which we can approximate the waves as being coherent.

If you are nearsighted, a laser's speckle pattern appears to be nearer than the viewing surface because, lacking any clues about its true distance, the natural focusing of your eye makes the speckle seem closer. If you move your head in one direction, the speckle pattern seems to move in the *opposite* direction past the viewing screen. A similar illusion of motion occurs if you hold a finger between an eye and a lamp. Move your head to one side, keeping the finger stationary. Because you know that lamps don't move but a finger can, you have a strong illusion that the nearby finger moves in the direction opposite your head's motion. If you are farsighted, then the speckle pattern appears to be farther from you than the viewing surface because your eye naturally adjusts for distant viewing. If you move your head in one direction, the distant pattern seems to move in the *same* direction.

If you have normal sight, the apparent motion of the pattern depends on the color of the light because different colors refract by different amounts upon entering the eye and thereby may be judged to originate from different distances. Some researchers have suggested that laser speckle should be used in eye examinations when a patient, such as a young child, is unable to read the letters on the usual vision chart.

The kinetic display in milk is probably due to two types of motion: (1) Evaporation along the thin edge of the milk sets up circulation of the fluid. (2) Even without that circulation, the molecules undergo *Brownian motion* in which they randomly collide with one another and with the milk's proteins and fat globules. Both types of motion continuously alter the way light scatters from the milk. The layer of milk must be thin or otherwise the light scatters multiple times before leaving the milk, the phases of the waves become jumbled, and no interference display can be seen. In that case, the milk is simply white.

The *kinetic speckle* on the side of an apple or tomato is believed to be due to slight motion of the pigmented bodies (plastids) in the skin. As they move, their distance from you changes, altering the interference of the light waves scattered to you and so also the resulting speckle pattern.

6.125 • Colors in fluorescent light

If you spin an object such as a coin in fluorescent light, you may find that the object displays faint colors such as blue and yellow. The demonstration works best if the object is on a dark background while illuminated with a single fluorescent tube. Similar colors can be seen if a string is made to oscillate in fluorescent light: Portions of the blur created by the string's motion are faintly tinted. Colors can also be seen in the thin layer of flowing water surrounding the spot at which the water from a faucet hits the underlying sink basin. What produces the colors?

Answer A fluorescent tube is excited by a burst of electrons (current) that runs through the mercury vapor atoms within the tube. The electrons collide with and excite the atoms, which then quickly de-excite, emitting blue and green light and also ultraviolet light. The ultraviolet light is absorbed by a phosphor coating on the interior of the tube, which then *fluoresces* for a short time (that is, the phosphor glows). If the overall light is meant to be white, the phosphor is chosen to emit largely red and yellow to complement the light from the mercury atoms, so that you perceive white.

To the eye, the tube continuously emits white light. However, the mercury's emission of blue and green light actually fluctuates many times per second because the current through the tube is controlled by the AC electrical supply, which varies in strength and direction many times per second. Between these bursts of mercury emissions, the tube emits only the phosphor's red and yellow. So, if a spinning object is illuminated by fluorescent light, its reflections at various orientations fluctuate between being whitish and being red and yellow—you see periodically appearing colors. The situation is similar for reflections from a vibrating string and from small waves moving over the water surface near the impact point below a faucet.

6.126 • Polarizing sunglasses

How do polarizing sunglasses more effectively block glare from the roadway than sunglasses that are only tinted? Why can polarizing sunglasses improve your view of the region below water so that, for example, you can better spot a fish while fishing?

Take off the glasses, hold one of the filters in front of one eye, close the other eye, and look along a slanted line at a puddle of water. Rotate the filter around your line of sight. Why does the puddle disappear for some orientations of the filter?

Answer Light is a wave of oscillating electric and magnetic fields. These fields are always directed perpendicular to the light's direction of travel, and if we represented them by short arrows, the drawing would look like a rose stem with thorns sticking out on the sides. The *polarization* of light (or the lack of it) refers to the orientation of the electric field—

that is, the short arrows representing the electric field. If light is *unpolarized*, those arrows can point in any direction perpendicular to the light's direction of travel. If the light is *polarized*, those arrows point along a single line, either in one direction along the line or in the opposite direction. Such polarized light is special, because light from almost all common sources, including the Sun, is unpolarized.

One way that unpolarized light can become polarized is by reflecting from certain types of surfaces. For example, if (unpolarized) sunlight reflects from pavement or water, it becomes *horizontally polarized*. This means that the electric fields in the light are horizontal (the short arrows are horizontal). If your eyes intercept this light, you see a bright spot on the pavement or water at the point of reflection, and the light is described as *glare*. Such light fatigues the eye and decreases visibility during many activities, such as driving.

You can decrease the brightness of the glare by wearing common sunglasses, which contain tinted plastic discs, but they also dim the rest of your view, which may be undesirable if you are watching traffic coming your way. Polarizing sunglasses are different: They consist of polarizing filters that block (absorb) horizontally polarized light, and thus they block glare off pavement and water. Because they are slightly tinted, they also dim your entire view but not by very much. So, they give you a clear, relatively bright view without the normal glare. You now might be able to see underwater fish that were previously masked by the bright glare off the water.

If you rotate one of the polarizing disks around your line of sight of a puddle, the puddle disappears when the filter blocks the horizontally polarized light reflecting from it. While flying in search of water, some aquatic insects detect water by detecting the horizontally polarized light reflected from it. Such light is brightest when the Sun is about 40° above the horizon, which may be one reason why aquatic insects search for water more often early in the morning or late in the afternoon than during midday.

6.127 • Sky polarization

Why is the light from much of the clear sky polarized? Why are some regions unpolarized?

Why is light from a cloud usually unpolarized? (This fact aids cloud photography: Place a polarizing filter over the lens and rotate it until the clouds stand out against the background sky.) Why is the light from clouds polarized if the clouds are in the east near sunset or in the west near sunrise? If you look down on such clouds while flying and while wearing polarizing sunglasses, you might see a bright path running over cloud tops off toward the part of the horizon opposite the Sun. If you turn the sunglasses around your line of sight, the path disappears. What produces the path?

Answer Although direct sunlight is unpolarized, the light from most of the sky is polarized because it scatters from air molecules. As an example, suppose that the Sun is low in the west and that light scatters to you from an overhead molecule. The electric field of the light you intercept is directed north–south, and we say that the light is horizontally polarized north and south. Most of the light from the rest of the sky is polarized north and south when the Sun is low. However, scattering from air molecules leaves parts of the sky near the northern and southern horizons vertically polarized.

The light from the region of the sky almost opposite the Sun (the eastern horizon for the low Sun in the west) should be polarized north–south (Fig. 6-46*a*). However, light from that region is actually vertically polarized because it is illuminated more by light from the rest of the horizon sky (which sends vertically polarized light) than by direct sunlight. There is a spot, somewhat above the region opposite the Sun, where there is no polarization—it marks the changeover between the lower vertically polarized region of the sky and the higher north–south polarized region.

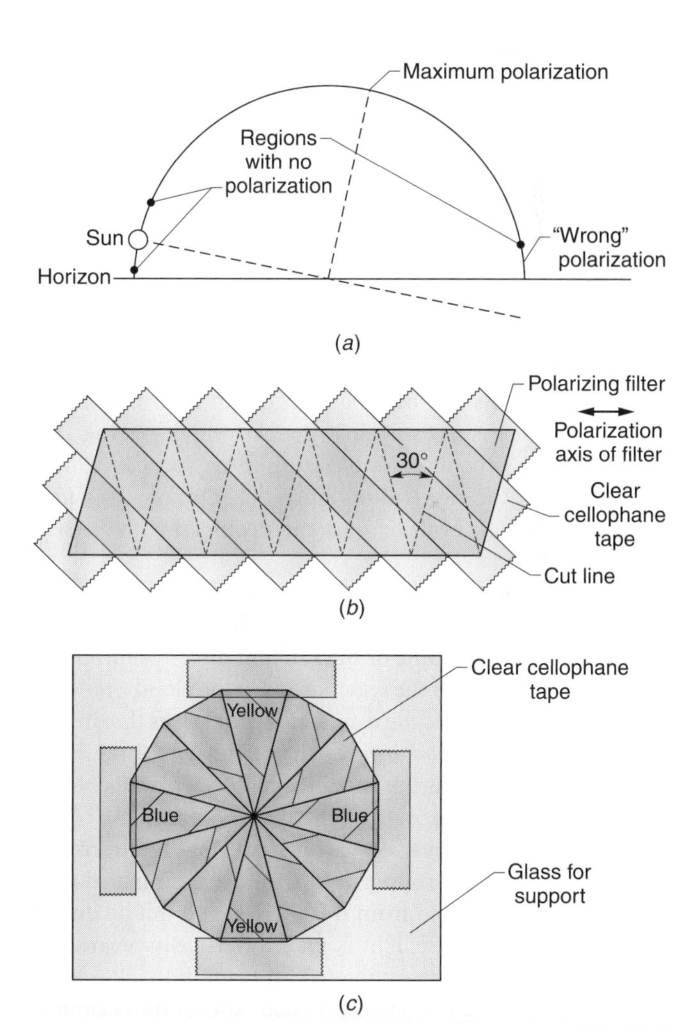

Figure 6-46 / *Item 6.127* (*a*) Polarizations of the sky light. Device for monitoring sky polarization is made by (*b*) attaching clear tape to a polarizing sheet and (*c*) then arranging the triangles in a circle.

Here is a simple device that colorfully reveals the polarization of the sky. Press parallel strips of clear cellophane tape onto a polarizing sheet, with the strips running 45° to the polarizing direction of the sheet. (If the polarizing direction is not indicated on the sheet, look through the sheet at a puddle. Rotate the sheet until the puddle disappears. The polarizing direction of the sheet is then vertical.) Add a second layer of the strips. Then cut the sheet into 12 isosceles triangles with apex angles of 30° and bases parallel to the sheet's polarizing direction (Fig. 6-46*b*). Next, on a clear piece of glass arrange the triangles in a circle with the sharp corners touching and with the sides holding the tape facing you (Fig. 6-46*c*). Now use as little tape as possible to attach the base of each triangle to the glass plate, and put a very small section of tape over the adjacent sharp corners. When you observe the sky through this device, with the triangles on the opposite side of the glass, the polarization of the sky light can result in the triangles being colored blue or yellow.

Although the colors are not easy to explain, the setup is straightforward. When the polarized sky light enters the device, it first travels through a certain thickness of tape, which causes the direction of the light's electric field to rotate around the light's direction of travel. Different colors rotate by different amounts. When the light emerges from the tape, some colors have the correct polarization to pass through a polarizing filter on the glass, and so you see those colors. The other colors, with incorrect polarizations, are absorbed by the filter, and you don't see them.

Sunlight scattering through clouds is normally unpolarized because the light undergoes multiple scattering that eliminates any history of polarization the light may have had. If you view clouds against the sky while looking (or photographing) through a polarizing filter, you can rotate the filter until the sky is dim. Since the rotation does not alter the brightness of the clouds, their contrast with the sky is enhanced.

However, when the Sun is low, clouds that are from 30° to 40° above the point directly opposite the Sun can display horizontal polarization. Some of the sunlight directly illuminating those clouds enters the water drops in the clouds, reflects from the rear surface of the drops, and then leaves the drops. This light forms a *cloudbow*, which is similar to a rainbow except that the water drops are too small to separate colors. Still, the process polarizes the light as in a rainbow.

When the Sun is near or somewhat below the horizon, clouds that are near the opposite horizon are illuminated primarily by light scattered from the horizon and not by direct sunlight. That scattered light is mainly vertically polarized, and thus so is the light coming to you from those clouds. If you wear polarizing sunglasses while looking at those clouds from an airplane, the glasses pass such vertically polarized light and those clouds are relatively bright. Clouds to either side, which send unpolarized light to you, are dimmer. So, you see bright clouds that seem to form a path running off to the part of the horizon opposite the Sun.

According to legend, Vikings located the Sun when it was not directly visible by means of a magic stone that is now believed to be cordierite. When light passes through such a stone, the stone's color depends on the direction of polarization of the light. The Viking observer looked through the stone at part of a clear sky while rotating the stone about his line of sight. During the rotation, the colors of the stone varied between being faint yellow and dark blue. With experience and after examining several other parts of the sky, he could then determine the location of the Sun even when the Sun was below the horizon as it often was at the high latitudes explored by the Vikings.

6.128 • Ant navigation

The desert ant *Cataglyphis fortis* lives in the plains of the Sahara desert. When one of the ants forages for food, it travels from its home nest along a haphazard search path that may be 500 meters long, with hundreds of abrupt turns. Moreover, the path can lie on flat sand that contains no landmarks. Yet, when the ant decides to return, it turns to face home and then runs directly there along a straight path. How does the ant know the way without any guiding clues on the desert plain?

Answer As one of the desert ants moves away from the nest, it keeps track of the distance traveled (it functions as an odometer) and in what directions it turns. It can determine the direction because its eyes are sensitive to polarized light. Thus, it can monitor the polarization of light from the sky and determine its body orientation relative to the direction of the polarized sunlight.

The truly remarkable feature of the ant's brain is that it continuously updates the information about distance and direction so that it knows almost exactly the direction back to the nest. Each part of its travel is treated as a vector (with length and direction), and the ant effectively performs a vector summation. It may also use landmarks if available, but in experiments in which landmarks are moved for the return-home trip, about half the ants choose to follow the vector-summation method rather than be misguided by the landmarks. Vector summation is a challenge for many students; the desert ant with a brain mass of only 0.1 milligram can do it automatically.

6.129 • Colors and spots and polarization

On a morning after a cold night, search for some thin layers of ice (called *frost flowers*) on a window facing the Sun. Wait for some of the ice to melt so that a pool of water lies on the windowsill, or add the water yourself (Fig. 6-47). Look into

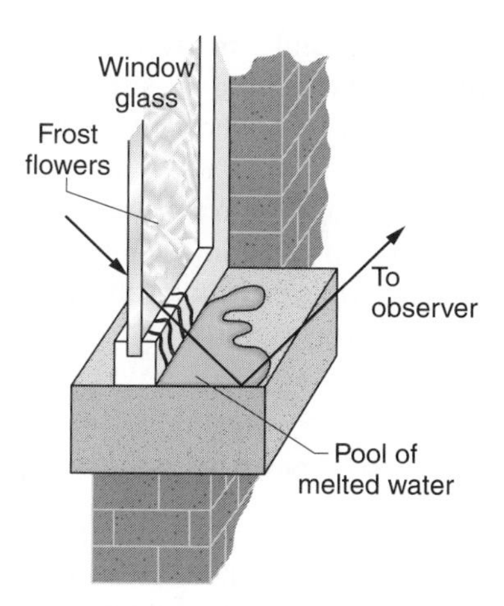

Figure 6-47 */ Item 6.129* The reflection of ice crystals via the pool of water produces colors.

the pool so that you see the layers of ice in reflection. Why is the ice colored in the reflection?

Plastic food wrap is colorless but if you insert a stretched length of the food wrap between two polarizing filters, it becomes covered with various colors. If you rotate one of the filters around its center, the colors change. Multiple layers of clear plastic tape can be substituted for the food wrap. What causes these colors?

Novel art pieces have been based on the colors produced by plastic in polarized light. In some pieces, a mosaic of different colors has been constructed by using different thicknesses and orientations of stretched cellophane layers. The mosaic is illuminated by light sent through a polarizing filter taped to the front of a light projector. Observers view the mosaic through another polarizing filter.

Some artists have also constructed three-dimensional art pieces from plastics and polarizing filters. The light passing through the plastic has been polarized either by a filter bonded to the plastic or by the scatter of the light from molecules in the sky. You might sometimes see a similar display of colors if you wear polarizing sunglasses while looking through an airplane window during a flight.

If you wear polarizing sunglasses while driving, you may notice large spots, usually arranged in patterns, on the rear windows of cars in front of you. What causes those spots?

Answer Ice, stretched plastic wrap, clear cellophane tape, and the stressed rear window of a car are said to be *birefringent materials.* When polarized light travels through a birefringent material, the direction of the polarization rotates around the path of the light. The extent of rotation differs for different colors: Red emerges from the material with a particular polarization, yellow emerges with a different polarization, and so on. If a polarizing filter lies in the path of the emerging light, it transmits the colors with the "right" polarization and blocks the colors with the "wrong" polarization. Although we start with white light, we end up with colored light.

In the situation with ice flowers, the initial light is polarized sky light and the ice is the birefringent material. However, the light emerging from the ice does not go through a polarizing filter. Instead, it reflects from the pool of water. That reflection selects only horizontally polarized light, and colors emerging from the ice with that direction of polarization reach you.

In the situation with a car's rear window, sunlight becomes horizontally polarized when it reflects from either the exterior or interior surface of the tilted window. The polarizing sunglasses on a trailing driver blocks light with that polarization, and the window should look relatively dim. However, some of the light reflecting from the interior surface travels through birefringent regions in the glass and the polarization changes. Those regions are where stress was embedded in the glass during the manufacturing process for safety reasons. The glass is cooled rapidly from the molten state by air jets so that it is put under stress. If, later, the glass shatters, the embedded stress causes it to break into relatively harmless small pieces rather than dangerous shards. The regions hit by the cooling air jets are where polarization is changed in light reflecting from the interior surface. The polarizing sunglasses on the trailing driver pass some of that light, and so those spots are somewhat brighter than the rest of the rear window.

The stress pattern might be faintly visible even without polarizing sunglasses if the light illuminating the rear window is primarily polarized sky light instead of unpolarized direct sunlight. As the polarized light passes through the window to reflect on the interior surface, some of it passes through the stress regions and its polarization is changed. The reflections of that light and the unaltered light from the rest of the interior surface differ in intensity, and so the trailing driver can see a pattern on the rear window. Similar patterns can be seen in an airplane window.

6.130 • Colorless foam and ground powder

Why is the foam of beer white instead of the amber or dark color of the bulk liquid in the glass? Why do most colorful materials lose their color when they are ground into powder? Why is clear glass transparent but not crushed glass? Why is a grain of salt transparent but not a layer of salt more than a few grains deep?

Answer Light passing through yellow beer loses some of its nonyellow colors to absorption by the molecules in the beer. In dark beer, even more absorption occurs. The foam of either beer is white because much of the light reflects from the multitude of surfaces on the bubbles and avoids the absorption within the liquid itself. (Although the thin films forming those surfaces may show faint interference colors when closely inspected, those colors normally overlap to produce white light when the beer is seen at a normal distance.)

The color of most solid materials is due to selective absorption of other colors within the bulk of the material. When the material is ground up, the incident light scatters from the resulting multitude of surfaces, and little of the light you receive has penetrated the powder grains to undergo selective absorption. If white light illuminates the grains, white light is scattered to you.

A pane of clear glass is transparent because of the arrangement of molecules within the glass. At both surfaces of the glass, some of the light is reflected, but most of the light passes through the interior of the glass. Each molecule in the interior scatters the passing light in all directions, but only the light scattered in the forward direction undergoes constructive interference (the waves reinforce one another). The light scattered in other directions undergoes destructive interference (the waves tend to cancel one another).

If the glass is crushed and the small pieces are gathered in a pile, light reflects many times from the multitude of surfaces of the pieces, and much of it ends up traveling back out of the pile. The light that does manage to pass through the pile has been reflected and refracted many times in random directions. So, if your eye intercepts this light, it cannot form an image of the original source of light. Thus, crushed glass is not transparent.

For similar reasons, a single grain of salt is transparent, but a jumble of salt grains provides too many surfaces to be transparent.

6.131 • Glossy black velvet, glossy varnish

Why does black velvet have a shiny side and a dull side? Since a material that is black absorbs all colors, how can black velvet be glossy? Why is varnish glossy? Why is a mirror made with a metallic coating instead of, say, paper?

Answer The shiny side of black velvet has a regular pattern of parallel furrows. If you view the cloth along a line perpendicular to the length of the furrows, light reflects from their sides to you. The cloth is brightest when you view it in this orientation while facing the light source. Although pigments within the threads absorb some of the light, enough light is reflected by the uniform arrangement of furrows to make the cloth shiny. The dull side lacks this arrangement of furrows. Light falling on it is scattered by the threads into many directions, eliminating the possibility of a bright reflection in any one direction.

The gloss of varnish and paint is due to the bright, mirror-like reflection of light by the outer surface. In a semi-glossy paint, some of the light passes into the paint layer where it scatters in many directions from the pigments contained in the paint. The combination of this scattered light with the light reflected from the outer surface decreases the glare from the paint.

Both a paper surface and a metallic coating have irregularities, but those on the coating are small compared to the wavelength of light. As a result, the light scattered by the coating can form an image in your eye. The irregularities on a paper surface are larger and scatter the rays in so many directions that they do not form an image in your eye. Thus, the mirror you look into while brushing your hair has a metallic coating rather than a paper layer.

6.132 • Colors of green glass and green velvet

If you look through a layer of green glass at the filament in an incandescent bulb, you will, of course, see green light. What do you see if you look at the filament through three or more layers of the green glass?

If you spread out green velvet in sunlight, you will see that the velvet is green. What do you see if you gather the velvet into folds? Why does the edge of a fold (the outer portion of a fold) brightly glisten without any color?

If the velvet is part of someone's clothing, why do some of the edges appear to be green (as normal) but other edges appear to be white?

Answer For many (but perhaps not all) types of green glass, the color you see while looking through it depends on its thickness. Through a single layer, you see primarily green light, but a certain amount of red light also passes through the glass. As you increase the thickness of the glass by adding more layers, the transmitted intensity of both green and red light decreases, but that of the green light decreases more rapidly than the red. With three layers you might find that the light is whitish, because the intensity of the green and red is then about equal and the combination is perceived as being white. With even more layers, the intensity of the red light begins to dominate and you see a reddish tint to the light.

A similar result occurs with green velvet, which reflects primarily green light but also a certain amount of red light. When the velvet is flat, the light that reaches you reflects once from the velvet and you see green. But when the velvet is gathered in folds, the light that reaches you reflects two or more times within the folds. With each reflection, the intensity of both the green and red light decreases, but that of the green decreases more rapidly. If the light reflects enough within the folds, it becomes reddish.

When you look at velvet clothing illuminated by white light, you intercept a lot of light that is scattered from the outer tips of the velvet fibers. That light loses little to absorption because it is scattered only once, and so it remains bright and white. The light that penetrates the velvet and then emerges becomes green and darker. In a painting of velvet clothing or draperies, the edges are painted with a bright white line to represent the glistening seen by the artist, in contrast to the darker color seen in the rest of the velvet.

6.133 • Peachy skin and apparent softness

You can tell whether skin on a person, such as a baby, is soft because of a visual clue, especially when a light source somewhat behind the person glances some of the light off the skin to you. A computer simulation of a person might lack this clue, and the skin would then look hard, telegraphing that it is only a simulation. You get the same type of visual clue of softness when you look at a ripe peach against a background light—it is noticeably different from, say, a nectarine and, as a result, you judge the peach to be softer. What is the visual clue?

Answer The clue about skin is the small amount of light that is scattered at the outer edge of your view of a person, either in the top layer of skin or (better) in hairs that extend from the skin. Even when skin looks hairless, it probably has short, downy hairs that are nearly imperceptible when you look directly down at it. However, when a light source somewhat behind a person glances light off the skin to you, some of the light from the outer edge of your view is scattered from those hairs, giving a soft glow to that edge. You interpret that glow as meaning "soft." An edge without the glow is sharper and better defined and thus looks "hard."

The short hairs that extend from a peach give a soft glow around the edge of the peach in proper illumination, making the peach look soft. A nectarine, with no such hairs, looks hard. The scatter by downy hairs on the skin of some young women gives such a soft appearance that the women are said to have *peachy skin.*

In a related effect, if someone stands in front of a light source, such as the Sun, you can see a dramatic glow in the outer layer of the hair on the top of the head.

6.134 • Twinkies and Vaseline parties

At MIT, some student parties are reportedly conducted in black light (ultraviolet light) from ultraviolet lamps. Participants smear their bodies with Vaseline petroleum jelly or the cream filling of Twinkies snack food. They also may be drinking tonic water, which has an eerie blue glow. Why do these products glow in ultraviolet light?

Answer Some component in the petroleum jelly and the Twinkies cream filling fluoresces by absorbing the ultraviolet light and emitting visible light at the blue end of the visible spectrum. I suspect the component is an aromatic hydrocarbon. Vaseline is used as a binder for holding certain other aromatic hydrocarbons to objects that police want to track as the objects pass through criminal hands. The passage is marked by chance smudges of the Vaseline, and the aromatic hydrocarbons can be identified later when the smudges are illuminated with ultraviolet light. Tonic water glows blue in black light because it contains quinine, which absorbs ultraviolet light and then emits blue light.

6.135 • The colors of meat

Why is the outside surface of fresh beef bright red while the interior is purplish red? Why does the interior of a roast turn bright red when cooked rare and brown when cooked well-done? Why are packages of bacon, ham, and corned beef partially covered with cardboard or turned over in the store so that the meat is not illuminated? Why does cured meat sometimes form fluorescent yellow or green layers? What produces the iridescence sometimes seen on meat?

Answer Most of the color of meat is due to the pigment myoglobin that, in the living animal, is responsible for taking up the oxygen transported into the meat from the lungs via the hemoglobin in the blood. When the animal is slaughtered, the oxygen supply is eliminated and the myoglobin turns purplish red. Once the meat is cut and exposed to the air, oxygen in the air combines with the myoglobin on the exposed surface, forming oxymyoglobin, which is bright red. Somewhat below the surface, where oxygen is present but less abundant, the oxymyoglobin dissociates and the iron within it oxidizes to the ferric state. The resulting complex, called metmyoglobin, is brownish red. Deeper in the meat, where there is little oxygen, the myoglobin remains purplish red. Butchers commonly pack meat in air-permeable wrappings so that the surface of the meat is bright red, which presumably suggests freshness to the customer.

When the meat is heated, the myoglobin within the bulk of the meat begins to combine with the oxygen there to form bright-red oxymyoglobin. (Thus, if you slice open a roast that is cooked rare, the interior of the meat is bright red because of this conversion.) Meanwhile, the myoglobin on the exterior of the meat begins to denature and its iron component begins to oxidize, leaving the exterior brown. If heating continues, this change in color extends into the meat's interior.

In cured meats such as ham, bacon, and corned beef, nitric oxide combines with the myoglobin to form the complex nitric-oxide-myoglobin, a pink pigment. If the meat is illuminated while exposed to oxygen, the light dissociates the nitric oxide from the myoglobin, and then the iron component oxidizes to form brownish red metmyoglobin. Sometimes another part of the complex oxidizes, forming yellow or green pigments that will fluoresce. To avoid this unpleasant change in color, the packages of cured meat are turned over to shield the meat from the light in the store. They can, instead, be sealed in vacuum-tight packages to eliminate the oxygen.

The iridescence on either fresh or cooked meat is due to the interference in light scattered by the myofibrils (muscle fibers) at or just below the meat surface. When the light is incident perpendicular to the myofibrils, the light scattered at one point can interfere with the light scattered at another point, with constructive interference typically occurring for green light. The iridescence, primarily in the green portion of the visible spectrum, occurs as the surface dehydrates. It is not necessarily associated with bacterial contamination of the meat nor does it necessarily mean that the meat is too old for consumption.

6.136 • A short beer

Why do beer mugs often have thick, tapered walls with a thick bottom? The design may be partially for the feel of a "good heft," but how can it give the visual illusion that the mug holds more beer than it actually does?

Answer The thick walls of the mug create the illusion because of the refraction the light undergoes when passing from the beer through the glass and then into the air. For example, a ray from the leftmost side of the beer is bent toward the center of your view of the mug (Fig. 6-48). When you intercept the ray, you mentally extend it straight back into the mug and conclude that it originated farther to the left than it actually did, and so there seems to be more beer in the mug than there actually is. The thickness and curvature of the glass can also alter the apparent depth of the beer. In an extreme case the actual contents of a mug may be only half the apparent contents.

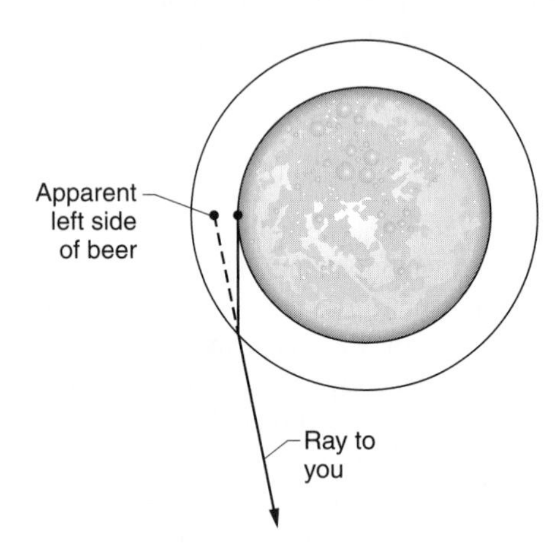

Figure 6-48 / *Item 6.136* Overhead view of a beer mug giving the illusion of extra contents.

6.137 • "Whiter than white"

A once popular commercial for a laundry product claimed that the product made white clothes "whiter than white." What does the saying mean?

Answer The laundry product leaves a fluorescent brightener in the clothing that converts some of the ultraviolet light of sunlight into blue light, thus increasing the amount of visible light coming from the clothing. The clothing is "whiter than white" in the sense that the clothing emits an additional amount of the visible spectrum.

The brightener is needed because the laundry product also leaves a yellow dye on the clothing. The dye absorbs blue light. Without the presence of the brightener, white clothing would be yellowish once washed. Before laundry products had brighteners, they contained a blue dye in addition to the yellow dye. The presence of both dyes in the clothing resulted in a near elimination of color in white clothing but also left an undesirable faint gray.

6.138 • Disappearing coin

Place a coin in a jar filled with water and arrange your line of sight so that you see an image of the coin through the top surface of the water as shown in Fig. 6-49. If you place your hand on the far side of the jar, the image is unchanged. However, if your hand is wet, the image disappears. Why does the wetness eliminate the image?

Answer The coin is initially visible because you see some of the light from it via reflection on the external surface on the far side of the jar. If you place your dry hand on that reflection area, the reflection is hardly altered for two reasons: (1) Your hand makes few contacts with the glass and (2) only a little light passes from the glass into your skin. However, if your hand is wet, the contact of the water with the glass is extensive. Moreover, where contact is made, much of the light can pass from the glass into the water because the optical properties of the two materials are close. Thus, most of the light that previously reflected on the far side of the jar is now lost into the water on your hand, and the coin's image has been eliminated.

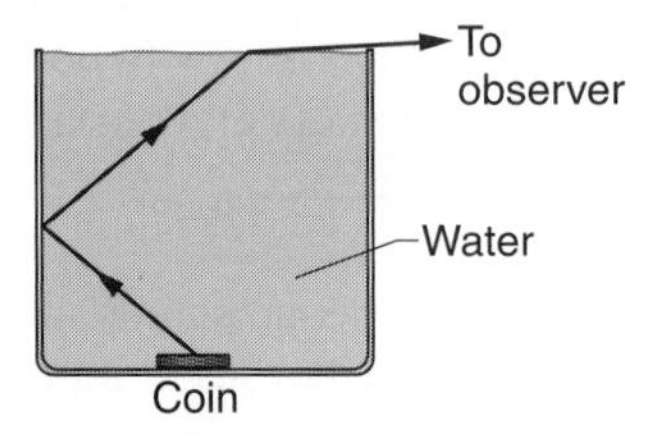

Figure 6-49 / *Item 6.138* Coin seen through the water surface in a jar.

6.139 • Sunglasses and smog

If you wear polarizing sunglasses while viewing a smog-covered mountain lying against a clear sky, you may be able to make the mountain appear and disappear by tilting your head and the sunglasses. Many features on the mountain may also appear and disappear. What accounts for these changes?

Answer When you view a smog-covered mountain against a clear background sky, you intercept three sources of light: (1) light scattered by air molecules in the background sky (this light is polarized), (2) light scattered by the smog particles between you and the mountain (if the particle diameters are in the range of 0.5 to 5.0 microns, then this light is partially polarized in the same orientation as the background sky light), and (3) light reflected from surface features on the mountain (this light is weakly polarized in a myriad of orientations). The extent of the light you see from these sources depends on the orientation of your polarizing sunglasses.

Suppose that the smog particles are in the diameter range quoted. If you tilt your head and the sunglasses to the left or right until the background sky is brightest, the sunglasses pass light from all three sources. However, the partially polarized light scattered by the smog particles is only partially passed by the sunglasses. What you see is a dim outline of the mountain against a bright sky. The contrast is enough for you to perceive the mountain, but the dim smog-scattered light is still sufficiently bright to mask the surface features on the mountain.

If you now tilt your head and the sunglasses through 90° (either from your left to your right, or vice versa, whichever you can do comfortably), the sunglasses will block all the light from the background sky and most of the light from the smog particles. You might then be able to distinguish surface features that happen to be oriented so as to reflect light with a polarization that can pass through the sunglasses.

6.140 • Brightness of the ocean

Suppose the sky is clear and you ride in a boat over ocean water that has only moderate waves. Where is the water brightest, directly below you as you look down on it or off toward the horizon? Where do you primarily see reflections of direct sunlight or the sky? Where might you be able to see submerged objects?

Answer There are three sources of light to compare: the glitter of sunlight reflecting from the water, the sky light reflecting from the water, and the upwelling of light from beneath the water surface. The ocean is always brightest toward the horizon because the sky-light reflection there is at such a glancing angle that only little of the light is lost into the water. As you bring your gaze inward, progressively more of the reflected sky light enters the water due to the steeper angles.

The next brightest region is that of the glitter, which is determined by how high the Sun is in the sky. The glitter region slides toward you from the eastern horizon during sunrise; it is below you or at least near you when the Sun is at its highest point; and it slides away from you toward the western horizon during sunset.

Away from the glitter region and closer to you than the distant water, upwelling might dominate the light that you intercept from certain regions. In those regions, you might be able to see submerged objects.

6.141 • Blue ribbon on sea horizon

The portion of the sea just below the horizon often appears to be a much deeper blue or darker gray than the rest of the sea or the portion of the sky just above the horizon. On a clear day that portion of the sea resembles a bright blue ribbon. The blue ribbon disappears if you lie on the beach or climb well above it. What produces the blue ribbon?

Answer When you look at the water surface just below the horizon, you usually see a reflection of the sky that is about 30° above the horizon (see preceding answer). When the sky is clear, that portion of the sky is bluer than the sky at the horizon, which is usually white, and the contrast between the white horizon and blue reflection makes the blue noticeable. If the reflected portion of the sky is gray, then the band is gray, which still contrasts with the adjacent white horizon. The blue or gray band disappears when you lie on the beach because incoming waves block your view of the sea just below the horizon. It also disappears if you climb too high from the beach, but I am not certain why.

6.142 • Darkness falls with a bang

Why is sunset more rapid in the tropical zones than at higher latitudes?

Answer The duration of twilight is the time between the beginning of a sunset and when the center of the Sun's image is a certain angle below the horizon. (For *civil twilight*, the angle is 6°; for *nautical twilight*, it is 12°; and for *astronomical twilight*, it is 18°.) How rapidly the Sun moves between these two positions depends on how its apparent path across the sky intercepts the horizon. At low latitudes the path may be vertical or nearly so, and thus the Sun moves quickly between the two positions, yielding a short twilight. At high latitudes the Sun's path intercepts the horizon at a small angle, and thus the Sun moves between the two positions along a slant, yielding a long twilight.

6.143 • Colorful contrail

Contrails (the trails of water condensation left by airplanes) are usually white. Why are they sometimes colorful? How can the colored portion of the contrails move along just behind the airplane? After all, the drops that comprise a contrail do not move that fast.

Answer As they form and then grow by absorbing water from the air, the drops can go through a certain size range where they separate colors in the sunlight because of diffraction (a type of scattering). Once the drops grow too large, the colors disappear. Thus, there can be a certain position behind the airplane that marks where the growth gives a colorful contrail. The drops closer to the airplane are too small; the drops farther from the airplane are too large.

6.144 • Nacreous clouds

Nacreous clouds (also called *mother-of-pearl clouds*) have beautiful, delicate colors. They are rare, normally seen only at high latitudes and only somewhat after sunset or before sunrise. This timing of their appearance means that they must be at high altitudes where they are still illuminated by sunlight while the underlying ground is in darkness. Normal, lower clouds show colors only along their fringes or when they lie in front of the Sun or Moon, but a nacreous cloud gives a grand display of colors even when it is as much as 40° from the Sun.

What accounts for the colors of nacreous clouds? If the colors are due to water drops, how can there be liquid water in the very cold environment (about −80°C) at the altitudes of the clouds?

Answer The colors of nacreous clouds are due to the scatter of sunlight and separation of colors by tiny drops of water and sulfur trioxide within the clouds. Tiny drops of pure water can remain liquid down to a temperature of about −40°C. But when sulfur trioxide is included, the drops can remain liquid at much lower temperatures, such as those found at altitudes of 18 to 22 kilometers where nacreous clouds form. If a drop in a nacreous cloud gradually collects more water and thus grows, freezing eventually begins, and the drop no longer contributes to the color separation by the cloud. Such frozen drops are believed to form the white tail that extends from a nacreous cloud.

6.145 • Twilight purple light

What produces the *twilight purple light* that appears briefly in the western sky above a setting Sun about 15 to 40 minutes after sunset, after the other sunset colors are fading? Is the same mechanism responsible for the unusual second purple light that may appear up to two hours later? How could the Sun still provide light to the western sky so long after sunset?

Answer The twilight purple light is a combination of red light scattered by low-altitude air and blue light scattered by high-altitude air. The red part is sunlight that skirts the curvature of Earth, passing through the dense, low-altitude air. Along this route the light loses much of its blue because of scattering by the air. When the light finally scatters to you, it is dominated by red. The blue part is sunlight that takes a long route through the high-altitude ozone layer. Along this route, the light loses much of its red to absorption by the ozone molecules. When the light finally scatters to you, it is dominated by blue. Thus, when you look at the sky above where the Sun set, you get a combination of red and blue that is perceived as purple. In some locations this purple light can illuminate mountain peaks after the Sun has set, in what is known as either *alpenglow* or *Alpine glow*.

The rare second twilight purple light is not well understood. It may be due to light that scatters from a high-altitude particle layer that is still illuminated by sunlight well after sunset. Such a tenuous particle layer lies at an altitude of about 85 kilometers and consists of comet and asteroid debris intercepted by Earth.

6.146 • Ripples in the sky

Several observers have reported dark and bright ripples moving across clouds. In some cases the ripples appeared to be random while in others they appear wave-like.

During fighting near the Siegfried Line in World War II, American troops spotted dark shadows crossing over white cirrus clouds. These shadows were arcs with centers located on the German side.

What would produce such displays?

Answer The ripples are not understood but are believed to be due to sound waves moving through ice crystals in the clouds. Since sound was not heard by the American troops, the sound sources must have been distant, perhaps heavy artillery or large explosions behind enemy lines. A sound wave would briefly reorient the crystals, changing the brightness of the reflection from them. (According to an early theory, the sound waves may, instead, momentarily alter the extent of condensation within a cloud and thereby change the brightness of the cloud along a moving band.)

6.147 • Line across distant rain

When you see distant rain falling in direct sunlight, you may notice a horizontal line separating higher, brighter precipitation from lower, dimmer precipitation. What accounts for the line?

Answer Generally, the temperature of the air decreases with altitude. The line marks the altitude at which falling ice crystals melt and become raindrops. The crystals reflect more sunlight than the raindrops, and so the region above the line is brighter than the region below it.

6.148 • Bright nights

If you live in a region that is unpolluted at night by stray light, you might find that some nights are unusually bright, even when the Moon is not present. Such nights seem to be associated with meteor showers, as if the meteors lend a glow to the sky. However, the burning up of a meteor is too brief to account for the extra light. Is there some other factor?

Answer When a meteor travels through the atmosphere at an altitude of about 90 kilometers, it heats the air along its trail. In the heated region nitrogen oxide is produced, which then reacts with oxygen to give nitrogen dioxide. The process emits light in the green, yellow, and red range. A dark-adapted eye is sensitive to such light. So, a meteor shower could give an extra glow to the night sky.

6.149 • Zodiacal light, gegenschein, and other nocturnal lights

In a dark, moonless sky that is well away from city lights, you may spot two curious patches of light. The *zodiacal light* is a milky triangle that can be seen in the west for a few hours after sunset or in the east before sunrise. The evening observations are best near the spring equinox, and the morning observations are best near the autumn equinox. The triangle is nearly as bright as the Milky Way and is oriented along the plane in which Earth orbits the Sun.

The *gegenschein* is a faint light that some people can see at the *antisolar point* in the sky (the point directly opposite the Sun's position). The light is so faint that you will need ideal conditions of darkness for the observation, with dark-adapted eyes. Even then you probably can see the light only with *averted vision*, in which you sweep your eyes to and fro across the sky rather than fixing them on the location of the gegenschein. In the Northern Hemisphere, viewing is probably best in October because then the background of stars is relatively dark.

On occasion, large-scale, moving, luminous regions have been seen in the night sky but they are unrelated to aurora.

What accounts for these night lights?

Answer The zodiacal light and the gegenschein are due to the scattering of sunlight by interplanetary dust, probably derived from comets. The scattering is brightest in the forward direction, weaker in the backward direction, and weaker still at other angles. The dust responsible for the zodiacal light lies inside Earth's orbit. You can see the light scattered from it when it is still within your view just after sunset or just before sunrise; in both situations you see light that is approximately scattered forward.

In the middle of the night you might be able to see sunlight scattered backward by dust outside Earth's orbit. This display is the gegenschein. The moving luminous displays are thought to be airglow, probably from high-altitude hydroxyl (OH) molecules that are excited by *density waves* (wave-like variations in pressure and temperature, often called gravity waves) traveling through the region.

6.150 • Reflections from sea horizon

While standing on a seashore, examine the reflections from the sea just below the horizon when the sea has plenty of waves. If the sea were flat, it would reflect the sky like a large, horizontal mirror. However, since the waves have tilted surfaces, they reflect various portions of the sky. The portion that is reflected by each wave is angled above the horizontal by twice the tilt of the wave.

Strangely, the reflection you see just below the horizon is usually the portion of the sky that is about 30° above the horizon. This angle suggests that the average slope of the waves is 15°. However, measurements reveal that waves infrequently have a slope that large. Why then is the reflection usually at 30°?

Answer Consider the waves on the sea just below the horizon. Small waves, having shallow tilts, reflect the portion of the sky near the horizon. Although these waves are numerous, their contribution to the reflection you see is small because each has only a small surface area (they are like small mirrors). Somewhat larger waves with intermediate tilts reflect higher portions of the sky. Although those somewhat larger waves are less numerous than the small waves, their contribution to the reflection you see is greater because each wave has a large surface area. Much larger waves are so few that not even their large reflecting areas matter. The net result is that most of the reflection you see is as if the water were covered with waves having an intermediate tilt of 15°, thus giving a reflection of the sky that is 30° above the horizon.

6.151 • Using a solid metal ball to focus light

In 1818, Augustin Jean Fresnel submitted a wave model of light to a competition at the French Academy. Simeon D. Poisson, one of the members of the judging committee, argued strongly against the model, attempting to reduce it to absurdity with this thought experiment: If an opaque object with a circular cross section (such as a coin or a ball) is illuminated with a beam of light, Fresnel's wave model predicted that a bright spot should appear at the center of the shadow cast by the object on a distant viewing screen.

Dominique F. Arago, another committee member, arranged to test the prediction in spite of the absurd conclusion. Surprisingly, he found the bright central spot. Through a quirk of history, the spot is now known as either the Poisson spot or the Arago spot, although neither man initially believed in its existence.

Since the spot's discovery, several researchers have employed opaque objects, such as a small metal ball from a bearing, to serve as a lens to form images. If the image forms on film, you can make a photograph of the image just as with a camera. Why does the Poisson spot form, and how can something like a solid ball force light to form an image?

Answer Suppose that the source of light is a distant, bright point and that the imaging is done with a solid ball. When the light waves reach the ball, they diffract around its sides, spreading radially outward and also into the shadow region of the ball. If a screen is placed well behind the ball, the light forms a small diffraction pattern of bright and dark concentric circles on it. The center of the pattern is a bright point because waves passing on one side of the ball travel the same distance to the center as waves passing on the opposite side of the ball, and so the waves arrive in phase and undergo constructive interference.

The first dark circle is due to destructive interference. Consider the top of the circle. Waves passing the bottom of the ball must travel a longer distance to reach that point than waves passing the top of the ball. The extra distance amounts to half a wavelength, and so the two sets of waves (one from the bottom and one from the top) undergo destructive interference when they reach the point on the screen.

The rest of the pattern is due to similar constructive or destructive interference. In some places, waves from opposite sides of the ball differ in the distance traveled by an integer number of wavelengths, which results in constructive interference, since that puts the waves in phase. In other places the difference in the distance traveled amounts to an odd number of half wavelengths, which results in destructive interference since that puts the waves out of phase.

When a ball forms an image of an object, each bright part of the object serves as a point source of light and creates a slightly displaced bright point near the center of the diffraction pattern. The composite of these bright points roughly reproduces the shape of the object and thus provides an image of the object.

6.152 • A fast spin in a curved mirror

Curve a sheet of shiny Mylar so that it forms part of a cylinder. Hold this curved mirror so that the long axis of the cylinder is horizontal and then look into the curved interior. Adjust the curvature of the mirror and your distance from it so that your image is inverted. (You are then outside the focal point of the mirror.) If you rotate the cylinder through 90° so that its axis is then vertical, your image is right-side up; that is, it has rotated through 180°. Why is the rotation of the image twice the rotation of the cylinder?

Answer Consider the cylinder as being two mirrors: one is straight and parallel to the cylinder's axis while the other curves around the mirror, making a perpendicular intersection with the first one. An image produced along the straight mirror is not inverted (the same as with a flat mirror), but any image produced along the curved mirror *is* inverted.

For any orientation of the cylinder you actually see both types of images: a real inverted image in front of the mirror (formed by the curved mirror) and a weaker noninverted virtual image behind the mirror (formed by the straight mirror). For one cylinder orientation, one type of image dominates your view. For the other orientation, the other types dominates. Thus, when you rotate the cylinder between the two orientations, the image you perceive switches from one type to the other, giving the illusion that a single image rotates 180°.

6.153 • Color of cigarette smoke

Why is the smoke rising directly from the burning end of a cigarette tinted blue while the smoke exhaled by the smoker is white?

Answer The smoke particles rising from the burning end are small enough that they scatter more blue light to you from the incident room light. When the smoke is inhaled, condensation of water onto the particles increases their size enough that, once exhaled, the particles scatter the different colors of the incident light about equally well, giving a whitish appearance.

6.154 • If you could see in the UV

The limit of your vision at the far blue end of the spectrum is partially due to the increased absorption of light by the cornea and lens in your eye. If you undergo cataract surgery and have an artificial lens implanted, you may be able to see light somewhat farther into the ultraviolet range. Suppose that is all you could see. Would the world look different?

Answer Cities would be dark even when their lights were on because the glass coverings on lamps and the windows on buildings absorb ultraviolet. So, you would not see light from those familiar sources. For the same reason, corrective eyeglasses would resemble dark sunglasses. Shadows would be faint, and you would not see as far even on a clear day, because air molecules more strongly scatter ultraviolet light than visible light. So, distant objects would look murky, and any shadow would be partially filled in by the scatter of light from the air.

6.155 • Diffracted alphabet

If a laser beam is sent through a small opening in the shape of a letter, can you predict the diffraction pattern that will result? Or, if given a photograph of the pattern, can you tell which letter was used? If you have a laser, you might like to produce a puzzle of different diffraction patterns, each created by a letter. To solve the puzzle, someone would have to determine a message spelled out by those letters.

Answer You can often guess the letter by considering the direction in which light diffracts. When light is diffracted by an edge, the light spreads perpendicularly to the edge. For example, the letter O diffracts light in all directions because of its roughly circular edges, while the letter Z diffracts light up and down and also along a line that is perpendicular to the central line segment of the letter.

6.156 • A game of reflection

Find two small rectangular mirrors, attach them with tape on the back so that they can pivot about a common length, and then place the assembly over a penciled sketch. The composite of the original sketch and its reflections can give a wild design.

Suppose that the sketch consists of only one or two straight line segments that run fully from mirror to mirror so that the segments and their images connect. If you adjust the position of the mirrors and their angle, you can produce regular geometric figures.

Keeping the angle less than 180°, find the minimal number of straight line segments needed to create a square, an octagon, and a six-point star. How about a star within a star, with the points of the two stars either aligned or exactly misaligned? What minimal number is needed to make a square with a small square tucked into each corner?

C • H • A • P • T • E • R 7

Armadillos Dancing Against a Swollen Moon

Figure 7-1 / Item 7.1

7.1 • Enlarging the Moon

The most striking illusion in the natural landscape is the apparent enlargement of the Moon when it is near the horizon. Is the enlargement produced by the refraction (bending) of light rays by the atmosphere, a change in the distance to the Moon, or a misinterpretation by you?

Answer The Moon appears to be about 50% larger when it is near the horizon than when it is overhead because of a misinterpretation. Actually, the Moon always occupies an angle of about 0.5° in your view regardless of whether it is high or low. If refraction of the light by the atmosphere is appreciable, it tends to reduce the vertical width of the Moon, not enlarge it. Also, the distance between Earth and the Moon does not change appreciably during the few hours it takes the Moon to rise or set.

The misinterpretation that leads to the apparent enlargement probably has several simultaneous causes. The primary cause seems to be that you associate a low Moon with the terrain in front of you—on the basis of the terrain, the Moon looks bigger. You can easily defeat the effect: Turn around, bend over, and look at the Moon through your legs—it looks its normal size, presumably because the terrain, being in the top half of your view, is no longer used for scaling. Other possible causes involve the tilt of the eyes to see the Moon and the simultaneous lack of convergence required of the eyes when viewing a very distant object.

7.2 • Shape of the sky

Does the sky appear to be hemispherical? Most people see it as an inverted soup bowl with the overhead section closer than the sections near the horizon. Try the following observations. When there is a crescent Moon in the daytime sky, mentally halve it into symmetrical parts with a line. Since the crescent is produced by sunlight, the line should point directly toward the Sun. However, it does not because your perception of the shape of the sky distorts the line as you mentally trace it across the sky. Searchlight beams are straight, but when viewed from the side they appear to bend because of the apparent shape of the sky. Why doesn't the sky appear to be hemispherical?

Answer The shape of the sky probably has many causes. Here are a few: Because you see a broad horizon, you probably attribute a large distance to the sky just above the horizon. Because you see nothing overhead, you probably attribute a shorter distance to the overhead sky because your eyes naturally relax.

The assignment of shape to the sky can be so strong that the searchlight appears to bend along the curvature of the sky and the mentally extrapolated line from the Moon misses the Sun. Both views are illusions.

7.3 • Decapitation with the blind spot

Each eye has a *blind spot* in which nothing is perceived. The spot lies in your field of view about 15° off the center of your gaze and toward the temple. You can find it by viewing with one eye as you bring a small object (say, a pencil's eraser held at arm's length) across your field of view. Hold your gaze steady. As the object passes through the blind spot, it disappears.

When the famous physiological psychologist Karl S. Lashley was forced to endure an irritating dinner guest, he amused himself by bringing his blind spot over the person's head, neatly decapitating the person. An old story (apparently untrue) relates how King Charles II of England also visually beheaded his guests, ironic in that his own father was truly beheaded.

How large is the region and why is it blind? Why is the blind spot normally unnoticed?

Answer The retina is covered with *rod and cone photoreceptors* except in the region where the nerve pathways leave the retina on their way to the brain. Lacking photoreceptors, that region is blind. The blind spot is normally unnoticed for several reasons: Usually both eyes are open, allowing one eye to see objects lying in the blind spot of the other eye. Also, you concentrate on the center of vision, which falls on the *fovea* (the pit in the retina holding the maximum packing of cones) and not the blind spot. In addition, some detail is filled into the blind spot by the natural tiny motions (*saccades*) of the eye in which the eye rotates about a degree. The eye also slowly drifts and suffers from tremor. Thus, an image that first falls on the blind spot soon falls elsewhere on the retina because of these motions. Even without the motions, the blind spot is often filled in because the brain can correlate the scenes on each side of the blind spot and then create an image straddling the blind spot to connect those scenes.

7.4 • Gray networks in the morning, dashing specks in the daylight

If you stare at a sunlit room immediately upon opening your eyes in the morning, your field of view will be covered with a gray network. The network quickly fades but can be generated at will with a penlight or an illuminated pinhole. (Do not damage your eyes with a bright light.) Slowly move the penlight around in your field of view in an otherwise dark room. Parts of the network should appear. What is the network, and why does it fade so quickly?

A related observation can be made on a bright day. While staring at the clear blue sky, I find my field of view filled with many floating dots (which are discussed in the next item) and moving specks. The specks are bright with dim tails. I can correlate them with my pulse: They move quickly during the systolic (contraction) phase of the pulse and slower during the diastolic (dilation) phase. Blue light enhances their visibility. They are seen everywhere except directly along the line of sight (that line intersects the retina at the fovea). What are the specks? Why is blue light best for seeing them? And why are they absent from the fovea?

Answer The network is formed by the shadows of the blood vessels in the retina as they block light from reaching the photoreceptors deeper in the retina. The specks are the white blood cells that move through the vessels. Blue light is best for contrast because red blood cells absorb light at a wavelength of about 415 nanometers (blue), and white blood cells do not. Thus, the motion of the white blood cells becomes more apparent with a blue background. Neither the network nor the specks pass through the fovea because that region lacks blood vessels.

Because any pattern that is stabilized on the retina loses its perceived contrast within seconds, the network fades quickly. If a small light is moved across your view, the shadows cast by the blood vessels vary enough that the network remains visible.

Making the retinal network visible with an illuminated pinhole may account for the otherwise puzzling observations of Venus made by the astronomer Percival Lowell. He consistently observed "spokes" on the surface on Venus (this was long before anyone knew that the surface could not be seen because of persistent cloud cover). Moreover, the spokes were in the same locations, which implied that Venus always held the same face toward Earth—very strange indeed. Those spokes Lowell saw were probably the retinal network in his viewing eye. He viewed Venus by using only a small section of the large refractor in his telescope, at high magnifying power. The situation was equivalent to viewing through a pinhole—he could see Venus, but the retinal network was superimposed on it.

7.5 • Floaters and other spots in your eye

When I view a bright featureless background such as the clear sky, my field of view is littered with small floating dots and moving specks. The moving specks are described in the preceding item. The small floating dots each consist of concentric circles, but I also see larger, elongated structures. A large structure in my right eye often interferes with my ability to read with that eye.

Floaters, as they are called, can be seen more clearly if the eye is illuminated by a small source of light. For example, I normally use an illuminated pinhole in opaque cardboard. However, any tiny bright source of light, such as a brightly reflecting paper clip, can also work. (I am very careful about bringing any object close to an eye.)

When I use the pinhole arrangement, I see several other curious features. There are bright specks that seem to lack the concentric circles of the more common floaters. Sometimes I see dark specks and a stationary pattern of dark lines extending from the center of my field of view. Just after I blink, I see bright spots and a pattern of horizontal bright and dark lines. Sometimes I also see stationary bright patches or floating wrinkled patches. Upon opening my eyes in the morning, I might see one or more spots that are much darker or (more rarely) much brighter than the rest of my view.

What causes these various appearances?

Answer The common floaters are probably due to irregularities in the *vitreous humor* (the transparent material filling most of the eyeball). You cannot see the irregularity itself or even its shadow on the retina. Rather, you see the diffraction pattern the irregularity casts on the retina. Diffraction is a type of interference that light waves undergo when they pass through a small opening or pass by a small obstacle. Here, when light from a pinhole passes an irregularity in the vitreous humor, the light is diffracted into an interference pattern on the retina. The pattern consists of concentric bright bands (where light waves happen to reinforce one another) and dark bands (where they tend to cancel one another).

If the irregularity is roughly circular, the pattern is also, with a bright central spot. An elongated irregularity produces an elongated pattern. The floater you normally see is a blurry diffraction pattern. If you look through a pinhole, you see the pattern more sharply and can distinguish individual bright and dark bands. Floaters drift through your line of sight because the vitreous humor is not rigid and can shift.

Some of the floaters may be due to bits of the vitreous humor that have broken off and that float in the liquid layer in front of the *fovea*, the pit-like structure where your line of sight falls. They can also be due to blood cells that leak into that liquid layer, but in that case the field of view is likely to be tinted red.

Everyone has floaters, and their presence is not

necessarily an indication of a medical problem. As you age, you probably will see more floaters.

The bright specks and the pattern of bright and dark lines that follow a blink are due to the liquid (tear) layer left on the cornea. Irregularities in that layer can slightly focus light rays to give the brighter regions. The lines radiating from the center of the view may be from the radial structure of the eye lens. Dark specks may be due to tiny opaque regions in the lens. The dark and bright spots seen by some people just after opening their eyes in the morning are not understood.

7.6 • Streetlight halos, candle glow, star images

At night, many people see rings (halos) surrounding bright light sources when the sources are viewed directly. (Viewing them through a window covered with condensation produces a different set of rings.) The diameters of the first four rings (as measured in the number of degrees of arc they occupy in your field of view) are approximately 2.5, 4.5, 5.5, 6.0, and 9.0. The rings are larger in red light than in blue light. So, if the source provides white light, the rings may be colored with red on the outside and blue on the inside. Why do the rings appear?

Some of the paintings by Vincent van Gogh show rings around light sources, such as the Sun in *Red Vineyard at Arles* and stars in *Starry Night.* He painted the rings partly for style because they give an impression of radiance on the canvas. However, he reportedly saw such rings around light sources because his vision was altered by his medical use of digitalis, which was at a toxic level.

Why is a candle flame surrounded by a faint glow when the flame is seen in an otherwise dark room? Stars shimmy because of the changes in the atmosphere, but what causes the typical star image with radial spikes or lines?

Answer The rings around bright lights, called *entoptic halos*, result from the diffraction of light as it passes small structures (nonuniform regions) in the eye on the way to the retina. Diffraction is a form of scattering in which light waves spread around an obstacle to produce a pattern of bright and dark bands concentric with a bright center spot. The bright regions are where light waves reinforce one another; the dark regions are where they tend to cancel one another. The central bright spot is not perceived because it coincides with the (much brighter) direct view of the light source, but the first of the bright rings can be perceived. Its angular size depends on the size of the structure diffracting the light and the distance between the structure and the retina: A smaller structure produces a larger ring; a greater distance also produces a larger ring.

When several rings are perceived, diffraction is produced by several structures with different sizes and distances from the retina. No one is certain what structures are responsible for the diffraction. Candidates include the cells of the corneal epithelium (between 10 and 40 micrometers in size), corneal endothelial cells, corneal striations, and lens fibers.

The diffraction of light within the eye is also responsible for the faint glow seen surrounding a candle flame and for the points you see in a typical spiked image of a star or any other bright, small source of light distant from you. The star spikes are probably caused by the irregularities of the suture lines (the junctions of the fibers) on the front surface of the eye lens.

7.7 • Phosphenes—psychedelic displays

Prisoners confined to dark cells sometimes see brilliant light displays called *phosphenes* that can be colored or speckled with colored dots. Truck drivers also see such displays after staring at snow-covered roads for long periods. In fact, whenever visual stimuli are lacking, these displays appear.

Migraine headaches and some hallucinogenic drugs (such as LSD) can produce magnificent phosphene displays. So can rapid accelerations of the head as experienced by pilots and astronauts. They also can be made at will by pressing lightly against a closed eye. Different displays appear as the finger moves over the eyelid. Increasing the pressure produces more complex patterns. (Do not press enough to hurt your eyes, and do not press at all if contact lenses are on the eyes.) Geometric designs appear if both eyes are pressed simultaneously.

They also appear when an observer looks at a flashing light such as a stroboscope in a rock concert or dance club. When I view a strobe light flashing at a rate between 10 and 30 times a second, vividly colored geometric arrays appear. (For safety, I close my eyes while facing the strobe. The light is bright enough to pass through my closed eyelids.) Sometimes I perceive a chessboard array of squares, sometimes hexagons or triangles. For slow flashes, the phosphenes are swirls. They disappear for rapid flashes. The complex geometric patterns require illumination of both eyes. With only one eye illuminated, I see simple patterns of lines and swirls.

Phosphenes can also be produced when a mild electrical current passes through the observer's head. (I would never do such a dangerous thing, and neither should you.) Phosphene parties were high fashion in the 18th century (even Benjamin Franklin once took part). People holding hands in a circle were shocked by a high voltage, low current electrostatic generator. Each time the current passed through them, they saw phosphene displays.

Even more bizarre (and dangerously stupid) were the 1819 experiments by the physiologist Johannes Purkinje. He placed one electrode on his forehead and another in his mouth and then repeatedly broke contact with one of them, so that there were pulses of current through his head. The pulses produced stable phosphene images.

What causes phosphenes?

Answer When you press against a closed eye, the vitreous humor filling the eye presses against the retina, triggering the photoreceptors or the nerve pathways to send signals to the brain as they would were the eye illuminated. Thus, I perceive light even without light entering the eye.

Phosphenes are also created if I look at a flashing light. The more complex, geometric patterns require stimulation of both eyes, implying that they are interpretations imposed by the brain on the signals arriving from the eyes. The geometric shapes appear because the nerve signals activate the line and shape detectors in the brain. Colors are created when the color detectors are activated. (Thus this perception of color is not directly due to the color detection by the cone photoreceptors on the retina.) Perhaps the flashing lights happen to match the brain's code for colors. If so, a flashing white light may result in a perception of brilliantly colored arrays. More likely the colors arise from mutual interference of the nerve pathways in the retina and those leading to the brain.

The electrically generated phosphenes can result when the brain is directly stimulated. With them some sightless persons may be given vision. A miniature video camera, mounted on a frame (such as used for glasses), sends signals to a microprocessor, which then generates phosphenes by sending a low-current signal directly to the brain. For example, when the video camera finds an object in the person's left field of view, the brain is stimulated so that the person perceives a phosphene in the left field of view. Thus the person's surroundings are represented by phosphenes and in a sense the person can "see."

Drug-induced phosphenes appear to be drawn in the Paleolithic art found on cliffs and in caves. The phosphenes could have been part of the visual experience of someone (perhaps a shaman) undergoing a trance, and they may be symbols of the underlying magic the people believed to rule the world.

7.8 • Humming becomes a stroboscope

By humming at the proper frequency, you can stroboscopically freeze the rotation of an airplane propeller or a fan blade. If you hum at a slightly lower frequency, the strobe pattern rotates slowly in the direction in which the object is turning. At a slightly higher frequency, the pattern rotates in the opposite direction.

A similar strobe-freezing can be produced if you hum while watching television from a sufficiently large distance. The humming generates lines on the screen that are stationary for one frequency but migrate up or down for other frequencies.

To study how humming alters my vision, I prepared a paper pattern to be mounted on a turntable. The pattern consisted of black and white sectors, each one degree wide, extending from the center point outward. I allowed sunlight to illuminate the pattern as it rotated at $33\frac{1}{3}$ revolutions per minute. (A lightbulb provides flickering light.) Since I cannot hold a note, I forced my chin against a small speaker that oscillated at 100 times a second (it was driven by an audio oscillator). The rotating pattern on the turntable stroboscopically froze into a dim, blurry pattern. In this and the other examples, why does the oscillation (from the speaker or the humming) freeze the rotation?

Answer Humming, purring, or oscillating my head with a small speaker results in vertical oscillation of my eyes. If the oscillation has the proper frequency, the pattern passing through my view is kept at the same position on the retina throughout most of the oscillation cycle of an eye.

Suppose I stare at a section of the pattern on a turntable as the section descends through my view. When my eye descends during a cycle of oscillation, the pattern continues to illuminate the same portion of my retina and appears stationary. When my eye begins to move upward, the illumination pattern shifts on the retina but only briefly. Quickly the original pattern of black and white falls on the same places on the retina as it did previously. My visual system averages the brightness during the cycle of oscillation. Those places receiving a bright image during most of the oscillation are perceived to be bright. Those places receiving a dark image during most of the oscillation are dark in comparison. Thus the black-and-white pattern appears to be stationary.

A television image is created by horizontal sweeps, line by line, from top to bottom of the screen. The fading of each line is masked by the speed of the sweep and my persistence of vision. When I hum at the appropriate frequency, the oscillation of my eye stroboscopically freezes the sweep. Throughout most of a cycle of oscillation, there is a horizontal line across my retina that is an image of the line on the screen where the old image has faded and a new image has yet to be made. Thus I continuously see a dark line on the screen.

7.9 • Keeping your eye on the baseball

The skilled baseball player Ted Williams claimed that he could see a pitched ball hit his bat. Several players have claimed that they could see the stitches and spin on the ball as it passed in front of them. Can a player truly make such observations? Does a player visually follow the ball from the moment of its release by the pitcher until it either passes behind home plate or is hit with the bat?

Must a player have two functioning eyes in order to play baseball? Apparently not. How then do players having a single functioning eye determine the distance and trajectory of the ball? Similarly, how does a person with sight in only one eye determine depth in the visual field when driving a car or flying an airplane? For example, landing an airplane surely requires depth perception, yet the vision of the famous pilot Wiley Post was limited to one eye.

Answer Suppose that a professional baseball player bats right-handed at home plate. If the player is to track a ball thrown across the plate, the player must rotate the line of sight rightward from the pitcher. Most good players can do this until the ball is about 5.5 feet away from the plate, but thereafter the required rotation is just too fast. However, a player *can* see the ball hit the bat if the player correctly anticipates where the ball–bat collision will be and jumps the line of sight to that point. Ted Williams probably used such a vision jump, called a *saccade*, to see a ball hit his bat.

Another factor may also be involved in the visual tracking of a ball. Apparently the visual system can perceive depth in the motion of an object even if it cannot perceive the location of the object. This capability has an obvious survival factor: You can tell whether an object is headed toward you even if you cannot tell exactly where the object is at any instant. The perception of depth in the motion of an object can be performed with a single eye. Thus people having only one functioning eye can play sports and fly airplanes. When both eyes function, the brain may compare the relative motion seen by each eye. For example, if your right eye sees an object with motion to the left and your left eye sees it with motion to the right, then the object is headed directly for you.

7.10 • Impressionism

In the style of painting known as impressionism, objects and their background are painted with only general shapes instead of their details. Claude Monet, for example, is renowned for his impressionistic paintings of outdoor scenes. As he aged, he retained the impressionistic style, but his work took on a "warmer color" of red and yellow, losing the opposite end of the visible spectrum. Although impressionism is an appealing artistic style, could it have begun for some physical or physiological reason? What accounts for the color change in Monet's art?

Answer Many of the artists in the impressionism era suffered from vision defects. Some were shortsighted and thus saw the objects they painted as vague and blurry—just right for the impressionistic style of vague and blurry images. At least one artist would paint a canvas at arm's length to also put the canvas out of focus. Others, such as Monet, suffered from cataracts that prevented sight farther than a few meters. Monet probably had *nuclear cataracts*, which absorbs the blue end of the spectrum and leaves the yellow–red end, thus explaining the yellow–red dominance in his later work. After undergoing cataract surgery late in life, he was furious with his earlier yellow–red works and threatened to destroy them or overpaint them.

7.11 • Pointillistic paintings

A pointillistic painting, such as *Sunday Afternoon on the Island of La Grande Jatte* by Georges Seurat, is painted not by using brush strokes in the usual sense but rather by using a myriad of small colored dots. You can see the dots if you stand close enough to the painting, but as you move away from it, they eventually blend and cannot be distinguished. Moreover, the color that you see at any given place on the painting can change as you move away—which is why a pointillistic painting contains the dots. What causes this change in color?

Answer When light passes through the circular iris of an eye, it diffracts; that is, it spreads and forms an interference pattern. If you view a point source of light, the diffraction forms a circular image of that source on your retina. If you view two adjacent sources of light, they each tend to form their own circular image, but if they are too close, the images, overlap and you can distinguish only a single, merged image. Thus, the onset of overlap of such images sets your limit of resolving the two light sources as separate points.

Two adjacent dots of paint in a pointillistic painting serve as two light sources. Suppose that the dots have different colors. If you stand just in front of the painting, the dots are far enough apart in your view to form separate images on your retina, and thus you see the true color of the dots. As you move away from the painting, the dots eventually produce overlapping images and then you can no longer distinguish them. The color that your brain brings to consciousness may not be the color of either dot or even a simplistic blend of their colors—it may be a color fabricated by the brain. For example, suppose that a magenta (blue and red) point is adjacent to a yellow point. The combination of the two colors is perceived as pink. Thus, a pointillistic painter uses your visual system to create the colors of the art.

A traditional oil painting is usually darker than a pointillistic painting because its colors depend on a mixture of paints within an oil layer. Light must cross through the layer, reflect and then re-cross the layer to reach you. When more pigment is included in the layer, the light emerging from the painting is dimmer. Because a pointillistic painting accomplishes the mixing of colors within your brain, not on the canvas, it does not dim the light as much.

Many colored surfaces, such as mosaics, woven fabrics, colored printing, and the screens of color monitors, are arrays of points having different colors. In the traditional color theory, three basic colors (red, blue, and green) are required to generate all possible colors. Thus a color monitor contains arrays of dots of those three colors. Any desired color is generated by controlling how brightly lit each dot is.

7.12 • Moiré patterns

When a fine-scale grid overlays a pattern of similar design, you can see a larger-scale pattern called a *moiré pattern.* I see moiré patterns when silk overlays silk or when one picket fence lies behind a parallel picket fence. I also find them with arrays of circular holes. When one of the hole arrays lies several centimeters in front of the other array, I see a circular moiré pattern in the composite. What produces moiré patterns?

Answer Moiré patterns are seen because of the periodic arrangement of the patterns that are overlaid. For example, consider two parallel picket fences that are somewhat distant, with an illuminated background. At certain points, empty spaces in the fences are lined up in your view, and so you see bright regions. At other points, pickets are lined up, and so you see dark regions. Where the alignment is incomplete, you see narrow regions of illumination. The composite of these bright and dark regions is the moiré pattern seen in the fences—a repeated pattern in the variation of brightness occurs along the fence length. If one of the fences were moved by less than the picket-to-picket spacing, the moiré pattern would noticeably shift, as if it amplified the actual fence motion.

The best explanation for why moiré patterns are so captivating is that your visual system is especially sensitive to intersections of lines and searches for them. Because of this sensitivity, even small displacements of an array producing a moiré pattern are apparent.

7.13 • Op art

When you view an example of *op art* (a phrase coined in 1964 for *optical art*), a static arrangement of lines, blocks, or spots creates illusions of motion, as if sections of the art shimmy, rotate, or flash on and off. The arrangement can also create illusions of colors that seem to bleed from one section to another. What causes these illusions?

Answer No one can explain the op art illusions. They are still being discovered, catalogued, and compared. From an artist's point of view, new discoveries can lead to novel art pieces. From a physiologist's point of view, they can lead to insights about how the visual system and brain work.

The visual system can briefly retain an image as an *afterimage*. The eye undergoes small jumps called *saccades*, in which the view changes slightly. When you examine op art having a geometric design, afterimages from successive saccades can be mentally superimposed. However, because the afterimages are slightly different, the design seems to have moved from one afterimage to the next. The illusion is subtle and you may not be aware of it—you simply know that the op art differs from other static pictures of lines.

Some geometric arrays create the illusion of bright or dark spots where there are none. For example, the grid in Fig. 7-2 can create the illusion of fleeting dark spots in the intersections of the white lanes—the dark spots are said to be *induced*. The illusion is not well understood but is probably caused by the photoreceptors in one region of the eye interfering with the photoreceptors in adjacent regions, as explained below about Mach bands. In some colored arrays, induced lines or spots of color (said to be *neon spreading*) can occur, indicating that the interference involves the messages about color sent from the eye to the brain.

Figure 7-2 / *Item 7.13* Grid with fleeting black dots appearing in the intersections.

7.14 • Depth in oil paintings

The 15th-century Flemish school of painting achieved an extra illusion of depth in oil painting by applying thin layers of translucent paint (*glaze*) over a white background. Why did portions of the paintings seem to be in front of other portions, and why did colors seem to originate within the painting rather than on the surface?

Answer Some of the light intercepted by one of these paintings reflects from the front of the glaze, and the rest travels through the layer (Fig. 7-3). Pigments suspended in the glaze scatter light outward and also toward the rear. Any light, whether direct or scattered, that reaches the rear reflects from the white (opaque) background layer. As the light travels through the glaze again, it can again be scattered by the pigments. When you intercept light from the painting, you perceive the part of it that has reflected from the outer surface of the glaze. However, points of color appear to originate behind that outer surface, especially when the painting is viewed with both eyes so that their convergence on a colored spot allows you to assign depth to the spot.

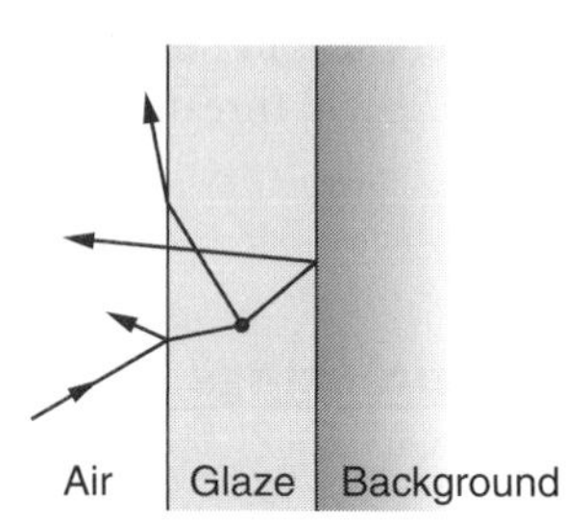

Figure 7-3 / *Item 7.14* Oil painting with glaze; light scatters from front and back and from a pigment within the glaze.

The painter can control the saturation (or brilliance) of a color by applying more than one layer of glaze with the same pigments. Each additional layer sharpens the color you see from the pigments because it causes additional scattering. For example, if a pigment scatters a particular blue wavelength more than other wavelengths, then additional layers containing the pigment bring out more blue in the painting.

A layer of varnish is often applied to a painting to protect it. This type of layer does not contribute to the sensation of depth in a painting. Rather, its partial absorption of light usually detracts from the painting by dulling the colors, perhaps even hiding certain tints.

7.15 • Reading in the dark

The most startling afterimage I have seen is produced by a bright flash in a dark room. I place an open magazine in front of me and then turn off the lights. After my eyes adjust to the darkness (about 10 to 15 minutes), I hold my camera's flash attachment next to my head and illuminate my view with a single flash of light. The flash is too bright for me to distinguish the magazine.

If I hold my gaze steady as the glare fades, a detailed image of the magazine appears, as if the magazine is illuminated with a steady bright light. What I see is said to be a *positive afterimage* because the bright areas of the magazine appear white and the dark areas appear black. Photographs, drawings, and paragraph formats are easily distinguished, although none of them could be seen during the flash of light. I can even read words. After about 15 seconds, the positive afterimage fades to a *negative afterimage* in which the black and white areas are exchanged.

If I flash the light twice, I perceive two superimposed positive afterimages. If I drop a coin during the flashes, I see two images of the coin during its fall, as if I am viewing a stroboscopic photograph. Sometimes the positive afterimage can be strange. If I flash the light while my hand is in front of me and then place my hand on my back, I perceive an image of my hand in its former position, yet feel its presence on my back. If I flash the light while standing and looking toward the floor and then squat, I perceive the floor as being distant, yet I know from my body orientation that it must be closer.

I have also seen a positive afterimage at a ballet. During intermission I kept my eyes closed. As the second half began, I briefly opened my eyes to see the flood of light on the stage. Once they were again closed, I perceived an afterimage of the dancers on stage.

I can sees similar afterimages when awakening in a well-lit room. As I lie with my eyes closed, I can see only red light leaking through my eyelids. I hold my hand in front of my face and then briefly open my eyes. With my eyes again closed, I perceive a vivid afterimage that is first negative and then positive. If I open my eyes for several minutes so that they adjust to the light in the room, I can no longer repeat the demonstration.

These different positive afterimages appear provided I hold my gaze steady but are immediately erased if I move my eyes relative to my head. What causes them? What produces the negative afterimages?

Answer The cause of the positive afterimage is not understood. The brief flash of light momentarily saturates the rod photoreceptors so that details of the magazine cannot be distinguished. However, the image of the magazine produces a substance or an effect along the visual pathway that lasts longer than the saturation. As the saturation fades, the image can be distinguished. The subsequent negative afterimage is probably due to a fatigue of the visual system. Those parts strongly stimulated by the bright areas in the image fatigue, leaving darker images than the parts weakly stimulated by the dark areas.

7.16 • Trailing ghost light

An afterimage that may be related to the positive afterimage in the preceding item involves the motion of a small point of light in a dark room. After adapting your eyes to the darkness for a few minutes, move the stimulus light around in front of your open eyes. A ghost light follows behind it, slightly delayed. Behind the ghost light, a dim trail of light remains.

If the stimulus light is deep red, the ghost and its trail are not produced. If the stimulus light is yellow or yellow-red, the ghost (and perhaps its trail) may be faintly blue. However, if your eyes are fully adapted to the dark (the adaptation requires 10 to 15 minutes), the ghost and its trail are always gray. Why does the ghost and its trail appear, and what accounts for the color? Why are they not produced by a deep-red stimulus light?

Answer The ghost light and its trail are probably afterimages generated by the rod photoreceptors once they are illuminated by the passage of the light source. A short time is needed for the rods, their nerve pathways, or the brain to generate a second perception of light. Thus your perception of the ghost light is delayed. (If the ghost light were due to persistence of vision, there would be no such delay.) The trail is the composite of the afterimages as they slowly fade. A deep-red light does not trigger the afterimage because it cannot activate rods.

The source of the ghost's color is not understood and is rarely mentioned. I believe that it results from the rods interfering with the message about color being sent by the cones lying along the illuminated path on the retina. Although rods are not believed to be able to send color information to the brain, they seem able to inhibit color information from the cones. When a yellow or yellow-red light stimulates the retina, the inhibition by the rods forces a perception of blue, the complementary color of yellow. This inhibition disappears once the eye becomes fully dark adapted. Then the ghost is gray.

7.17 • Reflecting eyes

Your flashlight throws a narrow beam through the thick darkness. Suddenly a pair of glowing eyes appears in the light, but your fear disappears when the animal meows softly.

Why do the eyes of a cat seem to glow when it faces directly into your flashlight beam but not when it looks somewhat away from that direction? Why can a person that is photographed end up with red eyes in the photograph?

Scallops have eyes that consist of a lens, a thick retina, and (behind the retina) a concave mirror. The lens is so weak that it hardly refracts (bends) the light rays and thus cannot be responsible for any formation of an image. Besides, unlike your eye, the scallop's eye has a lens that lies against the retina, giving no room for the refracted rays to cross through one another to form an image. How then does an image form in the scallop's eye? The mirror is an excellent reflector. How can a biological system have a reflecting surface that rivals (or even exceeds) modern metallic mirrors?

Answer Behind the photoreceptors in a cat's retina lies a layer that reflects some of the light back through the photoreceptors so that they have a second chance to absorb the light. This increased efficiency may be useful to night prowlers. When you shine a flashlight into the cat's eyes as the cat looks toward you, you see some of the light reflected from the back of the cat's retina.

In a human, the layer behind the photoreceptors is not as reflecting and does not result in especially bright eyes when illuminated with a flashlight at night. Still, some reflected light can be seen in a photograph if the person looks directly at the camera to which the flash unit is attached.

The formation of an image in your eye derives from the refraction of light by the cornea and eye lens. In the scallop's eye, the formation derives from the reflection of light by the concave mirror behind the retina. The rays enter the scallop's eye, pass through the lens and the retina, reflect from the mirror, and then focus (because of the reflection) to form an image while still within the retina.

The mirror is not a single layer of reflecting material such as your bathroom mirror. It consists of alternating layers of cytoplasm (which has a low index of refraction) and guanine crystals (which have a high index of refraction). (The index of refraction is a measure of the speed of light in a material. Light travels slowly through a material with a high index.) The thickness of each layer is about one quarter of the wavelength of light. Because of this thickness and the alternating values of the index of refraction, the reflected light waves are in step with one another, producing a much brighter reflection than can be obtained with a single reflecting layer.

7.18 • Underwater vision of humans, penguins, and crocodiles

Why do we lose most of our focusing ability when we are underwater? Why does a nearsighted person see better than others when underwater? Why can a mask restore the vision for anyone? Why can some people, such as the Moken in Burma and the West Coast of Thailand, see well underwater without the use of a mask?

Penguins live in the air but hunt underwater. How can they see in both air and water?

Answer In air, most of the focusing of light rays by a human eye occurs at the cornea; the rest of the focusing is done by the eye lens, which can be controlled by muscles. When you are submerged, you lose all the focusing by the cornea because the optical properties of the eye material almost match the optical properties of the water outside the eye. Thus, there is no bending of the light rays as they enter the eye. That leaves only the lens to focus the light rays, but most of us cannot change the shape of the lens enough to bring clear images onto the retina. However, the Moken have trained themselves to see underwater: They narrow the pupil to decrease the spread of light rays entering the eye. They also force the eye lens to be as curved as possible. These two measures produce retinal images that are reasonably clear. (Reportedly, anyone can learn this procedure.)

A nearsighted person has too much focusing by the cornea and eye lens, which tends to produce images in front of the retina. When light rays from a distant object reach the retina, their spread produces blurry images. When a shortsighted person is underwater, the elimination of the focusing by the cornea shifts the location of the focused image back toward the retina, perhaps even onto the retina. Thus, a nearsighted person can see more clearly underwater than a farsighted person or a person with normal vision.

If a diver wears a mask, air lies next to the cornea, and so the bending of light rays at the surface of the cornea is normal.

The cornea of a penguin is almost flat. So, when a penguin changes between vision in water and vision in air, the focusing by the cornea is hardly affected. The penguin is adapted for vision in water, because there it finds its food. Thus, its eye lens is highly curved so as to bring light rays into focus on the retina. When it is in air, it can relax some of the curvature of the lens but the lens probably still focuses too much for clear images to form on the retina. Thus, the penguin is probably very nearsighted in air. However, it can reduce the blurriness of images on the retina by narrowing the pupil to a small square pinhole. Such a narrow aperture restricts the angular extent of the rays coming from an object, making the object's image on the retina sharper.

Crocodiles can see well in air but not underwater. Like us, they cannot change the shape of their eye lens enough to compensate for the loss of focusing by the cornea. Still, they are skilled underwater stalkers because they use other means to locate prey.

7.19 • Underwater vision of "four-eyed fish"

A peculiar fish, the *Anableps anableps*, swims with its eyes partially extending above the water surface so that it can see simultaneously above and below water. How can its eyes focus in both air and water?

Answer The eye lens in the fish is egg-shaped to compensate for the comparatively small refraction suffered by the light arriving from an underwater scene. The light coming from above water is well refracted upon entering the eye and is refracted further by the lens, so that it is in focus when it reaches the retina in the lower half of the eye.

The light coming from below water is poorly refracted upon entering the eye. However, the extreme curvature of the eye lens creates enough bending that an image of the underwater scene is focused onto a separate retina in the upper half of the eye. The focusing is also aided by the relatively long distance between the lens and the top retina.

7.20 • Cheshire cat effect

Arrange for a mirror to reflect a scene to one eye while the other eye sees a second scene directly. You might be able to fuse the two scenes, see them alternately (in what is called *binocular rivalry*), or see only one of them for most of the time. But if you run your hand through either scene without following the motion with your eyes, the second (other) scene either partially or totally disappears. When there is partial disappearance, the erased portion in the second scene corresponds to the region through which the hand moves. If the second scene is the face of a person, you might be able to erase part of the face, perhaps leaving only a freely floating mouth that resembles the Cheshire cat in Lewis Carroll's *Alice's Adventures in Wonderland.* What accounts for the erasure?

Answer Presumably as a survival mechanism, the visual system concentrates on any motion seen by either eye and brings to consciousness the scene concerning that motion. Usually both eyes detect motion and the fused view appears to be normal. But when a mirror gives a very different scene to one eye (an unnatural situation), the concentration on the scene with motion disallows all or part of the other scene to reach consciousness. If your eyes were on opposite sides of your head (like the situation with some fish), this illusion would be more commonplace.

7.21 • Rhino-optical effect

Close your left eye, stare straight ahead with your right eye, extend your left arm leftward with one finger on the hand uplifted, and then rotate that arm until the finger just barely comes into view. Next, rotate your right eye toward the finger. The finger will probably disappear. It is visible when you look straight ahead but not when you attempt to look toward it. What makes the finger disappear?

Answer For you to see something, light from it must pass through the opening of your eye. If you stare straight ahead and bring your finger barely into view as described, light from the finger skims your nose before it enters the eye's opening. When you turn the eye toward the finger, you move the opening into the "shadow" of the nose. Thus, light from the finger is blocked by the nose and cannot enter the eye's opening, an effect dubbed the *rhino-optical effect* after the Greek word *rhino* for nose. However, this effect does not appear if you have a shallow nose.

The nose, forehead, and cheek always block part of the view from an eye. Your brain, however, concentrates on the information available in the region around your direct gaze and suppresses the missing sections that are well off to the side.

7.22 • Flying clouds and Blue Meanies

Thorne Shipley of the University of Miami School of Medicine once described a novel visual illusion. While flying in an airplane at high altitude he spotted two layers of clouds. One layer, which seemed to be the more distant one, moved rapidly toward the rear. The other one appeared to be rigidly connected to the airplane, following exactly along with it. Later during the flight Shipley was able to see the surface of the sea far below. Immediately the cloud layers traded distances, and he saw the distant one as stationary. Which interpretation of the distance and motion relative to the airplane was correct?

In the movie *The Yellow Submarine* (which was based on a song by the Beatles), a Blue Meanie undergoes a remarkable transformation. When he is distant he is large and ferocious. As he nears he becomes smaller and less fearsome. Why is such a transformation strange?

Answer One layer of clouds was distant enough that it did not appear to move in Shipley's view. The closer layer moved through his view because of the airplane's motion past it. Since the scene was almost devoid of other clues about distance and motion, the lack of motion by the first layer was interpreted as evidence that it was locked in flight with the airplane and thus must be nearby. (Such a perception is called *visual capture.*) The other layer was interpreted as lacking any connection with the airplane and thus as being distant. Only when Shipley was able to see the ocean surface were there sufficient clues about depth and motion to correct the illusion.

The Blue Meanie seems strange because he should occupy a small angle in your view when he is distant and a larger one as he approaches. Shipley also argues that nearer objects are more likely to be considered dangerous than distant ones, just the reverse of how the Blue Meanie is portrayed.

7.23 • Pulfrich illusion

Arrange to view a pendulum swinging across your field of view while one eye is covered with a dark (but not opaque) filter. A lens from a pair of sunglasses will serve. Although the pendulum swings in a vertical plane, you will perceive it as swinging in an ellipse. The apparent depth of its motion increases if you also see a vertical reference (such as a rod or a hanging string) or if you switch to a darker filter. With the filter over the left eye the pendulum seems to orbit clockwise (from an overhead sense). Its motion is reversed when the filter is moved to the right eye.

The motion is subtly perplexing if you arrange for two pendulums to swing near each other. Each appears to orbit around the other, which would require that the strings wrap up on each other. Since they don't become entangled, their motion is eerie.

If you ride in a car while wearing a dark filter over one eye, the speed at which an object in the landscape seems to pass the car is altered. Objects on one side of the car pass too slowly; objects on the other side pass too quickly. The filter also alters the apparent distances to the objects.

How does the dark filter alter normal perceptions of depths and speeds?

Answer The decreased light in the eye covered with the dark filter delays the signal sent from the eye to the brain. (The delay is called *visual latency.*) Thus, the true location of the pendulum is seen by the uncovered eye while a previous location is seen by the other eye. Your brain fuses the two views, and you perceive the pendulum as being either closer or farther than it truly is. Although the pendulum's swing is actually along a straight line, your brain forces depth into the motion, making the pendulum appear to swing in an ellipse.

For example, suppose the left eye is covered with the filter and the pendulum is moving toward the right (Fig. 7-4*a*). The delayed view of the pendulum lies to the left of its actual position. After the brain fuses the two views, you perceive the pendulum to be farther away than it truly is. Later when the pendulum moves toward the left, the delayed view is to the right of the true location (Fig. 7-4*b*). This time you perceive the pendulum to be nearer than it truly is.

The distorted views from the moving car also involve visual latency. Suppose that the filter covers the left eye while you look at an object on the right side of the car. The discrepancy between the object's true location (as seen by the right eye) and the location where it was a short time earlier (as seen by the left eye) forces you to perceive the object as being farther from the car than it truly is. Since the time the object takes to pass through your field of view is unaltered, you conclude that the seemingly distant object must be traveling quickly. On the other side of the car, visual latency

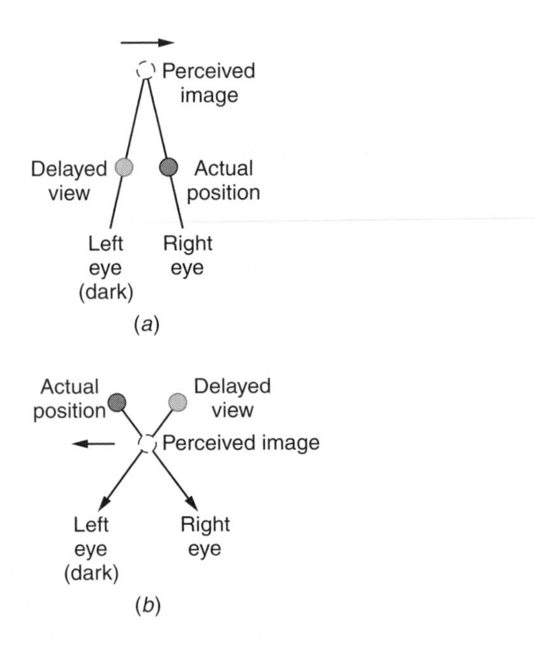

Figure 7-4 / Items 7.23 and 7.30 Pulfrich illusion with pendulum swinging to (*a*) the right and (*b*) the left.

forces you to perceive an object as being nearer than it truly is. This object seems to pass too slowly.

Jerry Lerner told me of a novel variation of the Pulfrich illusion. Replace the pendulum with an object rotating horizontally. You can arrange the filter and the rotation so that the apparent rotation is in the opposite direction of the true rotation and at twice the rate. The object appears to shrink and expand mysteriously.

The delay in the visual signal introduced by wearing dark sunglasses (or having a tinted windshield) might increase the distance required to stop a car. Suppose that the sunglasses are dark enough to delay the visual signal by about 0.1 second (an extreme case). At a speed of 55 miles per hour (about 90 kilometers per hour), the delay adds about 2.5 meters to the stopping distance.

7.24 • Streetlight delay sequence

When streetlights turn on simultaneously at dusk, the nearer ones may seem to turn on sooner, giving the illusion that the lighting of the streetlights proceeds down the street away from you. Streetlights clustered at intersections may seem to turn on slightly earlier than more widely spaced streetlights. What is responsible for these illusions? They cannot depend on the speed at which current flows from one streetlight to another because that speed is too great. Besides, the sequence proceeds away from you no matter where you are.

Answer The apparent streetlight sequence may be due to visual latency as discussed in the preceding item. The nearby lights are brighter than the more distant ones, and thus the visual system responds quicker to them. The progressively delayed response to the dimmer light arriving from more distant streetlights produces the illusion that the lights are turning on in sequence. This explanation may be incomplete. The response time may also depend on where the image of a streetlight falls on the retina.

7.25 • Mach bands

Suppose that only one strong light source (such as the Sun) illuminates an object to create a shadow. I would expect the shadow to have an edge across which brightness gradually fades to darkness. However, in many cases I find two mysterious bands running parallel to the edge. A dark band lies just within the shadow, and a bright band lies just outside the shadow (Fig. 7-5). If I take a photograph of the shadow and examine the edge in the photograph, I still see the bands. They are called *Mach bands* after Ernest Mach, an Austrian physicist and psychologist who first studied them.

Mach bands are almost always present along the edges of a shadow but are normally ignored. However, Paul Signac, a neo-impressionist painter born in the 19th century, studiously incorporated them in the shadows in his painting *Le Petit Dejeuner.* You can see them along the sides of your own shadow in sunlight. If the shadow is near you (perhaps it is on a wall), the shadow is probably too sharply defined for Mach bands to form. The bands are more apparent if your shadow lies on a sidewalk, especially while you move.

What produces the bands? Do they derive from some curious optics at the edge of the object casting the shadow, or are they created in the visual system of the observer?

Answer Mach bands are produced by the visual system and cannot be detected with instruments measuring the intensity of light at a shadow's edge. The visual system creates the bands as readily with a photograph of the shadow as with the shadow itself. No one presently has a full explanation of why the bands appear. Here I shall give only a partial explanation.

The bands are due to a mutual interference (called *lateral inhibition*) between groups of photoreceptors in the eye, their nerve pathways, and sections of the brain. A group that is activated by illumination decreases the strength of the signals sent to the brain by adjacent groups.

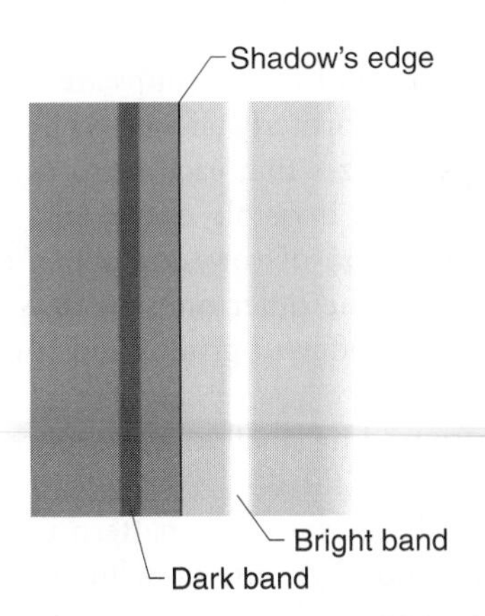

Figure 7-5 */ Item 7.25* Mach bands along a shadow's edge.

Consider groups whose photoreceptors lie reasonably near the shadow's edge on the retina. The groups far from the shadow inhibit one another and thus send a moderate signal, which comes to consciousness as an illuminated region. The groups well within the shadow are weakly activated, only weakly inhibit one another, and send a weak signal, which comes to consciousness as a dim region.

The curious bands appear from intermediate groups that straddle the shadow's edge. Consider a group having neighbors on one side that are strongly illuminated and neighbors on the other side that are weakly illuminated. The first set of neighbors inhibits the group but the second set does so only poorly. Because this group suffers only moderate inhibition, it sends more signal to the brain than does a group far from the shadow. Such groups near the shadow's edge form the bright Mach band.

Next consider a group that has neighbors on one side that are in the shadow and ones on the other side that are weakly illuminated at the shadow's edge. The group results in darkness because it suffers less inhibition than a group well within the shadow. Such groups form the dark Mach band.

Mach bands are prominent when the edge of the shadow (the transition region from well-lit to dimly lit groups) occupies an angle of about 0.2° in your view. They do not appear when the edge is more sharply defined.

7.26 • An upside-down world

Since the eye functions like a convex lens, it produces a real image of the world inverted on your retina. (For example, the ground appears at the top of the retina while the sky appears at the bottom.) Why then do you perceive the world as having proper orientation? If you wore special glasses (prisms) that inverted the image of the world, would you perceive the world as upside down?

Answer In normal viewing, the brain interprets the inverted image of the world as having proper orientation because of experience. For example, when you reach upward in front of your face, the image of your hand actually stretches downward on your retina. Still, the brain makes sense of the image by interpreting the stretch as being upward motion. If special glasses invert the image of the world, the brain takes several hours, perhaps days, to adjust its interpretation. Until then the world seems inverted. However, after adjusting, the brain again perceives the world in its correct orientation. When the glasses are removed, the brain requires another adjustment before the world again looks right-side up.

7.27 • Inverted shadows, the blister effect

Punch a pinhole in an opaque sheet of paper, hold the paper a few centimeters from one eye, close the other eye, and then hold a thin nail between the pinhole and your open eye.

Move the nail around until a shadowy figure of the nail appears in the circle of light from the pinhole. Why is the figure inverted from the nail's orientation? Why does it seem to be behind the paper?

With one eye, look at a distant light through the gap formed by your thumb and a finger, with your finger somewhat farther from you than your thumb. A "blister" seems to form on your finger along the portion within the gap. The narrower the gap, the more pronounced the blister is, until it fills the gap. What produces this *blister effect*?

The blister effect was seen by Captain James Cook in 1769 during the transit of Venus across the Sun. As Venus moved across Cook's view of the Sun, it formed a black dot. However, when it neared the edge of the Sun, a black bar appeared between the dot and the edge, as if the dot had developed a blister.

Answer The eye functions like a convex lens, producing an inverted image on the retina. Suppose that the nail has its head near the center of your view and extends downward. The image is upside down and extends through the upper half of your retina. Out of experience, your brain inverts the image so that you perceive the nail in its proper orientation.

The nail also casts a shadow on your retina because it blocks some of the light from the pinhole. Since the nail is the lower half of your view, the shadow falls across the lower half of your retina. However, since the brain inverts what is on the retina, you perceive the shadow in the upper half of your view. Thus you see the nail in its correct orientation, but its shadow is upside down.

In the blister effect, as you narrow the gap you see between thumb and finger, the thumb begins to block some of the light that passes the finger. That decrease in illumination produces a dim region on the retina immediately next to the finger image, which appears to extend into the gap as if it has a blister. The narrower the gap is, the more light is blocked and the larger the extension appears to be.

7.28 • Peculiar reflection from a Christmas tree ball

A shiny Christmas tree ball can reflect nearly the entire room to you. Suppose that you place a small source of light (such as an illuminated pinhole) in front of a ball and then view the ball's reflection from about 10 centimeters. If the room lights are on, the reflection of the source will be a point of light. If the room lights are then turned off, the reflection gradually stretches into a line. It quickly shrinks to a point again if the room lights are turned back on. What is responsible for the distortion when the room is dark?

Answer Rays of light from the point source of light reflect into many directions from the spherical surface of the ball. If your eye is close enough, it will intercept many of those diverging rays and focus them onto the retina. What you perceive is an image of the source that appears to lie behind the near surface of the ball. When the room is well lit, the pupil of your eye is small enough that only a narrow expanse of the reflected rays can enter it. Then the image of the source is a small point. When the room is dark, the pupil is larger, allowing a greater expanse of the reflected rays to enter the eye, and you see an extended image of the source.

7.29 • Rotated random-dot patterns

Sprinkle ink or paint onto a sheet of paper and then photocopy it onto a transparency. Place the transparency over the original copy so that corresponding dots are aligned. Use a finger to hold a point on the transparency in place while you rotate the rest of the transparency around the finger. For large rotations the dots remain random, but for small rotations many of the dots seem to lie on invisible circles.

With more rotation, this ordering of the dots progresses inward to your finger until all ordering disappears. Other designs can be introduced with distortions and other motions of the transparency. If the transparency is produced on a copy machine that reduces the copy somewhat, then when the transparency is placed over the original copy, spirals and other shapes appear. What is responsible for the perceived order in these various arrangements, collectively called *Glass patterns* after Leon Glass of McGill University, who discovered the arrangements in 1969? The order disappears if the second pattern is a negative one of the first and has a background of halftone gray. Does this result imply that the perception of rotation in the pattern depends on the contrast of dots and background in each of the patterns?

Answer The perception of rotation in the random-dot displays is not understood in detail. Somehow the visual system scans the whole pattern to correlate the pairs of dots that were formerly aligned. If you look at only a small section, the correlation fails and you do not perceive any pattern. Correlation in the whole scene may be due to organization of the visual system into groups that are designed to detect lines and edges.

Suppose that many of these groups are each excited by a pair of dots that were formerly aligned. Then the brain compares the groups, realizes that each has a correlated pair of dots connected by an invisible circle, and brings to consciousness a perception of rotation. When one pattern is rotated too far from the other one, the dots that were formerly aligned excite different groups and you are unable to perceive rotation.

In addition to needing small rotations of the patterns, the brain may require that the dots have the same contrast with the background. Thus when one pattern is the negative of the other and has a background of halftone gray, no pattern is perceived even when there is a small rotation of the dots.

7.30 • Patterns in television "snow"

Tune a television to a channel having no signal so that the screen is filled with the random noise called snow. If you lay a circle over the screen, the snow seems to flow along the boundaries of the circle. If you lay a grid of radial lines over the screen, the dots seem to move perpendicular to the lines and thus in a swirl. A grid of concentric circles forces the dots to stream radially outward from the center of the grid. Why is there such apparent organization of the snow?

When you view the screen of snow with one eye covered with a dark (but not opaque) filter, the snow breaks up into a startling organization. (A filter from a pair of sunglasses will serve.) The white dots on the screen appear to be in two planes, one in front of the screen and one behind it. The dots on one plane move uniformly to the left while those on the other move uniformly to the right. What is responsible for this apparent organization, direction of motion, and change of depth?

Answer The apparent motion of the dots induced by a pattern or grid on the screen is certainly illusionary because the dots are appearing and disappearing randomly. Still, the visual system insists on perceiving a sequence of dots along one direction as being the motion of a single dot. No one knows for certain why there is a preferred direction for the perceived motion. However, the preferred direction is apparently induced by the orientation of the edges in the grid or pattern.

When one eye is covered with a dark (but not opaque) filter, the visual signal from that eye is delayed. Thus while the uncovered eye sees a dot presently on the screen, the covered eye sees a dot formerly on the screen. As in the Pulfrich illusion (discussed in item 7.23), the brain attempts to identify these two images as being a single dot. To do so, the dot is perceived as being in front of or behind the screen (see Fig. 7-4). Thus, the illusion of depth to the dots derives from the delay in the visual signal from the covered eye.

Suppose that in the next instant another pair of dots is seen just to the right of the first pair. Again the brain fuses the two images to perceive a single dot. It also can interpret this new dot as being the former one that is now displaced to the right. Further fusing and interpretation gives the illusion that a dot travels to the right across the screen. Other dots appear to travel to the left.

My explanation may be wrong. The brain may be interested first in finding apparent motion and then in assigning depth. The net result is the same: The brain brings to consciousness an illusion of dots moving in opposite directions on planes at different depths.

7.31 • Mona Lisa's smile

One of the most enchanting smiles in the world is the one on the Mona Lisa painting by Leonardo da Vinci. What is so captivating about that smile?

Answer Your vision may seem to be reasonably consistent, but it is continuously altered by a subtle *random noise*—that is, fluctuations in signal and signal processing from the retina all the way to the conscious level of the brain. Photoreceptors and neurons fire spontaneously or fail to fire when triggered, the absorption of light in the photoreceptors fluctuates, and lines and shapes are misinterpretated or fluctuate between alternative interpretations. These and other variations subtly alter the corners of the Mona Lisa smile, randomly lifting or dropping the corners, changing the apparent mood of Mona Lisa. You are unaware of the changes but are captured by the ambiguous smile.

7.32 • Floating, ghostly images of a television screen

While watching television in an otherwise dark room, quickly look from about a meter to the left of the screen to about a meter to the right. During the sweep, off to the right side of the screen, you will see one or more bright, detailed, ghostly reproductions of the image then on the screen. Each reproduction is tilted toward the right.

What forms these reproductions? Which way do they tilt if you sweep your view in the opposite direction? Why does increasing your distance from the screen increase the skewness and the separation between the reproductions? Does a vertical sweep yield ghostly images? Can the images form when you sweep your view across a movie projected onto a screen?

Answer The picture on a television screen is created rapidly by horizontal traces, beginning at the top and descending line by line to the bottom. Suppose that your view sweeps rightward past the screen while the beam creates the top line. The line remains visible for a short time because of the persistence of vision. Since your view moves rightward, the persisting image of the top line is shifted to the right when the second line is created on the screen. In fact, as each new line is created on the screen, the preceding lines can still be perceived, but they are shifted to the right because of the motion of your view. The shift is greatest for the top line and least for the bottom line. The composite of the lines gives a skewed reproduction of what is shown on the screen.

A downward sweep of your view yields a squashed reproduction. An upward sweep gives a vertically stretched reproduction. Since a movie is projected as a series of whole frames, it does not yield the ghostly reproductions.

7.33 • Reading through pinholes

Can you read with less eyestrain if you read through pinholes placed in front of the eyes? Reading glasses consisting of pinholes are sold with a promise of easing muscle strain in the eyes because the need for accommodation (focusing) normally required by those muscles is reduced or eliminated.

The argument is that when you see through a pinhole, you increase your *depth of focus* (the range over which objects are in focus) and no longer need to accommodate the eyes. Is the argument correct?

Geckos, which live in desert regions, don't read through pinholes, but they do use them to see during bright light. During the night, two membranes over the opening of the eye open to form a vertical slit through which the lizard sees. During the day, the membranes close the slit, but four small openings remain where notches lie on the edges of the membranes. How does the gecko use those four openings to see prey? Wouldn't a single opening work better?

Answer The argument about pinhole reading is not correct. Accommodation is coupled to the convergence of the eyes, which is the action of angling them to bring their lines of sight onto the same object. When you read this page at a distance of, say, 25 centimeters, your eyes must converge at a certain angle so that you can mentally fuse the images from the eyes to form a single perceived image. The convergence automatically forces each eye to accommodate, even if they are viewing through pinholes. Thus the pinholes do not ease eyestrain.

The multiple openings in the closed eye of a gecko reduce the depth of focus. Then the gecko can adjust the eye so that the four openings produce a single sharp image of an object of prey while they produce four somewhat-overlapping images of any other object. The gecko's attention goes to the sharp image and not the other, blurry images.

7.34 • Finger colors

While in a dim room and viewing with only one eye, stretch out an arm and hold one finger upright in front of a distant, bright window. Focus on the window (or even more distantly), not on your finger. The sides of the finger are bordered with faintly colored light: Red is on one side, blue is on the other. What causes the colors?

Answer First consider a single ray of white light passing just to the left of the finger. The light is said to be white because it consists of approximately equal amounts of the colors in the visible spectrum. Once the light enters and then passes through the eye, the colors end up spread over a small region on the retina, with blue on one side and red on the other, with intermediate colors spread between red and blue. Generally, you cannot see these colors because of the overlap of colors on the retina. However, in the arrangement of a finger held up against a brightly lit window, the finger casts a dark shadow on the retina, eliminating the overlap of colors next to the shadow. So, along that shadow you can see color. Depending on the angle at which the light enters the eye, red appears on one side of the shadow and blue appears on the other side.

7.35 • Stars seen through a shaft during the daytime

Ever since Aristotle started the idea, people have believed that stars can be seen in the daytime if they are viewed through a long shaft such as a tall chimney. The shaft eliminates much of the sky glare, leaving only a small patch of sky at the top of the shaft. The decreased light also allows partial dark adaptation by the observer. Still, are these measures sufficient to permit a star to be seen in that patch of sky?

Answer A star cannot be seen through a long shaft in the manner described because the patch of sky surrounding the star is just as bright as it would be without the shaft. Although viewing through a shaft would reduce the overall glare into the eye, it does not alter the lack of contrast between a star and its surroundings. The shaft may even decrease the star's visibility: If you view a small luminous area surrounded by darkness, the luminous area must be brighter than a certain threshold amount if it is to be perceived. The threshold decreases if the surrounding region is brightened somewhat.

7.36 • A stargazer's eye sweep

Why do you have a better chance of seeing a dim star neighboring a bright star if you sweep your eyes to one side of the stars? When you are in partial darkness, why can you see a dim light source better if you avoid looking directly at it? Aristotle employed this technique to argue that comets were not planets having long periods between appearances because when the eyes are swept to one side of a comet, a tail can be distinguished.

Answer In low illumination, as when you look at a star on a dark night, the cone photoreceptors are insensitive and only the rod photoreceptors can detect light. If you look directly at the star, its image falls on the fovea, which contains only cones, and so you cannot see the star. If you sweep your direct view across the star, the image falls on other parts of the retina where rods are located. You might then be able to see the star.

7.37 • Resolution of Earth objects by astronauts

What are the smallest objects that orbiting astronauts can distinguish on Earth's surface without resorting to instruments? Can they see large cities or structures like the pyramids? The early Mars flybys were disappointing to some people because the photographs revealed no signs of intelligent life. What signs of intelligence would be seen in such photographs of Earth if the resolution were limited to about a kilometer?

Answer An orbiting astronaut can distinguish almost no signs of intelligent life when Earth's surface is viewed with unaided eyes during the daytime. The limit of resolution of a person's sight is set by the diffraction (spreading) of the light as it passes through the pupil. From orbit (say, 800 kilometers above the surface) the diffraction is enough to blur the details of nearly all human structures. Structures about a kilometer wide are just at the threshold of resolution. However, when an astronaut views Earth during the night, dramatic evidence for intelligent life abounds because the lights from large cities can be seen.

7.38 • Honeybees, desert ants, and polarized light

A light ray reaching you directly from the Sun is said to be *unpolarized* because its electric fields oscillate at every possible direction perpendicular to the ray. However, a light ray reaching you after scattering from an air molecule is said to be *polarized* because the electric fields oscillate along a single axis perpendicular to the ray. You may not notice the difference, but certain animals, such as honeybees and desert ants, navigate by monitoring the pattern of polarized light they see in the sky. For example, when a desert ant sets out in search of food, it notes the angles between its body direction and the polarization direction of the sky light. Then, when it wants to return to its nest, it calculates the direction of the nest by effectively combining all the angle information. The calculation is remarkable because many hundreds of angles can be involved.

How do these insects detect the polarization of the light?

Answer The eye of the honeybee and desert ant is said to be compound because it consists of a thousand or more light receptors called *ommatidia.* In each receptor, light passes through a front lens and a cone of transparent crystalline material and then into an elongated structure called the *rhabdom.* That structure is subdivided into nine sections arranged around a central axis running along its length. The sections join by overlapping the regions that have a light-sensitive pigment (rhodopsin). An ommatidium works by sending light along the centerline so that the pigment can absorb it and send a signal to the insect's brain.

One of the sections in an ommatidium can detect the polarization of the light. In some ommatidia the twist of this section is clockwise around the central axis, while in others it is counterclockwise. The insect detects the orientation and strength of the polarization of light by employing three signals. Two of them come from the polarization-sensitive sections, one each from the two directions of twist. The third signal comes from the ultraviolet detectors in the ommatidia that indicate the brightness of the ultraviolet light. On the basis of these three signals from a group of the ommatidia in part of the eye, the insect can monitor the sky polarization.

No one presently understands how the insect's brain employs this information. However, we do know that bees communicate the information between one another by performing a *waggle dance*, a special looping flight path in which the information is symbolized.

The ability to detect polarized light can be used in other ways besides navigation. For example, some aquatic insects detect bodies of water by the polarization of light reflecting from the water. Such reflected light is strongly polarized horizontally. People wear polarizing sunglasses because the lenses absorb horizontally polarized light, and thus they eliminate glare off water. More important to a motorist, the light reflecting from asphalt is also horizontally polarized and thus is eliminated by polarizing sunglasses. Mayflies sometimes misinterpret the polarized light from asphalt as indicating water. They then swarm onto the asphalt and lay their eggs, which soon perish.

7.39 • Haidinger's brush

Most people are able to detect the polarization of light with their eyes. View a bright, featureless background through a polarizing filter. (One of the filters in polarizing sunglasses will serve.) Along your line of sight and for a few seconds, you should see a small, faint, yellow figure with an hourglass shape, surrounded by blue regions (Fig. 7-6). The figure is called *Haidinger's brush* after Wilhelm Karl von Haidinger, who discovered it in 1844.

To keep Haidinger's brush visible, rotate the filter around your line of sight so that the polarization direction of the light entering your eye changes. The hourglass figure rotates also (maintaining the short axis of the yellow portion parallel to the polarization direction of the light). The figure is enhanced if the illumination has more blue than the other colors; a blue sky serves nicely.

Not everyone can see Haidinger's brush, and its appearance seems to dim with age. When I was younger, I could spot it without a filter by looking at sky light, which is polarized. What in the eye is responsible for the figure and the eye's sensitivity to the polarization of light?

Answer Traditionally the cause of Haidinger's brush is attributed to the *macula lutea,* which is a region overlapping the fovea. The sensitivity to the direction of polarization of the light was initially thought to be due to the arrangement of the pigment molecules giving a yellow color to that region.

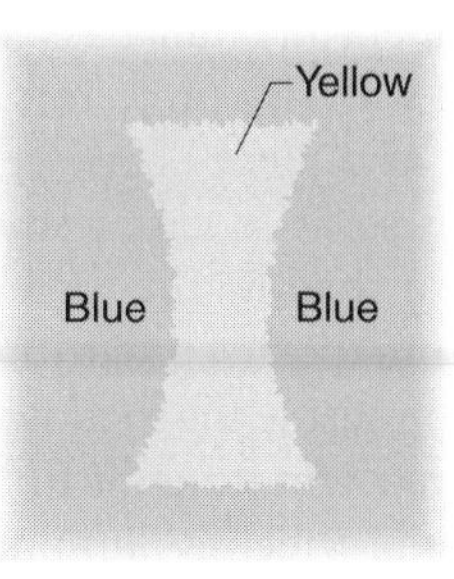

Figure 7-6 / *Item 7.39* Pattern seen in polarized light.

These molecules absorb blue light of a certain polarization. They were believed to be oriented along radial lines having a common center. A more recent model suggests that the molecules need not themselves be oriented. Instead, they might be collected in regions whose orientation with one another provides the selective absorption of one direction of polarization of the blue light.

To understand either of these models, I imagine that the pigments covering the macula lutea lie along two intersecting lines, one horizontal and the other vertical. If vertically polarized, blue light enters the eye, the vertical line passes the light to underlying cones but the horizontal line absorbs the light, preventing it from reaching underlying cones. If, instead, horizontally polarized light enters the eye, then the horizontal line passes the light and the vertical line does not.

Suppose that vertically polarized light enters the eye and that the light is nearly white but has more blue than the other colors. Then the cones behind the vertical line are excited and you perceive blue light along the vertical. However, the horizontal line absorbs the blue, and the cones behind that line receive only the rest of the colors in the light sent into the eye. The subtraction of blue from white or nearly white light is perceived as yellow. Thus you perceive a horizontal line of yellow that is the hourglass figure of the brush. The vertical line of blue is seen as the blue regions on the sides of the hourglass.

If this explanation were complete, you would not see the blue regions when you see the brush against the sky or any other broad source of predominately blue light because the blue regions could not be distinguished from the background. To complete the explanation, we apparently must suppose that the brain perceives *extra* blue in those regions. Presumably, this *subjective coloring* is provoked by the yellow of the hourglass that is adjacent to the regions.

7.40 • Colors of shadows

In 1810, Johann Wolfgang von Goethe, one of the pioneer explorers of color vision, described the following experiment: "Let a short, lighted candle be placed at twilight on a sheet of white paper. Between it and the declining daylight let a pencil be placed upright, so that its shadow thrown by the candle may be lighted, but not overcome, by the weak daylight: The shadow will appear of the most beautiful blue."

You can experiment in similar fashion. In a dark room, illuminate a screen with two projectors. In one beam place a colored filter such as a piece of red cellophane. Hold your hand in front of it so as to cast a small shadow onto the screen. Outside the shadow the screen is pink because it receives red light from the first projector and white light from the second projector. Inside the shadow the screen should be white, because the red light from the first projector is blocked by your hand and the screen is illuminated by only the second projector. However, inside the shadow the screen is blue-green. Why is the shadow region colored?

Answer I shall explain the projector experiment of the problem but leave Goethe's experiment for you to explore. The images of the screen and the shadow of your hand excite the three types of cone photoreceptors on the retina. The image of the pink screen strongly excites the red cones and less strongly the green and blue cones.

The image of the shadow should be white because the shadow region is illuminated by only the second, unfiltered projector. Thus this image should excite all the cones. However, the red cones excited by the pink screen inhibit the signal from the red cones excited within the image of the shadow. This inhibition is interpreted by the visual system as a signal of blue-green, the complementary color of red. How the inhibition is carried out and why the complementary color is perceived are unsolved puzzles.

7.41 • Safety of sunglasses

Sunglasses reduce the intensity of the visible and ultraviolet light entering the eye by absorbing the light, but the darkening also causes the pupil to widen. Is it possible that, because of the widening, there is *more* ultraviolet light entering the eye and therefore sunglasses should not be worn?

Why did native people in what is now Canada and Alaska commonly cover their eyes with pieces of bone or wood through which narrow slits had been cut for viewing? Why do athletes (especially American football players) put black paint or grease on their upper cheeks when playing in bright sunlight?

Answer Sunglasses reduce the net ultraviolet light entering the eye in spite of the widening of the pupil. The conclusion is based on a study of over 400 types of sunglasses—even the least expensive pair reduces the net UV rays.

The native people of Canada and Alaska sought to reduce the glare they faced in brightly lit snow and ice fields. The slits through which they looked greatly reduced not only the visible and ultraviolet light entering their eyes but also the infrared light that produces discomfort in the eyes. The black paint or grease on the cheeks of American football players decreases the reflection of light from the cheeks into the eyes, which can interfere with a player's view. The glare off the cheeks is especially troublesome when the cheek is wet with sweat and the game is being played below a high Sun during the day or bright lights at night.

7.42 • Fish lens

We can see because the eye bends (refracts) the light rays so that they form a sharp image on the retina. About two-thirds of this bending occurs at the curved surface of the cornea; the rest occurs as the rays pass through the eye lens, which is somewhat behind the cornea. A fish is different because its eye is immersed in water, which has about the same optical properties as the eye, and so only the lens can bend the light rays. Moreover, because the lens must bend the rays sharply to bring them to focus just behind the lens, the lens tends to be spherical. However, a spherical lens suffers from *spherical aberration* because rays passing through the periphery enter at such a large angle to the lens surface that they are bent significantly. The rays passing along the central axis of the lens enter at smaller angles and thus are bent much less. The result is that the rays are focused in a broad range behind the lens and thus do not produce a sharp image (Fig. 7-7*a*). In fact, the spherical lens of a fish should leave the fish effectively blind. How, then, does a fish manage to see?

Answer The extent of bending as the light rays pass into and out of a lens depends on the change in the index of refraction of the materials. If a ray passes from the water–protein mixture in the bulk of the eye into a lens with a large index of refraction, the rays tend to bend a lot. If the lens has a smaller index of refraction, they bend less. The lens in a fish eye does not have a single value of the index of refraction. Rather, it has a large index along the central axis and a smaller index toward the periphery. The result is that the focusing along the central axis and the focusing by the periphery produce an image at the same point behind the lens (Fig. 7-7*b*).

Thus, the fish can see. The variation in the index of refraction, said to be a *gradient index of refraction*, is due to a variation in the water–protein mixture in the eye. You can detect the mixture variation by examining either a fresh or cooked fish eye: The texture is firmer near the central axis.

The eye lens in a human also has a gradient index of refraction (varying from larger to smaller values toward the periphery). However, because we live in air instead of water, the human eye corrects for spherical aberration primarily at the surface of the cornea: The cornea is not spherical but is shaped so as to offset spherical aberration.

The horseshoe crab *Limulus* also uses a gradient index of refraction but in a more complicated way. It has compound eyes consisting of many transparent facets, each having a smooth, flat surface. Light passes through a facet to reach the visual system at the end of a channel. The lack of curvature should eliminate any image formation, yet behind each facet an image forms. Along the centerline of the channel (running from front to rear), the index is high. Toward the side walls of the channel, the index decreases. Thus rays of light passing near the center of the facet are bent more than rays passing near the walls. The unequal bending forces the rays to cross one another to form an image behind the channel.

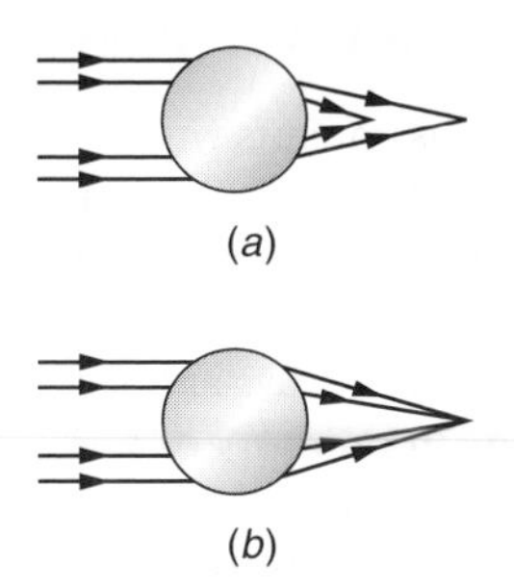

Figure 7-7 / *Item 7.42* Focusing of light rays by a spherical lens with (*a*) a uniform index of refraction and (*b*) a gradient index of refraction.

7.43 • Depth in red and blue signs

Under bright illumination the red sections of a red and blue sign appear to be in front of adjacent blue sections. Under dimmer conditions the apparent depth is reversed. What is responsible for the illusion of depth and its reversal under dimmer light?

Answer First, consider viewing three objects lying at different distances in front of you. If you focus on the intermediate object, each eye forms a sharp image of it at the intersection between the line of sight and the retina. The farther object forms a blurry image on the retina that is somewhat closer to the nose than the sharply focused image. The nearer object forms a blurry image on the retina that is somewhat closer to the temple than the sharply focused image. The brain compares the positions of these images and attributes the proper distances to the objects creating them.

A similar comparison of images accounts for the illusion of depth in red and blue signs. Suppose that under bright illumination you view two adjacent points, one red and one blue. Light rays from the points enter each eye and are refracted (bent) so as to form images on the retina. However, the blue rays are bent more than the red rays, and so both points cannot simultaneously be in sharp focus. Suppose that you look directly at and focus on the red point. In each eye its image forms at the intersection of the line of sight and the retina. The blue point forms a blurry, larger image on the retina (Fig. 7-8).

The depth that is attributed to these two images depends on where they form on the retina relative to the line of sight. Usually the line of sight does not pass through the center of the pupil. In bright illumination the line is on the nasal side of the pupil's center. With this geometry the center of the blurry image of the blue point is slightly displaced toward the nose from the line of sight. Out of experience in perceiving depth, the brain interprets this image as being produced by an object (the blue point) that is more distant than the one producing the sharp image on the line of sight (the red point). Thus, you perceive the blue point to be more distant than the red one.

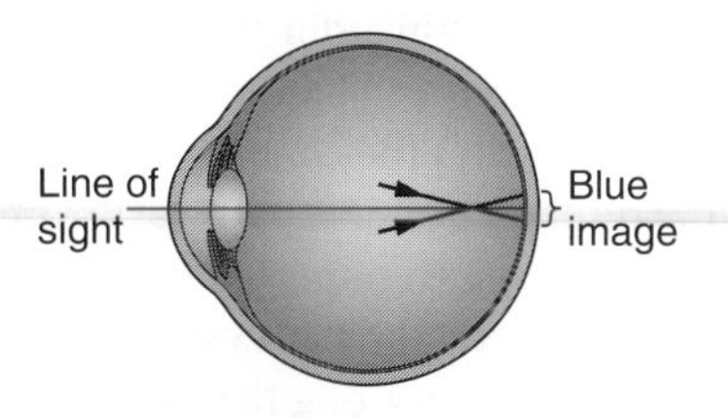

Figure 7-8 / *Item 7.43* When the eye focuses red sharply, it focuses blue in front of the retina.

In dim light, the pupil expands and the line of sight shifts to the temple side of the pupil's center. This arrangement moves the blurry image of the blue point across the retina until its center lies somewhat toward the temple from the line of sight. Your brain interprets this new position as meaning that the blue point is now closer than the red one.

Depth can be seen in maps that are color coded with red and blue if the maps are examined with a large convex lens, such as a large magnifying lens. Here, the separation of colors occurs within the convex lens because blue light is refracted more than red light.

7.44 • Purkinje's blue arcs

Outdoors on a dark night, the 19th century physiologist Johannes Purkinje noticed that a small ember excited two blue arcs in his field of view. Although they faded quickly, he could recreate them by wiggling the ember.

To see them, follow this procedure: Turn off the room lights for about two minutes. With one eye open, turn on a small red light. The best source of light is a narrow rectangle that occupies no more than 0.25° of arc in your view. With the open eye you should perceive a faint blue arc or spike for about a second. The shape of the arc depends on where the red light lies in your view. To recreate an arc, turn on the room lights for about two minutes and then repeat the procedure.

Dim arcs can also be seen just after the stimulus light is turned off. In either case, if you allow the eye to become fully adapted to the dark, the arcs are gray (colorless).

Why does the arc or spike appear, and why does the shape of the blue region depend on the position of the stimulus light? How can a small stimulus light create an arc that extends through a fairly broad region of your field of view? Why are the arcs blue when the eye is partially light-adapted but gray when it is fully dark-adapted?

Answer Where the image of the red light falls on the retina, the light activates the cones responsible for detecting red light. The nerve pathways leading from those cones lie next to the pathways connected to rods lying elsewhere on the retina. Apparently the excitation of the cone pathways stimulates the rod pathways, and the brain is misled into thinking that the rods are also illuminated. Since those rods are laid out in an arc on the retina, the brain perceives an illuminated arc.

The arcs are blue if some of the cones are still sending a signal of yellow to the brain from the previous exposure to room light. The perception of blue develops as follows: The red stimulus light activates the cones where the image of the light falls on the retina. The nerve pathways from those cones activate the nerve pathways connected to the rods in the perceived arc. Those activated rod pathways inhibit the signal of yellow from the cones lying along that arc.

Yellow and blue are said to be *opponent colors* because when a message of yellow is inhibited, the brain perceives blue. Thus when the nerve pathways of the rods inhibit the message of yellow from the cones in the arc, the brain perceives the arcs to be blue. Later when those cones become inactive (they adapt to the darkness), the rods have nothing to inhibit and the perceived arc is gray.

7.45 • Maxwell's spot

View a white sheet of paper through a yellow filter. Then quickly replace the filter with a blue one. You might momentarily see Maxwell's spot, a small dark or yellow spot lying on your line of sight. You can use other pairs of colored filters as long as the second filter in the pair passes more blue light than the first filter. What causes Maxwell's spot?

Answer One explanation of Maxwell's spot is that the rod photoreceptors interfere with the color information being sent to the brain by the cone photoreceptors. When you first view the white paper through a yellow filter, the yellow light entering the eye activates the cones and the rest of the visual system that is responsible for detecting yellow light.

Just after you switch to the blue filter, those cones are still active. With blue light now coming into the eye, other cones begin to send signals of blue to the brain. However, the rods respond to the blue light also (more than they did to the yellow light). Although they cannot send signals about color to the brain (they send only signals about brightness), their activity can inhibit the signal of yellow from the cones that are still active because of the previous exposure to yellow light.

Yellow and blue are said to be *opponent colors* because when a message of yellow is inhibited, the brain perceives blue. Thus when the rods inhibit the message of yellow from the cones, the brain perceives blue. Since it is also receiving a signal of blue from other cones that are activated by the blue light now entering the eye, the blue light seems to be brighter than it truly is.

Since there are no rods in the fovea (where the line of sight intercepts the retina), the additional blue is not perceived there. By contrast to the rest of the retina, the fovea looks yellow because of the blue-yellow opposition in the perception of colors. This perceived coloring of the fovea is Maxwell's spot.

7.46 • Visual sensations from radiation

While in deep space, astronauts with dark-adapted eyes have reported seeing flashes of light that formed points, stars, or double stars, or that would fill up much of the field of view. The displays are due to cosmic rays passing through the eyes of the astronauts. (Cosmic rays are particles, usually at high speed, that originate in outer space.)

Similar displays have been seen in research labs when fast particles have been directed through a subject's eye. How do the particles create the displays? Do they collide directly with the photoreceptors of the retina, forcing them to fire signals to the brain, or do they create light inside the eye that is then intercepted by the photoreceptors? Can the displays be seen at high altitudes by mountain climbers or airplane occupants?

Answer The displays seen by astronauts may result from light produced by extremely fast particles as they pass through the vitreous humor (the transparent material filling the eyeball) and the retina. The speeds of the particles exceed the effective speed of light in the eye. (Light's speed is effectively decreased because of its interaction with the molecules in the vitreous humor.) A resulting *shock wave* of light (called *Cerenkov radiation*) can be produced in the vitreous humor and detected by the photoreceptors in the retina.

Such displays of light have been observed in experiments where high-speed *muons* (particles similar to electrons) were sent through the eye of a subject. Particles (even slow ones) can also generate visual displays when they collide directly with the photoreceptors on the retina. A different sort of display results from x rays: With them an observer sees a uniform flood of light rather than the discrete lights described by astronauts. No one has reported observing any of the displays while traveling in an airplane, even on the high-altitude, high-latitude polar routes that involve enough radiation exposure to require radiation detectors.

7.47 • Red light for control boards

At night, why is the control board in a ship's bridge usually illuminated with far-red light—that is, light at the red edge of the visible spectrum?

Answer Although cone photoreceptors do not function well in dim light, rod photoreceptors can detect such light. However, to see in very dim light you need to allow the rods to adapt to dark conditions—that is, to turn off for at least 10 minutes. Then they will be the most sensitive to a dim light source. Because rods are not activated by light at the red edge of the visible spectrum, control boards are generally illuminated by such light at night. As you view the control board, the cones may be turned on but the rods are turned off and thus are ready when you look out into a dark night.

7.48 • Superman's x-ray vision

According to the stories, Superman can see through a solid wall by emitting x rays from his eyes. Let's ignore the obviously complex question about how eyes could possibly emit x rays. Instead, let's focus on an easier question: Could something on the opposite side of a wall be detected via x rays?

Answer If Superman is to intercept rays returned by, say, a criminal on the far side of a wall, the criminal will have to reflect the rays. But that means that the wall must also reflect the rays. You might argue that any material could partially transmit and partially reflect the rays. Then a portion of the rays would penetrate the wall; a fraction of those rays would reflect from the criminal; and finally a few surviving rays would reach Superman after passing back through the wall. The trouble is that the surviving rays would be so few that they would be masked by the overall glare of rays reflected by the wall and all the objects lying beyond the criminal. Even if Superman can somehow mentally process all the rays and construct an image of just the criminal, we still have the following problem: How can Superman's eyes absorb the rays if they are so readily reflected and transmitted? (I know, comics should be read, not studied.)

7.49 • Fireworks illusion

When a flare is shot directly upward in a windless, dark night, the flaming debris from its explosion should spread out horizontally and uniformly. Why then does all the debris seem to come in your general direction?

Answer The illusion has not been explained in detail. However, you might guess that a lifetime of experience sets the stage for the illusion: When you look at virtually every (three-dimensional) object, you see details on the near side, rarely on the far side. So, on a dark night without any depth clues available (from, say, background clouds), the expanding display of burning debris will be interpreted as being details on the near side of an expanding, unseen object.

7.50 • Looking at the ceiling

Lie on your back in the center of a room with ceiling fixtures and door frames. If you look at the ceiling in the direction of your feet, the ceiling fixtures, and door frames look normal. But if you tilt your head back so that you look at the opposite side of the ceiling, you will have an eerie sensation that you are looking down on the ceiling, as if you could walk on it. A ceiling lamp in that region will seem to sprout upward toward you, and a doorframe will appear to be an obstacle over which you would have to step. What produces this illusion? Do you see it if you look at the ceiling while standing on your head?

Answer The published research about the illusion suggests that when you are lying on your back, you then reverse up for down because you lack the normal clues from gravity about upward and downward. Normally, a region that is down is in the lower half of your view and a region that is up is in the upper half. If you mentally reverse up for down when you are on your back, the ceiling toward your feet is in the lower half of your view and looks like it is "up," and the opposite region of ceiling is in the upper half of your view and seems to be "down." The illusion does not appear if you stand on your head, presumably because you then have strong clues from gravity about what is up and down.

I • N • D • E • X